OF TRANSMISSION (1000–1500)

- Florence, Italy outlaws Hindu-Arabic numerals (1299)
- Chu Shih-Chieh — binomial coefficients (1303)
- Ockham — formal logic (1320)
- Burley — formal logic (1325)
- Oresme — fractional exponents (1360)
- Regiomontanus — trigonometry (1464)
- Hindu-Arabic numerals take present form (1479)
- Chuquet — notation for +, − of fractions (1484)
- Widmann — addition (+), subtraction (−) signs (1489)
- Calandri — long division process (1491)

EARLY MODERN PERIOD (1450 TO 1800)

- Rudoff — square root symbol (√) (1525)
- Apianus — notation for ×, ÷ of fractions (1527)
- Cardano — cubic equations (1545)
- Rheticus — right triangle trigonometry (1551)
- Diez Feile — writes 1st math book in Mexico (1556)
- Recorde — symbol for the equal sign (=) (1557)
- Bombelli — imaginary numbers (1572)
- Clavius — the dot (·) for multiplication (1583)
- Stevin — decimal fractions (1585)
- Napier — logarithms (1614)
- Oughtred — multiplication sign (×), slide rule (1621)
- Kepler and Briggs — table of logarithms (1624)
- Cavalieri — indivisibles in calculus (1635)
- Descartes — analytic geometry (1637)
- Desargues — projective geometry (1640)
- Fermat — probability, analytic geometry, number theory, calculus (1629–1654)
- Pascal — projective geometry, calculating machine, probability, binomial coefficients (1642–1654)
- Wallis — negative exponents, infinity symbol (∞) (1655)
- Rahn — division sign (÷) (1659)
- Isomura — magic circles (1660)
- Barrow — calculus, differentiation (1663)
- Leibniz — symbolic logic, calculating machine, calculus (1666–1684)
- Seki Kōwa — determinants magic squares (1685)
- Newton — calculus (1687)
- William Jones — symbol for pi (π) (1706)
- Jakob Bernoulli — theory of probability (1713)
- First math book printed in America (1719)
- DeMoivre — actuary math, complex numbers (1720)
- Greenwood — writes 1st American math book (1729)
- Maria Agnesi — popular math text book (1748)
- Achenwall — the word *statistik* (1749)
- Euler — uses the symbols i, e, and Σ (1750)
- Gabriel Cramer — systems of equations (1750)
- Bayes — statistics, origin of polls (1763)
- First Bank of the U.S. (1791)

MODERN PERIOD (1800 TO PRESENT)

- Gauss — statistics, normal curve, the term *complex number* (1809–1832)
- Laplace — probability theory (1814)
- Babbage — steam powered calculator (1823)
- Lobachevsky — non-Euclidean geometry (1829)
- Carl Jacobi — theory of determinants (1841)
- Ada Byron — computer programming (1843)
- George Boole — logic, Boolean algebra (1847)
- Georg Riemann — non-Euclidean geometry, calculus (1850–1854)
- Weierstrass — absolute value symbol (||), foundations o calculus (1841–1874)
- Möbius — topology (1865)
- John Venn — logic, Venn diagrams (1880)
- Hollerith — electrical tabulating device (1880)
- Burroughs — practical adding machine (1894)
- Markov — probability theory, Markov Chains (1907)
- Whitehead, Russell — Principia Mathematica (1910)
- Federal Reserve Act — U.S. banking system (1913)
- Lukasiewicz, Post, Wittgenstein — truth tables (1920)
- Ronald Fisher — statistics sampling techniques (1920)
- Thomas Watson forms IBM (1924)
- Mauchley and Eckert — ENIAC computer (1945)
- Gertrude Cox — statistical design (1950)
- Kilby, Texas Instruments — integrated circuit (1958)
- Mandelbrot — fractal geometry (1967)
- Gilbert Hyatt — computer microprocessor chip (1968)
- Edward Roberts — first personal computer (1971)
- Hewlett Packard — programmable calculator (1974)
- Proof of Fermat's Last Theorem by Andrew Wiles (1993)
- Team of Michael Cameron, George Woltman, Scott Kur(et al. discovered a new record prime $2^{13466917} - 1$ (2001)
- Dr. Kanada of University of Tokyo calculates Pi to 1.241 trillion digits (2002)

The Mathematical Palette

The Mathematical Palette

THIRD EDITION

RONALD STASZKOW
Ohlone College

ROBERT BRADSHAW
Ohlone College

THOMSON
BROOKS/COLE

Australia • Canada • Mexico • Singapore • Spain
United Kingdom • United States

Publisher: Bob Pirtle
Sponsoring Editor: John-Paul Ramin
Assistant Editor: Lisa Chow
Editorial Assistant: Darlene Amidon-Brent
Project Manager, Editorial Production: Janet Hill
Technology Project Manager: Christopher Delgado
Marketing Manager: Karin Sandberg
Marketing Assistant: Jennifer Gee
Advertising Project Manager: Bryan Vann
Print/Media Buyer: Kris Waller
Production Service: Hearthside Production Services/Laura Horowitz
Text Designer: Geri Davis
Art Editor: Hearthside Production Services
Photo Researcher: Sue Howard
Copy Editor: Barbara Willette
Illustrator: Hearthside Production Services
Cover Designer: Irene Morris
Cover Image: Matsu Illustration
Cover/Interior Printer: Quebecor/Versailles
Compositor: Better Graphics

Printed in the United States of America

2 3 4 5 6 7 07 06 05

For more information about our products contact us at:
Thomson Learning Academic Resource Center
1-800-423-0563
For permission to use material from this text, contact us by:
Phone: 1-800-730-2214
Fax: 1-800-730-2215
Web: http://www.thomsonrights.com

Library of Congress Control Number: 2003096133

Student Edition: ISBN 0-534-40365-4

Instructor's Edition: ISBN 0-534-40371-9

Brooks/Cole–Thomson Learning
10 Davis Drive
Belmont, CA 94002
USA

Asia
Thomson Learning
5 Shenton Way #01-01
UIC Building
Singapore 068808

Australia/New Zealand
Thomson Learning
102 Dodds Street
Southbank, Victoria 3006
Australia

Canada
Nelson
1120 Birchmount Road
Toronto, Ontario M1K 5G4
Canada

Europe/Middle East/Africa
Thomson Learning
High Holborn House
50/51 Bedford Row
London WC1R 4LR
United Kingdom

Latin America
Thomson Learning
Seneca, 53
Colonia Polanco
11560 Mexico D.F.
Mexico

Spain/Portugal
Paraninfo
Calle/Magallanes, 25
28015 Madrid, Spain

CONTENTS

Preface xi

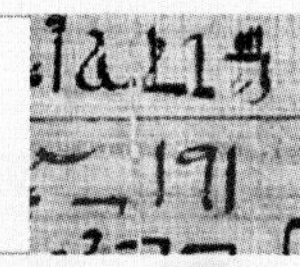

chapter 1 **Numbers—Old and New**

OVERVIEW 2
A SHORT HISTORY OF NUMBERS AND NUMERALS 2
PROJECTS 4
1.1 Ancient Systems of Numeration 6
1.2 The Hindu-Arabic System and Fractions 17
1.3 Numeration Systems with Other Bases 23
1.4 The Numbers of Technology 31
1.5 Types of Numbers 39
Summary 48
Review 50
Test 51

chapter 2 **Logical Thinking**

OVERVIEW 54
A SHORT HISTORY OF LOGIC 54
PROJECTS 55
2.1 Logic, Statements, and Definitions 56
2.2 Inductive and Deductive Reasoning 64
2.3 Symbolic Logic and Truth Tables 74
2.4 Logic and Flowcharts 85
2.5 Logic and Puzzles 93
Summary 103
Review 105
Test 106

chapter 3

Sets and Counting

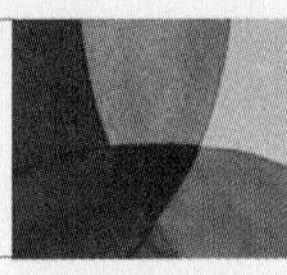

OVERVIEW 110
A SHORT HISTORY OF SETS 110
PROJECTS 112
3.1 Sets: Finite and Infinite 113
3.2 Set Operations and Venn Diagrams 121
3.3 Applications of Sets 128
3.4 Introduction to Counting 135

Summary 146
Review 147
Test 149

chapter 4

Probability

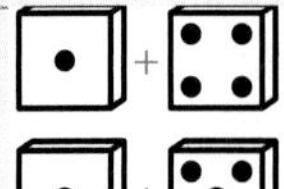

OVERVIEW 152
A SHORT HISTORY OF PROBABILITY 152
PROJECTS 154
4.1 Intuitive Concepts of Probability 155
4.2 Calculating Probabilities 163
4.3 Probability and Odds 172

4.4 Probability of Compound Events 176
4.5 Conditional Probability 187
4.6 Expected Value 195

Summary 205
Review 207
Test 209

chapter 5

Statistics and the Consumer

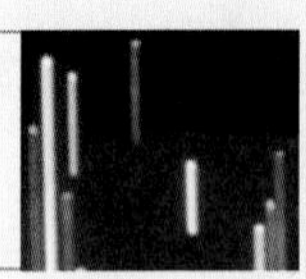

OVERVIEW 212
A SHORT HISTORY OF STATISTICS 212
PROJECTS 213
5.1 Arranging Information 215
5.2 Measures of Central Tendency 226
5.3 Measures of Dispersion 236
5.4 The Normal Distribution 245
5.5 Polls and the Margin of Error 257
5.6 Regression and Forecasting 264

Summary 274
Review 275
Test 277

chapter 6

Modeling with Algebra

OVERVIEW 280
A SHORT HISTORY OF ALGEBRA 280
PROJECTS 282
6.1 Linear Models 284
6.2 Quadratic Models 294
6.3 Exponential Models 304
6.4 Logarithmic Models 316
Summary 326
Review 327
Test 329

chapter 7

Geometry and Art

OVERVIEW 334
A SHORT HISTORY OF GEOMETRY 334
PROJECTS 337
7.1 Euclidean and Non-Euclidean Geometry 338
7.2 Perspective 351
7.3 Golden Ratios and Rectangles 368
7.4 Polygons and Stars 378
7.5 Tessellations 390
7.6 Fractals 402
Summary 417
Review 419
Test 421

chapter 8

Trigonometry: A Door to the Unmeasurable

OVERVIEW 424
A SHORT HISTORY OF TRIGONOMETRY 424
PROJECTS 426
8.1 Right Triangles, Sine, Cosine, Tangent 427
8.2 Solving Right Triangles 434

8.3 Right Triangle Applications 441
8.4 The Laws of Sines and Cosines 449
8.5 Acute Triangle Applications 458
8.6 The Motion of a Projectile 467

Summary 473
Review 475
Test 477

chapter 9 Finance Matters

OVERVIEW 480
A SHORT HISTORY OF INTEREST 480
PROJECTS 482
9.1 Percents 484
9.2 Simple Interest 492
9.3 Compound Interest 497
9.4 Annuities 506
9.5 Loans 511

Summary 520
Review 521
Test 523

chapter 10 Math from Other Vistas

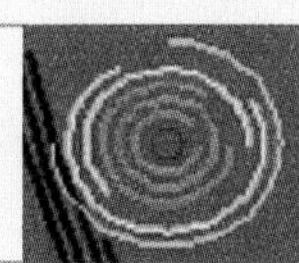

OVERVIEW 526
10.1 Differential Calculus 527
10.2 Integral Calculus 539
10.3 The Pascal–Yang Hui Triangle 548
10.4 Voting Systems 556
10.5 Apportionment 565
10.6 Linear Programming 574

Appendix 585
Selected Answers 589
Credits 629
Index 633

PREFACE

The Mathematical Palette, Third Edition, in its attempt to stimulate the creativity of the liberal arts student, makes mathematics understandable and enjoyable while being practical and informative. Just as an artist mixes paints on a palette, so *The Mathematical Palette* mixes the history of mathematics, its mathematicians, and its problems with a variety of real-life applications. The text presents this sampling of mathematics in a straightforward, interesting manner and is designed expressly for the liberal arts student, not for a mathematician.

The Mathematical Palette, Third Edition, is intended for the liberal arts student who has a background in high school algebra or who has successfully completed intermediate algebra in college. The text is designed not only to meet college general education requirements but also to help generate a positive attitude toward and an interest in mathematics. The text stresses learning mathematics rather than just learning about mathematical ideas. The intent is the development of problem-solving skills by actually solving problems. We believe that a greater appreciation of the beauty and power of mathematics is gained when students become active participants.

The Mathematical Palette, Third Edition, provides additional material that is not ordinarily included in mathematics texts. The Brief Histories and Research Projects provide students the opportunity to write about mathematics and mathematicians. The Math Projects give students a chance to do some creative problem solving. The exercises at the end of each section are divided into three categories, Explain, Apply, and Explore. The Explain questions ask the student to give written answers explaining the what, how, or why of topics covered in the section. The Apply questions ask the student to solve problems with a direct application of the methods presented in the section. The Explore problems require that the student go beyond the basic problems by exploring other applications or implications of the concepts of the section. In general, the Explore problems have a higher degree of difficulty than the Apply problems and ask the student to look a little deeper into the material.

FEATURES

Instructors will find that *The Mathematical Palette,* Third Edition, is a very teachable text. Students will find that *The Mathematical Palette,* Third Edition, is a very readable text. Its organization, style, and format have been developed with both the student and the instructor in mind. Each chapter is a self-contained unit and may be taught in any order according to the instructor's or student's interests. The history section along with the Research and Math Projects at the beginning of each chapter present material that is ideal for classroom discussions, research papers, group projects, speeches, presentations, and reports. The text's readable style, clear explanations, numerous solved examples, Explain-Apply-Explore problem sets, chapter summaries, chapter reviews, chapter tests and accurate answer section are valuable resources for the instructor and the student. In particular, the following features of each chapter are real assets to teaching and learning:

- Overviews that describe and introduce the chapter.
- Short history sections present the development of the mathematics and introduce students to important dates, mathematicians, and events.
- Research projects at the end of each history section go beyond the material and serve as an ideal source for written papers and various types of reports.
- Math projects present problems for students to use the concepts of the chapter in some creative problem solving. These math projects are ideal for group cooperative learning experiences.
- Clear explanations along with worked examples and illustrations are found throughout the text.
- Explain-Apply-Explore problem sets include a variety of writing and skill-building exercises.
- End-of-chapter summaries present the terminology, formulas, and objectives used in the chapter.
- End-of-chapter reviews and tests give material that serve as a comprehensive review of the chapter.
- A full-color format highlights important concepts and formulas and provides students with visual guideposts.
- Artwork, photographs, and cartoons that relate to the mathematics being discussed make text visually stimulating.
- Consistent use of calculators eliminates the time spent on paper-and-pencil computation.

CHANGES IN THE THIRD EDITION

Suggestions from the reviewers and users of the second edition of *The Mathematical Palette* have prompted the following changes in the third edition. The content has been reorganized, topics have been added, and sections have been eliminated to make the text even more alive to the liberal arts student. The third edition continues to emphasize not only problem solving but also writing and critical thinking.

- History sections have been shortened to emphasize general developments in mathematics rather than detailed names and dates.
- More real-life problems and data are used throughout the text.
- Projects at the beginning of each chapter are divided into two groups: Research Projects and Math Projects.
- Review problems, keyed to individual sections, have been added to the end-of-chapter materials.
- The numbers of technology are included in Chapter 1 ("Numbers—Old and New").
- A section on logic and puzzles has been added to Chapter 2 ("Logical Thinking").
- There is a separate chapter on sets and counting the number of items in a set—Chapter 3 ("Sets and Counting").
- Sections on the probability of compound events and conditional probability have been added to Chapter 4 ("Probability").
- The emphasis of Chapter 5 is statistics and the consumer.
- A section on regression and forecasting has been added to Chapter 5 ("Statistics and the Consumer").
- Chapter 6 ("Modeling with Algebra") includes many examples and problems from sports.
- A section on the motion of a projectile has been added to Chapter 8 ("Trigonometry: A Door to the Unmeasurable").
- Chapter 9 ("Finance Matters") begins with a section on percents that includes markups, markdowns, and commissions.
- The chapter on computers has been eliminated.
- The chapter on calculus has been become Chapter 10 ("Math from Other Vistas"), which has views from the past (differential calculus, integral calculus, the Pascal–Yang Hui triangle) and views of the present (voting systems, apportionment, linear programming).

ORGANIZATION

Each chapter is self-contained and can be treated as a separate unit. The ten chapters contain ample material for a one-semester three-credit course.

Chapter 1 ("Numbers—Old and New") introduces the basic building blocks of mathematics: numbers. It begins with ancient systems of numeration, progresses through the Hindu-Arabic system and the number systems of technology, and ends with a look at types of numbers.

Chapter 2 ("Logical Thinking") investigates logic and some of its applications. It begins with basic statements and types of reasoning and proceeds to use of syllogisms and truth tables to analyze arguments. The chapter concludes by looking at flow charts as a means of arriving at logical decisions and using logic to solve puzzles.

Chapter 3 ("Sets and Counting") examines sets, set operations, and Venn diagrams. It investigates finite and infinite sets and discusses techniques for counting the number of elements in a set.

Chapter 4 ("Probability") investigates probability, odds, and expected values as applied to games of chance. It also examines the probability of compound and conditional events.

Chapter 5 ("Statistics and the Consumer") looks at basic statistics as a means of arranging and reporting data, especially those connected with the consumer. It begins with simple statistical graphs, examines measures of central tendency, and investigates the normal distribution. The chapter ends by looking at the statistics used in analyzing polls and forecasting.

Chapter 6 ("Modeling with Algebra") examines the use of algebraic functions as a model in various situations. Linear, quadratic, exponential, and logarithmic functions are used to model real-life situations, especially those involving sports.

Chapter 7 ("Geometry and Art") begins with a review of some of the postulates, definitions, and theorems of Euclidean geometry and an introduction to non-Euclidean geometry. With this as a base, it introduces some of the geometry in art such as perspective, golden ratios, polygons, stars, tessellations, and fractals.

Chapter 8 ("Trigonometry: A Door to the Unmeasurable") focuses on some of the practical applications found in the study of trigonometry. It examines right triangles, acute triangles, and the motion of a projectile.

Chapter 9 ("Finance Matters") uses mathematics as a practical tool by examining the world of finance through a realistic study of percents, simple and compound interest, annuities, and loans.

Chapter 10 ("Math from Other Vistas") presents views from the past by looking at differential calculus, integral calculus, and the Pascal-Yang Hui triangle and views of the present by introducing voting systems, apportionment, and linear programming.

ANCILLARY MATERIALS

The following ancillary materials are available to all adopters of *The Mathematical Palette*, Third Edition:

- *Student Study Guide* is available for purchase by students. It includes summaries that emphasize important concepts and techniques, complete solutions to problems, and other material that will assist students in mastering the concepts discussed in each chapter.

- *Test Bank* written by Jay Domnitch of Palm Beach Community College. It includes 6 tests per chapter as well as 3 final exams. The tests are made up of a combination of multiple-choice, free-response, true/false, and fill-in-the-blank questions.

- *CNN Today for Liberal Arts Mathematics*
  Volumes I and II (Vol. I: 0-534-39641-0; Vol. II: 0-534-40066-3).
  Instructors can launch their lectures with footage from CNN, the world's leading 24-hour global news television network. CNN Today Videos allow instructors to integrate the newsgathering and programming power of CNN into the classroom to show students the relevance of course topics to their everyday lives. Organized by topics covered in a typical course, these videos are divided into short segments allowing instructors a way to introduce key concepts. Projects built around the CNN video clips will be accessible through <http://mathematics.brookscole.com>. A Thomson Brooks/Cole exclusive.

- *Instructor's Resource Manual* that contains answers and worked-out solutions to all Explain-Apply-Explore problems along with other teaching aids.

- The text includes access to the **BCA/iLrn Testing, Tutorial, and Course Management System**—a dynamic suite of teaching and learning tools for instructors and students. **BCA/iLrn** features a diagnostic tool that enables instructors to assess and place each student in any course, as well as provide a personalized study plan. At the same time, **BCA/iLrn Tutorial** gives students unlimited practice problems, instant analysis and feedback, and streaming video to illustrate key concepts, as well as live, online, text-specific tutorial help from **vMentor™.**

ACKNOWLEDGMENTS

We would like to thank those who were instrumental in developing this text:

Our wives, Dianne and Theresa, for their continual encouragement.

David McLaughlin for his contributions to Chapter 7.

Kelly and Jeff O'Connell for their projects, and Kelly for performing an accuracy check.

Our Ohlone College students for their corrections and suggestions.

Janet Hill, Brooks/Cole, at Thomson Learning and Laura Horowitz of Hearthside Publishing Services for their concern for every detail in producing our book.

We wish to thank the following instructors for their insightful reviews:

Steven H. Heath, Southern Utah University

Joseph D. Lakey, New Mexico State University

Adele J. Miller, Cabrillo College

Charles E. Mitchell, Tarleton State University

ViAnn E. Olson, Rochester Community & Technical College

Victoria J. Young, Motlow State Community College

Ronald Staszkow

Robert Bradshaw

The Mathematical Palette

CHAPTER ONE

NUMBERS— Old and New

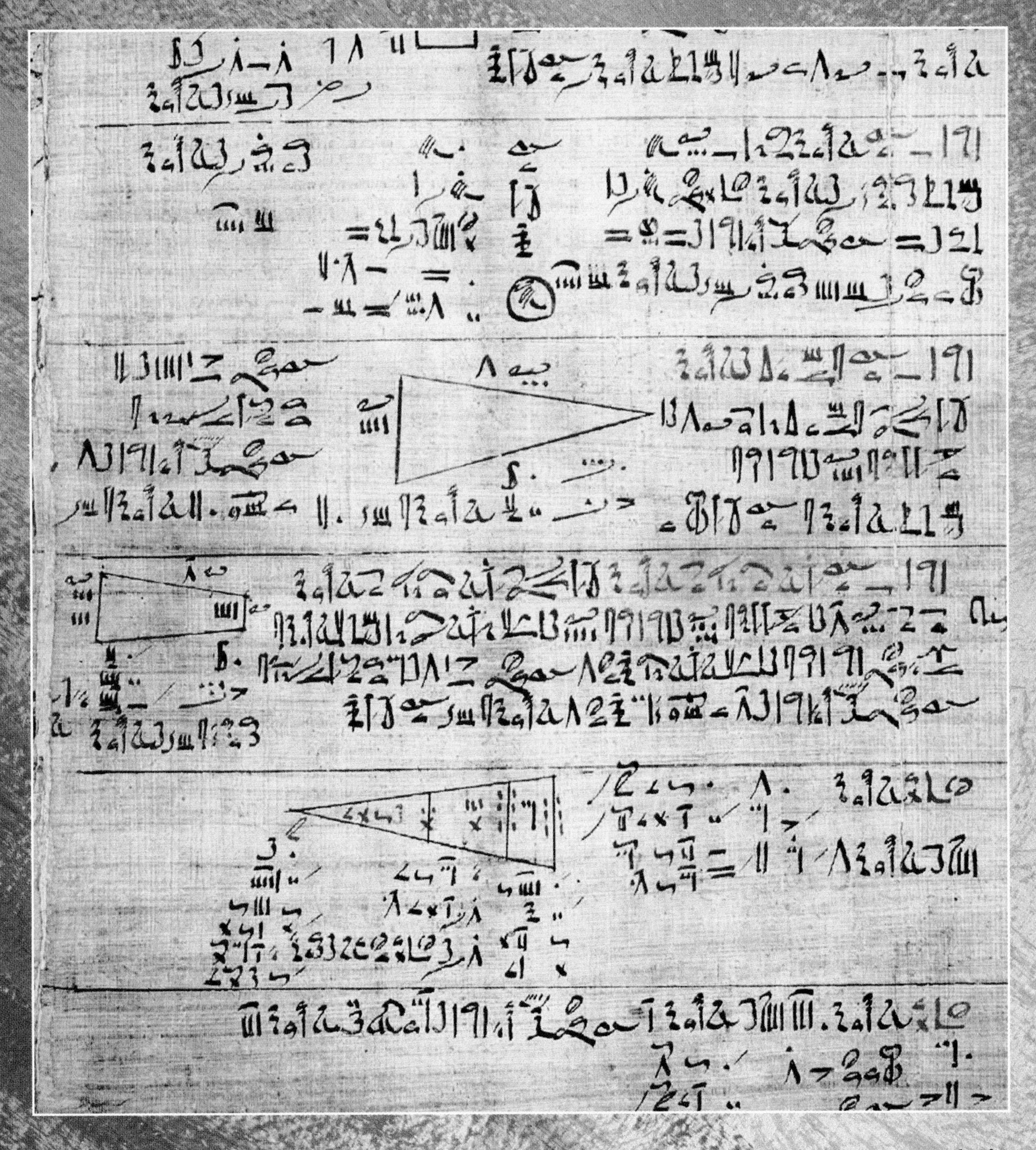

The formal study of mathematics has intrigued humans for centuries. An example is shown on this mathematical treatise written on Rhind papyrus dating from the middle of the 16th century B.C. (Art Resources, Inc.)

OVERVIEW

The numerals 0 through 9 are an intrinsic part of our society and are accepted and used throughout the world. Has this always been the case? Have all cultures always used these numerals? Were these the first numerals created? In this chapter, we answer those questions and give you insights into the lore of numbers. You will see that many cultures developed entirely different systems using symbols that might seem entirely foreign to us. You will discover how different systems from the past work and will be introduced to systems that are used today. You will learn about special types of numbers that have names such as *complex numbers, perfect numbers, friendly numbers,* and *pentagonal numbers*. You will investigate the numbers of computers: binary, octal, and hexadecimal.

one

A SHORT HISTORY OF NUMBERS AND NUMERALS

It seems appropriate that early in our journey through mathematics, we should study math's basic building blocks: numbers. Ever since prehistoric times, human beings have concerned themselves with numbers. During the Paleolithic period (1,000,000 B.C.–10,000 B.C.) men and women were almost totally consumed with survival. However, it is believed that they possessed a basic sense of numbers and had the ability to distinguish between more, less, and equal. Their language might have been lacking in the words to represent numbers, but anthropologists agree that, at the very least, they possessed a visual number sense and an awareness of form.

This stamp issued by Nicaragua in 1971 under the title "Mathematical Equations Which Changed the World" shows the basic process of using fingers to count and developing mathematical symbolism using the brain.

As humans progressed through the Neolithic period (8000 B.C.), they became civilized. They grew crops, domesticated animals, wove cloth, made pottery, and lived in villages. Their number sense also grew. Though we have no written records, archaeological findings and opinions of anthropologists suggest that by the beginning of recorded history (3000 B.C.), human beings had developed the ability to tally and count.

A **tally** is a mark that represents the object being counted. This process of tallying took the form of scratches on the ground or on cave walls, knots on ropes or vines, piles of pebbles or sticks, and notches on pieces of bone or wood. For example, to count the number of days between full moons, you could make a mark each evening on the wall of your bedroom until the next full moon appeared. Such a tally might look like this:

||||| ||||| ||||| ||||| ||||| |||

However, if the number of objects to be counted is very large, the tally method becomes very cumbersome. For example, if you wanted to record the population of the United States (281.4 million in 2000) using the tally system shown above and the tallies were typed on both sides of standard letter-sized paper, you would end up with a pile of more than 21,000 pieces of paper.

As society became more involved in measurement, commerce, and taxes, more efficient means of representing numbers were needed, so different systems of numeration were developed.

What Is a System of Numeration?

A **number** is a quantity that answers the question "How much?" or "How many?" Numbers are given a name in words and are represented by symbols. The symbols that are used to represent numbers are called **numerals.** In common usage, the terms number and numeral are used interchangeably, but the number is really the abstract concept, the amount or value, and the numeral is a symbolic representation of that amount or value. For example, the quantity of trees shown below can be represented by the numeral **10** in the Hindu-Arabic system, **=** in the Mayan system, ∩ in the Egyptian system, **X** in Roman numerals, **Δ** in the Attic Greek system, **1010** in the binary system, **+** in the traditional Chinese system, and so on. The word used for that amount of trees is *ten* in English, *zehn* in German, *diez* in Spanish, *i'wes* in Ohlone Indian of California, *decem* in Latin, *'umi* in Hawaiian, *dix* in French, *shyr* in Chinese, *tiz* in Hungarian, *daca* in Sanskrit, *desiat* in Russian, and so on.

No matter what symbols or words are used, the amount of trees is understood.

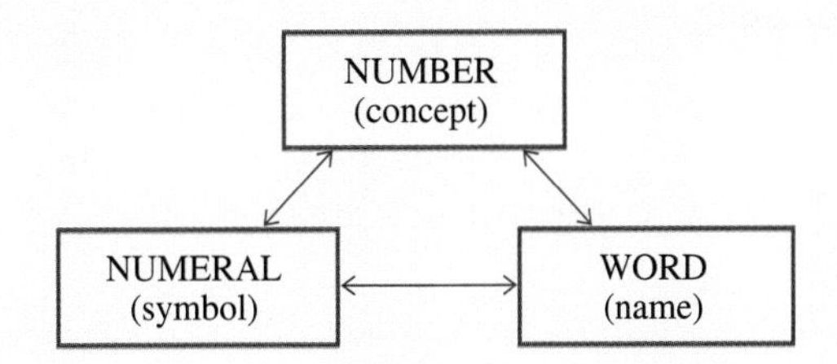

A system of numeration consists of a set of symbols and a method for combining those symbols to represent numbers. This chapter examines the major systems of numeration that humankind has created: Egyptian (c. 3400 B.C.), Babylonian (c. 2300 B.C.), Ionic Greek (c. 450 B.C.), Mayan (c. 300 B.C.), Attic Greek (c. 300 B.C.) Roman (c. 200 B.C.), Chinese (c. 200 B.C.), and Hindu-Arabic (c. 825). As we study these different systems, our goal is to develop a richer understanding of the system of numeration that we currently use, the Hindu-Arabic system, and introduce the systems used by computers.

CHAPTER 1 PROJECTS

Research Projects

1. One of the important advantages of the Hindu-Arabic system is that it allowed the common person to perform arithmetic calculations. From 1400 to 1700, the symbols, $+, -, \times, \div, \sqrt{\ }$, and techniques for performing the operations they symbolize were developed. Who were some of the people involved, what were their contributions, and when were they made?
2. Methods for creating magic squares are introduced in the text. Discuss at least two methods for creating magic squares of various sizes. Who developed these methods? Show how to generate magic squares using these methods.
3. What is numerology? How does the system of numerology work? Using your name and birth date, perform some numerological calculations. What do you think about the results?
4. Perfect and amicable numbers are introduced in Section 1.5. With the use of computers many more of these numbers have been found. What are some of these recently discovered perfect and amicable numbers? Who discovered them? When were they discovered?

Math Projects

1. **Fun with 1944**

   Using the four digits 1, 9, 4, and 4, in that order, create problems that would give all the integer answers from 1 to 44. You can use the addition, subtraction, multiplication, division, powers, and square roots along with parentheses and the correct order of operations to create the problems but you must use all the four digits (1, 9, 4, 4) in that order. For example,

$$
\begin{aligned}
1 &= 1^{944} \\
2 &= (-1 + 9 - 4) \div \sqrt{4} \\
3 &= 19 - 4 \cdot 4 \\
&\vdots \\
44 &= (1 + 9) \cdot 4 + 4
\end{aligned}
$$

   There might be more than one way to get each answer. Your objective is to create at least one problem for each integer from 1 to 44.

2. In this chapter, you will be exposed to various systems of numeration. With that background, design your own system of numeration using your own symbols and scheme for representing numbers. How would you represent fractions in your system? Give examples and explanations of the components of your system.

3. In Section 1.4, you are shown how binary addition and multiplication can be accomplished. How would you perform subtraction and division with binary numerals? Develop a technique for doing this and perform some calculations using the technique. Use decimal numerals to check whether your results are correct.

Ancient Systems of Numeration

In this section, we examine the four basic types of ancient systems of numeration. Each system uses symbols to represent numbers or a set of tallies. We examine how different systems of numeration use symbols to represent numbers. We do not expect you to be an expert in ancient systems of numeration. We merely want to expose you to a very important area of mathematics. Furthermore, we do not expect you to memorize all the symbols in this section. As you read this section and solve the problems, you will need to refer frequently to the lists of symbols for a given system of numeration.

A copy of a wall painting from the tomb of Menna, Thebes, XVIII Dynasty (c. 1420 B.C.), shows ancient Egyptians using tallying to record the amount of wheat harvested. (Egyptian Expedition of The Metropolitan Museum of Art, Rogers Fund, 1930. (30.4.44) Photograph "1979 The Metropolitan Museum of Art)

SYSTEMS USING ADDITION

Egyptian Hieroglyphic System

The oldest type of numeration system using addition is the Egyptian hieroglyphic system (c. 3400 B.C.). The numerals used in this system are shown in the table. The value of a number is simply obtained by finding the sum of the values of its numerals.

Value	Symbol	Name	Value	Symbol	Name
1	|	Staff	10,000		Pointing finger
10	∩	Heel bone	100,000		Tadpole
100		Spiral	1,000,000		Astonished man
1000		Lotus blossom	10,000,000		Sun

EXAMPLE 1

Find the number represented by [astonished man, pointing finger, pointing finger, heel bone, heel bone, heel bone, staff, staff]

Solution: Using the table, add the value of each of the numerals.

$$1{,}000{,}000 + 10{,}000 + 10{,}000 + 10 + 10 + 10 + 1 + 1 = 1{,}020{,}032$$

EXAMPLE 2

Write the number 1753 as a numeral in the Egyptian hieroglyphic system.

Solution: 1753 = 1000 + 700 + 50 + 3, which in Egyptian hieroglyphics is

𓆼𓍢𓍢𓍢𓍢𓍢𓍢𓍢𓎆𓎆𓎆𓎆𓎆𓏺𓏺𓏺.

In such an additive system, the symbols can be placed in any order.

123 = 𓏺𓏺𓏺𓎆𓎆𓍢 = 𓍢𓎆𓎆𓏺𓏺𓏺 = 𓏺𓏺𓎆𓎆𓍢𓏺

Attic System

Another example of a system of numeration that uses an additive grouping scheme was found in records in Athens, Greece, around 300 B.C. The Attic system uses the following numerals.

1	5	10	50	100	500	1000	5000	10,000	50,000
Ι	𐅃	Δ	𐅄	Η	𐅅	Χ	𐅆	Μ	𐅇

EXAMPLE 3

Find the number represented by the Attic numeral

ΜΜ𐅅ΗΗΗΔ𐅃ΙΙΙΙ.

Solution: Using the chart above, add the value of each numeral.

10,000 + 10,000 + 500 + 100 + 100 + 100 + 10 + 5 + 1 + 1 + 1 + 1 = 20,819

EXAMPLE 4

Write 6376 in the Attic system of numeration.

Solution: Using the chart above, we get

6376 = 5000 + 1000 + 100 + 100 + 100 + 50 + 10 + 10 + 5 + 1

= 𐅆ΧΗΗΗ𐅄ΔΔ𐅃Ι.

SYSTEMS USING ADDITION AND SUBTRACTION

Roman System

Around 200 B.C., the Romans also developed a system of numeration that used grouping symbols in conjunction with addition. This system also utilized an abacus to perform computation. In the following table, you will find the standardized numerals used in the Roman system.

1	5	10	50	100	500	1,000
I	V	X	L	C	D	M
	5000	10,000	50,000	100,000	500,000	1,000,000
	$\overline{V}$	$\overline{X}$	$\overline{L}$	$\overline{C}$	$\overline{D}$	$\overline{M}$

In this system, the value of a numeral was originally obtained by adding the value of each symbol. Later, Roman numerals also included subtraction. Subtraction was used when the numerals were written with "4's" and "9's"—that is, 4(IV), 9 (IX), 40 (XL), 90 (XC), 400 (CD), 900 (CM), and so on. With these numbers, the numeral representing a smaller number is placed before the numeral that represents a larger number. This indicates that the smaller numeral is to be subtracted from the larger one.

EXAMPLE 5

What number is represented by $\overline{D}M\overline{V}CCLIV$ in Roman numerals?

Solution: Using the table above, we get

$$\begin{aligned} \overline{D} &= 500{,}000 \\ M\overline{V} &= 5000 - 1000 = 4000 \\ CC &= 100 + 100 = 200 \\ L &= 50 \\ IV &= 5 - 1 = 4. \end{aligned}$$

Thus,

$$\overline{D}M\overline{V}CCLIV = 500{,}000 + 4000 + 200 + 50 + 4 = 504{,}254.$$

EXAMPLE 6

Write 1989 in the Roman numeration system.

Solution:

$$\begin{aligned} 1989 &= 1000 + 900 + 80 + 9 \\ 1000 &= M \\ 900 &= CM \\ 80 &= LXXX \\ 9 &= IX \end{aligned}$$

This gives 1989 = MCMLXXXIX.

SYSTEMS USING ADDITION AND MULTIPLICATION

Traditional Chinese System

The traditional Chinese system of numeration appearing in the Han dynasty around 200 B.C. also used grouping symbols and addition of numerals to represent numbers. However, instead of repeating a symbol when many of the same symbols are needed, multiplication factors are placed above the numeral. Numerals are written vertically with the following symbols.

1	2	3	4	5	6	7	8	9
一	二	三	四	五	六	七	八	九

10	100	1000	10,000
十	百	千	万

EXAMPLE 7

Find the number represented by the following.

三
千
六
百
一
十
八

Solution:

三 千 $3 \times 1000 = 3000$

六 百 $6 \times 100 = 600$

一 十 $1 \times 10 = 10$

八 8

This gives a total of 3618.

Notice that each digit of the numeral 3618, except for the units digit, is represented by two characters in the Chinese system.

EXAMPLE 8

Write 453 in the traditional Chinese system.

Solution:

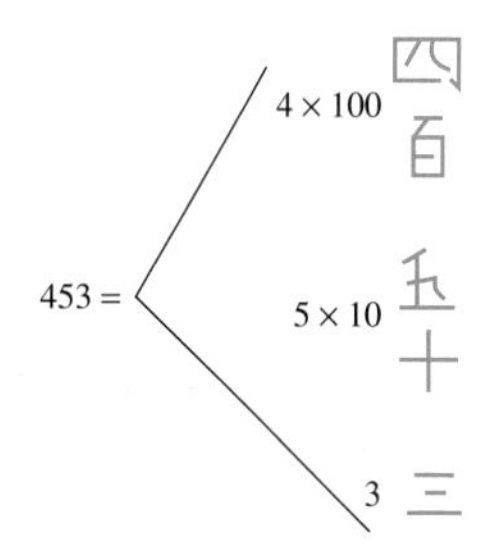

Ionic Greek System

The Greeks (c. 450 B.C.) used the 24 letters of their alphabet along with three ancient Phoenician letters for 6, 90, and 900, to represent numbers. Originally, the capital Greek letters were used, but they gave way to the lowercase letters. The symbols used to represent numbers in the Ionic Greek system are shown in the chart below.

1	α	10	ι	100	ρ
2	β	20	κ	200	σ
3	γ	30	λ	300	τ
4	δ	40	μ	400	υ
5	ϵ	50	ν	500	ϕ
6	ς	60	ξ	600	χ
7	ζ	70	o	700	ψ
8	η	80	π	800	ω
9	θ	90	Q	900	T

To represent a number from 1 to 999, write the appropriate symbols next to each other; for example, $\pi\delta = 84$ and $\omega\kappa\gamma = 823$. To obtain numerals for the multiples of 1000, place a prime to the left of the symbols for 1 to 9 to signify that it is multiplied by 1000; for example,

$$'\alpha = 1000 \qquad '\zeta = 7000 \qquad '\theta = 9000$$

One way to obtain the numerals that represent the multiples of 10,000 is to place the symbols for 1 to 9 above the letter M; for example,

$$\overset{\alpha}{\mathrm{M}} = 10{,}000 \qquad \overset{\beta}{\mathrm{M}} = 20{,}000 \qquad \overset{\theta}{\mathrm{M}} = 90{,}000$$

A system in which many different symbols are used to represent the digits is called a ciphered system. This type of system allows numbers to be written in a simple compact form but requires memorization of many different symbols.

EXAMPLE 9

What number is represented by $\overset{\eta}{\mathrm{M}}\,'\theta\phi\xi\epsilon$?

Solution:

$$\begin{aligned} \overset{\eta}{\mathrm{M}} &= 80{,}000 \\ '\theta &= 9000 \\ \phi &= 500 \\ \xi &= 60 \\ \epsilon &= 5 \end{aligned}$$

This gives 80,000 + 9000 + 500 + 60 + 5 = 89,565.

EXAMPLE 10

Write 2734 in the Ionic Greek system.

Solution:

$$2734 = 2000 + 700 + 30 + 4 = {}'\beta\psi\lambda\delta$$

SYSTEMS USING PLACE VALUES

Babylonian System

The most advanced numeration systems are those that not only use symbols, addition, and multiplication, but also give a certain value to the position a numeral occupies. The Babylonian system (c. 2300 B.C.) is an example of this type of system. The sexagesimal system, based on 60, uses only two symbols, formed by making marks on wet clay with a stick.

Groups of numerals separated from each other by a space signify that each group is associated with a different place value. Groups of symbols are given the place values from right to left. The place values are

$$60^0 = 1, \quad 60^1 = 60, \quad 60^2 = 3600, \quad 60^3 = 216{,}000, \quad \ldots$$

For example, means

$$\begin{array}{ccccccc} 12 \times 60^2 & + & 31 \times 60^1 & + & 23 \times 60^0 & & \\ 12 \times 3600 & + & 31 \times 60 & + & 23 \times 1 & & \\ 43{,}200 & + & 1860 & + & 23 & = & 45{,}083 \end{array}$$

The Babylonian system, however, did not contain a symbol for zero to indicate the absence of a particular place value. Some Babylonian tablets have a larger gap between numerals or the insertion of the symbol , which indicates a missing place value.

EXAMPLE 11

Find the number represented by

Solution: The groupings represent 2, 11, 0, and 34, which gives

$$\begin{array}{ccccccccc} 2 \times 60^3 & + & 11 \times 60^2 & + & 0 \times 60^1 & + & 34 \times 60^0 & & \\ 2 \times 216{,}000 & + & 11 \times 3600 & + & 0 \times 60 & + & 34 \times 1 & & \\ 432{,}000 & + & 39{,}600 & + & 0 & + & 34 & = & 471{,}634 \end{array}$$

EXAMPLE 12

Represent 4507 in the Babylonian numeration system.

Solution: The place values of the Babylonian system that are less than 4507 are 3600, 60, and 1. To determine how many groups of each place value are contained in 4507, we can use the division scheme shown below.

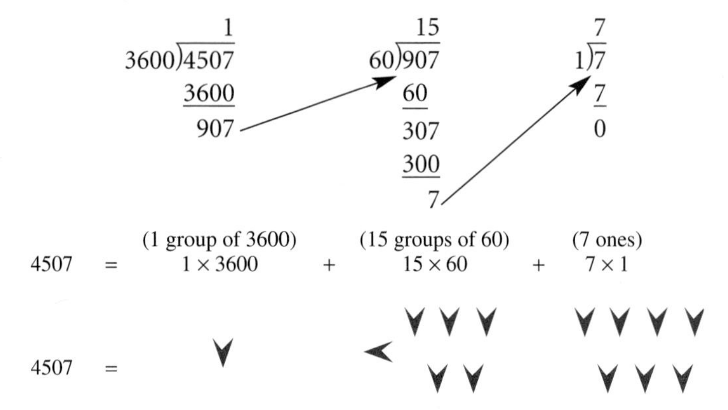

Mayan System

In about 300 B.C., the Mayan priests of Central America also developed a placevalue numeration system. Their system was an improvement on the Babylonian system because it was the first to have a symbol for zero. The Mayan system is based on 20 and 18. It uses the numerals shown below.

0	5	10	15
1	6	11	16
2	7	12	17
3	8	13	18
4	9	14	19

The numerals are written vertically, with the place value assigned from the bottom of the numeral to the top of the numeral. The positional values are

$$20^0 = 1, \quad 20^1 = 20, \quad 18 \times 20^1 = 360,$$
$$18 \times 20^2 = 7200, \quad 18 \times 20^3 = 144{,}000, \quad \ldots$$

Instead of the third position having a place value of 20^2, the Mayans gave it a value of 18×20^1. This was probably done so that the approximate number of days in a year, 360 days, would be a basic part of the numeration system.

For example, in the Mayan system, 168,599 is written as

$$\rightarrow 1 \times (18 \times 20^3) = 1 \times 144{,}000 = 144{,}000$$
$$\rightarrow 3 \times (18 \times 20^2) = 3 \times 7200 = 21{,}600$$
$$\rightarrow 8 \times (18 \times 20^1) = 8 \times 360 = 2{,}880$$
$$\rightarrow 5 \times 20^1 = 5 \times 20 = 100$$
$$\rightarrow 19 \times 20^0 = 19 \times 1 = 19$$

168,599

The Mayan system does have a zero, but because of the use of 18 in the third position, its place-value feature is irregular.

EXAMPLE 13

Find the number represented by

Solution:

$$= 4 \times 7200 = 28{,}800$$
$$= 0 \times 360 = 0$$
$$= 7 \times 20 = 140$$
$$= 12 \times 1 = 12$$

This gives a total of 28,952.

EXAMPLE 14

Write 17,525 in the Mayan numeration system.

Solution: The place values less than 17,525 in the Mayan system are 7200, 360, 20, and 1. To determine how many groups of each place value are contained in 17,525, we use the division scheme below.

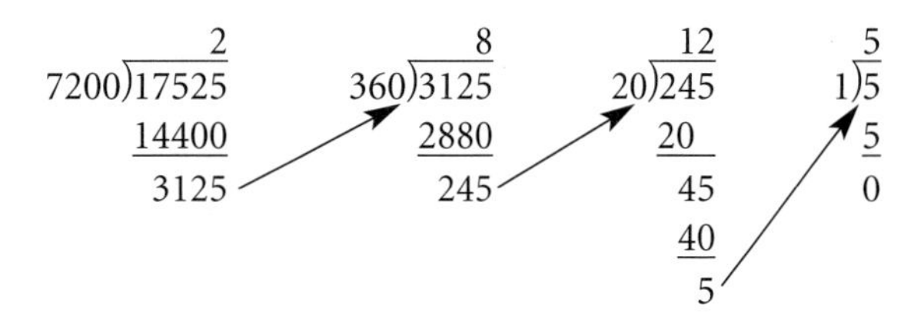

(2 groups of 7200) (8 groups of 360) (12 groups of 20) (5 ones)

$2 \times 7200 \rightarrow$

$8 \times 360 \rightarrow$

$12 \times 20 \rightarrow$

$5 \times 1 \rightarrow$

Our system of numeration originated with the Hindus in India in about 150 B.C., with its base 10 place-value feature and zero placeholder being introduced before 628. By 900, this Hindu-Arabic numeration system had reached Spain, and by 1210, it had been spread to the rest of Europe by traders on the Mediterranean Sea and scholars who attended universities in Spain. By 1479, the digits appeared in the form that is used today.

The Hindu-Arabic system of numeration made computation a reasonable task. However, the "abacists," those that used Roman numerals and calculated with their counting board abacus, were opposed to the "algorists," who used and computed with the Hindu-Arabic numerals. In fact, in 1299, the city-state of Florence in Italy outlawed the use of the Hindu-Arabic system. Many banks in Europe also forbade the use of these numerals because they were easily forged and/or altered on bank drafts. However, by 1500 the Hindu-Arabic system had won the battle and became the prevalent numeration system. In the next section, we look more closely at the Hindu-Arabic system of numeration.

section 1.1

PROBLEMS ○ Explain ○ Apply ○ Explore

○ Explain

1. What is meant by a system of numeration that uses addition and subtraction? Give some examples.
2. What is meant by a system of numeration that uses addition and multiplication? Give some examples.
3. Compare numeration systems that use addition and subtraction with those that use addition and multiplication. What are the advantages and disadvantages of each type of system?
4. What is meant by a system that uses place values? Give some examples.
5. Compare numeration systems that use addition and multiplication with those that use place values. What are the advantages and disadvantages of each type of system?
6. What are some advantages of a system that uses place values over a system that does not use place values?
7. If used today, which ancient systems of numeration would require that you invent new symbols for larger numbers? Which systems would not? Explain.
8. After studying seven ancient systems of numeration, you can better understand the advantages of the Hindu-Arabic system we currently use. What are some of these advantages?
9. Explain the statement "For any number, there are many numerals."
10. Which of the systems of numeration described in this section would you prefer to use? Give reasons for your choice.

○ Apply

What number is represented by each Egyptian hieroglyphic numeral?

11. 𓂭 𓂭 𓆐 𓁨
12. 𓂭 𓍢 𓍢 𓍢
13. 𓆼 𓂭 𓍢 𓏺

What number is represented by each Attic numeral?

14. Μ 𐅅 𐅄
15. Χ Η 𐅃 Ι
16. 𐅇 𐅇 Μ Η Η

What number is represented by each Roman numeral?

17. DCCXXXIV

18. $\overline{\text{C}}$MCM

19. $\overline{\text{XL}}$MMCDLVII

What number is represented by each traditional Chinese numeral

What number is represented by each Ionic Greek numeral?

23. $\lambda\zeta$

24. $\phi\mu\delta$

25. $\overset{\beta}{\text{M}}\,{}'\epsilon\xi\beta$

What number is represented by each Babylonian numeral?

26. ≺ ∨

27. ≺ ∨

28. ≺≺∨∨∨ ≺∨∨∨∨ ≺≺∨

What number is represented by each Mayan numeral?

29. •
⊖
•••

30. $\overset{\bullet}{—}$
$\overset{\bullet\bullet\bullet}{—}$
$\overset{}{=}$
$\overset{\bullet\bullet\bullet}{=}$
$\overset{\bullet}{=}$

31. $\overset{\bullet\bullet\bullet\bullet}{\equiv}$
$\overset{\bullet}{\equiv}$
$\overset{\bullet}{—}$
••••
⊖
⊖

32. Represent the speed of sound (750 mph) in each of the following numeration-systems:

a) Egyptian hieroglyphic

b) Roman

c) Ionic Greek

d) Mayan

33. Represent the number of minutes in a day (1440 min) in each of the following numeration systems:

a) Attic

b) Traditional Chinese

c) Babylonian

Explore

34. a) Represent the number of chairs shown below in each of the systems of numeration of Problem 33.

b) Represent the number of legs on the chairs shown below in each of the systems of numeration of Problem 32.

35. Suppose that the Egyptian numeral 𓂭𓁨𓁨𓍢𓍢𓍢𓎆||||| were converted by people from different cultures into a numeral in their own systems of numeration. How would the Egyptian numeral be represented in the following systems of numeration?

a) Roman system

b) Traditional Chinese system

c) Ionic Greek system

d) Babylonian system

e) Mayan system

36. The birth date of one of the authors is 11-8-1944. Written in Roman numerals, that would be XI-VIII-MCMXLIV.

a) Write that birth date in Mayan, traditional Chinese, and Egyptian.

b) Write your birth date in three of the systems of numeration described in this section.

37. Suppose ancient cultures had telephones. How would you write the phone number 1-800-YEA-MATH in three of the systems of numeration described in this section?

38. Using the symbols * for 1, / for 5, ¥ for 10, ∇ for 50, $ for 100, and □ for 1000, design a multiplicative system to represent numbers. Explain how the system works and represent 324 and 1995 in the system.

39. Using the symbols → for 1, ← for 9, ↓ for 81, and ↑ for 729, design an additive system of numeration. Use that system to represent the

a) number of chairs in Problem 34.

b) number of days in a leap year.

c) number of feet in a mile.

d) number of grams in a kilogram.

40. Using the existing symbols, modify the Roman system so that it is a multiplicative system. Explain your modifications and give three examples of numerals using your system.

41. Modify the existing Mayan system of numeration to create a base 10 system that uses the same format and symbols. Explain your modifications and give three examples of numerals using your system.

42. Using the existing symbols, modify the Attic Greek system so that it is a multiplicative system. Explain your modifications and give three examples of numerals using your system.

43. When adding in the Hindu-Arabic system, we use a "carrying" procedure. For example, to find 36 + 48 we add the digits in the one's place (6 + 8) and get 14. We write the 4 in the one's place of the sum and carry the 1 to the ten's place as shown below.

$$\begin{array}{r} {}^{1} \\ 36 \\ +\,48 \\ \hline 84 \end{array}$$

For each ancient system of numeration listed below, explain how this carrying procedure would work and use the procedure to add 178 + 47.

a) Egyptian

b) Roman

c) Traditional Chinese

section 1.2

The Hindu-Arabic System and Fractions

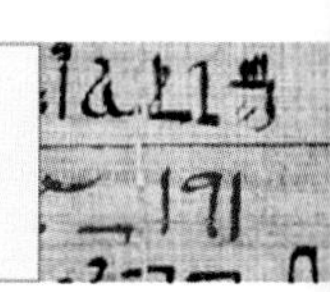

The Hindu-Arabic system is the numeration system that is used in the United States and many other parts of the world. It is also called the decimal system, from the Latin *deci*, meaning tenth. (Note that the Chinese system of numeration could also be called a decimal system.) The base of the Hindu-Arabic system is 10, and the symbols used in the system are the digits 0, 1, 2, 3, 4, 5, 6, 7, 8, and 9. The position a digit holds in a numeral gives it a certain value. The numeral 1,389,260,547 has place values as follows:

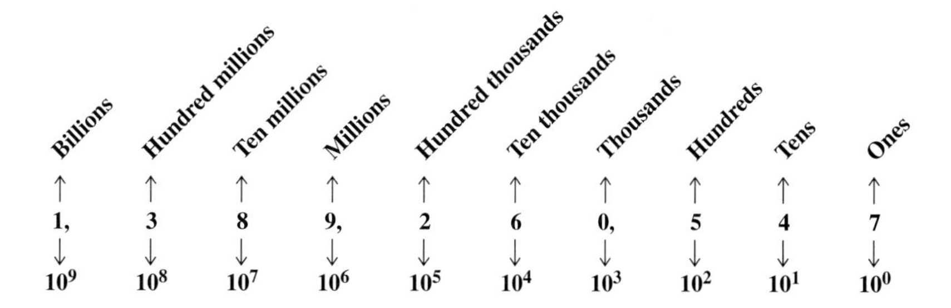

In expanded form,

$$\begin{aligned} 1{,}389{,}260{,}547 = {} & 1 \times 10^9 + 3 \times 10^8 + 8 \times 10^7 + 9 \times 10^6 \\ & + 2 \times 10^5 + 6 \times 10^4 + 0 \times 10^3 + 5 \times 10^2 \\ & + 4 \times 10^1 + 7 \times 10^0 \end{aligned}$$

I Saw the Figure 5 in Gold by Charles Henry Demuth (The Metropolitan Museum of Art, Alfred Stieglitz Collection, 1949. (49.59.1) Photograph "1986 The Metropolitan Museum of Art) gives a modern artist's visual interpretations of the Hindu-Arabic digit five.

The position of a digit tells us what it really represents. The 9 represents 9 millions (9,000,000), the 6 represents 6 ten thousands (60,000), the 4 represents 4 tens (40), and so on.

EXAMPLE 1

Write 4,175,280 in expanded form.

Solution:

$$\begin{aligned}4{,}175{,}280 &= 4{,}000{,}000 + 100{,}000 + 70{,}000 + 5000 + 200 + 80\\ &= 4 \times 10^6 + 1 \times 10^5 + 7 \times 10^4 + 5 \times 10^3\\ &\quad + 2 \times 10^2 + 8 \times 10^1\end{aligned}$$

EXAMPLE 2

In the numeral 576,239, what do the 5, 6, and 2 represent?

Solution:

5 represents 5 hundred thousands (500,000).

6 represents 6 thousands (6000).

2 represents 2 hundreds (200).

DECIMAL FRACTIONS

The decimal system also gives an efficient means of representing numbers that are less than a whole, numbers that fall between 0 and 1. Such numbers can be represented in two different forms, as a fraction or as a decimal (decimal fraction). For example, the shaded portion of the block shown below can be represented by the fraction 1/2 or the decimal 0.5.

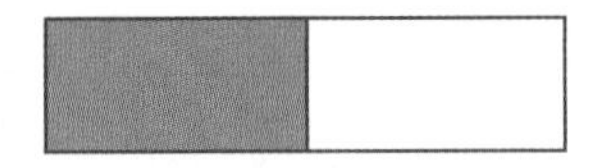

The decimal numbers simply extend the place-value system by using negative powers of 10. For example, 0.943271 means

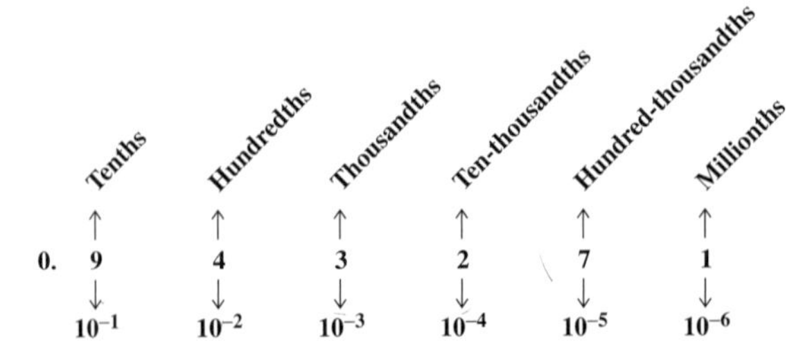

In expanded form,

$$\begin{aligned}0.943271 &= 9 \times 10^{-1} + 4 \times 10^{-2} + 3 \times 10^{-3} + 2 \times 10^{-4}\\ &\quad + 7 \times 10^{-5} + 1 \times 10^{-6}\\ &= \frac{9}{10} + \frac{4}{100} + \frac{3}{1000} + \frac{2}{10{,}000} + \frac{7}{100{,}000} + \frac{1}{1{,}000{,}000}\end{aligned}$$

The position a digit holds to the right of the decimal point tells us what value it represents. The 9 represents 9 tenths (9/10), the 3 represents 3 thousandths (3/1000), the 1 represents 1 millionth (1/1,000,000), and so on.

EXAMPLE 3

In the numeral 1.7312, what do the digits 7, 3, and 2 represent?

Solution:

The 7 represents 7 tenths (7/10).
The 3 represents 3 hundredths (3/100).
The 2 represents 2 ten-thousandths (2/10,000).

EXAMPLE 4

Write 75.324 in expanded form.

Solution:

$$\begin{aligned} 75.324 &= 70 + 5 + 3/10 + 2/100 + 4/1000 \\ &= 7 \times 10^1 + 5 \times 10^0 + 3 \times 10^{-1} \\ &\quad + 2 \times 10^{-2} + 4 \times 10^{-3} \end{aligned}$$

Some of the ancient systems of numeration discussed in the previous section also had methods of representing fractions. Although these systems were not as advanced as the decimal system, a look at them should prove interesting. You will have to refer to the numeration systems in the previous section to obtain the symbols used in ancient fractions.

EGYPTIAN HIEROGLYPHIC FRACTIONS

In this system, the sign for a mouth, $\circledcirc$ which in this context meant a part, was placed above a numeral to allow the Egyptians to represent fractions with a numerator of 1.

$$\overset{\circledcirc}{\text{III}} = \frac{1}{3} \qquad \overset{\circledcirc}{\cap\text{II}} = \frac{1}{12} \qquad \overset{\circledcirc}{\varrho\varrho\cap} = \frac{1}{210}$$

Any fraction whose numerator was not equal to 1 was represented as the sum of distinct fractions whose numerators were equal to 1.

$$\overset{\circledcirc}{\cap}\;\overset{\circledcirc}{\text{II}} = \frac{1}{10} + \frac{1}{2} = \frac{3}{5} \qquad \overset{\circledcirc}{\text{III}}\;\overset{\circledcirc}{\text{IIIII}}\;\overset{\circledcirc}{\cap} = \frac{1}{3} + \frac{1}{5} + \frac{1}{10} = \frac{19}{30}$$

BABYLONIAN FRACTIONS

The use of the symbols ⸜⸜ in the initial position of a numeral indicated that the number being represented was a fraction. A group of symbols following the ⸜⸜ became the numerator of a fraction, and successive powers of 60 (60, 3600, 216,000, etc.) were understood to be the denominator for each group of symbols. The Babylonian fraction being represented was the sum of the individual fractions.

⸌⸌ Y $= \frac{1}{60}$ ⸌⸌ < Y Y Y < Y $= \frac{13}{60} + \frac{11}{3600} = \frac{791}{3600}$

ROMAN FRACTIONS

For the Romans, fractions were mainly used in connection with their system of weights: 1 *as* (pound) = 12 *unciae* (ounces). Hence, their fractions were limited to parts of 12 (12ths). The table below gives the fractions of the Romans.

.	or	–	→ 1/12	S .	or	S –	→ 7/12
. .	or	=	→ 2/12 or 1/6	S . .	or	S =	→ 8/12 or 2/3
. . .	or	= –	→ 3/12 or 1/4	S . . .	or	S = –	→ 9/12 or 3/4
. . . .	or	= =	→ 4/12 or 1/3	S. . . .	or	S = =	→ 10/12 or 5/6
. . . . .	or	= = –	→ 5/12	S	or	S = = –	→ 11/12
		S	→ 6/12 or 1/2			I	→ 12/12 or 1

IONIC GREEK FRACTIONS

The Greek scheme for representing fractions involved using a prime (′) on the right side of the numeral. Fractions with a numerator of 1 were written with a single prime to the right of the numeral.

$$\theta' = \frac{1}{9} \qquad \pi\epsilon' = \frac{1}{85} \qquad \rho' = \frac{1}{100}$$

Fractions with numerators greater than 1 were represented with the numerator written as a normal numeral and the denominator written twice with a prime to the right of each.

$$\epsilon\eta'\eta' = \frac{5}{8} \qquad \iota\alpha\lambda\epsilon'\lambda\epsilon' = \frac{11}{35}$$

EXAMPLE 5

For each fraction below, write its equivalent fraction in the Hindu-Arabic system.

a) Egyptian: $\overset{\circ}{\text{III}}$ $\overset{\circ}{\cap\text{I}}$ **b)** Babylonian: ⸌⸌ Y Y <<<

c) Roman: S . . **d)** Greek: $\theta\pi\epsilon'\pi\epsilon'$

Solution:

a) 1/3 + 1/11 = 11/33 + 3/33 = 14/33
b) 2/60 + 30/3600 = 4/120 + 1/120 = 5/120 = 1/24
c) 8/12 = 2/3
d) $\theta = 9$, $\pi\epsilon = 85$, so the fraction is 9/85

section 1.2

PROBLEMS • Explain • Apply • Explore

Explain

1. What is a decimal fraction? How is it different from a fraction?
2. How are exponents used in the Hindu-Arabic system of numeration?
3. What is the expanded form of a Hindu-Arabic numeral?
4. What is the major difficulty in writing a fraction such as 3/8 in Egyptian fractions?
5. How is the Babylonian system of fractions similar to the decimal system of fractions?
6. Compare Egyptian, Babylonian, Roman, and Ionic Greek fractions. Which system is easiest to use? Explain.

Apply

In Problems 7–18, write each number in expanded form.

7. 139
8. 0.53
9. 437.15
10. 1,032,742
11. 0.314
12. 5.23
13. 543,867
14. 0.03874
15. 53.171
16. 5083
17. 0.62193
18. 1.043

19. In Problems 7–18, what does the digit 3 represent?

In Problems 20–23, write the equivalent Hindu-Arabic fraction for each Egyptian fraction.

20. 𓂋|| 𓂋|||| 𓂋∩
21. 𓂋∩ 𓂋||||
22. 𓂋𓍢∩ 𓂋𓍢
23. 𓂋𓆼 𓂋𓍢𓍢𓍢 𓂋∩∩

In Problems 24–27, write the equivalent Hindu-Arabic fraction for each Babylonian fraction given.

24. \\ <<<<
25. \\ < ∨ ∨
26. \\ ∨ ∨ ∨ ∨ <<
27. \\ < < <

In Problems 28–31, write the equivalent Hindu-Arabic fraction for each Roman fraction.

28. . .
29. = =
30. S .
31. S = –

In Problems 32–35, write the equivalent Hindu-Arabic fraction for each Greek fraction.

32. $\kappa\delta'$

33. $\xi\gamma'$

34. $\iota\lambda\beta'\lambda\beta'$

35. $\mu\theta\phi\alpha'\phi\alpha'$

Explore

The sum of a whole number and a fraction is called a mixed number. In the Hindu-Arabic system, $5 + 2/3$ is written as $5\frac{2}{3}$. In Problems 36–39, find the mixed numbers represented by each ancient numeral.

36. Egyptian: 𓆼𓍢 𓂋 over 𓎆 𓂋 over 𓏽

37. Babylonian: < ∨ ∨ < < ∨ ∨ ∨ ∖∖< ∨ ∨

38. Roman: X X I V S =

39. Greek: $\rho\alpha\lambda\delta'$

Changing a Hindu-Arabic fraction into the corresponding Egyptian fraction requires that you represent the Hindu-Arabic fraction as the sum of distinct fractions that have a numerator of 1. In Problems 40–44, find the Egyptian fractions.

40. 3/4

41. 5/12

42. 7/10

43. 8/15

44. 31/100

In Problems 45–48, write $1/3 + 1/2 = 5/6$ in the ancient system of numeration.

45. Egyptian

46. Babylonian

47. Roman

48. Greek

49. How would a Roman merchant write the following calculation?

$$12 \times 3\frac{3}{4} = 45$$

50. How would an ancient Egyptian merchant write the following calculation?

$$20 \times 3\frac{3}{5} = 72$$

51. How would an Ionic Greek merchant write the following calculation?

$$120 \times 3\frac{3}{4} = 450$$

52. How would a Babylonian merchant write the following calculation?

$$70 \times 3\frac{3}{5} = 252$$

53. Devise a system that would allow the Romans to write fractions other than 12ths. Give three examples and explain how your system works.
54. Devise a system that would allow the Mayans to write fractions. Your method should be consistent with the existing Mayan system. Give three examples and explain how your system works.
55. The traditional Chinese system of numeration has a way to represent fractions. Do some research to find out how this is done. Give three examples and explain how the system works.

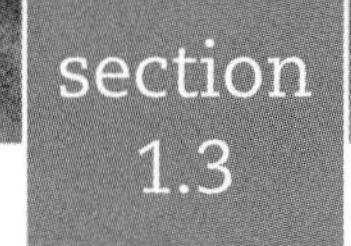

Numeration Systems with Other Bases

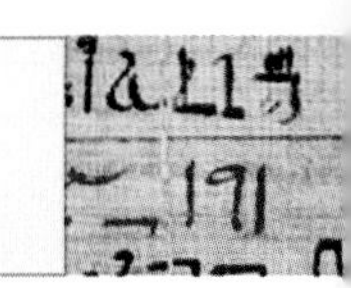

The decimal system uses the powers of 10 to determine the place value of each digit used in a numeral. The base of 10 is the result of human beings having ten fingers. However, as we saw in Section 1.1, other place-value systems did not use a base of 10. The Babylonians had a system based on 60, whereas the Mayan system was based on 20. Primitive tribes have been discovered that had a system of numeration based on 5, the number of fingers on one hand. The Duodecimal Society of America in the 1960s advocated a change to a base of 12. Computers, on the other hand, use a base of 2, since an electric pulse is in one of two states—on or off. If animals could develop a system of numeration, a horse might use a base 4 system, an octopus a base 8, an ant a base 6, and so on. In this section, we investigate how to write numbers in different bases and how to convert a numeral written in one base to the corresponding numeral in another base.

THE PLACE-VALUE SYSTEM FOR ANY BASE

The first component of any place-value system is its base. The base of a numeration system is a whole number that is larger than 1. The integer powers of the base give each position in a numeral its **place value**. If we let a dot separate the whole number part and the fractional part of a number, for any base b to the left of the dot the place values are the nonnegative integer powers of the base and to the right of the dot the place values are the negative powers of the base.

Whole-number Part					Fractional Part			
#	#	#	#		#	#	#	#
↓	↓	↓	↓		↓	↓	↓	↓
$\ldots b^3$	b^2	b^1	b^0	.	b^{-1}	b^{-2}	b^{-3}	$b^{-4}\ldots$

The second component of a place-value system is the set of digits. The digits are symbols that represent the quantities from 0 to 1 less than the base. The base determines the number of symbols that are in the system. In the decimal system, the base is 10, and it has 10 digits (0, 1, 2, 3, 4, 5, 6, 7, 8, and 9). The following is a summary of some numeration systems with various bases. It includes the base, the digits, and the place value for whole numbers from the right to the left of the numeral. Notice that base 12 and base 16 require the use of letters to represent some numbers greater than 9. For example, the letter A represents the number 10.

System	Base	Digits	Place Values
Binary	2	0, 1	1, 2, 4, 8, 16, . . .
Quintary	5	0, 1, 2, 3, 4	1, 5, 25, 125, 625, . . .
Octal	8	0, 1, 2, 3, 4, 5, 6, 7	1, 8, 64, 512, 4096, 32,768, 262,144, . . .
Duodecimal	12	0, 1, 2, 3, 4, 5, 6, 7, 8, 9, A(10), B(11)	1, 12, 144, 1728, 20,736, 248,832, . . .
Hexadecimal	16	0, 1, 2, 3, 4, 5, 6, 7, 8, 9, A(10), B(11), C(12), D(13), E(14), F(15)	1, 16, 256, 4096, 65,536, 1,048,576, . . .

When you read a numeral in a base other than 10, read each digit separately. For example, the numeral 123_5 is read "one two three, base five" and not "one hundred twenty-three, base five." The reason is that the concept of "hundreds" is part of the base 10 system and therefore should not be used in other bases.

CONVERTING TO A DECIMAL NUMERAL

To understand what number or amount is being represented by a numeral in a base other than 10, we need to convert it to the system we are familiar with: the decimal system. For example, the numeral 23014 in base 5, written 23014_5, represents an amount. To understand what amount that is, we will convert it to a decimal numeral.

EXAMPLE 1

Write 23014_5 in base 10.

Solution: Use the powers of 5 for the place value for each digit in the numeral.

$2\ 3\ 0\ 1\ 4_5$

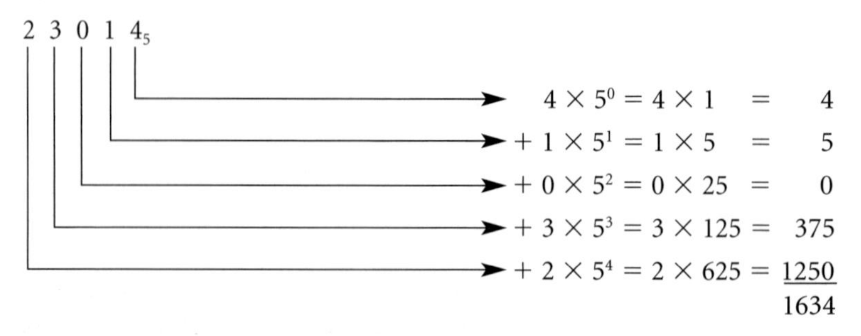

Thus, $23014_5 = 1634_{10}$.

EXAMPLE 2

Write $17A6_{12}$ in base 10.

Solution: Use the powers of 12 for the place value of each digit in the numeral.

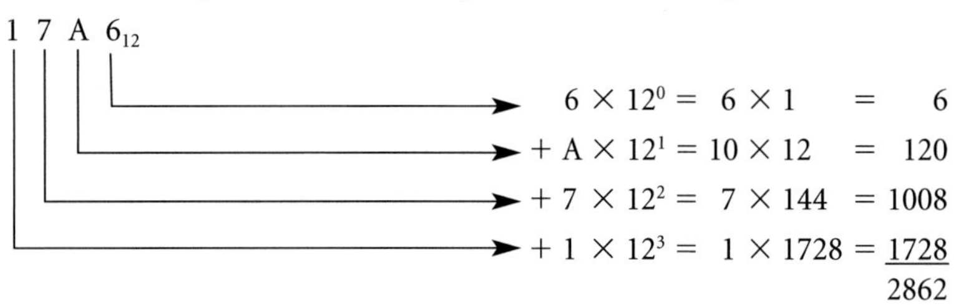

Thus, $17A6_{12} = 2862_{10}$.

EXAMPLE 3

Write $101\ 101_2$ in base 10.

Solution: Use the powers of 2 for the place values of each digit in the numeral.

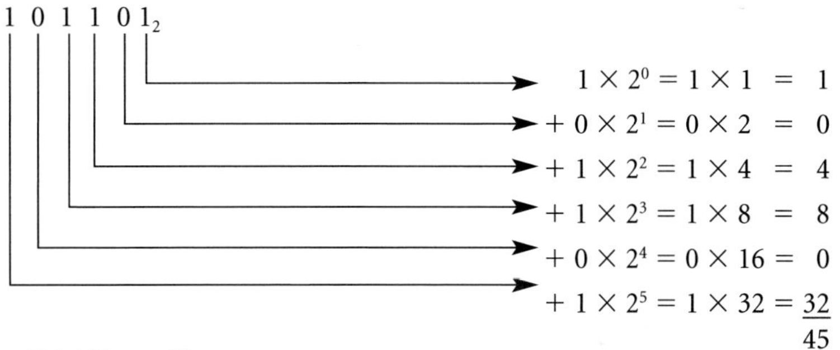

Thus, $101\ 101_2 = 45_{10}$.

EXAMPLE 4

Write $86DC0_{16}$ in base 10.

Solution: Use the powers of 16 for the place values of each digit in the number.

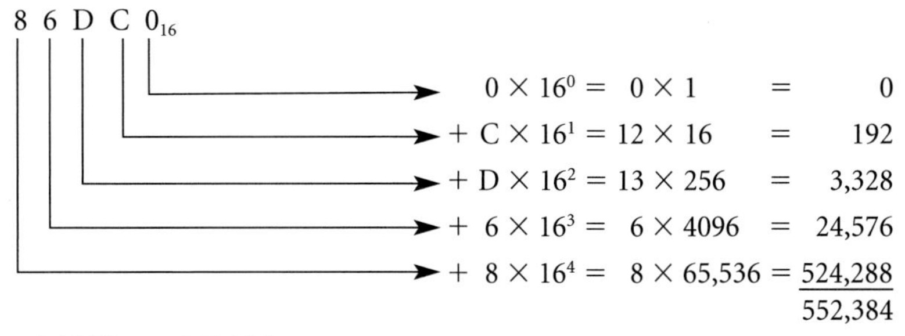

Thus, $86DC0_{16} = 552{,}384_{10}$.

CONVERTING DECIMAL NUMERALS TO OTHER BASES

To convert a decimal number to another base, we need to find how many groups of each appropriate place value are contained in the decimal number. The division scheme shown in the examples that follow will allow you to convert a base 10 numeral into another base.

EXAMPLE 5

Write 89 in base 5.

Solution: The powers of 5 that are less than 89 are 25, 5, 1. We need to find how many groups of each of those place values are contained in 89. We can do that by using the division scheme shown below.

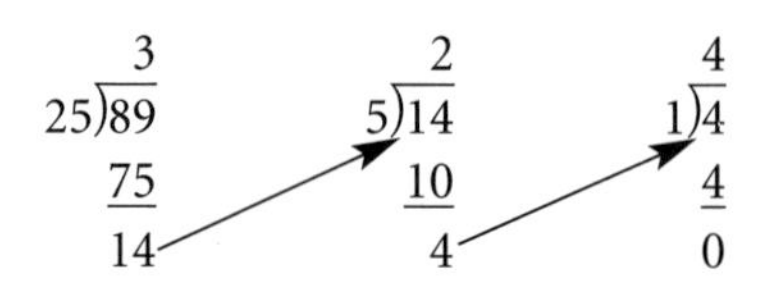

The number 89 contains 3 groups of 25, 2 groups of 5, and 4 groups of 1.

$$
\begin{aligned}
89_{10} &= 3 \times 25 + 2 \times 5 + 4 \times 1 \\
&= 3 \times 5^2 + 2 \times 5^1 + 4 \times 5^0 \\
&= 324_5
\end{aligned}
$$

EXAMPLE 6

Write 204 in binary.

Solution: The powers of 2 that are less than 204 are 128, 64, 32, 16, 8, 4, 2, and 1. Use the division scheme of Example 5.

1	1	0	0	1	1	0	0
128)204	64)76	32)12	16)12	8)12	4)4	2)0	1)0
128	64	0	0	8	4	0	0
76	12	12	12	4	0	0	0

$$
\begin{aligned}
204_{10} &= 1 \times 128 + 1 \times 64 + 0 \times 32 + 0 \times 16 \\
&\quad + 1 \times 8 + 1 \times 4 + 0 \times 2 + 0 \times 1 \\
&= 1 \times 2^7 + 1 \times 2^6 + 0 \times 2^5 + 0 \times 2^4 + 1 \times 2^3 + 1 \times 2^2 \\
&\quad + 0 \times 2^1 + 0 \times 2^0 \\
&= 11\,001\,100_2
\end{aligned}
$$

EXAMPLE 7

Write the decimal number 10,000 in base 8.

Solution: The powers of 8 that are less than 10,000 are 4096, 512, 64, 8, and 1. Use the division scheme of previous examples.

2	3	4	2	0
4096)10000	512)1808	64)272	8)16	1)0
8192	1536	256	16	0
1808	272	16	0	

$$
\begin{aligned}
10{,}000 &= 2 \times 4096 + 3 \times 512 + 4 \times 64 + 2 \times 8 + 0 \times 1 \\
&= 2 \times 8^4 + 3 \times 8^3 + 4 \times 8^2 + 2 \times 8^1 + 0 \times 8^0 \\
&= 23{,}420_8
\end{aligned}
$$

EXAMPLE 8

Write the decimal number 40,600 in base 16.

Solution: The powers of 16 that are less than 40,600 are 4096, 256, 16, and 1. Use the division scheme again.

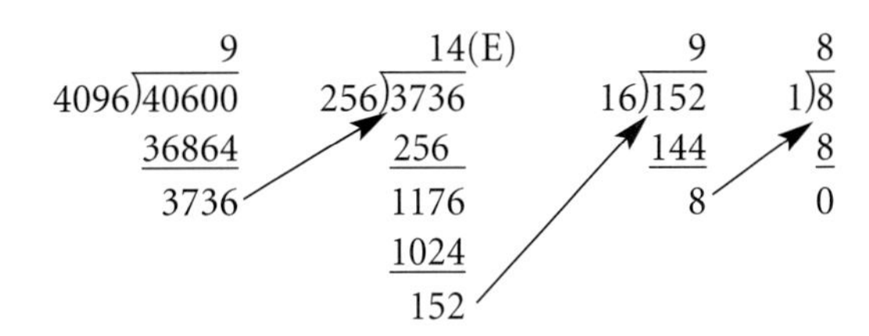

$$\begin{aligned}40{,}600 &= 9 \times 4096 + E \times 256 + 9 \times 16 + 8 \times 1\\ &= 9 \times 16^3 + E \times 16^2 + 9 \times 16^1 + 8 \times 16^0\\ &= 9E98_{16}\end{aligned}$$

FRACTIONS IN OTHER BASES

A base 10 number less than 1 can be represented by using a decimal point or negative integer powers of 10. For example,

$$\begin{aligned}0.2358 &= 2 \times 10^{-1} + 3 \times 10^{-2} + 5 \times 10^{-3} + 8 \times 10^{-4}\\ &= 2 \times \frac{1}{10} + 3 \times \frac{1}{100} + 5 \times \frac{1}{1000} + 8 \times \frac{1}{10{,}000}\end{aligned}$$

In other bases, the dot that is used to separate the whole number part of a number and the fractional part has different names. It is called the quintary point in base 5, the binary point in base 2, the octal point in base 8, and so on. However, the generic term used for the dot in any base is the **basimal** point. Fractional numbers can be represented in any base by using methods similar to those used with decimal numbers.

EXAMPLE 9

Write 0.2314_5 as a base 10 numeral.

Solution:

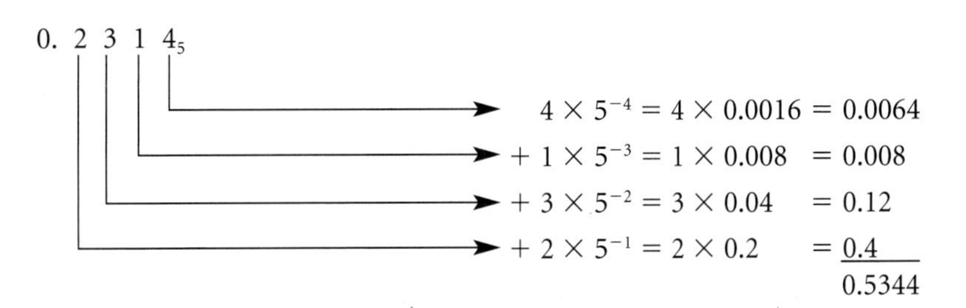

Thus, $0.2314_5 = 0.5344_{10}$.

EXAMPLE 10

Write 53.72_8 as a base 10 numeral.

Solution:

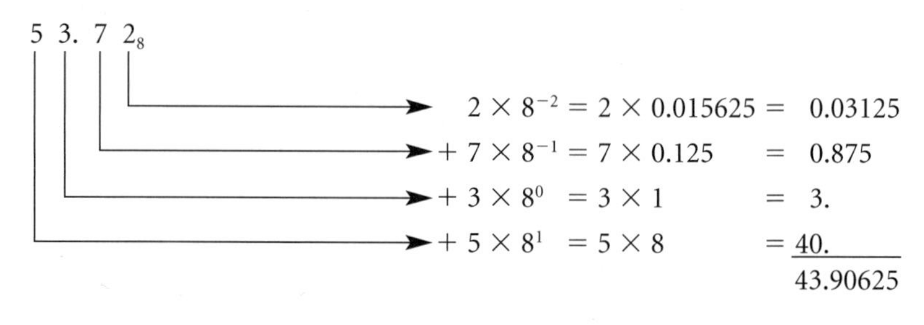

Thus, $53.72_8 = 43.90625_{10}$.

Looking back at the survey of the different systems of numeration, you can appreciate the advantages of the Hindu-Arabic system. First, the Hindu-Arabic system uses a small number of uncomplicated symbols. After memorizing these ten digits, a person can write any numeral. This is an improvement over the large number of symbols used by the Ionic Greeks and the elaborate characters of the Egyptian system.

Using too few characters also has drawbacks. While the Babylonian system and base 2 use only two characters, both methods require using the symbols many times to express a relatively small number. For example,

59 = ≺≺≺≺≺ YYYYY YYYY $= 111\,011_2$

The Hindu-Arabic system strikes a balance between using a large number of symbols and requiring a large number of digits to write a relatively small number.

A second advantage of the Hindu-Arabic system is its use of place value. Although initially more difficult to learn, a place-value system is advantageous for writing most large numbers. The advantages of a place-value system become clear when writing 888 in a system such as the Roman system:

$$888 = \text{DCCCLXXXVIII}$$

Finally, we leave it to your imagination to envision a long division problem such as $10{,}488 \div 23$ in any system other than the Hindu-Arabic system.

PROBLEMS ● Explain ● Apply ● Explore

● Explain

1. How many symbols are used to create numerals in base 10?
2. How many symbols would be needed to create numerals in base 17? What are they?

3. What is the common belief as to why modern cultures use a base 10 system?
4. Why should the number 135_7 be read "one three five base seven," rather than "one hundred thirty-five base 7"?
5. Describe in words the method for converting the numeral 12_8 into a decimal numeral.
6. Describe in words the method for converting the Hindu-Arabic numeral 12 into a numeral in base 8.

Apply

Write each of the following as a decimal numeral.

7. 302_5
8. 140.32_5
9. 101.101_2
10. $100\ 111\ 001_2$
11. 5610_8
12. $70\ 037.2_8$
13. $7A4.3_{12}$
14. $123ABC.D_{16}$
15. 253_7
16. 1068_9
17. 1111_2
18. 1111_5

Write the following decimal numerals in the specified base.

19. Write 401 as a numeral in base 5.
20. Write 186 as a numeral in base 4.
21. Write 990 as a numeral in base 8.
22. Write 88,888 as a numeral in base 8.
23. Write 777 as a numeral in base 7.
24. Write 7777 as a numeral in base 7.
25. Write 1860 as a numeral in base 12.
26. Write 4235 as a numeral in base 12.
27. Write 129 as a numeral in base 16.
28. Write 16,016 as a numeral in base 16.
29. Write 186,000 as a numeral in base 9.
30. Write 32 as a numeral in base 2.
31. Write 222 as a numeral in base 2.
32. Write 2222 as a numeral in base 2.

Represent the number of days in a leap year (366) in each of the following systems of numeration.

33. Quintary
34. Binary
35. Octal

Represent the number of pounds in a ton (2000) in each of the following systems of numeration.

36. Duodecimal
37. Hexadecimal
38. Base 6

Represent the number 1,000,000 in each of the following systems of numeration.

39. Quintary

40. Binary

41. Octal

42. Duodecimal

43. Hexadecimal

44. Base 6

Explore

In each pair of numerals, determine which one has a larger value.

45. 254_9 or 12202_3

46. $6C_{16}$ or 253_6

47. $101\ 101_2$ or 3033_4

48. 13.421_5 or 1.421_8

49. Solve the following $45_6 + 67_8 =$ ______ $_7$

50. Solve the following $23_4 + 23_5 =$ ______ $_6$

51. Suppose $200_B = 128$. Find B.

52. Suppose $301_B = 76$. Find B.

53. **a)** Write the numbers 45, 100, 200 in base 3 and base 9.
b) What do you notice about the number of digits in the numerals as the base increases?
c) Write each of the base 3 numerals in groups of two digits, starting at the right most digit. Compare these with the base 9 numerals. Do you notice the pattern? Explain the pattern.

54. **a)** Write the numbers 6, 18, and 45 in base 2, base 4 and base 8.
b) What do you notice about the number of digits in the numerals as the base increases?
c) Write each of the base 2 numerals in groups of two digits, starting at the right most digit. Compare these with the base 4 numerals. Do you notice the pattern? Explain the pattern.
d) Write each of the base 2 numerals in groups of three digits, starting at the right most digit. Compare these with the base 8 numerals. Do you notice the pattern? Explain the pattern.

55. Create a system of numeration for base 32. What are its digits? Write two 4-digit numerals in base 32 and determine the equivalent Hindu-Arabic numerals.

56. Create a system of numeration for base 64. What are its digits? Write two 4-digit numerals in base 64 and determine the equivalent Hindu-Arabic numerals.

57. Consider your telephone number as a 3-digit number followed by a 4-digit number. Write your telephone number in at least two different bases greater than 10. What do you notice about the number of digits in the telephone number when written in these bases?

58. Consider your telephone number as a 3-digit number followed by a 4-digit number. Write your telephone number in at least two different bases less than 10. What do you notice about the number of digits in the telephone number when written in these bases?

section 1.4

The Numbers of Technology

Binary numbers are intrinsic to the technology of the 21st century. Computers read data as electric pulses having low or high voltage levels. Disks and tapes store information using magnetic fields pointing forward or backward. Compact disks store data using tiny pits and non-pits on the surface of the disk. Recordable CDs store audio signals as dark marks and non-dark marks on the CD. In all these instances, the two states are associated with the digits of the binary system of numeration. The digit 1 is used for one state and the digit 0 is used for the other. Each 1 or 0 is called a **bit**. The number of bits that can be processed at one time varies from device to device. For example, an audio compact disk uses 16-bits to digitize the music. An inkjet printer uses 8-bits for each symbol on the keyboard. Present day 32-bit computers read and store data as sequence of thirty-two 1's or 0's. In this section, we will look more closely at binary numbers as they apply to computers.

Computers and their peripheral devices use binary digits to read, store, and write information. For example, the binary sequence 0101 1010 is used for the letter Z. The 0's and l's represent the two different states.

In Lester Lefkowitz's *Global Pathways*, the computer age has put the world on a path controlled by binary numbers.

Each letter of the alphabet, digit, and special character, such as a period, comma, or hyphen, has a binary code. This code is called **ASCII**, **A**merican **S**tandard **C**ode for **I**nformation **I**nterchange. When a computer reads or sends information, it is processed in ASCII code. When it sends data to a device such as a printer, the printer decodes the binary information and converts it to the corresponding character.

Since sequences of binary numerals are difficult to read, write, and remember, codes are usually written as base 8 or base 16 numerals. The ASCII chart in Figure 1.4.1 uses HEX codes. The abbreviation HEX is for hexadecimal or base 16. The ability to convert between binary, octal, and hexadecimal forms is useful in working with and understanding computer codes. The following examples will show you how to convert between binary, octal, and hexadecimal numerals.

ASCII (American Standard Code for Information Interchange)

Chr	Hex	Chr	Hex	Chr	Hex	Chr	Hex	Chr	Hex	Chr	Hex
space	20	0	30	@	40	P	50	`	60	p	70
!	21	1	31	A	41	Q	51	a	61	q	71
"	22	2	32	B	42	R	52	b	62	r	72
#	23	3	33	C	43	S	53	c	63	s	73
$	24	4	34	D	44	T	54	d	64	t	74
%	25	5	35	E	45	U	55	e	65	u	75
&	26	6	36	F	46	V	56	f	66	v	76
'	27	7	37	G	47	W	57	g	67	w	77
(	28	8	38	H	48	X	58	h	68	x	78
)	29	9	39	I	49	Y	59	i	69	y	79
*	2A	:	3A	J	4A	Z	5A	j	6A	z	7A
+	2B	;	3B	K	4B	[	5B	k	6B	{	7B
,	2C	<	3C	L	4C	\	5C	l	6C	|	7C
-	2D	=	3D	M	4D	]	5D	m	6D	}	7D
.	2E	>	3E	N	4E	^	5E	n	6E	~	7E
/	2F	?	3F	O	4F	_	5F	o	6F	delete	7F

FIGURE 1.4.1

CONVERTING BINARY, OCTAL, AND HEXADECIMAL NUMERALS

In Section 1.3. the place-value system for any base was discussed. For example, the binary number, $10\ 110_2$, means you have the following:

Furthermore, since $8 = 2^3$, an octal digit can be represented by a three-digit binary numeral, and since $16 = 2^4$, a hexadecimal digit can be represented by a four-digit binary numeral. If the digits of a binary numeral are written in groups of three or four digits, the binary numeral can be easily converted into an octal numeral or a hexadecimal numeral. If you add up the place values for each nonzero binary digit given in Figure 1.4.2, you get the corresponding octal or hexadecimal numeral.

Binary (place values) 4 2 1	Octal
0 0 1	1
0 1 0	2
0 1 1	3
1 0 0	4
1 0 1	5
1 1 0	6
1 1 1	7

Binary (place values) 8 4 2 1	Hex
0 0 0 1	1
0 0 1 0	2
0 0 1 1	3
0 1 0 0	4
0 1 0 1	5
0 1 1 0	6
0 1 1 1	7
1 0 0 0	8
1 0 0 1	9
1 0 1 0	A(10)
1 0 1 1	B(11)
1 1 0 0	C(12)
1 1 0 1	D(13)
1 1 1 0	E(14)
1 1 1 1	F(15)

FIGURE 1.4.2

EXAMPLE 1

Write 10 110 111_2 as (a) an octal numeral and (b) a hexadecimal numeral.

Solution:

(a) Octal: Separate the binary numeral into groups of three digits starting from the right of the numeral. Find the octal digit for each group of three digits by adding the nonzero place values shown in Figure 1.4.2.

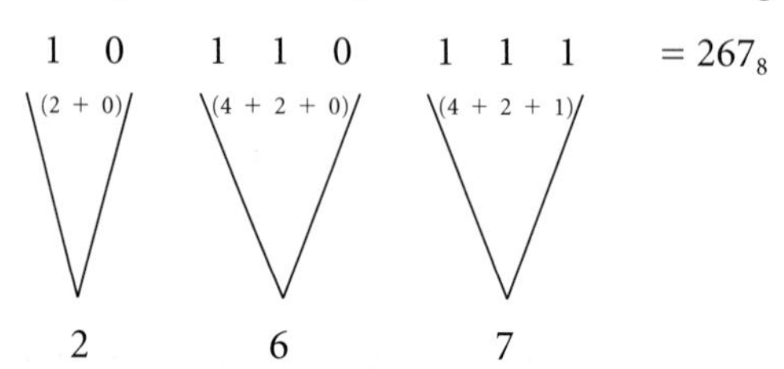

(b) Hexadecimal: Separate the binary numeral into groups of four digits and find the hexadecimal digit for each group of four digits adding the nonzero place values shown in Figure 1.4.2.

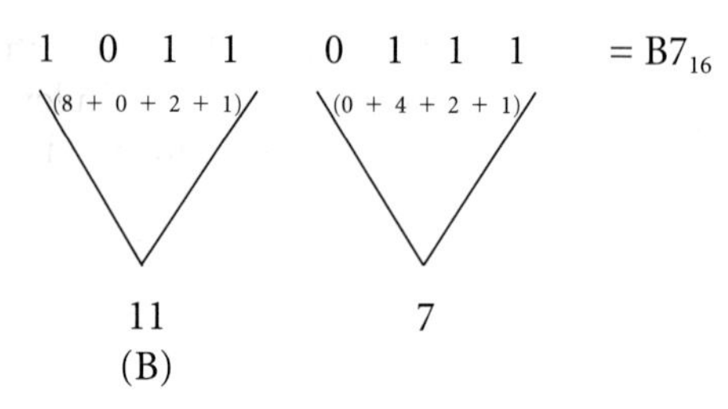

EXAMPLE 2

Using the ASCII chart, find the eight-digit binary code for A, Z, and >.

Solution: Since every hexadecimal digit can be represented by four binary digits, find the four-digit binary numeral for each digit in the hexadecimal representation of A, Z, and >.

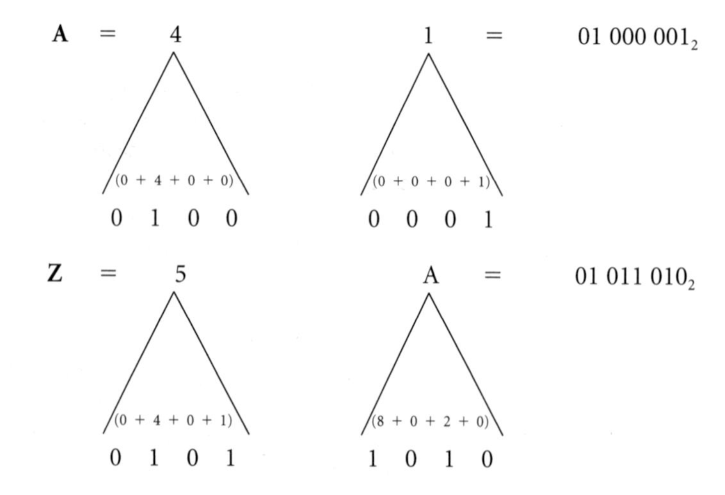

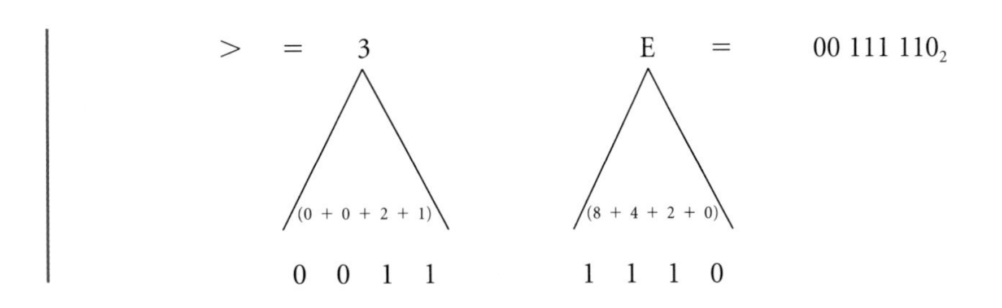

NOTE: By convention, binary numerals are written in groups of three digits.

HOW ARE CALCULATIONS DONE BY A COMPUTER?

The last aspect of computers we will investigate is how computers do calculations, particularly the operations of addition and multiplication. In our base-10 system, to add or multiply efficiently, you need to know the 100 addition facts from $0 + 0$ to $9 + 9$ and the 100 multiplication facts from 0×0 to 9×9. These facts along with the process of "carrying" allow you to add and multiply. However, in the binary system, there are only four addition facts and four multiplication facts.

Binary Arithmetic Facts

Addition	Multiplication
$0 + 0 = 0$	$0 \times 0 = 0$
$0 + 1 = 1$	$0 \times 1 = 0$
$1 + 0 = 1$	$1 \times 0 = 0$
$1 + 1 = 10$	$1 \times 1 = 1$

Because we are accustomed to thinking in base 10, there is an entry in the binary arithmetic facts chart that seems strange: $1 + 1 = 10$. You must understand that the "10" is in base 2 and 10_2 means $1 \times 2 + 0 \times 1 = 2$. Thus, the chart entry actually corresponds to our understanding of addition. With these binary arithmetic facts and the process of "carrying," we can explain how a computer performs addition and multiplication quickly and efficiently.

Binary Addition and Multiplication

To add or multiply decimal numerals, the computer first converts them into binary numerals. The addition or multiplication is done electronically by the computer using the binary numerals. Before the answer is displayed, it is converted back into its decimal form. Let's show how this is done by finding $28 + 13$ and 28×13.

1. The computer converts the 28 and 13 into binary numerals. Paper-and-pencil methods for doing that are given in Section 1.3.

$$28 = 11\,100_2 \qquad \text{and} \qquad 13 = 1\,101_2$$

2. Addition and multiplication are done electronically by using the basic binary arithmetic facts and a "carrying" procedure. We will use the process for addition and multiplication that we use with decimal numerals to demonstrate the process.

Addition:

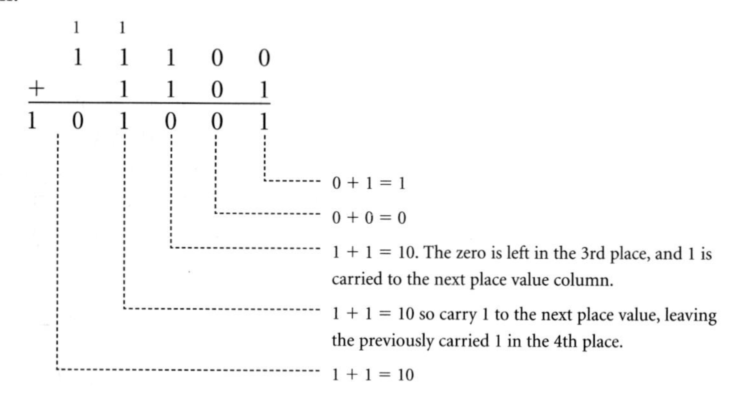

Check: $101\,001_2 = 1 \times 32 + 0 \times 16 + 1 \times 8 + 0 \times 4 + 0 \times 2 + 1 \times 1 = 41$.
Using decimal numerals, $28 + 13 = 41$.

Multiplication:

```
              1 1 1 0 0
           ×    1 1 0 1
              1 1 1 0 0 ------ Using the multiplier in the "1's" place, we get 1 × 11100 = 11100.
            0 0 0 0 0 -------- Using the multiplier in the "2's" place, we get 0 × 11100 = 00000.
          1 1 1 0 0 ---------- Using the multiplier in the "4's" place, we get 1 × 11100 = 11100.
        1 1 1 0 0 ------------ Using the multiplier in the "8's" place, we get 1 × 11100 = 11100.
      1 0 1 1 0 1 1 0 0 ------ Add the digits in each place value column. Remember to carry "1"
                               to the next column whenever you get 1 + 1.
```

Check: $101\,101\,100_2 = 1 \times 256 + 0 \times 128 + 1 \times 64 + 1 \times 32 + 0 \times 16 + 1 \times 8 + 1 \times 4 + 0 \times 2 + 0 \times 1 = 364$.
Using decimal numerals, $28 \times 13 = 364$.

3. To output the answers, the computer converts them back into decimal form. Paper-and-pencil methods for doing that are given in Section 1.3. Thus, we would see $28 + 13 = 41$ and $28 \times 13 = 364$.

Using only 1's and 0's simplifies computation. In the integrated circuits of a computer, the l's and 0's are manipulated at great speed. If you are not concerned about the decimal answer to a problem and simply work in binary, addition and multiplication can be done quite easily.

EXAMPLE 3

Find the sum of the binary numbers, 101 111 101 and 10 011 010.

Solution: The binary numbers are positioned above each other, and the digits in each place-value column are added by using the binary arithmetic facts and the normal "carrying" procedure as noted. Proceeding from the l's place and working to the left give the following result.

```
 1 111 11
   101 111 101
+   10 011 010
 1 000 010 111 = 1 000 010 111₂
```

Note: Writing the numbers in decimal form, we have 381 + 154 = 535.

EXAMPLE 4

Find the product of the binary numbers, 10 011 and 1 110.

Solution: The binary numbers are positioned above each other, and the first factor is multiplied by each digit of the second factor. The process starts with the 1's place and moves one place-value column to the left with each new multiplication. The digits in each place-value column are added using the binary arithmetic facts and the normal "carrying" procedure.

```
         1 0 0 1 1
      ×    1 1 1 0
         0 0 0 0 0
       1 0 0 1 1
     1 0 0 1 1
   1 0 0 1 1
 1 0 0 0 0 1 0 1 0 = 100 001 010₂
```

Note: In decimal form, we have 19 × 14 = 266.

PROBLEMS ⊳ Explain ⊳ Apply ⊳ Explore

⊳ Explain

1. What do the binary digits 1 and 0 represent to a computer?
2. What is a bit? What is a 16-bit machine?
3. What is ASCII? What is it used for?
4. How is information stored by a CD?
5. Why is 1 + 1 = 10 in the binary system?
6. How does a computer use binary numbers in addition?
7. How does a computer use binary numbers in multiplication?

⊳ Apply

In Problems 8–13, write each binary numeral in (a) octal and (b) hexadecimal.

8. $11\ 011_2$
9. $10\ 101_2$
10. $111\ 111_2$
11. $1\ 100\ 100_2$

12. $111\ 110\ 101\ 110_2$

13. $101\ 110\ 101\ 110_2$

In Problems 14–19, use the ASCII chart in Figure 1.4.1 to find the 8-bit binary code for each character.

14. The letter M

15. The letter e

16. The letter n

17. The percent sign (%)

18. The equals sign (=)

19. The plus sign (+)

In Problems 20–27, give each answer as a binary numeral and check your results using decimal numerals.

20. $1\ 010_2 + 1\ 110_2$

21. $110\ 011_2 + 10\ 110_2$

22. $10\ 011\ 111_2 + 111\ 011_2$

23. $111\ 000\ 101_2 + 101\ 011\ 001_2$

24. $110_2 \times 101_2$

25. $10\ 100_2 \times 1\ 101_2$

26. $101\ 111_2 \times 1\ 011_2$

27. $111\ 000\ 111_2 \times 111_2$

Explore

To transfer data, it can be stored and read in binary ASCII format. If each line of binary code represents one character, change each line to hexadecimal and find what characters were being transferred in Problems 28–29?

28. 0100 0111
0110 1111
0110 1111
0110 0100
0010 0000
0100 0100
0110 1111
0110 0111

29. 0101 0100
0110 1111
0010 0000
0110 0010
0110 0101
0010 0000
0110 1111
0111 0010
0010 0000
0110 1110
0110 1111
0111 0100
0011 1111

In Problems 30–34, use the ASCII code to write each word in binary.

30. LOVE

31. HOPE

32. M*A*S*H

33. Mom

34. YES!

35. This section gives a method for converting binary numerals into octal (base 8) and hexadecimal (base 16). How could you convert binary numerals into base 32 numerals or base 32 numerals into binary numerals? Give some examples of your method. (Hint: $2^5 = 32$.)

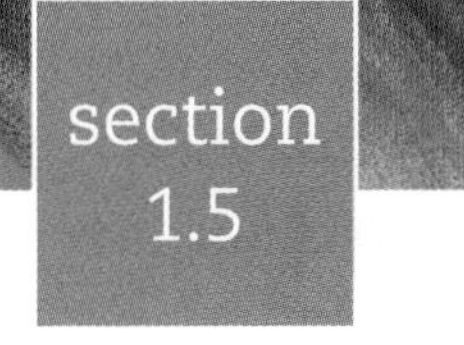

Types of Numbers

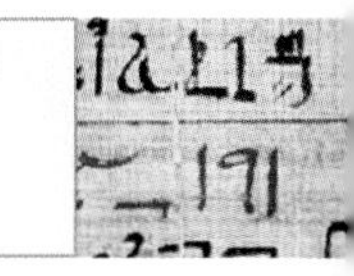

Numbers are a basic part of our daily lives. They are all around us. As we have seen in this chapter, they have been of interest to human beings from earliest times. Just as humans classified the animals, insects, plants, and objects around them, they also classified their numbers. So as we continue our study of mathematics, it would be appropriate to study the many types of numbers that have been classified.

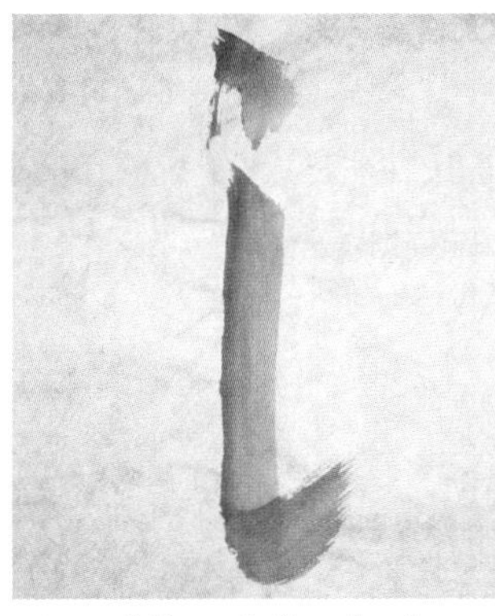

0, e, and *i* by David McLaughlin (courtesy of the artist) and π by Tom Marioni (Courtesy of Crown Point Press): artists' representation of four historically significant numerical constants.

REAL NUMBERS

The **natural numbers**, also called counting numbers, consist of the numbers $\{1, 2, 3, 4, \ldots\}$. The **whole numbers** consist of the natural numbers and zero $\{0, 1, 2, 3, 4, \ldots\}$. The **integers** consist of the whole numbers and the negatives of the whole numbers, $\{\ldots -4, -3, -2, -1, 0, 1, 2, 3, 4, \ldots\}$.

The **rational numbers** are numbers that can be represented by the quotient a/b, where a and b are integers and $b \neq 0$. It can be shown that when these quotients are converted into decimals, they are either terminating or repeating decimals. For example, the rational number 3/5 is the terminating decimal 0.6, $-7/4$ is -1.75, 53 is 53.0, 156 17/25 is 156.68, 2/3 is the repeating decimal 0.666. . . , $-36/11$ is $-3.2727\ldots$, and 43 1/7 is 43.142857142857 . . . .

The **irrational numbers** are decimal numbers that do not terminate and do not repeat (as the rational numbers do). It can be shown that an irrational number cannot he written as a ratio of two integers. Included in this group of numbers are radical numbers, such as $\sqrt{2}$, $\sqrt{19}$, $\sqrt[4]{101}$, $\sqrt[3]{34}$, and mathematical constants, such as π and e.

The **real numbers** consist of all the rational and irrational numbers. They can be visualized by using a horizontal number line where a zero point (the origin) and a unit length are marked off. Each real number corresponds to exactly one point on the line, and each point on the line corresponds to exactly one real number. Numbers that are larger than zero, the positive numbers, are placed to the right of zero, and numbers that are less than zero, the negative numbers, are placed to the left of zero.

The Real Number Line

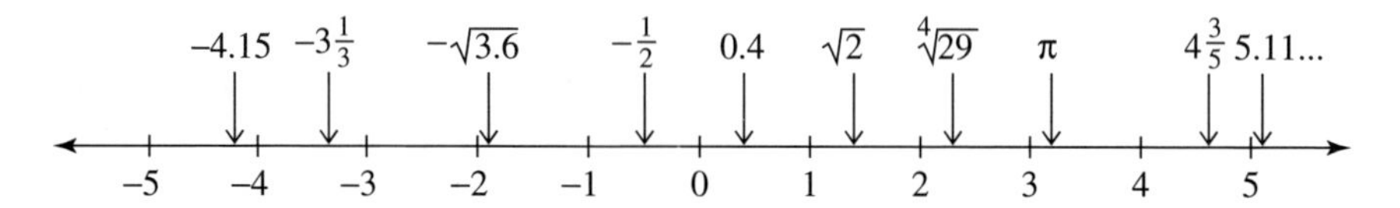

The following chart shows the classification of the real numbers.

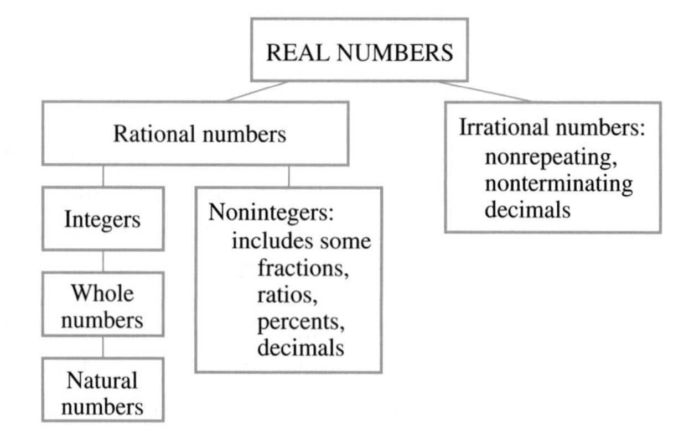

EXAMPLE 1

$-10, -4.67, -\sqrt{17}, -3\frac{5}{8}, -\sqrt[5]{2.9}, -1, -\frac{2}{3}, 0, 0.333\ldots, \sqrt[3]{8.41}, \frac{5}{2}, \sqrt{25}, \sqrt{27},$ $7.87, 27$

Determine which of the numbers in the list above are (a) integers, (b) irrational numbers, (c) natural numbers, (d) rational numbers, (e) real numbers, and (f) whole numbers.

Solution:

a) Integers: $-10, -1, 0, \sqrt{25}, 27$

b) Irrational numbers: $-\sqrt{17}, -\sqrt[5]{2.9}, \sqrt[3]{8.41}, \sqrt{27}$

c) Natural numbers: $\sqrt{25}, 27$

d) Rational numbers: all except those in part (b)

e) Real numbers: all numbers in the list

f) Whole numbers: $0, \sqrt{25}, 27$

EXAMPLE 2

Explain why $\sqrt{36}$ can be classified as a rational number, a natural number, a whole number, and an integer but not an irrational number.

Solution: Since $\sqrt{36} = 6$, it is a rational number (6/1), a natural number (1, 2, 3, 4, ...), a whole number (0, 1, 2, 3, 4, . . .), and an integer

(. . . , −3, −2, −1, 0, 1, 2, 3, . . .). It is not an irrational number because it can be expressed as a quotient of integers (6/1) and because, as a decimal, it terminates (6.0).

ZERO AND NEGATIVE NUMBERS

Before we continue our investigation of other types of numbers, we will look at the origins of zero and negative numbers. Though sometimes taken for granted, zero frequently finds its way into our daily activities. The symbol for zero, 0, is used quite often when we write numerals and perform computations. Negative numbers have also gained acceptance today. We use a negative number for an amount less than zero, such as the temperature at the North Pole or a checking account balance after writing a check for more money than is in the account. Surprisingly, this modern use and acceptance of zero and negative numbers did not occur overnight; it took many centuries for these mathematical concepts to be accepted by the scientific community.

Let's examine some facts about zero. The concept of zero, which indicates the absence of a quantity, has most likely been understood since prehistoric times. Each time a hunter came home without any game, the number zero was experienced. Though the early Egyptian, Greek, and Roman civilizations understood the concept of zero, they had no symbol for zero. Their systems of numeration did not need a symbol for zero. Zero was, however, represented on ancient Babylonian tablets and Chinese counting boards as a blank space used for a missing place value in a numeral. In the fourth century B.C., the symbol was used by the Babylonians, and the symbol was used by the Mayans of South America in place of the blank space. The symbol 0 is believed to have originated in India years after forms of the other nine numerals appeared. This symbol for zero developed before A.D. 870, since it was contained on an A.D. 870 inscription in Gwalior, India. However, some experts believe that this symbol might have come to India by way of Indochina, since inscriptions in Cambodia and Sumatra (A.D. 683) used this same symbol for zero.

In Western culture, the use of the symbol 0 is a fairly recent development. It became well established in Europe during the late 1400s when the Hindu-Arabic system replaced Roman numerals. The word *zero* comes from the Latin word *ziphrum. Ziphrum* is a translation from the Arabic word *sifr,* which came from the Hindu word *sunya,* meaning void or empty.

As with the number zero, negative numbers have an interesting history. There is no trace of negative numbers in ancient Egyptian, Babylonian, or Greek writings. In A.D. 270, when negative numbers occurred as solutions to equations, the Greek mathematician Diophantus dismissed them as being absurd. This had such an effect that it was not until the Renaissance (14th and 15th centuries) that mathematicians began to be more receptive to negative numbers.

Negative numbers gained acceptance in Europe in the 16th century through the work *Ars Magna* by Girolamo Cardano (1545), who used negative numbers as solutions to equations, and the work of Michael Stifel (1544), who described negative numbers as those numbers that are less than zero. The terms *positive* and *affirmative* were used to indicate positive numbers, and the terms *privative, negative,* and *minus* were used for negative numbers.

In non-Western cultures, such as China, India, and Arabia, negative numbers were readily accepted. In the second century B.C., Chinese counting boards used red or triangular rods to represent positive numbers and black or square rods to represent negative numbers. In India. c. A.D. 628, Brahmagupta mentioned negative numbers, and the Hindu and Arabian mathematicians that followed continued to use negative numbers in their arithmetic and algebra.

COMPLEX NUMBERS

Besides the real numbers described previously, there is another set of numbers based on $\sqrt{-1}$. These numbers are called **imaginary numbers** and use the letter i, where $i = \sqrt{-1}$. For example, $\sqrt{-16} = 4\sqrt{-1} = 4i$, $\sqrt{-27} = 3\sqrt{3}\sqrt{-1} = 3i\sqrt{3}$ and $\sqrt{-93} = i\sqrt{93}$. If a real number and an imaginary number are added together, the sum is called a **complex number** and is written in the form $a + bi$, where a and b are real numbers. Thus,

$$5 + \sqrt{-9} = 5 + 3i \quad \text{and} \quad -7.2 - \sqrt{-20} = -7.2 - 2i\sqrt{5}.$$

Imaginary numbers are a recent mathematical development. Until the 1500s, square roots of negative numbers were considered an impossibility. The work of Girolamo Cardano (1545) and Rafael Bombelli (1572) introduced imaginary numbers as roots of equations. René Descartes (1637) called them "imaginary," and Leonhard Euler (1748) used i to represent $\sqrt{-1}$. Though imaginary numbers are not used in everyday transactions, they are used to solve problems in mathematics, electronic circuit design, vibration analysis, and other branches of science and engineering.

NUMBERS BASED ON GEOMETRIC SHAPES

Polygonal numbers are numbers that were devised to conform to basic geometric shapes. These geometric-based numbers were of interest especially to the Greeks because of their simplistic geometric beauty and the many different mathematical patterns found between the terms of each number and between different polygonal numbers.

Triangular numbers take the shape of triangles, as pictured below.

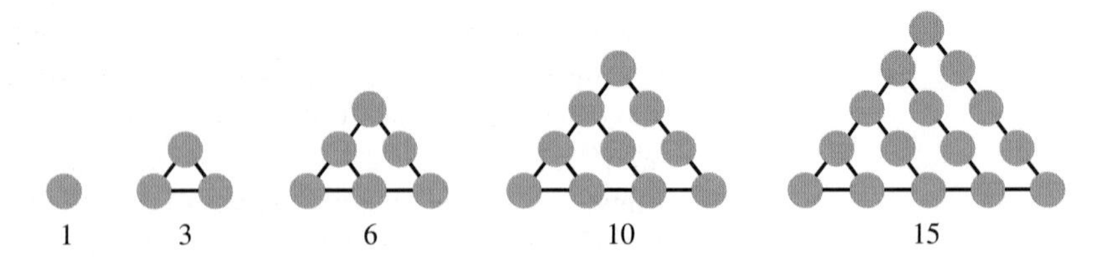

Square numbers take the shape of squares, as pictured below.

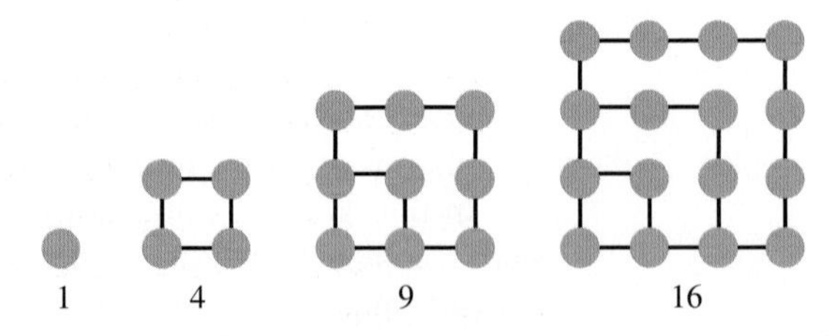

Pentagonal numbers take the shape of pentagons, as pictured below.

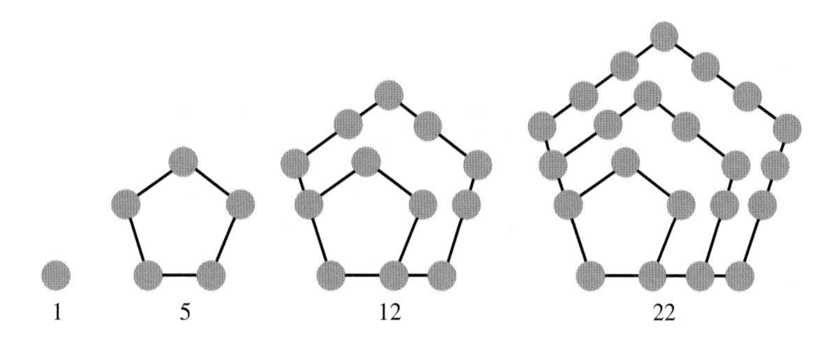

Other polygonal numbers can be formed by using other regular geometric shapes such as hexagons, octagons, and decagons.

EXAMPLE 3

Find the next two triangular numbers and describe the pattern that exists in going from one triangular number to the next.

Solution: The next two triangular numbers are 21 and 28. If you list the triangular numbers and find the difference between successive terms, the differences are the whole numbers 2, 3, 4, 5, 6, ... Thus, to get from one triangular number to the next, just continue adding consecutive integers.

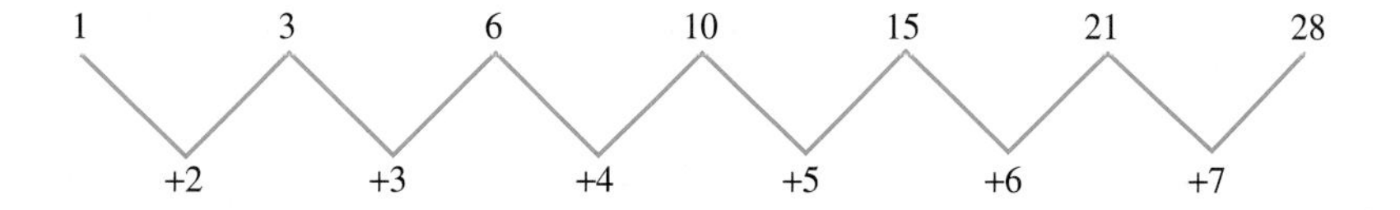

NUMBERS BASED ON FACTORS

A **proper factor of a natural number** is a natural number less than the number that divides evenly into the number. For example, the proper factors of 8 are 1, 2, and 4. A **prime number** is a natural number whose only proper factor is 1, and a **composite number** is a natural number that has proper factors greater than 1. According to these definitions, the number 1 is not a prime number. The first ten prime numbers are 2, 3, 5, 7, 11, 13, 17, 19, 23, and 29. It has been proved that there are an infinite number of primes and that every composite number can be represented as a product of prime numbers. The Greek scholar Eratosthenes (274–194 B.C.). invented an arithmetical sieve for finding prime numbers. For example, to find all the prime numbers less than 102, write down all the natural numbers from 2 to 101 (Fig. 1.5.1). The first number, 2, is a prime. Draw a box around it and cross out all other multiples of 2 (every second number: 4, 6, 8, 10, . . .). The next number that is not crossed, 3, is a prime. Draw a box around it and cross out all other multiples of 3 (every third number: 6, 9, 12, 15, 18, . . .). Continuing in this manner, you can see that there are 26 prime numbers less than 102.

Since every composite number has proper factors, the properties of these factors have also been studied. The following types of numbers are a result of these investigations.

FIGURE 1.5.1

2	3	~~4~~	5	~~6~~	7	~~8~~	~~9~~	~~10~~	11
~~12~~	13	~~14~~	~~15~~	~~16~~	17	~~18~~	19	~~20~~	~~21~~
~~22~~	23	~~24~~	~~25~~	~~26~~	~~27~~	~~28~~	29	~~30~~	31
~~32~~	~~33~~	~~34~~	~~35~~	~~36~~	37	~~38~~	~~39~~	~~40~~	41
~~42~~	43	~~44~~	~~45~~	~~46~~	47	~~48~~	~~49~~	~~50~~	~~51~~
~~52~~	53	~~54~~	~~55~~	~~56~~	~~57~~	~~58~~	59	~~60~~	61
~~62~~	~~63~~	~~64~~	~~65~~	~~66~~	67	~~68~~	~~69~~	~~70~~	71
~~72~~	73	~~74~~	~~75~~	~~76~~	~~77~~	~~78~~	79	~~80~~	~~81~~
~~82~~	83	~~84~~	~~85~~	~~86~~	~~87~~	~~88~~	89	~~90~~	~~91~~
~~92~~	~~93~~	~~94~~	~~95~~	~~96~~	97	~~98~~	~~99~~	~~100~~	101

A **perfect number** is a natural number with the property that the sum of its proper factors equals the number. Six is the first perfect number because its proper factors, 1, 2, and 3, have a sum of 6 ($1 + 2 + 3 = 6$). Another perfect number is 496 because the sum of its proper factors, 1, 2, 4, 8, 16, 31, 62, 124, and 248 equals 496. In 1952, there were only 12 known perfect numbers. Since then, with the aid of computers, more perfect numbers have been found. Some of these have more than 100,000 digits.

An **abundant number** is a natural number with the property that the sum of its proper factors is greater than the number. Twelve is an abundant number because its proper factors 1, 2, 3, 4, and 6 have a sum of 16. One hundred is an abundant number because its proper factors 1, 2, 4, 5, 10, 20, 25, and 50 have a sum of 117.

A **deficient number** is a natural number with the property that the sum of its proper factors is less than the number. Twenty-seven is a deficient number because the sum of its proper factors 1, 3, and 9 equals 13. Forty-three is a deficient number because its only proper factor is 1.

Amicable numbers are two natural numbers such that the sum of the proper factors of one number equals the other number, and vice versa. The numbers 220 and 284 are amicable because the sum of the proper factors of 220 (1, 2, 4, 5, 10, 11, 20, 22, 44, 55, 110) is 284 and the sum of the proper factors of 284 (1, 2, 4, 71, 142) is 220. This pair of numbers has been ascribed to the Greek mathematician Pythagoras (c. 540 B.C.). The fact that each amicable number generates the other gives them an intimate relationship that played a role in mysticism and superstition through the ages. In 1636, the French mathematician Pierre Fermat discovered a second amicable pair, 17,296 and 18,416. In 1638, René Descartes discovered a third pair, and in 1750, Leonhard Euler found 60 other pairs. In 1866, a 16-year-old Italian, Nicolo Paganini, astounded the mathematical world when he discovered the small amicable pair of 1184 and 1210. There are more than 900 known amicable pairs of numbers, and they are still intriguing to mathematicians and computer enthusiasts.

EXAMPLE 4

Classify the numbers 18, 28, 31, and 45 as (a) composite or prime and (b) perfect, abundant, or deficient.

Solution:

18: **a)** Composite number because 1 is not the only proper factor.
b) Abundant number because the sum of its proper factors (1, 2, 3, 6, 9) equals 21, and $21 > 18$.

28: **a)** Composite number because 1 is not the only proper factor.
b) Perfect number because the sum of its proper factors (1, 2, 4, 7, 14) equals 28.

31: **a)** Prime number because its only proper factor is 1.
b) Deficient number because its only proper factor is 1, and $1 < 31$.

45: **a)** Composite number because 1 is not its only proper factor.
b) Deficient number because the sum of its proper factors (1, 3, 5, 9, 15) equals 33, and $33 < 45$.

EXAMPLE 5

Show that the numbers 17,296 and 18,416 found by Pierre Fermat are amicable numbers.

Solution: The proper factors of 17,296 are 1, 2, 4, 8, 16, 23, 46, 47, 92, 94, 184, 188, 368, 376, 752, 1081, 2162, 4324, and 8648. The sum of these factors is 18,416. The proper factors of 18,416 are 1, 2, 4, 8, 16, 1151, 2302, 4604, and 9208. The sum of these factors is 17,296. Thus, 17,296 and 18,416 are amicable numbers.

NOTES ON π

Another number that has been of interest since ancient times is the ratio of the circumference (C) of a circle to its diameter (D), given by C/D. No matter what size circle is considered, this ratio has the same value. By about 2000 B.C., the Babylonians used a value of 25/8 for this ratio. Many brilliant minds have worked on obtaining approximate values for this ratio. Here are some examples.

Archimedes of Syracuse (200 B.C.): $211875/67441 \approx 3.141635$

Astronomer Ptolemy (160): $377/120 \approx 3.141667$

Liu Hui (263): $157/50 = 3.14$

Āryabhata (499): $626832/200{,}000 = 3.13416$

Valentin Otho (1573): $355/113 \approx 3.141593$

In 1706, William Jones used the symbol $\boldsymbol{\pi}$ (the Greek letter **pi**) to represent the ratio. With usage of this symbol in 1736 by Leonhard Euler, π became a standard. Between 1500 and 1800, others used trigonometry and calculus to approximate π to more than 500 decimal places. In 1766, Johann Lambert proved that π could not be represented by the ratio of two natural numbers and was therefore an irrational number. In the 20th century, calculators and computers were used to determine π to thousands of decimal places. In September 2002, Dr. Kanada of the University of Tokyo and his team, calculated 1.241 trillion digits of π (over six times their 1999 record). However, evidence of accurate estimations of π without the use of computers has been noted. For example, in the Great Pyramid of Gizeh in Egypt (2600 B.C.), the ratio of twice the width of the pyramid ($w = 230.364$ m) to the height ($h = 146.599$ m) of the pyramid gives π accurate to the hundredths place ($2w/h \approx 3.14278$).

The first 501 digits of π are as follows:

3.14159 26535 89793 23846 26433 83279 50288 41971 69399 37510 58209 74944
59230 78164 06286 20899 86280 34825 34211 70679 82148 08651 32823 06647
09384 46095 50582 23172 53594 08128 48111 74502 84102 70193 85211 05559
64462 29489 54930 38196 44288 10975 66593 34461 28475 64823 37867 83165
27102 19091 45648 56692 34603 48610 45432 66482 13393 60726 02491 41273
72458 70066 06315 58817 48815 20920 96282 92540 91715 36436 78925 90360
01133 05305 48820 46652 13841 46951 94151 16094 33057 27036 57595 91953
09218 61173 81932 61179 31051 18548 07446 23799 62749 56735 18857 52724
89122 79381 83011 94921 (never stops or repeats)

PROBLEMS ○ Explain ○ Apply ○ Explore

Explain

1. What are prime numbers?
2. What are composite numbers?
3. What are abundant numbers?
4. What are perfect numbers?
5. Is it possible to have a real number that is not rational? Explain.
6. What are square numbers?
7. Trace the history of negative numbers and zero.
8. Answer true or false for each statement. If a statement is false, explain why and give an example to show that it is false.
   a) All rational numbers are real numbers.
   b) All real numbers are rational numbers.
   c) All irrational numbers are real numbers.
   d) All real numbers are irrational numbers.
   e) All integers are whole numbers.
   f) All whole numbers are integers.
   g) All rational numbers are irrational numbers.
   h) All irrational numbers are rational numbers.
   i) All imaginary numbers are irrational numbers.
   j) All irrational numbers are imaginary numbers.
   k) All radicals are irrational.
   l) Complex numbers of the form $a + bi$ are real numbers when $b = 0$.
9. Even numbers are integers that end in 0, 2, 4, 6, or 8, and odd numbers end in 1, 3, 5, 7, or 9. How can even numbers be defined by using the concept of a factor? Explain.
10. Why is 2 the only even prime number?
11. Why is every prime number a deficient number?

Apply

12. From the list below, choose the numbers belonging to each category.

$-11, \quad -9.4, \quad -8\frac{2}{9}, \quad -\sqrt{50}, \quad -4, \quad 1, \quad -\sqrt[3]{7.3}, \quad 0, \quad \frac{3}{4}, \quad \sqrt{-2},$
$\sqrt{2}, \quad 6.1212\ldots, \quad \sqrt{49}, \quad 9, \quad 10.12$

a) Natural numbers
b) Whole numbers
c) Integers
d) Rational numbers
e) Irrational numbers
f) Real numbers
g) Imaginary numbers
h) Noninteger rational numbers

13. From the list below, choose the numbers belonging to each category.

-14.785, -7, $-\sqrt{64}$, $-\sqrt{-25}$, $\frac{-5}{16}$, 0, $\sqrt[5]{19}$, $\sqrt{-8}$, $\sqrt{8}$, π, $5\frac{7}{8}$, $9.76555\ldots$, $\sqrt{100}$, 19

a) Natural numbers
b) Whole numbers
c) Integers
d) Rational numbers
e) Irrational numbers
f) Real numbers
g) Imaginary numbers
h) Nonradical irrational numbers

14. List five numbers that satisfy each description.

a) Real numbers that are not rational
b) Rational numbers that are not integers
c) Irrational numbers that are not square roots
d) Real numbers that are not irrational
e) Integers that are not natural numbers
f) Noninteger rational numbers

15. What are the fifth and sixth pentagonal numbers? Make a sketch of each number.

16. The first three hexagonal numbers are 1, 6, 15. Make a sketch of these three polygonal numbers.

17. Classify each of the following numbers as (a) prime or composite and (b) perfect, abundant, or deficient.

a) 31
b) 77
c) 145
d) 1988
e) 8128
f) 9000

Explore

18. Show that the pair of numbers 1184 and 1210, found by 16-year-old Nicolo Paganini in 1866, is an amicable pair of numbers.

19. Example 3 showed that there is a pattern going from term to term in the triangular numbers. Find the pattern for the square numbers and use the pattern to find the first twelve square numbers.

20. Example 3 showed that there is a pattern going from term to term in the triangular numbers. Find the pattern for the pentagonal numbers and use the pattern to find the first twelve pentagonal numbers.

21. Explain why there is only one prime number with a last digit of five.

22. A basic math text stated that $\pi = \frac{22}{7}$. Why is this statement incorrect? What is the correct relationship between π and $\frac{22}{7}$?

23. Explain why it is incorrect to say, $\sqrt{2} = 1.414$. What is the correct relationship between $\sqrt{2}$ and 1.414?

CHAPTER 1 SUMMARY

Key Terms

The important terms in this chapter are:	Section
Abundant number: A natural number with the property that the sum of its proper factors is greater than the number.	1.4
Amicable numbers: Two natural numbers with the property that the sum of the proper factors of one number equals the other number and vice versa.	1.4
ASCII: American Standard Code for Information Interchange.	1.4
Basimal point: The generic term for the dot in a number in any base. In base ten, this is the decimal point.	1.3
Binary: A number system with a base of 2 using the digits 0 and 1.	1.3, 1.4
Bit: A 1 or a 0 that can be read electronically *on* or *off* or as a positive $(+)$ or negative $(-)$ charge.	1.4
Complex number: The sum of a real number and an imaginary number.	1.5
Composite number: A whole number that has proper factors greater than 1.	1.5
Deficient number: A natural number with the property that the sum of its proper factors is less than the number.	1.5
Duodecimal: A number system with a base of 12 using the digits 0, l, 2, 3, 4, 5, 6, 7, 8, 9, A, and B.	1.3
Hexadecimal: A number system with a base of 16 using the digits 0, 1, 2, 3, 4, 5, 6, 7, 8, 9, A, B, C, D, E, and F.	1.3, 1.4
Imaginary number: A number involving $\sqrt{-1} = i$.	1.5
Integer: A number from the set, $\{\ldots, -4, -3, -2, -1, 0, 1, 2, 3, 4, \ldots\}$.	1.5
Irrational number: A number that cannot be represented as the ratio of two integers; as a decimal, it does not terminate or repeat.	1.5
Natural number: A number from the set $\{1, 2, 3, 4, 5, \ldots\}$.	1.5
Number: A measure of a quantity or amount.	HISTORY
Numeral: A symbol used to represent a number.	HISTORY
Octal: A number system with a base of 8 using the digits 0, 1, 2, 3, 4, 5, 6, and 7.	1.3,1.4
Pentagonal numbers: A type of polygonal number based on the shape of a pentagon.	1.5
Perfect number: A whole number with the property that the sum of its proper factors equals itself.	1.5

Term	Section
Pi (π): The ratio of the circumference of a circle to its diameter; $\pi \approx 3.14159$.	1.5
Place value: The value given to the position a digit holds in a numeral.	1.1
Polygonal numbers: Numbers based on geometric shapes, such as triangular, square, pentagonal, and hexagonal numbers.	1.5
Prime number: A whole number that is divisible only by 1 and itself.	1.5
Proper factor of a natural number *N*: A natural number less than *N* that divides evenly into *N*.	1.5
Quintary: A number system with a base of 5 using the digits 0, 1, 2, 3, and 4.	1.3
Rational number: A number that can he represented as the ratio of two integers; as a decimal it terminates or repeats.	1.5
Real number: A number that is either rational and irrational; each real number corresponds to a point on a number line.	1.5
Square numbers: A type of polygonal number based on the shape of a square.	1.5
Tally: A mark used to represent objects being counted.	HISTORY
Triangular numbers: A type of polygonal number based on the shape of a triangle.	1.5
Whole number: A number from the set {0, 1, 2, 3, 4, 5, . . .}.	1.5

After completing this chapter, you should be able to:	Section
1. Explain the difference between a number, a tally, a numeral, and the word used to verbalize a quantity or amount.	HISTORY
2. Represent numbers in ancient systems of numeration that use grouping symbols along with:	
a) Addition—Egyptian hieroglyphic system, Attic system	1.1
b) Addition and subtraction—Roman numeral system	1.1
c) Addition and multiplication—Traditional Chinese system, Ionic Greek system	1.1
d) Place values—Babylonian system, Mayan system, Attic system	1.1
3. Represent numbers in the Hindu-Arabic system (decimal system), a place-value system using a base of 10.	1.2
4. Show how fractions are written in some systems of numeration.	1.2
5. Represent numbers in a place-value system that has any base and be able to convert between decimal numerals and numerals in other bases.	1.3
6. Convert between binary, octal, and hexadecimal numerals.	1.4
7. Perform addition and multiplication with binary numerals.	1.4
8. Classify different types of numbers, such as real numbers, complex numbers, numbers based on geometric shapes, and numbers based on factors.	1.5

CHAPTER 1 REVIEW

Section 1.1

1. Represent the tally as a numeral in each of the following systems of numeration:

   √√√√ √√√√ √√√√ √√√√ √√√√
   √√√√ √√√√ √√√√ √√√√ √√√√
   √√√√ √√√√ √√√√ √√√√ √√√√
   √√√√ √√√√ √√√√ √√√√ √√√√
   √√√√ √√√√ √√√√ √√√√ √√√√
   √√√√ √√√√ √√√√ √√√√ √√√√
   √√√

   a) Hindu-Arabic
   b) Egyptian
   c) Roman
   d) Traditional Chinese
   e) Ionic Greek
   f) Attic Greek
   g) Babylonian
   h) Mayan

2. By using the letters of the alphabet—o for 0 (zero), a for 1, b for 2, c for 3, d for 4, e for 5, f for 6, g for 7, h for 8, i for 9, and dots placed to the right of a letter to indicate that a certain digit is being multiplied by a power of 10 (. placed to the right of a letter indicates it is being multiplied by 10; . . . placed to the right of a letter indicates it is being multiplied by 100, and so on)—a system of numeration that uses both addition and multiplication can be formed that is similar to the Chinese system. What is the amount represented by the following numerals?

   a) a. d
   b) f.. a. d e...
   c) h... i.. d. e
   d) i. h.. g... g.... f
   e) b... e... e.. f.
   f) c a.... d..

Section 1.2

3. Write the following in expanded form using powers of 10.

   a) 25.47 **b)** 120.045

4. Write the fraction 3/4 in the following systems of numeration:

   a) Decimal
   b) Roman
   c) Egyptian
   d) Babylonian
   e) Ionic Greek

5. Devise a method for representing fractions in the system of numeration in Problem 2. Give examples showing how various fractions can be written by using the system.

Section 1.3

6. Find the decimal equivalent of the following:

   a) 1234_5
   b) 12.34_8
   c) $12A4_{12}$
   d) $C2.04_{16}$

7. Represent the tally in Problem 1 as a numeral in the following systems of numeration:

   a) Binary
   b) Base 5
   c) Octal
   d) Duodecimal
   e) Hexadecimal

Section 1.4

8. Explain how binary numbers are associated with storing information?

9. Use the ASCII chart to find 8-bit binary code for the phrase "Do it!"

10. Convert 101 110 011_2 into both octal and hexadecimal.
11. Convert 67 into binary, octal, and hexadecimal.
12. Find the answer for the following in binary and check the results using decimal numerals.
    a) 101 110 011_2 + 1 100 111_2
    b) 10 $101_2 \times 1\ 011_2$

Section 1.5

13. Explain:
    a) why 6.55 and 6.555. . . are both classified as a rational numbers.
    b) why 3.14 is a rational number but π is an irrational number.
    c) the difference between $-\sqrt{9}$ and $\sqrt{-9}$.
14. From the list below, choose the numbers belonging to each category.
    $-7.6,\quad -7,\quad -5\frac{2}{3},\quad -\sqrt{16},\quad \sqrt{-16},\quad 0,\quad 2,\quad 8.333\ldots,\quad 10\frac{3}{5},\quad \sqrt[3]{125},\quad \sqrt[3]{127}$
    a) Natural numbers
    b) Whole numbers
    c) Integers
    d) Rational numbers
    e) Irrational numbers
    f) Real numbers
    g) Imaginary numbers
    h) Noninteger rational numbers
15. Classify the four numbers, 400, 461, 496, and 512, as (a) composite or prime and (b) as perfect, abundant, or deficient.
16. The algebraic expression, $n^2 - n + 41$, generates prime numbers for whole numbers, $0 \le n \le 40$. Find the first nine prime numbers generated by this expression. Show that this expression generates a composite number when $n = 41$. Find another value for n for which this expression generates a composite number.
17. Find the sixth triangular, square, and pentagonal numbers.

CHAPTER 1 TEST

one

In Problems 1–10, write the year 2003 in the indicated system.

1. Egyptian
2. Roman
3. Traditional Chinese
4. Ionic Greek
5. Attic Greek
6. Babylonian
7. Mayan
8. Binary
9. Base 5
10. Hexadecimal

In Problems 11–15, write the mixed number, $10\frac{2}{3}$, in the indicated system.

11. Decimal
12. Roman
13. Egyptian
14. Babylonian
15. Ionic Greek
16. Find the decimal equivalent of $70B3.6_{12}$.

17. Convert $111\ 101\ 011_2$ to octal and hexadecimal.

In Problems 18–19, find the answer for the following in binary and check the results using decimal numerals.

18. $111\ 101_2 + 10\ 111_2$

19. $11\ 101_2 \times 1\ 101_2$

20. $-7.45, -\sqrt{40}, -5, \sqrt[3]{-27}, \frac{-5}{9}, 0, \sqrt[4]{19}, \pi,$ $3.1416, 4.6161, 7\frac{2}{3}, 13, \sqrt{-400}$

From that list, determine which are

a) Imaginary.

b) Integers.

c) Irrational.

d) Natural.

e) Rational.

f) Real.

g) Whole.

CHAPTER TWO

Logical Thinking

A modern interpretation of Auguste Rodin's "The Thinker." (Computer Art Image, Courtesy of CORBIS)

OVERVIEW

Logical thinking is a key to making sound decisions and solving complex problems. You use logic in everyday events such as determining why your car won't start, planning the route you are going to take when shopping in the city, or filling out your 1040 tax form in April. In this chapter, we will examine a few of the many facets of logic in the hope of helping you become a better thinker and problem solver. In this journey, we will discuss some basic ideas about premises and conclusions, explore methods of logical argument, analyze truth tables and flowcharts, and use logic to help solve puzzles.

2 A SHORT HISTORY OF LOGIC

The ability to think logically is an innate human ability. From earliest times, human beings have used this power. However, the first systematic study of logic is credited to the Greek philosopher Aristotle (384–322 B.C.). In his work *Organon,* Aristotle systematized principles of reasoning and laws of logic. He is known for his work on arguments consisting of two statements and a conclusion, called syllogisms. The Stoic and Megarian schools in Greece (c. 300 B.C.) continued his study of logic. They created a logic of propositions and valid inference schemes and were very interested in paradoxes. The most productive of the Stoic-Megarian logicians was Chrysippus.

Logic in the Middle Ages was marked by several different schools of thought. The Roman philosopher Boëthius (c. 480–520) was instrumental in passing on the logical traditions of the Greeks to European monks. These monks sought to preserve the logic found in classical Greek texts. Meanwhile, the Nyaya schools in India also made contributions to the field of logic during this period. As the Middle Ages came to an end, scholastic logic made its entrance. Scholastic logic was characterized by its use of Latin and the influence of Christian theology. It was developed for discussions of theological questions. Noted contributors in the development of logic during this period were Peter Abelard (c. 1130), Robert Grosseteste (c. 1240), St. Thomas Aquinas (c. 1250), and Petrus Hispanus (c. 1260).

Seki Kowa, pictured on this 1992 stamp issued by Japan to commemorate his 350th birthday, shows some of the logic he used in developing magic squares.

With the Renaissance period in Europe, advances in formal logic again appeared. William of Ockham (c. 1320), in *Summa Logicae,* and Walter Burley (c. 1325), in *De puritate artis logicae,* improved on Aristotle's logic. Gottfried Leibniz

(1666) initiated the study of symbolic and mathematical logic in his essay *De arte combinatoria.* In this work and others over the next 25 years, Leibniz implied that mathematics can be derived from the principles of logic. His work began what is considered the period of mathematical logic. Leonhard Euler (c. 1770) adopted a method of visually checking syllogisms by means of circle diagrams. John Stuart Mill (c. 1843) made contributions to the development of inductive logic. Augustus De Morgan (c. 1850), in *Formal Logic,* and George Boole, in his works *The Mathematical Analysis of Logic* (1847) and *An Investigation of the Laws of Thought* (1854), applied algebraic operations to logic and placed logic on a mathematical basis. John Venn (1880) introduced the Venn diagram, which is a modification of the circle diagrams of Euler. Alfred North Whitehead and Bertrand Russell (1910–1913) published the three volumes of *Principia Mathematica,* which attempted to develop mathematics from only undefined concepts and principles of logic. Between 1920 and 1921, Jan Lukasiewicz, Emil Post, and Ludwig Wittgenstein independently introduced truth tables as a means of reaching logical conclusions. Developments in logic continue to the present day. In this chapter, we will not look at the details of formal logic, but we will investigate some of the principles of logic and applications.

CHAPTER 2 PROJECTS

Research Projects

1. Who is Bertrand Russell? What is Russell's Paradox? Explain the reasoning behind Russell's Paradox?
2. Investigate the history of magic squares.
3. Advertisements in magazines make use of logic to persuade us to buy certain products. Examine some ads and determine the "logic" used in the ad.

Math Projects

1. Recreational magazines such as Dell's *Math Puzzles and Logic Problems* contain logic problems. Find and solve one of these problems. Include a detailed description of how you solved the problem, where logic was used, and where you had to resort to guessing.
2. Present solutions to other recreational math puzzles. Those by H. E. Dudeney, Sam Lloyd, or Martin Gardener would be appropriate.
3. Who was Zeno? What are Zeno's paradoxes? Explain the mathematics and the reasoning behind each of Zeno's paradoxes?

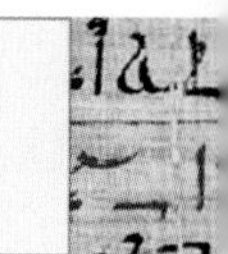

section 2.1 Logic, Statements, and Definitions

LOGIC

Even though humans have the ability to think, they do not always reason correctly. The science of correct reasoning is called **logic.** An understanding of logic will help you correctly arrange supporting evidence that leads to a conclusion. It will also help you understand the process of proving mathematical facts. This process of proving mathematical propositons is probably one of the most challenging aspects of mathematics. The mathematician wants to be sure that a certain proposition actually follows logically from what has already been accepted or, just as important, that a proposition does not follow logically from what has been accepted. To a mathematician, creating an original proof is as gratifying as finishing a painting is to an artist, as setting a record is to an athlete, or as finding a cure to a disease is to a scientist.

STATEMENTS

The basic components of logic are its statements. Statements in logic must have a clear meaning and be either true or false. Statements cannot be both true and false at the same time. They must have only one truth value—that is, either true or false. In general, questions, commands, or vague sentences cannot be used as statements in logic because they cannot be judged to be true or false. The following list gives examples of some sentences that are statements and some that are not.

Statements T/F	Nonstatements
Today is a holiday.	Are we having fun yet?
Pigeons fly.	Do your homework!
The square of 7 is 49.	Don't worry, be happy!
This book contains history sections.	It smells like whatchamacallit.
I did my homework.	This statement is true.

In the study of logic, letters are used to represent statements just as letters are used to represent numbers in algebra. For example, the letter Q can be used to represent the statement, "All the players on this year's team are over 6 feet tall."

EXAMPLE 1

For each of the following, classify S as a statement or nonstatement.

a) S: All men are mortal.
b) S: Yea, team!
c) S: Euclid did not study geometry.
d) S: Finish your dinner.

Solution:

a) S is a statement.
b) S is a nonstatement.
c) S is a statement.
d) S is a nonstatement

Negation of Statements

If you change a statement to one that has the opposite meaning, you form the negation of the statement. The negation of a statement has a truth value that is the opposite of the truth value of the given statement. That is, if a statement is true, then its negation is false. Similarly, if a statement is false, its negation is true. If S represents a statement, then $\sim S$ represents the negation of the statement and is read "not S." For example, if S represents the statement "A triangle has three sides," then $\sim S$ represents the statement "A triangle does not have three sides." Notice that S is true, but its negation $\sim S$ is false. In most statements, forming the negation is simply a matter of changing the action in the statement by adding or deleting the word "not." For example:

Statement	Negation
Kai is running.	Kai is not running.
Kristi does not smile.	Kristi does smile.
The two amounts are equal.	The two amounts are not equal.
Logic is not a five-letter word.	Logic is a five-letter word.
He is a guitar player.	He is not a guitar player.

In statements that involve the words "all," "every," "some," "none," or "no," forming the negation is not as easy as in the previous examples. For example:

Statement	Negation
All men are mortal.	Some men are not mortal.
Some of the numbers are not positive.	All of the numbers are positive.
No birds are fish.	Some birds are fish.
Some women can swim.	No women can swim.
None of the flashlights worked.	Some of the flashlights worked.

The basic forms for negative statements that involve "all," "every," "some," "none," or "no" can be summarized as follows:

Statement	Negation
All/every	Some . . . not
Some . . . not	All/every
None/no	Some
Some	None/no

EXAMPLE 2

Write the negation of each of the following statements.

a) My car did not start.
b) Some of the cars did not start.
c) None of the cars started.

d) Every car started.
e) Some of the cars started.

Solution:

a) My car did start.
b) All of the cars started.
c) Some of the cars started.
d) Some cars did not start.
e) No car started.

Conditional Statements

A very important type of statement used in logic is the **conditional statement.** It is a statement formed by two individual statements joined by the words "if . . . then . . . ." In the conditional statement "If A, then B," the letter A represents the "if" clause or the **hypothesis** or **antecedent,** and the letter B represents the "then" clause or the **conclusion** or **consequent.** For example, in the statement "If you are a student, then you should study," "you are a student" is the hypothesis and "you should study" is the conclusion. We will see in the next section that conditional statements are used extensively in formulating logical arguments.

The same conditional statement can be written in different ways.

If A, then B	If you are a student, then you should study.
A implies B	Being a student implies that you should study.
$A \rightarrow B$	Being a student $\rightarrow$ one should study.
All A are B	All students should study.

Related to the conditional statement $A \rightarrow B$ are three other basic types of statements.

Converse:	$B \rightarrow A$
Inverse:	$\sim A \rightarrow \sim B$
Contrapositive:	$\sim B \rightarrow \sim A$

These statements are important because they are used in creating valid arguments. It can be shown that if a statement is true, then its contrapositive is always true, and if a statement is false, then its contrapositive is also false. That is, a conditional statement and its contrapositive are logically equivalent. However, if a statement is true, its inverse and converse may be either true or false. A conditional statement and its converse or its inverse are not logically equivalent.

NOTE: A conditional statement is logically equivalent to its contrapositive. A conditional statement can be replaced with its contrapositive and keep its same truth value. ($A \rightarrow B$ is logically equivalent to $\sim B \rightarrow \sim A$.)

This interrelationship between these statements can be seen by studying the following examples:

EXAMPLE 3

Write the converse, inverse, and contrapositive of the true conditional statement below. Determine whether each of the statements is true or false.

If it is an IBM PC, then it is a computer.

Solution: Let

$A =$ It is an IBM PC.

$B =$ It is a computer.

$A \rightarrow B$ the conditional statement

Converse:	$B \rightarrow A$	If it is a computer, then it is an IBM PC. (false)
Inverse:	$\sim A \rightarrow \sim B$	If it is not an IBM PC, then it is not a computer. (false)
Contrapositive:	$\sim B \rightarrow \sim A$	If it is not a computer, then it is not an IBM PC. (true)

EXAMPLE 4

Write the converse, inverse, and contrapositive of the false conditional statement below. Determine whether each of the statements is true or false.

If x is an even number, then the last digit of x is 2.

Solution: Let

$A = x$ is an even number.

$B =$ The last digit of x is 2.

$A \rightarrow B$ the conditional statement

Converse:	$B \rightarrow A$	If the last digit of x is 2, then x is an even number. (true)
Inverse:	$\sim A \rightarrow \sim B$	If x is not an even number, then the last digit of x is not 2. (true)
Contrapositive:	$\sim B \rightarrow \sim A$	If the last digit of x is not 2, then x is not an even number. (false)

EXAMPLE 5

Write the converse, inverse, and contrapositive of the true conditional statement below. Determine whether each of the statements is true or false.

If two lines are perpendicular, then the two lines form a right angle.

Solution: Let

$A =$ Two lines are perpendicular.

$B =$ Two lines form a right angle.

$A \rightarrow B$ the conditional statement

Converse	$B \rightarrow A$	If two lines form a right angle, then the two lines are perpendicular. (true)

Inverse:	$\sim A \rightarrow \sim B$	If two lines are not perpendicular, then the two lines do not form a right angle. (true)
Contrapositive:	$\sim B \rightarrow \sim A$	If two lines do not form a right angle, then the two lines are not perpendicular. (true)

In Example 5, notice that the statement and its converse are both true statements; that is, $A \rightarrow B$ and $B \rightarrow A$ are both true. In such a situation, the two statements are combined into a **biconditional statement,** which is written in the following ways:

A if and only if B

A iff B

$A \leftrightarrow B$

The above box introduces the abbreviation "iff," which stands for "if and only if." We will continue to use this abbreviation throughout the text. Thus, the results of Example 5 could be written as follows:

Two lines are perpendicular if and only if the two lines form a right angle.

Two lines are perpendicular iff the two lines form a right angle.

Two lines are perpendicular $\leftrightarrow$ the two lines form a right angle.

DEFINITIONS

Besides conditional statements, definitions of terms are used as basic building blocks of a mathematical system. A definition states properties of the term being defined, gives us a way of recognizing what is defined, and provides a way of distinguishing what is being defined from other objects. A definition must:

1. Name the term being defined.
2. Use words that have already been defined or already understood.
3. Be biconditional. The statement of the definition and its converse must both be true.

Besides those three necessary properties, a good definition should also:

1. Place the term in the smallest or nearest group to which it belongs.
2. Use the minimum information needed to distinguish the object from other objects.

EXAMPLE 6

Explain why the following statements are not examples of definitions.

a) It is a place where tennis matches are played.

b) Charity is the act of being eleemosynary.

c) A mother is a parent of a child.
d) A social insect is a bee.

Solution:

a) The term that is being defined is not included.
b) Charity is being defined by a word that is more difficult to understand and is probably not previously understood.
c) The statement is not biconditional. The converse statement "A parent of a child is a mother" is false.
d) The definition is not biconditional. The statement is false because a social insect could also be an ant.

EXAMPLE 7

Even though the following statements satisfy the three properties of a definition, they are not good definitions. Explain why.

a) A treasurer is in charge of finances of an organization.
b) A triangle is a plane figure with three sides, three angles, and three vertices.
c) A Chevy is a Chevrolet automobile.

Solution:

a) The group to which a treasurer belongs is not included. A treasurer is a *person* in charge of finances of an organization and not a machine, report, or computer program.
b) More information is given than is needed to distinguish the object from all other plane figures.
c) The definition does not place the term into the group to which it belongs. "Chevy" is a *slang* term for a Chevrolet automobile.

In this section, we have looked at the various kinds of statements and two ways of combining statements. We should pay particular attention to these concepts:

1. If a statement is true, its negation is false, and vice versa.
2. If a conditional statement is true, then its contrapositive is also true.
3. A definition must be biconditional.

PROBLEMS ▶ Explain ▶ Apply ▶ Explore

▶ Explain

1. Explain the difference between a statement and a nonstatement.
2. Explain how to form the converse of a statement.
3. Explain how to form the inverse of a statement.
4. Explain how to form the contrapositive of a statement.
5. Explain how to form the negation of a statement.

6. What is meant by a *biconditional* statement?

In Problems 7–14, explain why each statement is not an example of a definition.

7. A skean is a falchion.

8. To gasconade is to vaunt.

9. It is used to remove the skin of a potato.

10. We call it the period from noon to sunset.

11. An integer is a positive or negative number.

12. An obtuse angle is not a 90° angle.

13. A Toyota is an automobile.

14. A pencil is a writing implement.

In Problems 15–18, show how each statement satisfies the three necessary conditions of a definition.

15. A puppy is a young dog.

16. A crook is a person who steals or cheats.

17. A quadratic equation is an equation of the form $ax^2 + bx + c = 0$, where a, b, and c are real numbers and a is not equal to zero.

18. A rational number is a number that can be represented as the ratio of two integers a/b, where b is not equal to zero.

Apply

In Problems 19–26, determine which of the following are considered statements and which are not considered statements. Be sure to give your reasoning.

19. Don't eat the daisies!

20. My dog is a dalmation.

21. Do you enjoy reading novels?

22. The jokes are great.

23. Mozart composed classical music.

24. The camera is not a Kodak™.

25. This statement is false.

26. Use the quadratic formula on that one.

In Problems 27–40, write the negation of the statements.

27. My car is in the shop.

28. Fred did not do his research paper.

29. I hate sitting around doing nothing.

30. The two lines are parallel.

31. This is an example of an exponential equation.

32. No rational number is irrational.

33. All fish can live under water.

34. Every chef knows how to boil water.

35. Some numbers are not prime numbers.

36. Some dogs do not have long tails.
37. Some trees are always green.
38. Some TV shows are boring.
39. None of the numbers are positive.
40. My uncle did not like what you did to his lawn.

In Problems 41–48, write the converse, inverse, and contrapositive of each conditional statement.

41. If you get a busy signal, the phone is in use.
42. If there is a leak in the tube, it will become flat.
43. If it is a point on the circle, then it will be 16 in. from the center of the circle.
44. If a whole number ends in 3, then it is an odd number.
45. When G. H. Mutton speaks, I listen.
46. When I am asleep, nothing bothers me.
47. If the figure has five sides, it is not a hexagon.
48. If it is an ellipse, then its graph is not a circle.

Explore

In Problems 49–60, if possible, find an example of a conditional statement (P) that satisfies the stated condition.

49. P is true and its inverse is false.
50. P is true and its inverse is true.
51. P is true and its converse is true.
52. P is true and its converse is false.
53. P is true and its contrapositive is false.
54. P is true and its contrapositive is true.
55. P is false and its inverse is false.
56. P is false and its inverse is true.
57. P is false and its converse is true.
58. P is false and its converse is false.
59. P is false and its contrapositive is false.
60. P is false and its contrapositive is true.

For the terms in Problems 61–64, write definitions that satisfy the three necessary conditions of a definition.

61. Dog
62. Table
63. Book
64. Shoe
65. Using a standard dictionary, find three words that have definitions that satisfy the three necessary conditions of a definition.
66. Using a standard dictionary, find three words that have definitions that do not satisfy the three necessary conditions of a definition.

67. Construct the contrapositive of the inverse of $A \rightarrow B$. What is this equivalent to?

68. Construct the inverse of the contrapositive of $A \rightarrow B$. What is this equivalent to?

Inductive and Deductive Reasoning

We can now use the statements discussed in Section 2.1 to formulate arguments. An argument consists of statements of supporting evidence organized to show that a conclusion is true. Two of the reasoning processes used in creating an argument are induction and deduction.

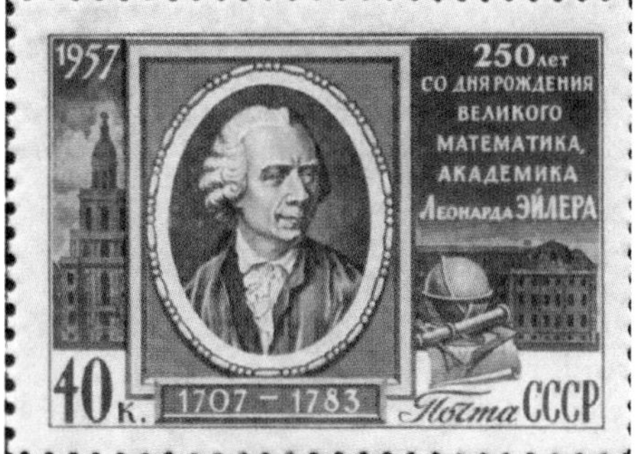

Famed mathematician Leonhard Euler (c. 1770) made contributions to the formal study of logic.

INDUCTION

Induction is the process of reasoning in which conclusions are based on experimentation or experience. When using induction, we make a conclusion about a situation after observing results, analyzing experiences, citing authorities, or presenting statistics. We predict future experiences by extending patterns seen in present experiences.

EXAMPLE 1

On a cold winter night, there is a fire burning in the fireplace. A baby crawls up to the fireplace and touches the screen covering the fireplace. The baby burns his little hand and cries. A few weeks later, the baby does the same thing and burns his hand again. Because of these experiences, the baby stays away from the fireplace when he sees a fire burning. He has used the process of induction. On the basis of his experience, the baby has made the conclusion that touching an object heated by a fire will cause his hand to hurt.

EXAMPLE 2

In the triangles, use a ruler to measure each side and use a protractor to measure each angle. Is there a relationship between the length of a side and the size of the angle that is opposite that side, that is, between $\angle A$ and side a, $\angle B$ and side b, and $\angle C$ and side c?

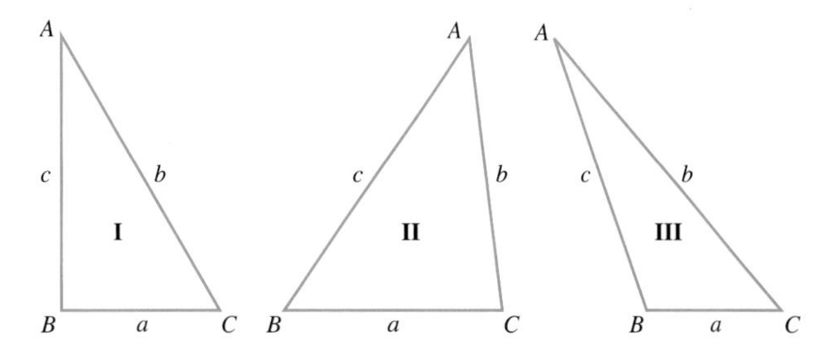

Solution:

Triangle I	Triangle II	Triangle III
$a = 2$ cm, $\angle A = 30°$	$a = 2.7$ cm, $\angle A = 42°$	$a = 1.7$ cm, $\angle A = 20°$
$b = 3.9$ cm, $\angle B = 90°$	$b = 3.4$ cm, $\angle B = 56°$	$b = 4.4$ cm, $\angle B = 111°$
$c = 3.3$ cm, $\angle C = 60°$	$c = 4.1$ cm, $\angle C = 82°$	$c = 3.6$ cm, $\angle C = 49°$

From studying the results, we can conclude that in each triangle, opposite the longest side is the largest angle and opposite the shortest side is the smallest angle.

EXAMPLE 3

Consider the expression $n^2 - n + 11$. It seems to generate prime numbers when whole number values are substituted for n.

$$n = 0 \quad n^2 - n + 11 = 0 - 0 + 11 = 11$$
$$n = 1 \quad n^2 - n + 11 = 1 - 1 + 11 = 11$$
$$n = 2 \quad n^2 - n + 11 = 4 - 2 + 11 = 13$$
$$n = 3 \quad n^2 - n + 11 = 9 - 3 + 11 = 17$$
$$n = 4 \quad n^2 - n + 11 = 16 - 4 + 11 = 23$$
$$\vdots$$

Using inductive reasoning, you might conclude that the expression $n^2 - n + 11$ always generates prime numbers for whole numbers, n. This, however, is not true. When $n = 11$, the value of the expression is 121, and 121 is not a prime number. By not investigating a sufficient number of cases, we could have made a false conclusion.

Inductive reasoning is the process of determining a general conclusion by examining individual cases or particular facts. It can show us that there is a good chance that our conclusion is true, but we will not be absolutely certain. For example, if the first ten people you meet at a new school are very helpful and friendly, you might generalize that the people at the school are really nice. However, the next person you meet might be extremely hostile. On the basis of the first ten people, your conclusion seemed true. However, the 11th person disproved your conjecture. When reasoning inductively, one has to make sure that there is a sufficient number of facts or specific cases to warrant a conclusion. Scientists, for example, repeat experiments many times before making conclusions. Again, a good inductive argument only gives a high probability that a statement is true or an action should be performed.

EXAMPLE 4

Give an inductive argument to persuade your friend Maria to vacation in Hawaii.

Solution: The following list of premises gives an example of an inductive argument.

1. The weather is great in Hawaii, and the beaches are fantastic.
2. My friend has a condo you can rent for only $150 a week.
3. The airlines are having a special on fares to Hawaii this month.
4. I went there last year and had a wonderful time.
5. The people there were friendly and treated me kindly.
6. There are lots of nice guys vacationing there. You'll have a great time and are bound to meet that special man you have been looking for.
7. In a recent travel magazine, 95% of the vacationers polled said that they enjoyed their vacation in Hawaii.
8. Anna Holiday, worldwide traveler and economist, in her book *Travels to Paradise,* states that a vacation in Hawaii is the best bet for your travel dollar.

Such an argument shows that vacationing in Hawaii makes good sense. It implies that there is a good chance that Maria would enjoy a vacation in Hawaii based on past experience, statistics, and the opinions of experts. However, even if all the premises of this argument were true, Maria could still have a miserable time in Hawaii.

Even though inductive reasoning does not ensure certainty, it is the basis of many everyday decisions and is used to extend our knowledge by making suppositions based on experimentation. If certainty is desired, we can in some situations use a second reasoning process: deduction.

DEDUCTION

Deduction is the process of reasoning in which conclusions are based on accepted premises. These premises are usually articles of faith, laws, rules, definitions, assumptions, or commonly accepted facts. The conclusions we reach are either explicitly or implicitly contained in the premises.

A deductive argument is a series of statements consisting of premises and a conclusion. The premises are the statements of evidence from which the conclusion is drawn. In deductive arguments, the premises are usually written as conditional statements. Arguments may take many different forms. One of the common forms is the syllogism. The basic syllogism consists of two statements or premises, and a logical conclusion drawn from them. According to Aristotle, "A syllogism is a discourse in which, certain things being posited, something else follows from them by necessity." In this chapter, we discuss three types of syllogisms: hypothetical syllogisms, affirming the antecedent, and denying the consequent.

Hypothetical Syllogism

If A, B, and C represent statements, a **hypothetical syllogism** is constructed from the statements, the first two lines being the premises and the third being the conclusion. The hypothetical syllogism can be written in three different ways:

If A, then B.	A implies B.	$A \to B$
If B, then C.	B implies C.	$B \to C$
$\therefore$ If A, then C.	Therefore, A implies C.	$\therefore A \to C$

Note: $\therefore$ means therefore.

In the hypothetical syllogism, the argument is correct even when one or both of the premises are false. The truth or falsehood of the premises does not affect the logic of the argument. Logic deals with the relationship between premises and conclusion, not the truth of the premises. To say that a deductive argument is correct means that the premises are related to the conclusion in such a way that, if the premises are true, the conclusion must be true. A conclusion cannot be false if the logical form is correct and the premises are true.

NOTE: To reason deductively toward a true conclusion using a syllogism, you must have the correct form and true premises.

The following are examples of correct deductive arguments that use hypothetical syllogisms and lead to true conclusions.

If you live in Palolo, then you live on Oahu.
If you live on Oahu, then you live in Hawaii.
Therefore, if you live in Palolo, then you live in Hawaii.

If a triangle is isosceles, then it has two equal sides.
If a triangle has two equal sides, then it has two equal angles.
Therefore, an isosceles triangle has two equal angles.

EXAMPLE 5

Is the following argument a hypothetical syllogism? Why or why not?

If you have a party, you should invite your friends.
If you are graduating from college, you should invite your friends.
Therefore, if you are having a party, you are graduating from college.

Solution: The argument is not a hypothetical syllogism. The premises do not link properly. The conclusion of the first premise should be the hypothesis of the second premise, and no logical rearrangement can accomplish the proper linking of the statements.

EXAMPLE 6

Even though the conclusion of this argument is true, explain why the following argument is a poor one.

If you are over 18 years old, then you can read.

If you can read, you can vote.

Therefore, if you are over 18 years old, then you can vote.

Solution: The argument has the form of a hypothetical syllogism, so it is a correct argument. However, it is a poor argument, since neither of the premises is true. The argument does not actually prove its conclusion.

Affirming the Antecedent

If A and B represent statements, an argument that affirms the antecedent has the following form.

Major premise:	$A \rightarrow B$
Minor premise:	A
Conclusion:	$\therefore B$

The major premise is a conditional statement. The minor premise states that the hypothesis of the major premise is true or has occurred. This is called **affirming the antecedent.** An example of an argument of this form is as follows.

If I study for 6 hours, I will pass the exam.

I studied for 6 hours.

Therefore, I will pass the exam.

This classical argument is another example of affirming the antecedent.

All men are mortal.

Socrates is a man.

Therefore, Socrates is mortal.

This can be written so that the correct form is apparent.

If one is a man, then one is mortal.

Socrates is a man.

Therefore, Socrates is mortal.

If an argument has the correct form, it is a logical argument. However, if it is to be a convincing argument with a true conclusion, its premises must also be true. You can affirm the antecedent to reason deductively if the argument has the correct form and true premises.

EXAMPLE 7

Is the following argument a good one? Explain.

> If you want to run a marathon, then you should train for the race.
>
> Kerry wants to run a marathon.
>
> Therefore, Kerry should train for the race.

Solution: The argument has the correct form for affirming the antecedent. If we take its first premise as true because of commonly accepted notions about the physical stamina needed to run a marathon (26.2 mi), the argument is a good one.

EXAMPLE 8

What is wrong with the following argument?

> All good chess players wear glasses.
>
> Sylvia is a good chess player.
>
> Therefore, Sylvia wears glasses.

Solution: Rewriting the argument into conditional statements, we get the following:

> If one is a good chess player, then one wears glasses.
>
> Sylvia is a good chess player.
>
> Therefore, Sylvia wears glasses.

Even though the argument has the correct form of an argument using the technique of affirming the antecedent, the major premise is not true. Thus, it is a correct argument but it does not arrive at a true conclusion. You need both the correct form and true premises to ensure true conclusions.

Denying the Consequent

If A and B represent statements, an argument that denies the consequent has the following form.

Major premise:	$A \rightarrow B$
Minor premise:	$\sim B$
Conclusion:	$\therefore \sim A$

Examples of arguments of this form are as follows:

> If John is at the beach, then he wears sun screen on his nose.
>
> John does not have sun screen on his nose.
>
> Therefore, John is not at the beach.

If you pay the bill on time, then you are not charged a penalty.

You are charged a penalty.

Therefore, you did not pay the bill on time.

The major premise is a conditional statement. The minor premise is a denial (negation) of the consequent (conclusion) of the conditional statement. For this reason, this argument is called **denying the consequent.** This form of argument is based on the contrapositive principle in which the statement $A \to B$ is logically equivalent to $\sim B \to \sim A$. We can see that this form of argument is correct by observing that it is really an application of affirming the antecedent.

Major premise:	$A \to B$	is equivalent to	$\sim B \to \sim A$
Minor premise:	$\sim B$		$\sim B$
Conclusion:	$\therefore \sim A$		$\therefore \sim A$

EXAMPLE 9

Is the following argument a good one? Explain.

If a number is not positive, then the number is negative.

Zero is not negative.

Therefore, zero is positive.

Solution: The argument has the form of an argument using denying the consequent, so it is a correct argument. However, its first premise is not true, since if a number is not positive, it could be either negative or zero. Thus, the argument is faulty.

We can also make correct arguments from premises that do not at first glance seem to be one of our standard logical forms, as in the next example.

EXAMPLE 10

Construct a logically correct argument from the following premises:

If P, then $\sim Q$.

If $\sim R$, then Q.

If R, then $\sim S$.

Solution: To have the correct form of the hypothetical syllogism, the conclusion of one statement must be the hypothesis of the next statement. Since we know that if a statement is true, its contrapositive is true, we can use that principle on the second premise, that is, "If $\sim R$, then Q" implies "If $\sim Q$, then R."

If P, then $\sim Q$.		If P, then $\sim Q$.
If $\sim R$, then Q.	$\to$	If $\sim Q$, then R.
If R, then $\sim S$.		If R, then $\sim S$.
		$\therefore$ If P, then $\sim S$.

Thus, the conclusion of this argument is: If P, then $\sim S$.

While you may use the contrapositive statement in a logical argument, do not use the inverse or converse.

Watch out for the following:

Converse: $A \rightarrow B$ does *not* necessarily imply $B \rightarrow A$.

Inverse: $A \rightarrow B$ does *not* imply $\sim A \rightarrow \sim B$

Although there are other forms of syllogisms and methods of reasoning deductively, the three forms of syllogisms explained in this section will be adequate for our brief excursion into logic.

THE ROLES OF INDUCTION AND DEDUCTION

You have been introduced to two methods of reaching reasonable conclusions: induction and deduction. Both processes play a role in the formulation of a mathematical system of geometry. Induction is used to conjecture facts about geometry. Early Egyptians, Babylonians, and Greeks used experience and experimentation to hypothesize relationships about geometric objects. So, too, modern mathematicians discover mathematical relationships by looking at specific cases and examples and generalizing their findings. The deductive process is used to determine if the findings follow logically from what has been previously accepted as true in the mathematical system. This process is sometimes very difficult. For example, it took nearly 2000 years to prove that, in general, it is impossible to trisect an angle with only a straight edge and a compass.

section 2.2

PROBLEMS ○ Explain ○ Apply ○ Explore

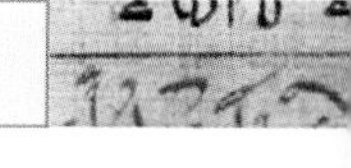

○ Explain

1. Explain what is meant by "inductive reasoning."
2. Explain what is meant by "deductive reasoning."
3. What is a syllogism?
4. Explain what is meant by "affirming the antecedent."
5. Explain what is meant by "denying the consequent."
6. Give an example in which a statement is true but the converse of the statement is not true.
7. Give an example in which a statement is true but the inverse of the statement is not true.
8. Explain why it is possible to have a correct argument even if the conclusion is false.

Apply

Determine whether or not the following arguments are correct. For those that are not correct, (a) explain what is wrong with the argument; (b) change the minor premise and make a correct argument.

9. When it is midnight, I am asleep.
I was asleep.
Therefore, it was midnight.

10. All NBA basketball players are over 5 ft tall.
Russell is 6 ft tall.
Therefore, Russell plays in the NBA.

11. If you are a farmer in Polt County, then you grow corn.
Farmer Ron does not farm in Polt County.
Therefore, Farmer Ron does not grow corn.

12. All Rhode Island Red hens lay brown eggs.
My hen, Marguerite, is a Rhode Island Red.
Therefore, Marguerite lays brown eggs.

13. If *ABCD* is a square, it has four sides.
If it has four sides, then it is a quadrilateral.
Therefore, if *ABCD* is a square, it is a quadrilateral.

14. If a triangle is equilateral, then it has three equal sides.
$\triangle ABC$ does not have three equal sides.
Therefore, $\triangle ABC$ is not equilateral.

Create valid deductive arguments for the following statements.

15. All even numbers greater than 2 are not prime numbers.

16. If you are a shoplifter, you are dishonest.

17. My teacher, Mrs. Santos, is a nice person.

The deductive arguments based on the hypothetical syllogism have two premises and a conclusion. An argument may, however, have many premises that logically lead to a conclusion. What statement is proven in each of Problems 18–20?

18.
$$3(x+4)=18 \rightarrow 3x+12=18$$
$$3x+12=18 \rightarrow 3x=6$$
$$3x=6 \rightarrow x=2$$

19. If you are serious about school, you should be a good student.
If you are a good student, you should study regularly.
If you study regularly, you will have less time to watch TV.

20. If *A*, then *B*.
If *B*, then *C*.
If *C*, then $\sim D$.
If $\sim D$, then $\sim E$.

21. In Problem 20, find statements for *A*, *B*, *C*, *D*, and *E* that will make the argument correct and the conclusion true.

Explore

22. Example 2 dealt with triangles with sides of unequal lengths. The triangles here have at least two sides with the same length. What conclusions you can induce from analyzing the lengths of their sides and the measures of their angles?

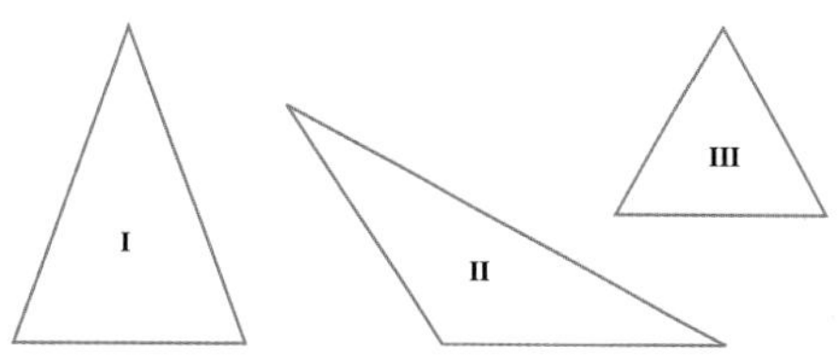

23. Combine the results of Example 2 and Problem 22 to formulate an accurate statement about the relationship between the lengths of the sides of triangles and the angles opposite those sides. Draw a few triangles that verify your statement.

24. Draw any triangle. Find the midpoints of two sides of the triangle and the length of the third side of the triangle. Draw the line segment connecting the two midpoints. Find the length of that segment. State the relationship that seems to exist between the length of the third side and the length of the segment joining the two midpoints. Test out your findings on a few other triangles.

25. Draw any quadrilateral (four-sided figure). Find the midpoint of each of the four sides. Use line segments to join the midpoints (in order) so that another quadrilateral is formed. Repeat this for several other quadrilaterals. What seems to be true about the quadrilaterals by connecting the midpoints?

26. Explain how Problems 22–25 are examples of inductive reasoning. Explain what role deduction would play in further analyzing the geometric principles involved in each problem.

27. Use a compass and straightedge to draw a large semicircle and its diameter. Mark any point on the semicircle and draw line segments joining each endpoint of the diameter to the point marked on the semicircle. Use a protractor to measure the angle formed by the two line segments. Repeat this for several other points marked off on the semicircle. What seems to be true about the angle formed by the line segments connecting the endpoints of the diameter to a point on a semicircle?

28. Show that the expression $n^2 - n + 17$ seems to generate prime numbers for whole number values of n. Explain how using induction might cause you to make a false conclusion in this case. What is the first value of n that causes the expression to produce a composite number?

29. Give an inductive argument to convince a friend that he should not vacation in Hawaii.

30. Give both an inductive argument and a deductive argument to convince a friend to quit smoking.

31. Explain what type of argument is used to show that a frying chicken purchased at a local grocery store weighs less than 4 lb.

a) Every frying chicken sold over the last 25 years has weighed less than 4 lb. Thus, the next chicken you purchase at a local grocery store will surely weigh less than 4 lb.

b) If a frying chicken weighs 4 lb or more, it is not sold in grocery stores.
I purchased this frying chicken in a local grocery store.
Therefore, it must weigh less than 4 lb.

The following premises are not in the exact syllogistic form shown in this section. However, they can be rearranged or logically rewritten so that they correspond to one of the types of syllogisms. What conclusion follows from these premises?

32. All mice are not birds.
Some pets are mice.

33. Every integer is a rational number.
Some numbers are integers.

34. If you can hear, you are not hearing impaired.
If you can't hear, I'll use sign language.

35. If the car is full of gasoline, we can drive 300 mi.
If we haven't reached Yosemite Park, we haven't driven 300 mi.
If we can't see Vernal Falls, we haven't reached Yosemite Park.

36. $P \leftrightarrow Q$
$P \rightarrow R$
$S \rightarrow \sim R$

37. $P \leftrightarrow Q$
$Q \rightarrow \sim R$
$S \rightarrow R$

38. Using your solution to Problem 36, find statements for P, Q, R, and S that make the conclusion true.

39. Using your solution to Problem 37, find statements for P, Q, R, and S that make the conclusion true.

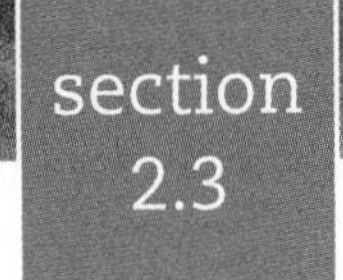

Symbolic Logic and Truth Tables

One of the common beliefs about mathematics is that everything can be turned into abstract symbols that must follow certain sets of rules. It is also believed that these rules always lead the mathematician to a correct solution of whatever problem is at hand and that there is no other correct answer. As we have begun to see, this is not always the case. Chapters 1 and 7 discuss developments in number systems and geometry systems that have led to different, and sometimes conflicting, systems of mathematics. Which system is best is not easy to say, particularly in the case of geometry. Still, mathematicians often try to systematize their work so that, as often as possible, a certain set of information always leads to one and only one solution.

In this chapter, we have seen several methods of logical thinking. This section covers what is, perhaps, the most mathematical method of logical thinking, **symbolic logic.** We have seen a small portion of symbolic logic in earlier sections. For example, the section covering statements, the conditional statement "If A, then B" was abbreviated by using the symbols $A \rightarrow B$. In this section, we provide a list of symbols and their meanings, explain how to translate English into these symbols, and show how proofs may be developed using these symbols.

One important note before we start concerns our philosophy about this material. You should treat this material as you did the material in Chapter 1 on ancient numeration systems. We do not expect you to be an expert in symbolic logic. We merely want to expose you to another area of mathematics, to give you a view of one more of the many facets of mathematics.

TRANSLATING ENGLISH TO SYMBOLS

The following is a partial list of the symbols used in symbolic logic. While looking through the chart, you might have recognized the symbols $\therefore$ and $\sim$. They are found in earlier sections of this chapter. The remaining symbols have not been used but each represents a common English word and each has a formal name. For example, the symbol "$\wedge$" represents the English word "and" and has the formal name "conjunction."

Symbols used in Logic

English	Symbol	Name
Not	$\sim$	Negation
Therefore	$\therefore$	
Implies: If . . . then . . .	$\rightarrow$	Conditional
Equivalent statements	$\equiv$	Logical equivalency
And	$\wedge$	Conjunction
Or	$\vee$	Inclusive disjunction

Our next step is to see how these symbols can be used to abbreviate English sentences.

EXAMPLE 1

Translate the following sentences into symbols.

a) If it is raining, the skies are cloudy.

b) If it is a dog and it can run quickly, then it has four legs.

c) If my grade is not a C and my grade was better than average, then my grade must be an A or a B.

Solution:

a) This is a conditional statement. If we let R = it is raining and C = the skies are cloudy, we can write

$$R \rightarrow C.$$

b) The premise of this statement involves two conditions: the dog and running quickly. To handle this, we need to use a conjunction (and) as well as the conditional statement. Letting D = dog, R = run quickly, and F = has four legs, we can write

$$(D \wedge R) \rightarrow F.$$

We use the parentheses to enclose the portion of the statement that must be considered first.

c) Like the earlier statements, this is a conditional. It contains a conjunction (and) in the antecedent and a disjunction (or) in the consequent. Letting G = better than average grade and A, B, and C represent those letter grades, we have

$$(\sim C \land G) \rightarrow (A \lor B).$$

In our next example, we want to see how an entire syllogism can be translated into symbolic logic.

EXAMPLE 2

Translate the following syllogism into symbols.

If I own chickens, I can have fresh eggs for breakfast.

If I have fresh eggs for breakfast, I am in a good mood until lunchtime.

Therefore, if I own chickens, I am in a good mood until lunchtime.

Solution: We will use C = owning chickens, E = having fresh eggs for breakfast, and G = being in a good mood until lunchtime. Using these letters and two of the symbols from the chart, we can write the syllogism as three statements:

$$C \rightarrow E$$

$$E \rightarrow G$$

$$\therefore C \rightarrow G.$$

EXAMPLE 3

Translate the following paragraph into a set of symbolic logic statements.

If the ground is rocky and the horse needs new shoes, its feet will be sore. However, if the horse does not need new shoes, its feet will not be sore. Therefore, either the ground is not rocky and the horse's feet are not sore, or the horse needs new shoes.

Solution: First, let R = the ground is rocky, N = the horse needs new shoes, and S = the horse's feet are sore. Using these letters, we can abbreviate the paragraph as

If R and N, then S.

If $\sim N$, then $\sim S$.

Therefore, $\sim R$ and $\sim S$, or N.

The problem now looks more approachable. The only difficult part remaining is that the first statement contains both a conjunction and a conditional statement, whereas the third statement contains both a conjunction and a disjunction. To enable us to handle complex statements of this type, we use parentheses.

If (R and N), then S.
If $\sim N$, then $\sim S$.
Therefore, ($\sim R$ and $\sim S$), or N.

Now, using our symbols for the conjunction and conditional, we have

$$(R \wedge N) \rightarrow S.$$
$$\sim N \rightarrow \sim S.$$
$$\therefore (\sim R \wedge \sim S) \vee N.$$

With the completion of this last example, we have seen how a complicated paragraph in English can be reduced to a few short sentences in logic. To complete this section, we show how logical arguments can be used to answer questions about complicated situations. We do this by using a mathematical structure called a truth table.

TRUTH TABLES

A **truth table** is a chart consisting of all the possible true and false combinations of the clauses in a statement. Certain sequences of logical statements, such as syllogisms, have been discussed in earlier sections, but we have not been able to analyze the situation presented in Example 3. With care and patience, a truth table can provide this analysis.

The following three truth tables are the basis for all truth tables. They handle the possible situations that can arise in the disjunction, conjunction, and conditional statements.

The Disjunction Statement

A **disjunction** is a logic statement used in connection with the English word "or." As a result, a disjunction is considered true whenever one or both of its clauses are true. Thus, for the disjunction $A \vee B$, we have the following truth table.

Disjunction Truth Table

A	B	$A \vee B$
T	T	T
T	F	T
F	T	T
F	F	F

Notice that the value of $A \lor B$ is true unless both A and B are false. This corresponds to our intuitive idea of the meaning of an *or* statement.

The Conjunction Statement

A **conjunction** is a logic statement used in connection with the English word "and." A conjunction is true only if *both* of its clauses are true. As a result, we have the following truth table for conjunctions.

Conjunction Truth Table

A	B	$A \land B$
T	T	T
T	F	F
F	T	F
F	F	F

The Conditional Statement

A conditional is a logic statement used when the statement is in the form "if . . . then . . . ." A conditional statement is logically true in three of the four possible cases, as shown here.

Conditional Truth Table

A	B	$A \to B$
T	T	T
T	F	F
F	T	T
F	F	T

At first, it might seem peculiar that the conditional statement is true even when the first clause (the antecedent) is false. To understand this, we need to realize what is meant by "logically true." Recall from our work on deductive reasoning that if the premise of a syllogism is false, we can prove any statement based on the false assumption. Similarly, if the antecedent of a conditional statement is false, any conclusion is considered logically true. Therefore, the only situation that causes the conditional statement to be false is when the first clause (the antecedent) is true and the second clause (the consequent) is false.

Now that we have these three truth tables, we have the basic building blocks for all truth tables. We can use these tables to determine when a statement is logically true.

EXAMPLE 4

Use a truth table to determine the truth values of A and B that make the statement $\sim A \wedge \sim B$ logically true.

Solution: As in the three previous truth tables, we have rows that give all the possible truth values for A and B.

A	B	$\sim A$	$\sim B$	$\sim A \wedge \sim B$
T	T	F	F	F
T	F	F	T	F
F	T	T	F	F
F	F	T	T	T

The first two columns of the truth table give all possible combinations of the truth values of A and B, and the second two columns of the truth table are merely negations of the first two columns. The statement $\sim A \wedge \sim B$ is the conjunction of the two statements $\sim A$ and $\sim B$. Therefore, the only case in which $\sim A \wedge \sim B$ is true is when both A and B are false.

EXAMPLE 5

Determine the truth values of A and B that make the statement $\sim(A \vee B)$ logically true.

Solution: As before, the truth table consists of four rows. The third column of the truth table gives the truth values for the disjunction $A \vee B$. The fourth column gives the truth values for the negation of the disjunction. Therefore, $\sim(A \vee B)$ is truly only when both A and B are false.

A	B	$A \vee B$	$\sim(A \vee B)$
T	T	T	F
T	F	T	F
F	T	T	F
F	F	F	T

Equivalent Statements

Notice that the truth values of $\sim(A \vee B)$ in Example 5 and $\sim A \wedge \sim B$ in Example 4 are identical. Whenever two statements have the same truth values, the statements are said to be **logically equivalent.** Referring to our chart used to convert English to symbols (found at the beginning of the section), we find that the symbol for equivalent statements is $\equiv$. Thus, we can write

$$\sim(A \vee B) \equiv \sim A \wedge \sim B.$$

EXAMPLE 6

Use a truth table to show that the conditional $A \rightarrow B$ and its contrapositive $\sim B \rightarrow \sim A$ are logically equivalent.

Solution: To do this, we construct a truth table containing both $A \rightarrow B$ and $\sim B \rightarrow \sim A$ and show that the two statements always have the same truth values. Being careful about the order of $\sim B$ and $\sim A$ in the fourth and fifth columns of the truth table shows us that the truth values of $A \rightarrow B$ and $\sim B \rightarrow \sim A$ are the same and, therefore, that $A \rightarrow B \equiv \sim B \rightarrow \sim A$.

A	B	$A \rightarrow B$	$\sim B$	$\sim A$	$\sim B \rightarrow \sim A$
T	T	**T**	F	F	**T**
T	F	**F**	T	F	**F**
F	T	**T**	F	T	**T**
F	F	**T**	T	T	**T**

Verifying Syllogisms

A second use of truth tables is to verify the conclusion of a logical argument. To do this, we form a conditional statement. The hypothesis of the conditional will consist of a conjunction of all the premises. The conclusion of the conditional will be the conclusion of the syllogism. A syllogism has a true conclusion if the conditional that is formed in this manner is always true.

EXAMPLE 7

Use a truth table to verify that the following syllogism has a true conclusion.

$$P \rightarrow Q$$
$$P$$
$$\therefore Q$$

Solution: To verify the syllogism, we form a conditional statement whose hypothesis is the conjunction of the premises and whose conclusion is the conclusion of the syllogism. Doing so gives us the statement $[(P \rightarrow Q) \wedge P] \rightarrow Q$. Next we construct a truth table with this statement as its last column. If the last column always has a "true" value, the syllogism has a true conclusion.

P	Q	$P \rightarrow Q$	$(P \rightarrow Q) \wedge P$	$[(P \rightarrow Q) \wedge P] \rightarrow Q$
T	T	T	T	T
T	F	F	F	T
F	T	T	F	T
F	F	T	F	T

Since the last column of this truth table is always true, the syllogism has a true conclusion.

EXAMPLE 8

Determine whether the following logical argument is correct:

$$P \to Q$$
$$\sim Q$$
$$\therefore P$$

Solution: To determine whether the argument is correct, we form a conditional statement from the premises and the conclusion of the argument. Doing so gives us the statement $[(P \to Q) \wedge \sim Q] \to P$. Next, we construct a truth table with this statement as its last column. If the last column always has a "true" value, the argument is correct.

P	Q	$P \to Q$	$\sim Q$	$(P \to Q) \wedge \sim Q$	$[(P \to Q) \wedge \sim Q] \to P$
T	T	T	F	F	T
T	F	F	T	F	T
F	T	T	F	F	T
F	F	T	T	T	F

Since the last column of this truth table is not always true, the argument is not correct.

EXAMPLE 9

Use a truth table to determine whether the following syllogism has a true conclusion.

$$S \to R$$
$$N \to S$$
$$\therefore N \to R$$

Solution: If $N \to R$ is the conclusion of $S \to R$ and $N \to S$, then the statement $[(S \to R) \wedge (N \to S)] \to (N \to R)$ must always be true.

Setting up a truth table for this statement requires eight lines because there are three variables: R, N. and S.

S	R	N	$S \to R$	$N \to S$	$(S \to R) \wedge (N \to S)$	$N \to R$	$[(S \to R) \wedge (N \to S)] \to (N \to R)$
T	T	T	T	T	T	T	T
T	T	F	T	T	T	T	T
T	F	T	F	T	F	F	T
T	F	F	F	T	F	T	T
F	T	T	T	F	F	T	T
F	T	F	T	T	T	T	T
F	F	T	T	F	F	F	T
F	F	F	T	T	T	T	T

Since the last column always has a true value, the conclusion must be true.

In summary, this section has presented a few topics from symbolic logic. The important topics are

1. The English-to-symbol dictionary

English	Symbol	Name
Not	$\sim$	Negation
Therefore	$\therefore$	
Implies, if . . . then . . .	$\rightarrow$	Conditional
Equivalent statements	$\equiv$	Logical equivalency
And	$\wedge$	Conjunction
Or	$\vee$	Inclusive disjunction

2. The three basic truth tables

Disjunction Truth Table

A	B	$A \vee B$
T	T	T
T	F	T
F	T	T
F	F	F

Conjunction Truth Table

A	B	$A \wedge B$
T	T	T
T	F	F
F	T	F
F	F	F

Conditional Truth Table

A	B	$A \rightarrow B$
T	T	T
T	F	F
F	T	T
F	F	T

3. Truth tables can be used to determine whether two statements are logically equivalent and to determine the correctness of a logical argument.

section 2.3

PROBLEMS ◆ Explain ◆ Apply ◆ Explore

Explain

1. What is a disjunction? What is the logic symbol used to represent a disjunction?
2. What is a conjunction? What is the logic symbol used to represent a conjunction?
3. What is a conditional? What is the logic symbol used to represent a conditional?
4. What does it mean for two statements to be logically equivalent? What is the logic symbol used for logical equivalence?
5. What is a truth table? What is one of the uses of a truth table?
6. What are the three basic truth tables? When is each used?

Apply

In Problems 7–16, translate the given sentence(s) into symbols. State what each symbol represents.

7. If it was midnight, I was asleep.
I was asleep.
Therefore, it was midnight.

8. If a person plays basketball in the NBA, he is over 5 ft tall.
Russell (a person) plays NBA basketball.
Therefore, Russell is over 5 ft tall.

9. If you are a farmer in Polt County, then you grow corn.
Farmer Ron does not grow corn.
Therefore, Farmer Ron does not farm in Polt County.

10. If *ABCD* is a square, it has four sides.
ABCD does not have four sides.
Therefore, *ABCD* is not a square.

11. If $\triangle ABC$ is not scalene, then it has either two or three equal sides.
$\triangle ABC$ does not have three equal sides.
Therefore, $\triangle ABC$ has two equal sides.

12. If the correct answer is not yes, then the correct answer is either no or maybe.
The correct answer is not no.
Therefore, the correct answer is maybe.

13. If you are a winning professional golfer, then you have good hand-eye coordination and a positive attitude. Therefore, if you have good hand-eye coordination and a positive attitude, then you are a winning professional golfer.

14. Bill will study hard and pass the test, or Bill will not study hard and not pass the test.

15. If you voted for Bush or Nader, then you did not vote for Gore.

16. If Joan plays sports or works out, then Joan will not be fat and will not be lazy.

In Problems 17–20, use a truth table to determine the truth values of *A* and *C* that make each statement true.

17. $(A \to C) \vee \sim C$

18. $A \to (C \vee \sim C)$

19. $(\sim A \to C) \to \sim C$

20. $(A \to C) \to \sim C$

In Problems 21–24, use a truth table to determine whether the two statements are equivalent.

21. $\sim(A \wedge \sim B)$ and $\sim A \vee B$

22. $\sim A \to \sim B$ and $B \to A$

23. $(A \vee B) \wedge (A \vee \sim B)$ and A

24. $(A \wedge B) \vee (A \wedge \sim B)$ and A

In Problems 25–28, use truth tables to determine whether the arguments are logically correct.

25. $P \rightarrow Q$
P
$\therefore Q$

26. $\sim Q \rightarrow \sim P$
P
$\therefore Q$

27. $\sim A$
$A \rightarrow Q$
$\therefore A$

28. B
$B \rightarrow \sim C$
$\therefore \sim C$

Explore

In Problems 29–32, translate each argument into symbolic form and determine whether each is logically correct by constructing a truth table.

29. If elections become TV popularity contests, then good-looking, smooth-talking candidates will get elected. Therefore, if elections do not become TV popularity contests, then good-looking, smooth-talking candidates will not get elected.

30. If affirmative action policies are adopted, then minorities will be hired. If minorities get hired, then discrimination will be addressed. Therefore, if affirmative action policies are adopted, then discrimination will be addressed.

31. If high school graduates have poor math skills, then they will be less able to get a job in the computer industry. If high school graduates have poor writing skills, then they will be less able to get a job in the computer industry. Therefore, if high school graduates have poor math skills, then they have poor writing skills.

32. Money causes all the world's troubles or money helps the poor. If money helps the poor, it is not the cause of all the world's troubles. Money is the cause of all the world's troubles. Therefore, money does not help the poor.

33. Determine whether the argument in Example 3 is correct.

34. Lewis Carroll, author of *Alice in Wonderland,* was also an accomplished logician. One of his logic problems was the following.

No ducks waltz.
No officers ever decline to waltz.
All my poultry are ducks.
Therefore, my poultry are not officers.

If we let D = ducks, P = my poultry, O = officers, and W = willing to waltz, the problem becomes

$$D \rightarrow \sim W$$
$$O \rightarrow W$$
$$P \rightarrow D$$
$$\therefore P \rightarrow \sim O$$

Construct a 16-row truth table to give all the possible true-false combinations of D, W, O, and P. Use the truth table to verify whether the conclusion of the argument is true.

section 2.4

Logic and Flowcharts

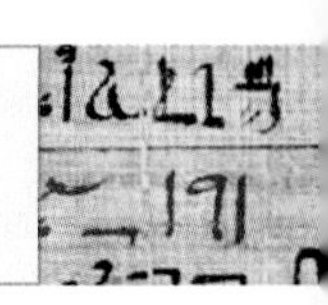

Logic is used in many areas. It is particularly evident in a diagram called a flowchart. A **flowchart** describes the logical path that is followed when a decision is made. Formal flowcharts are used in computer programming to outline the decisions and steps that are followed by a computer as it performs its operations. Industry uses flowcharts to indicate how a process will occur as it progresses through different levels. Corporations, schools, and government bureaus also use flowcharts to display their organizational or management structures. In this section, we will examine how flowcharts are used to outline the logical solution to a problem or the structure of an organization.

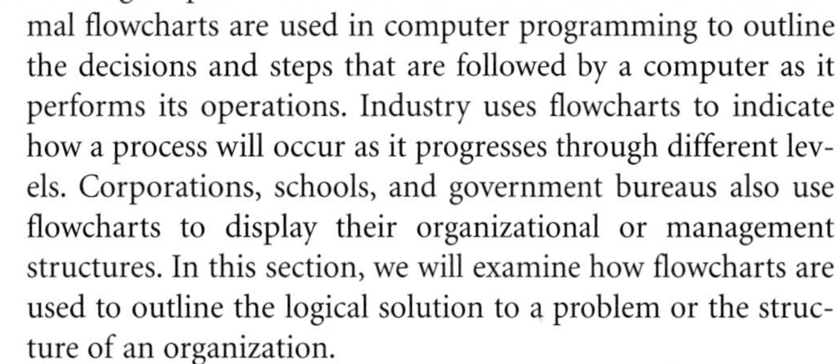

Rube Goldberg is famous for his flowchart-like cartoons.

SYMBOLS IN FLOWCHARTS

We will use four basic symbols: the start/stop symbol, the statement symbol, the decision symbol, and the flow line in constructing flowcharts. While a more thorough treatment of flowcharts will include more than these four symbols, they will be sufficient for our needs.

The **start/stop symbol** is an oval or circle and indicates the beginning or end of a line of logic.

The **statement symbol** is a rectangle and is used to indicate either an action, a person, or a result.

Statement Symbol

The **decision symbol** is a diamond-shaped object known as a rhombus. The decision symbol indicates a question in the flowchart.

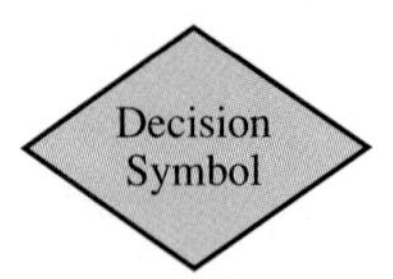

The final symbol, the **flow line,** is a directed line segment (a ray) that connects the other symbols and describes the path that will be followed as you progress through the flowchart.

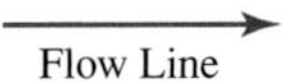

CONSTRUCTING A FLOWCHART

To construct a flowchart, it is first necessary to have a clear idea of what situation the flowchart is to describe and what information is necessary to completely describe the situation. The following two examples show how a flowchart can be used to describe increasingly complex situations.

EXAMPLE 1

A company wants to construct a flowchart that represents the method it will use in answering a customer's telephone call. The initial plan is to have the telephone receptionist route the incoming calls to either the business department or the technical staff, depending on the nature of the call.

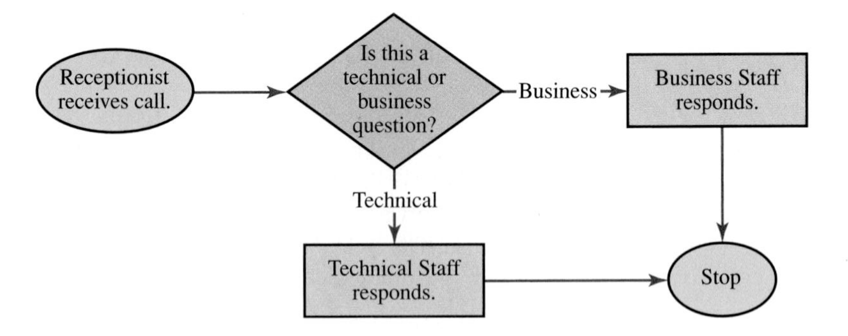

Solution: The flowchart drawn up by the company is shown in the figure. In a fairly straightforward process, the receptionist is expected to answer each call and forward it to the appropriate department. A member of that department will then respond to the call, and the process ends.

The process described in the preceding example might seem straightforward and logical. However, because of the simplicity of the flowchart, it might not describe the process actually followed by the people working in the company. In such an instance, the flowchart can be expanded to handle a more complicated situation.

EXAMPLE 2

Occasionally, a customer will call the company and have a question that both the technical and business staffs must respond to. The customer may also have more than one question. How can the company add the capability to handle these situations into the flowchart?

Solution: After discussions among the staff in both departments and the telephone receptionist, it is decided that the best approach is to allow either department to transfer a call to the other department. The resulting flowchart is shown in this figure.

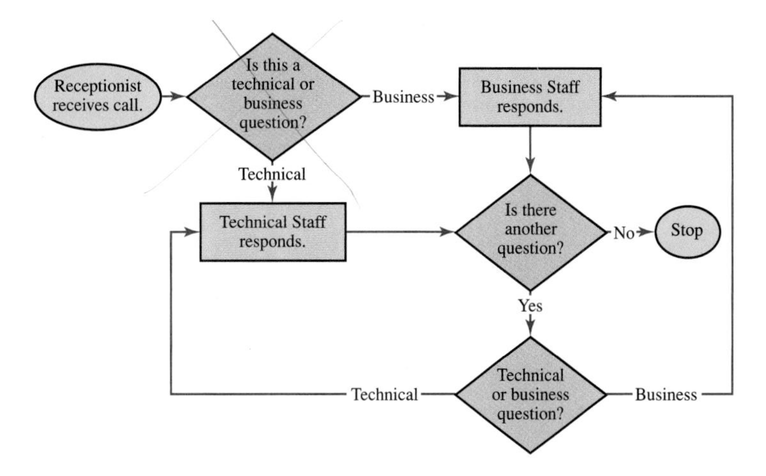

Notice that this new flowchart allows two new possibilities. The customer is allowed to ask more than one question in any department and also be transferred back and forth between the two departments more than once. Only when the customer has no more questions does the process reach an end.

Throughout this chapter, we have shown how logic can be used to solve problems. The process of solving problems is one of analyzing both the available information and the desired goal and then determining a logical method of reaching that goal. For a simple problem, this can be accomplished without the use of flowcharts. However, as the problem becomes more involved, a flowchart provides a visual map to guide you to a solution. In the following example, we show how to construct a flowchart to answer a particular question.

EXAMPLE 3

Early one morning, your car won't start. Before calling for a tow truck, you want to make sure that the cause is not a simple problem that you can fix yourself. What steps should be used in reaching the decision to call the tow truck?

Solution: Since the goal is to call the tow truck only after all the simple causes for the car to not start are eliminated, the plan is as follows:

1. List the possible simple causes.
2. Determine the appropriate action if one of these causes is at fault.
3. Call the tow truck only as a last resort.

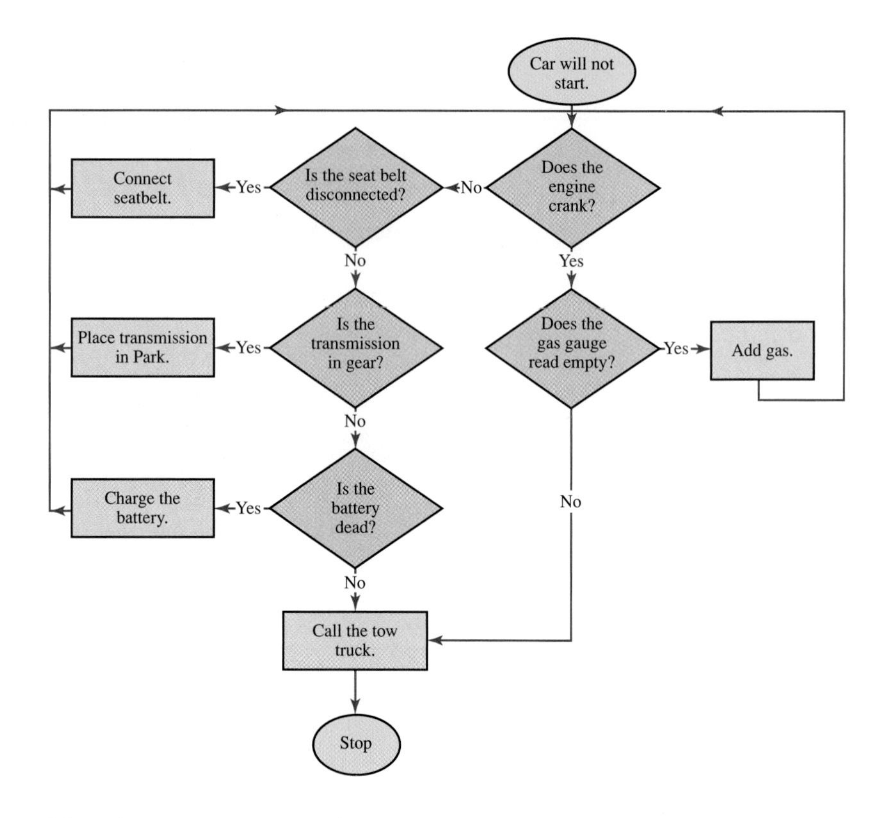

Although you are not a mechanic, you understand that if the engine is not receiving the correct amount of gasoline, the engine will crank but will not start. Alternatively, if the problem is electrical in nature, the engine will not crank. Thus, determining whether the engine will crank is the first decision.

If the engine cranks but does not start, the car has an empty gas tank or there is some other problem with the fuel system. If the tank is empty, it can be partially filled with gas from a spare can in the tool shed. If there is some other problem with the fuel system, you will call the tow truck.

If the engine will not crank, you know of three possible causes: the seat belt is not connected, the automatic transmission is in gear, or the battery is dead. If none of these is the cause, you decide to call the tow truck.

ORGANIZATIONAL CHARTS

Our final example of how flowcharts are used is to describe the organizational structure of a company. While the flowchart itself does not contain any decision blocks, it does provide a diagram of how decisions are made within the company.

EXAMPLE 4

You have been given the following list of names and positions for the Vallejo Manufacturing Company and have been asked to create an organizational chart.

President	P. Vallejo
V.P. Finance	P. Dang
Accounts Payable/Receivable	T. Tran
Controller	M. Yotter
V.P. Engineering	T. Hopkins
Engineering	T. Lee
Documentation	C. Wilson
V.P. Marketing	B. DeSousa
Sales	M. Yoshinobi
Customer Service	M. Sanchez
V.P. Manufacturing	M. Agredano
Purchasing	F. White
Assembly	J. O'Connor
Shipping	T. Root

Solution: The organizational chart for Vallejo Manufacturing is shown in the figure.

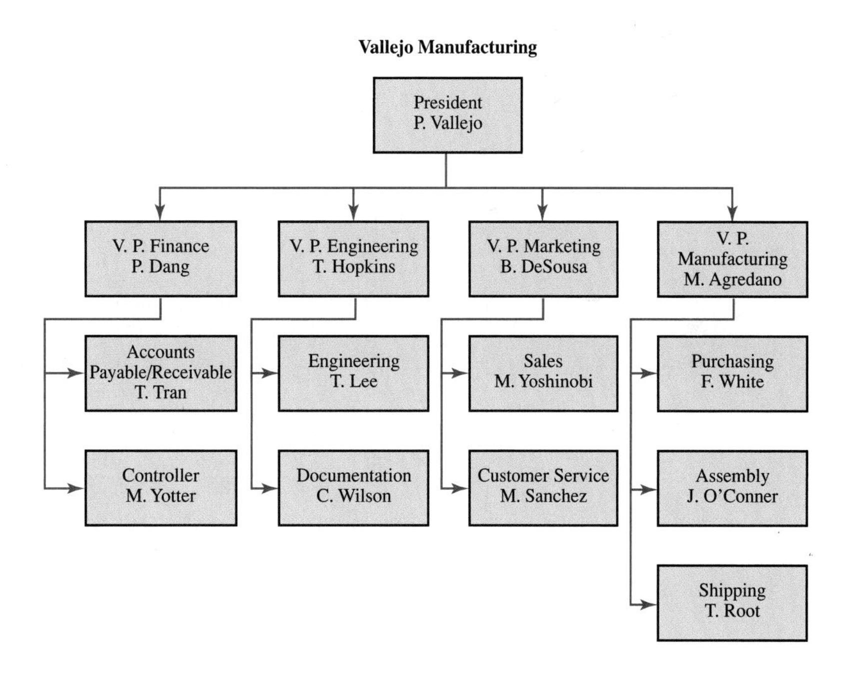

Based on the organizational chart, the chain of command in the company becomes clear. If a major policy change in the shipping department is being discussed, it must be approved first by T. Root, then by the V.P. of Manufacturing, M. Agredano, and finally by the President, P. Vallejo.

The three situations described in this section show how a flowchart can be an effective way to organize the logical path to the solution of a problem. The construction of a flowchart forces you to think clearly and organize an efficient solution to a problem. In essence, a flowchart is a picture of logic in use.

PROBLEMS ● Explain ● Apply ● Explore

● Explain

1. What is the purpose of a flowchart?
2. What is the purpose of a flow line in a flowchart? What symbol is used?
3. What is the decision symbol and how is it used in a flowchart?
4. What is the statement symbol and how is it used in a flowchart?
5. What is the purpose of an organizational chart?
6. Are decision statements used in organizational charts? Explain why or why not.
7. Explain the statement "A flowchart is a picture of logic in use."

● Apply

In Problems 8 and 9, give a written description of the flowchart.

8.

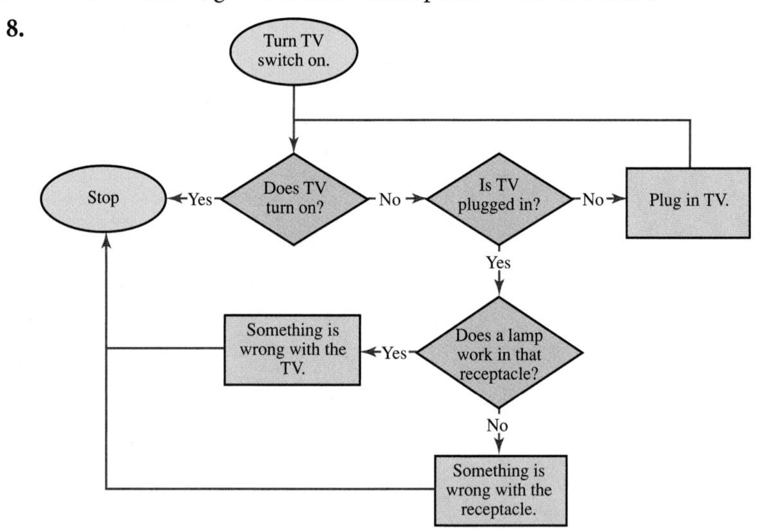

9.

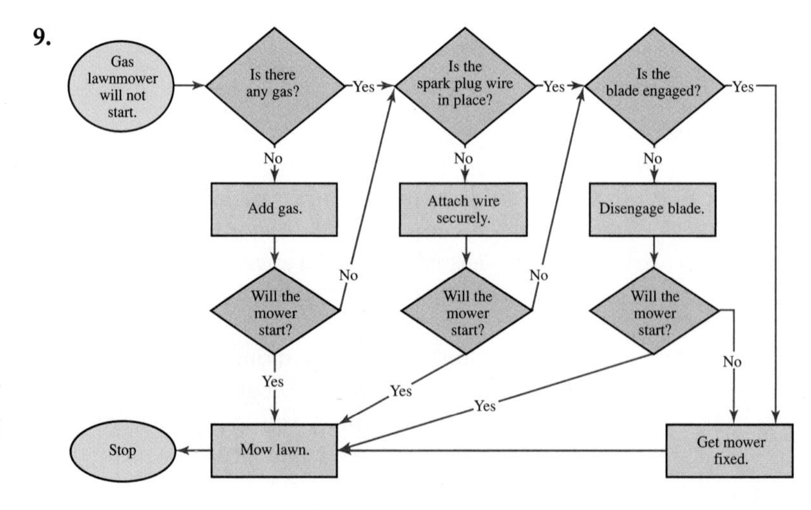

10. Given the following flowchart,

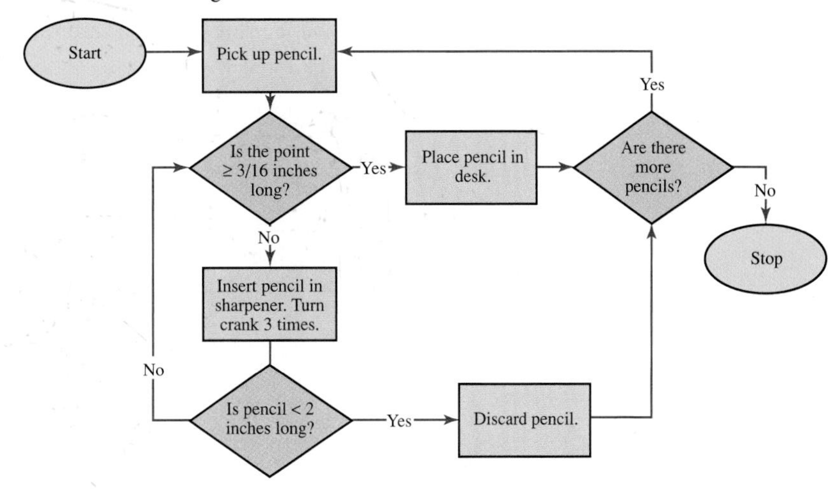

a) What happens to a 4-in.-long pencil with a point that is 1/4 in. long?

b) What happens to a 6-in.-long pencil with a point that is 1/8 in. long?

c) What happens to a 1-in.-long pencil?

d) What happens to a 5-in.-long pencil?

e) What happens to a pencil that has a point that keeps breaking as you sharpen it?

11. The following is a possible strategy for playing blackjack. You win the game when the total of your cards is less than or equal to 21 points and greater than the total held by the dealer. An ace may count as either 1 or 11 points, and jacks, queens, and kings count as 10 points. All other cards are counted at face value.

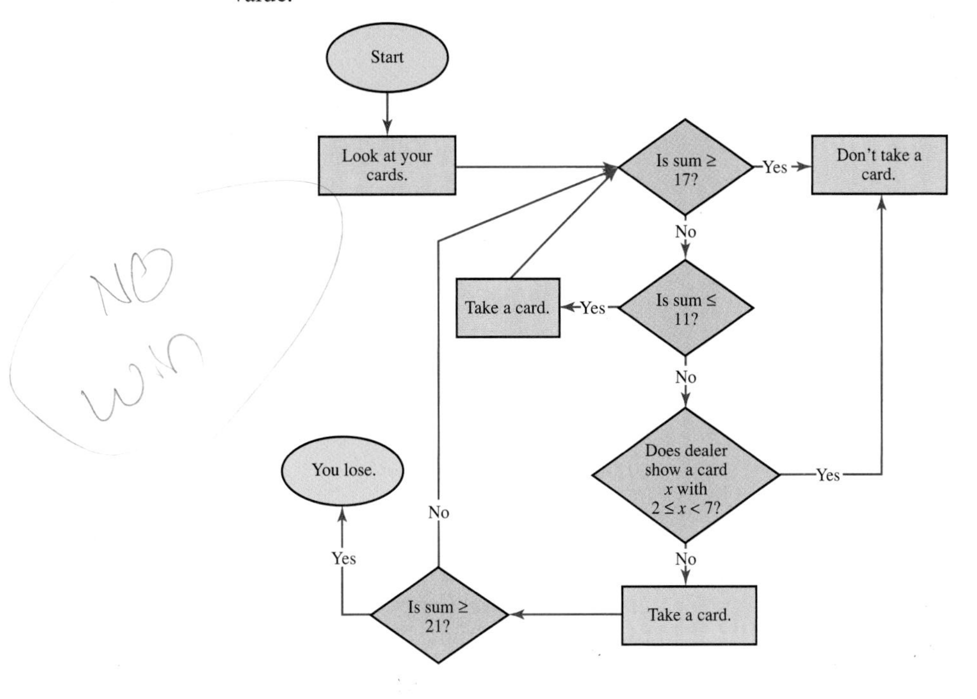

a) What happens if your two cards are a 10 and an 8?

b) What happens if your two cards are a 4 and an 8?

c) What happens if your two cards are a 4 and a 2?

d) What happens if your two cards are a 2 and a king when the dealer shows a queen?

e) What happens if your two cards are a 2 and a king when the dealer shows a 6?

f) Does this flowchart show how you can win the game? If so, explain. If not, modify the flowchart to show how the game can be won.

Explore

In Problems 12–13, suppose that the company used in Example 2 decides to add an additional step to its existing flowchart. In each problem, modify the flowchart in Example 2 to accommodate the additional step.

12. Sometimes when a customer calls the technical staff, all the telephones are busy. Rather than requiring the customer to remain on hold, the company has instituted the following policy: If all lines are busy, the telephone receptionist is to ask customers if they wish to remain on hold or if they want to have their call returned by the next available representative of the technical staff. Change the flowchart in Example 2 to accommodate this situation.

13. Sometimes when a customer calls the business staff, the question is too complex for the staff to answer immediately. Rather than requiring the customer to remain on hold, the company has instituted the following policy: In situations in which an immediate answer cannot be given, the business staff have been instructed to ask customers if they wish to be placed on hold or if they want to have their call returned after the answer has been determined. Change the flowchart in Example 2 to accommodate this situation.

14. A member of the quality assurance staff of a company that manufactures and sells prefabricated, unassembled cabinets is responsible for determining whether each carton contains the correct parts. If the carton contains the correct parts, it is sent on to shipping. If not, the carton is removed from the assembly line, has the necessary parts added to the carton, and then is sent to shipping. Write a flowchart for this process.

15. Write a flowchart that shows how to determine whether a positive integer is odd or even.

16. Write a flowchart that shows how to determine whether a positive integer has an integer square root.

17. A light on your desk no longer works. Describe in writing and construct a flowchart describing the process of determining whether the light bulb, the circuit breaker, or some other factor is the cause of the light not working.

18. A mechanical coin sorter puts coins in different bins depending on the size of the coins. The sorter is capable of handling quarters, dimes, nickels, and pennies. Write a flowchart that describes the process followed by the coin sorter. Assume that another machine has previously removed all foreign coins, U.S. coins other than those listed above, and all other objects.

19. Write a flowchart to tally the points scored by a basketball team during a game.
20. Write a flowchart to tally the points scored by a professional football team during a game.
21. Write a flowchart that describes a process to determine which new car you should buy.
22. Write a flowchart that a good student would use to determine whether she has time to go to the beach.
23. Write a flowchart for the process of signing up for classes at your school.
24. Write an organizational chart describing the administrative structure at your school.
25. Write an organizational chart describing the separation of powers of the three major branches of the federal government.
26. Write an organizational chart describing the federal judiciary system.

section 2.5

Logic and Puzzles

Math and logic puzzles are a popular hobby for many people, some of whom subscribe to monthly puzzle magazines.

The ability to think logically is critical in everyday living and very helpful in solving problems. In this section, we introduce you to puzzles in which the use of logic makes it easier to arrive at correct solutions.

MAGIC TRIANGLES

A magic triangle is a triangle in which an equal number of counting numbers {1, 2, 3, . . .} is placed on its sides so that the sum on the numbers on each side is the same.

EXAMPLE 1

Place the whole numbers from 1 to 6 in the circles to create a magic triangle.

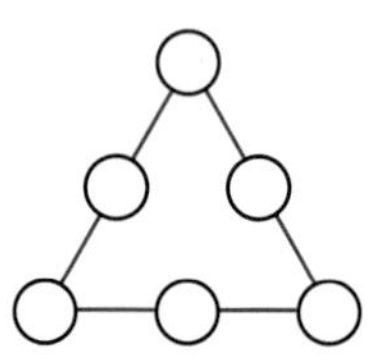

Solution: Since there are only six numbers to place in the circles, you could use trial and error: place numbers, check the sums, make adjustments, and arrive at a solution. However, using a little logic might make it easier. Here are some factors to consider.

1. Three numbers are odd, and three numbers are even. Maybe putting either the odd numbers or the even numbers at the vertex points and then using trial and error to place the other three numbers might work. Doing that, we get two solutions: a sum of 10 and a sum of 11.

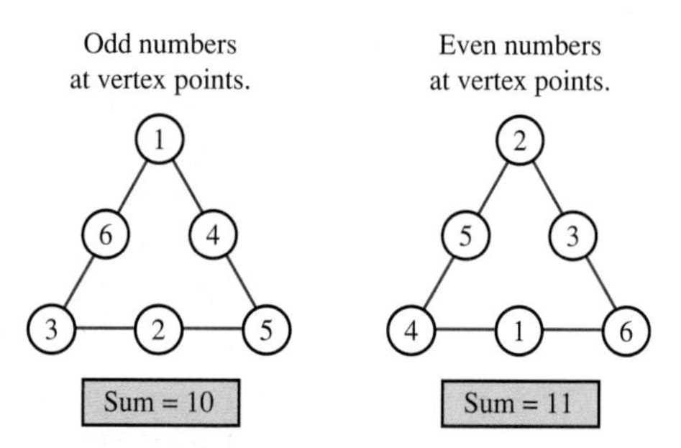

2. You could also split the numbers into two groups: the smaller numbers (1, 2, 3) and the larger numbers (4, 5, 6). Putting either group at the vertex points and using trial and error to place the other three numbers, we get a sum of 9 and 12.

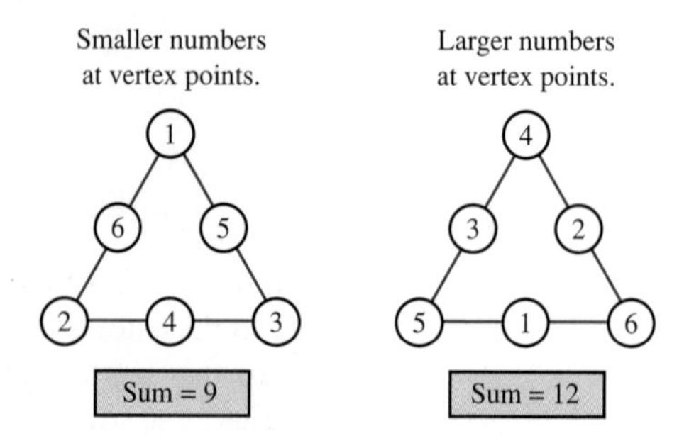

MAGIC SQUARES

A **magic square** is a square array of numbers {1, 2, 3, . . .} that has the same sum horizontally, vertically, and diagonally. Some ancient cultures believed that these square arrays of numbers showed mathematical harmony and contained mystical powers. Interest in magic squares dates back to 2200 B.C.

EXAMPLE 2

Consider the 3×3 square array shown below. Make a magic square by placing the whole numbers from 1 to 9 in the squares so that the sum of each row, column, and diagonal is the same.

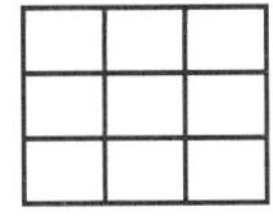

Solution: You could use trial and error and place numbers in the squares and hope that each row, column, and diagonal has the same sum. However, since there are 362,880 ways in which the numbers 1, 2, 3, 4, 5, 6, 7, 8, and 9 can be placed in the squares, it might be better to apply some logic before you start. Here are some factors you might consider.

1. *What is the sum of each row, column, and diagonal?* The sum of the whole numbers from 1 to 9 is 45. Since there are three rows, it seems logical that each row should have a sum of $45 \div 3 = 15$.
2. *What combinations of numbers are not possible?* With a sum of three numbers being 15, if two of the numbers already add up to 15 or more, those two numbers cannot be in the same row, column, or diagonal. Thus, $9 + 6$, $9 + 7$, $9 + 8$, and $8 + 7$ cannot be in any row, column, or diagonal. With a sum of three numbers being 15, if two numbers have a sum that is below 6, there is no way to get a total of 15. Thus, $1 + 2$, $1 + 3$, $1 + 4$, and $2 + 3$ cannot be in any row, column, or diagonal.
3. *What sets of three numbers have a sum of 15?* Using the above results, a careful listing will give the following sets with a sum of 15:

   (1, 5, 9), (1, 6, 8), (2, 4, 9), (2, 5, 8), (2, 6, 7), (3, 4, 8), (3, 5, 7), (4, 5, 6)
4. *Where should the numbers be placed?* In the magic square, the number in the center is used four times: once horizontally, once vertically, and twice diagonally. From the above list, the only number that appears in four sets is the number 5. Therefore, let's place 5 in the center.

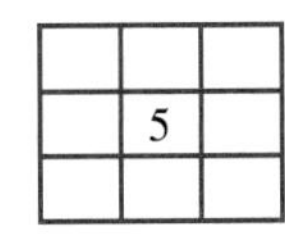

In the magic square, a number in a corner is used three times: once horizontally, once vertically, and once diagonally. From the above list, the numbers that appear in three sets are the numbers 2, 4, 6, and 8. Therefore, let's place 2, 4, 6, and 8 in the corners so that the sum of the diagonals is 15.

8		4
	5	
6		2

Now the remaining numbers, 1, 3, 5, and 9, can be placed so that the sums are all 15.

8	3	4
1	5	9
6	7	2

With the use of trial and error and some logic, the magic square was solved.

ALPHAMETIC PUZZLES

In an **alphametic,** short for alphabet arithmetic, each letter represents a unique whole number from 0 to 9. This means that no two letters are assigned the same digit value. In an alphametic, the value assigned to a letter that starts a "word" cannot be a zero. The first published alphametic was created by the well-known puzzler H. E. Dudeney in 1924. His alphametic, SEND + MORE = MONEY, is given as an exercise.

EXAMPLE 3

Find the value of each letter in the given alphametic puzzle.

$$\begin{array}{r} \text{A T D} \\ +\ \text{S I X} \\ \hline \text{P N T S} \end{array}$$

Solution: You could use trial and error to find the value assigned to each letter, or you could use some logic before you start. Are there some underlying principles that are used in adding two numbers?

1. *The effect of carrying:* If the sum of digits in a column is less than 10, you do not carry a one to the next column. If the sum of the digits is more than 9, you carry a one to the next column.

$$\begin{array}{r} \text{A B} \\ +\ \text{C} \\ \hline \text{A E} \end{array} \rightarrow \text{B + C must be less than 10, no carrying.}$$

$$\begin{array}{r} \text{A B} \\ +\ \text{C} \\ \hline \text{D E} \end{array} \rightarrow \text{B + C must be greater than 9, carrying occurred, D = A + 1.}$$

2. *The same letter in a column:* If the same letter appears in the ones column, the value of the letter is 0.

$$\begin{array}{r} \text{C A} \\ +\ \text{B A} \\ \hline \text{E A} \end{array} \rightarrow \text{A must be 0.}$$

If the same letter appears in another column, the value of the letter is 0 or 9.

$$\begin{array}{r} \text{E A B} \\ +\text{F A C} \\ \hline \text{G A D} \end{array}$$

→ A is 0 if B + C is less than 10.
→ A is 9 if B + C is greater than 9.

3. *The same letter in top and bottom of a column:* If the same letter appears in the top and bottom of a column, the value of the letter in the middle is 0 or 9.

$$\begin{array}{r} \text{C A} \\ +\text{B D} \\ \hline \text{T E A} \end{array}$$

→ D must be 0. (no carrying)

$$\begin{array}{r} \text{C A} \\ +\text{D R} \\ \hline \text{T C E} \end{array}$$

→ D must be 9. (carrying from ones column)

Armed with those principles, let's solve the alphametic.

$$\begin{array}{r} \text{A T D} \\ +\ \ \text{S I X} \\ \hline \text{P N T S} \end{array}$$

a) I = 0 (Numbers in the top and bottom are the same in the tens column.)
b) P = 1 (One more digit in the answer indicates carrying.)

Realizing that each letter represents a different digit, we logically place digits, check sums, and make adjustments with the remaining digits to get a solution.

$$\begin{array}{r} 962 \\ +\ 503 \\ \hline 1465 \end{array}$$

There are other possible solutions to this alphametic puzzle.

CROSS NUMBER PUZZLES

In a cross number puzzle, you fill in the blank squares using the numbers below the puzzle to obtain the answers given in the bottom row and the right hand column. You must use the correct order of operations. Each number from the list at the bottom of the puzzle is used only once.

EXAMPLE 4

Solve the following cross number puzzle.

50	÷		×		30
÷	■	−	■	−	■
	+		×		34
−	■	+	■	+	■
	+		−		4
19	■	8	■	4	■

2 3 4 5 6 7 8 9

Solution: Let's label the blank squares so that you can follow the solution.

50	÷	*a*	×	*b*	30
÷	■	−	■	−	■
c	+	*d*	×	*e*	34
−	■	+	■	+	■
f	+	*g*	−	*h*	4
19	■	8	■	4	■

2 3 4 5 6 7 8 9

In this puzzle, you must use your knowledge of the four operations of arithmetic and the order of operations. Here are some factors to consider.

1. Since the answers are whole numbers, divisors are probably factors of the dividend. Thus, in the first row, it is $50 \div 2 = 25$ or $50 \div 5 = 10$. Square a is either 2 or 5. Since the answer is multiplied by a number to get 30, $a = 5$ and, consequently, $b = 3$.
2. Since each number can be used only once, in the first column $c = 2$ and, consequently, $f = 6$.
3. With only the numbers 4, 7, 8, and 9 to work with, trial and error techniques lead us to the solution.

50	÷	5	×	3	30
÷		−		−	
2	+	4	×	8	34
−		+		+	
6	+	7	−	9	4
19		8		4	

The exercises that follow will give you practice in using logic to solve math puzzles. Have fun!

PROBLEMS ▶ Explain ▶ Apply ▶ Explore

▶ Explain

1. What is a magic square?
2. What is an alphametic puzzle?
3. Why is it necessary to use more than a trial-and-error technique in solving these math puzzles?
4. In a 3×3 magic square, why can you not place 6 and 9 in two consecutive squares?
5. Consider a subtraction alphametic puzzle with letters as shown below.

$$\begin{array}{r} \text{A B C} \\ -\ \text{D E} \\ \hline \text{G C} \end{array}$$

   What conclusions can you draw about the letters A and E?

6. Consider a subtraction alphametic puzzle with letters as shown below.

$$\begin{array}{r} \text{A B C} \\ -\ \text{D C} \\ \hline \text{E G C} \end{array}$$

   What conclusions can you draw about the letters C and E?

7. Consider a multiplication alphametic puzzle with letters as shown below.

```
   A B C
 ×   D E
 -------
   E E E
 A B C
 -------
 A B C E
```

What conclusions can you draw about the letters D and E?

8. Consider a division alphametic puzzle with letters as shown below.

```
          GA
ABC )ADEC
     ABC
      FEC
      FEC
```

What conclusions can you draw about the letters G and C?

Apply

9. Place the even integers 2, 4, 6, 8, 10, and 12 in the circles of the triangle in Example 1 so that the sum of the numbers on each side of the triangle is the same. Find all possible solutions.

10. Place the odd integers 1, 3, 5, 7, 9, and 11 in the circles of the triangle in Example 1 so that the sum of the numbers on each side of the triangle is the same. Find all possible solutions.

11. Place the odd integers 1, 3, 5, 7, 9, 11, 13, 15, and 17 in the 3×3 square in Example 2 so that the sum of each row, column, and diagonal is the same number.

12. Place the even integers 2, 4, 6, 8, 10, 12, 14, 16, and 18 in the 3×3 square in Example 2 so that the sum of each row, column, and diagonal is the same number.

13. A $3 \times 3 \times 3$ magic cube has the whole numbers from 1 to 27 placed in three layers with nine numbers each layer. The sum of the three numbers in each horizontal row, vertical column, and four diagonals of the cube has the same sum. Some of the numbers of the magic cube are given. Determine the missing numbers.

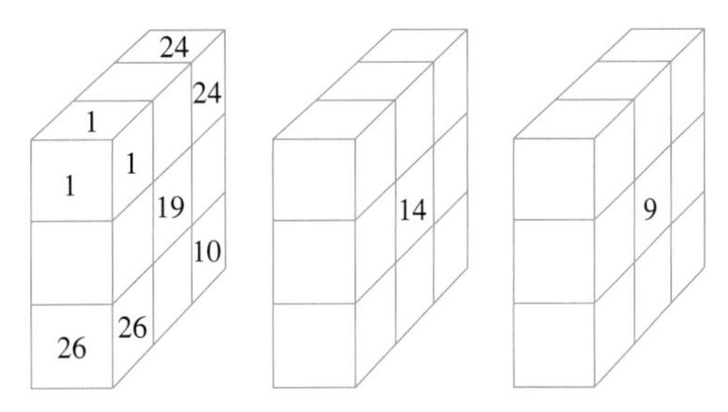

14. A figure is created by placing the whole numbers in circles at each intersection point on the lines of a star. If the sum of the numbers along each line of the star is the same, determine the missing values in the stars on the following page.

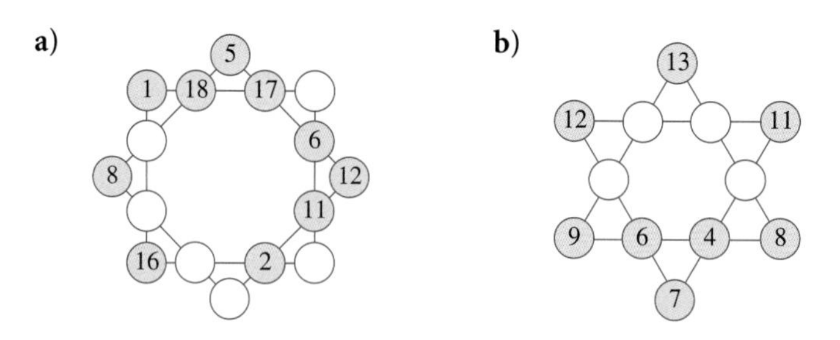

15. Find a solution to the alphametic puzzles in Problems 5 and 8.
16. Find three different solutions to the alphametic puzzles in Problems 6 and 7.

Explore

17. In the squares on the cross below, place the whole numbers from 1 to 9 so that the sum of the numbers on each leg is the same. Find all possible ways to do this.

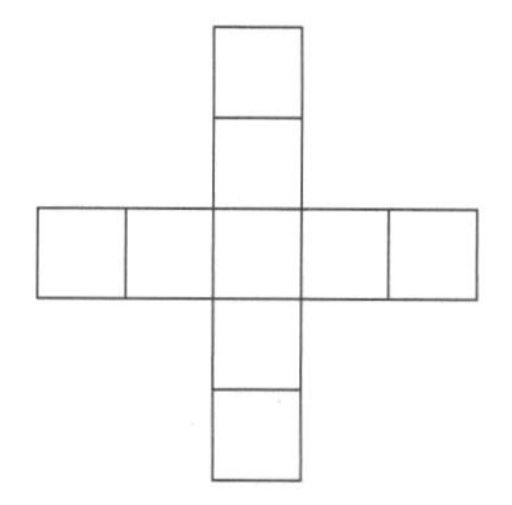

18. Magic letters are created by placing the consecutive whole numbers 1, 2, 3, 4, 5, . . . , in the circles at the intersection and the end of each line segment of a letter so that the sum of the numbers on each segment is the same. Determine the placement of the numbers in the magic letters A, F, T, E, and X.

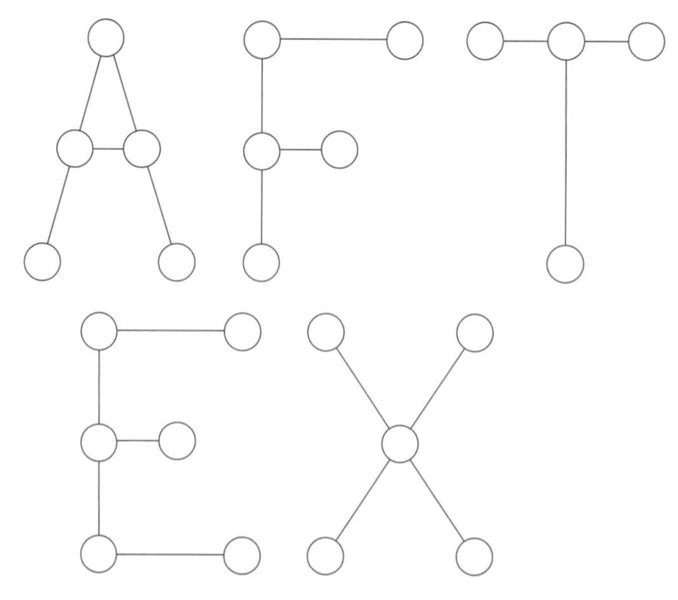

19. The number 4 is the only "magic number." Place the whole numbers 1, 2, 3, 4, and 5 in the circles so that the sum of the numbers on each segment is the same.

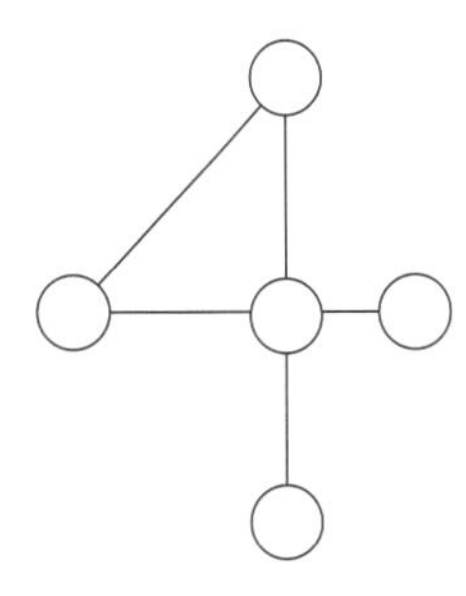

20. Consider a triangle with four circles on each side as shown below. Place the whole numbers from 1 to 9 in the circles so that the sum of the numbers on each side is the same.

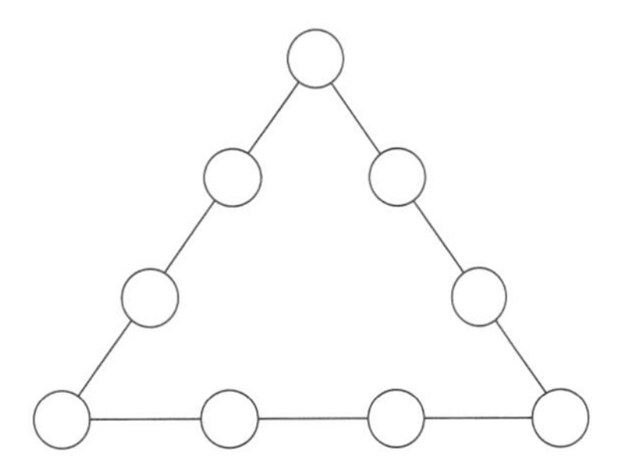

Solve the alphametic puzzles in Problems 20–30. There may be more than one solution to each puzzle.

21.
$$\begin{array}{r} \text{GOLF} \\ +\ \text{RAIN} \\ \hline \text{NOFUN} \end{array}$$

22.
$$\begin{array}{r} \text{WIN} \\ +\ \text{YEA} \\ \hline \text{TEAM} \end{array}$$

23.
$$\begin{array}{r} \text{FIVE} \\ -\ \text{FOUR} \\ \hline \text{ONE} \end{array}$$

24.
$$\begin{array}{r} \text{CAT} \\ \times\ \text{N} \\ \hline \text{HAT} \end{array}$$

25.
$$\begin{array}{r} \text{YOU} \\ \times\ \text{R} \\ \hline \text{COOL} \end{array}$$

26.
$$\begin{array}{r} \text{FUN} \\ \times\ \text{IN} \\ \hline \text{RIN} \\ \text{NOA}\phantom{\text{N}} \\ \hline \text{RAIN} \end{array}$$

27.

```
     BIG
×  MAC
     BIG
   AAA
YESA
YESBIG
```

28.

```
  BEST
 MATH
+TEST
 BLEST
```

29.

```
   SEND
+MORE
MONEY
```

In Problems 30–33, solve the cross number puzzle by using each number below the puzzle only once.

30.

	×		+		7
×		+		×	
	+	24	÷		7
−		÷		−	
	+		−		5
7		7		5	

1 2 3 4 5 6 7 8

31.

36	÷		×		6
÷		−		+	
	−		×		−6
−		×		−	
	−		×		0
−4		−6		4	

1 2 3 4 5 6 8 9

32.

	+		×		−	5	13
÷		×		−		+	
	−		+	13	+		23
+		−		+		−	
	×	15	−		−		30
−		+		+		+	
1	+		+		−		12
5		16		12		8	

2 3 4 6 7 8 9 10 11 12 14 16

33.

	÷	√	×		10
+		+		−	
	−	√	−		−8
−		×		×	
	+		+		13
−2		9		−8	

1 2 3 4 5 6 7 8 9

CHAPTER 2 SUMMARY

Key Terms, Concepts & Formulas

The important terms in this chapter are:	Section
Affirming the antecedent: A deductive argument of the form If A, then B. A. Therefore, B.	2.2
Alphametic: Alphabet arithmetic: Each letter represents a unique whole number from 0 to 9.	2.5
Biconditional statement: If A and B represent statements, the biconditional statement combines both the statement "If A, then B" and the statement "If B, then A" into one statement, "A if and only if B."	2.1
Conclusion: The "then" clause of a conditional statement, also called the consequent.	2.1
Conditional statement: A complex statement formed by two individual statements joined by the words "If . . . then . . . ."	2.1
Conjunction: A statement in which categories are connected with the word "and." A conjunction is true whenever a quality applies to both categories.	2.3
Contrapositive: If A and B represent statements, the contrapositive of the conditional statement "If A, then B" is the statement "If not B, then not A."	2.1
Converse: If A and B represent statements, the converse of the conditional statement "If A, then B" is the statement "If B, then A."	2.1
Decision symbol: A diamond-shaped object that indicates a question in a flowchart.	2.4
Deduction: The process of reasoning in which conclusions are based on general principles.	2.3
Denying the consequent: A deductive argument of the form If A, then B. Not B. Therefore, not A.	2.2
Disjunction: A statement in which categories are connected with the word "or." A disjunction is true whenever a quality applies to one or both categories.	2.3
Flowchart: A diagram that shows the logical path that is followed when a decision is made.	2.4
Flow line: A directed line segment (or ray) that connects the symbols in a flowchart.	2.4
Hypothesis: The "if" clause of a conditional statement; also called the antecedent.	2.1

Hypothetical syllogism: A deductive argument of the form — 2.2

If *A*, then *B*.
If *B*, then *C*.
Therefore, if *A*, then *C*.

Induction: The process of reasoning in which conclusions are based on experience or experimentation. — 2.2

Inverse: If *A* and *B* represent statements, the inverse of the conditional statement "If *A*, then *B*" is the statement "If not *A*, then not *B*." — 2.1

Logic: The science of correct reasoning. — 2.1

Logically equivalent: Two statements are logically equivalent when they have the same truth values. — 2.1

Magic square: A square array of numbers {1, 2, 3, . . .} that has the same sum horizontally, vertically, and diagonally. — 2.5

Start/Stop symbol: An oval or circle that is used to indicate the beginning or end of a flowchart. — 2.4

Statement symbol: A rectangle used in a flowchart to indicate an action, person, or result. — 2.4

Symbolic logic: A system of logic that uses symbols to represent statements. The symbols used in this chapter are: — 2.3

Symbol	English	Name
$\sim$	Not	Negation
$\therefore$	Therefore	
$\rightarrow$	Implies, if . . . then . . .	Conditional
$\equiv$	Equivalent statements	Logical equivalency
$\wedge$	And	Conjunction
$\vee$	Or	Inclusive disjunction

Truth table: A chart used to check the validity of a series of logic statements. — 2.3

After completing this chapter, you should be able to:	**Section**
1. Distinguish between different types of conditional statements and use the contrapositive as the logical equivalent of a given statement.	2.1
2. Recognize and write good definitions.	2.1
3. Distinguish between inductive and deductive reasoning.	2.2
4. Make deductive arguments using hypothetical syllogism, affirming the antecedent, and denying the consequent.	2.2
5. Use truth tables to determine the validity of an argument or determine the equivalency of statements.	2.3
6. Construct flowcharts to show a decision making process or organization structure.	2.4
7. Use logic to help solve math puzzles.	2.5

CHAPTER 2 REVIEW

Section 2.1

1. Write the negation of "All my relatives live in Argentina."
2. Write the contrapositive of "If it isn't broken, don't fix it."
3. Write a definition of "math" that satisfies the three necessary conditions of a definition.
4. In logic, what is meant by a "statement"?

Section 2.2

5. Explain what is wrong with the following argument. Change the minor premise to make the argument correct.

   If there is a drought, you don't water the lawn.

   You don't water the lawn.

   Therefore, there is a drought.
6. The expression $n^2 - n + 41$ seems to generate prime numbers for whole number values of n. Explain how using induction might cause you to make a false conclusion in this case. What is the first value of n that causes the expression to produce a composite number?
7. Give both an inductive argument and a deductive argument to convince a friend to not to drive after excessive drinking of alcoholic beverages.
8. The following premises are not in exact syllogistic form. Rearrange and/or logically rewrite the premises so that they correspond to one of the types of syllogisms. What conclusion follows from the premises?

   If you are photographing water, you use a polarizing filter.

   If there are no reflections in the scene, you do not use a polarizing filter.

Section 2.3

9. Translate the following into symbols. State what each symbol represents.

   If the dog is a Brittany, it likes to hunt birds.

   The dog is a Brittany.

   Therefore, the dog likes to hunt birds.
10. Use a truth table to determine the truth values of A and B that makes the following statement true: $(\sim A \rightarrow \sim B) \rightarrow B$.
11. Use a truth table to determine whether the following is logically correct.

$$\sim(A \wedge B) \equiv \sim A \vee \sim B$$

12. Translate the argument into symbolic form and determine whether it is logically correct by constructing a truth table.

    If there is no election reform, candidates will buy elections. Therefore, if there is election reform, candidates will not buy elections.

Section 2.4

13. Your computer printer is not working. Describe in writing and construct a flowchart describing the process of determining whether your printer is not plugged into a wall outlet, it is not connected to the computer, it is not turned on, or some other factor is the cause of the printer not working.

Section 2.5

14. The Japanese mathematician Kittoku Isomura (c. 1660) did a great deal of work on magic circles. In these circles, consecutive natural numbers, starting at 1, are placed on the diagram shown. If we add the numbers on any circle and the number in the center of the diagram,

we get the same result as the sum of the numbers on each of the diagonals of the circle.

a) Place the numbers 1 to 9 in the diagram so that you have created a magic circle.

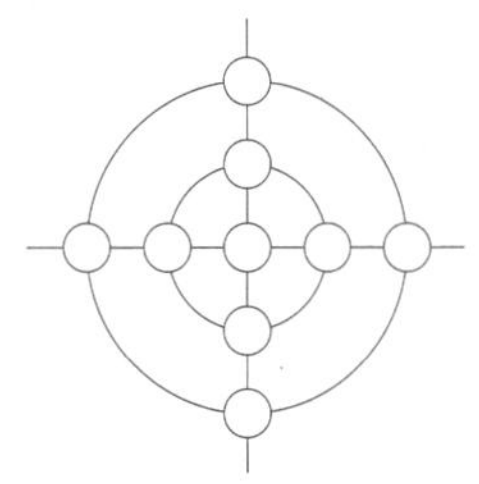

b) Place the numbers 1 to 19 in the diagram so that you have created a magic circle.

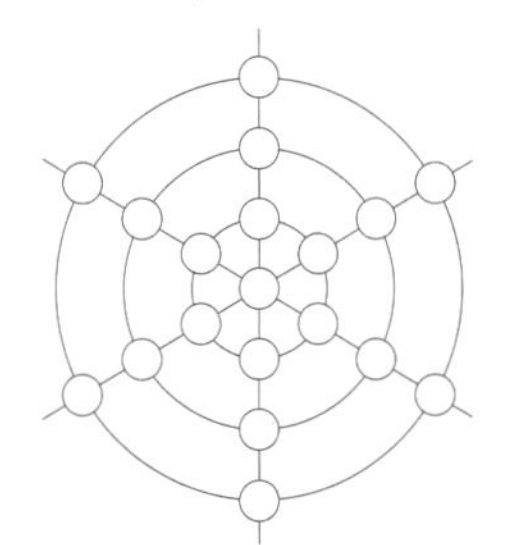

In Problems 15 and 16, solve the alphametic puzzles.

15.

```
  TEE
  HEE
+ HEE
-----
 JOKE
```

16.

```
 FUN
×  N
----
 SUN
```

17. Solve the cross number puzzle by using each number below the puzzle once.

36	÷		×		81
÷		+		×	
	+		−		1
÷		−		÷	
	×		÷		2
3		10		24	

1 2 3 4 6 7 8 9

CHAPTER 2 TEST

1. For the conditional statement "I am in bed if it is after midnight."
   a) State its hypothesis and conclusion.
   b) Write its converse.
   c) Write its inverse
   d) Write its contrapositive.
2. What are the three necessary components of a definition? What additional properties are needed for a good definition? For each of the following, state what is wrong with the "definition" and change it to make it a good definition.
   a) Perpendicular line segments are segments that intersect each other.
   b) Microscopic: that which can only be seen through a microscope.
   c) Natural numbers are numbers that are greater than zero.
3. Write a statement that is true but whose converse and inverse are false.
4. Explain the difference between an inductive argument and a deductive argument. Give both an inductive argument and a deductive argument to persuade a friend to help you build a fence.
5. Rewrite the following in correct syllogistic form and state a valid conclusion from the premises.

a) $X \rightarrow Y$
$Z \rightarrow \sim Y$
$\sim Z \rightarrow P$

b) If the geometry is Riemannian, then there are no parallel lines.
If the geometry is Euclidean, then parallel lines exist.
If the geometry is non-Euclidean, then at least one of Euclid's postulates is changed.

6. Translate the following statements into symbolic logic.

a) If John has a headache, then John is either grumpy or silent.

b) If the weather is not good, then we will not play baseball and we will not have a picnic.

c) If you do not study math and you want a good job, it will be harder to advance in business.

7. Use truth tables to show that $(A \vee B) \wedge (A \vee \sim B)$ is equivalent to A.

8. Use a truth table to determine whether the argument is logically valid.

Parrots can crack walnuts, and if it is my animal, it cannot crack walnuts.

Therefore, if it is my animal, it is not a parrot.

9. Write a flowchart that describes the process of using the Internet to find information on perfect numbers. Include directions on what to do if your initial search does not provide useful or sufficient information.

10. The following four layers contain some of the numbers that form a $4 \times 4 \times 4$ magic cube. Determine the missing numbers.

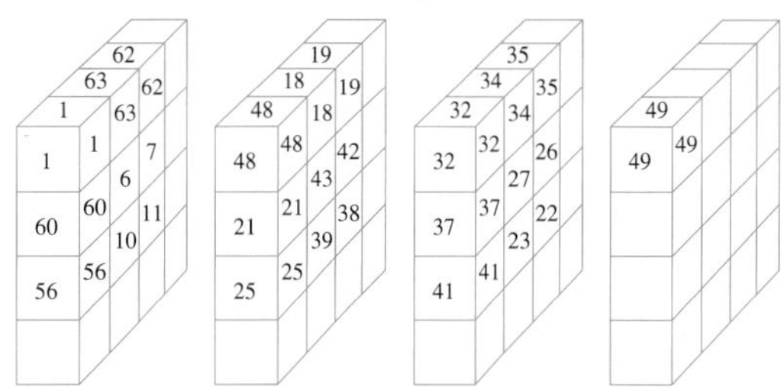

11. Solve the alphametic puzzle.

$$\begin{array}{r} \text{YEA} \\ \times \quad 4 \\ \hline \text{MATH} \end{array}$$

12. Solve the cross number puzzle using each number only once.

72	÷		×		9
÷		−		×	
	−		×		−3
−		−		+	
	×		−		10
6		0		4	

0 1 2 3 4 5 8 9

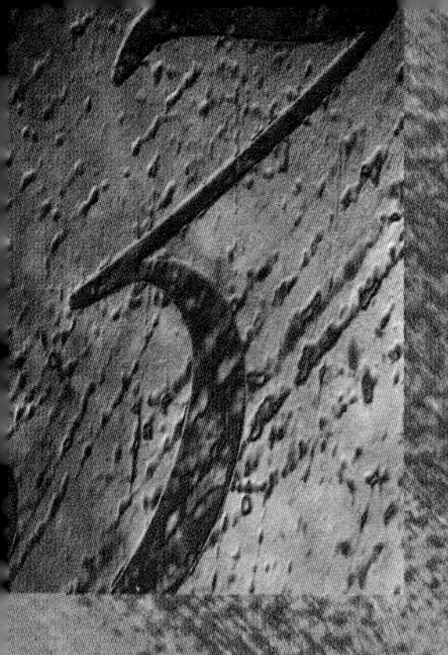

CHAPTER THREE

Sets and Counting

Golden Venn by David McLaughlin makes use of Golden Ratios and colorfully represents the Venn diagram of three intersecting sets. (Courtesy of the artist)

OVERVIEW

In this chapter, you will learn about sets and how the abstract mathematical concept of a set is applied to real-world problems. In particular, you will become familiar with the terminology and operations of sets such as unions and intersections. You will see how Venn diagrams give a graphical representation of sets. Finally, you will see how the number of items in a set is directly related to counting things such as the number of possible lottery tickets or the number of car license plates in the state of California.

A SHORT HISTORY OF SETS

This chapter introduces the concepts of sets and Venn diagrams. Unlike most of the material in this book, sets and Venn diagrams are a modern creation. While geometry has its roots in Greece in 600 B.C. and trigonometry can be traced to the Egyptians in 1550 B.C., sets and Venn diagrams do not have such a lengthy history. In addition, unlike the development of trigonometry for surveying and navigation, the development of sets was of a more technical, mathematical nature.

Although the intuitive concept of a set as simply a collection of objects is not a difficult idea, the formal mathematical use of sets did not occur until the middle of the 1800s. Beginning with Bernhard Bolzano (1781–1848) in the 1840s, set theory was used to investigate the concept of infinity. It was Bolzano who showed that an infinite set A could be put into a one-to-one correspondence with a subset of A (see Section 3.1).

Georg Cantor (1845–1918) is credited with the work that made set theory a major field of mathematical investigation. In a paper published in 1874, Cantor distinguished at least two different levels of infinity. His work showed that the set of whole numbers and the set of rational numbers had the same cardinality, that is, the same number of elements. Of course, there are an infinite number of elements in these sets. However, Cantor demonstrated that the set of real numbers has a greater cardinality—a larger infinity. This discovery caused controversy in the mathematics community. The development of set theory was an active field of study through the early 1900s, with work continuing through the 1960s.

At least one paradox arose from this work on set theory. As phrased by Bertrand Russell (1872–1970) in 1903:

> *A barber in a certain town has stated that he will cut the hair of all those persons and only those persons in the town who do not cut their own hair. Does the barber cut his own hair?*

In terms of sets, if S is the set of people whose hair is cut by the barber, is the barber an element of S?

Venn diagrams are named for the British logician John Venn (1834–1923) and are a refinement of Euler diagrams, named for Leonhard Euler (1707–1783). Venn specified that each new circle added to a diagram must intersect every other compartment. Venn diagrams were again refined by the logician Charles Dodgson (1832–1898). Dodgson's contribution was to add a rectangle around the diagram, representing the universal set. Dodgson is also known by his pen name, Lewis Carroll, and was the author of the well-known children's story *Alice's Adventures in Wonderland.*

In a card game such as this one, *The Card Players* by Paul Cezanne, we can determine the number of ways a particular hand can be dealt. (The Metropolitan Museum of Art, Bequest of Stephen C. Clark, 1960. (61.101.1) Photograph "1982 The Metropolitan Museum of Art)

CHAPTER 3 PROJECTS

Research Projects

1. What are DeMorgan's Laws and how are they used in real-world applications?
2. Describe how Georg Cantor (1845–1918) showed that the set of rational numbers is countably infinite and that the set of real numbers is infinite but not countable.
3. Compare Euler diagrams and Venn diagrams. What are the similarities and differences? How are Euler diagrams used?

Math Projects

1. In this chapter, it will be necessary to find the general formulas for the terms in a set of numbers. For example, for the set {2, 4, 8, 16, 32, . . .}, the general formula for the kth term is 2^k. Find the general formulas for the following sets.
   - **a)** $\{1, 7, 13, 19, 25, \ldots\}$
   - **b)** $\{-2, 6, -18, 54, -162, \ldots\}$
   - **c)** $\{2, 5, 10, 17, 26, 37, \ldots\}$
   - **d)** $\{1, -1, 1, -1, 1, -1, \ldots\}$
   - **e)** $\left\{\frac{1}{1}, \frac{3}{2}, \frac{5}{6}, \frac{7}{24}, \frac{9}{120}, \frac{11}{720}, \ldots\right\}$
   - **f)** $\left\{4, \frac{9}{4}, \frac{16}{9}, \frac{25}{16}, \frac{36}{25}, \ldots\right\}$
2. For the set $Q = \{*\}$, there are two subsets: { } and {*}.
   - **a)** How many subsets can you find for the set $A = \{*, \times\}$?
   - **b)** How many subsets can you find for the set $B = \{*, \times, \dagger\}$?
   - **c)** How many subsets can you find for the set $C = \{*. \times, \oplus, \nabla\}$
   - **d)** Suppose a set D contains 20 elements. Use the above results to conjecture the total number of possible subsets.

section 3.1

Sets: Finite and Infinite

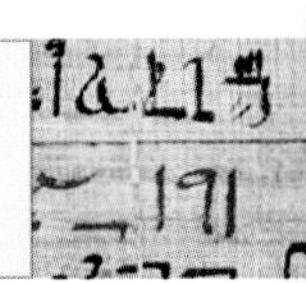

Philip Jaeger's *What a Great Pair, We Two* can be viewed as a Venn Diagram of two nonintersecting sets. (CORBIS)

A **set** is simply a collection of items. Although we have not previously defined sets and their use, we have used and will continue to use the idea of a set throughout the book. For example, our system of numeration uses a set of ten symbols to represent numbers. When we graph an equation, the points on the graph are the set of points that satisfy the equation. In statistics, a collection of data is called a data set. In this section, we want to examine the rules for working with sets and some of the uses of sets.

SET NOTATION AND MEMBERS OF A SET

To begin our study of sets, we examine the notation that is used with sets. Since a set can contain numbers, words, equations, or even other sets, we need a very general term to describe the items in a set. In mathematics, we use the word **element.** Each item in a set is an element of the set.

Suppose we want to have a set consisting of all the even integers between 1 and 9. Using the symbols { and } (called braces) to form a set, we can write this in two different ways. Since the even integers between 1 and 9 are 2, 4, 6, and 8, we can write the set by enclosing these numbers inside the braces,

$$\{2, 4, 6, 8\}.$$

This method is called the **listing method.** Though it is easy to use, it is not convenient if a set contains a large number of elements.

A second method of writing a set is to use **set builder notation,** also called the **descriptive method.** When using set builder notation, we use mathematical notation to describe the contents of the set rather than listing every element in the set. Instead of $\{2, 4, 6, 8\}$, we can write

$$\{x \mid x \text{ is an even integer with } 1 < x < 9\}.$$

The vertical bar between the x's means "such that." Therefore, the notation is read, "the set of x such that x is an even integer with $1 < x < 9$." Using set builder notation allows us to write sets with a large number of elements without resorting to the tedium involved in writing out each element. For example, the set of odd whole numbers less than 1001 would contain 500 elements but could be written as

$$\{x \mid x \text{ is an odd whole number with } x < 1001\}.$$

To indicate that a certain item is an element of a set or is not an element of a set, we use the symbols

$\in$ for "is an element of" and

$\notin$ for "is not an element of".

Thus, $6 \in 0, 3, 6, 9, 12, 15, 18\}$, but $8 \notin \{0, 3, 6, 9, 12, 15, 18\}$.

EXAMPLE 1

Use set notation to write all the whole numbers less than 13 that are divisible by 3.

Solution: We can do this in two ways.

a) The descriptive method: $\{x \mid x \in \text{whole numbers}, x < 13, \text{and } x \text{ is divisible by } 3\}$.

b) The listing method: $\{0, 3, 6, 9, 12\}$.

CARDINAL NUMBER OF A SET

One of the characteristics of a set is the number of elements in the set. This number is called the **cardinal number** of a set. For a set S, the cardinal number is written $n(S)$. If the cardinal number of a set is a whole number, the set is a **finite set.** For example, the set $D = \{\text{the days of the week}\}$ is a finite set with $n(D) = 7$. If a set has an unlimited number of elements, it is an **infinite set.** For example, the set $W = \{\text{natural numbers}\}$ is an infinite set. We will finish this section by first examining finite sets and then taking a brief look at infinite sets.

EXAMPLE 2

Find the cardinal number of each of the following sets.

a) $A = \{0, 3, 6, 9, 12\}$

b) $B = \{x \mid x \in \text{whole numbers}, x < 18, \text{and } x \text{ is divisible by } 5\}$

Solution:

a) Since there are five elements in the set A, the cardinal number $n(A) = 5$.

b) To find $n(B)$, we first write out the elements of B, $B = \{5, 10, 15\}$. We can now see that $n(B) = 3$.

SUBSETS AND PROPER SUBSETS

Set B is called a **subset** of a set A if every element in the set B is also an element of the set A. The symbol used to indicate subsets is $\subseteq$. Thus, to indicate that B is a subset of A, write $B \subseteq A$. For example, if we let $X = \{\text{ingredients in pizza}\}$ and $Y = \{\text{ingredients in Italian cooking}\}$, then we can say $X \subseteq Y$ because every element in the set X is a member of the set Y.

Set B is called a **proper subset** of a set A if every element in the set B is an element of the set A and $n(B) < n(A)$. The symbol used to indicate a proper subset is $\subset$. The subtle distinction is not too difficult. It really means that the set B has fewer elements than the set A.

To indicate that a set is not a subset of another set, we use the symbol $\not\subset$. Thus, for the sets X and Y, since there are some ingredients in Italian cooking that are not used in pizza, Y is not a subset of X. This may be represented as $Y \not\subset X$.

EXAMPLE 3

Let $A = \{1, 2, 3\}$, $B = \{1\}$, and $C = \{1, 2, 3\}$, and decide whether each of the following statements is true or false.

a) $B \subset A$

b) $B \subseteq A$

c) $C \subset A$

d) $C \subseteq A$

e) $A \subset B$

Solution:

a) $B \subset A$ states that B is a proper subset of A. It is a true statement, since every element of B is also an element of A and $n(B) < n(A)$, $(1 < 3)$, is also true.

b) $B \subseteq A$ states that B is a subset of A. It is a true statement, since every element of B is also an element of A.

c) $C \subset A$ states that C is a proper subset of A. This is a false statement, since $n(C) = n(A)$, $(3 = 3)$.

d) $C \subseteq A$ is a true statement, since every element of C is also an element of A.

e) $A \subset B$ is a false statement, since there are elements in A that are not in B. We could write $A \not\subset B$.

An important technical note about the notation for sets is that the subset and element symbols serve different purposes. The subset symbol can occur only between two sets, and the element symbol can occur only between an element and a set. The following example gives correct and incorrect usage for each symbol.

EXAMPLE 4

Let

$$A = \{\text{all the planets in our solar system}\}$$
$$B = \{\text{all the celestial objects}\}$$

Which of the following statements make correct use of the subset and element symbols?

a) $A \in B$ **b)** The earth $\in B$ **c)** $B \subset A$ **d)** The earth $\subset A$

Solution:

a) $A \in B$ is an incorrect usage of the element symbol, since A is a subset of B, not an element of B. The correct statement is $A \subset B$.

b) The earth $\in B$ is a correct usage of the element symbol, since the earth is one element in the set consisting of all celestial objects.

c) $B \subset A$ is a correct usage of the subset symbol, since both B and A are sets. However, the statement is not true, since not every celestial object is a planet. A correct statement is $B \not\subset A$.

d) The earth $\subset A$ is not a correct usage of the subset symbol, since the earth is an element of A, not a subset of it. A correct statement is "The earth $\in A$."

UNIVERSAL SETS

A **universal set** is a set that includes all the items that are under consideration in a particular situation. If we are talking about the set of people in a classroom who are wearing jeans, the people in the classroom are the universal set. Similarly, if we are talking about the set of people in a school who are wearing jeans, the people in the school are the universal set.

EMPTY SETS

An **empty set** is a set that does not contain any elements. It can be expressed in two ways. The first way uses the standard set notation of braces. Since the empty set does not contain any elements, the empty set is written as a pair of braces that do not enclose any symbols:

$$\text{empty set} = \{\ \ \}.$$

The second way to indicate an empty set is to use the symbol $\varnothing$. Thus,

$$\text{empty set} = \varnothing.$$

Empty sets can also be used to describe impossible events, such as a square circle or a natural number that is even and odd at the same time.

EQUAL VERSUS EQUIVALENT SETS

Two sets are said to be **equal** if they have exactly the same elements. Thus, the sets $A = \{x, y, z\}$ and $B = \{z, y, x\}$ are equal. Two sets are said to be **equivalent** if they have the same *number* of elements, that is $n(A) = n(B)$. Thus, the sets $A = \{1, 2, 3\}$ and $B = \{\text{apple, orange, banana}\}$ are considered equivalent, since both sets have three elements.

INFINITE SETS

Since you first learned to count, your mathematical world has continuously expanded. In the beginning, there were the numbers $\{1, 2, 3, \ldots\}$. This set of numbers is called the **natural numbers** or counting numbers. Next came zero (0). When added to the set of natural numbers, you had the set of **whole numbers**, $\{0, 1, 2, 3, \ldots\}$. At some point in elementary school, fractions were added to your mathematical world. By the time you began your first algebra course, negative numbers were included, and you had the set of **integers**, $(\ldots, -3, -2, -1, 0, 1, 2, 3, \ldots)$ and the set of **rational numbers**, (numbers that can be written as fractions). Since these sets have an unlimited number of elements, they are infinite sets.

SIZES OF SETS

If we have the sets $A = \{1, 2, 3\}$, and $B = \{5, 6\}$, since there are three elements in A and only two elements in B, it certainly makes sense to say that the size of A is larger than the size of B, $n(A) > n(B)$. Similarly, if $A = \{1, 2, 3\}$ and $B = \{5, 6, 7\}$, since both sets contain three elements, it makes sense to say that the sizes of the sets are equal or equivalent, $n(A) = n(B)$. How does this discussion work with infinite sets? Consider the natural numbers $\{1, 2, 3, \ldots\}$ and the rational numbers $\{x \mid x \text{ can be written as a fraction}\}$. Since both sets are unlimited, there is an infinite number of both natural numbers and rational numbers. However, there are certainly some rational numbers, such as 0.25, that are not natural numbers. So is the infinite set of rational numbers larger than the infinite set of natural numbers? The concept of a "larger" set that seemed reasonable only two paragraphs ago has now become complicated.

The concept of an *infinite* set is at the root of our difficulty. Can there be two sets that are both "infinite" but of different "size"? This difficulty was addressed by Georg Cantor in the late 1800s. His methods put the concept of infinity on a firm mathematical foundation. In looking at the size of sets, Cantor used the concept of equivalence. Two sets A and B are said to be equivalent if you can set up some relationship between them so that for every element of A, there is exactly one element in B and for each element in B, there is exactly one element in A. This type of relationship is called a **one-to-one** relationship.

EXAMPLE 5

For the two sets $A = \{1, 2, 3\}$ and $B = \{5, 6, 7\}$, we can set up the relationship below.

$$\begin{array}{ccc} 1 & 2 & 3 \\ \updownarrow & \updownarrow & \updownarrow \\ 5 & 6 & 7 \end{array}$$

Therefore, A and B are equivalent.

EXAMPLE 6

For the sets $A = \{1, 2, 3\}$ and $B = \{5, 6\}$, there is no one-to-one relationship. Thus, for sets that have a finite size, the concept of equivalent sets is the same as sets having the same size, $n(A) = n(B)$.

Now let's extend this concept to infinite sets.

EXAMPLE 7

For infinite sets, look at the set $A = \{1, 2, 3, 4, 5, \ldots\}$ and $B = \{2, 4, 6, 8, 10, \ldots\}$. We can set up the relationship below.

$$\begin{array}{ccccccc} 1 & 2 & 3 & 4 & 5 & \ldots & k \\ \updownarrow & \updownarrow & \updownarrow & \updownarrow & \updownarrow & & \updownarrow \\ 2 & 4 & 6 & 8 & 10 & \ldots & 2k \end{array}$$

Therefore, for each natural number in A, there is exactly one even natural number in B. As was true in Example 5, A and B are said to be equivalent sets. Thus, for each $k \in A$, $2k \in B$. This time, however, we can see that A contains all the elements of B (the even numbers) plus the odd numbers.

Example 7 provides us with an interesting situation. The infinite sets A and B are said to be equivalent, but there seem to be twice as many elements in A as there are in B. This is the fundamental difficulty with working with infinite sets: *Infinite sets can be equivalent but contain a seemingly unequal number of elements.*

COUNTABLY INFINITE SETS

A set is called **countably infinite** if you can put the elements into one-to-one correspondence with the natural numbers. Cantor showed that the rational numbers are countably infinite. However, he proved that there is no way to make a one-to-one correspondence between natural numbers and the real numbers. The set of real numbers is infinite but cannot be counted. His investigations opened the doors to the study of different levels of infinity.

EXAMPLE 8

Show that the set $A = \{2, 4, 8, 16, 32, 64, \ldots\}$ is countably infinite. What is the mathematical relationship that creates the one-to-one correspondence between set A and the natural numbers?

Solution: To show that a set is countably infinite, we need to find a one-to-one correspondence with the set and the natural numbers, N. By examining the elements of set A, you will notice this relationship.

$$\begin{array}{cccccccc} N: & 1 & 2 & 3 & 4 & \ldots & k \\ & \updownarrow & \updownarrow & \updownarrow & \updownarrow & & \updownarrow \\ A: & 2^1 = 2 & 2^2 = 4 & 2^3 = 8 & 2^4 = 16 & \ldots & 2^k \end{array}$$

The set A is countably infinite, and the mathematical relationship is for each natural number k, 2^k is in set A, that is, for $k \in N$, $2^k \in A$.

EXAMPLE 9

Show that the sets, $R = \{1, 1/2, 1/4, 1/8, \ldots\}$ and $S = \{1, 1/4, 1/16, 1/64, \ldots\}$, are equivalent.

Solution: To show the sets are equivalent, we need to find a one-to-one correspondence between the sets, as we did in Example 8.

$$\begin{array}{cccccccc} N: & 1 & 2 & 3 & 4 & \ldots & k \\ & \updownarrow & \updownarrow & \updownarrow & \updownarrow & & \updownarrow \\ R: & \frac{1}{2^{1-1}} = 1 & \frac{1}{2^{2-1}} = \frac{1}{2} & \frac{1}{2^{3-1}} = \frac{1}{4} & \frac{1}{2^{4-1}} = \frac{1}{8} & \ldots & \frac{1}{2^{k-1}} \\ & \updownarrow & \updownarrow & \updownarrow & \updownarrow & & \updownarrow \\ S: & \frac{1}{4^{1-1}} = 1 & \frac{1}{4^{2-1}} = \frac{1}{4} & \frac{1}{4^{3-1}} = \frac{1}{16} & \frac{1}{4^{4-1}} = \frac{1}{64} & \ldots & \frac{1}{4^{k-1}} \end{array}$$

section 3.1

PROBLEMS ➤ Explain ➤ Apply ➤ Explore

➤ Explain

1. What is a set?
2. What is the listing method of showing the items in a set? Give an example.
3. What is the descriptive method of showing the items in a set? Give an example.
4. What is meant by cardinality of a set?
5. What does it mean if a set A is a subset of set B?
6. If $A \subseteq B$ and $B \subseteq A$, what can you say about sets A and B? Give an example.
7. What is meant if two finite sets are equivalent?
8. What is meant if two infinite sets are equivalent?
9. What does it mean to say that a set is countably infinite?
10. What is meant by a one-to-one relationship?

➤ Apply

11. Write the set of even whole numbers less than or equal to 10, using the
    a) Descriptive method.
    b) Listing method.
12. Write the set of odd whole numbers less than or equal to 12, using the
    a) Descriptive method.
    b) Listing method.

In Problems 13–24, use the sets $T = \{3, 6, 9, 12\}$, $F = \{5, 10, 15, 20, 25\}$, and $I =$ {all the integers} to determine whether the subset and element symbols are used correctly. If the symbol is being used correctly, is the statement true? If not, write a true statement. If the symbol is being used incorrectly, write a true statement using the correct symbol.

13. $F \subset I$
14. $I \subset F$
15. $T \subset F$
16. $T \subset I$
17. $5 \subset I$
18. $5 \subset F$
19. $5 \notin I$
20. $5 \notin F$
21. $5 \in I$
22. $5 \notin T$
23. $F \in I$
24. $T \notin I$
25. Show that the set $F = \{5, 10, 15, 20, 25, \ldots\}$ is countably infinite. What is the mathematical relationship that shows the one-to-one correspondence?

26. Show that the set $F = \{1, 5, 9, 13, 17, \ldots\}$ is countably infinite. What is the mathematical relationship that shows the one-to-one correspondence?

27. Show that the set $F = \{1, 4, 9, 16, 25, 36, \ldots\}$ is countably infinite. What is the mathematical relationship that shows the one-to-one correspondence?

28. Show that the set $F = \{1, 3, 9, 27, 81, \ldots\}$ is countably infinite. What is the mathematical relationship that shows the one-to-one correspondence?

Explore

29. Use the listing method to create a set A of the states that border Indiana.

30. Use the listing method to create a set E of the countries that border France.

31. Use the listing method to create a set N of the countries that border South Africa and are north of the equator.

32. Use the listing method to create a set S of the states that border California and are also touching the Atlantic Ocean.

33. Let P be the set of U.S. Presidents in the years 1940–2003. Find $n(P)$.

34. Let C be the set of state capitals in the United States. Find $n(C)$.

35. Let B be the set of major league baseball teams as of 2003. Find $n(B)$.

36. Let J be the set of current U.S. Supreme Court justices. Find $n(J)$.

37. Show that the sets $R = \{3, 6, 9, 12, 15, \ldots\}$ and $S = \{1, 3, 9, 27, 81, \ldots\}$ are equivalent.

38. Show that the sets $R = \{1, 2, 6, 24, 120, 720, \ldots\}$ and $S = \{1, 3, 5, 7, 9, 11, \ldots\}$ are equivalent.

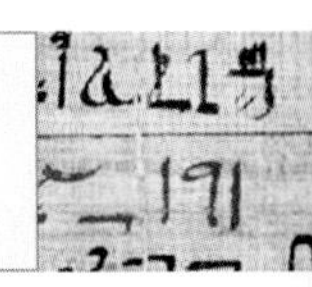

section 3.2

Set Operations and Venn Diagrams

The Jockey (*Le Jockey*) by Henri de Toulouse-Lautrec shows the age-old fascination with horse racing. (National Gallery of Art)

OPERATIONS WITH SETS

There are three set operations that will be used in this text: the operations of complement, union, and intersection.

The **complement** of a set A is the set of items in a universal set that are not contained in the set A. The complement is denoted as $\overline{A}$ and is read "*not A*" or "A bar." Using the complement of a set can often simplify the description of a set, and, more importantly, simplify finding the cardinality of a set.

EXAMPLE 1

Suppose that a math class of 37 students has taken a quiz in which the scores are in the set $\{0, 1, 2, 3, \ldots, 100\}$ and the six students who scored 90 or above received a grade of A.

a) If we are interested in the actual scores earned by the students, what is the universal set, U, for these scores?

b) On the basis of the above information, do we know $n(U)$?

c) If A is the set of students who earned a grade of A, find $n(A)$.

d) Describe the elements of $\overline{A}$ and find $n(\overline{A})$.

Solution:

a) $U = \{\text{the set of all test scores}\}$

b) We cannot know $n(U)$, since there might be some duplicate scores; however, we do know $n(U) \leq 37$.

c) Since there were six students who earned a grade of A, $n(A) = 6$.

d) $\overline{A}$ is the set of students who did not earn a grade of A, so $n(\overline{A}) = 37 - 6 = 31$.

The **union** of two sets creates a new set containing all the elements of the two sets. The symbol used to indicate the union of two sets is $\cup$.

EXAMPLE 2

Given the sets $A = \{0, 3, 6, 9\}$ and $B = \{1, 3, 5, 7, 9\}$, find the union of A and B.

Solution: The union of A and B is given by $A \cup B$. Since the union must contain any element that is in either of the original sets, we have

$$A \cup B = \{0, 1, 3, 5, 6, 7, 9\}.$$

Notice that the elements in the union can be in either of the original sets or in both of the original sets. If there is an element that is contained in both of the original sets, that element is written only once in the union of the two sets. For example, the number 3 was in both of the original sets but was written only once in $A \cup B = \{0, 1, 3, 5, 6, 7, 9\}$.

The third set operation is the **intersection** operation. The intersection of two sets is a set consisting of all the elements that are in both sets. The symbol representing the intersecting operation is $\cap$.

EXAMPLE 3

Find the intersection of the sets $A = \{0, 3, 6, 9\}$ and $B = \{1, 3, 5, 7, 9\}$.

Solution: The intersection of A and B is given by $A \cap B$. Since the intersection contains any element that is in both of the orignal sets, we have

$$A \cap B = \{3, 9\}.$$

In summary, we will be using the following symbols when we discuss sets.

English	Symbol
Is an element of	$\in$
Is not an element of	$\notin$
Is a proper subset of	$\subset$
Is a subset of	$\subseteq$
Union	$\cup$
Intersection	$\cap$
Empty set	$\varnothing$ or $\{\ \}$
Complement of a set	$\overline{A}$

TRANSLATING WORDS INTO SET OPERATIONS

In algebra, certain key words are translated into mathematical symbols or operations. For example, the word "is" often is represented by the symbol "=" when a word problem is translated into mathematical symbols. The same is true for set operations. Certain key words can be translated into set operations. When the word *or* is used, it may be represented in set operations by the union symbol.

EXAMPLE 4

Find the set of whole numbers less than 13 that are divisible by 3 or 2.

Solution: The whole numbers less than 13 that are divisible by 3 can be written as the set $A = \{0, 3, 6, 9, 12\}$. Similarly, the set of whole numbers less than 13 that are divisible by 2 is the set $B = \{0, 2, 4, 6, 8, 10, 12\}$. Since the question asks for those numbers that are divisible by 3 or 2, we find the union of A and B.

$$A \cup B = \{0, 2, 3, 4, 6, 8, 9, 10, 12\}$$

In a similar fashion, the word *and* is used to indicate the intersection of two sets.

EXAMPLE 5

Find the set of whole numbers less than 13 that are divisible by 3 and 2.

Solution: As before, we let $A = \{0, 3, 6, 9, 12\}$ and $B = \{0, 2, 4, 6, 8, 10, 12\}$. Since the question asks for those numbers that are divisible by 3 and 2, we want the intersection of A and B.

$$A \cap B = \{0, 6, 12\}$$

NOTE: It is important to remember that *and* and *or* have very specific meanings when used in connection with sets. In particular, *or* indicates that an element can be in either of the original sets or in both of the original sets. This contradicts the common usage of *or* meaning one set or the other but not both.

SETS AND VENN DIAGRAMS

Venn diagrams (named after John Venn, a 19th century English mathematician) are illustrations depicting sets. Typically, Venn diagrams consist of a rectangular region containing several circles. Shown here is a Venn diagram consisting of two intersecting sets, A and B, depicted by circles. The boundary rectangle is labeled U, for universal set. As we shall see in the following examples, Venn diagrams can be used to help us interpret mathematical problems.

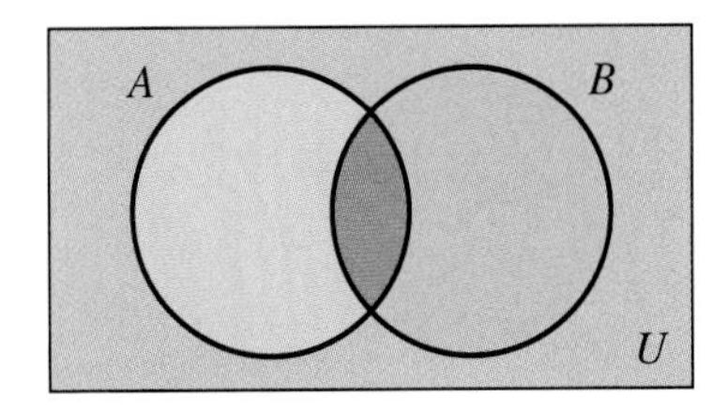

EXAMPLE 6

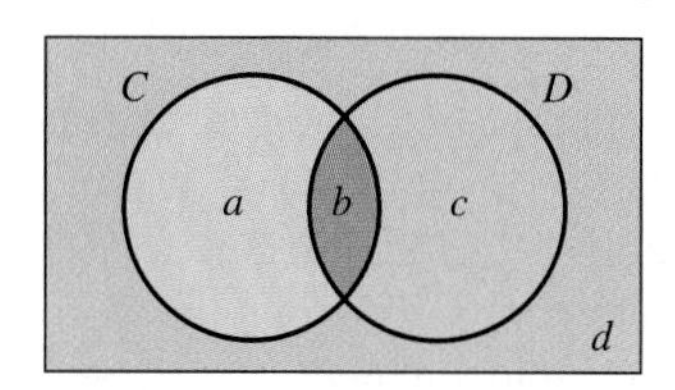

Examine the Venn diagram shown in which U is the set of all music, the circle C represents music available on cassettes, and the circle D represents music available on compact discs (CDs). Describe what is represented by each of the regions a, b, c, and d, and express each region in terms of the sets C and D.

Solution: Region a is inside the circle C but outside the circle D. Therefore, region a represents the music that is available on cassettes but not CDs. In terms of set notation, this is represented as $C \cap \overline{D}$.

Similarly, region c is inside the circle D but outside the circle C. Therefore, region c represents the music that is available on CDs but not cassettes. In terms of set notation, this is represented as $\overline{C} \cap D$.

Region b is inside both the circle C and the circle D. Therefore, region b represents the music that is available on both cassettes and CDs. In terms of set notation, this is represented as $C \cap D$.

Region d is outside of both circles. Therefore, region d represents the music that is not available on either cassettes or CDs. In terms of set notation, this is represented as $\overline{C} \cap \overline{D}$ or $(\overline{C \cap D})$.

Venn diagrams can help in determining the cardinality of a set. In fact, as the following example shows, Venn diagrams are useful in developing formulas for cardinality.

EXAMPLE 7

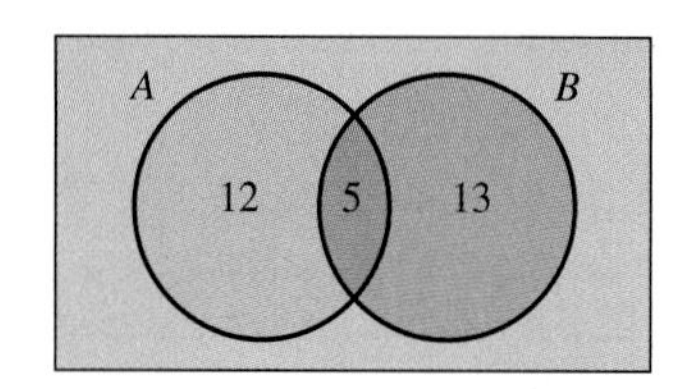

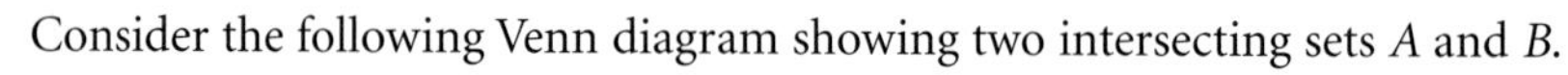

Consider the following Venn diagram showing two intersecting sets A and B.

a) Find $n(A)$.

b) Find $n(B)$.

c) Find $n(A \cap B)$.

d) Find $n(A \cup B)$.

Solution:

a) The elements contained in A are the elements in the shaded region in the picture at the right. Therefore, $n(A) = 12 + 5 = 17$.

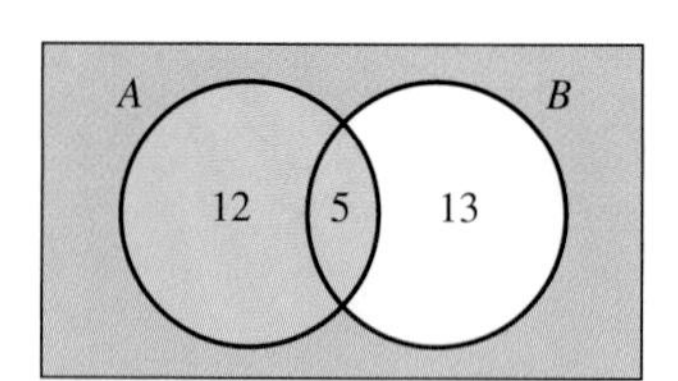

b) Similarly, $n(B) = 5 + 13 = 18$.

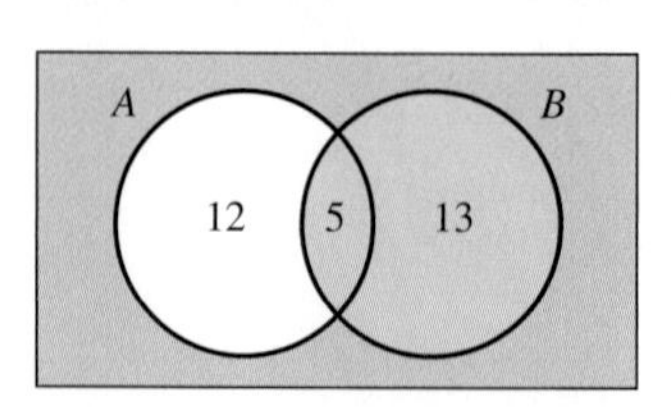

c) The intersection is the region contained in both circles. Therefore, $n(A \cap B) = 5$.

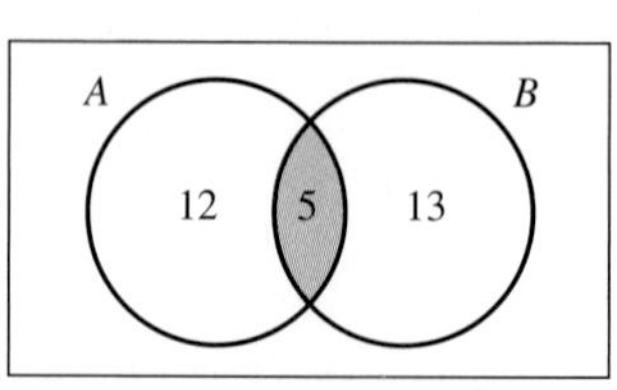

d) The union is the region that contains the elements of A or the elements of B. Therefore, $n(A \cup B) = 12 + 5 + 13 = 30$.

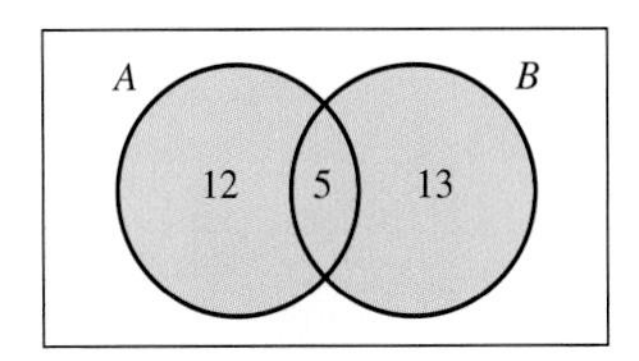

In Example 7, notice that $n(A \cup B) = 30$ is less than the sum of $n(A) = 17$ and $n(B) = 18$. This is because adding $n(A)$ and $n(B)$ includes the intersection twice. Therefore, if we want a formula for the cardinality of a union, the cardinality of the intersection should be subtracted. This results in the following formula.

Cardinality of a Union

$$n(A \cup B) = n(A) + n(B) - n(A \cap B)$$

EXAMPLE 8

In a group of 100 people, there are 18 people with red hair, 12 people with blue eyes, and 10 people with both red hair and blue eyes. Find the number of people in this group having either red hair or blue eyes.

Solution: Using R for the set of people with red hair and B for the set of people with blue eyes, the above formula gives the following.

$$\begin{aligned} n(R \cup B) &= n(R) + n(B) - n(R \cap B) \\ &= 18 + 12 - 10 \\ &= 20 \end{aligned}$$

The final example for this section is one in which the diagram does not contain the standard two intersecting circles that we have been studying.

EXAMPLE 9

Let

$J =$ the set of irrational numbers

$N =$ the set of natural numbers $= \{1, 2, 3, 4, 5, \ldots\}$

$I =$ the set of integers $= \{\ldots, -3, -2, -1, 0, 1, 2, 3, \ldots\}$

Draw the Venn diagram for these sets. Use $U =$ the set of real numbers.

Solution: In this example, the set N is a subset of I. Therefore, when we draw the diagram, the set N should be completely enclosed inside the set I. In addition, I and J have no common element, that is, $I \cap J = \varnothing$. Therefore, they are shown as nonoverlapping circles. When two sets have no intersection, they are called **disjoint sets.**

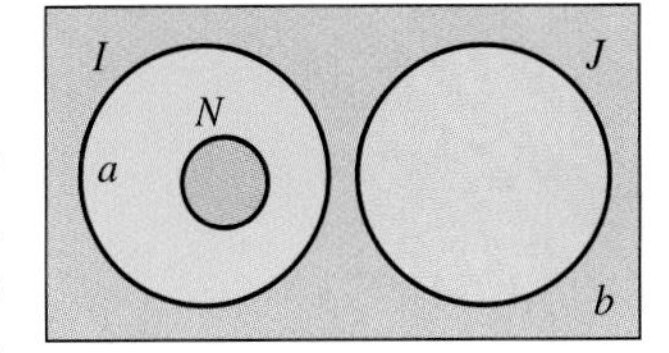

The region inside the circle N is the set of natural numbers. The region labeled a is the set consisting of zero and negative integers. The region b contains all noninteger, rational numbers.

PROBLEMS ▶ Explain ▶ Apply ▶ Explore

Explain

1. What is meant by the intersection of two sets?
2. What is meant by the union of two sets?
3. What is meant if two sets are described as disjoint?
4. If A and B are disjoint sets, why can the formula $n(A \cup B) = n(A) + n(B) - n(A \cap B)$ be simplified to $n(A \cup B) = n(A) + n(B)$?
5. Explain why the formula $n(A \cup B) = n(A) + n(B) - n(A \cap B)$ requires the subtraction of the term $n(A \cap B)$.
6. If A and B are finite sets, what does it mean for $n(A \cup B) = n(B)$?
7. If A and B are finite sets, what does it mean for $n(A \cap B) = n(B)$?
8. What does it mean for $A \cap B = \varnothing$? Draw a picture to illustrate your reasoning.

Apply

9. Let M = {apples, bananas, peaches, tomatoes} and T = {beans, peas, sprouts, tomatoes}. Determine the following.
   a) $M \cup T$
   b) $M \cap T$
10. Let F = {oranges, apples, apricots, peaches} and N = {coconuts, filberts, almonds}. Determine the following.
    a) $F \cup N$
    b) $F \cap N$
11. Let $n(A) = 25$, $n(B) = 30$ and $n(A \cap B) = 7$. Find $n(A \cup B)$.
12. Let $n(A) = 32$, $n(B) = 38$, and $n(A \cap B) = 7$. Find $n(A \cup B)$.
13. Let $n(A) = 25$, $n(B) = 30$, and $n(A \cup B) = 40$. Find $n(A \cap B)$.
14. Let $n(A) = 32$, $n(B) = 20$, and $n(A \cup B) = 40$. Find $n(A \cap B)$.
15. Show that it is not possible for $n(A) = 25$, $n(B) = 30$, and $n(A \cup B) = 60$.
16. Show that it is not possible for $n(A) = 32$, $n(B) = 20$, and $n(A \cap B) = 25$.

In Problems 17–24, use the Venn diagram to find the cardinality of each set.

A B
51 25 17
7

17. $n(A)$
18. $n(B)$
19. $n(A \cup B)$
20. $n(A \cap B)$
21. $n(A \cap \overline{B})$
22. $n(\overline{A} \cap B)$
23. $n(\overline{A} \cap \overline{B})$
24. $n(\overline{A} \cup \overline{B})$

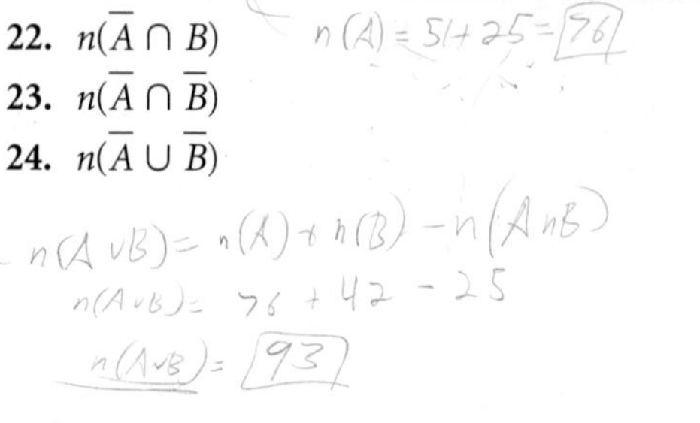

Explore

25. Let U = the set of all people, R = the set of all people with red hair, and W = the set of all women. Draw a diagram for these sets and describe the type of people contained in each region of the diagram.

26. Let U = the set of all books, N = the set of all novels, and P = the set of all books of poetry. Draw a diagram for these sets and describe the type of books contained in each region of the diagram.

27. Let U = the set of all items in the library, N = the set of novels in the library, B = the set of books in the library, and V = the set of all videos in the library. Draw a diagram for these sets and describe the type of item contained in each region of the diagram.

28. To construct the budget for English as a Second Language teachers, the principal at Elizabeth Haddon Elementary School needs to figure out how many new students are not fluent in English. Let U = the set of all new students at the school, S = the set of all new students whose first language is Spanish, E = the set of new students whose first language is English, and V = the set of new students whose first language is Vietnamese. Draw a Venn diagram for these sets and describe the type of people contained in each region of the diagram.

29. The Venn diagram shown gives the result of a survey in which 300 people were asked whether they drink coffee (C) or orange juice (O) for breakfast. Determine the number of people who drink both coffee and orange juice for breakfast.

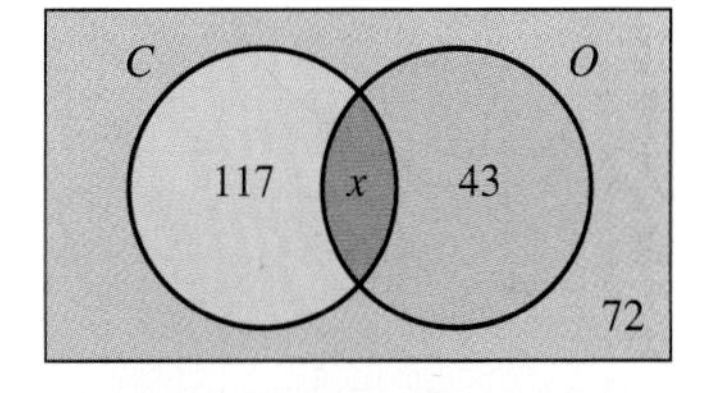

30. The Venn diagram shown gives the result of a survey in which 500 people were asked whether they drink coffee (C) or orange juice (O) for breakfast. Determine the number of people who drink neither coffee nor orange juice for breakfast.

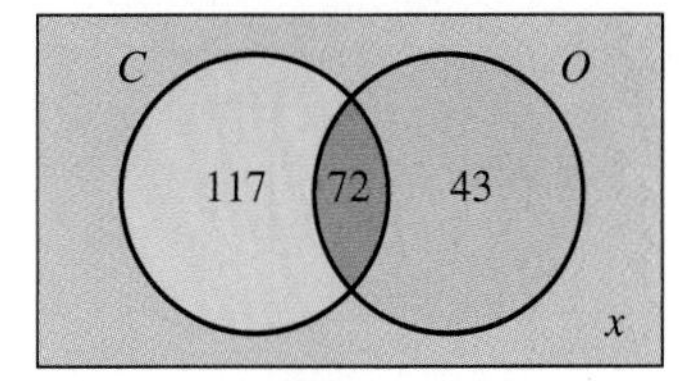

31. Given the diagram shown, what can be said about $A \cap B$?

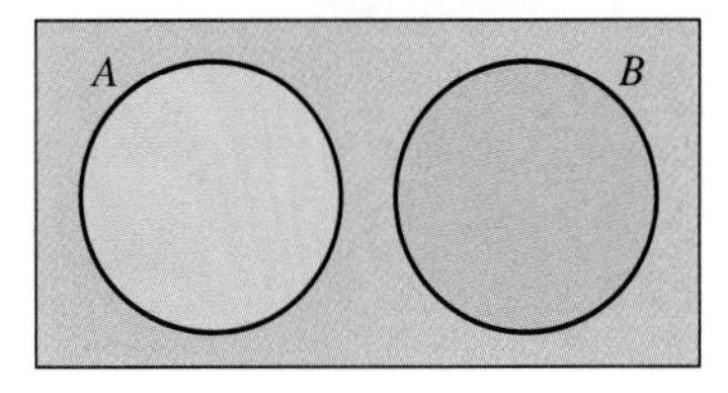

32. Draw a Venn diagram in which A contains a total of 30 elements, B contains a total of 50 elements, and U contains a total of 80 elements. (There are many possible answers.)

33. For marketing purposes, the maker of CoffeeMate™ wants to find out how many people put milk in their coffee or tea. A survey of 1000 coffee and tea drinkers has found that 625 people drink coffee with milk, 370 drink tea with milk, and 130 do not drink coffee or tea with milk. Draw a Venn diagram to represent this information and determine the number of people who drink both coffee and tea with milk.

34. Five hundred apples were examined for traces of pesticides. Forty of the apples showed traces of only Malathion, 35 showed traces of both Malathion and Diazinon, and 420 showed traces of neither Malathion nor Diazinon. Draw a Venn diagram to represent this information and determine the number of apples that showed traces of only Diazinon.

section 3.3

Applications of Sets

In this section, we continue our discussion of sets by examining two topics: using sets to investigate tables of information and the use of Venn diagrams for more complicated situations.

READING TABLES

When we look at real-world information, it is often arranged in tables. Each column or row of the table is some subset of this information. Consider the following example.

EXAMPLE 1

Based on statistics published by the U.S. Department of Labor, the employment classifications of 1200 men and 1300 women are given in the following table.

Occupation	Number of Males	Number of Females	Total
Executive, administrative, management	180	185	365
Professional specialty	163	234	397
Technical, sales, administrative support	236	520	756
Service occupations	119	227	346
Precision production, crafts	223	27	250
Operators, fabricators, laborers	232	92	324
Farming, forestry, fishing	47	15	62
Totals	1200	1300	2500

a) If F is the set of females in the survey, what is $n(F)$?

b) If T is the set of people in the survey who are working in technical, sales, or administrative support, what is $n(T)$?

c) Describe in words the set $T \cap F$ and find the value of $n(T \cap F)$.
d) Describe in words the set $T \cup F$ and find the value of $n(T \cup F)$.

Solution:

a) Since there are 1300 females in the survey, $n(F) = 1300$.
b) Since there are a total of 756 people working in technical, sales, or administrative support, $n(T) = 756$.
c) $T \cap F$ is the set of females who work in technical sales, or administrative support, so $n(T \cap F) = 520$.
d) $T \cup F$ is the set of people who are female or who work in technical, sales, or administrative support. Using the rule

$$n(A \cup B) = n(A) + n(B) - n(A \cap B).$$

that was developed in Section 3.2, we have

$$\begin{aligned} n(T \cup F) &= n(T) + n(F) - n(T \cap F) \\ &= 756 + 1300 - 520 \\ &= 1536. \end{aligned}$$

USING VENN DIAGRAMS

Venn diagrams can answer specific questions about a situation. For each example, each region of a Venn diagram can be used to represent one of the possible outcomes of a problem. Because this method involves reading and drawing pictures, it is easy to use. It is limited, however, by the complexity of the pictures that can be drawn. First, we build on the work from Section 3.2 by examining another Venn diagram with two sets.

EXAMPLE 2

A survey of 200 employees determined that 22 of the employees do not have any medical insurance, while 43 of the employees are covered by the company's policy and an additional policy. If 160 employees are covered by the company's insurance policy, how many employees are covered only by some other policy?

Solution: To determine the solution of this problem, we will use three sets, U, C, and O. U is the set of the 200 employees in the survey. Let C be the set of employees covered by the company's policy, and let O be the set of employees covered by some other policy. Since 43 employees were covered by more than one policy, there are 43 people in the set $C \cap O$. Since a total of 160 employees are covered by the company's policy, there must be $160 - 43 = 117$ employees covered only by the company's policy. Also, we know that there are 22 people who do not have any medical insurance. This information is summarized in the Venn diagram shown here.

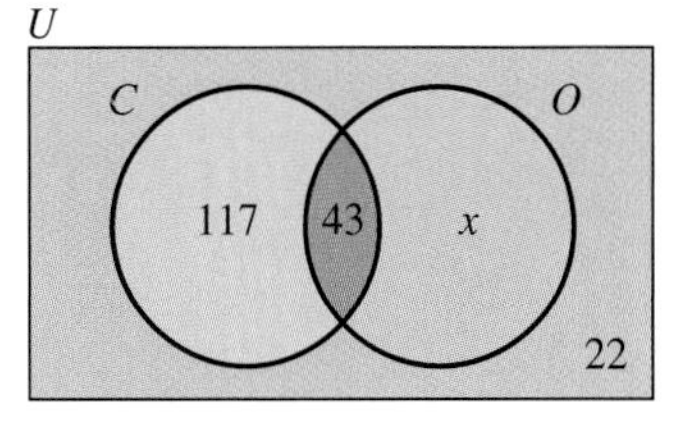

The intersection contains 43 people, the remainder of set C contains 117 people, and there are 22 people not contained in either circle. The number of people who are insured only by some policy other than the company's are represented by x. Since the total number of people surveyed is 200, we have the following equation and solution:

$$22 + 117 + 43 + x = 200$$
$$182 + x = 200$$
$$x = 18$$

Therefore, there are 18 people whose only insurance is not through the company policy.

So far, the examples of Venn diagrams have been restricted to those using at most two intersecting circles. In the following example, we look at a situation containing three intersecting circles.

EXAMPLE 3

Out of 600 doctors, 255 accept children as patients, 360 perform surgery, and 370 are general practitioners. If 175 of the general practitioners accept children as patients, 185 of the doctors performing surgery will operate on children, 260 of the general practitioners perform surgery, and there are 135 general practitioners who will perform surgery on children, how many of the 600 doctors are general practitioners who do not perform surgery and do not accept children?

Solution:

Let

C = the set of doctors who accept children as patients

S = the set of doctors who perform surgery

G = the set of doctors who are general practitioners

Using the given information in conjunction with the intersection symbol, gives the following.

The number of doctors in $C = 255$.

The number of doctors in $S = 360$.

The number of doctors in $G = 370$.

The number of doctors in $G \cap C = 175$.

The number of doctors in $S \cap C = 185$.

The number of doctors in $G \cap S = 260$.

The number of doctors in $G \cap S \cap C = 135$.

At this point, the only information that can be entered in the Venn diagram is that 135 people are contained in the intersection of the three circles. This gives the Venn diagram.

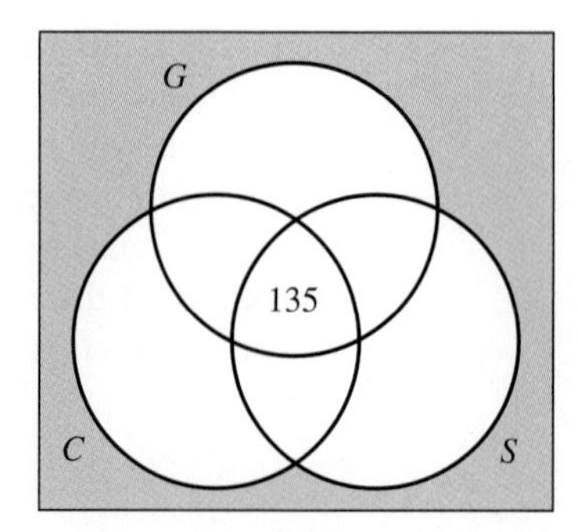

Since we know there are 135 general practitioners who perform surgery on children and there are a total of 260 general practitioners who perform surgery, there must be $260 - 135 = 125$ general practitioners who perform surgery but not on children.

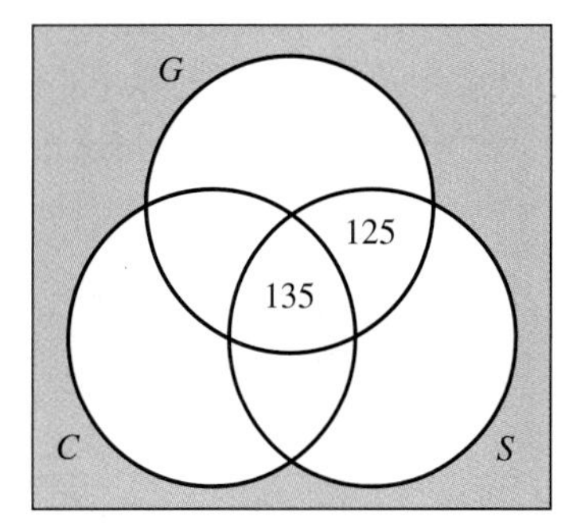

Performing similar operations with the intersections $S \cap C = 185$ and $G \cap C = 175$ gives us the third version of the Venn diagram. The question is to determine how many general practitioners do not perform surgery and do not accept children as patients. This is given by the region labeled x.

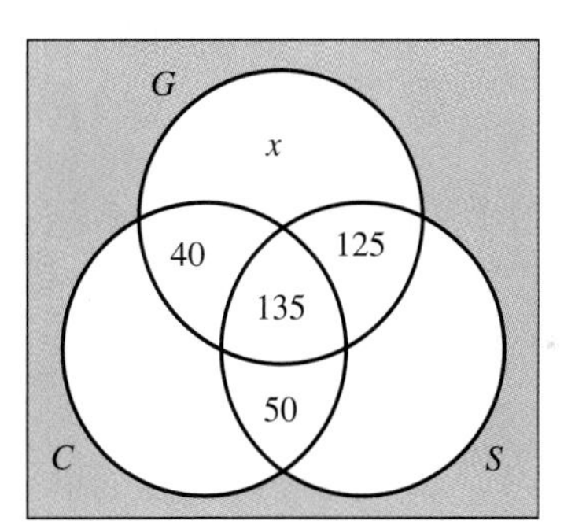

Since there are a total of 370 general practitioners, the number of doctors in the four regions of set G must add to 370. Therefore, we have the following algebraic equation and solution.

$$40 + 135 + 125 + x = 370$$
$$300 + x = 370$$
$$x = 70$$

PROBLEMS ● Explain ● Apply ● Explore

● Explain

1. Explain the difference between $T \cup F$ and $n(T \cup F)$.
2. Explain the difference between $T \cap F$ and $n(T \cap F)$.

Use the following table for Problems 3–6.

	Males	Females	Total
Golf	100	85	a
Fish	65	b	130
Total	c	d	e

3. Explain how to find the values a and b.
4. Explain how to find the values c, d, and e.
5. Using golfers as the universal set, draw a Venn diagram with exactly one circle for the above information.
6. Using all the people in the table as the universal set, draw a Venn diagram with exactly two circles for the above information using
   a) Circles G for golfers and F for females.
   b) Circle P for fish and M for males.

● Apply

7. Use the information in Example 3 to find the number of doctors who accept children as patients but do not perform surgery and are not general practitioners.
8. Use the information in Example 3 to find the number of doctors who perform surgery but are not general practitioners and do not accept children as patients.

Use the table in Example 1 to answer the following questions. If M is the set of males in the survey, F is the set of females in the survey, and A is the set of people in the survey who are working in farming, forestry, or fishing,

9. Find the value of $n(A)$.
10. Describe in words the set $M \cap A$ and find the value of $n(M \cap A)$.
11. Describe in words the set $M \cup A$ and find the value of $n(M \cup A)$.
12. Describe in words the set $M \cap F$ and find the value of $n(M \cap F)$.
13. Describe in words the set $M \cup F$ and find the value of $n(M \cup F)$.
14. Describe in words the set $M \cap \overline{A}$ and find the cardinality of $M \cap \overline{A}$.
15. Describe in words the set $\overline{M} \cap \overline{A}$ and find the cardinality of $\overline{M} \cap \overline{A}$.
16. Describe in words the set $(M \cup F) \cap A$ and find the cardinality of $(M \cup F) \cap A$.

For each of the following, use the Venn diagram to find the cardinality of the indicated set.

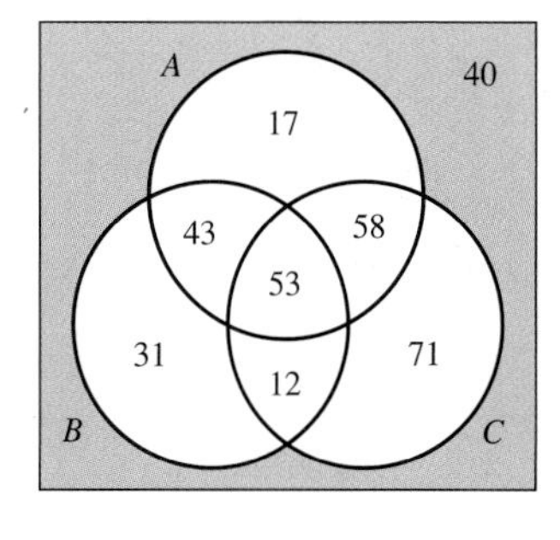

17. $A \cap B \cap C$
18. $\overline{A} \cap B \cap \overline{C}$
19. $\overline{A} \cup B \cup \overline{C}$
20. $A \cup B$
21. $A \cup C$
22. $A \cup B \cup C$
23. The given Venn diagram has each of its regions marked with a Roman numeral. Match each of the following set representations with the correct Roman numeral.

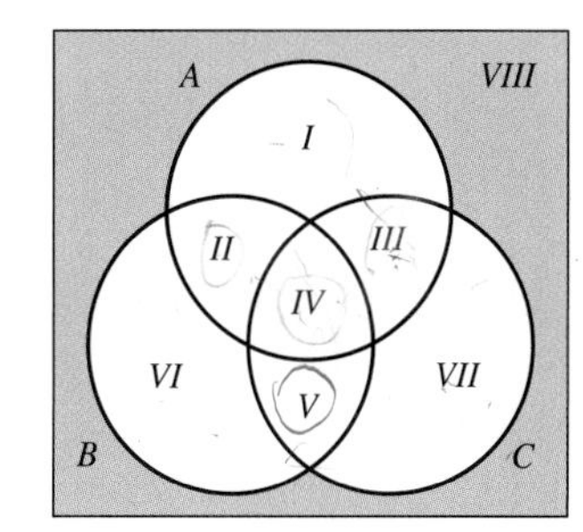

a) $A \cap B \cap C$
b) $A \cap B \cap \overline{C}$
c) $A \cap \overline{B} \cap C$
d) $\overline{A} \cap B \cap C$
e) $A \cap \overline{B} \cap \overline{C}$
f) $\overline{A} \cap \overline{B} \cap C$
g) $\overline{A} \cap B \cap \overline{C}$
h) $\overline{A} \cap \overline{B} \cap \overline{C}$

Explore

24. A survey of 1000 people determined the following results about their radio listening habits:

In the morning, between 5:00 and 7:00 A.M., 620 people listen to the radio.

During the evening, between 4:00 and 7:00 P.M., 640 people listen to the radio.

During the day, between 7:00 A.M. and 4:00 P.M., 450 people listen to the radio.

During all three periods, 210 people listen.

During the morning and the evening only, 220 people listen.

During the morning and the day only, 70 people listen.

During the day and the evening only, 130 people listen.

a) How many people listen only during the morning?
b) How many people listen only during the day?
c) How many people listen only during the evening?
d) How many people do not listen to the radio?

25. A survey of 500 people determined the following results about their exercise habits.

Two hundred forty-two people participate in aerobics.

Two hundred seventy-eight people participate in weight lifting.

Two hundred ninety-eight people participate in running.

Forty-three people participate only in aerobics and weight lifting.

Fifty people participate only in aerobics and running.

Ninety-two people participate only in weight lifting and running.

Seventy-five participate in all three activities.

a) How many participate only in aerobics?

b) How many participate only in weight lifting?

c) How many participate only in running?

d) How many participate in none of these three activities?

26. Given the Venn diagram with the two intersecting circles A and B,

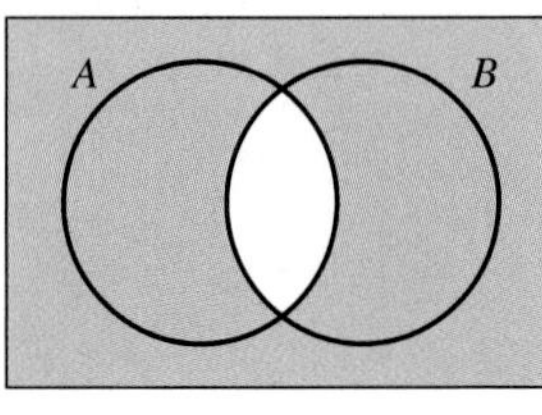

a) Shade the region $\overline{A \cap B}$.

b) Shade the region $\overline{A} \cup \overline{B}$.

Since the shaded regions are the same, we can make the statement $\overline{A \cap B} = \overline{A} \cup \overline{B}$. This is one

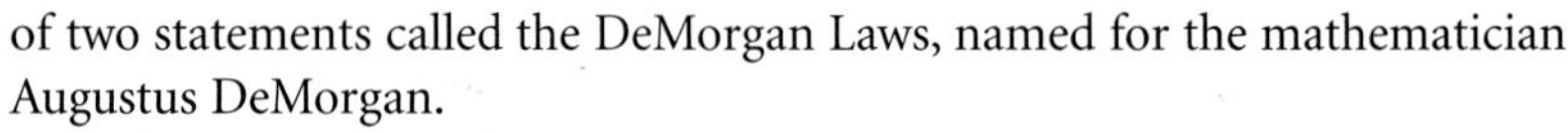

of two statements called the DeMorgan Laws, named for the mathematician Augustus DeMorgan.

27. Given the Venn diagram with the two intersecting circles A and B,

a) Shade the region $\overline{A \cup B}$.

b) Shade the region $\overline{A} \cap \overline{B}$.

Since the shaded regions are the same, we can make the statement $\overline{A \cup B}. = \overline{A} \cap \overline{B}$. This is the second of two statements called the DeMorgan Laws, named for the mathematician Augustus DeMorgan.

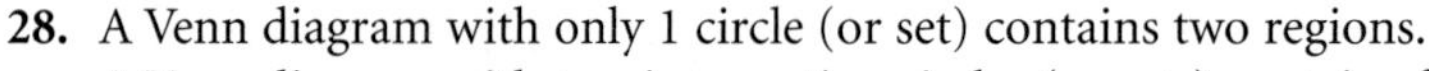

28. A Venn diagram with only 1 circle (or set) contains two regions.
A Venn diagram with two intersecting circles (or sets) contains four regions.
A Venn diagram with three intersecting circles (or sets) contains eight regions.

a) A Venn diagram with four intersecting sets contains 16 regions. Explain why this should be true.

b) Draw a Venn diagram containing four intersecting sets and 16 distinct regions. *Hint:* Use three circles and one set that is not represented by a circle.

29. When we referred to a Venn diagram with three intersecting circles, the formula for the cardinality of a union was given by $n(A \cup B) = n(A) + n(B) - n(A \cap B)$. Devise a similar formula for $n(A \cup B \cup C)$ based on the three circles in the following Venn diagram.

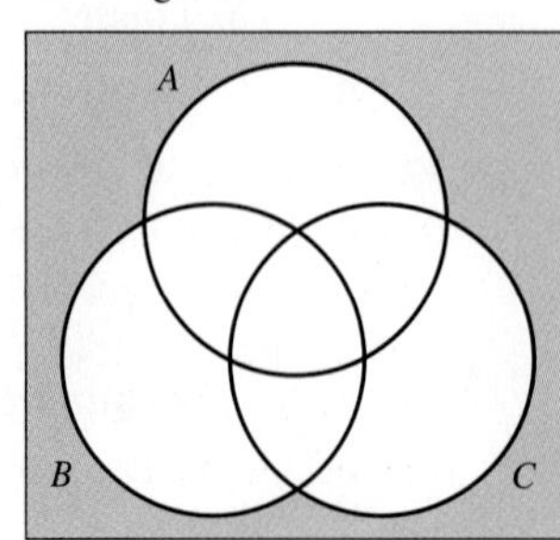

section 3.4

Introduction to Counting

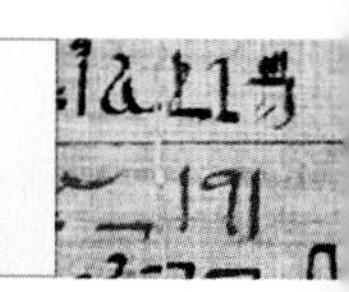

Every day people use the process of counting in many activities. Most of the time, the counting is fairly simple: "How many people are coming to dinner?" or "How many shopping days are left until Christmas?" When a more difficult problem arises, simple methods either fail or are overpowered by the problem. For example, in how many different ways can the spark plug wires be connected to the distributor cap of your car? Although only one arrangement is correct, there are 720 ways for a six-cylinder car. This section will introduce you to some basic principles that can make counting an easier task.

COUNTING BY LISTING POSSIBLE OUTCOMES

The most straightforward method for determining the number of ways in which an event can occur is to simply list all the possible outcomes and count them. For example, if four coins are tossed, in how many different ways could they land with two heads and two tails facing up? If we list all the possible ways this could happen, we see that there are six possibilities.

1st Coin	2nd Coin	3rd Coin	4th Coin
H	H	T	T
H	T	T	H
H	T	H	T
T	H	H	T
T	H	T	H
T	T	H	H

The listing of all desired outcomes is a legitimate way to count. However, it becomes very tedious if there are many possibilities. If you were to list all the possible five-card hands you could deal from a standard deck of 52 cards, you would have to list 2,598,960 different hands. To avoid this kind of task, we will investigate other methods of counting the number of ways in which an event can occur. However, the listing method is always a possibility in a given problem.

With or Without Replacement?

Before examining other methods of counting, there is a concern that could affect the answer to a counting problem. Is the counting done with items being replaced or without being replaced? Counting is done **with replacement** if the same object can be used more than once. For example, suppose a person is buying a candy bar from a well-stocked candy machine. If the machine is working correctly and is well stocked, choosing a particular brand of candy bar does not prohibit the next person using the machine from choosing the same brand of candy. The candy bar chosen by the first person was replaced by another bar of the same brand. If the machine had only one bar of each brand, candy bars would be chosen **without replacement.**

A standard deck of cards can be used to show the difference between an event done with replacement and without replacement. Suppose a player is to draw two cards from a standard deck of cards and the queen of hearts is drawn on the first card. If the second card is drawn without replacement (that is, the first card is not returned to the deck), it is not possible to draw another queen of hearts. On the other hand, suppose the second card is drawn after the first card has been replaced in the deck. It is now possible to draw the queen of hearts on the second card. When counting the number of ways an event can occur, it makes a difference whether the event happens with or without replacement.

COUNTING USING THE BASIC COUNTING LAW

Besides listing outcomes, the number of ways an event can occur can be determined by the basic counting law. According to the **Basic Counting Law**, if there are n choices for the first item and m choices for the second item, there are $n \times m$ ways in which to pick a set consisting of two items.

EXAMPLE 1

At the Produce Market Restaurant, the light lunch special consists of a choice of one of the four salads and one of the six types of fruit. How many different lunches are available?

Solution: Since there are four choices of salad and six choices of fruit, the Basic Counting Law gives a result of $4 \times 6 = 24$ different light lunch specials.

EXAMPLE 2

A group of three people decide to have the light lunch special at the Produce Market Restaurant. In how many different ways can the group of three order lunches?

Solution: Assuming that there are enough supplies at the restaurant to accommodate many orders of the same item, this is a situation involving replacement. This means that all the people can order the same item. Using the results of Example 1, each of the people has 24 choices. Thus, by the Basic Counting Law, there are $24 \times 24 \times 24 = 13{,}824$ different orders.

EXAMPLE 3

For those whose appetites are not satisfied by the light lunch special, the restaurant has a dessert cart. There are ten different desserts on the cart but only one of each type. In how many ways can our group of three each order a dessert?

Solution: In this case, it is not possible for the same dessert to be chosen by more than one person. The first person will have ten choices for dessert, but the second person will have only nine choices, and the third person only eight choices. Thus, by the Basic Counting Law, there are $10 \times 9 \times 8 = 720$ different possibilities.

In Examples 1–3, the Basic Counting Law was used. In Examples 1 and 2, the selections were made with replacement, whereas in Example 3, and selections were made without replacement.

COUNTING TOOLS

To continue our investigation into counting, we first introduce three tools: factorials, combinations, and permutations. All three of these functions are included on most scientific calculators.

Factorials

The expressions $n!$ is read n **factorial.** When you see an expression such as 5!, it means

$$5! = 5 \times 4 \times 3 \times 2 \times 1 = 120.$$

Factorials are defined only for whole numbers, with $0! = 1$ being a special case. Since factorials are often very large, you will frequently see the results expressed in scientific notation. Thus,

$$20! = 20 \times 19 \times 18 \times \cdots \times 3 \times 2 \times 1 = 2{,}432{,}902{,}008{,}176{,}640{,}000$$

would be expressed as

$$20! = 2.43 \times 10^{18}.$$

Permutations

A **permutation** is one of the arrangements of a set of items. For example, if we want to pick two letters from the set $\{a, b, c, d\}$, they can be arranged in twelve ways. Notice that the different arrangements of the letters count as different permutations.

$$\begin{matrix} ab & ac & ad & bc & bd & cd \\ ba & ca & da & cb & db & dc \end{matrix} \qquad (P_{4,2} = 12)$$

For larger groups of items, it is awkward to list all the permutations. However, we can count the number of permutations using the formula

$$P_{n,r} = \frac{n!}{(n-r)!}.$$

Thus,

$$P_{15,3} = \frac{15!}{(15-3)!} = \frac{15!}{12!} = 2730.$$

Combinations

A **combination** is a set of objects in which the order of the objects does not matter. For example, if we want to pick two letters from the set $\{a, b, c, d\}$ but we are not concerned about the order, there are six ways. Notice that only one ordering of each pair of letters is included.

$$ab \quad ac \quad ad \quad bc \quad bd \quad cd \qquad (C_{4,2} = 6)$$

For larger groups of items, we can count the number of combinations using the formula

$$C_{n,r} = \frac{n!}{r!(n-r)!}.$$

Thus,

$$C_{15,3} = \frac{15!}{3!(15-3)!} = \frac{15!}{3! \times 12!} = 455.$$

You will find it easier to compute factorials, permutations, and combinations using a calculator. For help, read your calculator manual or ask your instructor. Using these three tools, we can now continue with our discussion of counting.

COUNTING USING PERMUTATIONS AND COMBINATIONS

In some counting problems, you may be selecting a group of items from a larger set of items, such as dealing five cards from a deck of 52 cards or picking two horses from a field of 12 horses or selecting three desserts from a tray containing ten desserts. Counting in situations such as these can be accomplished by using permutations or combinations. The number of permutations determined by $P_{n,r}$ indicates how many arrangements of r items can be made from a set of n items. The number of combinations determined by $C_{n,r}$ indicates how many groupings of r items can be made from a set of n items without regard for the order of the items. If we are selecting items without replacement, permutations are used if the order of the items matters, and combinations are used if the order does not matter. With this in mind, let us take another look at the problem posed in Example 3. It can be reworded into the statement formulated in the next example.

EXAMPLE 4

In how many ways can three people select a dessert from a dessert tray containing ten different desserts?

Solution: Since they are selecting desserts without replacement and the order of the selections gives each person a different dessert, we can use permutations to determine the number of ways in which this can be done. Using $n = 10$ and $r = 3$, the number of permutations of ten items, choosing three at a time, is

$$P_{10,3} = \frac{10!}{(10-3)!} = \frac{10!}{7!} = \frac{10 \times 9 \times 8 \times 7 \times \cdots \times 1}{7 \times 6 \times \cdots \times 1}$$
$$= 10 \times 9 \times 8 = 720.$$

EXAMPLE 5

The lock on a safe is a combination lock with 50 numbers on the dial. Four numbers are needed to unlock the safe.

a) How many four-number sequences are possible if a number can be used only once?

b) How many four-number sequences are possible if a number can be used more than once?

Solution:

a) If the numbers can be used only once, the selection of numbers is done without replacement. Since the order of the numbers is essential for opening the lock, permutations should be used. Thus, the number of four number sequences is

$$P_{50,4} = \frac{50!}{(50-4)!} = \frac{50!}{46!} = \frac{50 \times 49 \times 48 \times 47 \times 46 \times \cdots \times 1}{46 \times 45 \times \cdots \times 1}$$

$$= 50 \times 49 \times 48 \times 47 = 5{,}527{,}200.$$

b) If the numbers can be used more than once, the numbers are chosen with replacement. There are 50 choices for each of the four numbers in the combination. Therefore, by the Basic Counting Law, the number of possible four number sequences is

$$50 \times 50 \times 50 \times 50 = 6{,}250{,}000.$$

EXAMPLE 6

A researcher has to select three out of a group of five rabbits to test a new dietary supplement. How many different sets of three rabbits can be selected?

Solution: Since she needs to select three rabbits from a group of five, the selection is done without replacement. Since the order in which they are selected does not matter, she can count the number of ways in which this can be done using combinations with $n = 5$ and $r = 3$.

$$C_{5,3} = \frac{5!}{3!(5-3)!} = \frac{5!}{3!2!} = \frac{5 \times 4 \times 3 \times 2 \times 1}{3 \times 2 \times 1 \times 2 \times 1} = \frac{5 \times 4}{2 \times 1} = 10$$

EXAMPLE 7

In the game of lotto, each player picks some numbers in the hope of matching the winning numbers. For example, in a state lotto game, seven different numbers between 1 and 49 are chosen. How many different sets of seven numbers are possible?

Solution: Because the problem states that seven different numbers are drawn, we cannot use the same number twice. Therefore, the sampling is done without replacement. Since we are concerned only about a set of numbers and not about the order in which they are drawn, we should use combinations. Thus, the number of combinations of 49 numbers, taken seven at a time is

$$C_{49,7} = \frac{49!}{7!(49-7)!} = \frac{49!}{7!42!} = 85{,}900{,}584.$$

EXAMPLE 8

A jar contains 75 balls, numbered 1 through 75. If balls are selected without replacement and the order of selection does not matter, are there more ways to select groups of two balls from the jar or ways to select groups of 73 balls?

Solution: Since we are selecting a group of balls from the 75 balls without replacement and the order in which the balls are selected does not matter, we compute the number of combinations in both cases.

$$C_{75,2} = \frac{75!}{2!(75-2)!} = \frac{75!}{2!73!} = 2775$$

$$C_{75,73} = \frac{75!}{73!(75-73)!} = \frac{75!}{73!2!} = 2775$$

The results are the same! After arriving at this answer, it is not hard to see why it is true. Imagine being told to select 73 items from a jar containing 75 items. Not wanting to go through the tedious chore of selecting 73 items, it would be simpler to select and discard two items while keeping the remaining 73 items. This means that there are as many groups of two as groups of 73 that can be chosen from the set of 75 items.

The preceding examples have introduced you to the following four methods used in counting the number of ways an event can occur.

1. Counting by listing all possibilities.
2. Counting with replacement using the Basic Counting Law.
3. Counting without replacement where the order of the objects matters. This can be done using permutations.
4. Counting without replacement where the order of the objects does not matter. This can be done using combinations.

If you are selecting a group of items and listing all possible outcomes is not feasible, the following chart can help you decide how to approach a given counting problem.

Summary of Counting Principles

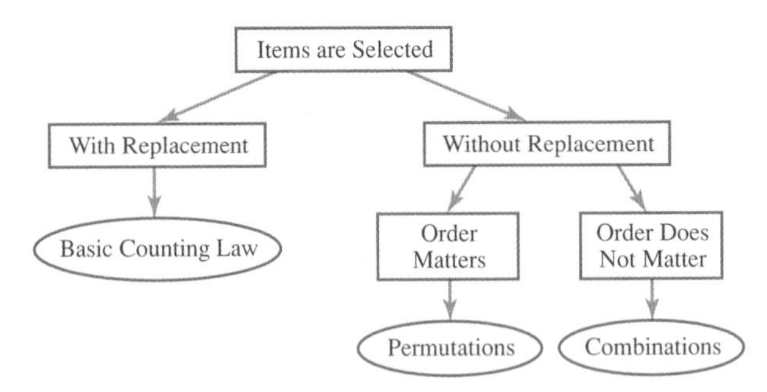

With these basic counting principles, we can examine applications of a more complex nature. We must state here that although these basic techniques will allow you to examine a large variety of applications, there are many counting problems that are beyond the scope of this text.

EXAMPLE 9

A standard deck of cards has 52 cards consisting of four suits of 13 cards each. In five-card poker, a hand with three cards of one rank and two cards of another rank is called a full house. How many ways are there to deal a five-card hand consisting of three 8's and two 7's?

Solution: Since the order of selection does not matter and the cards are dealt without replacement, we will use combinations. Since we want to get three of the four 8's in the deck, the number of ways to get three 8's is given by

$$C_{4,3} = \frac{4!}{3!(4-3)!} = 4.$$

Similarly, the number of ways to get two of the four 7's in the deck is given by

$$C_{4,2} = \frac{4!}{2!(4-2)!} = 6.$$

Now, by the Basic Counting Law, the number of ways to get three 8's and two 7's is given by

$$C_{4,3} \times C_{4,2} = 4 \times 6 = 24.$$

An actual listing of the 24 possible ways to get a full house of three 8's and two 7's is shown below.

8♥ 8♦ 8♣ 7♥ 7♦
8♥ 8♦ 8♣ 7♥ 7♣
8♥ 8♦ 8♣ 7♥ 7♠
8♥ 8♦ 8♣ 7♦ 7♣
8♥ 8♦ 8♣ 7♦ 7♠
8♥ 8♦ 8♣ 7♣ 7♠
8♥ 8♦ 8♠ 7♥ 7♦
8♥ 8♦ 8♠ 7♥ 7♣
8♥ 8♦ 8♠ 7♥ 7♠
8♥ 8♦ 8♠ 7♦ 7♣
8♥ 8♦ 8♠ 7♦ 7♠
8♥ 8♦ 8♠ 7♣ 7♠

8♥ 8♣ 8♠ 7♥ 7♦
8♥ 8♣ 8♠ 7♥ 7♣
8♥ 8♣ 8♠ 7♥ 7♠
8♥ 8♣ 8♠ 7♦ 7♣
8♥ 8♣ 8♠ 7♦ 7♠
8♥ 8♣ 8♠ 7♣ 7♠
8♦ 8♣ 8♠ 7♥ 7♦
8♦ 8♣ 8♠ 7♥ 7♣
8♦ 8♣ 8♠ 7♥ 7♠
8♦ 8♣ 8♠ 7♦ 7♣
8♦ 8♣ 8♠ 7♦ 7♠
8♦ 8♣ 8♠ 7♣ 7♠

EXAMPLE 10

How many ways are there to deal a full house of any type?

Solution: This problem is similar to Example 9 except that there are fewer conditions. Instead of having three 8's, we now want three cards of any rank. Since there are 13 different ranks, there are 13 ways to choose the set of three of a kind. Similarly, there will be only 12 ranks from which to get the pair. Therefore, the number of ways to deal a full house is

$$13 \times C_{4,3} \times 12 \times C_{4,2} = 13 \times 4 \times 12 \times 6 = 3744.$$

EXAMPLE 11

a) How many different five-card hands contain four aces?
b) How many different five-card hands contain four cards of the same rank?

Solution:

a) To get four aces, it seems that it is necessary only to compute $C_{4,4}$, but it is not that simple. Because we have a five-card hand, we must also include the number of ways to select the fifth card. Since there are 48 cards in the deck that are not aces, the number of ways to select the fifth card is given by $C_{48,1}$. Thus, the Basic Counting Law gives

$$C_{4,4} \times C_{48,1} = 1 \times 48 = 48.$$

b) This is the same problem as part (a) except that we now have 13 choices for the rank (2's through aces). Therefore, the number of possible hands with four cards of the same rank is

$$13 \times C_{4,4} \times C_{48,1} = 13 \times 1 \times 48 = 624.$$

EXAMPLE 12

In the game blackjack (also called twenty-one or vingt-et-un), a player can win the hand by receiving a blackjack: an ace and a ten-point card in the first two cards. A ten-point card is a king, queen, jack, or 10. Find the number of ways in which a player can get a blackjack when two cards are dealt from a standard deck of 52 cards.

Solution: Since we want to get one of the four aces and one of the 16 ten-point cards, the number of ways to get a blackjack is

$$C_{4,1} \times C_{16,1} = 4 \times 16 = 64.$$

EXAMPLE 13

In casinos, blackjack is played with several decks of cards. Find the number of ways in which a player can draw a blackjack from three standard decks of 52 cards.

Solution: Since there are three decks of cards, each containing four aces, there is a total of 12 aces. This means that the number of ways to get one ace out of the 12 possible aces can be determined by $C_{12,1}$. Similarly, since there is a total of 48 ten-point cards in the three decks, the number of ways to get a ten-point card is given by $C_{48,1}$. Thus, the number of ways to get a blackjack is

$$C_{12,1} \times C_{48,1} = 12 \times 48 = 576.$$

EXAMPLE 14

KENO

1	2	3	4	5	6	7	8	9	10
11	12	13	14	15	16	17	18	19	20
21	22	23	24	25	26	27	28	29	30
31	32	33	34	35	36	37	38	39	40
41	42	43	44	45	46	47	48	49	50
51	52	53	54	55	56	57	58	59	60
61	62	63	64	65	66	67	68	69	70
71	72	73	74	75	76	77	78	79	80

In the game of keno, 80 numbers are displayed on a board. Twenty of these numbers are randomly drawn and are considered the winning numbers. Suppose a person has selected nine numbers on a playing card. In how many ways can the person get exactly six winning numbers?

Solution: This problem is similar to the card problems. We are playing a game with 80 possibilities, 20 of which are considered the winning values. Since the order that the numbers are selected does not matter and we want to select six winning numbers from these 20, there are $C_{20,6}$ ways to pick the winning numbers. Since we are picking a total of nine numbers, we must also account for the three losing numbers. Since there are $80 - 20 = 60$ losing values, the three losing numbers can be chosen in $C_{60,3}$ ways.

Therefore, when playing a keno card with nine numbers, the number of ways to pick exactly six of the 20 winning numbers is

$$C_{20,6} \times C_{60,3} = 38{,}760 \times 34{,}220 = 1{,}326{,}367{,}200.$$

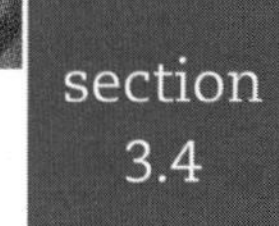

section 3.4

PROBLEMS ○ Explain ○ Apply ○ Explore

○ Explain

1. What does counting by listing involve?
2. Why would you use other methods besides the listing method in determining the number of ways an event can occur?
3. What is meant by counting with replacement?
4. What is meant by counting without replacement?
5. What does the Basic Counting Law tell us about counting?
6. Under what conditions would combinations be used to count the number of ways an event can happen?
7. Under what conditions would permutations be used to count the number of ways an event can happen?
8. Under what conditions would the listing method be used to count the number of ways an event can happen?

○ Apply

In Problems 9–21, determine whether the number of ways for the event can be found by using the Basic Counting Law, permutations, or combinations. Explain the reason for your choice. (There may be more than one correct answer.)

9. The number of ways in which eight runners in a 100-meter dash can finish first, second, and third.
10. The number of ways in which three cards can be dealt from a standard deck of cards.
11. The number of ways to answer six true/false questions on a test.
12. The number of ways to pick five baseballs from a box that contains 50 baseballs.
13. The number of ways in which a group of five people can be selected from a class of 30 people.
14. The number of serial numbers that consist of a letter of the alphabet followed by five digits.
15. The number of ways in which you can dress if you have five pairs of shorts, six T-shirts, and two pairs of shoes to choose from.
16. The number of ways in which you can press three different keys on a basic calculator that has keys for the 10 digits and the $+$, $-$, $\times$, $\div$, $=$, and . keys.

17. The number of different four-letter code words that can be made from consonants if no letter can be used more than once.
18. The number of different code words that can be made from consonants if a letter can be used more than once.
19. The number of ways to plug in a black cord and a gray cord into an electrical strip that has six outlets in a row.
20. The number of ways to answer eight multiple-choice questions on a test if each question has five choices.
21. The number of ways in which you can choose 4 of your 15 shirts to take on vacation.

Explore

22. A slot machine consists of three wheels with 12 different objects on a wheel. How many different outcomes are possible?
23. A computer chip consists of four different switches. Each switch can be in either the off or the on position. What is the total number of arrangements of the chip's switches? Make a list of all the possible outcomes.
24. A VISA® credit card has a first digit of 4 followed by 15 other digits. How many different VISA accounts does this allow?
25. Social Security numbers have nine digits. How many different people can have distinct social security numbers?
26. How many seven-digit phone numbers are possible if the phone number does not begin with a 1 or 0?
27. A young couple has decided to have three children. If all three births are single births, in how many different ways could the couple have exactly two boys? Make a list of all cases.
28. An eight-cylinder car has eight wires running from the spark plugs to the distributor cap. A prankster has removed all the wires from the distributor. In how many possible ways can the wires be reattached?
29. Seven candidates are running for three positions on the local board of supervisors. The candidate with the most votes will be the board president while the second and third place finishers will have correspondingly lesser positions. How many different outcomes can the election have?
30. A pitcher knows how to throw five different pitches. He threw four pitches before the last batter was called out.
    a) Assuming that he threw each pitch at most once, determine the number of possible arrangements of his pitches.
    b) Assuming that he can use any pitch an unlimited number of times, determine the number of possible arrangements of his pitches.
31. Determine the number of ways in which a ten-question true/false test can be answered.
32. Determine the number of ways in which a ten-question multiple-choice test can be answered if there are five possible answers to each question.
33. Determine the total number of five-card hands that can be dealt from a deck of 52 cards.
34. Determine the total number of 13-card bridge hands that can be dealt from a deck of 52 cards.

35. In poker, a straight is five cards in consecutive numerical order. The suit of the card does not matter.

   a) Find the number of ways to get a straight beginning with a 5 and ending with a 9.

   b) Find the number of ways to get any straight. (*Note:* An ace can be part of either A, 2, 3, 4, 5 or part of 10, J, Q, K, A but cannot be used in K, A, 2, 3, 4.)

36. In poker, a straight flush is five cards of the same suit in consecutive numerical order. If five cards are dealt,

   a) Find the number of ways to get a straight flush beginning with the 5 of hearts and ending with the 9 of hearts.

   b) Find the number of ways to get a straight flush beginning with a 5 and ending with a 9.

   c) Find the number of ways to get any straight flush. (*Note:* An ace can be either part of A, 2, 3, 4, 5 or part of 10, J, Q, K, A but cannot be used in K, A, 2, 3, 4.)

37. In bridge, a yarborough is a 13-card hand containing only cards numbered 2 through 9. Find the number of ways in which to get a yarborough.

38. A basketball league intends to add two more teams to the league. If 18 cities have applied for franchises, in how many ways can the league add two more cities?

39. In a dog show, a German Shepherd is supposed to pick the correct two objects from a set of 20 objects. In how many ways can the dog pick any two objects?

40. There are eight people on a committee. How many different subcommittees of two people can be formed?

41. An auditorium has scheduled three basketball games, two concerts, and four poetry readings. You have a ticket allowing you to attend any three of the events. In how many ways can you go to two of the poetry readings and one of the other events.

42. In a seven-card poker hand, find the number of hands containing four aces.

43. Find the total number of license plates that can be printed by the state of California using the format of a digit, followed by three letters, followed by three digits. Assume that all arrangements of letters can be used.

44. Twenty people are at a party. If everyone at the party shakes the hand of everyone else at the party, determine the total number of handshakes.

45. Ten married couples are at a party. If each person at the party shakes the hand of everyone else except his or her spouse, determine the total number of handshakes at the party.

46. In the game of keno, 80 numbers are displayed on a board. Twenty of these numbers are chosen to be the winning numbers. Suppose a person has selected seven numbers on a playing card. In how many ways can the person select exactly five of the winning numbers?

47. In the game of keno, 80 numbers are displayed on a board. Twenty of these numbers are chosen to be the winning numbers. Suppose a person has selected six numbers on a playing card. In how many ways can the person select six of the winning numbers?

CHAPTER 3 SUMMARY

Key Terms, Concepts, and Formulas

The important terms in this chapter are:	Section
Basic Counting Law: States that if there are n choices for one item and m choices for a second item, there are $n \times m$ ways to pick both items.	3.4
Cardinal number: The number of elements in a set.	3.1
Combination: A group of objects chosen from a set in which the order of the objects does not matter.	3.4
Complement ($\overline{A}$): The set of items in a universal set that are not contained in the set A.	3.2
Countably infinite: A set A is countably infinite if there is a one-to-one relationship between A and the set of natural numbers.	3.1
Descriptive method: The technique of representing a set by giving a rule for its elements.	3.1
Disjoint sets: Sets that have no intersection.	3.2
Element ($\in$): An item in a set.	3.1
Empty set ($\emptyset$ or $\{\ \}$): A set that does not contain any elements.	3.1
Equal sets: Two sets that have exactly the same elements.	3.1
Equivalent sets: Two sets that have the same number of elements.	3.1
Factorials: The mathematical expression given by $n! = n \times (n-1) \times (n-2) \times \cdots \times 2 \times 1.$	3.4
Finite Set: A set that has a whole number for its cardinal number.	3.1
Infinite set: A set with an unlimited number of elements.	3.1
Integers: The set of numbers $\{\ldots, -3, -2, -1, 0, 1, 2, 3, \ldots\}$.	3.1
Intersection ($\cap$): An operation on two sets that creates a set containing elements common to both sets.	3.2
Listing method: The technique of representing a set by writing all its elements in braces.	3.1
Natural numbers: The set of numbers $\{1, 2, 3, \ldots\}$.	3.1
One-to-one: The relationship between two sets A and B if for every element of A, there is exactly one element in B and for each element in B, there is exactly one element in A.	3.1
Permutation: A group of objects chosen from a set in which the order of the objects matters.	3.4
Proper subset ($\subset$): A set A whose elements are contained in another set B and $n(A) < n(B)$.	3.1
Rational numbers: The set of numbers that can be written as fractions.	3.1
Set: A collection of items.	3.1

Set builder notation: The technique of representing a set by giving a rule for its elements. 3.1

Subset (⊆): A set A whose elements are contained in another set B. 3.1

Union (∪): An operation on sets that creates a set containing all the elements of both sets. 3.2

Universal set: A set that contains all the elements of a certain category. 3.1

Venn diagrams: A picture used to depict sets. 3.2

Whole numbers: The set of numbers {0, 1, 2, 3, . . .}. 3.1

With replacement: When an object can be used more than once. 3.4

Without replacement: When an object cannot be used more than once. 3.4

After completing this chapter, you should be able to:	Section
1. Use sets and Venn diagrams to analyze a situation.	3.1
2. Use the notation of sets ($\cup, \cap, \subset, \subseteq, \not\subset, \in, \notin, \varnothing$) to convert worded statements into set notation.	3.1
3. Explain the difference between combinations and permutations.	3.4
4. Explain the difference between using "with replacement" and "without replacement."	3.4
5. Determine the number of ways an event can occur by applying the Basic Counting Law, listing possibilities, and using formulas for combinations, $C_{n,r} = \frac{n!}{r!(n-r)!}$, and permutations, $P_{n,r} = \frac{n!}{(n-r)!}$.	3.4

CHAPTER 3 REVIEW

Section 3.1

1. Write the set of whole numbers less than or equal to 20 that are divisible by 3, using the
   a) Descriptive method.
   b) Listing method.
2. Explain the difference between equivalent sets and equal sets.
3. What is meant by the cardinality of a set?
4. Show that the set {12, 23, 34, 45, 56, . . .} is countably infinite. What is the mathematical relationship that shows the one-to-one correspondence?

Section 3.2

5. Let U be the set of all dogs, S be the set of all dogs that like to swim, and H be the set of all dogs with black hair. Draw a Venn diagram

for these sets and describe the type of dog contained in each region of the diagram.

6. Let $T = \{\text{all integers divisible by } 3\}$ and let $S = \{4, 8, 12, 16, 20, \ldots\}$. Write a set listing the first four numbers in the set $T \cap S$.

In Problems 7 and 8, use the Venn diagram to find the indicated values.

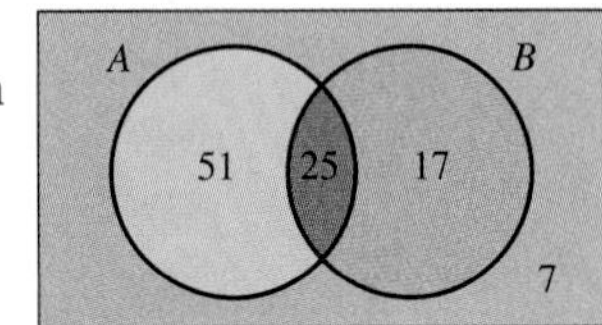

7. $n(A \cup B)$
8. $n(A \cap \overline{B})$

Section 3.3

Use the following information to answer the questions in Problems 9 and 10. A teacher took a survey of the 24 students in her eighth grade class. The students were asked whether they preferred to play video games or watch a movie. The results are shown in the table. Let G = the set of girls, B = the set of boys, M = the set of students who prefer to watch movies, and V = the set of students who prefer to play video games.

	Movies	Games
Girls	6	5
Boys	3	10

9. $n(G \cap \overline{M})$
10. $n(B \cup \overline{V})$

Use the information in the Venn diagram to answer the questions in Problems 11 and 12.

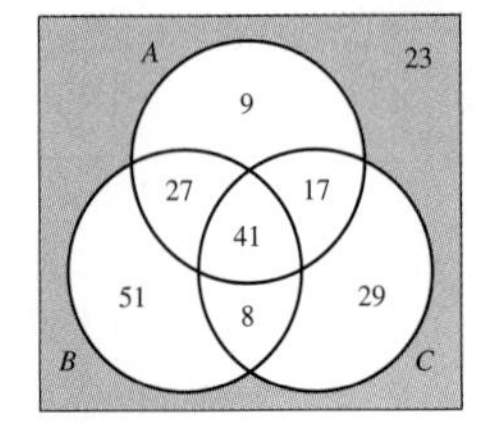

11. $n(A \cap B \cap \overline{C})$
12. $n(\overline{A} \cap \overline{B})$

Section 3.4

13. In what situations should you use
    a) the Basic Counting Law?
    b) permutations?
    c) combinations?
14. A delicatessen offers a lunch combination consisting of one salad, one sandwich, and one drink. There are three different types of salads, six types of sandwiches, and four different drinks. How many different lunch combinations are there?
15. A lottery has a total of 20 different numbers, and two will be chosen as the winning numbers. How many different pairs of winning numbers are possible?
16. A set of three standard dice is rolled. How many ways are there to have the sum of the three dice to equal 5?
17. A combination lock has three wheels, each with 16 numbers. How many different lock combinations are possible if no number is used more than once?

CHAPTER 3 TEST

1. Give an example of two events whose intersection is the empty set.
2. Give an example of two sets A and B whose intersection is the same as set A.
3. Find the value of the following.

   a) $C_{17,4}$ b) $P_{17,4}$

4. Show that the set $F = \{7, 11, 15, 19, 23, \ldots\}$ is countably infinite. What is the mathematical relationship that shows the one-to-one correspondence?
5. Of 400 students, 360 students are taking a math course or an English course. Eighty-five of the students in math classes are also taking an English class. If a total of 190 students are taking a math course, how many students are taking an English course but not a math course?
6. Let $F = \{1, 3, 5, 7, 9, 11, 13, 15\}$ and let I be the set of all integers. For each of the following statements, if the statement uses correct notation and is true, write "True": if not, write a true statement using correct notation.

   a) $5 \subset I$

   b) $5 \in F$

   c) $F \in I$

   d) $4 \notin F$

7. If there are six candidates for the president of an organization, three candidates for secretary, and four candidates for treasurer, how many different executive teams of a president, secretary, and treasurer are possible?
8. a) Find the number of ways in which three aces and two kings can be drawn in a six-card hand from a standard 52-card deck.

   b) Find the number of ways in which three aces and a pair of any rank can be drawn in a six-card hand from a standard 52-card deck.

   c) Find the number of ways in which three cards of one rank and two cards of another rank can be drawn in a six-card hand from a standard 52-card deck.

The table below shows the number of students with a GPA greater than 3.2 depending on the number of hours they work per week and the average number of units they take per semester. Use this information to answer Problems 9–12.

	Working ≤ 10 hours	Working > 10 hours	
Enrolled in ≤ 6 units	4	7	11
Enrolled in > 6 units	41	29	70
	45	36	81

9. Take the information from the table and write it in a Venn diagram with two circles. Let W be the set of students who are working more than 10 hours and E be the students enrolled in more than 6 units.
10. Of these 81 students, how many are working more than 10 hours per week or are enrolled in more than 6 units?
11. Of these 81 students, how many are working 10 or fewer hours per week or are enrolled in 6 or fewer units?
12. Of these 81 students, how many are working 10 or fewer hours per week and are enrolled in 6 or fewer units?

CHAPTER FOUR

Probability

At the races by Edouard Manet. Horse racing is just one of the many sporting events in which the study of probability and odds plays an important role. (National Gallery of Art)

OVERVIEW

In this chapter, you will study probability from the point of view of its intuitive concepts. You will learn how these intuitive ideas can be connected to mathematical rules through the use of both sets and common sense. Once these rules have been established, we can use them to answer probability questions that are beyond the solutions provided by your intuition. These questions are drawn from a wide variety of areas, including games of chance, employment statistics, consumer buying habits, and medicine.

4 A SHORT HISTORY OF PROBABILITY

Detail from *The Card Players* by Lucas van Leyden shows a game in which probability theory has relevant application. (National Gallery of Art)

Probability is the science of determining the likelihood, or chance, that an event will occur. Combinatorics is the mathematical tool that is used to find the number of ways in which an event can occur. The histories of these two topics are interwoven, developing with people's interest in games.

Writers from several civilizations contributed early work on combinatorics. In China in 1100 B.C., permutations were mentioned in *I-Ching* [Book of Changes] concerning the possible number of trigrams. The Latin writer Boethius (c. A.D. 510) gave a rule for selecting items two at a time from a large set. The Hindu mathematician Bhāskara gave rules for calculating permutations and combinations and discussed them as they related to such varied topics as medicine, music, and architecture. Although the Hebrew writer Rabbi ben Ezra did not provide a formula, he used combinations to discuss possible arrangements of Saturn and the other planets.

Although probability was mentioned as early as 1477 in a commentary on Dante's *Divine Comedy,* it is said to have had its origins in an unfinished dice game. Two gamblers were unable to complete a game of chance. They agreed to divide the stakes according to their respective chances of winning the game but could not decide what those chances were. The mathematician Blaise Pascal received a letter from his friend Chevalier de Méré requesting a solution to the problem (c. 1654). Pascal sent the problem to another French mathematician, Pierre de Fermat. Working together, they are credited with developing modern probability theory.

As is always true in science, probability did not spring forth as a completely developed theory. Of the many who have made substantial contributions to probability, the Spanish alchemist Raymond Lulle (1234–1315) is credited as being the father of combinatorics. He wanted to find the symbols for all the chemical elements and then to write down all possible arrangements of these symbols. By doing so, he believed, he would be able to construct every possible thing.

It was during the 16th and 17th centuries that European mathematicians developed combination theory and applied it to games of chance. In 1663, *Liber de ludo aleane* [The book on games of chance] was published. Written by the Italian Girolamo Cardano (1501–1576) nearly a century earlier, it was published after the works of Pascal and Fermat.

The Dutch astronomer, Christiaan Huygens (1629–1695), wrote an introduction to dice games in *De ratiociniis in ludo aleane* [On reasoning in games of dice] in 1657. This treatise included the concept of mathematical expectation (see Section 4.6). Huygens was also one of the first to study probability from what is now considered the classical viewpoint. Whereas the investigations of Fermat and Pascal started with games, Huygens considered probability as the ratio of the number of successful outcomes of an event to the total number of possible outcomes. The classical use of probability theory and combinatorics in the study of games of chance continued to develop throughout the 18th and 19th centuries.

Probability is applied to many different fields. One of these fields is in the area of games and gambling. The study of probability and combinatorics is critical to state lottery commissions and gambling casinos in determining the payoffs for their games. These organizations are in the business of acquiring your money. By tempting you with the prospect of a lucky win, they draw your attention.

CHAPTER 4 PROJECTS

Research Projects

1. What does a standard deck of playing cards consist of? What are the origins of playing cards? Give the origins and history of at least three card games, such as poker, blackjack, whist, bridge, pinochle, and euchre.
2. Who was Hoyle and what is *Book of Hoyle?* Give the origins and history of at least three games contained in *Hoyle.* Be sure to include card games and noncard games.
3. Research the profitability of gambling casinos and the effects that such casinos have on the economy of their local communities.

Math Projects

1. Examine one of the state-run lotteries. How is the lottery played? What are the odds or probabilities of winning in the lottery? What is the expected value for this lottery?
2. The games of chance in gambling casinos make use of probability. Write a short paper on one of the games listed below. Include a history of the game, a description of how the game is played, and an analysis of the probabilities, odds and expected values involved in the game.

   a) Keno **b)** Roulette
   c) Craps **d)** Slot machines

3. The calculations for the payoffs in horse and dog races involve the use of probability. Write a short paper on horse or dog racing. Include a history of the event, a description of how the gambling supports the horse owners and the tracks, and an analysis of the probabilities or odds involved in the payoffs. Also include a discussion of the pari-mutuel betting and trifectas.
4. Find the expected value of an actual three-reel slot machine. The website www.simslots.com/slot12.htm simulates a slot machine and displays its pay table.

   a) Determine what is on each reel.

   b) Determine the probability of each winning combination.

   c) Determine the probability of losing.

   d) Determine the expected value of the machine.

   e) Write a paragraph explaining the reasons for such an expected value.

section 4.1

Intuitive Concepts of Probability

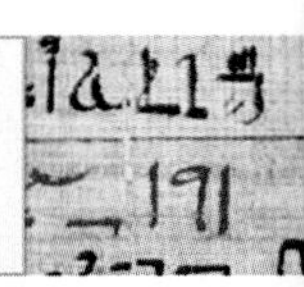

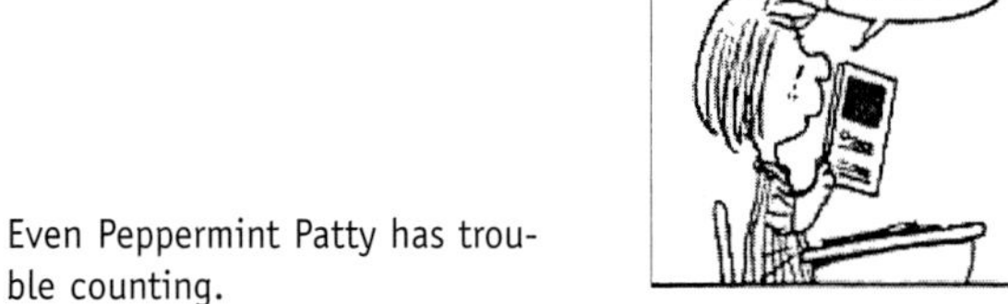

Even Peppermint Patty has trouble counting.

In everyday conversations, we frequently ask questions such as "What's my chance of getting an *A* in this class?" "What's the probability of my winning the drawing for the trip to Hawaii?" "What is the chance that it will rain today?" What we are looking for is some kind of a measure of the chance that the event will occur. I might say that there is a 90% chance that I will get an *A* in the class, or the probability of my winning the drawing for the Hawaiian vacation is 1 in 20,000, or there is a 50% chance of rain today. The numbers measure the likelihood that the event will occur. For example, the probability of flipping a coin and having it land with the head side up is 1/2. Since the coin could land on either heads or tails, the head outcome is one of the two possible outcomes. In general, the **probability** of an event is found by dividing the number of ways that an event can occur (number of *desired* outcomes) by the *total* number of possible outcomes. If we will let $P(E)$ represent the probability of an event E, the basic formula for determining the probability is

$$P(E) = \frac{\text{number of ways an event can occur } (\textit{desired})}{\text{total number of possible outcomes } (\textit{total})}.$$

EXAMPLE 1

Find the probability of drawing an ace from a standard deck of 52 cards.

Solution: In this case, the event is drawing an ace. Because there are four aces in the standard deck, there are four ways to draw an ace out of a total of 52 possibilities.

$$P(\text{ace}) = \frac{4}{52} = \frac{1}{13}$$

The probability of an event can be represented as a fraction, decimal, or percent. Each form gives a numerical way to analyze the chance that the event occurs. Thus, the probability of drawing an ace in Example 1 can be written

$$P(\text{ace}) = \frac{1}{13} \approx 0.077 \approx 7.7\%.$$

EXAMPLE 2

A question on a multiple-choice test has five answers. What is the probability that you guess the correct answer to the question?

Solution: There is only one correct (*desired*) answer out of the five possible (*total*) answers. Therefore,

$$P(\text{correct answer}) = \frac{1}{5} = 0.2 = 20\%.$$

EXAMPLE 3

Suppose you draw an ace from a standard deck of cards and do not return the ace to the deck before you draw from the deck a second time. What is the probability that you draw an ace on this second draw from the deck?

Solution: Since one ace was removed, there are now three aces in the deck. Similarly, since one card was removed, there are a total of 51 cards in the deck. Thus,

$$P(\text{ace on 2nd card}) = \frac{3}{51} = \frac{1}{17}.$$

TWO RESTRICTIONS ON PROBABILITY

There are two restrictions on the value of a probability. First, in the formula for probability,

$$P(E) = \frac{\text{number of ways an event can occur } (\textit{desired})}{\text{total number of possible outcomes } (\textit{total})},$$

the number on the bottom of the fraction must be greater than or equal to the number on the top of the fraction. Therefore, $P(E)$ must be less than or equal to 1.

The second restriction is that since the number of ways in which an event occurs cannot be negative, $P(E)$ must be greater than or equal to zero. Thus, we have

$$0 \leq P(E) \leq 1.$$

The closer the probability is to zero, the less of a chance there is that the event will occur. The closer the probability is to 1, the more of a chance there is that the event will occur. If it is *impossible* for an event to occur, the probability of the event is 0. If an event is *certain* to happen, the probability of the event is 1. The next example demonstrates these concepts.

EXAMPLE 4

Two standard dice are tossed. (a) What is the probability that the sum of the two dice is 14? (b) What is the probability that the sum of the two dice is less than 13?

Solution:

a) Since the maximum number of dots on a die is six, it is impossible to get a sum of 14 with two standard dice. Therefore, $P(\text{sum of } 14) = 0$.

b) Since the sum of two standard dice is always less than 13, the event is a certain event. The probability of an event that is certain is 1. Therefore, $P(\text{sum less than } 13) = 1$.

WHAT DOES PROBABILITY REALLY TELL US?

In the previous examples, we showed how to determine the probability of an event, but what does a given probability actually tell us? When the word *probability* is used, it can have either of two meanings: experimental probability or theoretical probability.

If we flip ten coins, we might find that we get seven heads. To say that the probability of getting a head is $7/10$ is an example of **experimental probability.** This means that we arrived at the value through an experiment rather than through mathematical calculations. Experimental probability can be used to analyze events even if there is no mathematical way to determine the probability. For example, there is no mathematical way to determine the probability that the next person who walks into a restaurant will be wearing a red dress. However, if for an entire month we counted the number of people who entered the restaurant and the number who wore red dresses, we could give an approximation to the probability of that event.

Theoretical probability is what is predicted by mathematics. For example, we say that the probability of a coin landing heads is 1/2, since the coin has two sides and only one of the sides is a head. When an event has a theoretical probability, let's say 1/2, it means that in a large number of trials, 1/2 of the trials will result in the given event. The larger the number of trials, the closer the ratio of desired to total outcomes will be to 1/2. For example, if 1000 coins were tossed with 523 landing on heads, the experimental probability would be $523/1000 = 0.523$. As we increase the number of coin tosses, the experimental probability will become close to the theoretical probability. The fact that the experimental probability is close to the theoretical probability when an experiment includes a large number of trials is known as the **Law of Large Numbers.** Probabilities cannot be used to predict individual events. Probabilities can only determine long-range outcomes or trends.

DETERMINING PROBABILITIES OF SINGLE EVENTS

Theoretical Method

Suppose we want to determine the possible outcomes for the sum of two dice and determine the number of ways in which the sum could be seven. To use theoretical probability, we list all possible outcomes of two dice as shown below.

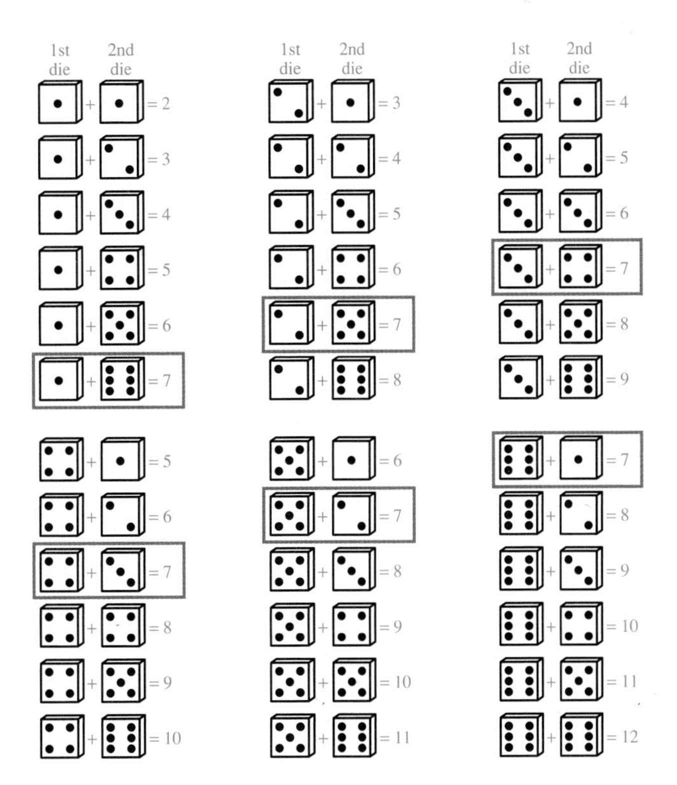

There are 36 different ways to roll the dice, and six of these ways have a sum of 7. Therefore,

$$P(\text{sum of } 7) = \frac{6}{36} = \frac{1}{6}.$$

This means that if we were to roll the dice many times, about 1/6 of them would have a sum of 7. It does not mean that one of every six rolls will give a sum of 7.

Experimental Method

If we wanted to test the effectiveness of a new drug in relieving migraine pain, we could administer the drug to 100 volunteers with migraines and observe the results. If 91 of the volunteers receiving the drug get some pain relief, we could say that the probability that a person will get some pain relief is

$$P(\text{pain relief}) = \frac{91}{100}.$$

In our advertising for the medication, we could state that 91% of those tested received relief for migraine pain.

Experimental probability is very important in any type of quality control process for a new product, whether it be a new painkiller or a new type of computer chip. However, experimental probability requires looking at the results of hundreds of experiments. Therefore, we will focus most of our attention on determining the theoretical probability of various events.

COMPLEMENTARY EVENTS

The **complement** of an event E, denoted by the symbol $\overline{E}$, is the nonoccurrence of E. For example, if E is the event that it is raining, $\overline{E}$ is the event that it does not rain. The probability of $\overline{E}$ is always given by the formula

$$P(\overline{E}) = 1 - P(E).$$

EXAMPLE 5

A German shepherd is being tested for its ability to choose a specific toy from a group of 20 toys.

a) Find the probability that the dog picks the correct toy at random.

b) Find the probability that the dog does not pick the correct toy.

Solution:

a) Let T be the event that the dog picks the correct toy at random. Since there is only one correct toy and there is a total of 20 toys, $P(T) = \frac{1}{20}$.

b) Let $\overline{T}$ be the event that the dog does not pick the correct toy. Therefore,

$$P(\overline{T}) = 1 - P(T) = 1 - \frac{1}{20} = \frac{19}{20}.$$

EXAMPLE 6

An American roulette wheel has 38 compartments around its circumference. Thirty-six compartments are numbered from 1 to 36, half of them being colored

red and half colored black. The remaining two compartments, numbered 0 and 00, are colored green. A ball is spun and lands in one of the compartments.

a) What is the probability that the ball lands on the number 27?

b) What is the probability that the ball does not land on the number 27?

c) What is the probability that on the tenth spin, the ball lands on the number 27?

d) What is the probability that the ball lands on an odd number?

e) What is the probability that the ball lands on a green-colored compartment?

Solution: In this problem, we can determine the number of desired outcomes and the number of total outcomes. Therefore, we will not need to use an experimental approach.

a) There is one desired outcome out of the 38 total outcomes,

$$P = \frac{1}{38}.$$

b) Since this is the complementary event of part (a),

$$P = 1 - \frac{1}{38} = \frac{37}{38}.$$

c) The probability for the ball landing on the number 27 for the tenth or any other spin is the same,

$$P = \frac{1}{38}.$$

d) There are 18 desired outcomes (odd-numbered compartments) out of the 38 total outcomes,

$$P = \frac{18}{38} = \frac{9}{19}.$$

e) There are two desired outcomes (green-colored compartments) out of the 38 total outcomes,

$$P = \frac{2}{38} = \frac{1}{19}.$$

EXAMPLE 7

In the National Football League's Super Bowl XXXIV, Kurt Warner of the St. Louis Rams had a completion rate of about 53.3%. If Warner threw 45 passes during the game, how many of them were completed?

Solution: In sporting events, the percents given are arrived at by actually tallying the results. They are examples of experimental probability. Warner's 53.3% completion rate means that the probability that any given pass will be completed is

$$\frac{53.3}{100} = 0.533.$$

Since the probability that a pass is completed is 0.533 and 45 passes were thrown, the number of completed passes is $0.533 \times 45 = 23.985 \approx 24$.

PROBLEMS ● Explain ● Apply ● Explore

Explain

1. What is probability?
2. Explain what $P(E) = 0$ means and describe an event that has a probability of zero.
3. Explain what $P(E) = 1$ means and describe an event that has a probability of 1.
4. Explain the difference between theoretical and experimental probability.
5. Would you find the probability of the following events using experimental or theoretical methods? Explain why you think the method is appropriate.
   - **a)** The probability that a given serve by a tennis player will hit the net
   - **b)** The probability that three dice are thrown and the sum is 18
   - **c)** The probability that a senior citizen will be riding a certain city bus on Friday
   - **d)** The probability that the six numbers you picked for a lottery are all winning numbers
6. Explain how you could determine the probability of randomly selecting a word from Shakespeare's play *Romeo and Juliet* and having it be the word "the."
7. Explain how you could determine the probability that a given household in your community is watching the 10:00 P.M. TV news on a Wednesday night?
8. What does the Law of Large Numbers tell us when $P(E) = 1/20$?
9. What are complementary events and how is the probability of complementary events expressed? Give an example of complementary events and their probabilities.
10. What is the meaning of a weather forecast that says there is a 40% chance of rain?
11. What is the meaning of a political poll that says there is a 63% chance that the candidate will win reelection?

Apply

12. A card is drawn from a standard deck of 52 cards.
    - **a)** What is the probability that the seven of spades is drawn?
    - **b)** What is the probability that a seven is drawn?
    - **c)** What is the probability that a face card (king, queen, or jack) is drawn?
    - **d)** What is the probability that a heart is drawn?
    - **e)** What is the probability that a red card is drawn?
    - **f)** What is the probability that either an ace or an eight is drawn?
13. A card is drawn from a standard deck of 52 cards.
    - **a)** What is the probability that the ten of diamonds is drawn?
    - **b)** What is the probability that a ten is drawn?
    - **c)** What is the probability that a nonface card (ace, 2, 3, 4, . . . , 10) is drawn?

d) What is the probability that a spade is drawn?

e) What is the probability that a black card is drawn?

f) What is the probability that either a queen or a king is drawn?

14. Two six-sided dice, with sides numbered 1 through 6, are rolled.

a) What is the probability that the sum of the two dice is 8?

b) What is the probability that the sum of the two dice is 1?

c) What is the probability that exactly one of the two dice shows a 3?

d) What is the probability that the sum of the two dice is 13?

e) What is the probability that the sum of the two dice is less than 13?

f) If the dice are rolled 9000 times, about how many times would you expect the dice to have a sum of 8?

15. In baseball, if a player has a batting average of 0.279, it means that for every 1,000 official at bats, the player gets a hit 279 times. Ty Cobb of the Detroit Tigers had a lifetime batting average of 0.367. How many hits would Ty expect in a season if he had 530 official at bats?

16. In the course of a professional baseball season, the probability that a good leadoff hitter will get on base is about 0.435. If a player comes to bat 620 times during the season, how many times would you expect him to reach base?

17. If the probability of your making a three-point shot in basketball is 0.15, what is the probability that you will not make a three-point shot?

18. If there is a 75% chance that you will get up when the alarm rings, what is the probability that you will not get up when the alarm rings?

19. Martina believes that the probability of her passing her next test is 60%. What is the probability that Martina does not pass?

20. If the probability of a gene having a defect that causes a certain disease is 1/25, what is the probability of the gene not having that defect?

Explore

21. Suppose five standard decks of cards are combined.

a) What is the probability that the seven of spades is the first card drawn?

b) What is the probability that a seven is the first card drawn?

c) What is the probability that a face card (king, queen, or jack) is the first card drawn?

d) What is the probability that a heart is the first card drawn?

e) What is the probability that either an ace or an eight is the first card drawn?

f) Suppose that the first card drawn is an ace. What is the probability that the second card drawn is an ace?

g) Suppose the first and second cards drawn are aces. What is the probability that the third card drawn is an ace?

22. The game Dungeons and Dragons uses two regular dodecahedral dice. Such dice are solids with 12 congruent hexagon shaped faces. A pair of these dodecahedral dice with the numbers 1 to 12 on the faces are tossed. A possible result when such dice are tossed is shown.

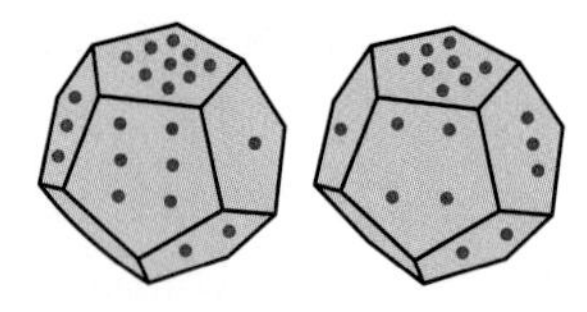

a) List all the possible outcomes for tossing the two dice.

b) What is the probability that the sum of the two dice is 24?

c) What is the probability that the sum of the two dice is greater than 24?

d) What is the probability that the sum of the two dice is 17?

e) What is the probability that the sum of the two dice is greater than 17?

f) What is the probability that the sum of the two dice is greater than 1?

23. The probability of gene mutation under certain conditions is 0.00006. What is the probability of a gene under these conditions not mutating?

24. A pinochle (*pē-nuk′-l*) deck of cards consists of 48 cards: eight aces, eight kings, eight queens, eight jacks, eight tens, and eight nines. Suppose you are cutting cards from a pinochle deck.

a) What is the probability that you draw a king on the first card?

b) Suppose you draw a king on the first card and do not return the king to the deck before you draw a second card. What is the probability of drawing a king?

c) Suppose you draw another king on the second card and again you do not return the king to the deck before you draw a third time. What is the probability that you now draw a king?

25. A BINGO card consists of a square containing 25 smaller squares, as shown in the figure. Five numbers from 1 to 15 are placed under the letter B, five numbers from 16 to 30 are placed under the letter I, four numbers from 31 to 45 along with the "free space" are placed under the letter N, five numbers from 46 to 60 are placed under the letter G, and five numbers from 61 to 75 are placed under the letter O.

B	*I*	*N*	*G*	*O*
3	17	40	52	73
8	21	42	56	63
1	20	FREE	47	69
13	19	45	59	71
10	26	34	46	68

When a regular Bingo game is played, the numbers from 1 to 75 are randomly called until someone's card has five numbers in a horizontal, vertical, or diagonal row. Suppose you are playing Bingo with the card in the figure.

a) What is the probability that the first number called is on your card?

b) What is the probability that G-59 is the first number called?

c) What is the probability that the first number called is in your *N* row?

d) Suppose the first number called is not on your card. What is the probability that the second number called is on your card?

e) Suppose the first and second numbers called are not on your card. What is the probability that the third number called is on your card?

f) After 20 numbers (7 from the *B* row) are called, you notice that you have B-3, B-13, B-8, and B-10 covered. You are one number away from getting a BINGO (five numbers in a vertical row). What is the probability that the next number called will give you a BINGO?

26. This problem repeats the experiment performed by French naturalist Georges Buffon in 1777. Take a piece of paper with regularly spaced lines in one direction and measure the distance between the lines. Call this distance *d*. Obtain a pin or a short nail and measure its length. Call this length *L*. (The length of the pin should be less than the distance between two lines. If not, cut the pin with a pair of pliers.) Toss the pin onto the paper 100 times and count how many times the pin rests on top of a line. Calculate the probability *p* that the pin crossed a line. Finally, perform the calculation $\frac{2L}{pd}$ and express it as a decimal. What value does this seem close to?

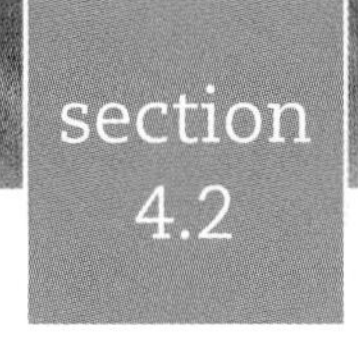

Calculating Probabilities

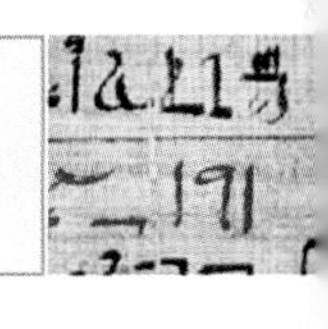

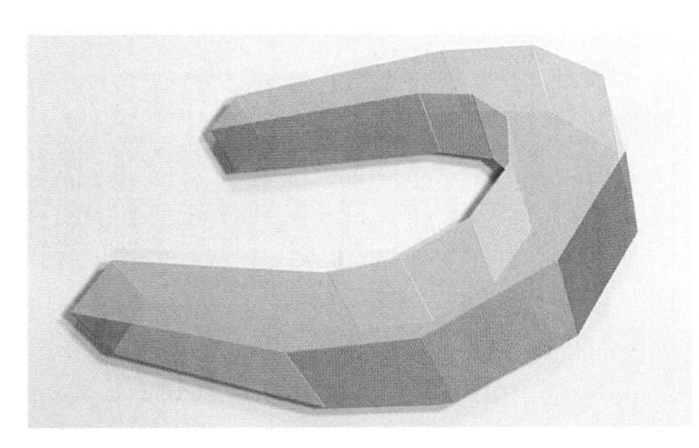

Horseshoe by Ron Davis shows the symbol of good luck valued by many a gambler. (Courtesy of the artist and The John Berggruen Gallery)

In Section 4.1, we introduced the basic terminology and formulas used in the study of probability. We are now ready to combine those concepts with the counting principles studied in Section 3.4 to further examine probabilities involved in real-world situations such as games of chance. We will address the problems using the techniques of theoretical probability. Some problems will involve using permutations and combinations, whereas others will require actually listing the outcomes of the event or using the Basic Counting Law.

EXAMPLE 1

If an automobile's license plate consists of any three letters of the alphabet and three digits as shown, what is the probability that the license plate assigned to your car has the first three letters of your name on it?

Solution: To find the probability, determine (a) the number of license plates with the first three letters of your name on it and (b) the total number of license plates consisting of three letters and three digits.

a) Since the letters of your name must be in the first three positions of the license plate, each letter can occur in one way. Since any of the ten digits can be used in the three digits of the license plate, each digit could occur in ten ways. Thus, by the Basic Counting Law, there are $1 \times 1 \times 1 \times 10 \times 10 \times 10 = 1000$ license plates with the first three letters of your name on it.

b) Since the license plate consists of three letters and three digits, by the Basic Counting Law, there are $26 \times 26 \times 26 \times 10 \times 10 \times 10 = 17{,}576{,}000$ total license plates. Therefore,

$$P(\text{3 letters of your name}) = \frac{1000}{17{,}576{,}000} = \frac{1}{17{,}576}.$$

EXAMPLE 2

If three standard dice are tossed, what is the probability that the sum of the numbers on the three dice is 15?

Solution: To determine the probability of the sum being 15, we must determine (a) the number of ways in which we can get a sum of 15 and (b) the number of total outcomes for tossing three dice.

a) By listing the ways in which the sum of three dice can be 15, we can count the number of ways that event can occur. There are ten different ways for three dice to have a sum of 15.

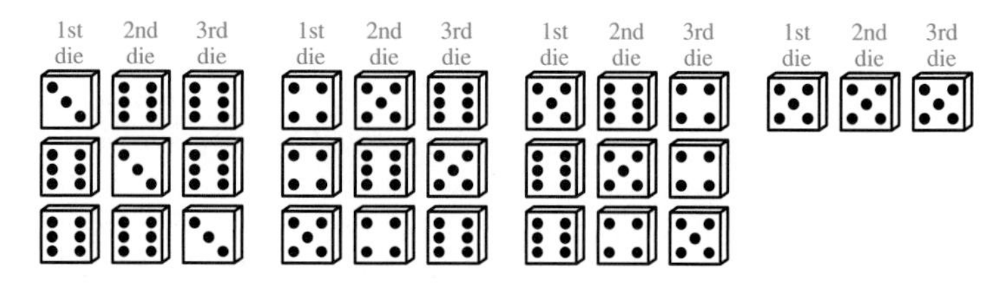

b) Since the first die can land in six ways, the second die can land in six ways, and the third die can land in six ways, by the Basic Counting Law, the total number of outcomes for the three dice is $6 \times 6 \times 6 = 216$. Therefore,

$$P(\text{sum of } 15) = \frac{10}{216} = \frac{5}{108} \approx \frac{1}{21.6} \approx 0.0463.$$

(*Note:* Answers to a probability problem are often given as a fraction with 1 as its numerator. This simply makes the answer easier to read and understand.)

EXAMPLE 3

You discover that the six spark plug wires were removed from the distributor cap of your car. What is the probability of randomly reconnecting the wires in the correct order?

Solution: We must connect all the wires in the correct order, and each wire can be used only once. Since the order in which we connect the wires matters, we use permutations to find the number of total ways this can be done. Using $n = 6$ and $r = 6$, the number of different arrangements of the wires is given by $P_{6,6} = 6! = 720$. Since only one of these arrangements is correct, the probability of randomly selecting the correct order is $1/720$.

EXAMPLE 4

Find the probability of having four queens when dealing five cards from a standard deck of 52 cards.

Solution: There are four queens and 48 other cards in the deck. Since we want to get all four of the queens and one of the other 48 cards, the number of ways to draw the hand is given by

$$C_{4,4} \times C_{48,1} = 1 \times 48 = 48.$$

The number of ways to draw five cards from a 52-card deck is given by

$$\begin{aligned} C_{52,5} &= \frac{52!}{5!(52-5)!} = \frac{52!}{5!47!} \\ &= \frac{52 \times 51 \times 50 \times 49 \times 48 \times 47 \times \cdots \times 1}{5 \times 4 \times 3 \times 2 \times 1 \times 47 \times 46 \times 45 \times \cdots \times 1} \\ &= \frac{52 \times 51 \times 50 \times 49 \times 48}{5 \times 4 \times 3 \times 2 \times 1} \\ &= 52 \times 51 \times 10 \times 49 \times 2 = 2{,}598{,}960. \end{aligned}$$

Thus, the probability of drawing four queens in five cards from a standard 52-card deck is

$$P(4 \text{ queens}) = \frac{48}{2{,}598{,}960} = \frac{1}{54{,}145} \approx 0.00001847.$$

This means that, on the average, four queens will appear in a five-card poker hand once every 54,145 hands.

In the examples that follow, we will not show all the computations. Instead, we wil give the answers in the following form.

$$\frac{C_{4,4} \times C_{48,1}}{C_{52,5}} = \frac{1 \times 48}{2{,}598{,}960} = \frac{1}{54{,}145}$$

In this way, the solutions and explanations can be presented without the page being filled with calculations. It is hoped that you have a calculator with combination and permutation keys to help simplify your computations.

EXAMPLE 5

a) Find the probability of getting three queens and two aces (a full house of queens over aces) from a standard deck of 52 cards when seven cards are dealt.

b) Find the probability of getting three cards of one rank and two cards of another rank (any full house) from a standard deck of 52 cards when seven cards are dealt.

Solution:

a) Since the order in which the cards are received does not affect the results, we can use combinations to determine the number of ways the events can occur. The number of ways to get three queens out of the four queens in the deck is $C_{4,3}$. Similarly, there are $C_{4,2}$ ways to get two of the four aces. Since we are receiving a total of seven cards, we get two more cards from the 44 cards remaining in the deck that are not queens or aces. There are $C_{44,2}$ ways to do this. We can now compute the desired probability.

$$\begin{aligned} P(\text{full house of queens over aces}) &= \frac{C_{4,3} \times C_{4,2} \times C_{44,2}}{C_{52,7}} \\ &= \frac{4 \times 6 \times 946}{133{,}784{,}560} \\ &= \frac{22{,}704}{133{,}784{,}560} \approx \frac{1}{5893} \approx 0.00017. \end{aligned}$$

b) Since there are 13 choices for the set of three cards and 12 choices for the set of two cards, the probability of getting a full house when being dealt seven cards is

$$\begin{aligned} P(\text{any full house}) &= \frac{(13 \times C_{4,3}) \times (12 \times C_{4,2}) \times C_{44,2}}{C_{52,7}} \\ &= \frac{(13 \times 4) \times (12 \times 6) \times 946}{133{,}784{,}560} \\ &= \frac{3{,}541{,}824}{133{,}784{,}560} \approx \frac{1}{37.8} \approx 0.0265. \end{aligned}$$

In the game twenty-one or blackjack, casinos like to use several decks of cards at once. In the next example, we will investigate whether using more than one deck changes the probability of getting a blackjack (21-point total consisting of two cards—an ace along with a ten-point card, king, queen, jack, or ten).

EXAMPLE 6

Find the probability of getting a blackjack when drawing two cards from (a) one standard deck and (b) two standard decks.

Solution:

a) Since the order in which the cards are received does not affect the results, we use combinations to determine the number of ways the events can occur. We want to get one of the four aces and one of the 16 ten-point cards that are in the deck. Thus, the number of desired hands is given by $C_{4,1} \times C_{16,1}$. The total number of possible two-card hands is given by $C_{52,2}$. Thus, the probability of getting a blackjack is

$$P(\text{blackjack}) = \frac{C_{4,1} \times C_{16,1}}{C_{52,2}} = \frac{4 \times 16}{1326} = \frac{64}{1326} \approx \frac{1}{20.7} \approx 0.04827.$$

b) When playing with two decks of cards, there are 104 cards with eight aces and 32 ten-point cards. Thus, the probability of drawing a blackjack is

$$P(\text{blackjack}) = \frac{C_{8,1} \times C_{32,1}}{C_{104,2}} = \frac{8 \times 32}{5356}$$
$$= \frac{256}{5356} \approx \frac{1}{20.9} \approx 0.04780.$$

When more than one deck is used, the probability of getting a blackjack decreases. It also makes it more difficult for the players to keep track of which cards have been played.

EXAMPLE 7

KENO

1	2	3	4	5	6	7	8	9	10
11	12	13	14	15	16	17	18	19	20
21	22	23	24	25	26	27	28	29	30
31	32	33	34	35	36	37	38	39	40

41	42	43	44	45	46	47	48	49	50
51	52	53	54	55	56	57	58	59	60
61	62	63	64	65	66	67	68	69	70
71	72	73	74	75	76	77	78	79	80

In the game of keno, 80 numbers are displayed on a board. A player marks from 1 to 20 numbers on a keno card. Twenty of the 80 numbers are randomly drawn as the winning numbers. Suppose a person selects seven numbers on a keno card. What is the probability that the person gets exactly five winning numbers out of the seven marked on the keno card?

Solution: This is a game with 80 numbers: 20 winning ones and 60 losing ones. Since the order in which the numbers are selected does not affect the results, use combinations to determine the number of ways the events can occur. The number of ways to select five out of the 20 winning numbers is $C_{20,5}$. The number of ways to select two out of the 60 losing numbers is $C_{60,2}$. Finally, there are $C_{80,7}$ total ways in which any seven numbers can be selected out of the 80 numbers. This gives the probability of picking exactly five winning numbers from a keno card of seven marked numbers as

$$P(5 \text{ of } 7 \text{ marked}) = \frac{C_{20,5} \times C_{60,2}}{C_{80,7}} = \frac{15{,}504 \times 1770}{3{,}176{,}716{,}400} \approx \frac{1}{115.8} \approx 0.00864.$$

Lotteries are held in many states of the United States. Millions of people wager a dollar each week hoping that they will become millionaires. What are their chances of "striking it rich"? The following example will show you that the probability of becoming a millionaire in a state-run lottery is very small. In fact, there is a greater chance of a person being struck by lightning than winning the State of California SuperLotto.

EXAMPLE 8

In a state lottery, a player selects six numbers (from 1 to 51) and wagers $1. Every Wednesday and Saturday evening at 7:56 P.M., six winning numbers are selected for the lottery. If the numbers a player chooses match the six winning numbers, the player wins the lottery grand prize for that game. What is the probability that the player selects the six winning numbers?

Solution: The game has 51 numbers, six of which are the winning numbers. Since the order in which the numbers are selected does not matter, use combinations to determine the number of ways the events can occur. There are $C_{6,6}$ ways to pick the winning numbers. There are $C_{51,6}$ total ways in which any six winning numbers can be drawn. This gives the probability of picking six numbers correctly in the lottery as

$$P(6 \text{ out of } 6) = \frac{C_{6,6}}{C_{51,6}} = \frac{1}{18{,}009{,}460} \approx 0.0000000555.$$

EXAMPLE 9

A certain jackpot-only $1 slot machine has three wheels on it as shown below. Each wheel rotates and lands on the pay line with either a BAR symbol or a blank.

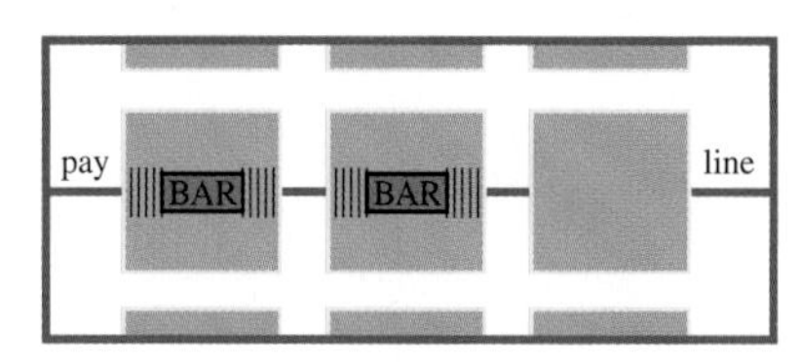

The first wheel contains six BAR symbols and 14 blanks, the second wheel contains five BAR symbols and 15 blanks, and the third wheel contains three BAR symbols and 17 blanks. After depositing $1 and spinning the wheels, you win the jackpot when each wheel lands on the pay line with a BAR symbol showing. What is the probability of winning the jackpot?

Solution: The desired outcome is to have a BAR symbol on each wheel. There are six BAR symbols on the first wheel, five on the second wheel, and three on the third wheel. By the Basic Counting Law, a BAR symbol on all three wheels can occur in $6 \times 5 \times 3 = 90$ ways. Since there are 20 symbols on each wheel, by the Basic Counting Law, the total number of possible ways in which the wheels can land on the pay line is $20 \times 20 \times 20 = 8000$. Thus, the probability of the jackpot is

$$P(\text{jackpot}) = \frac{6 \times 5 \times 3}{20 \times 20 \times 20} = \frac{90}{8000} = \frac{9}{800} \approx 0.01125.$$

In this section, we have determined the probability for various events using four methods: listing possibilities, the Basic Counting Law, combinations, and permutations. Armed with these principles, you should be able to determine probability in many different situations.

PROBLEMS ◐ Explain ◐ Apply ◐ Explore

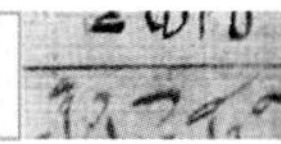

◐ Explain

1. What are four methods used in determining the number of ways an event can occur? Under what conditions is each one used?
2. In the card game of blackjack, what constitutes a blackjack?
3. In the game of keno, why are you not allowed to select more than 20 numbers?
4. What does it mean to have a full house in playing cards?

◐ Apply

5. Three standard dice are tossed. Find the probability for obtaining
   - **a)** a sum of 18 on the three dice.
   - **b)** a sum of 4 on the three dice.
   - **c)** a sum of 16 on the three dice.
6. Five cards are dealt from a standard deck of 52 cards.
   - **a)** What is the probability of being dealt three kings?
   - **b)** What is the probability of being dealt three of any rank?
7. Five cards are dealt from a standard deck of 52 cards.
   - **a)** What is the probability of being dealt an ace and four kings?
   - **b)** What is the probability of being dealt an ace and four more cards of another rank?
8. Five cards are dealt from two standard decks mixed together.
   - **a)** What is the probability of being dealt an ace and four kings?
   - **b)** What is the probability of being dealt an ace and four more cards of another rank?
9. Five cards are dealt from two standard decks mixed together.
   - **a)** What is the probability of being dealt three kings?
   - **b)** What is the probability of being dealt three cards of any rank?
10. In poker, a full house is a five-card hand with three cards of one rank and two cards of another rank. Suppose you are dealt five cards from a standard deck of 52 cards.
    - **a)** What is the probability of drawing a full house with three kings and two queens?
    - **b)** What is the probability of drawing a full house with three kings and two cards of another rank?
    - **c)** What is the probability of drawing any full house?
11. In horse racing, an *exacta* occurs when you correctly pick the horses that finish first and second. It is important to pick the correct order. If eight horses are in the race, what is the probability of correctly guessing the exacta?
12. In horse racing, a *trifecta* occurs when you correctly pick the order of the first

three horses. If six horses are in the race, what is the probability of correctly picking a trifecta by just guessing?

13. In the game of keno, 80 numbers are displayed on a board. Twenty of these numbers are chosen to be the winning numbers. Suppose a person has selected five numbers on a keno card.

a) What is the probability of selecting no correct numbers?

b) What is the probability of selecting exactly five correct numbers?

c) What is the probability of selecting exactly four correct numbers?

14. In the game of keno, 80 numbers are displayed on a board. Twenty of these numbers are chosen to be the winning numbers. Suppose a person has selected ten numbers on a keno card.

a) What is the probability of selecting no correct numbers?

b) What is the probability of selecting exactly six correct numbers?

15. A certain lottery has 49 numbers, six of which are the winning numbers for a particular game. To play the game, each participant chooses six numbers. What is the probability of choosing exactly

a) six correct numbers?

b) five correct numbers?

c) four correct numbers?

16. A certain lottery has 49 numbers, six of which are the winning numbers for a particular game. To play the game each participant chooses six numbers. What is the probability of choosing exactly

a) three correct numbers?

b) two correct numbers?

c) one correct number?

d) zero correct numbers?

17. A Wild 7's slot machine has three wheels on it. On each wheel there are four 7 symbols and six blanks. Find the probability and the odds of hitting the jackpot by getting a 7 symbol on the pay line on each wheel of the slot machine.

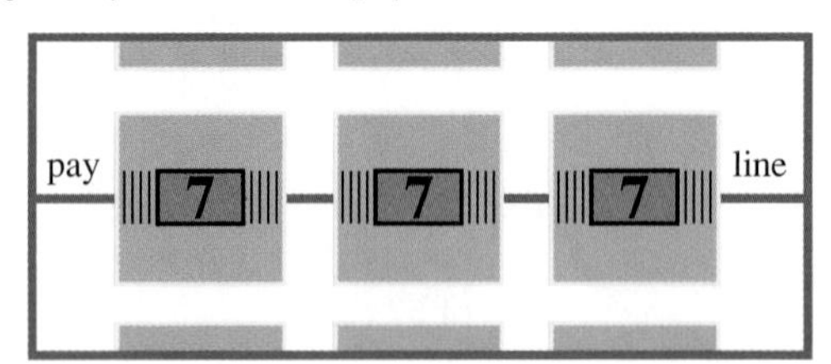

Explore

18. Suppose the NBA (National Basketball Association) has decided to add two more teams to the league. If 18 cities have applied for franchises, what is the probability that in a random drawing, the league selects the two westernmost cities?

19. In a dog show, a beagle is supposed to correctly pick two objects from a set of 20 objects. What is the probability that the dog picks the two objects?

20. Eight people on a committee are randomly selecting a subcommittee of two people. What is the probability that the two oldest committee members are selected?

21. An auditorium has scheduled three basketball games, two concerts, and four poetry readings. You have a ticket allowing you to attend any three of the events. What is the probability that by selecting three events at random, you attend two of the poetry readings and one of the other events?

22. Fourteen paintings are to be picked at random and placed along a wall. What is the probability that the paintings will be placed on the wall in alphabetical order, according to their titles? Assume that no two paintings have the same title.

23. License plates are printed by the State of California using the format of a digit, followed by three letters, followed by three digits. Assume that all arrangements of letters can be used. What is the probability that the last digit of a plate is 5?

24. Ten people are on a bus. There are three stops until the end of the line. Assume that it is equally likely for a person to get off at any of the three stops. What is the probability that all the people get off the bus at the last of the three stops?

25. There are 30 people in your speech class. For the last ten days of the semester, three students are randomly selected to give their final speech for the semester.

a) What is the probability that you are selected to give your speech on the first of the ten days?

b) If you were not selected to give your speech during the first five days, what is the probability that you are selected to give it on the sixth day?

c) If you were not selected to give your speech during the first nine days, what is the probability that you are selected to give it on the tenth day?

26. A regular octahedron is a solid with eight faces that are equilateral triangles. Suppose the numbers from 1 to 8 were placed on the faces of the octahedron. If three octahedrons are tossed, what is the probability that the sum of the numbers on the octahedrons is

a) 3.

b) 15.

c) 24.

27. In horse racing, a "Pick Six" bet requires that you pick the first-place horse in six designated races. Suppose that on a certain day at Churchill Downs in Louisville, Kentucky, there are nine horses in the first, ten horses in the second, 12 horses in the third, 15 horses in the fourth, eight horses in the fifth, and seven horses in the sixth Pick Six race. What is the probability of randomly guessing the winner in all of the six races?

28. Find the probability of getting a BINGO in the B column of a card when only five numbers have been called. (See Section 4.1, Problem 25, for an explanation of the game of bingo.)

29. Find the probability of getting a BINGO in the N column of a card when only four numbers have been called. (See Section 4.1, Problem 25, for an explanation of the game of bingo.)

30. There are 40 people in your math class. Four people are chosen at random to write their solution to Problem 28 on the chalkboard. What is the probability that you are one of those selected?

Probability and Odds

In many situations, rather than probability, the odds of an event are given. The **odds** for any event can be thought of as a ratio of the number of *desired* outcomes to the number of outcomes that are *not desired,* such as wins to losses or successes to failures.

EXAMPLE 1

The odds of War Emblem winning the Kentucky Derby in 2002 were 11 to 10. Explain what this means and give the probability that War Emblem would win.

Solution: Odds of 11 to 10 indicate that for every 11 times War Emblem wins, he loses 10 times. Since probability is the ratio of the number of successes to the total number, finding the probability requires knowing the total. On the basis of odds, we can think of War Emblem running 21 (11 + 10) times. Thus, the probability of War Emblem winning was $\frac{11}{21}$.

As seen in Example 1, like probability, odds are also used to measure the likelihood of an event occurring. We were able to convert from odds to probability by thinking about the meaning of both odds and probability. While we feel that this is a straight-forward way to convert from one to the other, there are also formulas to do the conversions.

To convert from odds to probability:
If the odds for an event E are a to b (a:b), then

$$P(E) = \frac{a}{a+b}.$$

To convert from probability to odds:
If the probability of an event E is $P(E)$, then the odds, $O(E)$, are given by

$$O(E) = \frac{\text{number of ways an event can occur } (\textit{desired})}{\text{number of ways an event fails to occur } (\textit{not desired})} = \frac{P(E)}{P(\overline{E})}.$$

EXAMPLE 2

Suppose the San Francisco Giants have a 10% chance of winning the National League pennant. What are the odds of the Giants winning the pennant? What are the odds of the Giants not winning the pennant?

Solution: Since the probability is 10%, $P(E) = 0.10$ and $P(\overline{E}) = 1 - 0.10 = 0.90$. The odds of the Giants winning the pennant are

$$O(E) = \frac{0.10}{0.90} = \frac{1}{9} = 1{:}9.$$

The odds of the Giants not winning the pennant are 9:1.

EXAMPLE 3

The chairman of a corporation states in a press conference that the odds are 10 to 1 that the company will reach $1,000,000,000 (one billion) in sales in the coming year. What is the probability that the corporation will reach one billion dollars in sales this year?

Solution: Since the odds are 10 to 1, use $a = 10$ and $b = 1$ in the formula that converts odds to probability. Thus,

$$P = \frac{a}{a+b} = \frac{10}{10+1} = \frac{10}{11} \approx 91\%.$$

There is a 91% chance of reaching one billion dollars in sales.

HOUSE ODDS

The odds for various events as given by casinos, racetracks, lotteries, and the like are called **house odds.** They give the odds against an event occurring. For example, before the 2002 baseball season, a Las Vegas casino stated that the odds for the New York Yankees to win the World Series were 5:2. This really meant that the odds against the Yankees winning the World Series were 5 to 2 or the probability of the Yankees winning was 2/7. House odds are given in such a manner as to facilitate the betting of money that might accompany the event. Odds of 5 to 2 would mean that for every $2 you bet on the Yankees, you would win $5 if the Yankees won the World Series. The odds for the Tampa Bay Devil Rays to win the World Series in the same year were 200:1. This would mean that if Tampa Bay won the World Series, you would win $200 for every $1 you bet. Odds are established to balance the amount won with the likelihood that the event occurs. Less probable events have higher house odds and greater potential winnings if the event occurs.

EXAMPLE 4

The oddsmakers in Atlantic City have stated that Jhun's odds are 7 to 1 in the title fight with Chavez.

a) What is the probability that Jhun wins the match?

b) If you bet $50 on Jhun, and Jhun beats Chavez, how much money will you win?

Solution:

a) The house odds of 7 to 1 indicate that Jhun's odds of losing are 7 to 1. This means that the odds of Jhun winning are 1 to 7. Using $a = 1$ and $b = 7$ in the formula that converts odds to probability, we find that the probability that Jhun wins the fight is

$$P = \frac{a}{a+b} = \frac{1}{1+7} = \frac{1}{8}.$$

b) The house odds for Jhun of 7 to 1 indicate that if Jhun wins, each \$1 bet on Jhun will generate \$7 in winnings. Thus, if you bet \$50, you will win $7 \times \$50 = \$350$.

EXAMPLE 5

An American roulette wheel has thirty-eight compartments around its circumference. Thirty-six compartments are numbered from 1 to 36, half of them being colored red and half colored black. The remaining two compartments, numbered 0 and 00, are colored green. A ball is spun and lands in one of the compartments. The odds for betting on a single number on an American roulette wheel are given by casinos as 35:1.

a) What do the odds mean in terms of the amount won when \$1 is bet on a single number?

b) On the basis of the actual roulette table and not the odds, what are the odds of winning a single bet?

c) Are the odds posted by the casino fair to the players?

Solution:

a) House odds of 35:1 indicate that for every dollar bet on a single number, the bettor wins \$35.

b) If E is the event that the single number comes up,

$$P(E) = \frac{1}{38}, \qquad P(\overline{E}) = 1 - \frac{1}{38} = \frac{37}{38}, \qquad \text{and} \qquad O(E) = \frac{1/38}{37/38} = \frac{1}{37}.$$

c) The odds against the single number coming up are 37:1. The house odds are not correct. Since the casino is paying out \$2 less than is mathematically correct, the odds are not fair to a player. The discrepancy in the odds gives the casino a margin of profit.

The concept of *fair* extends far beyond house odds. It is important in many areas of business, especially in determining insurance premiums. Calculating whether a game is fair and how much you should expect to win or lose in a game is the subject of Section 4.6.

section 4.3

PROBLEMS ○ Explain ○ Apply ○ Explore

Explain

1. What are odds?
2. What are house odds?
3. What do house odds tell you about money being bet on an event?
4. How do you convert from probability to odds?
5. How do you convert from odds to probability?
6. What is the relationship between odds and house odds?

Apply

7. The odds of an event are 3 to 2. What is the probability of the event?
8. The odds of an event are 1 to 3. What is the probability of the event?
9. The probability of an event is 25%. What are the odds of the event?
10. The probability of an event is 1/10. What are the odds of the event?
11. The house odds of a game are 10 to 1. What is the probability that the house wins? What is the probability that you win?
12. The house odds of a game are 20 to 1. What is the probability the house wins? What is the probability you win?

Explore

13. If the probability of you making a three-point shot in basketball is 0.15,
    a) What is the probability that you will not make a three-point shot?
    b) What are the odds of you making a three-point shot?
    c) What are the odds of you not making a three-point shot?
14. If there is 75% chance that you will get up when the alarm rings,
    a) What is the probability that you will not get up when the alarm rings?
    b) What are the odds of you not getting up when the alarm rings?
    c) What are the odds of you getting up when the alarm rings?
15. If the house odds for winning the prize in an Iowa lottery are posted as 28,560:1,
    a) What is the probability of winning the lottery?
    b) What are the odds of winning the lottery?
    c) How much should you win if a lottery ticket costs $2?
16. If the probability of a given gene having a defect that causes a certain disease is 1/25,
    a) What are the odds of the gene having that defect?
    b) What is the probability of the gene not having that defect?

17. If a casino's house odds on Serena Williams in a tennis match are posted at 3 to 2,
    a) What is the probability that Serena wins the match?
    b) What is the probability that Serena loses the match?
    c) How much should you win if you bet \$10 on Serena to win the match?
18. If the probability of Harvard winning the Ivy League championship is $\frac{1}{25}$.
    a) What are the odds of Harvard winning the championship?
    b) What is the probability of Harvard not winning the championship?
    c) What are the odds of Harvard not winning the championship?
    d) What odds would a casino post for betting purposes?
19. The probability of gene mutation under certain conditions is 0.00006.
    a) What is the probability of a gene under these conditions not mutating?
    b) What are the odds of a gene under these conditions mutating?
    c) What are the odds of a gene under these conditions not mutating?

Probability of Compound Events

Playing games can sometimes be very addicting. (©Hilary Price. Reprinted with special permission of King Features Syndicate)

In the previous section, we introduced the concept of probability for single events. In this section, we continue our investigation of probability by extending it to compound events. A **compound event** is an event that consists of more than one single event. An example of a compound event is "you drove to school and it was raining," which consists of the two single events "you drove to school" and "it was raining." As we shall see over the next two sections, there are several methods that can be used in determining the probability of compound events.

TWO KEY WORDS

There are key words that are used to understand the probability of compound events. The first word is ***and.*** It is used specifically to mean that all parts of the compound event are true. For the event "you drove to school and it was raining" to be true, you must have driven to school and it also must have been raining.

The symbol that will be used to represent *and* is ∩, which is called the *intersection* symbol. This is the same symbol that was used in Section 3.2. Letting D be the event that you drove and R be the event that it was raining, the event "you drove to school and it was raining" could be represented as $D \cap R$.

The second word is ***or.*** It is used to indicate that one or more of the parts of the compound event is true. Thus, the statement "you drove to school or it was raining" is true in three different situations:

if you drove and it was raining,

if you drove and is wasn't raining, or

if you didn't drive but it was raining.

The symbol that will be used to represent *or* is ∪, which is called the *union* symbol. The event "you drove to school or it was raining" could be represented as $D \cup R$. To summarize, we have placed the above information in the following table.

Situation	Name	Symbol
and	intersection	∩
or	union	∪

As we proceed through the sections, you will see how *and* and *or* are used in determining probabilities.

EXAMPLE 1

Write the following using an intersection symbol.

> Venus and Serena both won a Grand Slam tennis tournament in 2002.

Solution: If we let V be the event that Venus won a Grand Slam tennis tournament in 2002 and S be the event that Serena won a Grand Slam tennis tournament in 2002, then both of them winning in 2002 can be written as $V \cap S$.

EXAMPLE 2

Write the following using a union symbol.

> Summer weather conditions in the Rockies usually implies that you have either a thunderstorm or crystal clear blue skies.

Solution: If we let T be the event that the weather on a summer afternoon in the Rockies includes a thunderstorm and B be the event that the weather on a summer afternoon in the Rockies has crystal clear blue skies, then the summer afternoon weather in the Rockies can be given by $T \cup B$.

PROBABILITY OF COMPOUND EVENTS

Let us now turn our attention to determining the probability of compound events. We will present four ways to do this.

Reading Tables

The first method applies when the information is either supplied in a table or can be arranged in a table.

EXAMPLE 3

Based on statistics published by the U.S. Department of Labor, the employment classifications of 1200 men and 1300 women are given in the following table.

Occupation	Number of Males	Number of Females	Totals
Executive, administrative, management	180	185	365
Professional specialty	163	234	397
Technical, sales, administrative support	236	520	756
Service occupations	119	227	346
Precision production, crafts	223	27	250
Operators, fabricators, laborers	232	92	324
Farming, forestry, fishing	47	15	62
Totals	1200	1300	2500

a) What is the probability of a person chosen at random being female?

b) What is the probability of a person chosen at random being in a service occupation?

c) What is the probability of a person chosen at random being female in a service occupation?

d) What is the probability of a person chosen at random being female or in a service occupation?

Solution:

Let F = a person being female and S = a person in a service occupation.

a) Since there are 1300 females in the survey out of a total of 2500 people,

$$P(F) = \frac{1300}{2500} = \frac{13}{25} = 52\%.$$

b) Since there are a total of 346 people out of 2500 employed in a service occupation,

$$P(S) = \frac{346}{2500} = 13.84\%.$$

c) "Being a female in a service occupation" means that the person chosen must be both female and in a service occupation, so this is an instance in which we should use an intersection. There are 227 females in service occupations, so

$$P(F \cap S) = \frac{227}{2500} = 9.08\%.$$

d) "Being a female or in a service occupation" means that the person chosen can be either a female or in a service occupation or both, so this is an instance in which we should use a union. There are 1300 females, including those in a service occupation. There are also 346 people in a service occupation, including females. If we add these two groups together, we would be counting the 227 females in a service occupation twice. Therefore, we should subtract 227, which gives

$$P(F \cup S) = \frac{1300}{2500} + \frac{346}{2500} - \frac{227}{2500} = \frac{1419}{2500} = 56.76\%.$$

Formula for the Probability of a Union

Part (d) in the previous example leads to the formula for the probability of the union of two events A and B.

Union Formula

$$P(A \cup B) = P(A) + P(B) - P(A \cap B)$$

This formula gives us a second way to find the probability of certain compound events.

EXAMPLE 4

If the probability of having red hair is 18%, the probability of having blue eyes is 12%, and the probability of having red hair and blue eyes is 10%, find the probability of having red hair or blue eyes.

Solution: If we let R be the event that a person has red hair and B be the event that a person has blue eyes, we have $P(R) = 0.18$, $P(B) = 0.12$, $P(R \cap B) = 0.10$. Therefore,

$$\begin{aligned} P(R \cup B) &= P(R) + P(B) - P(R \cap B) \\ &= 0.18 + 0.12 - 0.10 = 0.20 = 20\%. \end{aligned}$$

Venn Diagrams

A third way of finding the probability of compound events is to use Venn diagrams, which were introduced in Section 3.2. In addition to being able to solve the problem in Example 4, Venn diagrams can help to answer several other questions. Let's expand the previous example, this time using numbers of people rather than percentages.

EXAMPLE 5

In a group of 100 people, there are 18 people with red hair, 12 people with blue eyes, and 10 people with both red hair and blue eyes.

a) Find the probability of a person in this group having either red hair or blue eyes.

b) Find the probability of a person having red hair but not having blue eyes.

c) Find the probability of a person having blue eyes but not having red hair.

d) Find the probability of a person with red hair or blue eyes but not both.

Solution: Since there are ten people with both red hair and blue eyes, we can represent this as the intersection of two circles, R and B.

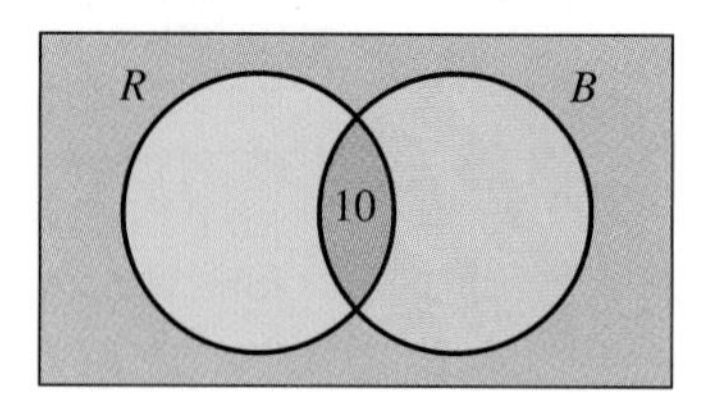

Since there are a total of 18 people with red hair and there are ten people in the intersection, there must be an additional eight people inside the circle R but outside of the intersection.

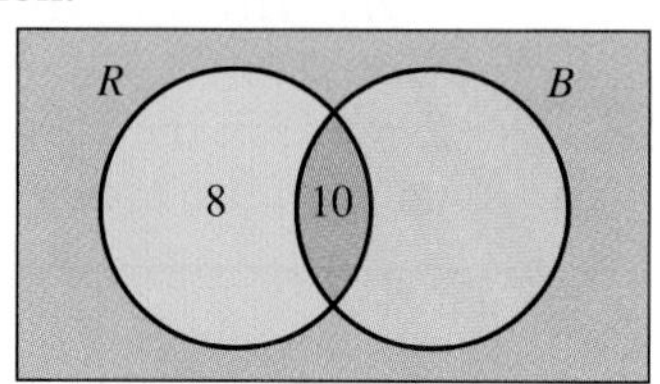

Since there are a total of 12 people with blue eyes, there also must be two more people inside the circle B but outside of the intersection.

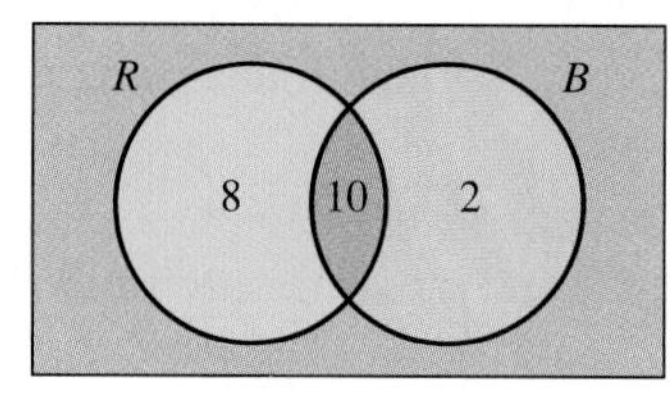

Since there are 100 people total and we have already placed $8 + 10 + 2 = 20$ of them, there must be $100 - 20 = 80$ people outside of the circles.

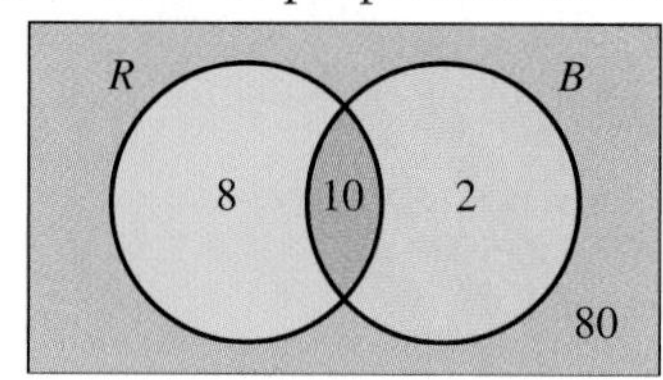

Once we have set up the Venn diagram, answering the questions requires reading only the correct numbers.

a) $P(R \text{ or } B) = P(R \cup B) = \dfrac{8 + 10 + 2}{100} = \dfrac{20}{100} = 20\%$

b) Using the notation $\overline{B}$ (Section 3.2) for people who do not have blue eyes, the group of people who have red hair and do not have blue eyes is denoted as $R \cap \overline{B}$. Thus, $P(\text{red hair and not blue eyes}) =$

$$P(R \cap \overline{B}) = \frac{8}{100} = 8\%.$$

c) Similarly,

$$P(\overline{R} \cap B) = \frac{2}{100} = 2\%.$$

d) The people who have either red hair or blue eyes but not both is a combination of the answers in parts (b) and (c). Therefore,

$$P = \frac{8}{100} + \frac{2}{100} = \frac{10}{100} = 10\%.$$

Tree Diagrams

The fourth method we will discuss in this section is called a tree diagram. A **tree diagram** is a listing of all the possible solutions presented in an organized fashion and then written in a structure resembling a tree.

EXAMPLE 6

Suppose that a migraine sufferer has a history that indicates the following probabilities. The probability that she will have a migraine on a random day is 25%. If she has a migraine one day, the chance that she will have a migraine the next day is 40%. If she does not have a migraine, the chance that she will have a migraine the next day is 20%. What is the probability that she has a migraine on one of two days?

Solution: If we let M be the event that she has a migraine, there are four possibilities of what can happen on two consecutive days:

MM = migraine on the first day and the second day
$M\overline{M}$ = migraine on the first day but not on the second
$\overline{M}M$ = no migraine on the first day and a migraine on the second day
$\overline{MM}$ = no migraine on the either day

We can draw a tree diagram of this information.

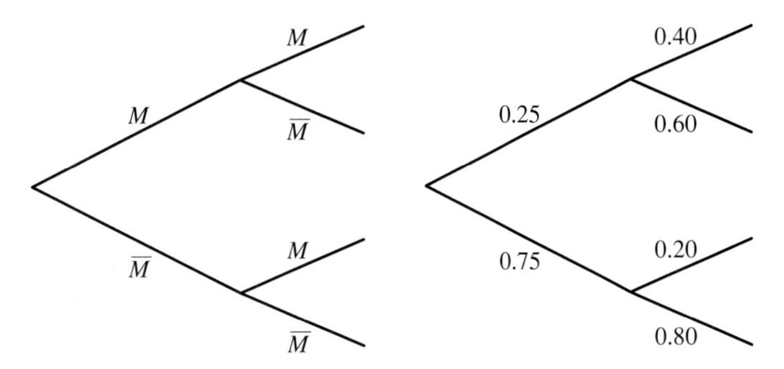

To find the probability of getting a migraine on the first day and not on the second, we multiply the corresponding numbers on the tree diagram. Thus,

$$P(M\overline{M}) = 0.25 \times 0.60 = 0.15.$$

Similarly,

$$P(\overline{M}M) = 0.75 \times 0.20 = 0.15.$$

Therefore, the probability of having a headache on only one of two days is

$$0.15 + 0.15 = 0.30 = 30\%.$$

EXAMPLE 7

In the league playoffs for major league baseball, two teams play until one team has won three games. Suppose that the New York Yankees are playing the Boston Red Sox in the playoffs and that in any single game, the Yankees have a 60% chance of winning.

a) What is the chance that the Yankees win in three straight games?

b) What is the chance that the Yankees win in four games?

c) What is the chance that the Yankees win in five games?

d) What is the chance that the Yankees win the series?

Solution: First we draw the tree diagram for the status of the playoffs after three games. In this picture, Y indicates that the Yankees win, and R indicates that the Red Sox win.

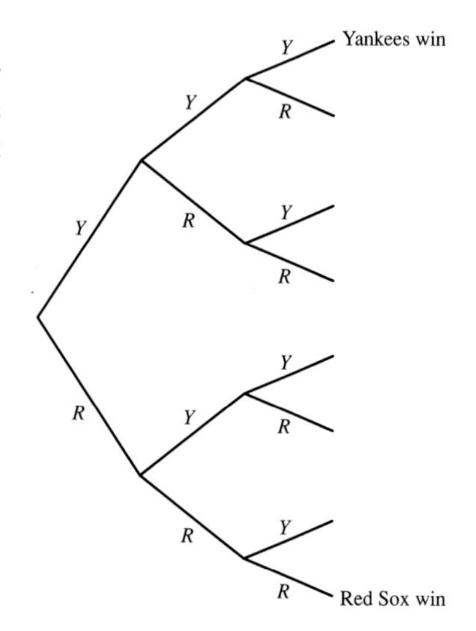

At this point, the tree diagram gives only two of the possibilities of one team winning three games. To finish, each branch of the tree must show either the Yankees or the Red Sox winning three games.

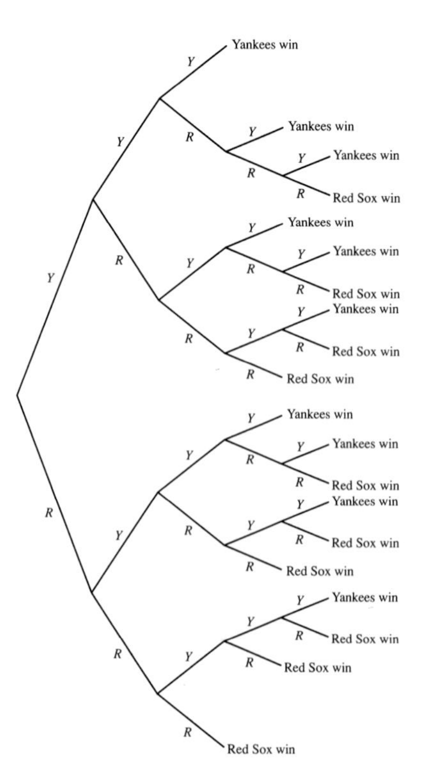

To calculate the probability that the Yankees win the playoffs, we use the probability that the Yankees will win a particular game (0.6) and the probability that the Red Sox will win (0.4) for each part of the tree. Finally, multiply the numbers along a particular path to get that probability. For example, the chance that the Yankees win three straight games is

$$(0.6)(0.6)(0.6) = 0.216.$$

As another example, the chance that the Yankees win three games in the order win, win, lose, win is

$$(0.6)(0.6)(0.4)(0.6) = 0.0864.$$

Filling out the tree in this fashion for the ways that the Yankees win gives the following:

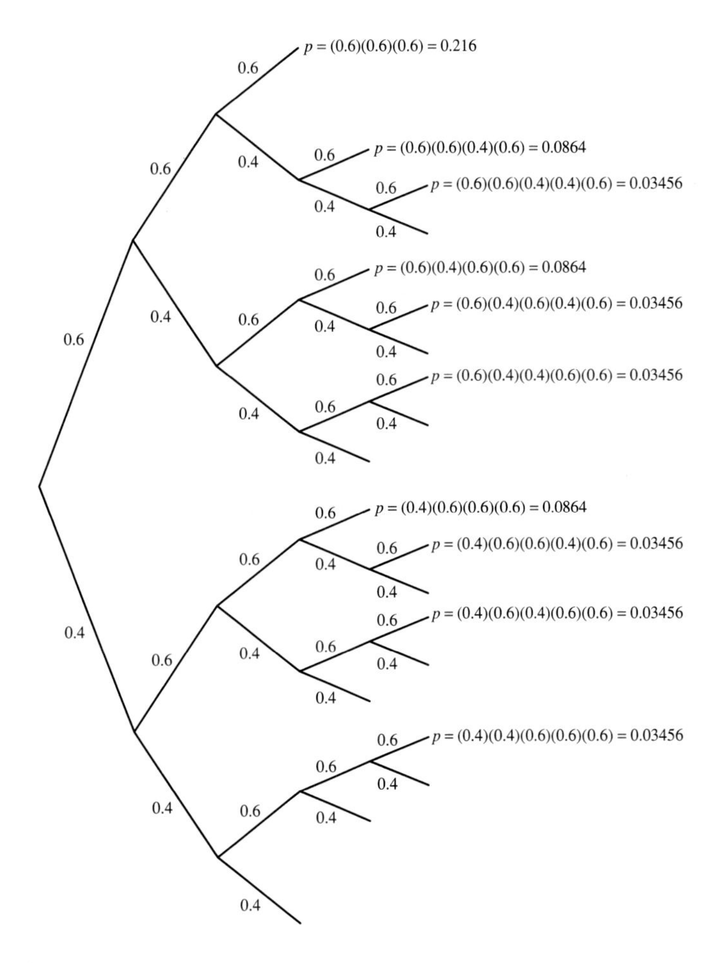

We can now use this information to answer the questions.

a) $P(\text{Yankees in 3 straight}) = 0.216 = 21.6\%$

b) $P(\text{Yankees win in 4 games}) = 0.0864 + 0.0864 + 0.0864$
$= 3 \times 0.0864 = 0.2592 \approx 25.9\%$

c) $P(\text{Yankees win in 5 games}) = 6 \times 0.03456 = 0.20736 \approx 20.7\%$

d) $P(\text{Yankees win series}) \approx 21.6\% + 25.9\% + 20.7\% = 68.2\%$

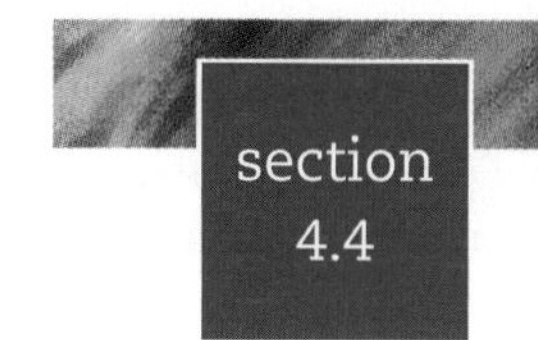

PROBLEMS ● Explain ● Apply ● Explore

● Explain

1. What is a compound event?
2. In probability, what is meant when you are finding $P(A \text{ and } B)$?
3. In probability, what is meant when you are finding $P(A \text{ or } B)$?
4. When finding the union of two events, what must be true about the two events?
5. When finding the intersection of two events, what must be true about the two events?
6. What are the four methods used in this section to find the probability of compound events?

● Apply

7. If $P(A) = 0.56$, $P(B) = 0.48$, and $P(A \cap B) = 0.12$, find $P(A \cup B)$.
8. If $P(A) = 0.78$, $P(B) = 0.29$, and $P(A \cap B) = 0.21$, find $P(A \cup B)$.
9. If $P(A) = 0.56$, $P(B) = 0.48$, and $P(A \cup B) = 0.86$, find $P(A \cap B)$.
10. If $P(A) = 0.78$, $P(B) = 0.29$, and $P(A \cup B) = 0.86$, find $P(A \cap B)$.
11. It is not possible to have $P(A) = 0.48$, $P(B) = 0.24$, and $P(A \cap B) = 0.29$. Explain.
12. It is not possible to have $P(A) = 0.58$, $P(B) = 0.32$, and $P(A \cap B) = 0.35$. Explain.
13. It is not possible to have $P(A) = 0.48$, $P(B) = 0.36$, and $P(A \cup B) = 0.29$. Explain.
14. It is not possible to have $P(A) = 0.58$, $P(B) = 0.32$, and $P(A \cap B) = 0.92$. Explain.

There are 20 envelopes in a hat. Each envelope contains a piece of paper in the shape of a plane or a boat. On each slip of paper is written Hawaii or Mexico. You have been chosen to pick an envelope from the hat, and then you get to go on that type of vacation. The following table shows the number of each of the four types of vacations that are in the hat. Use this information to solve Problems 15–24.

	Hawaii	Mexico	
Plane	7	8	15
Boat	2	3	5
	9	11	20

15. What is the probability that you fly to Mexico?
16. What is the probability that you take a cruise to Hawaii?
17. What is the probability that you take a cruise?

18. What is the probability that you go to Hawaii?

19. What is the probability that you fly or go to Hawaii?

20. What is the probability that you take a cruise or go to Hawaii?

21. What is the probability that you take a cruise or go to Mexico?

22. What is the probability that you fly or go to Mexico?

23. What is the probability that you do not go on a cruise to Mexico?

24. What is the probability that you do not fly to Hawaii?

25. Use the information in Example 6 to find the probability of having a migraine two days in a row.

26. Use the information in Example 6 to find the probability of not having a migraine on two consecutive days.

Use the information in Example 7 to answer the following questions.

27. What is the probability of the Red Sox winning the series in three games?

28. What is the probability of the Red Sox winning the series in four games?

29. What is the probability of the Red Sox winning the series in five games?

30. What is the probability of the Red Sox winning the series?

31. What is the probability of the Yankees not winning the series in three games?

32. What is the probability of the Yankees not winning the series in four games?

33. What is the probability that the series ends in exactly three games?

34. What is the probability that the series ends in exactly four games?

35. What is the probability that the series takes more than four games?

36. What is the probability that the series take fewer than five games?

Explore

37. In a sampling of 100 golden retrievers, a total of 16 dogs had a serious hip problem and a total of seven dogs had epilepsy. Two dogs had both a serious hip problem and epilepsy.

a) Draw a Venn diagram for this information.

b) What is the probability that a dog had a serious hip problem but not epilepsy?

c) What is the probability that a dog had epilepsy but not a serious hip problem?

d) What is the probability that a dog had neither a serious hip problem nor epilepsy?

38. In a sampling of 200 cats, a total of 12 cats had leukemia, and a total of nine cats had cataracts. Three cats had both leukemia and cataracts.

a) Draw a Venn diagram for this information.

b) What is the probability that a cat had leukemia but not cataracts?

c) What is the probability that a cat had cataracts but not leukemia?

d) What is the probability that a cat had neither leukemia nor cataracts?

39. A marketing firm is interested in what type of electronic devices college students use daily. In a survey of 450 college students, 373 said that they used only a cellular telephone, while 39 students used both a cellular telephone and a pager. If 15 students in the survey used neither a cellular telephone nor a pager,

a) Draw a Venn diagram for this information.

b) Find the probability that a student uses only a pager.

c) Find the probability that a student uses a pager.

d) Find the probability that a student uses a cellular telephone.

40. A marketing firm is interested in what people are drinking with breakfast. In a survey of 300 people, 117 people said that they drank only coffee with breakfast, 43 said that they drank only orange juice with breakfast, and 72 said that they drank neither coffee nor orange juice with breakfast. Assume that the rest of the people in the survey drank both coffee and orange juice.

a) Draw a Venn diagram for this information.

b) Find the probability that a person drinks orange juice for breakfast.

c) Find the probability that a person drinks coffee for breakfast.

d) Find the probability that a person drinks coffee or orange juice for breakfast.

A survey of 730 students has the following results.

	Enrolled in		
Working	**Fewer than 6 Units**	**Between 6 and 13 Units**	**More than 13 Units**
Fewer than 10 Hours/Week	28	72	85
Between 10 and 20 Hours/Week	29	150	72
More than 20 Hours/Week	150	82	62

Use this table to answer the following questions.

41. What is the probability that a student is working less than 10 hours per week?

42. What is the probability that a student is working more than 20 hours per week?

43. What is the probability that a student is enrolled in fewer than six units and working more than 20 hours per week?

44. What is the probability that a student is enrolled in more than 13 units and working fewer than 10 hours per week?

45. What is the probability that a student is enrolled in fewer than six units or working less than 10 hours per week?

46. What is the probability that a student is enrolled in more than 13 units or working more than 20 hours per week?

If the probability that a newborn child is a boy is 49% and the probability that the child is a girl is 51%, answer the following questions.

47. What is the probability of having a two-child family consisting of two girls?

48. What is the probability of having a two-child family consisting of two boys?

49. What is the probability of having a two-child family consisting of one boy and one girl?
50. What is the probability of having a two-child family consisting of at least one boy?
51. What is the probability of having a two-child family consisting of at least one girl?
52. What is the probability of having a two-child family consisting of two children of the same sex?

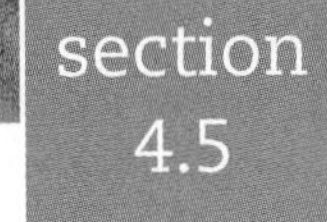

section 4.5

Conditional Probability

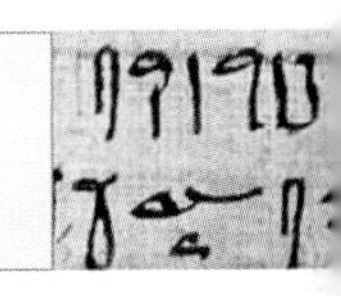

In this section, we continue the discussion of the probability of compound events, this time emphasizing the concept of conditional probability. **Conditional probability** is the probability of an event occurring if some other condition has already occurred. The knowledge of this other condition often drastically changes the probability. For example, the probability that you are more than six feet tall if your parents are both more than six feet tall is much higher than the probability that a person selected at random is more than six feet tall.

A NEW SYMBOL

Before we look at some further examples, we give you the symbol that is used for conditional probability. A conditional probability is written as $P(A \mid B)$.

$$P(A \mid B) = \text{the probability of event } A \text{ if event } B \text{ is true}$$

It is read as "probability of A given B." Combining this with the symbols from the last section, we have the following table of symbols used in probability.

Situation	Name	Symbol
and	intersection	$\cap$
or	union	$\cup$
if	conditional	$\mid$

AN OLD CONCEPT

Although we have a new word (*conditional*) and a new symbol $P(A \mid B)$, the concept of conditional probability was used in the previous section whenever we discussed tree diagrams. For instance, in Example 6 of Section 4.4, we discussed the probability of a migraine headache. The probability of having a migraine was greater if you had a migraine the day before (40%) than if you hadn't had a migraine the day before (20%).

INTUITIVE EXAMPLES

As was true in the preceding section, there are several methods for working with conditional probabilities. Tables of information, formulas, Venn diagrams, and tree diagrams will be used in the same fashion as they were in the previous section. However, one of the easiest methods (an often neglected one) is simply to read the problem and use your intuition.

EXAMPLE 1

A box contained three glazed doughnuts and five chocolate doughnuts and then your friend ate one of the chocolate doughnuts. If you now pick a doughnut from the box at random, what is the probability that you pick a glazed doughnut?

Solution: If we let G be the event that you pick a glazed doughnut and C be the event that your friend ate a chocolate doughnut, then we are looking for $P(G \mid C)$. Since you know that your friend ate a chocolate doughnut, there still must be three glazed doughnuts left in the box out of a total of seven. Therefore, $P(G \mid C) = \frac{3}{7}$. Notice that this answer is different than if you simply found $P(G) = \frac{3}{8}$.

EXAMPLE 2

A standard deck of cards contains 52 cards, including four queens. You are going to pick and keep two of the cards.

a) What is the probability that the second card is a queen if the first card is a queen?

b) What is the probability that the second card is a queen if the first card is not a queen?

Solution: If we let Q mean that we pick a queen and $\overline{Q}$ mean that we did not pick a queen,

a) After you pick the first queen, there are still three queens in the deck out of a total of 51 cards. Therefore, $P(\text{second card } Q \mid \text{first card } Q) = \frac{3}{51} = \frac{1}{17}$.

b) Since the first card you picked is not a queen, there are still four queens left in the deck out of a total of 51 cards. Therefore, $P(\text{second card } Q \mid \text{first card } \overline{Q}) = \frac{4}{51}$.

USING TABLES

EXAMPLE 3

Based on statistics published by the U.S. Department of Labor, the employment classifications of 1200 men and 1300 women are given in the following table.

Occupation	Number of Males	Number of Females	Total
Executive, administrative, management	180	185	365
Professional specialty	163	234	397
Technical, sales, administrative support	236	520	756
Service occupations	119	227	346
Precision production, crafts	223	27	250
Operators, fabricators, laborers	232	92	324
Farming, forestry, fishing	47	15	62
Totals	1200	1300	2500

a) What is the probability of a person chosen at random being male if that person works in a service occupation?

b) What is the probability of a person chosen at random working in a service occupation if that person is male?

Solution: We will let M indicate males and S indicate service occupations. Notice that the wording of the two questions is reversed.

a) In part (a), it is known that the person works in a service occupation. Therefore, we want $P(M \mid S)$. Since there are a total of 346 workers in S and 119 of them are male, $P(M \mid S) = \frac{119}{346}$.

b) In part (b), it is known that the person is male. Therefore, we want $P(S \mid M)$. Since there is a total of 1200 male workers and 119 of them are in service occupations, $P(S \mid M) = \frac{119}{1200}$.

Notice how the wording of the two parts is different and that this drastically affects the answer. It is important to read carefully when you are answering this type of question.

FORMULA FOR CONDITIONAL PROBABILITY

We are now going to examine the information in the preceding table and use the results to develop a formula for calculating conditional probability.

Based on the information in the above table, the probability that a worker is a male and in a service occupation is given by $P(M \cap S) = \frac{119}{2500}$, and the probability that any worker is in a service occupation is given by $P(S) = \frac{346}{2500}$. Notice that if we divide these two results, we have

$$\frac{P(M \cap S)}{P(S)} = \frac{119/2500}{346/2500} = \frac{119}{346}.$$

This is exactly the result we get from calculating $P(M \mid S)$ directly from the table in Example 3. This gives us the general formula for calculating conditional probability.

Conditional Probability Formula

$$P(A \mid B) = \frac{P(A \cap B)}{P(B)} \quad \text{or} \quad P(A \cap B) = P(A \mid B)P(B)$$

EXAMPLE 4

Use the above formula to calculate the probability of $P(S \mid M)$.

Solution: First, we write out the formula

$$P(S \mid M) = \frac{P(S \cap M)}{P(M)}.$$

Next, notice that the intersection of S and M must be the same as the intersection of M and S. Thus,

$$P(S \mid M) = \frac{P(S \cap M)}{P(M)} = \frac{119/2500}{1200/2500} = \frac{119}{1200}.$$

USING VENN DIAGRAMS

If you have a Venn diagram of some information, calculating conditional probability is sometimes simply a matter of reading the information from the diagram.

EXAMPLE 5

In a group of 100 people, there are 18 people with red hair, 12 people with blue eyes, and 10 people with both red hair and blue eyes. This information is represented by the Venn diagram first shown in Example 5 of Section 4.4.

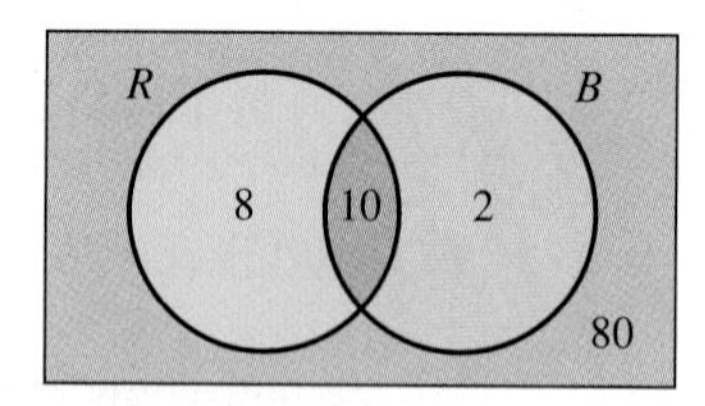

Find the probability that a person has blue eyes if that person has red hair by

a) Reading the information from the Venn diagram.

b) Using the previous formula.

Solution:

a) Since we know that the person must have red hair, we are looking at a total number of $8 + 10 = 18$ people. Since ten of these people have blue eyes, we have

$$P(B \mid R) = \frac{10}{18} = \frac{5}{9}.$$

b) Using the formula for conditional probability, we have

$$P(B \mid R) = \frac{P(B \cap R)}{P(R)} = \frac{10/100}{18/100} = \frac{10}{18} = \frac{5}{9}.$$

EXAMPLE 6

Continue with Example 5. This time, find the probability of a person having red hair if you know that she has blue eyes.

Solution: Since we know that she has blue eyes, we are looking at a total number of $10 + 2 = 12$ people. Since ten of these people have blue eyes,

$$P(R \mid B) = \frac{10}{12} = \frac{5}{6}.$$

USING TREE DIAGRAMS

When we worked with tree diagrams in the preceding section, even though it was never mentioned, conditional probability was an essential part of the discussion. Let's look at the following example.

EXAMPLE 7

A box of doughnuts originally had nine chocolate doughnuts and three glazed doughnuts. If your friend ate one of the doughnuts, what is the chance of you picking a chocolate doughnut when you randomly select a doughnut from the box?

Solution: We will let *YC* be the event that you pick a chocolate doughnut and *FC* be the event that your friend picks a chocolate doughnut. The chance that you select a chocolate doughnut depends on what your friend chose. If your friend chose a chocolate doughnut, there are eight chocolate doughnuts remaining in a box of 11 doughnuts, so

$$P(\text{you pick chocolate if friend picks chocolate}) = P(YC \mid FC) = \frac{8}{11}.$$

If your friend did not choose a chocolate doughnut, there are still nine chocolate doughnuts remaining in a box of 11 doughnuts, so

$$P(\text{you pick chocolate if friend does not pick chocolate}) = P(YC \mid \overline{FC}) = \frac{9}{11}.$$

This is represented by the following tree diagrams.

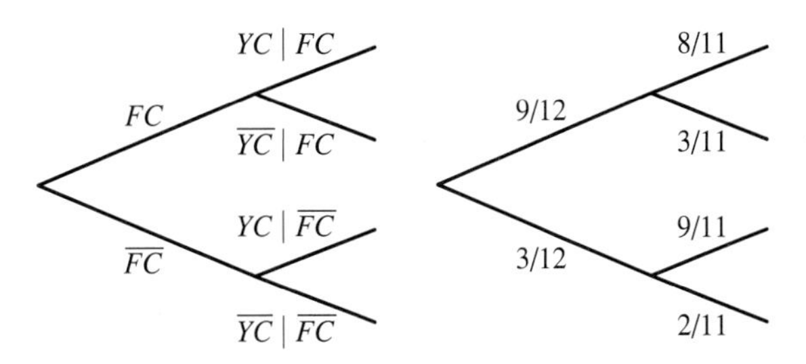

section 4.5

PROBLEMS ⊙ Explain ⊙ Apply ⊙ Explore

⊙ Explain

1. What is meant by conditional probability?
2. If you are investigating $P(A \mid B)$, what is assumed about B?
3. If A is the event that you like to eat apples and B is the event that you like to go bike riding, explain in words the meaning of $P(A \mid B)$.
4. If A is the event that you like to eat apples and B is the event that you like to go bike riding, explain in words the meaning of $P(B \mid A)$.
5. If A is the event that you like to eat apples and B is the event that you like to go bike riding, explain in words the difference between $P(A \mid B)$ and $P(A \cap B)$.
6. If event B is a sure thing, explain in words why $P(A \mid B)$ and $P(A \cap B)$ are equal.

⊙ Apply

7. If $P(B) = 0.54$ and $P(A \cap B) = 0.12$, find $P(A \mid B)$.
8. If $P(B) = 0.63$ and $P(A \cap B) = 0.21$, find $P(A \mid B)$.
9. If $P(B) = 0.56$ and $P(A \mid B) = 0.12$, find $P(A \cap B)$.
10. If $P(B) = 0.29$ and $P(A \mid B) = 0.21$, find $P(A \cap B)$.
11. If $P(A) = 0.56$, $P(B) = 0.48$, and $P(A \cup B) = 0.86$, find $P(A \cap B)$ and $P(B \mid A)$.
12. If $P(A) = 0.78$, $P(B) = 0.29$, and $P(A \cup B) = 0.86$, find $P(A \cap B)$ and $P(B \mid A)$.

There are 20 envelopes in a hat. Each envelope contains a piece of paper in the shape of a plane or a boat. On each slip of paper is written Hawaii or Mexico. You have been chosen to pick an envelope from the hat, and then you get to go on that type of vacation. The following table shows the number of each of the four types of vacations that are in the hat. Use this information to solve Problems 13–16.

	Hawaii	Mexico	
Plane	7	8	15
Boat	2	3	5
	9	11	20

13. What is the probability that you won a trip to Mexico if you know you won a plane flight?
14. What is the probability that you won a trip to Hawaii if you know you won a plane flight?
15. What is the probability that you won a cruise if you know you won a trip to Hawaii?
16. What is the probability that you won a plane trip if you know you won a trip to Mexico?

Use the information in Example 3 to solve the Problems 17–20.

17. What is the probability that you are employed in a professional specialty if you are female?

18. What is the probability that you are employed in a professional specialty if you are male?

19. What is the probability that you are male if you are employed in a professional specialty?

20. What is the probability that you are female if you are employed in a professional specialty?

21. A jar has 10 pieces of chocolate candy and 15 pieces of butterscotch candy. You pick and eat two pieces of candy from the jar at random, one piece at a time.

a) Write two tree diagrams for this information. One diagram should have the formulas such as $P(B \mid C)$, and the other diagram should have the numerical probabilities (as in Example 7).

b) Use the tree diagram to find the probability that you picked first a chocolate and then a butterscotch.

22. A jar has 20 pieces of chocolate candy and 30 pieces of butterscotch candy. You pick and eat two pieces of candy from the jar at random, one piece at a time.

a) Write two tree diagrams for this information. One diagram should have the formulas such as $P(B \mid C)$, and the other diagram should have the numerical probabilities (as in Example 7).

b) Use the tree diagram to find the probability that you picked first a butterscotch and then a chocolate.

Explore

23. In a sampling of 100 golden retrievers, a total of 16 dogs had a serious hip problem, and a total of seven dogs had epilepsy. Two dogs had both a serious hip problem and epilepsy.

a) Draw a Venn diagram for this information.

b) What is the probability that a dog had a serious hip problem if it had epilepsy?

c) What is the probability that a dog had epilepsy if it had a serious hip problem?

24. In a sampling of 200 cats, a total of 12 cats had leukemia, and a total of nine cats had cataracts. Three cats had both leukemia and cataracts.

a) Draw a Venn diagram for this information.

b) What is the probability that a cat had leukemia if it had cataracts?

c) What is the probability that a cat had cataracts if it had leukemia?

25. A marketing firm is interested in what type of electronic devices college students use daily. In a survey of 450 college students, 373 said that they used only a cellular telephone, while 39 students used both a cellular telephone and a pager. If 15 students in the survey used neither a cellular telephone nor a pager,

a) Draw a Venn diagram for this information.

b) Find the probability that a student uses a pager if she uses a cellular telephone.

c) Find the probability that a student uses a cellular telephone if she does not use a pager.

26. A marketing firm is interested in what people are drinking with breakfast. In a survey of 300 people, 117 people said that they drank only coffee with breakfast, 43 said that they drank only orange juice with breakfast, and 72 said that they drank neither coffee nor orange juice with breakfast. Assume that the rest of the people in the survey drank both coffee and orange juice.

a) Draw a Venn diagram for this information.

b) Find the probability that a person drinks orange juice for breakfast if he drinks coffee.

c) Find the probability that a person drinks coffee for breakfast if he does not drink orange juice.

A survey of 730 students has the following results.

	Enrolled in		
Working	**Fewer Than 6 Units**	**Between 6 and 13 Units**	**More than 13 Units**
Fewer Than 10 Hours/Week	28	72	85
Between 10 and 20 Hours/Week	29	150	72
More than 20 Hours/Week	150	82	62

Use this table to answer the following questions.

27. What is the probability that if a student is working fewer than ten hours per week, the student is enrolled in more than 13 units?

28. What is the probability that if a student is working more than 20 hours per week, the student is enrolled in more than 13 units?

29. What is the probability that if a student is enrolled in more than 13 units, the student is working more than 20 hours per week?

30. What is the probability that if a student is enrolled in more than 13 units, the student is working fewer than 10 hours per week?

If the probability that a newborn child is a boy is 49% and the probability the child is a girl is 51%, answer the following questions.

31. What is the probability of having a two-child family consisting of two girls if the first child is a girl?

32. What is the probability of having a two-child family consisting of two boys if the first child is a boy?

33. What is the probability of having a two-child family consisting of one boy and one girl if the first child is a girl?

34. What is the probability of having a two-child family consisting of two children of the same sex if the first child is a boy?

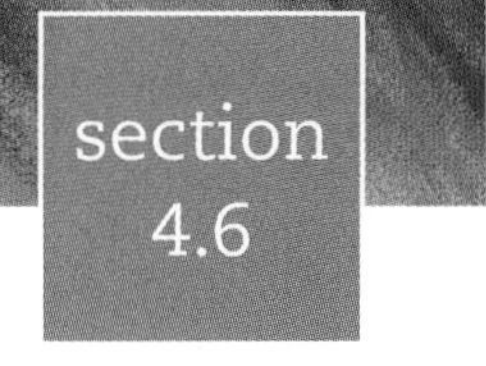

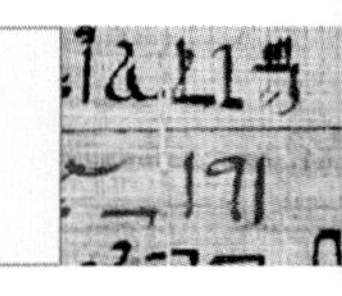

section 4.6 Expected Value

Expected value is the term used to describe the expected winnings from a contest or a game. Expected value is used throughout the world to determine prizes in contests and premiums on insurance policies. It is also used in the mathematical field called decision theory. In this section, we will show how an expected value can be calculated.

Suppose a game is played with one six-sided die. If the die is rolled and lands on 1, 2, or 3, the player wins nothing. If the die lands on 4 or 5, the player wins \$3. If the die lands on 6, the player wins \$12. The following table summarizes this information, including the probability of each situation.

Event	P(event)	Winnings
1, 2 or 3	3/6	\$0
4 or 5	2/6	\$3
6	1/6	\$12

If you play this game, how much can you expect to win? What is the expected value of this game?

Expected Value

To compute the expected value of a game, find the sum of the products of the probability of each event and the amount won or lost if that event occurs.

For this game, the expected value (E.V.) is given by

$$\text{E.V.} = \frac{3}{6} \times 0 + \frac{2}{6} \times 3 + \frac{1}{6} \times 12 = \$3.$$

An expected value of \$3 means that we would expect to win an average of \$3 for each game played. This means that if we played 1000 games, we would expect to win \$3000. However, as is true for probability, this is true only for a large number of games. If we played three games, although the expected winnings are \$9, we could win between \$0 and \$36.

Suppose the operator of the game charges \$1 to play the game. The amount won by the customer would then be reduced by \$1. By using the following table, we can compute the expected winnings for a player, including the cost of the game.

Event	P(event)	Winnings
1, 2, or 3	3/6	−$1
4 or 5	2/6	$2
6	1/6	$11

For this game, the expected value is

$$\text{E.V.} = \frac{3}{6}(-1) + \frac{2}{6}(2) + \frac{1}{6}(11) = \$2.$$

As might be expected, a charge of $1 to play the game reduces the expected value by $1, from $3 down to $2.

EXAMPLE 1

A state lottery has 49 numbers, six of which are the winning numbers for a particular game. The cost of playing the lottery is $1. To play the game, the player must pick six numbers. If a player picks three winning numbers, the payment is $20. Similarly, picking four numbers pays $100, picking five numbers pays $10,000, and picking six winning numbers pays $1,000,000. Find the expected value of this game, including the cost of playing.

Solution: First, we calculate the probability of correctly guessing the winning numbers. Using the methods of the previous sections, this is summarized in the following table along with the winnings at each level. As an example, the probability of selecting three numbers correctly is given by

$$P(3) = \frac{C_{6,3} \times C_{43,3}}{C_{49,6}} = \frac{20 \times 12{,}341}{13{,}983{,}816} = 0.0176504.$$

Similar computations give the following table. Notice that the amounts paid are reduced by $1 to account for the cost of playing the game.

Event (number correct)	P(event)	Winnings
0	$\frac{C_{6,0} \times C_{43,6}}{C_{49,6}} = 0.4359650$	−$1
1	$\frac{C_{6,1} \times C_{43,5}}{C_{49,6}} = 0.4130195$	−$1
2	$\frac{C_{6,2} \times C_{43,4}}{C_{49,6}} = 0.1323780$	−$1
3	$\frac{C_{6,3} \times C_{43,3}}{C_{49,6}} = 0.0176504$	$19
4	$\frac{C_{6,4} \times C_{43,2}}{C_{49,6}} = 0.0009686$	$99
5	$\frac{C_{6,5} \times C_{43,1}}{C_{49,6}} = 0.0000184$	$9999
6	$\frac{C_{6,6} \times C_{43,0}}{C_{49,6}} = 0.0000001$	$999,999

From this table, we can calculate the expected value.

$$\begin{aligned} \text{E.V.} &= (-1) \times 0.4359650 + (-1) \times 0.4130195 + (-1) \\ &\quad \times 0.1323780 + 19 \times 0.0176504 \\ &\quad + 99 \times 0.0009686 + 9999 \times 0.0000184 \\ &\quad + 999{,}999 \times 0.0000001 \\ &\approx -\$0.27 \end{aligned}$$

By doing this calculation, we see that we expect to lose an average of 27¢ every time we play the lottery. This means that the operators of the lottery expect to earn 27¢ every time someone buys a ticket.

EXAMPLE 2

A game is called **fair** if the expected value of the game is zero. Suppose a certain game is fair and costs \$3 to play. The probability of winning is 0.6, and the probability of losing is 0.4. How much should you win for the game to be fair?

Solution: Since the game is fair, we can set up an equation for the expected value of the game with E.V. = 0. We will use W to represent the prize for winning. Since it costs \$3 to play the game, the amount won if we win the game is $W - 3$ and the amount won if we lose the game is -3.

$$\text{E.V.} = \begin{pmatrix}\text{amount won if} \\ \text{you win game}\end{pmatrix} \times P(\text{winning}) + \begin{pmatrix}\text{amount won if} \\ \text{you lose game}\end{pmatrix} \times P(\text{losing})$$

$$\begin{aligned} 0 &= (W - 3) \times 0.6 + (-3) \times 0.4 \\ 0 &= 0.6W - 1.8 - 1.2 \\ 0 &= 0.6W - 3 \\ -0.6W &= -3 \\ W &= 5 \end{aligned}$$

Therefore, since the game is fair, a player should receive \$5 if he wins. Note that this means a real winnings of only \$2, since it costs \$3 to play the game.

EXAMPLE 3

Suppose a certain game is fair and costs \$3 if we lose and has a net payoff of \$10 if we win. The only possible outcomes of the game are winning and losing. What is the probability of winning?

Solution: Let p be the probability of winning. Because there are only two choices, winning and losing, the probability of losing must be $1 - p$. Since the game is fair, we can set up an equation for the expected value of the game with E.V. = 0.

$$\begin{aligned} \text{E.V.} &= (\text{amount won}) \times P(\text{winning}) + (\text{amount lost}) \times P(\text{losing}) \\ 0 &= 10p + (-3)(1 - p) \\ 0 &= 10p - 3 + 3p \\ 0 &= 13p - 3 \\ -13p &= -3 \\ p &= \frac{3}{13} \end{aligned}$$

Therefore, the probability of winning the game is 3/13.

Insurance companies determine the premiums for a policy by examining the risk involved. The risk is calculated by looking at statistics involving the situations covered by the policy. A person in a high-risk category pays more for insurance than someone in a low-risk category does.

EXAMPLE 4

The Lagomorph Insurance Company has a customer, Mr. Roger Abbit, who holds a \$250,000 fire insurance policy on his art collection. The company estimates that there is a 1% chance that the art will be destroyed by fire. If the insurance company tries to maintain an expected value of \$200 on each policy, what should Mr. Abbit's premiums be?

Solution: If a = amount of the premiums, then $a - 250{,}000$ is the amount the insurance company will lose if Roger's art collection is destroyed by fire. Since the company estimates that there is a 1% chance of loss by fire, there is a 99% chance that there will be no loss.

$$\text{E.V.} = \begin{pmatrix}\text{amount company}\\ \text{stands to lose}\end{pmatrix} \times P\begin{pmatrix}\text{losing}\\ \text{the art}\end{pmatrix} + \begin{pmatrix}\text{amount paid}\\ \text{in premiums}\end{pmatrix} \times P\begin{pmatrix}\text{not losing}\\ \text{the art}\end{pmatrix}$$

$$200 = (a - 250{,}000) \times 0.01 + a \times 0.99$$

$$200 = 0.01a - 2500 + 0.99a$$

$$200 = a - 2500$$

$$2700 = a$$

This means that the Lagomorph Insurance Company should charge \$2700 for this policy.

DECISION THEORY

Expected values can also be used to help make decisions that have financial implications. The process of weighing the risks versus the benefits of two or more alternatives is called **decision theory.** In decision theory, the expected value for each possible outcome is computed, and a decision is made by analyzing the results.

EXAMPLE 5

An engineering firm, Sasselli Satellites, is expanding its facilities and needs some electrical work done within a week. The firm has received three bids on the work. The first contractor, Acme Parts, says that it will charge \$10,000. The second company, Yablok Electric, will charge \$11,000 if it finishes within one week or \$9000 if it cannot finish the job within one week. The third company, Zak Communications, has the low bid of \$8500 but wants an extra \$4500 if it can complete the job in less than one week. Sasselli researches the history of all three contractors and finds that they all do very good work. Sasselli also finds that Yablok Electric completes its work as scheduled 85% of the time and that Zak Communications finishes ahead of schedule 20% of the time. Which contractor should Sasselli Satellites choose to do the electrical work if the primary concern is to keep costs low?

Solution: Since all three contractors have a reputation for high-quality workmanship and since cost is the primary consideration. Sasselli Satellites should choose the contractor with the lowest expected cost.

For Acme Parts, the cost will always be \$10,000.

For Yablok Electric, the cost will be \$11,000 with a probability of 0.85 and \$9000 with a probability of 0.15 ($1 - 0.85 = 0.15$). Therefore, the expected costs for Yablok Electric are

$$11{,}000 \times 0.85 + 9000 \times 0.15 = \$10{,}700.$$

For Zak Communications, the cost will be \$8500 with a probability of 0.80 and \$13,000 (\$8500 + \$4500 = \$13,000) with a probability of 0.20. Therefore, the expected costs for Zak Communications are

$$8500 \times 0.80 + 13{,}000 \times 0.20 = \$9400.$$

From this information, Sasselli Satellites should use Zak Communications.

EXAMPLE 6

A bank account is guaranteed to earn a fixed rate of 9% on a \$10,000 deposit over the next year. A speculative investment offers the possibility of 15% earnings on the \$10,000 if the investment succeeds and a loss of 5% of the \$10,000 if the investment fails. Determine the probability of success necessary for the speculative investment to be the better choice.

Solution: To choose the better way to invest the money, we find the expected gain from each investment over the next year. Since the bank account offers a 9% gain with no possibility of loss, the bank account has an expected value of

$$10{,}000 \times 0.09 = \$900.$$

To find the expected gain from the speculative investment, we must know the probability of a success. Since this is unknown, we will assign it a variable, p. The probability of failure is then $1 - p$. The expected gain from the speculative account is then given by

$$(10{,}000 \times 0.15) \times p + [(-10{,}000) \times 0.05] \times (1 - p).$$

Since we want to determine when the speculative account is the better investment, its expected value must be greater than the expected value of the bank account. If we solve the following inequality, we will find the value of p.

$$\begin{aligned}
(10{,}000 \times 0.15) \times p + [(-10{,}000) \times 0.05] \times (1 - p) &> 900 \\
1500p - 500(1 - p) &> 900 \\
1500p - 500 + 500p &> 900 \\
2000p &> 1400 \\
p &> 0.70
\end{aligned}$$

Thus, for the speculative venture to be the more lucrative investment, the probability of success must be greater than 70%.

PROBLEMS ▶ Explain ▶ Apply ▶ Explore

▶ Explain

1. What is expected value and how is it calculated?
2. In terms of expected value, when is a game considered to be fair?
3. What does it mean for the expected value of a \$1 slot machine to be −\$0.15?
4. What does it mean for the expected value of a \$1 lotto game to be \$0.15?
5. What is decision theory?
6. Why is your expected value for games of chance in a gambling casino negative? Explain.
7. Over a long period of time, what could you predict for the owners of a carnival game if the expected value for a player is a positive amount? Explain.

▶ Apply

8. In a certain game, the probability of winning is 0.3, and the probability of losing is 0.7. If a player wins, the player will collect \$50. If the player loses, the player will lose \$5. What is the expected value of this game? If the game is played 100 times, what are the expected winnings (or losses) of the player?
9. In a game of dice, the probability of rolling a 12 is 1/36. The probability of rolling a 9, 10, or 11 is 9/36. The probability of rolling any other number is 26/36. If the player rolls a 12, the player wins \$5. If the player rolls a 9, 10, or 11, the player wins \$1. Otherwise, the player loses \$1. What is the expected value of this game? If the game is played 100 times what are the expected winnings (or losses) of the player?
10. In a game of dice, the probability of rolling a 12 is 1/36. The probability of rolling a 9, 10, or 11 is 9/36. The probability of rolling any other number is 26/36. If the player rolls a 12, the player wins \$8. If the player rolls a 9, 10, or 11, the player wins \$2. If the game is fair, how much should the player lose when the player rolls any number less than 9?
11. In a certain game, the probability of winning is 0.2, and the probability of losing is 0.8. If the player loses, the player will lose \$5. If the game is fair, how much does the player collect when the player wins?
12. In the game of keno, 80 numbers are displayed on a board. Twenty of these numbers are chosen to be the winning numbers. Suppose a person has selected five numbers on a keno card. The probabilities and winnings are as follows.

Event	Probability	Amount Won
0 winning numbers	0.227184	−\$1
1 winning number	0.405686	−\$1
2 winning numbers	0.270457	−\$1
3 winning numbers	0.083935	\$1
4 winning numbers	0.012092	\$10
5 winning numbers	0.000645	\$250

a) What is the expected value of the game?

b) If the game costs $1, how much money should the player expect to win or lose after playing 1000 games?

13. Many charities and other organizations use lotteries or other similar marketing devices to acquire funds. (Publisher's Clearinghouse or *Reader's Digest* might come to mind.) Many states' laws require the odds of winning the various prizes to be posted on the back of the tickets or in some other conspicuous spot.

a) *Fancy That Poultry* magazine is running a contest with the following odds and prizes.

Odds	Prize
1 to 49	$10 (in back issues)
1 to 9999	$50 (in poultry feed)
1 to 99,999	$2000 (in rare ducks)

If tickets are free, find the expected value of a "winning ticket."

b) Is the drawing worth the price of a first-class stamp?

c) At what postage rate would the drawing be considered fair?

14. Dennis is in charge of designing a game for the school fund raiser. Participants will be paying $2 for each game. There will be three prizes. The lowest has a value of $0.50, the second has a value of $1, and the third has an undetermined value. The probability of winning the lowest prize is 0.35, the probability of winning the second prize is 0.15, and the probability of winning the third prize is 0.01. The probability of not winning any prize is 0.49. If the school wants an expected value of $1 per ticket, what should Dennis choose as the value of the third prize?

15. The batting average of a baseball player gives the probability that the player will get a hit in the next at bat. The table gives the batting averages for the San Francisco Giants during the 2001 season and the number of official at bats each player expects in a particular game. Determine the expected number of hits for the team during this game.

Player	Batting Average	At Bats
Rios cf	0.259	5
Kent 2b	0.298	5
Aurilia ss	0.324	4
Bonds lf	0.328	4
Benard rf	0.265	4
Santiago c	0.262	4
Feliz 3b	0.227	4
Snow 1b	0.246	4
Ortiz p	0.194	3

16. The following table gives a partial listing of mortality rates for guinea pigs. It gives the probability of a guinea pig living to a certain age. Assuming that all guinea pigs die by age 5, determine the life expectancy of guinea pigs.

Age	Probability
1	0.14
2	0.07
3	0.26
4	0.29
5	0.24

17. At many bingo parlors, the operators sell pull tabs, which are very similar to slot machines. Each pull-tab card has rows of symbols that are covered by paper tabs. If the paper tabs are removed and three of the same symbols are in a straight line, you win a designated amount. A summary of a typical pull-tab game in which each card costs $0.50 is shown below.

Symbols	Probability	Amount Won
Three diamonds	2/2783	$149.50
Three rubies	2/2783	74.50
Three pearls	4/2783	9.50
Three coins	20/2783	2.50
Three stars	250/2783	0.50
Three moons	400/2783	0
Any other combinations	2105/2783	−0.50

Determine the expected value of this pull-tab game.

Explore

18. In the game of roulette, players bet that a ball will land on a certain number. A player can choose any number from 1 through 36, 0, or 00. It costs $1 to play the game. If the player correctly selects the number, the $1 is returned, and the player receives an additional $35. What is the expected value of this game? Suppose a casino has 100 players, each of whom plays ten times each hour for 24 hours. Each player bets $1. What is the casino's expected profit?

19. In the game of roulette, a player can bet that a ball will land on any one of the numbers in a square of four numbers by placing a bet at the juncture of four numbers as shown below.

If a player bets \$1 on a square of four numbers and one of the numbers comes up, the \$1 is returned, and the player receives an additional \$8. What is the expected value of such a bet?

20. In the California State Lottery game of keno, 20 winning numbers are randomly selected from a total of 80 numbers. Suppose you play the 3-Spot Game, in which you choose three numbers on your keno card and bet \$2. If zero or one of your numbers is correct, you lose your bet. If two of your numbers are correct, you win \$2. If all three of your numbers are correct, how much should you win to make this a fair game? Why does the California State Lottery give you winnings of \$38 when you match three numbers on the 3-Spot Game?

21. In a 5-spot game of Keno at a casino, you select five numbers on your card. The number of winning numbers out of the 20 winners you get and the amount you bet determine the amount you win. The payoffs are shown in the chart that follows.

Number of Winners Picked	\$1 Ticket (\$)	\$2 Ticket (\$)	\$3 Ticket (\$)
5	820	1640	2460
4	10	20	30
3	1	2	3

If you pick fewer than three winning numbers, you lose the amount you bet. The amounts given in the chart are the actual amounts paid to you if you win. Deduct the amount bet to get the actual amount won in each case. Find the expected value for each of the three tickets shown in the chart. From your results, which is the best ticket to play?

22. A television game show contestant has current prizes worth \$12,500. If the contestant participates in the next round of competition, he will have \$50,000 if he wins and \$0 if he loses. The contestant uses expected values to determine that he should participate in the next round. What did the contestant determine his probability of winning the next round to be?

23. An engineer has provided a customer with the choice of two different procedures for extending the lifetime of a certain structure. The first procedure has a success rate of 93% and will extend the life of the building by eight years. The second procedure is still experimental and has not been perfected. It will extend the life of the structure by 15 years if the procedure works. However, there is only a 47% success rate. Assume that the customer can afford to use only one procedure and that if a procedure fails, the building will last two more years. Which procedure should be used to maximize the expected life of the building?

24. In planting a playing field, a park manager must decide between planting seed or sod. If seed is used, there is a 33% chance that the grass lawn will grow with one seeding and a 67% chance that it will need two seedings. If the lawn is seeded once, it will cost \$60. If the lawn needs two seedings, the cost will be \$400. Planting sod will cost \$300 and has a 100% success rate. Which method is more cost effective?

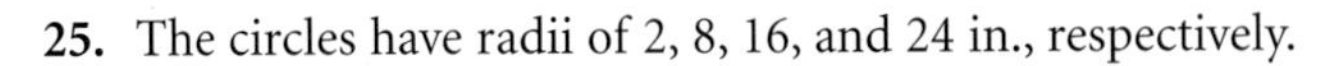

25. The circles have radii of 2, 8, 16, and 24 in., respectively.

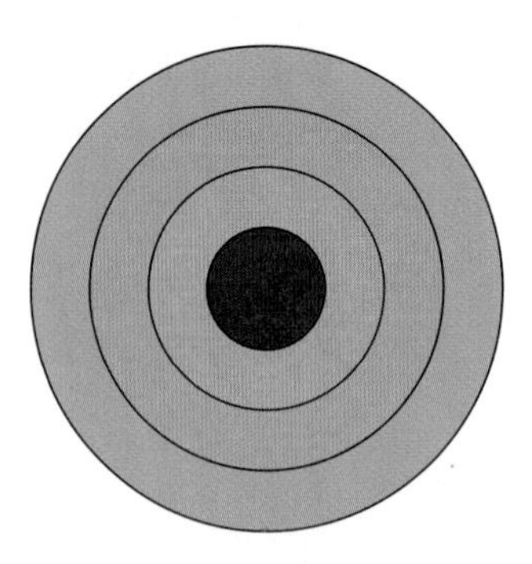

In the game of skeeball, a player rolls a ball up a ramp and wins \$10 if the ball lands in the center circle, wins \$2 if the ball lands in the second band, wins nothing if it lands in the third band, and loses \$1 if it lands in the outer band. If the ball misses the target, the ball is rolled again. The probability that the ball lands in a certain region is the same as the area of that region divided by 576π. The areas of the regions are given in the following table.

Region	Area
Center	4π sq in.
2nd band	60π sq in.
3rd band	192π sq in.
Outside band	320π sq in.

a) Find the expected value of the game.

b) Assuming that everything else is unchanged, what should the prize be for the inner circle so that the game is fair?

26. The squares have sides of 4, 8, 16, and 32 in., respectively.

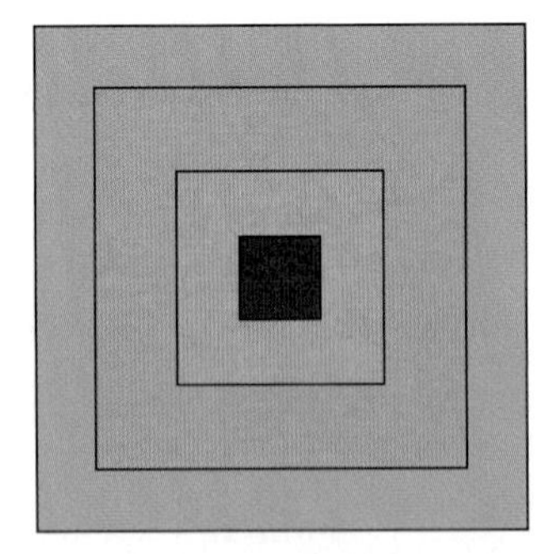

Suppose that in the game of skeeball, a ball is rolled up a ramp and wins \$10 if it lands in the center square, wins \$5 if it lands in the second band, wins \$1.50 if it lands in the third band, and loses \$1 if it lands in the outer band. If the ball misses the target, the ball is rolled again. The probability that the ball lands in a certain region is equal to the area of the region divided by 1024. The areas of the regions are given in the following table.

Region	Area
Center	16 sq in.
2nd band	48 sq in.
3rd band	192 sq in.
Outside band	768 sq in.

a) Find the expected value of the game.

b) Assuming that everything else is unchanged, what should the prize be for the inner square so that the game is fair?

CHAPTER 4 SUMMARY

Key Terms, Concepts, and Formulas

The important terms in this chapter are:	Section
And: The word used to indicate an intersection.	4.4
Complementary events: If an event E occurs, then its complementary event $\overline{E}$ is the nonoccurrence of E. The formula for complementary events is given by $P(\overline{E}) = 1 - P(E)$.	4.1
Compound event: A compound event is an event that consists of more than one single event.	4.4
Conditional probability: The probability of an event occurring if some other event has already occurred.	4.5
Decision theory: The use of mathematical expectation to determine choices.	4.6
Expected value: The average outcome of an event; this could be the expected amount of money won when playing a game or the anticipated cost of a project based on the associated risks involved in the project.	4.6
Experimental probability: The probability of an event that has been approximated by analyzing a number of trials rather than using mathematical calculations.	4.1
Fair game: A situation in which the expected value is zero.	4.6
House odds: The odds against an event occurring.	4.3
Law of Large Numbers: The principle that states that as the number of trials increases, the experimental probability approaches the theoretical probability.	4.1

Odds: Another way of expressing probability in which the probability of an event is $P(E)$ and the probability of its complement is $P(\overline{E})$, the odds that the event will occur are given by $O(E) = \frac{P(E)}{P(\overline{E})}$. 4.3

Or: The word used to indicate a union. 4.4

Probability: A measure of the chance that an event will occur. 4.1

Theoretical probability: The probability of an event that has been predicted through mathematical calculations rather than by an experiment. 4.1

Tree diagram: A listing of all the possible solutions presented in an organized fashion and then written in a structure that resembles a tree. 4.4

After completing this chapter, you should be able to:	**Section**
1. Apply intuitive concepts of probability.	4.1
2. Understand the difference between experimental and theoretical methods of calculating probability.	4.1
3. Apply the concepts of probability, odds, and expected value to games of chance.	4.1–4.6
4. Apply the formula for the probability of the union of two events.	4.4
5. Apply the formula for conditional probability.	4.5

Formula for item 4:

$$P(A \cup B) = P(A) + P(B) - P(A \cap B)$$

Formula for item 5:

$$P(A \mid B) = \frac{P(A \cap B)}{P(B)}$$

CHAPTER 4 REVIEW

Section 4.1

1. A jar contains 12 black balls and seven red balls. What is the probability that the first ball picked from the jar at random will be red?
2. Two standard dice are rolled. What is the probability that the total on the two dice is eight?
3. The weather forecast says that the probability of a storm arriving tonight is 75%. Is this an example of empirical probability or theoretical probability? Explain.
4. The probability that a Brittany spaniel has an arthritis-type disease is 0.024. What is the probability that a Brittany spaniel does not have an arthritis-type disease?

Section 4.2

5. If there are eight horses running in a race, what is the probability of correctly picking the first and second place finishers by just guessing?
6. When dealt a five-card hand from a standard 52-card deck, what is the probability that you will receive two kings and two queens?
7. When dealt a five-card hand from a standard 52-card deck, what is the probability that you will receive two different pairs (for example a pair of threes, a pair of jacks, and one card other than a three or a jack)?
8. When dealt a seven-card hand from a standard 52-card deck, what is the probability that you will receive five cards of the same suit?
9. In a state lottery, there are 50 numbers. Each Saturday, the seven winning numbers are drawn. If you buy a lottery ticket and choose a total of seven numbers, what is the probability that you have a ticket with exactly five winning numbers?
10. In keno, there are 80 total numbers, and 20 numbers are chosen as the winners in each game. If your keno ticket has ten numbers, what is the probability that your ticket has exactly eight winning numbers?

Section 4.3

11. What is the difference between the probability that an event occurs and the odds that an event occurs?
12. What are the odds that a spade is drawn from a deck of cards?
13. Suppose the odds of the San Francisco Forty-Niners winning the Superbowl are given as 1 to 12. What is the probability that they win the Superbowl?
14. What are the odds of rolling a pair of standard dice and having the sum of the dice equal 10?
15. If the probability of getting four aces in a five-card hand is $\frac{1}{54,145}$, what are the house odds of getting four aces in a five-card hand?

Section 4.4

16. A jar contains 12 black balls and seven red balls. Draw a tree diagram showing all the possible ways to pick two balls from the jar (without returning the balls to the jar). Include the probabilities for each branch.
17. A jar contains 12 black balls and seven red balls. Pick two balls without returning the balls to the jar. What is the probability that the first ball picked from the jar at random is red and the second ball is black?
18. A jar contains 12 black balls and seven red balls. Pick two balls without returning the balls to the jar. What is the probability that the first ball picked from the jar at random is red or the second ball is black?
19. In a survey of 290 high school students, a marketing firm finds that 120 have a portable CD player, 117 have an MP3 player, and 88 have neither. What is the probability that one of these 290 students has both a CD player and an MP3 player?

Section 4.5

20. A jar contains 12 black balls and seven red balls. If a black ball is removed from the jar, what is the probability that the next ball picked from the jar is also black?
21. A jar contains 12 black balls and seven red balls. If a black ball is removed from the jar and then returned to the jar, what is the probability that the next ball picked from the jar is also black?
22. If $P(B) = 0.73$ and $P(A \cap B) = 0.21$, find $P(A \mid B)$.
23. The numbers of flowers of two types and two colors being grown in a greenhouse are given in the following table.

	White	Red
Rose	82	122
Carnation	75	89

What is the probability that if a rose is chosen at random, it will be white?

Section 4.6

24. Explain what is meant by expected value.
25. The probabilities and net payoffs for a game are given in the following table. Find the expected value of the game.

payoff	−\$1	\$2	\$10	\$100	\$1000	\$100,000
probability	0.8789498	0.1	0.02	0.001	0.00005	0.0000002

26. Is the game described in Problem 25 fair? If so, explain why. If not, how should the \$100,000 prize be changed to make the game fair?
27. An insurance company has determined that it should charge \$150 to insure your car. The insurance policy covers only the case in which your car is completely destroyed, and in this case, the company will pay you \$4500. Assuming that this is fair, what does the insurance company believe is the probability that your car will be completely destroyed?

CHAPTER 4 TEST

1. Two special 12-sided dice have sides numbered 1 through 12. What is the probability of rolling the dice and having the sum of the two dice be 16?
2. What is wrong with the following argument? Since the probability of a flipped coin landing heads up is 0.5, flipping a fair coin ten times will result in five heads and five tails.
3. If the house odds for a team winning the championship are 99:1, what is the probability that the team will win the championship? What do the house odds tell you about wagering on that team?
4. If the probability of a baby being a boy is 0.51 and the probability of the baby being a girl is 0.49, what is the probability of two babies chosen at random both being boys?
5. A group of 12 women and seven men are candidates for a teaching job. If the finalists were to be chosen at random, what is the probability that both of the finalists would be women?
6. A group of 12 women and seven men are candidates for a teaching job. If the finalists were to be chosen at random, what is the probability that at least one of the finalists would be a woman?
7. If ten greyhounds are running in a race, what are the probability and odds of correctly picking the top three finishers by just guessing?
8. a) Find the probability of drawing three aces and four kings in a seven-card hand from a standard 52-card deck.

   b) Find the probability of drawing three aces and four cards of any rank in a seven-card hand from a standard 52-card deck.

   c) Find the probability of drawing three cards of one rank and four cards of another rank in a seven-card hand from a standard 52-card deck.
9. A certain game is played with two standard six-sided dice. The player will win an amount equal to the sum of the spots on both dice if the sum is greater than 7. The player will lose an amount equal to the sum if the sum is less than or equal to 7. What is the expected value of this game?
10. In the following picture are three equilateral triangles. The largest has sides of 2 ft, the next has sides of 1 ft, and the smallest has sides of 0.5 ft.

A ball will be rolled up a ramp toward the target and will win 5 points for hitting the center region, 2 points for hitting the middle region, and 1 point for hitting the outer region. If the ball completely misses the target, the ball will be thrown again. If the probability of hitting a certain region of the target is equal to the area of the region divided by $\sqrt{3}$ and we are given the following information about the areas of the three regions:

The area of the innermost triangle is $\dfrac{\sqrt{3}}{16}$.

The area of the middle band is $\dfrac{3\sqrt{3}}{16}$.

The area of the outer band is $\dfrac{3\sqrt{3}}{4}$.

   a) Find the expected value of the game.

   b) Find the total number of points a player will expect to accumulate in ten tosses that land in the target.
11. a) In the game of keno as described in this chapter, what are the probability and the odds of selecting four numbers on your playing card and having all four of the numbers be winning numbers?

   b) What are the probability and odds of not selecting all four winning numbers?

   c) If you wager \$1.00 and win \$250.00 if all four of the numbers you chose are winning numbers and lose your \$1.00 wager if you do not select all four winning numbers, what is the expected value of the game.

d) By examining the results of part (c), what can you say about this game? If you played this game for a long period of time, what can you expect?

12. The following is a famous problem known as the birthday problem.

a) In how many ways can 20 days of the year be selected if the same day may be used more than once? Leave your answer in exponential form.

b) In how many ways can 20 days be selected if the same day may be used only once? Leave your answer in terms of a formula.

c) Calculate the value of the result of part (b) divided by the result of part (a).

d) Explain why the answer to part (c) is the probability that no two people from a set of 20 are born on the same day of the year.

e) Explain why the probability that at least two people from a set of 20 are born on the same day is given by 1 minus the result from part (c).

CHAPTER FIVE

Statistics and the Consumer

The feeling of merely being a statistic in a future world can be sensed in Anton Brezezinski's painting for the *Story of Days to Come.* (CORBIS)

OVERVIEW

In this chapter, we will introduce you to some of the basic concepts of statistics. We will show you how data can be organized, displayed, and analyzed. Our focus in this chapter will be on the statistics that can be applied in areas associated with aspects of consumer affairs. You will compare salaries of professional athletes, examine the margin of error in Gallup polls, and predict gasoline prices.

5

A SHORT HISTORY OF STATISTICS

The word *statistik* was first used in 1749 by Gottfried Achenwall. It comes from the Latin word *statisticus,* meaning "of the state." This is appropriate because until the 1850s, statistics almost always referred to information about a country or other political body, such as political, social, or economic conditions. This information was usually presented in tables of numerical data.

John Graunt of England (1620–1674) is known as the father of vital statistics, primarily because of his analysis of England's mortality rates. By studying the English death records, Graunt found patterns in the number of deaths by suicide, disease, and accidents in English cities. He also found that the number of male births exceeded the number of female births. Graunt's work, along with that of William Petty, established the life expectancy tables used by life insurance companies to determine the premiums that the companies should charge their policyholders. By the mid-1700s, all Western nations were compiling information from periodic censuses, recording information such as age of death, the cause of death, and the male-to-female ratio of births. The availability and use of this information allowed life insurance rates to be determined by scientific methods. Adolphe Quetelet (1796–1874) of Belgium constructed the first statistical breakdown of a national census in 1829. He examined the census information to

Carl Friedrich Gauss and the normal (bell) curve are depicted on the 10 Deutschmark note.

determine possible connections between age of death and various other variables such as season of the year, occupation, age, and economic status.

In 1763, the Englishman Thomas Bayes' (1702–1761) posthumous publication *Essay Towards Solving a Problem in the Doctrine of Chances* became one of the first works to extrapolate from a small sample of information to what could be expected from the population as a whole. This was the forerunner of the activities of George Gallup and the Gallup poll.

Today, magazines and newspapers have statistical presentations every day about a myriad of facts. Stock averages, the latest cancer research, and the status of the national debt are some ever-present examples. The availability of computers allows us to accumulate and arrange numerical data in many ways. It is our hope that an understanding of basic statistics will help you interpret data and its presentation and to determine the validity of the position the data is being used to support.

CHAPTER 5 PROJECTS

Research Projects

1. Find some examples of statistics in a recent magazine or newspaper. Describe the situation and how statistics are being used.
2. What is the Dow Jones average? What does it represent? How is it calculated? What is its history?
3. What is demography? What are its uses? Cite some examples.
4. What is the meaning of the expression "lie with statistics"? Give some examples of this and explain how each of the examples is a "lie."

Math Projects

1. Pick a company listed in the stock market (e.g., General Motors, Apple Computer). Find the price of the stock at the end of the year for each year from 1990 to 1999. Plot this information on a graph. Compute the regression line for this information and draw the

line on the graph. Use the regression line to predict the price of the stock at the end of June 2001 and the end of December 2001. Compare the actual price on these days with the prediction of the regression line.

2. Find the world record time for running the 100-meter dash in the years 1950, 1955, . . . , 2000. Plot this information on a graph. Compute the regression line for this information and draw the line on the graph. Use the regression line to predict the world record time for the 100-meter dash in the year 2050. Does the regression line seem appropriate for the data? Explain.

3. You have been hired as a consultant by a strawberry farmer. After being awarded the contract, you receive the following letter.

```
To:       Recently Hired Statistician

From:     Morgan Hill Berry Growers

Please advise us on which company to use as our strawberry
distributor. Four highly recommended distributors have
provided us with statistical data on the weekly prices for
one load of strawberries per week for a ten-week period
last year. Prices fluctuate according to availability, and
We would like to use the company with the lowest overall
price and the least amount of fluctuation. We would like
your written report showing your results and a detailed
recommendation as to which company we should choose.

Thank you.
```

Week	FreshPicked, Inc.	FastNFresh	AlwaysRipe Fruit Company	BerryDelicious Deliveries
1	$355	$350	$350	$360
2	$350	$345	$350	$300
3	$310	$295	$320	$320
4	$330	$325	$320	$320
5	$340	$315	$330	$290
6	$290	$290	$300	$305
7	$305	$305	$310	$290
8	$315	$300	$315	$310
9	$325	$315	$345	$340
10	$350	$340	$290	$345

What is your recommendation to Morgan Hill Berry Growers? You will need to use the material in Sections 5.2 and 5.3 to determine an answer.

section 5.1 Arranging Information

Statistics are numerical data assembled in such a way as to present significant information about a subject. You come into contact with statistics on a daily basis. Credit bureaus use your payment history to determine your credit rating. Schools use standardized tests to determine student placement. Executives accumulate and organize data for a presentation. Government officials present findings regarding pollution, demographics, and economics. Magazines and newspapers contain articles with charts and data in almost every issue. In this section, we discuss methods of graphically presenting data.

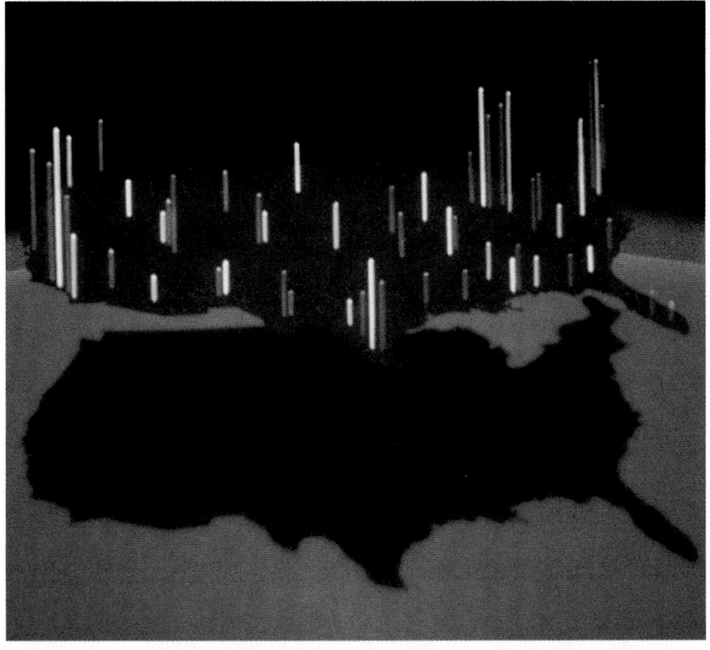

This population-density map of the United States gives a unique method of arranging information. (Chart of the U.S. courtesy of CORBIS)

THE DATA

When a study or a poll is done, much of the information is in numerical form. Merely listed, the results of a survey can be overwhelming. Suppose 100 people respond to a survey with ten questions, each with numerical answers. There will be 10 numbers from each survey, giving a total of 1000 numbers. Most people will not be able to make any pertinent observations when the data are in this form. The following will show how to overcome such a difficulty.

Displayed here are the test scores of 70 students who have taken a math placement test. The test has a possible low score of 0 and a possible high score of 25.

12	23	25	5	9	5	24
14	14	15	21	2	18	13
22	16	17	15	19	23	14
11	8	7	16	11	10	19
24	16	11	22	20	14	17
11	13	18	9	6	15	4
6	23	20	13	9	7	15
16	14	21	10	20	3	16
11	22	7	10	11	18	14
15	12	19	25	23	2	21

As we look at the data, there does not appear to be any pattern, nor can we come to any conclusions about the test scores. To examine the data more carefully, one technique is to put the data into categories.

When data are divided into categories, the different categories are called **classes.** The number and size of the classes we choose are determined by the goals of the investigation. Often, studies have predetermined classes. For example, many college courses use the standard groupings 90–100 for an A, 80–89 for a B, 70–79 for a C, 60–69 for a D, and 0–59 for an F. Economic studies concerned with household income have classes such as "below the poverty level (less than $18,400 income per year)," and "lower middle income ($31,000)."

A second criterion for choosing the classes is common sense. It would not be very wise to pick only one class because all the data will fall into that class. On the other hand, creating 30 classes for our data would mean that some of the classes would be empty and most of the classes would contain only a few scores. Finally, if we want the number of items in a class to be meaningful, the size of each class should be the same.

For our test data, we do not have any predetermined classes. Since the data vary from 0 through 25, it is convenient to pick five classes. This number was chosen because it allows us to create nearly equal size classes. The classes are 0 through 5, 6 through 10, 11 through 15, 16 through 20, and 21 through 25.

Now that we have chosen the classes, the next step is to determine the **frequency,** or the number of data values in each class. To do this, we have created the following chart, called a frequency distribution. The column labeled Tally is used to count the number of scores that fall into a class. The column labeled Frequency contains the number of tallies for that class. The entries in the column labeled Relative Frequency are found by dividing the frequency for each class by the total number of items, in this case 70. The final column is called Percentage Frequency. Its entries are found by multiplying the relative frequency by 100. They are the percentages of the data that fall into each category.

Frequency Distribution

Class	Tally	Frequency	Relative Frequency	Percentage Frequency
0–5	~~IIII~~ I	6	0.09	9%
6–10	~~IIII~~ ~~IIII~~ II	12	0.17	17%
11–15	~~IIII~~ ~~IIII~~ ~~IIII~~ ~~IIII~~ II	22	0.31	31%
16–20	~~IIII~~ ~~IIII~~ ~~IIII~~ I	16	0.23	23%
21–25	~~IIII~~ ~~IIII~~ IIII	14	0.20	20%

With the information presented in this way, we can see that of the 70 test scores, 22 were in the class 11–15. This group comprised 31% of those who took the test.

Having the data organized in this fashion is a great improvement, but it is still a bit intimidating. Many people do not like to look at tables of numbers. Information can often be presented with greater impact if it is in graphical form. For our discussion, we will examine three of the many types of graphical representations. The first of these is called a **bar graph.** A bar graph uses one axis to represent the different classes, while the other axis is used to indicate the frequency for each class.

Bar graphs may be oriented horizontally or vertically and may use either the frequencies or the relative frequencies as the lengths of the bars. The choice of display is left to the person creating the graphs. With a bar graph, it is easy to see which class contains the largest number of test scores. The numbers at the ends of the bar permit the reader to determine the actual number of scores in each class.

Three different bar graphs of our data are shown.

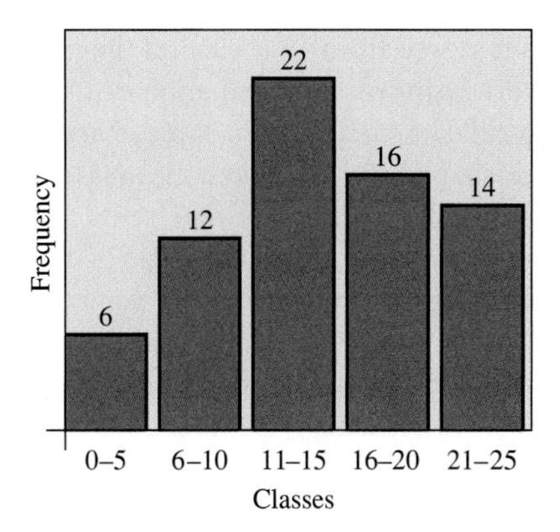

Vertical bar graph with frequencies

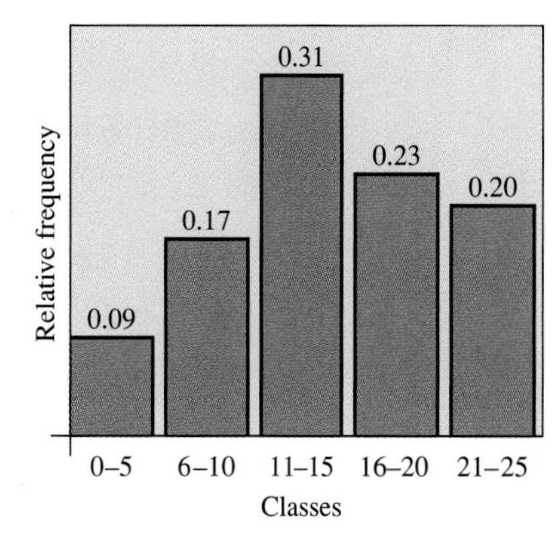

Vertical bar graph with relative frequencies

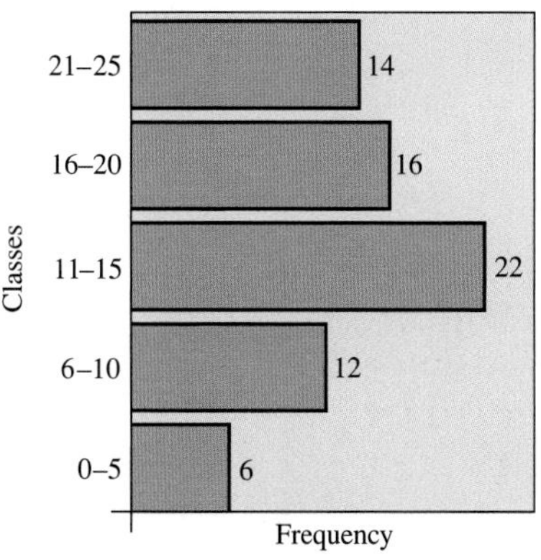

Horizontal bar graph with frequencies

A second way in which information can be displayed is with a **line graph,** also called a frequency polygon or broken line graph. In a line graph, the bars of a bar graph are replaced with dots that are connected by line segments. A line graph that represents the frequency of scores on the math placement test is shown in the following figure.

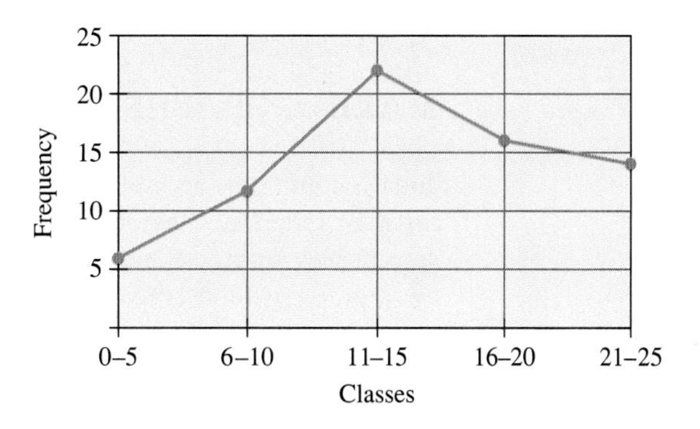

A line graph allows you to quickly see the increases and decreases in the number of scores in each class. For example, you can see that there is a large jump between the number that scored from 6 to 10 and the number that scored from 11 to 15. On the other hand, you can see that there is only a slight decrease from the number that scored from 16 to 20 and the number that scored from 21 to 25.

A third way in which information can be presented is through the use of a **pie chart,** also called a circle graph. A pie chart is a circular diagram divided into sectors (wedge-shaped pieces), where the areas of the sectors are used to represent percentage frequencies. Each sector represents a part or percentage of a whole. The percentage frequency determines the angle of the sector, the entire circle representing 100%. The percentage frequencies of the math placement test can be displayed in the pie chart that follows. From the pie chart, you can visually determine the relative size of each class. You can quickly see which class had the largest number of scores and which class had the smallest number of scores.

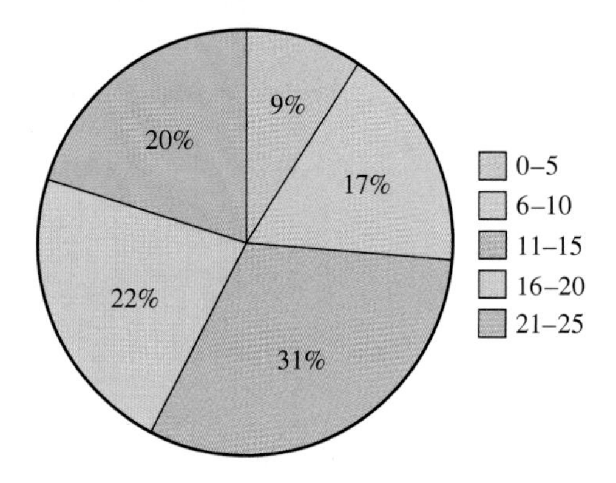

Which Type of Graph Should Be Used?

Since there are three basic types of statistical graphs, which one should be used for a given set of data? Depending on what you are trying to emphasize, the data can be displayed with any of these three statistical graphs. If you want to clearly show which category contains the largest or the least number of items, a bar graph is effective. If you want to emphasize the patterns of change (rise and fall) of various categories, a line graph is useful. If you want to show how a whole is divided into parts, then a pie chart is the best one to use.

MAKING GRAPHS

Today's computers accomplish the drawing of these graphs by using chart-drawing software. Often these abilities are incorporated into large software packages, such as spreadsheet programs and word processors. However, it might be necessary to create a graph without the aid of a computer. The remaining part of this section describes how to draw these basic statistical graphs.

Suppose we want to look at a breakdown of how people are employed in the United States according to occupation and sex. According to statistics published by the U.S. Department of Labor, the employment classification for each gender is given in the following table.

Occupation	% Males	% Females
Executive, administrative, management	15.0%	14.2%
Professional specialty	13.6%	18.0%
Technical, sales, administrative support	19.7%	40.0%
Service occupations	9.9%	17.4%
Precision production, crafts	18.6%	2.1%
Operators, fabricators, laborers	19.3%	7.1%
Farming, forestry, fishing	3.8%	1.1%

The data can be represented by either a bar graph or pie chart.

Bar Graphs

To make a bar graph, determine the following:

1. The categories to be placed along the vertical and horizontal axes
2. The scale used to encompass the numerical data
3. The style of bars used and the length of each bar

One possible way to draw the bar graph comparing occupation data is to have the vertical axis measure the percent and the horizontal axis contain the occupation. Since the data consists of percentages that range from 1% to 40%, by letting the vertical scale be 1/2 in. for every 10%, we could fit the graph in a reasonable area. Since each 10% will be represented by 0.5 in., to find the length of each bar, we could set up a proportion such as

$$\frac{0.5 \text{ in.}}{10\%} = \frac{L \text{ in.}}{P\%}.$$

For 15% we get:

$$\frac{0.5 \text{ in.}}{10\%} = \frac{L \text{ in.}}{15\%}$$

$$L = \frac{0.5}{10\%} \times 15\% = 0.75 \text{ in.}$$

Finally, since we are comparing the occupations of males versus females, we can use blue bars for males and pink bars for females. Using vertical bars that fit the data, we get the following statistical graph.

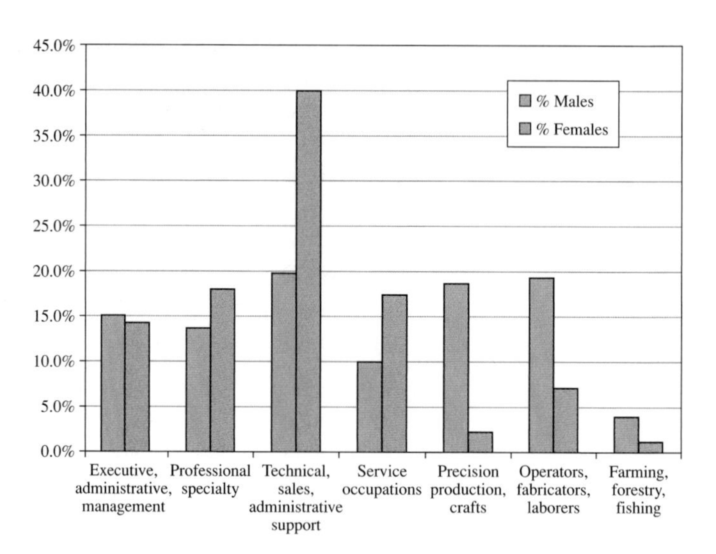

The bar graph allows us to visually analyze the data. By simply looking at the bar graph, we can make conclusions about percentage of a gender in a particular field. For example, 40% of women are employed as technical, sales, and administrative support staff, compared to only 20% of men.

Pie Charts

To make a pie chart, determine the size of each sector that will be used to represent the percentage frequency. Since there are 360 degrees around the center of a circle, the angle for each sector is found by multiplying the percentage by 360°. For example, to find the angle used to represent the 15% of men employed in management, multiply $0.15 \times 360° = 54°$.

Using a protractor to measure the desired angles, each circle is divided into seven sectors. Using a different color for each occupation and including labels, we get the following pie charts for the occupations of each gender.

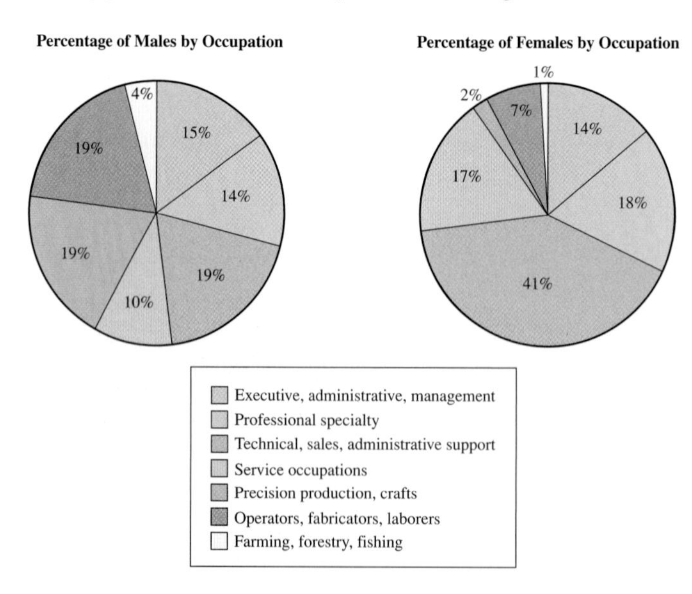

As we did with the bar graphs, we can draw conclusions about the occupations of each sex. For example, we can see that a large percentage of women are employed in technical, sales, and administrative support positions.

Line Graphs

Line graphs use data consisting of two parts, such as the x and y coordinates used in algebra. The points are plotted on a graph and then connected with line segments. For example, consider the following table of years and average prices of unleaded gasoline.

Year	Average Price
1980	$1.25
1982	$1.30
1984	$1.21
1986	$0.93
1988	$0.95
1990	$1.16
1992	$1.13
1994	$1.11
1996	$1.23
1998	$1.06
2000	$1.47

Plotting these points gives the following line graph.

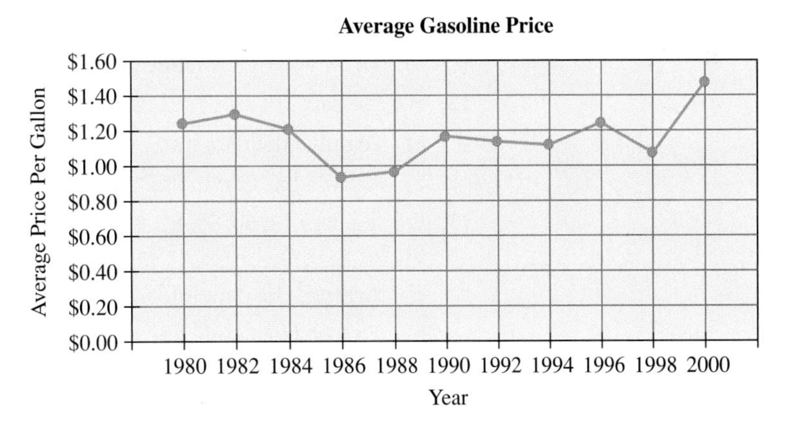

A quick look through magazines and newspapers will convince you that graphs are a common way of displaying information. While software packages often have many different types of graphs, the three types of graphs presented here give you a good basis for understanding all of them.

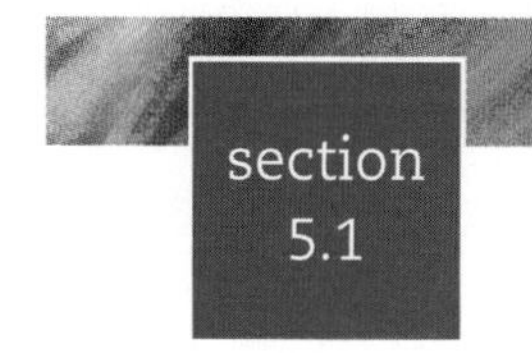

PROBLEMS ○ Explain ○ Apply ○ Explore

○ Explain

1. What are statistics?
2. What is a bar graph?
3. What is the difference between the frequency, relative frequency, and percentage frequency distribution?
4. What is a line graph?
5. What is a pie chart?
6. What are the three major steps in making a bar graph?
7. How is the angle of each sector of a pie chart determined?
8. When would it be effective to display data using a bar graph? line graph? pie chart?

○ Apply

9. For the data 2, 4, 6, 1, 7, 9, 5, 3, 7, 6:
   a) Arrange the data into a frequency distribution with three classes.
   b) Draw a bar graph, line graph, and pie chart.
10. For the data 2, 4, 3, 7, 2, 8, 6, 3, 7, 5:
    a) Arrange the data into a frequency distribution with four classes.
    b) Draw a bar graph, line graph, and pie chart.
11. For the data 10, 12, 14, 13, 17, 12, 18, 16, 13, 17, 15, 17:
    a) Arrange the data into a frequency distribution with five classes.
    b) Draw a bar graph, line graph, and pie chart.
12. For the data 20, 29, 21, 21, 28, 28, 23, 23, 23, 26, 26, 26, 25, 25, 25, 25, 25:
    a) Arrange the data into a frequency distribution with five classes.
    b) Draw a bar graph, line graph, and pie chart.
13. For the data 42, 64, 75, 82, 96, 93, 77, 82, 67, 78, 88, 90, 80, 72, 71, 80, 81, 98, 61, 75:
    a) Arrange the data into a frequency distribution with six classes.
    b) Draw a bar graph, line graph, and pie chart.
14. The bar graph shown gives the number of seeds of a particular type of cactus that germinated under greenhouse conditions within a specified number of weeks after planting. At the end of six weeks, the experiment was discontinued.
    a) Determine the total number of germinated seeds.
    b) What percentage of the germinated seeds sprouted during each week?
    c) Make use of the results of part (b) to create a line graph and pie chart displaying the percentage of seeds sprouting in each of the six weeks.

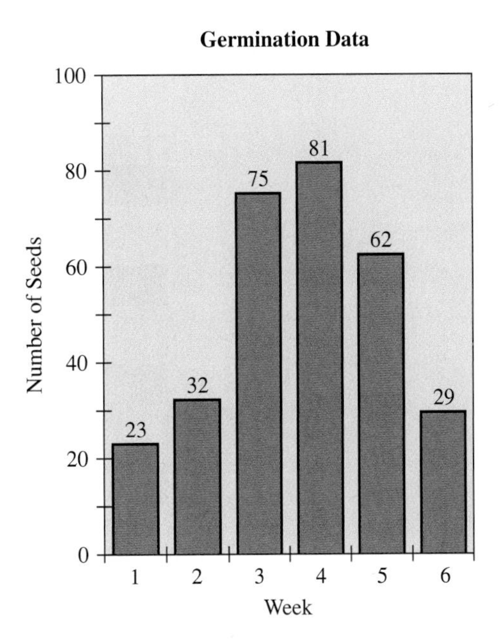

15. The bar graph (from statistics compiled by the U.S. Department of Labor) shows the U.S. Consumer Price Index (CPI). The CPI represents the relative costs of goods. Using 1967 as the base year, something that cost \$100 in 1967 cost \$30 in 1915 and \$512 in 2000.

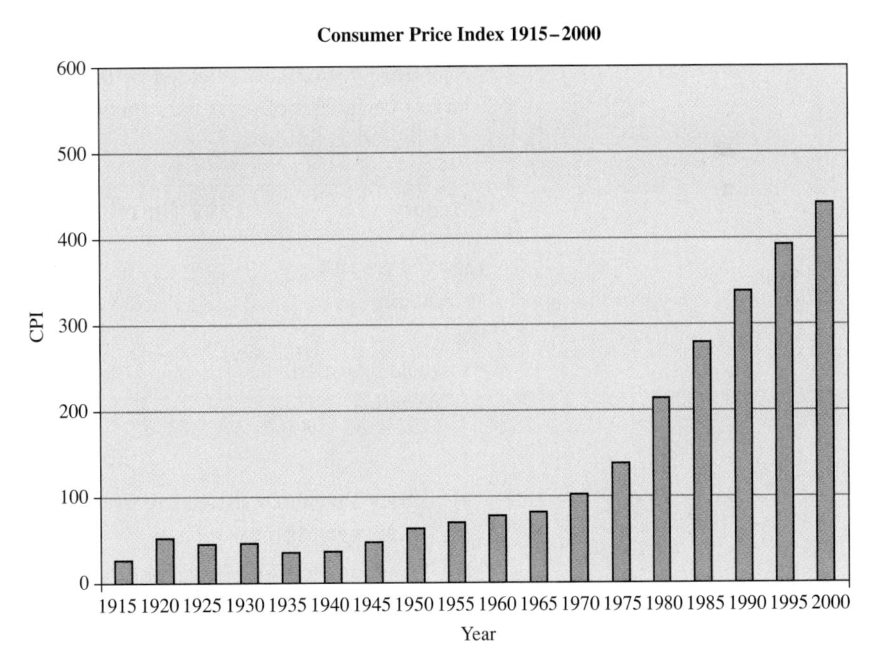

a) If a washing machine cost \$350 in 1967, what would the machine cost in 2000?

b) If a teacher earns \$40,000 per year in 2000, what would her pay have been in 1915?

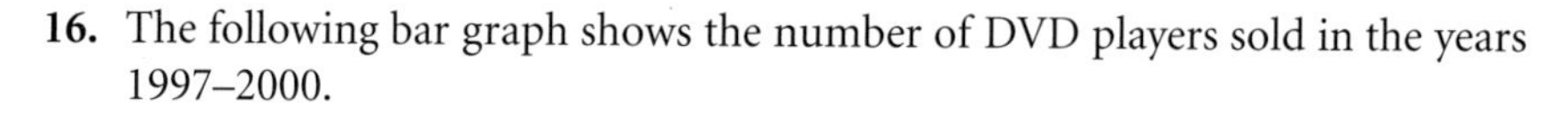

16. The following bar graph shows the number of DVD players sold in the years 1997–2000.

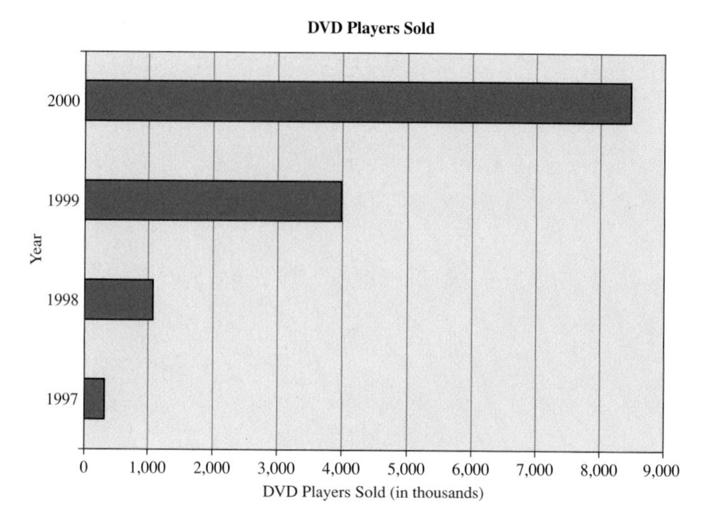

a) How many more DVD players were sold in 1999 than in 1998?

b) Find the percentage of the total DVD players sold in each of the years 1997–2000 and create a pie chart from the information. Use the sum of the DVD players sold in the four years for the total DVD players sold.

Explore

17. According to statistics released by the U.S. Department of Commerce, the five largest categories of personal expenditures in the United States for 1998 were distributed as follows.

Category	1992 (in billions of $)	1998 (in billions of $)
Food and tobacco	709.5	907.4
Medical care	733.2	1,032.3
Housing	646.8	855.9
Household operation	470.6	646.5
Transportation	471.5	647.4

a) Draw a pie chart for each year. Use the sum of the five categories as the total expenditures.

b) Draw one bar graph that shows the expenditures in each category for both years.

c) Compare the relative costs of medical care in the two years.

18. According to statistics released by the U.S. Department of Commerce, the five largest categories of personal expenditures in the United States for 1998 were distributed as follows.

Year	Medical Care Costs (in billions of $)	Food and Tobacco (in billions of $)
1992	733.2	709.5
1993	785.5	733.4
1994	826.1	761.7
1995	871.6	783.8
1996	912.4	805.2
1997	957.3	832.3
1998	1032.3	907.4

a) Draw one line graph showing both medical care and food/tobacco costs.

b) Draw one bar graph showing both medical care and food/tobacco costs.

c) Compare the relative increases in health care costs to that in food and tobacco.

d) In which year did food and tobacco have the greatest increase?

19. The fees collected for Vehicle Licenses and Registration in the State of California for the budget year 1993–94 were distributed as follows. (Source: California Department of Motor Vehicles Document, DMV 77 B REV 8/93.)

Distribution of License and Registration Fees (1993–94)	
City and County, Local Governments	$3044 million
California Highway Patrol	$ 661 million
Dept. of Motor Vehicles	$ 513 million
State Highways	$ 381 million
State's General Fund	$ 450 million
Various State Agencies*	$ 121 million
Total	$5170 million

*Includes Dept. of Justice, Air Resources Board, and Environmental Agencies.

a) Which type of graph would be effective in displaying the data? Explain.

b) Draw the graph.

20. Another graphical representation for data displays cumulative frequencies. The cumulative frequency for each class is determined by adding the frequency of the class to the combined frequency of the previous classes. In such a graph, the cumulative frequency for the last class must be the same as the total number of values in the survey. These graphs are called ogives (pronounced *oh'-jives*). Ogives can be used to compare data that use time as the variable for the classes. For example, two marathons, run in two different cities, have the finishing times presented in the two ogives shown.

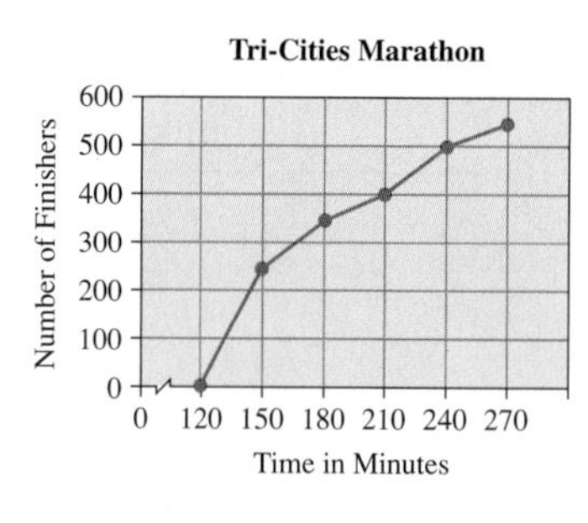

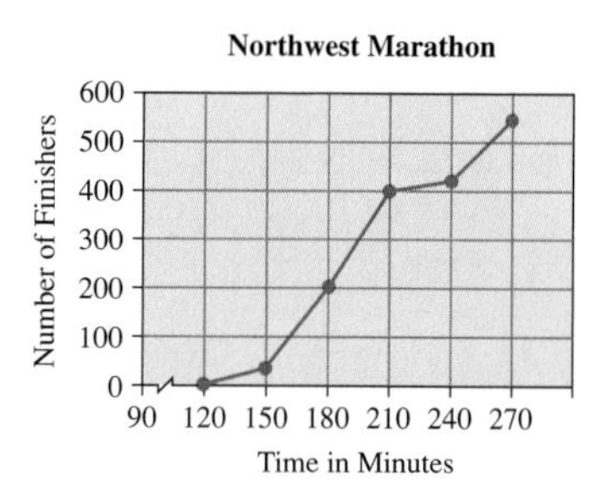

a) How many runners finished in 4 hours or less in each marathon?

b) How many runners finished in 3 hours or less in each marathon?

c) What percentage of the runners finished in 3 hours or less in each marathon?

d) If we assume that the abilities of the runners in each marathon were similar, which marathon was the easier race? Explain your reasoning.

section 5.2

Measures of Central Tendency

Finding an average is a computation that requires only arithmetic skills. It has, however, very important uses in real-world problems. An average allows us to find one value that represents the middle of a set of data. As we shall see, however, the middle of a set of data can be described in different ways. Collectively, these ways of describing the middle value are called measures of central tendency. In this section, we introduce three measures of central tendency and describe situations in which one of the three measures is the most appropriate to use.

Census forms are an efficient means of collecting the information that is used in the statistical analysis of many facets of American life.

THE MEAN

The **mean** or **arithmetic** (pronounced *ar-ith-met'-ic*) **mean** of a set of values is what most people are referring to when they say "average." It is found by adding up all the data and then dividing by the number of data values. Statisticians use the symbol μ (read *mu*) for the mean and define it with the following formula.

$$\text{mean} = \mu = \frac{\sum_{i=1}^{n} x_i}{n} \quad \text{where} \quad \begin{cases} n \text{ is the number of data values} \\ x_i \text{ represent the data values} \end{cases}$$

The symbol $\sum_{i=1}^{n} x_i$ is the mathematical notation for adding a sequence of numbers. The notation x_i means the ith number. Therefore, if you have three numbers, 10, 20, and 30, $x_1 = 10$, $x_2 = 20$, and $x_3 = 30$. This gives

$$\sum_{i=1}^{3} x_i = x_1 + x_2 + x_3 = 10 + 20 + 30.$$

EXAMPLE 1

Find the mean of the numbers 3, 6, 8, 5, 4.

Solution:

$$\mu = \frac{\sum_{i=1}^{5} x_i}{5} = \frac{3 + 6 + 8 + 5 + 4}{5} = \frac{26}{5} = 5.2$$

EXAMPLE 2

Find the mean of 1, 1, 1, 1, 3, 3, 5, 5, 6, 6, 6, 6, 6.

Solution: Rather than merely adding all these numbers, it is more convenient to multiply each value by its frequency and then add. This total is then divided by the total frequency. In other words, rather than doing the computation

$$\mu = \frac{1 + 1 + 1 + 1 + 3 + 3 + 5 + 5 + 6 + 6 + 6 + 6 + 6}{13}$$

$$= \frac{50}{13} = 3.85$$

it is easier to write

$$\mu = \frac{4(1) + 2(3) + 2(5) + 5(6)}{13} = \frac{50}{13} = 3.85.$$

In general, when a set of data has values that appear several times or when the data are grouped into classes, we can use the following formula.

$$\text{mean for grouped data} = \mu = \frac{\sum_{i=1}^{n} f_i x_i}{\sum_{i=1}^{n} f_i} \quad \text{where} \begin{cases} n \text{ is the number of classes} \\ f_i \text{ is the frequency of each class} \\ x_i \text{ represent the data values} \end{cases}$$

EXAMPLE 3

Find the mean of the following frequency distribution.

Class	Frequency	Class Midpoint x_i
1–5	6	3
6–10	12	8
11–15	22	13
16–20	16	18
21–25	14	23

Solution: In the table, there is a column that we have not previously used. The **class midpoint** is the mean of the boundary values of each class. For example, the boundary values for the first class are 1 and 5. The mean of these two numbers is 3. This gives the midpoint for the class. We use the midpoints because we do not know the actual values of the data in each class. For example, we know that there are six values in the 1–5 class, but we do not know what those values are. As an estimate of these values, we will use the class midpoints. Now, using the formula for the mean of grouped data, we get the following.

$$\mu = \frac{\sum_{i=1}^{5} f_i x_i}{\sum_{i=1}^{5} f_i}$$

$$= \frac{6(3) + 12(8) + 22(13) + 16(18) + 14(23)}{6 + 12 + 22 + 16 + 14}$$

$$= \frac{1010}{70}$$

$$= 14.43$$

Another use of the formula for the mean of grouped data is to compute a **weighted average.** A weighted average is the mean of a group of numbers in which certain values have more importance, or weight, than do other values. When computing a weighted average, we use the weights of each value as frequencies in the mean of grouped data formula. An example of this type of situation can be seen in the next problem.

EXAMPLE 4

A student is trying to calculate his grade in an English class. His total scores are as follows.

Midterms	82
Homework	87
Final	92

The course syllabus says that the midterms count for 60% of the grade, homework for 10% of the grade, and the final for 30% of the grade. Compute the student's total score in the class.

Solution: Using the percentages as the weights of the scores in each category, we have

$$\mu = \frac{0.60(82) + 0.10(87) + 0.30(92)}{0.60 + 0.10 + 0.30} = \frac{85.5}{1.00} = 85.5.$$

The remainder of this section will assume that you have a calculator that has a function to compute the mean of a set of data. For help with your calculator, read your calculator manual or ask your instructor for help.

The Median

EXAMPLE 5

Five houses are listed for sale in a real estate broker's advertisement. The prices are \$269,000, \$256,000, \$249,000, \$235,000, and \$749,000. Find the average price of the houses listed by the agent.

Solution:

$$\mu = \frac{\sum_{i=1}^{5} x_i}{5} = \frac{269{,}000 + 256{,}000 + 249{,}000 + 235{,}000 + 749{,}000}{5}$$

$$= \frac{1{,}758{,}000}{5}$$

$$\mu = \$351{,}600$$

Notice that the average is higher than all but one of the house prices. The reason is that the price of the most expensive house is distorting the results. The mean price of these five houses is \$351,600, but the mean does not accurately reflect the typical or "average" price.

To solve this difficulty, we will use a measure of central tendency called the **median.** The median is found by listing the data in increasing order and choosing the middle value. If there is an even number of items, the median is found by taking the mean of the two middle values. The median is frequently used when there are a few extreme values in the data that will greatly alter the value of the mean.

EXAMPLE 6

Find the median of the prices listed in Example 5.

Solution: Listing the data in increasing order gives

\$235,000, \$249,000, \$256,000, \$269,000, \$749,000.

The median price is \$256,000 because this is the middle value. This value is much closer to what we would call the "average" price of these five houses. Also notice that increasing the price of the most expensive house will have no effect on the value of the median.

EXAMPLE 7

Find the median of the values 0, 3, 4, 9, 16, 90.

Solution: The values are listed in order, so we need only find the middle value. Since there is an even number of values, the median is found by calculating the mean of the center pair of values.

$$\text{median} = \frac{4 + 9}{2} = 6.5$$

The Mode

The **mode** of a set of data is the value that occurs most frequently. It is the only measure of central tendency that can be used with nonnumerical as well as numerical data. For example, if a design consultant wanted to determine the most popular color for exterior house paint, he or she would conduct a survey to find out how many people liked each color. It would not be possible to use the mean or median.

EXAMPLE 8

Determine the mode of the following sets of values.

a) 1, 2, 2, 3, 4, 6, 6, 6, 8, 9

b) 1, 2, 2, 3, 4, 6, 6, 7, 8, 9

c) 1, 2, 2, 3, 4, 4, 6, 6, 8, 9

Solution:

a) In the first group of data, the number 6 appears three times, and all the other numbers appear at most twice. Therefore, the mode is 6.

b) In the second group, the numbers 2 and 6 both appear twice. No other number appears more than once. In a situation like this, there are two modes: 2 and 6. When a set of data has two modes, it is called bimodal.

c) In the third set of data, the numbers 2, 4, and 6 all appear twice. When more than two values have the highest frequency, we say that the set of data does not have a mode.

EXAMPLE 9

In 1990, there were 66,090,000 families in the United States. Of these families, 33,801,000 had no children, 13,530,000 families had one child, 12,263,000 had two children, 4,650,000 families had three children, and 1,846,000 families had four or more children.

a) Determine the modal number of children per family.

b) Determine the mean number of children per family.

Solution:

a) Since the class with the highest frequency is the class with no children, the modal number of children is 0.

b) Finding an accurate mean of the data will not be possible. Since the final category consists of families with four or more children, we do not know the actual number of children in these families. However, if we use 4 as the number of children for the last group, we can use the method of finding the mean for grouped data to give an estimate.

$$\mu = [33{,}801{,}000(0) + 13{,}530{,}000(1) + 12{,}263{,}000(2) + 4{,}650{,}000(3) + 1{,}846{,}000(4)] \div 66{,}090{,}000$$

$$= \frac{59{,}390{,}000}{66{,}090{,}000} \approx 0.90$$

Since the number of children in the last class may be greater than four per family, we can say that the mean number of children per family is greater than or equal to 0.90.

This final example might be of interest in a typical classroom.

EXAMPLE 10

An intermediate algebra class had test scores that are shown in the following frequency distribution. Find the mean score of the frequency distribution.

Class	Frequency
0–9	0
10–19	0
20–29	1
30–39	0
40–49	1
50–59	1
60–69	6
70–79	8
80–89	6
90–99	3

Solution: To find the mean of the frequency distribution, first find the midpoint of each class. Since some of the classes have a frequency of zero, those class are omitted.

Class	Midpoint	Frequency
20–29	24.5	1
40–49	44.5	1
50–59	54.5	1
60–69	64.5	6
70–79	74.5	8
80–89	84.5	6
90–99	94.5	3

Using the midpoints and frequencies for each class and the total number of students taking the test as 26, we find that the mean is given by

$$\mu = \frac{24.5(1) + 44.5(1) + 54.5(1) + 64.5(6) + 74.5(8) + 84.5(6) + 94.5(3)}{26} = 73.0.$$

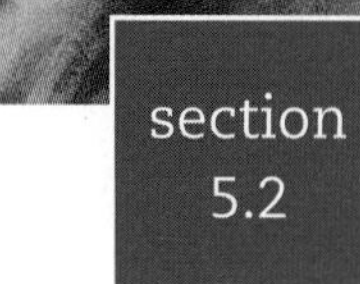
section 5.2

PROBLEMS ○ Explain ○ Apply ○ Explore

Explain

1. What are measures of central tendency?
2. What is the mean for a set of data and how is the mean determined?
3. What is the weighted average? When is it used?
4. What is the median for a set of data?
5. How is the median determined for an odd number of data? an even number of data?
6. What conditions in the data would make the median an effective measure of central tendency?
7. What is the mode for a set of data?
8. How is the mode determined?
9. For what kind of data would the mode be an effective measure of central tendency?
10. What does it mean for a set of data to be bimodal?

Apply

11. For the set of values 2, 4, 7, 2, 1, 8, 9, 10, 9, 6:
    - a) Find the mean.
    - b) Find the median.
    - c) Find the mode.
12. For the set of values 3, 8, 4, 2, 4, 6, 7, 1, 5, 0:
    - a) Find the mean.
    - b) Find the median.
    - c) Find the mode.
13. For the data 20, 29, 21, 21, 28, 28, 23, 23, 23, 26, 26, 26, 25, 25, 25, 25, 25:
    - a) Find the mean.
    - b) Find the median.
    - c) Find the mode.
14. For the data 42, 64, 75, 82, 96, 93, 77, 82, 67, 78, 88, 90, 80, 72, 71, 80, 81, 98, 61, 75:
    - a) Find the mean.
    - b) Find the median.
    - c) Find the mode.
15. For health reasons, many people watch their calorie and sodium intake. The following table lists a selection of items from two fast-food chains and their calorie and sodium contents.

	Calories (cal)	Sodium (mg)
McDonald's Hamburger	280	590
McDonald's Cheeseburger	330	830
McDonald's Fish Sandwich	470	890
McDonald's Grilled Chicken Sandwich	340	890
Wendy's Plain Single Hamburger	360	580
Wendy's Single with Everything Burger	420	920
Wendy's Grilled Chicken Sandwich	310	790

a) Find the mean and median amounts of calories and sodium for the McDonald's sandwiches.

b) Find the mean and median amounts of calories and sodium for the Wendy's sandwiches.

16. The top five scorers for the Los Angeles Lakers during the 1999–2000 NBA finals are given in the table. Included are the number of field goal attempts and the field goal percentage for each of the five players.

Player	Field Goal Attempts	Percentage
O'Neal	157	61%
Bryant	90	37%
Harper	56	46%
Rice	55	40%
Horry	43	51%

In parts (a) and (b), find the average field goal percentage by computing:

a) The weighted mean shooting percentages, using the number of attempts as the weights.

b) The mean of the field goal percentages.

c) Which of the averages computed gives a better representation of the team shooting percentage? Explain.

17. A student had seven 100-point tests during the semester. The scores on his tests were 90, 92, 59, 65, 94, 73, and 94. The instructor has given him the option of determining his semester grade by using the mean, median, or mode of his scores. Which would be the fairest selection? Explain.

18. A student is computing her cumulative grade point average (GPA). During six semesters, she has received 5 C's, 11 B's, and 8 A's. If a grade of C is worth 2 points, a B worth 3 points, and an A worth 4 points, and all classes have the same number of units, compute the student's GPA. (*Hint:* Use the number of grades as the weights.)

Explore

19. The top ten highest-paid athletes over the 1999–2000 season for the National Basketball Association (NBA), Major League Baseball (MLB), and the National Hockey League (NHL) and the 1999 season for the National Football League (NFL) were as follows. Figures are in millions of dollars. (Source: *Street & Smith's Sports Business Journal* and *USA Today.*)

National Football League

		Salary (in millions of dollars per year)
1	Troy Aikman	6.667
2	Drew Bledsoe	6.398
3	Brett Favre	6.249
4	Warren Sapp	6.231
5	Deion Sanders	6.200
6	Steve McNair	6.075
7	Steve Young	5.850
8	Joe Johnson	5.600
9	Dan Marino	5.447
10	Levon Kirkland	5.346

National Basketball Association

		Salary (in millions of dollars per year)
1	Shaquille O'Neal	17.140
2	Kevin Garnett	16.810
3	Juwan Howard	15.070
4	Patrick Ewing	15.000
5	Alonzo Mourning	15.000
6	Scottie Pippen	14.800
7	Hakeem Olajuwon	14.300
8	Karl Malone	14.000
9	Dikembe Mutombo	12.820
10	Jayson Williams	12.380

Major League Baseball

		Salary (in millions of dollars per year)
1	Kevin Brown	15.714
2	Randy Johnson	13.600
3	Albert Belle	13.000
4	Bernie Williams	12.357
5	Larry Walker	12.143
6	Mike Piazza	12.071
7	David Cone	12.000
8	Pedro Martinez	11.500
9	Mo Vaughn	11.167
10	Sammy Sosa	11.000

National Hockey League

		Salary (in millions of dollars per year)
1	Jaromir Jagr	10.360
2	Paul Kariya	10.000
3	Peter Forsberg	9.000
4	Theo Fleury	8.500
5	Eric Lindros	8.500
6	Pavel Bure	8.000
7	Patrick Roy	7.500
8	Dominik Hasek	7.000
9	Mats Sundin	7.000
10	Brian Leetch	6.680

a) Find the mean and median salaries of the top ten in each sport.

b) Find the mean and median salaries for the entire group of 40 athletes.

c) Is the median salary for the entire group of 40 athletes a realistic measure of the "average" salary of highest paid professional athletes? Explain.

d) Is the mean salary for the entire group of 40 athletes a realistic measure of the "average" salary of highest paid professional athletes? Explain.

Measures of Dispersion

Suppose the mean, mode, and median for two sets of data are identical. Does this suggest that the data are the same? Consider the two sets of data: 1, 1, 100, 100, 100, 199, 199 and 99, 99, 100, 100, 100, 101, 101. For each set, the mean, mode, and median all equal 100, yet the data are not the same. Not only are the data not the same, the first set has values that range between 1 and 199, whereas the second set is closely clustered around 100. From this, we can see that we need more tools to help describe a distribution of numbers. Collectively, the tools used to do this are called measures of dispersion. A measure of **dispersion** will provide a tool to determine the extent to which the data in a set differ from a central value. In this section, we will discuss two of these measures, range and standard deviation.

THE RANGE

The **range** of a set of values is the difference between the highest and lowest values in the set.

EXAMPLE 1

Find the range of each of the following sets of numbers.

a) 1, 1, 100, 100, 100, 199, 199

b) 99, 99, 100, 100, 100, 101, 101

Solution:

a) The range of the first set of data is $199 - 1 = 198$.

b) The range of the second set of data is $101 - 99 = 2$.

We can see that the range will provide some help in analyzing the difficulty mentioned in the introductory remarks. However, it will not completely solve the problem. Consider the following two sets of data: 1, 1, 1, 100, 199, 199, 199 and 1, 100, 100, 100, 100, 100, 199. In both sets, the range is 198, but the first set of data has most of its values at the extreme ends, whereas most of the data in the second set have the value of 100. Only two of the numbers are at the extremes. What is needed is a method to determine the average of the distance between each data value and the mean. If this average is high, then the data are spread out. If the average is low, then the data are clustered together.

STANDARD DEVIATION

The measure that we will use to determine how closely the data are clustered around the mean is called the **standard deviation.** The standard deviation is the square root of the average of the squares of the differences between the data values and the mean. The standard deviation is represented by the Greek letter σ (sigma) and can be determined by using the following formula.

Standard Deviation Formula

$$\sigma = \sqrt{\frac{\sum_{i=1}^{n} (x_i - \mu)^2}{n}} \quad \text{where} \quad \begin{cases} x_i \text{ are the data values} \\ \mu \text{ is the mean of the data} \\ n \text{ is the number of data} \end{cases}$$

EXAMPLE 2

Find the standard deviation of the numbers 4, 6, and 11.

Solution: To find the standard deviation, first find the mean.

$$\mu = \frac{4 + 6 + 11}{3} = 7$$

Next, subtract the mean from each of the data values and then square the results. Using the formula gives the following.

$$\begin{aligned} \sigma &= \sqrt{\frac{(4 - 7)^2 + (6 - 7)^2 + (11 - 7)^2}{3}} \\ &= \sqrt{\frac{9 + 1 + 16}{3}} \\ &= \sqrt{\frac{26}{3}} = 2.94 \end{aligned}$$

Although this process might not seem too difficult, performing the standard deviation calculation by hand is very tedious. Suppose you were asked to find the standard deviation of a set of data with mean 6.429. To find the standard deviation using the formula, 6.429 would be subtracted from each of the data values, and the result would then be squared. Although algebraic techniques exist to simplify this process, the ready availability of calculators eliminates the need for tedious calculations. As a result, for the remainder of this section we will assume that you have a calculator that has a standard deviation function. For help with your calculator, read your calculator manual or ask your instructor.

If you haven't done so already, use the standard deviation function on your calculator to find the standard deviation of the numbers given in Example 2.

STANDARD DEVIATION OF GROUPED DATA

As was true for the mean of data arranged into groups, there is a formula for the standard deviation of grouped data. The formula is as follows.

$$\sigma = \sqrt{\frac{\Sigma f_i(x_i - \mu)^2}{n}} \quad \text{where} \quad \begin{cases} x_i = \text{the midpoint of each group} \\ f_i = \text{the frequency of each group} \\ n = \text{the total of all the frequencies} \end{cases}$$

Since this formula is built into many calculators, we will not be doing the computations by hand. On most Texas Instruments calculators, this involves putting the midpoints of the classes in the first list of calculator data and the frequencies in the second list of calculator data. Do the following example on your calculator to verify that you are able to use this feature on your calculator.

EXAMPLE 3

Use the standard deviation button on your calculator to find the standard deviation of the grouped data.

Class	Midpoint	Frequency
1–5	3	2
6–10	8	5
11–15	13	7
15–19	17	3

Solution: After you have entered the midpoints and frequencies, your calculator should show that $\sigma \approx 4.26$.

HOW IS STANDARD DEVIATION USED?

The standard deviation of data is a measure of how much the data spreads out from the mean but, so far, we have no real feeling for what standard deviation indicates or how it can be used. In the remainder of this section, we discuss the meaning of standard deviation in terms of a frequency distribution.

EXAMPLE 4

Consider the following sets of test scores from two intermediate algebra classes.

Morning Class	Evening Class
78, 65, 83, 91, 98, 25, 67, 88, 81, 77,	34, 87, 81, 93, 99, 24, 77, 62, 98, 100,
53, 76, 80, 72, 75, 69, 64, 62, 85, 93,	57, 31, 81, 72, 61, 59, 68, 74, 77, 94,
70, 44, 85, 73, 75, 63	56, 71, 70, 81, 78, 83, 25, 94, 31

a) Find the mean and standard deviation of each class.

b) Construct the frequency distribution of each class.

c) Draw a bar chart for each class.

d) Use the bar chart to draw conclusions about the meaning of standard deviation.

Solution:

a) Using a calculator to find the mean and standard deviation of each class, we have the following.

Morning Class:

$$\mu = \frac{\sum_{i=1}^{26} x_i}{26} = \frac{1892}{26} = 72.77 \approx 73$$

$$\sigma = \sqrt{\frac{\sum_{i=1}^{26} (x_i - 72.77)^2}{26}} = \sqrt{\frac{6101.70}{26}} = 15.32$$

Evening Class:

$$\mu = \frac{\sum_{i=1}^{29} x_i}{29} = \frac{2021}{29} = 69.69 \approx 70$$

$$\sigma = \sqrt{\frac{\sum_{i=1}^{29} (x_i - 69.69)^2}{29}} = \sqrt{\frac{14{,}094.81}{29}} = 22.05$$

b) Completing the frequency distribution for each class gives the following.

Morning Class		Evening Class	
Class	**Frequency**	**Class**	**Frequency**
0–9	0	0–9	0
10–19	0	10–19	0
20–29	1	20–29	2
30–39	0	30–39	3
40–49	1	40–49	0
50–59	1	50–59	3
60–69	6	60–69	3
70–79	8	70–79	7
80–89	6	80–89	5
90–100	3	90–100	6

c) The bar charts for both classes are as follows.

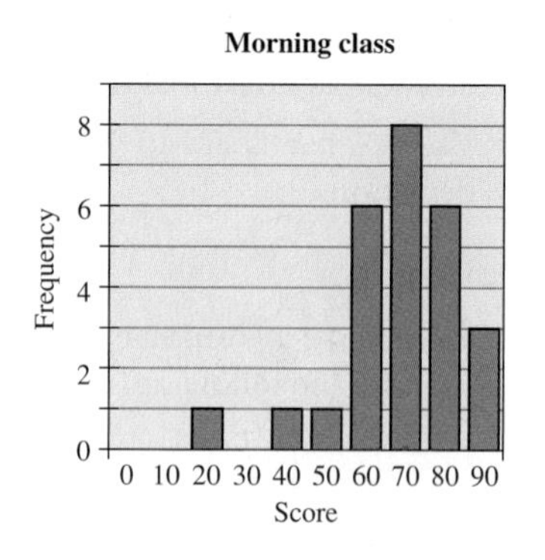

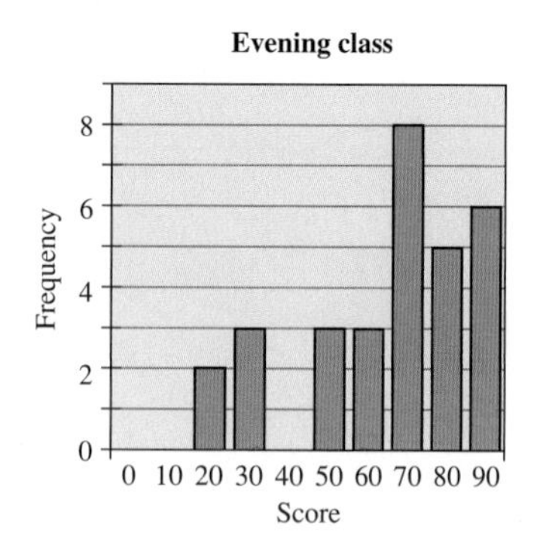

d) We now come to the important part of this problem. We have taken two sets of data, found the mean and standard deviation of each set, and drawn the bar graphs for each set of data. Both classes have means close to 70. This is seen in the bar charts by noting that most of the shading is clustered around the bar representing the 70–79 category. The standard deviation of the evening class is 22.05 versus 15.32 for the morning class. In the bar graphs, this can be seen by noticing that the evening class has more scores 20 units or more away from 70. This observation that the higher standard deviation corresponds to the bar graph showing the greater spread from the mean confirms the discussion at the beginning of this section.

section 5.3

PROBLEMS ➤ Explain ➤ Apply ➤ Explore

➤ Explain

1. What is a measure of dispersion?
2. What is the range of a set of data?
3. What is the standard deviation of a set of data?
4. The data 1, 3, 5, and 8 have a smaller standard deviation than the data 1, 3, 5, and 20. Explain.
5. In Example 3, how are the mid-points of each class determined and why do you need the midpoints?
6. You are comparing data from two similar experiments that have a mean of 34.5. The data in experiment A has a range of 150 and the data for experiment B has range of 20. What does that tell you about the data in the experiments?
7. You are comparing data from two similar experiments that have a mean of 34.5. Experiment A has a standard deviation of 10.1, and experiment B has a standard deviation of 5.5. What does this tell you about the data in the experiments?

8. Suppose two frequency distributions are represented by the following bar graphs. Which distribution has the greater standard deviation? Explain.

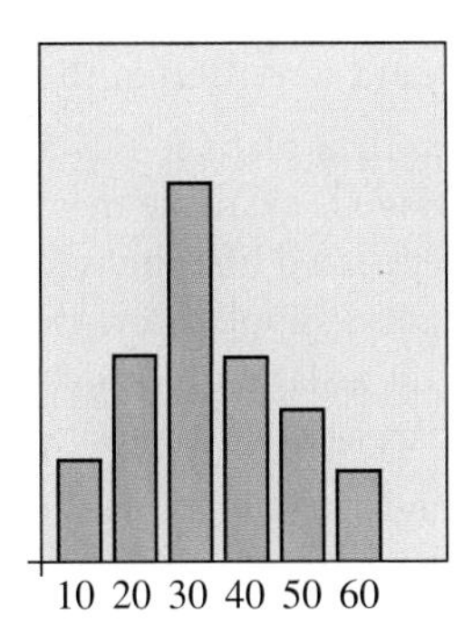

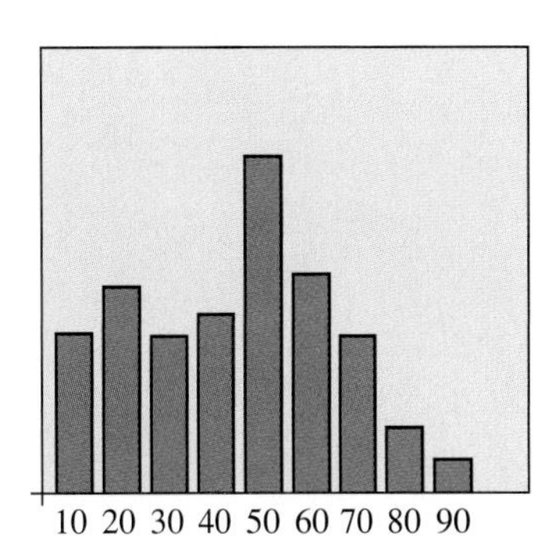

9. If a set of test scores has a standard deviation of zero, what can be said about the scores?

Apply

10. For the set of numbers 3, 6, 10, 14, 17:

a) Find the range.

b) Find the standard deviation.

c) Suppose that the original data are changed. The 3 is replaced by 2, and the 17 is replaced by 18. Without doing any calculations, what effect does this have on the standard deviation?

11. For the set of numbers 2, 6, 11, 12, 14:

a) Find the range.

b) Find the standard deviation.

c) Suppose that the original data are changed. The 2 is replaced by 5, and the 14 is replaced by 12. Without doing any calculations, what effect does this have on the standard deviation?

12. For the set of values 2, 4, 7, 2, 1, 8, 9, 10, 9, 6:

a) Find the range.

b) Find the standard deviation.

13. For the set of values 3, 8, 4, 2, 4, 6, 7, 1, 5, 0:

a) Find the range.

b) Find the standard deviation.

14. For the data 20, 29, 21, 21, 28, 28, 23, 23, 23, 26, 26, 26, 25, 25, 25, 25, 25:

a) Find the range.

b) Find the standard deviation.

15. For the data 42, 64, 75, 82, 96, 93, 77, 82, 67, 78, 88, 90, 80, 72, 71, 80, 81, 98, 61, 75:

a) Find the range.

b) Find the standard deviation.

Explore

16. Find the standard deviation of the calorie data in Problem 15 of Section 5.2.

17. Find the standard deviation of the sodium data in Problem 15 of Section 5.2.

18. An accounting firm plans to buy a large number of cartridges for laser printers. The cartridge is available from two different suppliers. The first supplier says that the expected lifetime is 3000 pages with a standard deviation of 100 pages. The second supplier says that its cartridge has an expected lifetime of 3000 pages with a standard deviation of 400 pages. If you are the buyer for the accounting firm, which supplier would you choose? Explain your reasoning.

19. The top ten highest-paid athletes over the 1999–2000 season for the National Basketball Association (NBA), Major League Baseball (MLB), and the National Hockey League (NHL) and the 1999 season for the National Football League (NFL) were as follows. Figures are in millions of dollars. (Source: *Street & Smith's Sports Business Journal* and *USA Today.*)

National Football League

		Salary (in millions of dollars per year)
1	Troy Aikman	6.667
2	Drew Bledsoe	6.398
3	Brett Favre	6.249
4	Warren Sapp	6.231
5	Deion Sanders	6.200
6	Steve McNair	6.075
7	Steve Young	5.850
8	Joe Johnson	5.600
9	Dan Marino	5.447
10	Levon Kirkland	5.346

National Basketball Association

		Salary (in millions of dollars per year)
1	Shaquille O'Neal	17.140
2	Kevin Garnett	16.810
3	Juwan Howard	15.070
4	Patrick Ewing	15.000
	Alonzo Mourning	15.000
6	Scottie Pippen	14.800
7	Hakeem Olajuwon	14.300
8	Karl Malone	14.000
9	Dikembe Mutombo	12.820
10	Jayson Williams	12.380

Major League Baseball

		Salary (in millions of dollars per year)
1	Kevin Brown	15.714
2	Randy Johnson	13.600
3	Albert Belle	13.000
4	Bernie Williams	12.357
5	Larry Walker	12.143
6	Mike Piazza	12.071
7	David Cone	12.000
8	Pedro Martinez	11.500
9	Mo Vaughn	11.167
10	Sammy Sosa	11.000

National Hockey League

		Salary (in millions of dollars per year)
1	Jaromir Jagr	10.360
2	Paul Kariya	10.000
3	Peter Forsberg	9.000
4	Theo Fleury	8.500
	Eric Lindros	8.500
6	Pavel Bure	8.000
7	Patrick Roy	7.500
8	Dominik Hasek	7.000
9	Mats Sundin	7.000
10	Brian Leetch	6.680

a) Find the range of the 40 salaries.

b) Find the median of the 40 salaries.

c) Arrange the data into a frequency distribution with four classes, the first one being

d) Make a bar graph of the grouped data.

e) Find the mean of the grouped data.

f) Find the standard deviation of the grouped data.

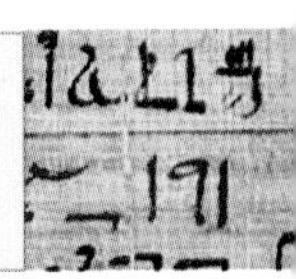

section 5.4

The Normal Distribution

Grading on the curve can both help and hurt students' grades. (PEANUTS reprinted by permission of United Features Syndicate, Inc.)

The following curve is the **normal distribution,** also called a **bell curve** or the **Gaussian distribution.** Since the total area under this curve is equal to 1, we can connect concepts of probability and statistics.

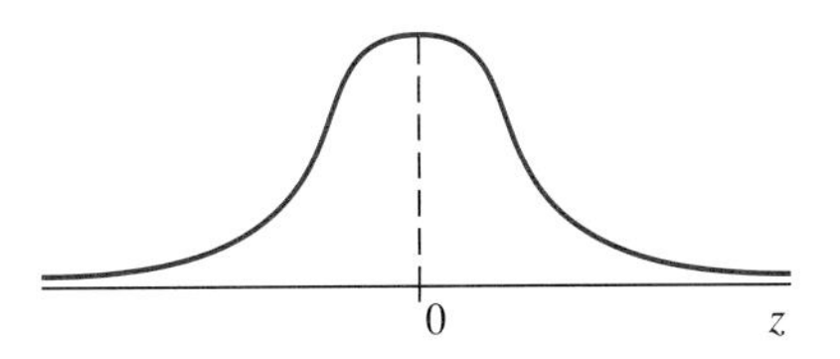

First, let's reexamine the results of the math placement scores from the first problem discussed in Section 5.1. The frequency distribution and the bar graph are given again here.

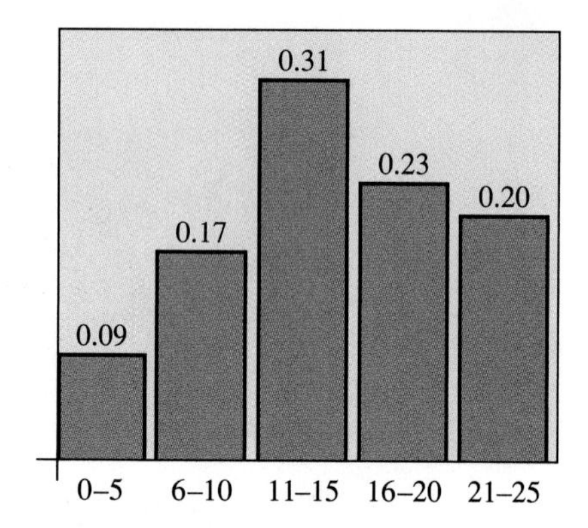

Frequency Distribution

Class	Relative Frequency
0–5	0.09
6–10	0.17
11–15	0.31
16–20	0.23
21–25	0.20

Notice that the height of each rectangle in the bar graph also represents the probability of a data value being in a certain class. For example, the probability that a randomly picked test score will be in the 6–10 class is 0.17. This means that, on the average, 17 out of 100 test scores will be in the class 6–10. This is true because both the probability and the relative frequency for a certain class are calculated by taking the number of items in a group and dividing by the total number of items. Thus, there is a connection between relative frequency, bar graphs, and probability.

A set of data is said to be normal if it follows a very definite set of mathematical criteria that is beyond our discussion. Informally, a set of data is said to be normal if 95% of all data values are within two standard deviations of the mean and 50% of the data is on each side of the mean. The normal distribution is used for many mathematical models, ranging from test scores to the number of defective items coming off a production line to the analysis of sociological surveys.

Using the normal distribution requires being able to find the area between the x axis and the graph of the normal curve. The equation of the normal curve is $y = (1 \div \sqrt{2\pi})e^{-x^2/2}$. Even using the powerful tools of calculus, this is a difficult problem. To avoid this difficulty, it is standard practice to use a table of values, called the normal table. The normal table is given in the Appendix. In a normal table, the letter z is used as the independent variable and as the label for the horizontal axis. To use the normal table, look up the desired value of z and read the corresponding probability in the adjacent column.

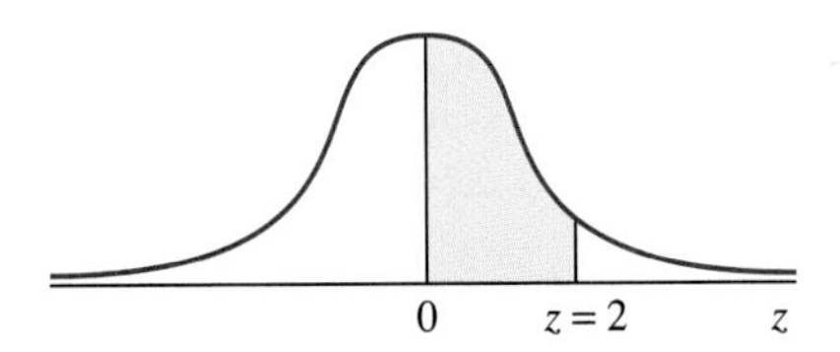

The values in the table give the area between the z axis and the normal curve, bounded on the left by the line $z = 0$ and bounded on the right by the value of z. For example, for $z = 2$ the table gives a value of 0.4772. This means that the area of the shaded region in the figure is 0.4772. It also means that the probability that z is between 0 and 2 is 0.4772. Symbolically, this is written as

$$P(0 < z < 2) = 0.4772.$$

This means that 47.72% of the area under the curve is between $z = 0$ and $z = 2$.

As you can see in the diagram, $z = 0$ is the middle value for the normal distribution. For most real-life problems, the middle or mean is usually a value other than 0. Also, the normal distribution has a standard deviation of 1, which is probably not the case in most situations. Therefore, to make the normal distribution useful, we must transform information that is given in a problem into z values. To do this, we use the following formula.

$$z = \frac{x - \mu}{\sigma} \quad \text{where} \quad \begin{cases} x \text{ is a data value} \\ \mu \text{ is the mean of the data} \\ \sigma \text{ is the standard deviation of the data} \end{cases}$$

At this introductory level, if the bar graph has a large number of rectangles and has a "bell shape," we will assume that the bar graph can be approximated by a normal distribution.

The z value gives the number of standard deviations between the mean and a particular data value.

EXAMPLE 1

The heights of 1000 students are measured and found to have a mean of 70 in. and a standard deviation of 3 in. Assuming that the heights of the students are normally distributed, what is the probability that a student, chosen at random, has a height between 70 and 73 in.?

Solution: The first step is to draw a normal distribution with a mean of 70 and shade the desired area. Since the mean is 70, that is the value at the center of the distribution. We want to find $P(70 < x < 73)$, where x represents the height of the student. Therefore, the region between 70 and 73 is shaded on the diagram.

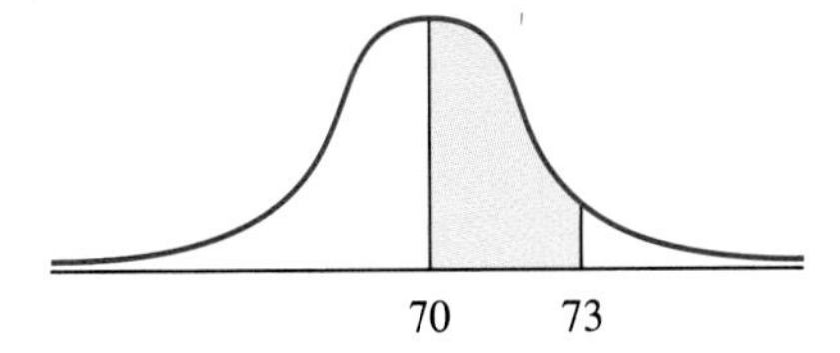

To find $P(70 < x < 73)$, convert 73 into a z value. The formula gives

$$z = \frac{73 - 70}{3} = 1.$$

In the normal table, for $z = 1$, the area is 0.3413. This can be interpreted in two ways. We can say that the probability that the student is between 70 in. and 73 in. tall is 0.3413, or we can say that approximately 34% of the students are between 70 and 73 in. tall.

EXAMPLE 2

Suppose the problem is similar to that in Example 1 except that the standard deviation is 4 in. rather than 3 in. Find $P(70 < x < 73)$.

Solution: Before doing the calculation, think about the physical meaning of standard deviation. We stated in the previous section that the standard deviation is a measure of how far apart the data are spread. Since the standard deviation in Example 2 is larger than the standard deviation in Example 1, the data are more spread out. Because of this, we should expect $P(70 < x < 73)$ to now be smaller.

We can verify this by doing the calculations. First, we need to find the z value. This is

$$z = \frac{73 - 70}{4} = 0.75.$$

Looking up $z = 0.75$ in the z table, we get 0.2734. Therefore, $P(70 < x < 73) = 0.2734$, which, as expected, gives a smaller value than in Example 1.

EXAMPLE 3

A study of the educational aspirations of military women under age 21 was undertaken. Results of the study showed that these women hoped to achieve, on the average, 15.301 years of schooling, with a standard deviation of 1.893 years. Assuming a normal distribution,

a) What percentage of the women hoped to have more than 15.301 years of school?

b) What percentage hoped to have between 12 and 15.301 years of school?

c) What percentage desired to have no more than 12 years of school?

d) What percentage of the women wanted to have between 16 and 18 years of school?

e) What percentage of the women wanted to have between 14 and 16 years of school?

Solution:

a) Since 15.301 is the mean, we expect that half of the respondents wanted to have more than 15.301 years of education while half did not want more. Therefore, 50% of the women desired to have more than 15.301 years of school.

For the remaining questions, we will sketch a graph of the normal distribution.

b) To find the percentage of women who hoped to have between 12 and 15.301 years of school, look at the graph.

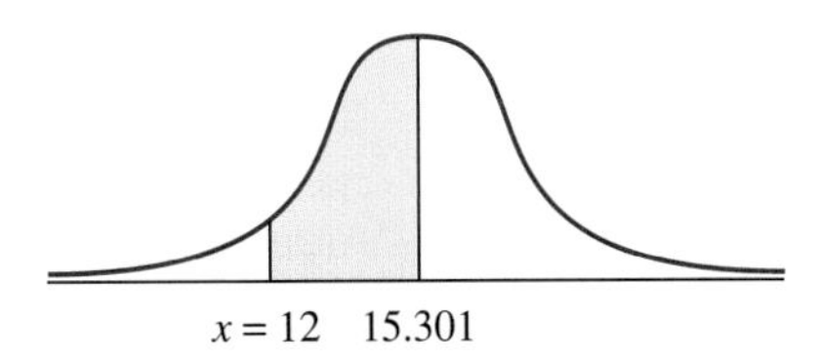

As we did in previous examples, we will find the z value:

$$z = \frac{12 - 15.301}{1.893} = -1.74.$$

This z value is negative, but the normal table includes only positive values. However, since the graph of the normal distribution is symmetric with respect to the mean, we can find the area between 12 and 15.301 by using $z = +1.74$. This gives the value $A = 0.4591$. This means that 45.91% of the women surveyed hoped to have between 12 and 15.301 years of school.

c) In this part of the problem, we are interested in the percentage of women who wanted 12 or fewer years of schooling, so we want to find the area to the left of 12.

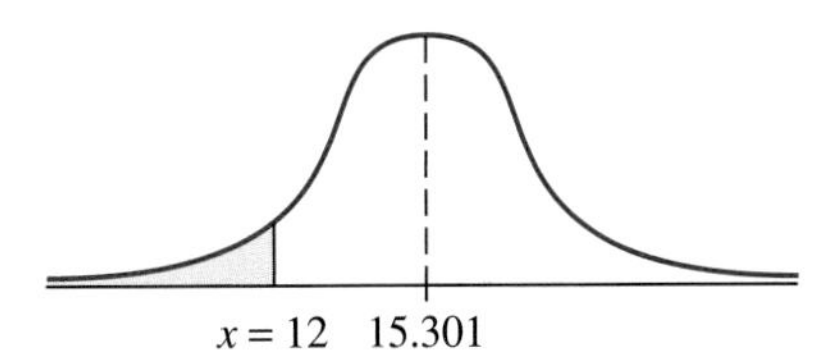

Since the normal table gives only areas that are adjacent to the center line of the distribution, we need to do this problem in another way. We use the fact that the area to the left of 15.301 is 0.5000 and the area between 12 and 15.301 is 0.4591. Subtracting these gives $A = 0.5000 - 0.4591 = 0.0409$. Therefore, only 4.09% of the women wanted the equivalent of a high school education or less.

d) To determine the percentage of women who wanted to have between 16 and 18 years of schooling, we again examine a picture.

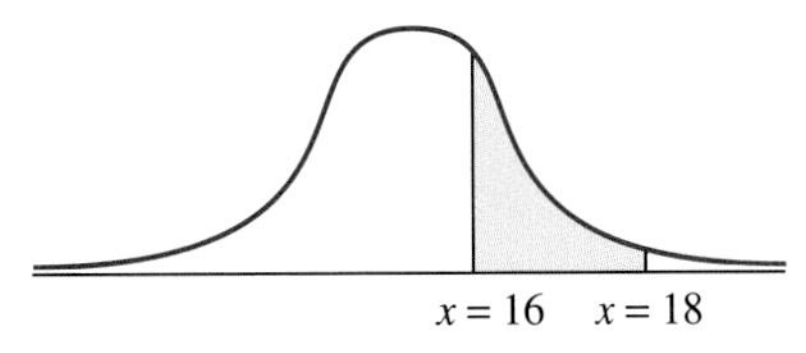

As was true in part (c), we cannot get the desired area directly, since the area is not adjacent to the middle of the distribution. As we did in part (c), we look at two separate areas and subtract the smaller from the larger to get the final answer.

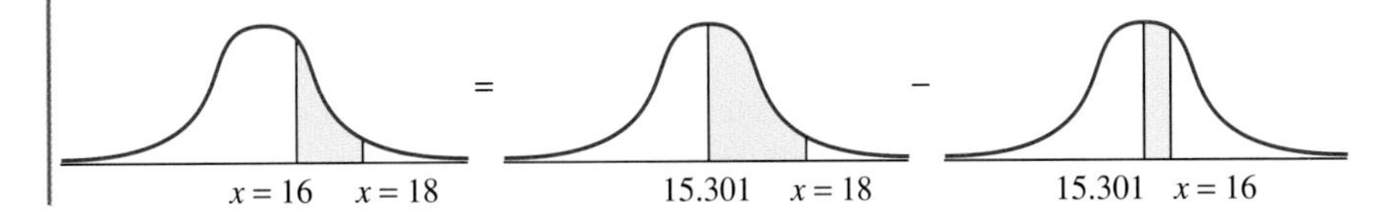

With the help of these pictures, we can see that if we want to find the area between 16 and 18, we can do it by finding the area between 15.301 and 18 and then subtracting the area between 15.301 and 16. We will use two z values. For $x = 18$,

$$z = \frac{18 - 15.301}{1.893} = 1.43$$

which gives an area of 0.4236. For $x = 16$,

$$z = \frac{16 - 15.301}{1.893} = 0.37$$

giving an area of 0.1443. Therefore, the probability that a randomly selected woman surveyed wanted between 16 and 18 years of education is $0.4236 - 0.1443 = 0.2793$. This means that 27.93% of the women hoped to have between 16 and 18 years of schooling.

e) For the final part of this problem, we want to determine the percentage of women who wanted to have between 14 and 16 years of schooing.

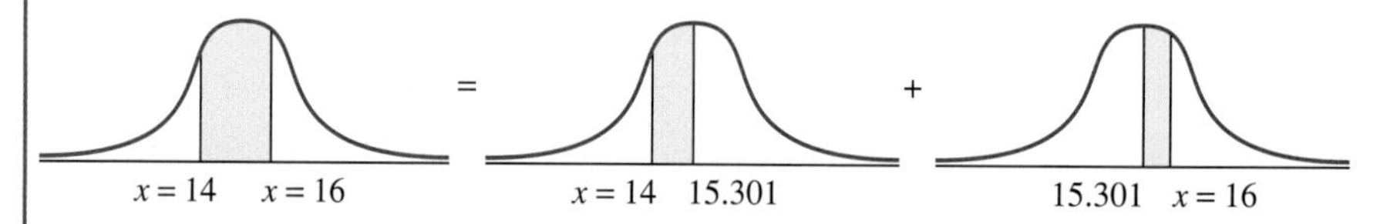

As we can see from the pictures, this problem is similar to part (d) except that since the mean is between the values of 14 and 16, we need to add the two areas to find the area of the combined regions. For $x = 14$,

$$z = \frac{14 - 15.301}{1.893} = -0.69$$

which gives an area of 0.2549. For $x = 16$,

$$z = \frac{16 - 15.301}{1.893} = 0.37$$

giving an area of 0.1443. Therefore, since $0.2549 + 0.1443 = 0.3992$, we can say that 39.92% of the women surveyed aspired to have between 14 and 16 years of education.

To summarize, we used the normal table to determine the probability of an event occurring. Even though the normal table gives only $P(0 < z < c)$, where c is some number, we found that using the formula

$$z = \frac{x - \mu}{\sigma}$$

allows us to determine the probability of an event if we know the mean and the standard deviation.

The final topic for this section is the use of the normal table to reverse the above process. In this type of problem, we know the percentage or probability but we want to determine the value of z or x that gives this probability. The next two examples will demonstrate what we mean.

EXAMPLE 4

Find the values of c that make the following statements true.

a) $P(0 < z < c) = 0.2580$ **b)** $P(c < z < 0) = 0.2580$

c) $P(0 < z < c) = 0.3000$ **d)** $P(z > c) = 0.0563$

Solution: As we did in previous examples, we will draw a picture for each problem. The picture will help us understand what we must do.

a) The shaded part of the following diagram represents the area or the probability. The problem is to determine the value of c that makes $P(0 < z < c) = 0.2580$. Look in the z table for the z value that gives a corresponding area of 0.2580. This gives $z = 0.70$. This means $P(0 < z < 0.70) = 0.2580$.

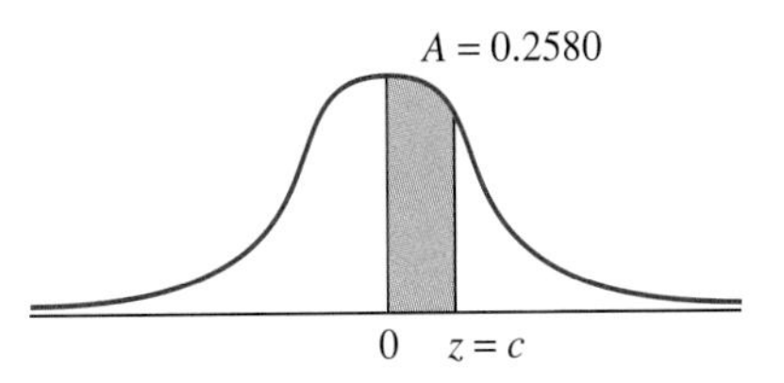

b) To find the solution to $P(c < z < 0) = 0.2580$, we know that the value of c must be negative because c is less than 0. Since the value of the area is the same as it was in part (a) and we know that c must be negative, we know that $P(-0.70 < z < 0) = 0.2580$.

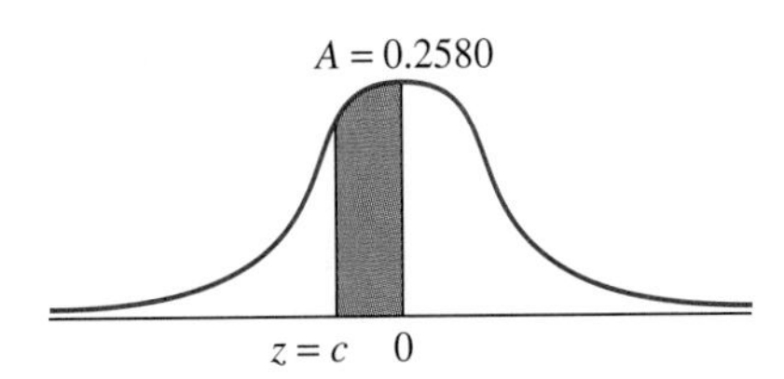

c) The problem $P(0 < z < c) = 0.3000$ is very similar to part (a). The difference arises when we try to find the value 0.3000 in the area column of the z table. For $z = 0.84$, $A = 0.2995$, and $z = 0.85$ gives $A = 0.3023$. Since neither of these values is 0.3000, we need to make some compromises. The choices are to find a table that has more digits of accauracy, use a mathematical technique called interpolation, or choose the value of A that is closest to 0.3000. Because this is only a survey course, we choose the easiest method, picking the closest value, which is $A = 0.2995$. Therefore, we will say $P(0 < z < 0.84) \approx 0.3000$.

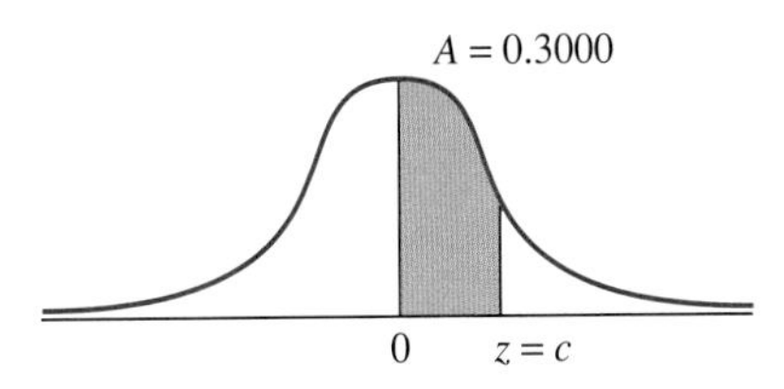

d) The problem $P(z > c) = 0.0563$ is different from parts (a), (b), and (c), since the specified area is given by $z > c$, which means that the area is at the far right tail (yellow region). Since the normal table only gives areas that are adjacent to the mean, using the blue region gives $0.5000 - 0.0563 = 0.4437$. The normal table now gives $P(z > 1.59) = 0.0563$.

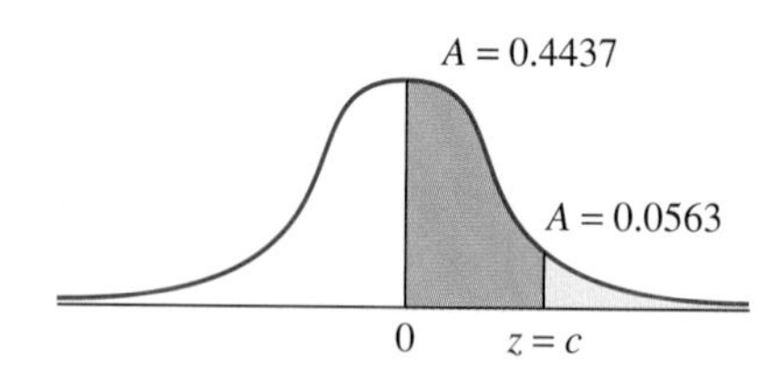

EXAMPLE 5

A certain college instructor has a fixed grading policy. He always gives the top 8% of the students A's; the next 15% receive B's, 54% receive C's, 15% receive D's, and 8% receive F's. Assuming that the scores are normally distributed with a mean of 72 and a standard deviation of 12, determine the scores that receive each grade.

Solution: Because there are five different grades, we must divide the normal distribution into five different regions, as shown in the diagram. Since the A's

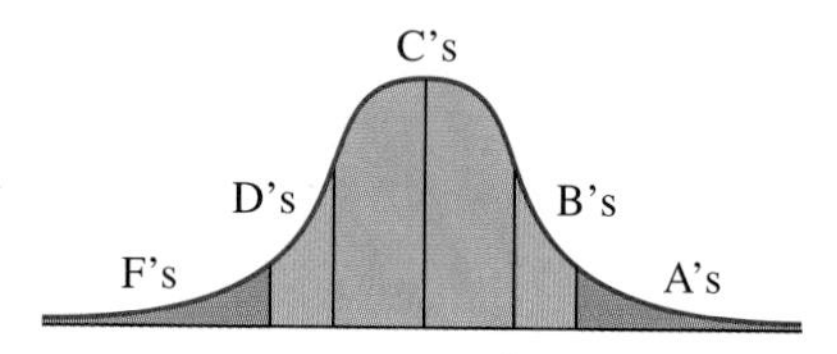

occupy the top 8%, we have the diagram shown. As we did in part (d) of the previous example, look at the area between the mean and some value of x, labeled c. Using $A = 0.5000 - 0.0800 = 0.4200$, we find $z = 1.41$.

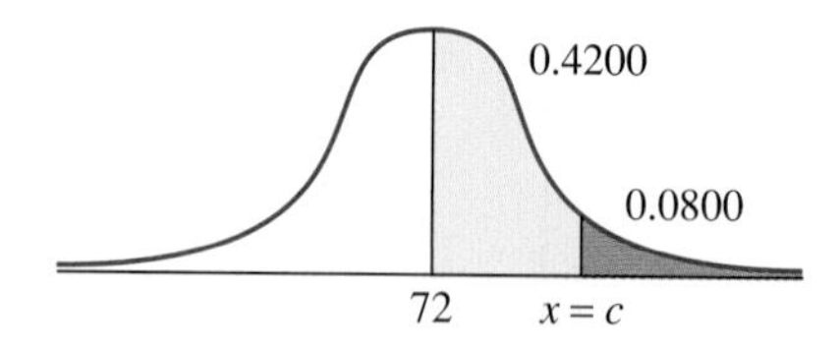

Since $\mu = 72$ and $\sigma = 12$, use the formula $z = (x - \mu)/\sigma$. This gives the following.

$$1.41 = \frac{x - 72}{12}$$

$$16.92 = x - 72$$

$$88.9 \approx x$$

This means that all students scoring 88.9 or higher will receive a grade of A.

Since 15% of the students will receive B's, the scores that will receive a grade of B are shown in the following diagram in the green region marked 0.1500. To find the area of the yellow region, we use the fact that 50% of the scores must be greater than the mean. Since 8% receive A's and 15% receive B's, we find the yellow region to be $0.5000 - 0.0800 - 0.1500 = 0.2700$.

Using $A = 0.2700$, we find $z = 0.74$. Using the formula $z = \dfrac{x - \mu}{\sigma}$, we get the following.

$$0.74 = \frac{x - 72}{12}$$

$$8.88 = x - 72$$

$$80.9 \approx x$$

Thus, students with scores between 80.9 and 88.9 will receive B's.

To find the scores for the other three grades, we use the fact that the normal distribution is symmetric about the mean.

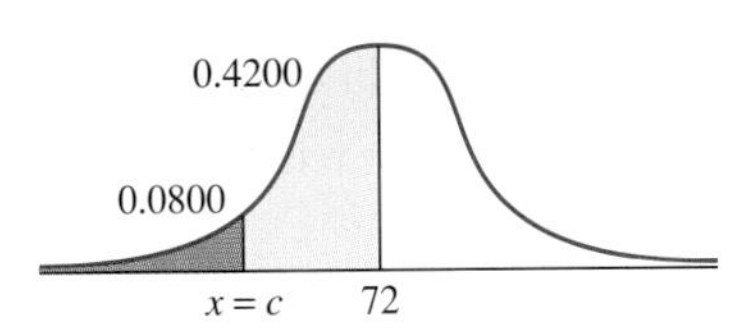

To find the scores receiving a grade of F, look at the lowest 8% of the scores. Since the diagram is the mirror image of the picture used to find the A grades, we know that $z = -1.41$.

$$-1.41 = \frac{x - 72}{12}$$

$$-16.92 = x - 72$$

$$55.1 \approx x$$

This means that all students scoring 55.1 or lower will receive a grade of F.

The scores earning D's are found in a similar way. From our work on the B's, we have $z = -0.74$. This gives the following.

$$-0.74 = \frac{x - 72}{12}$$

$$-8.88 = x - 72$$

$$63.1 \approx x$$

This means that all students scoring between 55.1 and 63.1 will receive a grade of D.

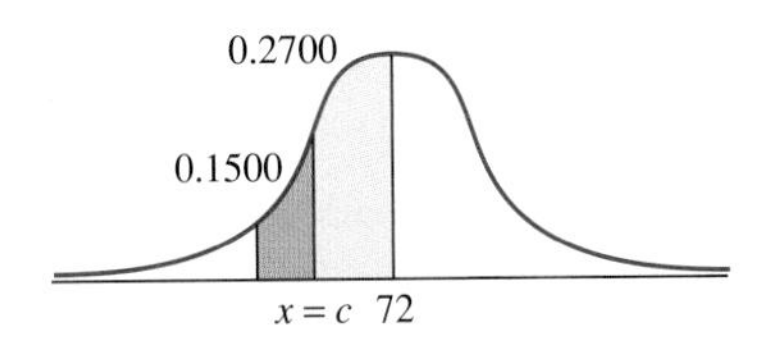

At this point, we have only to determine the scores that will earn a grade of C. Since the highest score that will earn a D is 63.1 and the lowest score that will earn a B is 80.9, all scores between 63.1 and 80.9 will receive a grade of C.

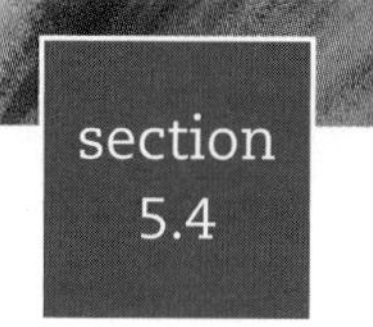

PROBLEMS ▸ Explain ▸ Apply ▸ Explore

▸ Explain

1. What is a normal distribution?
2. When is a set of data said to be normal?
3. What does a z value for a score give you?
4. In a normal distribution, if $z = 1.53$, what does this mean in terms of an area?
5. In a normal distribution, if $z = 1.53$, what does this mean in terms of a probability?
6. In a normal distribution, if $z = -0.76$, what does this mean in terms of probability?
7. In a normal distribution, if $z = -0.76$, what does this mean in terms of an area?
8. In a normal distribution with $\mu = 25$ and $\sigma = 5$, if a score x has a z value of 1.2, what does that tell you about x?
9. In a normal distribution with $\mu = 25$ and $\sigma = 5$, what does it mean if for score x, $P(x > 30) = 0.159$?
10. In a normal distribution with $\mu = 25$ and $\sigma = 5$, what does it mean for score x to have an area of $A = 0.4332$?

▸ Apply

In Problems 11–18, find the indicated area (probability) under the normal curve.

11.

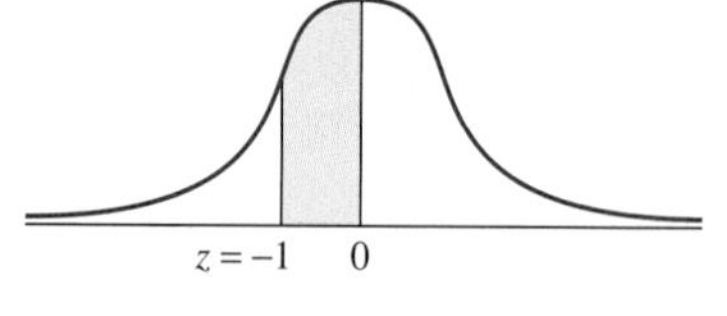

12.

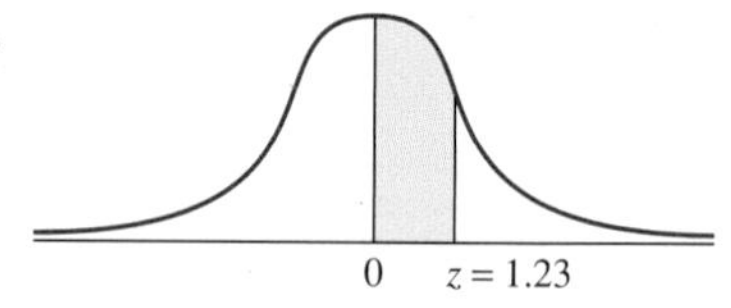

13.

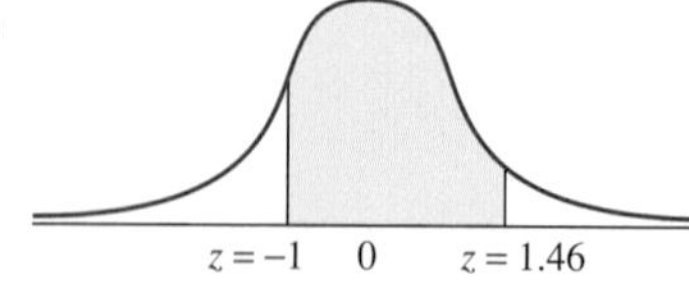

14.

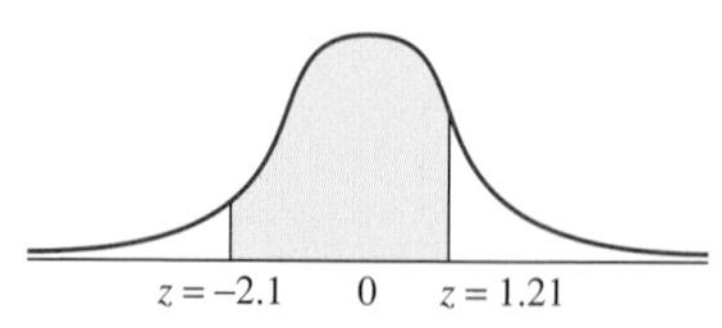

15.

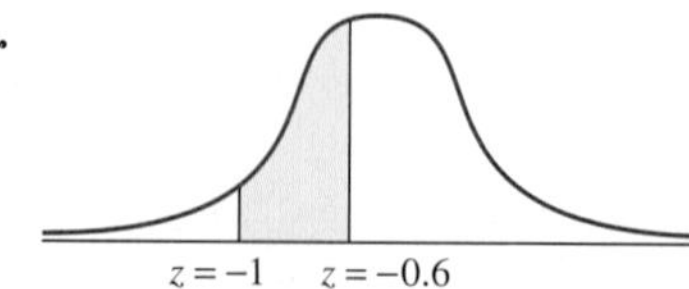

16.

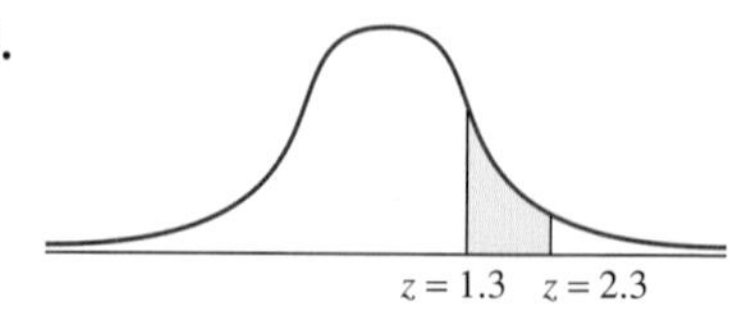

17.

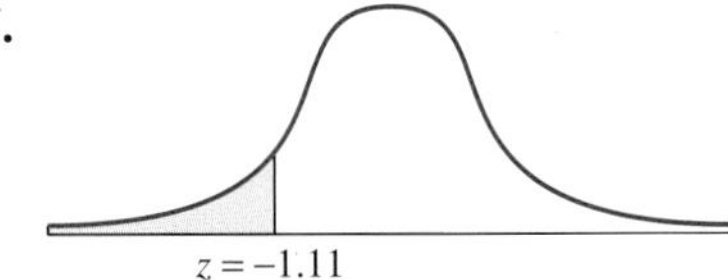

18.

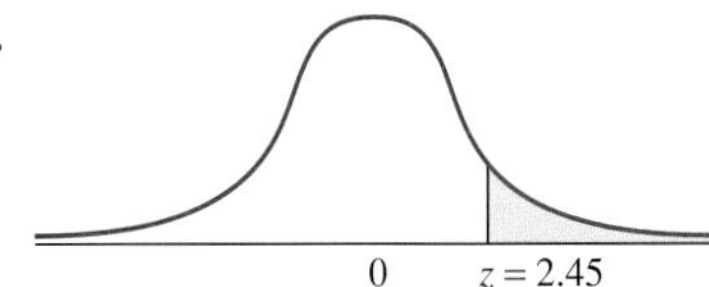

In Problems 19–24, determine the value of c that will give the indicated area (probability) under the normal curve.

19.

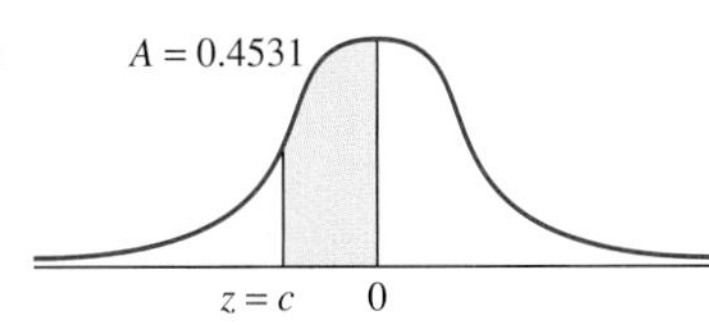

20.

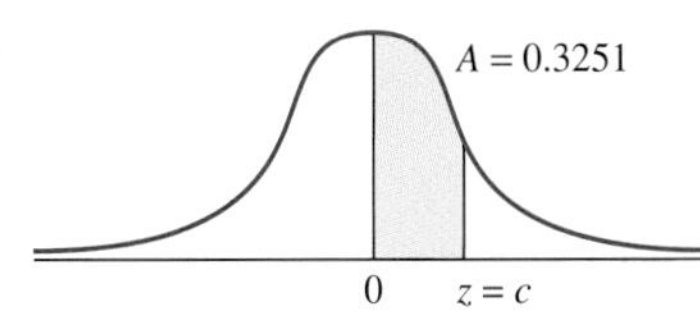

21.

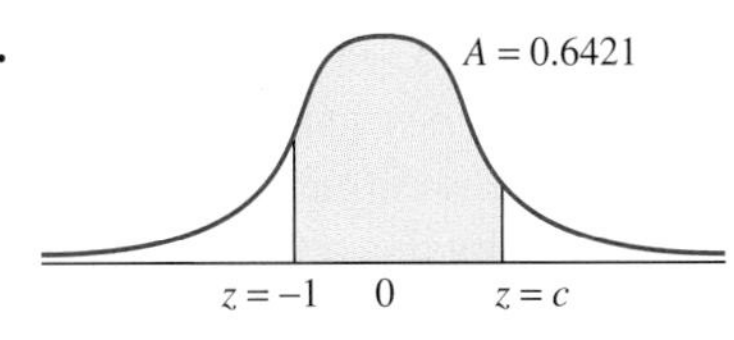

22.

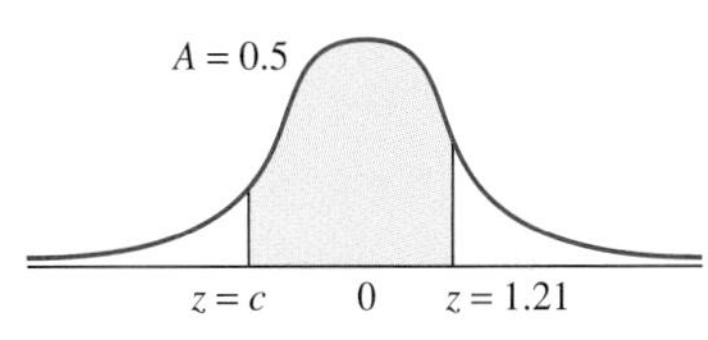

23.

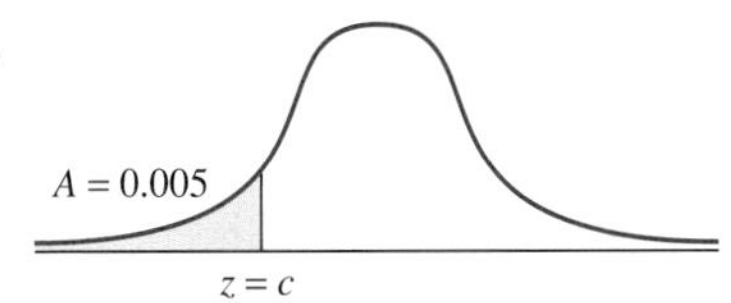

24.

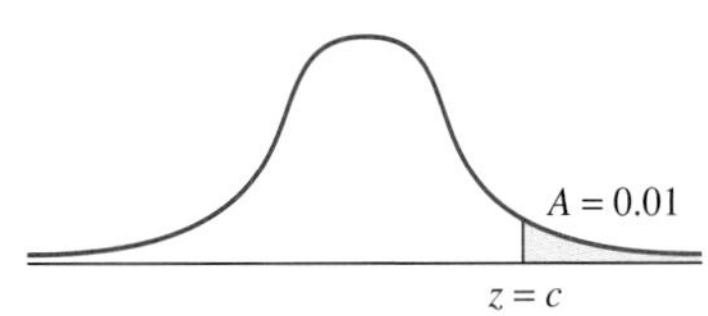

Explore

25. A survey of blue-collar workers found that the workers had a mean of 17.34 years of experience, with a standard deviation of 11.14 years. Assume a normal distribution.

- **a)** What percentage of the blue-collar workers in the survey had between 11 and 15 years of experience?
- **b)** Find the probability that a blue-collar worker in the survey, selected at random, had more than 15 years of experience.

26. A study was made of black families with a female as the head of the household. Among working mothers of age 30 or more, the mean number of hours worked per year was 1582.2, with a standard deviation of 728.5 hr. Assume a normal distribution.

- **a)** What percentage of the women in the study worked between 1000 and 1582.2 hr per year?
- **b)** What percentage of the women in the study worked less than 1000 hr per year?
- **c)** Assuming a 40-hr work week, how many full work weeks are represented by 1582.2 hr?

27. A survey of women aged 14–21 determined that the mean travel time to work was 17 min with a standard deviation of 13 min. Assume a normal distribution.

- **a)** What is the probability that a woman from this survey, chosen at random, traveled more than 20 min to reach work?

b) What percentage of the women traveled between 10 and 20 min to get to work?

c) What percentage of the women traveled between 5 and 10 min to get to work?

28. A survey of men aged 14–21 determined that the mean travel time to work was 19.7 min with a standard deviation of 20 min. Assume a normal distribution.

a) What is the probability that a man from this survey, chosen at random, traveled more than 30 min to reach work?

b) What percentage of the men traveled between 10 and 30 min to get to work?

c) What percentage of the men traveled between 5 and 10 min to get to work?

29. A study of men aged 35–57 at increased risk of coronary disease found that the average number of cigarettes smoked per day was 21.7 with a standard deviation of 20.5. Assuming a normal distribution, determine the number of cigarettes smoked by the 10% of the group who were the heaviest smokers.

30. The All American University accepts, unconditionally, students who score in the top 10% nationally on the SAT exam. For those students in the next 15% on the SAT exam, the university accepts the students on the basis of their grades in the senior year of high school. Assume that the SAT average is 950 with a standard deviation of 280.

a) Determine the lowest SAT score a student may have and be accepted to the All American University unconditionally.

b) Determine the SAT scores that will allow a student to enter All American University dependent on his or her senior-year grades.

31. Pick one of the following topics (or decide on one of your own) for a statistical investigation. Collect 40 or more data values for your topic. Arrange the data in a frequency distribution and draw a bar graph for the data. Find the mean and standard deviation for the data. Finally, determine the cutoff for the top 10% of the data.

a) Season-ending batting averages of baseball players.

b) Bowling scores of a bowling league.

c) Lengths of the most popular songs on the radio according to the *Billboard* charts.

d) Winning times in the 100-m dash at the collegiate level for track meets across the country.

e) Weights of chickens sold in the grocery store.

section 5.5

Polls and the Margin of Error

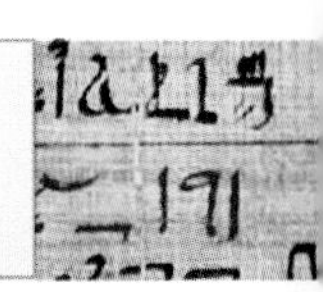

The Chicago Tribune, basing its conclusion on voting returns and statistics received too early in the night, picked the wrong winner of the 1948 Presidential election. (UPI/Bettman Newsphotos)

One of the most visible uses of statistics and the normal curve can be found in public opinion surveys. In these polls, a relatively small segment of the population is asked about a certain issue, and the responses are used to predict the views of the entire population. In this section, we will examine how the results of these polls can be interpreted and the accuracy of the predictions determined.

INFERENTIAL STATISTICS AND CONFIDENCE INTERVALS

Inferential statistics are the statements about a population that are derived from information about a small segment of a population. In other words, inferential statistics are exactly the types of information that can be derived from a poll or survey. A particular type of inferential statistic is the confidence interval. Before defining this term, consider the following example.

EXAMPLE 1

Two thousand registered voters have been asked to respond to a questionnaire about whether the state's members of the House of Representatives should be subject to term limitations of eight years. The results of the survey show that 1120 of the 2000 people think that the Representatives should be limited to a maximum of eight years in office. What does this survey say about the entire voting population of the state?

Solution: Since 1120 of the 2000 voters endorse term limitations, it is appropriate to say that approximately $1120 \div 2000 = 0.56$ or 56% of voters endorse term limitations. However, we cannot say that 56% of all the voters endorse the proposal. If we took another survey of a different group of 2000 people, we might find that 1160 or 1070 people endorse term limitations.

Now that we recognize that any survey will produce only an estimate of an actual value, the next step is to determine what factors contribute to the difference between the estimate and an actual value. One obvious factor that will affect the accuracy of the estimate is the size of the sample. If the survey asks a large number of people to respond to a question, the results of the survey should be more accurate than if the survey asks only a small number of people. With an increase in the size of the sample comes an increase in the accuracy of the estimate and an increase in the confidence we have in the validity of our estimate.

An additional consideration comes from the theory of statistics. Suppose that we conduct the poll described in Example 1 one hundred times, using a different sample of the population in each poll. Each poll will have a resulting estimate of the percentage of the people who are in favor of term limitations. By drawing a graph of these proportions, it can be shown that the percentages are distributed in the shape of a normal curve. As was true with the situations described in Section 5.4, the exact shape and position of the normal distribution will depend on the mean and the standard deviation of the sample percentages.

The above discussion leads to two important concepts in statistics, confidence intervals and confidence levels. A **confidence interval** is an interval centered around the estimate generated by the survey. For instance, in Example 1, the results of the survey stated that 56% of the voters endorsed term limitations. If we allow a margin of error of four percentage points, we anticipate that the true percentage in favor of term limitations could be as low as 52% (56 − 4) and as high as 60% (56 + 4). A confidence interval for this situation would then be 52% to 60%.

A **confidence level** is a statement of the probability that the actual percentage being studied is contained within the confidence interval. With all other factors remaining equal, the higher the confidence level, the larger the confidence interval. With these considerations in mind, it can be shown that the formula for the margin of error in a poll or survey is given as follows.

Margin of Error

$$M = \frac{z}{2\sqrt{n}} \quad \text{where} \quad \begin{cases} M = \text{margin of error} \\ z = \text{value determined by the confidence level and the normal distribution} \\ n = \text{sample size} \end{cases}$$

CONFIDENCE LEVELS AND THE NORMAL DISTRIBUTION

Before continuing with our discussion of estimates, it is important to understand what a confidence level means in terms of the normal distribution. Suppose that we have a 95% confidence level. This means that the probability that the confidence interval contains the actual percentage is 0.95. Using the methods of Section 5.4, we can now calculate the z value used to determine the margin of error.

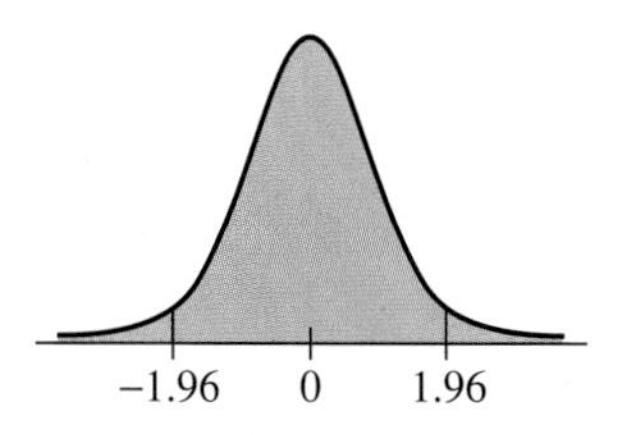

Since the confidence interval is centered on the estimate, the standard normal distribution is centered on zero. Therefore, dividing 0.95 by 2 gives 0.475. Using area $= 0.475$, we find $z = 1.96$. Notice that this calculation is done without consideration of the survey or the number of people responding to the survey. Therefore, any 95% confidence interval will use $z = 1.96$.

EXAMPLE 2

Determine the z value for a 90% confidence interval.

Solution: Dividing 0.90 by 2 gives area $= 0.45$. Looking up this area in the normal table gives $z = 1.645$. Thus, a 90% confidence interval will use $z = 1.645$.

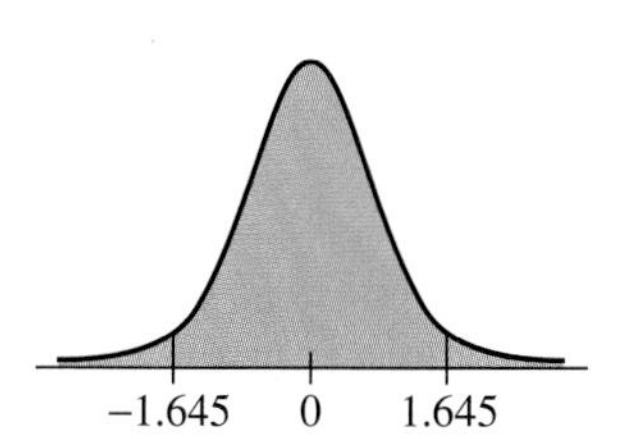

Since a particular confidence interval will result in a particular z value, we give the common confidence levels in the following table.

Confidence Level	z value
90%	1.645
95%	1.96
98%	2.327
99%	2.575

MARGIN OF ERROR AND SURVEYS

Now that we have completed the background material, we can apply the margin of error formula to a variety of situations.

EXAMPLE 3

An October 2001 Harris Poll found that 65% of Americans had Internet access. If the survey was conducted with a sample size of 1011 adults and had a 95% confidence level, what is the margin of error in the survey?

Solution: Since we have a 95% confidence level, $z = 1.96$. Using this z value and $n = 1011$, we have

$$M = \frac{1.96}{2\sqrt{1011}} = 0.03.$$

Thus, the margin of error is 3%. This means that the percentage of people who had Internet access is somewhere between $65 - 3 = 62\%$ and $65 + 3 = 68\%$.

EXAMPLE 4

Suppose that the pollster in Example 3 wants to decrease the margin of error from 3% to 1% while maintaining the same level of confidence. How large a sample should the pollster use?

Solution: Using $M = 0.01$ and $z = 1.96$, we have

$$\begin{aligned} 0.01 &= \frac{1.96}{2\sqrt{n}} \\ 0.02\sqrt{n} &= 1.96 \\ \sqrt{n} &= 98 \\ n &= 98^2 \\ n &= 9604. \end{aligned}$$

Notice that the reduction in the margin of error requires a very large increase in the sample size. This is an important consideration for pollsters. While it is desirable to decrease the margin of error, the resulting increase in the sample size could cause a large increase in the cost of the poll.

EXAMPLE 5

A sample of 2023 adults who use the Internet were asked their age group. Twenty-eight percent of those responding said they were in the age group 18–29. The margin of error in the survey was stated to be 3%.

a) What is the confidence level of this survey?

b) What is the confidence interval for the percentage of Internet users who are in the 18–29 age group?

Solution: Using $n = 2023$ and $M = 0.03$, we have

$$\begin{aligned} 0.03 &= \frac{z}{2\sqrt{2023}} \\ 0.03 &= \frac{z}{89.96} \\ z &= 0.03 \times 89.96 \\ z &\approx 2.70. \end{aligned}$$

a) Since this value is not in the table of common confidence levels, we must use the normal table. Using a z value of 2.70, the normal table returns a value of 0.4965. Because this is the area on only the right side of the normal distribution, we must double the result. This gives a value of approximately $0.4965 \times 2 = 0.993$. Thus, the confidence level is 99.3%.

b) To determine the confidence interval, we construct a confidence interval with 28% at its center and include all values within the margin of error. Since the margin of error is 3%, the confidence interval is

$$(28 - 3)\% \text{ to } (28 + 3)\% \quad \text{or} \quad 25\% \text{ to } 31\%.$$

Thus, we have a 99.3% assurance that the percentage of Internet users who are in the 18–29 age group is somewhere between 25% and 31%.

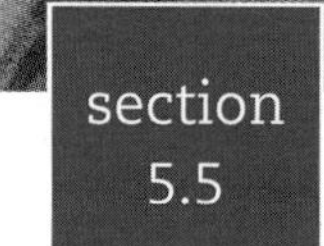

section 5.5

PROBLEMS ● Explain ● Apply ● Explore

● Explain

1. What is a confidence interval?
2. What is a confidence level?
3. What is a margin of error?
4. What is a sample size?
5. What happens to the margin of error if the sample size increases? Explain.
6. What happens to the margin of error if the confidence level increases? Explain.
7. A survey estimates that 25% of the viewing audience watched the Rose Bowl. If there is a 5% margin of error, what does this say about the percentage of the viewing audience that watched the Rose Bowl?
8. A poll estimates that 63% of the public wants to see an increase in handgun control laws. The margin of error is 4%. What does this say about the percentage of the public that wants to see an increase in the number of handgun control laws?

● Apply

9. Determine the z value for an 80% confidence level.
10. Determine the z value for a 70% confidence level.
11. Determine the z value for a 85% confidence level.
12. Determine the z value for a 75% confidence level.
13. A poll has a result of 63% with a margin of error of 4%. What is the confidence interval for this poll?
14. A poll has a result of 52% with a margin of error of 3%. What is the confidence interval for this poll?
15. A poll has a result of 47.6% with a margin of error of 2.4%. What is the confidence interval for this poll?
16. A poll has a result of 39.2% with a margin of error of 3.1%. What is the confidence interval for this poll?
17. A poll of 750 people is taken and the confidence level is 98%. What is the margin of error?
18. A poll of 1200 people is taken, and the confidence level is 95%. What is the margin of error?
19. A poll of 1000 people is taken, and the confidence level is 86%. What is the margin of error?
20. A poll of 800 people is taken, and the confidence level is 99.9%. What is the margin of error?
21. A poll of 2000 people has a margin of error of 4%. What is the confidence level for this poll?

22. A poll of 600 people has a margin of error of 2.5%. What is the confidence level for this poll?

23. A poll of 800 people has a margin of error of 3.5%. What is the confidence level for this poll?

24. A poll of 1000 people has a margin of error of 3%. What is the confidence level for this poll?

25. A customer requests that a polling company produce a survey with a 98% confidence level and a margin of error of 2%. How many people must respond to the survey?

26. A customer requests that a polling company produce a survey with a 99% confidence level and a margin of error of 1%. How many people must respond to the survey?

27. A customer requests that a polling company produce a survey with an 80% confidence level and a margin of error of 3%. How many people must respond to the survey?

28. A customer requests that a polling company produce a survey with an 85% confidence level and a margin of error of 2%. How many people must respond to the survey?

Explore

29. In a July 2001 Gallup poll of 266 smokers, 52% said that secondhand smoke was very harmful. If the margin of error was 7%,

a) What is the confidence level of the poll?

b) How large a sample would be needed to have a confidence level of 99% while maintaining a 7% margin of error?

c) How large a sample would be needed to decrease the margin of error to 2% while maintaining a 99% confidence level?

30. In a July 2001 Gallup poll of 266 smokers, 44% said that smoking should be completely banned from restaurants. If the margin of error was 7%,

a) What is the confidence level of the poll?

b) How large a sample would be needed to have a confidence level of 90% while maintaining a 7% margin of error?

c) How large a sample would be needed to decrease the margin of error to 2% while maintaining a 90% confidence level?

31. In a November 2001 Gallup poll of 1005 adults, 40% said that the economic conditions in their local economy were getting better. If the margin of error was 3%,

a) What is the confidence level?

b) What is the confidence interval for this survey?

c) Explain what the results of parts (a) and (b) indicate.

32. In a November 2001 Gallup poll of 1005 adults, 42% said that the economic conditions in their local economy were getting worse. If the margin of error was 3%,

a) What is the confidence level?

b) What is the confidence interval for this survey?

c) Explain what the results of parts (a) and (b) indicate.

33. In a March 2001 Gallup poll of 506 adults, 74% said that they would be willing to pay $100 more each year in higher prices so that industry could reduce air pollution.

a) If the margin of error is 5%, what is the confidence level?

b) How large a sample would be needed to have a confidence level of 99%?

c) How large a sample would be needed to decrease the margin of error to 4% while maintaining the original confidence level?

34. In a March 2001 Gallup poll of 554 adults, 63% said that they would be willing to pay $500 more each year in higher prices so that industry could reduce air pollution.

a) If the margin of error is 5%, what is the confidence level?

b) How large a sample would be needed to have a confidence level of 90%?

c) How large a sample would be needed to decrease the margin of error to 3% while maintaining the original confidence level?

35. In a November 2001 Gallup poll of adult men, 49% said that they were trying to lose weight.

a) If the margin of error is 3% and the confidence level is 95%, how many men were surveyed?

b) How large a sample would be needed to have a margin of error of 1% and a confidence level of 99%?

36. In a November 2001 Gallup poll of adult women, 68% said that they were trying to lose weight.

a) If the margin of error is 3% and the confidence level is 95%, how many women were surveyed?

b) How large a sample would be needed to have a margin of error of 1% and a confidence level of 99%?

Regression and Forecasting

While sometimes very useful, regression is not always an appropriate choice. (FOXTROT ©2002 Bill Amend. Reprinted with permission of UNIVERSAL PRESS SYNDICATE. All rights reserved.)

In algebra, you saw how to find the equation of the line that passes through two points. In this section, we expand that concept so that we can find the equation of a line (or other type of curve) that approximates a set of two-variable data. To see how this is done, we first look at an example with whole numbers. After that, we look at real-world data. Consider the table and graph below. There is a general upward trend to the data.

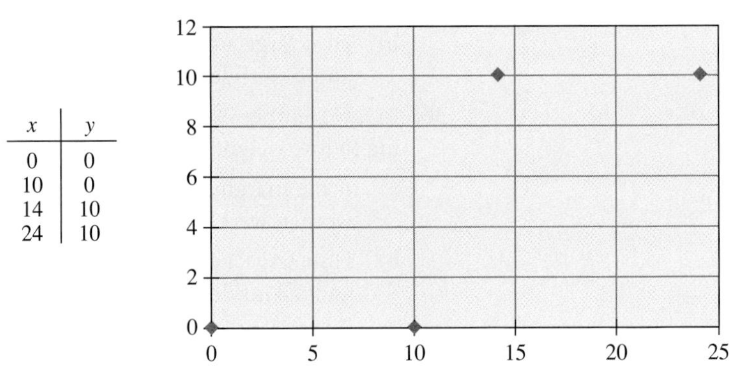

x	y
0	0
10	0
14	10
24	10

In the next three graphs, we have added different lines to the data. In the first, we chose a line passing through the first and last points. In the second graph, the slope is more than the slope in the first graph. In the third graph, the slope is less than the slope in the first graph. The goal of this section is to decide which line best describes the data.

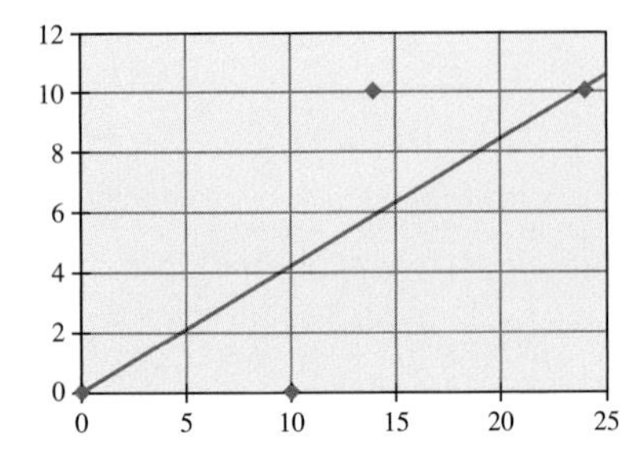

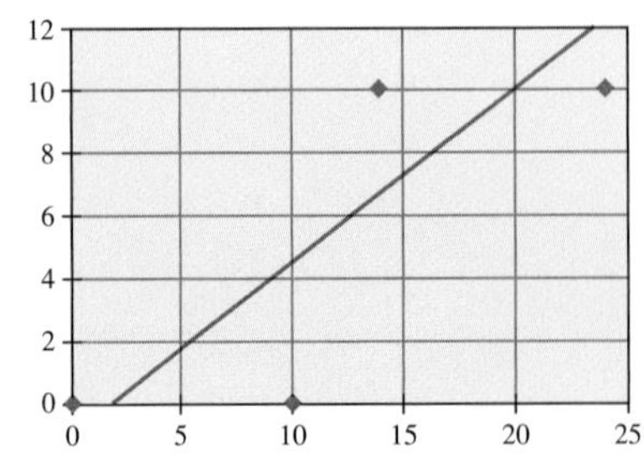

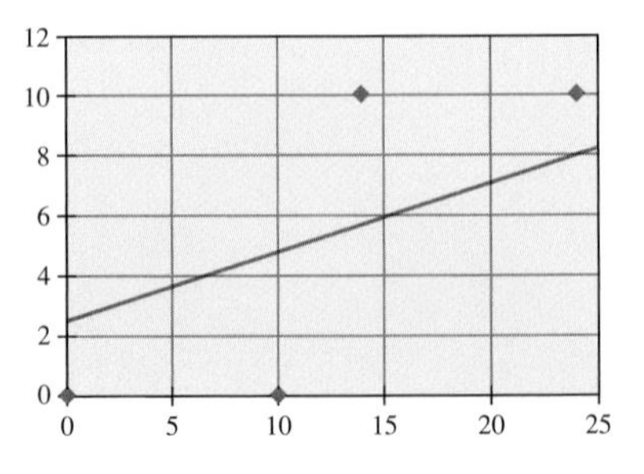

The mathematical method called the least-squares method is often used for determining the "best" line through a set of points and the resulting equation of the line is called a **regression line.** Using techniques from calculus, the regression line is chosen so that the total of the squares of the vertical distances from the points to the line is as small as possible. Note that if all the points were on the line, the line would be a perfect match with the data.

The equation of the regression line is given by the following:

$$y = a + bx \quad \text{where} \quad \begin{cases} b = \dfrac{n\Sigma(xy) - (\Sigma x)(\Sigma y)}{n\Sigma(x^2) - (\Sigma x)^2} \\ a = \dfrac{\Sigma y - b\Sigma x}{n} \\ n = \text{number of points} \end{cases}$$

We can calculate the regression line by hand, or we can use the regression feature that is built into many calculators. To do the calculations by hand, create a table with columns for x, y, x^2, and xy. Do the calculations for each point and then add up the columns.

	x	y	x^2	xy
	0	0	0	0
	10	0	100	0
	14	10	196	140
	24	10	576	240
Sum of each column	48	20	872	380

Now use the formulas to find the values of a and b. Using $n = 4$, $\Sigma(x) = 48$, $\Sigma(y) = 20$, $\Sigma(x^2) = 872$, and $\Sigma(xy) = 380$, we get the following.

$$b = \frac{n\Sigma(xy) - (\Sigma x)(\Sigma y)}{n\Sigma(x^2) - (\Sigma x)^2} = \frac{4(380) - (48)(20)}{4(872) - (48)^2} = \frac{560}{1184} \approx 0.473$$

$$a = \frac{\Sigma y - b\Sigma x}{n} \approx \frac{20 - 0.473(48)}{4} = -0.676$$

Thus, the equation of the line is $y = -0.676 + 0.473x$, and its graph appears below.

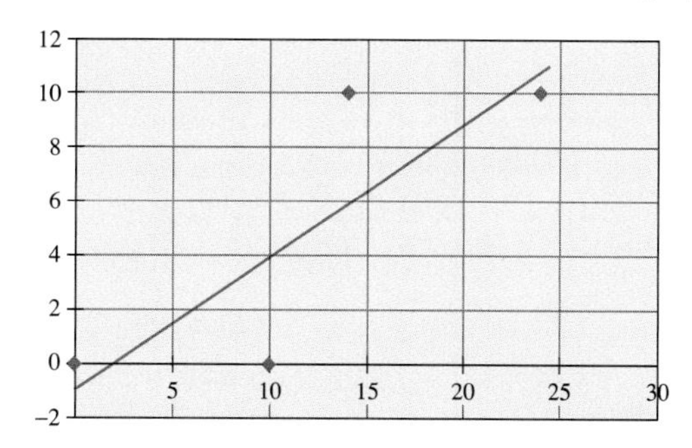

To do this on a calculator, ask your instructor or read your calculator manual.

Once you have the equation of the regression line, you can use it to predict what should happen for other values of x.

EXAMPLE 1

Using the line $y = -0.676 + 0.473x$, what y value occurs when $x = 20$?

Solution: Substituting $x = 20$ into the equation gives $y = -0.676 + 0.473(20) = 8.784$.

One use of regression lines is to estimate trends in prices over a period of time. In the following example, we look at the number of households with cable television.

EXAMPLE 2

According to statistics released by Nielsen Media Research, the following table gives the percentage of U.S. households with cable television.

Year	Percentage
1980	22.6%
1985	46.2%
1990	59.0%
1995	65.7%
1999	68.0%

Using $x = 0$ to represent 1980, $x = 5$ to represent 1985, and so on, find the regression line representing this data and draw a sketch of the data and regression line.

Solution: Using the information in the following table

x	y	x^2	xy
0	22.6	0	0
5	46.2	25	231
10	59.0	100	590
15	65.7	225	985.5
19	68.0	361	1292
49	261.5	711	3098.5

we have $\Sigma x = 49$, $\Sigma y = 261.5$, $\Sigma x^2 = 711$, and $\Sigma xy = 3098.5$, $n = 5$. This gives

$$b = \frac{n\Sigma(xy) - (\Sigma x)(\Sigma y)}{n\Sigma(x^2) - (\Sigma x)^2} = \frac{5(3098.5) - (49)(261.5)}{5(711) - (49)^2} = \frac{2679}{1154} \approx 2.32$$

$$a = \frac{\Sigma y - b\Sigma x}{n} \approx \frac{261.5 - 2.32(49)}{5} \approx 29.6.$$

Therefore, the equation of the regression line is $y = 29.6 + 2.32x$. Sketching the graph of the data and the regression line, we have the following.

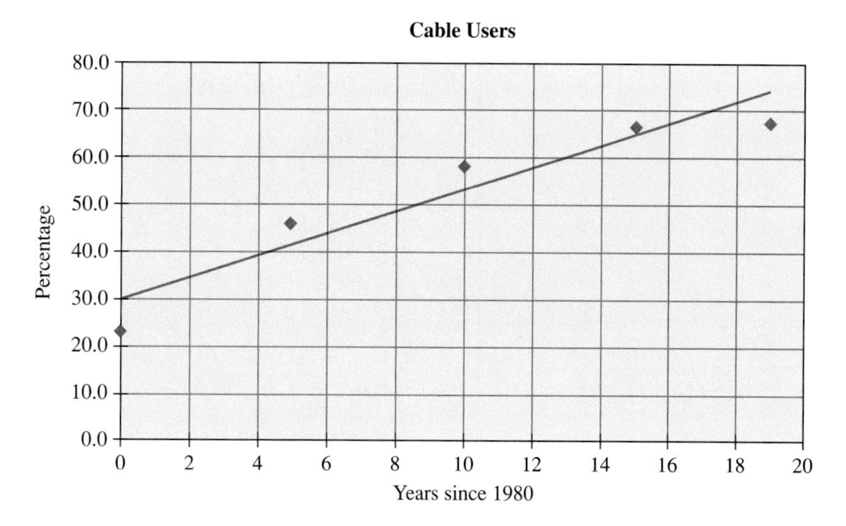

The regression line in Example 2 appears to be a good model of the data. However, as seen in the next example, we must be careful in using this model.

EXAMPLE 3

Use the regression line from Example 2 to estimate the proportion of people who will have cable television in 1997 and in 2020.

Solution: The year 1997 is represented by $x = 17$. Substituting this value into the equation gives $y = 29.6 + 2.32(17) = 69.0$, or 69%. This is close to the actual percentage of cable users in 1997 of 67.3%. The year 2020 is represented by $x = 40$. Substituting this value into the equation gives $y = 29.6 + 2.32(40) = 122.4$, or 122.4%. Since the percentage of viewers with cable cannot exceed 100%, the regression line cannot be used predict cable usage too far into the future.

We conclude this section with two examples that point out the importance of using all the available information and the importance of graphing your data.

EXAMPLE 4

According to statistics released by the U.S. Department of Energy, the following table gives the average price of unleaded regular gasoline in U.S. cities.

Year	Average Price
1980	$1.25
1985	$1.20
1990	$1.16
1995	$1.15
2000	$1.47

Use this information to find the equation of the regression line, predict the average price of unleaded gasoline in 2005, and draw the graph of the data and the regression line.

Solution: As we did in Example 2, we first convert the years into x values, with $x = 0$ representing 1980.

x	y
0	\$1.25
5	\$1.20
10	\$1.16
15	\$1.15
20	\$1.47

Using a calculator, we find that the equation of the line is $y = 1.168 + 0.0078x$. *Note:* If you are doing the calculations by hand,

$$\Sigma x = 50, \quad \Sigma y = 6.23, \quad \Sigma x^2 = 750, \quad \Sigma xy = 64.25, \quad \text{and} \quad n = 5.$$

According to this information, in the year 2005, unleaded gasoline prices should average $1.168 + 0.0078(25) = \$1.36$. Plotting the data and graphing the regression line gives the following.

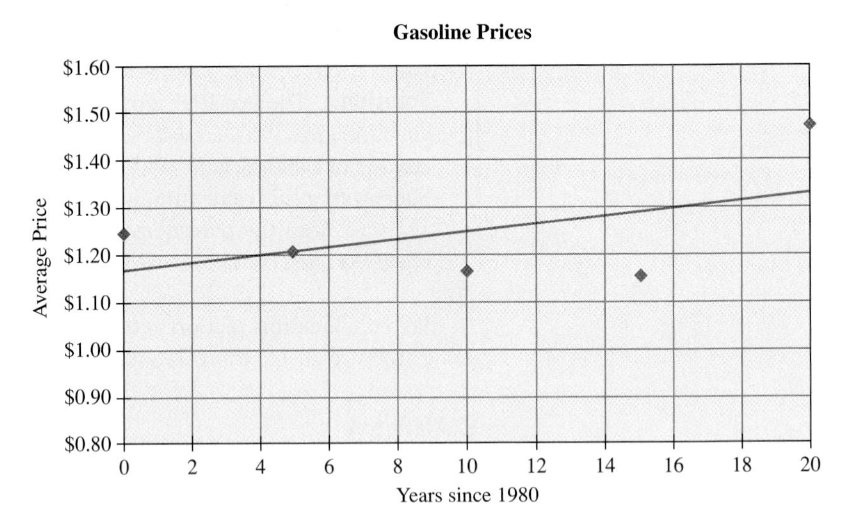

It is important to note that the regression line shows an upward trend in gasoline prices despite a 15-year decline in prices. This disparity provides a good example of why it is always important to be careful in using statistics. The example raises the question of whether the year 2000 was an error, an unusual price spike, or a real indication of future trends. To examine this situation more closely, we could use more data.

EXAMPLE 5

Suppose we repeat the previous example using the gasoline prices for every year starting in 1980. The data are as follows.

Year	Average Price	Year	Average Price
1980	$1.25	1991	$1.14
1981	$1.38	1992	$1.13
1982	$1.30	1993	$1.11
1983	$1.24	1994	$1.11
1984	$1.21	1995	$1.14
1985	$1.20	1996	$1.23
1986	$0.93	1997	$1.23
1987	$0.95	1998	$1.06
1988	$0.95	1999	$1.12
1989	$1.02	2000	$1.47
1990	$1.16		

Solution: Again, letting $x = 0$ represent 1980, we find that the equation of the regression line for the data is $y = 1.1667 - 0.0008x$. From the graph that follows, notice that the data has no simple pattern of linear increase or decrease. The points on the regression line indicate that the price of gasoline has stayed close to $1.16 over a 20-year period of time. Using the regression line to make a prediction of future prices would have little value.

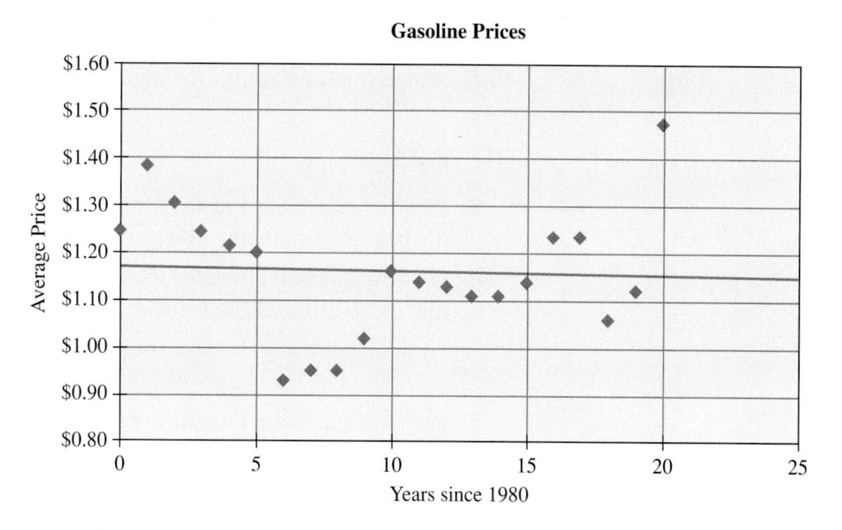

The above examples illustrate some of the uses and pitfalls of regression lines. You should know how to calculate a regression line and how to use a regression line to predict the value of y for a given value of x. You should also understand that it is important to graph the data and the regression line to check whether the regression line is a useful model for the data.

Correlation Coefficient

Most calculators that automatically compute regression lines also compute the correlation coefficient, r or r^2. The correlation coefficient is a statistical measure of how well the regression line models the data. Although a discussion of the correlation coefficient is beyond the scope of this text, users of regression lines will want to read more about this topic in a statistics book.

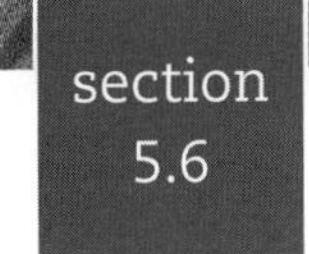

section 5.6

PROBLEMS ▶ Explain ▶ Apply ▶ Explore

▶ Explain

1. What is linear regression?
2. In the linear regression equation $y = a + bx$, which letter represents the slope of the equation?
3. Why is it important to look at a graph of the data and the regression line?
4. Why is it important for the data to be close to the regression line?

▶ Apply

5. Find the regression line for the points (2, 3), (4, 7), (12, 15), and (15, 17). Use the line to estimate the value of y if $x = 10$.
6. Find the regression line for the points (2, 30), (4, 17), (12, 5), and (15, 3). Use the line to estimate the value of y if $x = 10$.
7. Find the regression line for the points (20, 310), (42, 250), (60, 195), and (82, 146). Use the line to estimate the value of y if $x = 70$.
8. Find the regression line for the points (20, 31), (42, 75), (60, 115), and (82, 135). Use the line to estimate the value of y if $x = 70$.

▶ Explore

9. The following table shows the personal savings of Americans as a percentage of income, as given by the U.S. Department of Commerce.
   a) Find the regression line for the data. Use $x = 0$ to represent 1970.
   b) Estimate the percentage of personal income that will be saved in 2004.
   c) Draw a graph of the data and the regression line.

Year	Percentage
1970	8.4%
1975	9.0%
1980	8.2%
1985	6.9%
1990	5.0%
1995	3.4%
1999	2.2%

10. The following table shows the per capita public debt of U.S. citizens (total federal debt divided by the number of people in the United States—that is, your share of the debt), as given by the U.S. Department of Treasury, for years between 1960 and 1980.

a) Find the regression line for the data. Use $x = 0$ to represent 1960.

b) Estimate the per capita public debt in 1985.

c) Draw a graph of the data and the regression line. (Compare the estimated value with the actual value given in Problem 11 below.)

Year	Debt
1960	$9.20
1965	$11.30
1970	$19.00
1975	$32.70
1980	$74.90

11. The following table shows the per capita public debt of U.S. citizens, as given by the U.S. Department of Treasury, for years between 1980 and 2000.

a) Find the regression line for the data. Use $x = 0$ to represent 1980.

b) Estimate the per capita public debt in 2005.

c) Draw a graph of the data and the regression line.

Year	Debt
1980	$74.90
1985	$178.90
1990	$264.80
1995	$332.40
2000	$362.10

12. The following table shows the retail unleaded gasoline prices in Germany, as given by the U.S. Department of Energy.

a) Find the regression line for the data. Use $x = 0$ to represent 1990.

b) Estimate the price of gas in 2005.

c) Draw a graph of the data and the regression line.

d) In Example 5, the U.S. price of gas was shown to fluctuate in such a fashion that a regression line is not a good predictor of prices. Can a regression line be used to get accurate predictions of German gas prices?

Year	Price/gallon	Year	Price/gallon
1990	\$2.65	1995	\$3.88
1991	\$2.90	1996	\$3.94
1992	\$3.27	1997	\$3.54
1993	\$3.07	1998	\$3.34
1994	\$3.52	1999	\$3.42

13. The following table shows the retail unleaded gasoline prices in Japan, as given by the U.S. Department of Energy.

a) Find the regression line for the data. Use $x = 0$ to represent 1990.

b) Estimate the price of gas in 2005.

c) Draw a graph of the data and the regression line.

d) In Example 5, the U.S. price of gas was shown to fluctuate in such a fashion that a regression line is not a good predictor of prices. Can a regression line be used to get accurate predictions of Japanese gas prices?

Year	Price/gallon	Year	Price/gallon
1990	\$3.17	1995	\$4.43
1991	\$3.46	1996	\$3.65
1992	\$3.59	1997	\$3.27
1993	\$4.02	1998	\$2.82
1994	\$4.39	1999	\$3.27

14. According to statistics compiled by the U.S. Department of Labor, the values of the U.S. Consumer Price Index (CPI) are given in the following table. The CPI represents the relative costs of goods. Using 1967 as the base year, what cost \$100 in 1967, cost \$30 in 1915, and \$512 in 2000.

a) Find the regression line for the data. Use $x = 0$ to represent 1960.

b) Draw a graph of the data and the regression line.

c) Use the regression line to predict the CPI in 2005 and 2010.

Year	1960	1965	1970	1975	1980	1985	1990	1995	2000
CPI	89	95	116	161	249	322	391	457	512

15. The sales of DVD players for the years 1997–2000 are shown below. Find the regression line for the data and draw a graph of the data and the regression line. Use the regression line to estimate the number of DVD players to be sold in 2003. Use $x = 0$ to represent 1997.

Year	DVD Player Sales (in thousands)
1997	315
1998	1,089
1999	4,019
2000	8,499

16. If you have a calculator such as the Texas Instruments TI-83 that has built-in regression features, the calculator might be able to do other types of regression. That is, it can determine other mathematical functions that might approximate the data. If so, use your calculator to find the linear regression, exponential regression, and logarithmic regression for the data given in the following table. Sketch the data and all three regression curves. Discuss which curve best matches the data.

x	y
1	5
3	10
5	14
7	16
9	17
11	18
13	18.5

CHAPTER 5 SUMMARY

Key Terms, Concepts, and Formulas

The important terms in this chapter are:	Section
Bar graph: A graphical representation of grouped data formed with rectangles.	5.1
Central tendency: A general term describing the middle value of a set of data; usually refers to mean, mode, or median.	5.2
Class: A category into which data are distributed; used in frequency distributions.	5.1
Class midpoint: The mean of the boundary values for a class.	5.1
Confidence interval: An interval centered on the estimate of the desired statistic.	5.5
Confidence level: A statement of the probability that the actual percentage being studied is contained within the confidence interval.	5.5
Dispersion: A general term describing how far data are spread from a central value.	5.3
Frequency: The number of times a data value or a class value occurs.	5.1
Line graph: A graphical representation of data formed by dots connected by line segments.	5.1
Margin of error: The distance from the center of a confidence interval to each endpoint.	5.6
Mean (arithmetic mean): The sum of the data values divided by the number of data values.	5.2
Median: The middle value of a set; found by listing the data in increasing numerical order and selecting the middle value; if there are an even number of data values, it is the arithmetic mean of the middle pair.	5.2
Mode: The data value that occurs most frequently.	5.2
Normal distribution: A statistical distribution that models many real-world situations; also called the **bell curve** or the **Gaussian distribution.**	5.4
Pie chart: A graphical representation of percentage frequencies using sectors of a circle.	5.1
Range: The difference between the highest and lowest data values.	5.3
Regression line: The line that is used to model a set of points so that the total of the squares of the vertical distances from the points to the line is as small as possible.	5.6
Standard deviation: A method of measuring dispersion of data from the mean.	5.3
Weighted average: The mean of a group of numbers in which certain values have greater importance, or weight, than do other values.	5.2

After completing this chapter, you should be able to:	Section
1. Organize data into tables, charts, and graphs.	5.1
2. Use and explain the concepts of central tendency and dispersion.	5.2, 5.3
3. Know which measure of central tendency is most appropriate for a situation and how to calculate or find that value.	5.2
4. Find the standard deviation of either grouped or nongrouped data.	5.3
5. Use the mean and standard deviation of two sets of data to compare the distributions of the two sets.	5.3
6. Analyze a normal distribution using the normal table.	5.4
7. Interpret the meaning of margin of error and confidence interval in a survey.	5.5
8. Calculate the regression line for a set of data and use the regression line to make predictions about one of the variables.	5.6

CHAPTER 5 REVIEW

Section 5.1

In Problems 1–4, use the following data.

23	45	32	12	25	40	42
33	12	23	41	43	32	36
8	3	27	29	17	17	23
21	29	30	43	7	4	11

1. Use the data to create a frequency distribution with five classes.
2. Use the data and the frequency distribution to create a bar graph.
3. Use the data and the frequency distribution to create a pie chart.
4. Use the data and the frequency distribution to create a line graph.

Section 5.2

5. Find the mean, mode, and median of the original data in Problem 1.
6. Find the mean of the frequency distribution in Problem 1.
7. Describe a real-world situation in which the mode is the best measure of central tendency.
8. Describe a real-world situation in which the median is the best measure of central tendency.

Section 5.3

9. Find the standard deviation of the original data in Problem 1.
10. Find the standard deviation of the frequency distribution in Problem 1.

11. Find the range of the original data in Problem 1.

12. What does standard deviation tell you about a set of data?

Section 5.4

13. What is the area under the normal distribution on the interval $1.5 \leq z \leq 2.3$?

14. If the area under the normal distribution on the interval $z \geq c$ is 0.1222, what is the value of c?

15. The mean of a set of test scores is 73, and the standard deviation is 9.1. Assuming that the scores are normally distributed, find the probability that a particular test score is greater than 90.

16. The mean of a set of test scores is 73, and the standard deviation is 9.1. If scores are normally distributed and the top 12% of the scores are *A*'s, what is the lowest possible score that will receive a grade of *A*?

Section 5.5

17. Explain what is meant by *margin of error* and *confidence level.*

18. Explain why the margin of error should decrease if the sample size increases.

19. A customer requests that a polling company produce a survey with a 90% confidence level and a margin of error of 1%. How many people must respond to the survey?

20. A poll was taken with 2500 people. If the confidence level is 98%, what is the margin of error of the poll?

Section 5.6

21. Find the regression line for the following data.

x	5	6	8	10	10	11	13
y	14	20	19	24	26	30	28

22. Use the results from Problem 21 to estimate y if $x = 20$.

23. Use the results from Problem 21 to estimate x if $y = 40$.

24. Draw a graph showing the data and the regression line.

CHAPTER 5 TEST

1. Given the numbers 2, 4, 6, 8, 3, 6, 7, 9, 1, 3, find
   a) The mean.
   b) The median.
   c) The mode.
   d) The range.
   e) The standard deviation.
2. In 1998, there were 201.3 million people of age 18 or over in the United States. The age distribution is as follows. (Source: U.S. Bureau of the Census.)

Age group	People
18–24	25.7 million
25–44	83.1 million
45–64	58.0 million
65 years and over	34.5 million

   a) Draw a pie chart for this data.
   b) Draw a bar graph for the information. Why is this graph misleading?
3. The recording industry uses various recording media such as phonograph records, cassettes, and compact discs. The following data gives the net number of units shipped for various recording media from 1975 to 2000. (Source: Recording Industry Association of America, www.riaa.org.)
   a) Make a line graph that displays all three types of recording media on the same graph.
   b) What are some conclusions you can draw from the graph?

Recording Media Shipments in Millions

	1975	1980	1985	1990	1995	2000
Phono Records	421.0	487.1	287.7	39.3	12.4	7.0
Cassettes	16.2	110.2	339.1	442.2	272.6	76.0
Compact Discs	0	0	22.6	285.6	722.9	942.5

4. According to the 2000 *Current Population Survey*, published by the U.S. Bureau of the Census, the numbers of working adults age 25 and over, as classified by educational achievement and income level, are given in the following table.
   a) Find the mean income for each education level.
   b) What conclusion can you draw?

	High Diploma or Less	Some College, no Bachelors Degree	Bachelors Degree	Advanced Degree
$1 to $19,999	23,008	10,182	4531	1614
$20,000 to $39,999	19,342	14,123	7864	2562
$40,000 to $59,999	6616	6644	5719	3077
$60,000 to $79,999	1978	2433	3567	2239
$80,000 to $99,999	374	634	1403	1166
$100,000 and more	564	839	2077	2095

5. Suppose you need to purchase 500 resistors rated at 10 ohms. It is critical that the resistance does not vary substantially from the rated value. Two companies have told you that they can supply the part for the same low, low price. Both companies claim their resistors are rated at 10 ohms. However, company A says that the standard deviation of its resistors is 0.6 ohms, while company B says that the standard deviation of its resistors is 1.8 ohms. Which company should you choose as your supplier? Explain your answer.

6. A survey of white-collar workers found that the workers had a mean of 18.14 years of experience, with a standard deviation of 10.08 years. Assuming that the experience level is normally distributed,

   a) What percentage of the white-collar workers had between 10 and 20 years of experience?

   b) What percentage of the workers had between 5 and 10 years of experience?

   c) Find the probability that a blue-collar worker selected at random had more than 15 years of experience.

   d) Find the probability that a blue-collar worker selected at random had more than 25 years of experience.

7. An algebra class averaged 73 on the first midterm with a standard deviation of 14.2. If the top 10% received A's, the next 20% received B's, the next 40% received C's, the next 20% received D's, and the final 10% received F's, use the normal distribution to determine which scores receive which grades.

8. In a CNN poll of 1005 Americans, 89% said that it is morally unacceptable to clone humans. If the margin of error was 3%,

   a) What is the confidence level of the poll?

   b) How large a sample would be needed to have a confidence level of 99% while maintaining a 3% margin of error?

   c) How large a sample would be needed to decrease the margin of error to 1% while maintaining a 99% confidence level?

9. According to a November 2001 Harris poll, the percentage of adults that used the Internet for the years between 1995 and 2001 is given in the following table. Find the regression line for the data and graph the data and its regression line. Use $x = 0$ to represent 1995. What does the model predict for 2010? Explain why a linear regression is a not a good model for the data.

Year	Percentage
1995	9
1997	30
1998	56
2000	63
2001	64

CHAPTER SIX

Modeling with Algebra

Parabolas, curves, and slopes, figures commonly graphed in algebra, form the image of Joseph Stella's *The Brooklyn Bridge: Variation on an Old Theme.* (Whitney Museum of Art, New York)

OVERVIEW

In this chapter, you will investigate how the functions of algebra can act as models for real-life situations. You will see linear functions determine the profits at Carmen's Coffee Shop. You will find out how quadratic functions can predict the flight of a field goal in football. You will study exponential functions that estimate population growth and logarithmic functions that tell how long it takes for a vanilla latte to cool. You will see that algebra is alive in everyday events and can be used to make predictions about the real world.

6

A SHORT HISTORY OF ALGEBRA

As the men and women of ancient civilizations investigated the geometry of the world around them and used their number systems to count, measure, and calculate, they began to generalize the procedures of arithmetic and apply them to unknown quantities. Anthropologists believe that before 2000 B.C., the Chinese, Persians, Babylonians, and people of India might have begun this process and had some elementary knowledge of what we now call algebra. However, the first definite evidence of algebra is found in the Rhind papyrus (c. 1650 B.C.). In this work, the Egyptian mathematician Ahmes included problems such as this:

If a "heap" and a seventh of a "heap" are 19, what is the value of the "heap"?

The ancient Greeks (650 B.C.–A.D. 200) contributed much to the development of mathematics, but their main concern was geometry, not algebra. This left algebra in a stage at which its problems and solutions were stated only in words and used mainly in reference to geometric figures. A major advance in the development of algebra occurred around 250 A.D. The Greek mathematician Diophantus (210–290) worked out a system to solve problems by using symbols to replace numbers and operations. His work made a significant contribution to mathematical notation and expanded the scope of algebra. For this, he is considered by many to be the father of algebra.

The period after the disintegration of the Roman Empire in the fourth and fifth centuries become known as the Dark Ages. These were years of very little progress in the development of algebra in Europe. The main advances came from India and Arabia. Hindu mathematicians such as Brahmagupta (c. 625) continued to use symbols in the solution of mathematical problems. Around 825, Al-Khwârizmî, a teacher of mathematics in Baghdad used the word we know as algebra in his work *Ilm al-jabr walmuqabalah,* meaning "the science of transposition and cancellation." Through his writing, algebra became known as the study of solving equations. The interest in algebra also spread to Persia, where famed poet and mathematician Omar

Founders of algebra are celebrated on postage stamps. On the left, Al-Khwârizmî, the teacher in Baghdad who first used the word *algebra,* was commemorated on his 1200th birthday on a Soviet Union stamp in 1983. On the right, René Descartes, the French mathematician who developed the rectangular coordinate graphing system, was commemorated on his 400th birthday on an Albanian stamp in 1966.

Khayyam (1050–1123) wrote a book on algebra. The Arabian and Indian influence did much to improve number notation and the symbolism of algebra.

As Europe emerged from the Dark Ages, contributions to the development of algebra by Europeans again appeared. Italian merchant Leonardo de Pisa (1202), commonly known as Fibonacci, summarized Arabian algebra and introduced the Hindu-Arabic number system to Europe in his work *Liber Abaci.* Development of algebra also occurred in the East. In 1247, Ch'in Kiu-shao showed the high degree of sophistication of Chinese mathematics in his works on solving higher-degree equations by numerical methods, a discovery that was not made in Europe until 1819. In 1275, a Chinese official, Yang Hui, gave the first account of binomial coefficients some 250 years before they were published in Europe.

The Renaissance in Europe was a time of great progress and creativity in the development of algebra. From 1450 to 1750, most of the symbols, notation, and techniques of algebra that we use today were introduced and accepted. The most notable contributions to elementary algebra are credited to René Descartes (1627). In *La Geometrie,* he introduced notation similar to what we find in present-day algebra. He used x, y, and z for unknowns and the superscript (x^3) for cubes. But most significantly, he brought algebra and geometry together by creating the (x, y) rectangular coordinate system. This allowed the equations of algebra to be represented graphically, laid the foundation for algebraic geometry (analytic geometry), and made possible the development of the calculus.

Following Descartes, many other mathematicians contributed to the development of algebraic techniques, algebraic notation, and the underlying structure of algebra. Because of their work, we find algebra to be a mathematical system that

a) Has a rich history of development.

b) Uses symbols to represent numbers and operations.

c) Allows us to graphically represent and analyze mathematical concepts.

d) Has a logical foundation.

e) Gives us a powerful problem-solving tool.

CHAPTER 6 PROJECTS

Research Projects

1. What are conic sections? Discuss how they are produced, both geometrically and algebraically. What applications do they have? In particular, what are the uses of parabolas in the field of optics and satellite communication technology?
2. What is Fermat's Last Theorem? Why has it caused great interest over the centuries? Investigate claims to prove this theorem. What do these claims suggest about Fermat's Last Theorem?
3. Compare the Dark Ages to the Renaissance in Europe. What factors in the Dark Ages contributed to the stagnation of intellectual pursuits during this period? What factors in the Renaissance contributed to the rebirth of intellectual pursuits during this period? What contributions were made in algebra during these periods?
4. There is evidence that many of the developments in algebra that have been attributed to European mathematicians were developed independently by Asian mathematicians. Investigate this claim. Who were some of these Asian mathematicians and what were their contributions?

Math Projects

1. **Fun with your birth year (19"xy")**
Using the four digits of the year of your birth, represent the whole numbers from 1 to "xy" by creating problems that use all the four digits 1, 9, *x*, and *y*. You can use addition, subtraction, multiplication, division, powers, square roots, factorials, greatest integer functions, parentheses, and the order of operations to create the problems. You must use all the four digits (1, 9, *x*, *y*) in that order. For example, if you were born in 1983, you would represent all the whole numbers from 1 to 83. Here are some examples.

$$2 = -1^{98} + 3$$

$$20 = (1 + 9) \cdot [|8 \div 3|]$$

$$31 = 1 + (\sqrt{9})! + 8 \cdot 3$$

(*Note:* A factorial of a positive whole number is the product of all the positive whole numbers less than the number. For example, 4 factorial = $4! = 4 \cdot 3 \cdot 2 \cdot 1 = 24$, and $3! = 3 \cdot 2 \cdot 1 = 6$. Further, 0! is defined as 1, $0! = 1$. The greatest integer of a number is the closest integer that the number is greater than or equal to. For example, the greatest integer of $3.4 = [|3.4|] = 3$, and the greatest integer of $-3.4 = [|-3.4|] = -4$.)

2. Records in running and swimming events have improved over the last 100 years. Are the runners and swimmers improving at comparable rates? To answer this question, examine the changes in the world records for the 100 m run and freestyle swim every 10 years since 1900.
   a) Analyze the data graphically. Plot the times and the year the record was set on a rectangular coordinate system (100 m time on the *y*-axis and year on the *x*-axis). Are the graphs similar? Can the data be modeled with the functions of algebra studied in this chapter? Explain.
   b) Analyze the data numerically by determining percent decrease in the record times for each 10-year period. How do

the percent decreases compare between runners and swimmers?

c) Using the results of parts (a) and (b), what do you think the records will be in the year 2020? Justify your results.

d) Write a paragraph answering the question "Are the runners and swimmers improving at comparable rates?"

3. In this chapter, you will study exponential functions that model quantities which increase more and more rapidly as time progresses. However, there are phenomena that initially behave like exponential functions but have a slower rate of growth as time progresses and approach a limiting value. A model for such phenomena is a **logistic** (S-shaped) **curve.**

a) What is the algebraic form and graph of such a curve?

b) What kind of phenomena can be modeled with logistics curves?

c) Give some examples of such models, make some predictions using the models, and explain how saturation levels are reached in the models.

section 6.1 Linear Models

The Australian grass tree can be algebraically described by a set of lines emanating from the same point. (Photograph by David Hosking)

In algebra, we study a system in which symbols (usually letters) are used to represent numbers. We can use this system to create mathematical models for various situations. The first model we will look at is one in which the situation can be graphically displayed by a line or part of a line: a **linear model.** In such a model, we assume that the rate at which the quantities change (the slope) is constant. Before we look at some situations in which a linear function makes an appropriate model, let us review some basics about linear functions. For a line passing through the points (x_1, y_1) and (x_2, y_2).

Slope

$$m = \frac{\text{change in the } y \text{ values}}{\text{change in the } x \text{ values}} = \frac{y_2 - y_1}{x_2 - x_1}$$

Equation of a Line

$$y = mx + b \qquad \text{where} \begin{cases} m = \text{slope} \\ b = y \text{ intercept} \end{cases}$$

We will use these equations to find linear functions that can act as models in various situations.

EXAMPLE 1

The equation to convert Celsius (°C) to Fahrenheit (°F) temperature is the linear function

$$F = \frac{9}{5}C + 32.$$

a) Graph this equation.

b) Find its slope and explain what it tells us about the different temperature scales.

c) Find the Fahrenheit equivalent of 40°C.

Solution:

a) To graph a linear function, we find and plot two pairs of values that satisfy the equation. Since the value for C(the domain) can be any real number, we draw a continuous line.

°C	°F
0	32
10	50

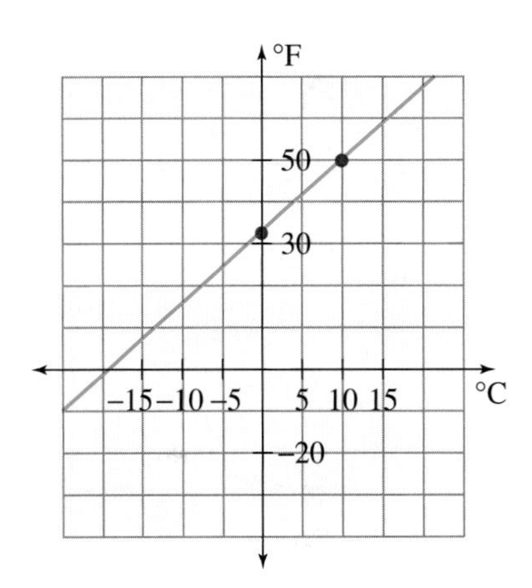

b) From our equation $F = \frac{9}{5}C + 32$, the slope, $m = 9/5$.

$$m = \frac{9}{5} = \frac{\text{change in Fahrenheit (°F)}}{\text{change in Celsius (°C)}}$$

Thus, a 9° change in F corresponds to a 5° change in C.

c) If $C = 40°$, $F = \frac{9}{5}(40) + 32 = 72 + 32 = 104°$.

EXAMPLE 2

Carmen's Coffee Shop had a net loss of \$300 in its first month of operation, January. In April, it had a net profit of \$240. If business continues to grow at this rate, how much profit will the coffee shop make in December? next April?

Solution: Carmen is looking for a way to predict her monthly profit, assuming that her profit will continue to increase at the present rate. A linear function would satisfy those conditions. Let

t = time in months (t is a whole number greater than 0)

p = net monthly profit

(t, p) = ordered pairs relating time and profit.

We are looking for the linear equation that gives us the profit p based on the time t, using the following facts.

$$\text{January, loss of \$300} \rightarrow (1, -300)$$
$$\text{April, profit of \$240} \rightarrow (4, 240)$$

The equation of a line is normally $y = mx + b$. The equation of the line, using the ordered pairs (t, p) instead of (x, y), becomes

$$p = mt + b.$$

We must now find the slope m and the p intercept b to get the linear function that determines profit based on time.

a) Find m: $m = \dfrac{p_2 - p_1}{t_2 - t_1}$. (Note: t and p are used instead of x and y.)

$$m = \frac{240 - (-300)}{4 - 1} = \frac{540}{3} = 180$$

b) Find b: $p = mt + b$. Since $m = 180$, $p = 180t + b$. To find b, substitute either known pair for (t, p); $(4, 240)$ is used here.

$$240 = 180(4) + b$$
$$240 = 720 + b$$
$$-480 = b$$

Thus, the linear function that determines profit based on time for Carmen's Coffee Shop for any month (t) is

$$p = 180t - 480.$$

In December, $t = 12$ and $p = 180(12) - 480 = \$1680$; next April, $t = 16$ and $p = 180(16) - 480 = \$2400$. The graph of Carmen's profit equation is as follows.

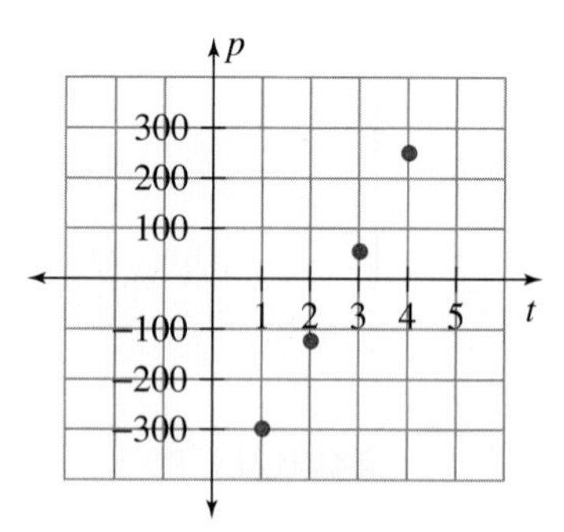

NOTE: Since profit is determined at the end of the month, the graph consists of a dot for each month. The graph is a discrete one; it shows that there are no values at fractional parts of months.

This linear model for Carmen's Coffee Shop is based on the assumption that profit will continue to increase at the same rate. Factors such as competition, prices, salaries, weather, and advertising, can affect profit but have not been included in our model. In any linear model, the approximations become less accurate as you move away from the ordered pairs used to establish the model. Our model does, however, give us a means of making approximations based on present facts.

EXAMPLE 3

The speed of sound has been calculated to be approximately 1090 ft/s, when the temperature is 32°F. However, as the temperature rises above 32°F, the speed at which sound travels increases at a constant rate. At 50°F, the speed of sound is about 1110 ft/s. Find the linear equation that relates the speed of sound to the Fahrenheit temperature and determine the speed of sound at 100°F.

Solution:

Let

$$T = \text{temperature in Fahrenheit where } T \geq 32°\text{F}$$
$$s = \text{speed of sound}$$
$$(T, s) = \text{ordered pairs relating temperature and speed}$$

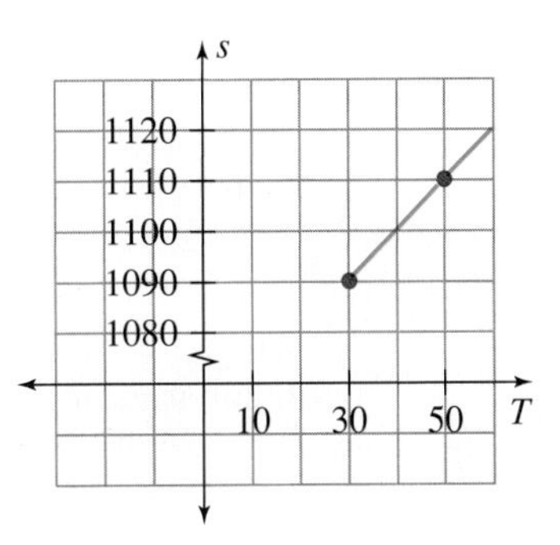

Equations using T and s instead of x and y are

$$s = mT + b \quad \text{where} \quad m = \frac{s_2 - s_1}{T_2 - T_1}.$$

Known pairs for (T, s) are (32, 1090) and (50, 1110).

a) Find m:

$$m = \frac{s_2 - s_1}{T_2 - T_1} = \frac{1110 - 1090}{50 - 32} = \frac{20}{18} = \frac{10}{9}$$

b) Find b in $s = mt + b$. Substitute $m = 10/9$ and either known pair for (T, s). (32, 1090) is used here.

$$1090 = \frac{10}{9}(32) + b$$
$$1090 = 35.6 + b$$
$$1054.4 = b$$

Thus the linear function that gives the approximate speed of sound based on the Fahrenheit temperature is

$$s = \frac{10}{9}T + 1054.4.$$

When $T = 100°F$,

$$s = \frac{10}{9}(100) + 1054.4 = 1165.5 \text{ ft/s}.$$

Linear functions can also be used in analyzing scoring in sports. Suppose a place-kicker scored a school record of 17 points with field goals (3 points each) and extra points (1 point each). In how many different ways could he have scored the 17 points? We could use some logic and determine the different ways to score 17 points—for example, four field goals and 5 extra points or five field goals and 2 extra points. However, if we are not careful in our analysis, we might miss some of the possible solutions. To help us be more accurate, we set up an equation to model the problem and then systematically find all solutions.

Let

x = the number of field goals scored

y = the number of extra points scored.

Since each field goal is 3 points and each extra point is 1 point, we get the following linear equation.

$$3x + 1y = 17$$

There are some logical restrictions on x and y. Both must be whole numbers; the maximum value for $x = 5$, since 6 field goals would be 18 points; and the maximum value for $y = 17$. Thus, the solution to the problem becomes solving the equation $3x + 1y = 17$, where x and y are whole numbers, $0 \leq x \leq 5$, and $0 \leq y \leq 17$. There is no simple algebraic way to find all the solutions to such an equation. The best thing we could do would be to set up a table for x and y and systematically find all solutions by assigning an integer value for x from 0 to 5 and calculating the value for y. (*Note:* In the table, the (3) and (1) next to the x and y were put there to remind us of the number of points for each x and y.)

x(3)	y(1)
0	17
1	14
2	11
3	8
4	5
5	2

There are six ways in which the kicker could score 17 points.

Such an equation with integer coefficients and integer solutions is called a **Diophantine equation** after the Greek mathematician Diophantus (c. A.D. 250).

EXAMPLE 4

A basketball player scored 38 points with only 2-point and 3-point shots. In how many ways could he do this?

Solution:

Let

$$x = \text{the number of 3-point shots}$$
$$y = \text{the number of 2-point shots}$$

Since the total points scored is 38, we get the following linear equation.

$$3x + 2y = 38$$

In that equation, both x and y must be whole numbers, the maximum value for $x = 12$, since thirteen 3-point shots would be 39 points, and the maximum value for $y = 19$. Setting up a table for x and y and systematically finding all solutions, we get the following.

x(3)	y(2)
0	19
2	16
4	13
6	10
8	7
10	4
12	1

There are seven ways in which the basketball player could score the 38 points.

You can see from Example 4 that even with the use of equations, the solution to the problem requires that you think logically and be systematic in your approach. The exercises that follow will let you experience linear functions that model real-world situations.

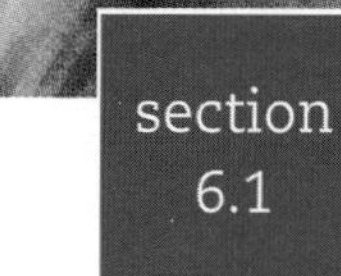

PROBLEMS ○ Explain ○ Apply ○ Explore

○ Explain

1. What assumption is made when a linear equation is used to model a situation?
2. Given the equation $C = mx + b$, where C is the cost of producing x items, what does the value of b represent? Explain.
3. Given the equation $C = mx + b$, where C is the cost of producing x items, what does the value of m represent? Explain.
4. Given the equation $V = mx + b$, where V is the value of a car after it has depreciated for x years, what does the value of m represent?
5. Given the equation $V = mx + b$, where V is the value of a car after it has depreciated for x years, what does the value of b represent?
6. Explain why $x = 2.5$ and $y = 2$ is not a solution to the Diophantine equation, $4x - 3y = 4$.
7. What are Diophantine equations?

○ Apply

In Problems 8–12, conversions from U.S. customary units to metric units are given by linear functions. Graph each equation if the number of U.S. customary units varies from 0 through 10.

8. $c = 2.54i$ c = centimeters, i = inches
9. $k = 2.2p$ k = kilograms, p = pounds
10. $g = 454p$ g = grams, p = pounds
11. $l = \frac{50}{53}q$ l = liters, q = quarts
12. $k = \frac{1000}{621}m$ k = kilometers, m = miles

In Problems 13–16, the profit (P) of a small business after time (t) in months is modeled by a linear function. Graph each equation for integer values of t with $1 \leq t \leq 12$.

13. $P = 750t + 1000$
14. $P = 750t - 1000$
15. $P = -750t + 1000$
16. $P = -750t - 1000$
17. For the four profit functions given in Problems 13–17, give a description of the profit outlook for each small business.
18. Find five integer solutions to the Diophantine equation $4x - 3y = 4$.
19. Find five integer solutions to the Diophantine equation $3x + 5y = 8$.
20. During the summer, as the temperature gets over 80°F, the chickens on a chicken farm drink more water. This behavior is modeled by the following equation.

$w = 25t - 1250$ where $\begin{cases} w = \text{number of gallons of water drunk per hour} \\ t = \text{Fahrenheit temperature } (t \geq 80°) \end{cases}$

a) Sketch a graph of this function.

b) What is the slope of the function and what does it tell us about the situation?

c) How many gallons of water are used in an hour when the temperature is 100°F?

21. Clover, a local department store, notices that there is a direct relationship between the gross revenue R in dollars on a given day and the number n of customers entering the store. It is determined that the equation $R = 2.3n$ approximates this revenue.

a) Sketch a graph of this equation.

b) What is its slope and what does it tell us about this situation?

c) If 1500 people enter the store on a given day, what is the approximate revenue?

22. The equation $W = 1.7x + 0.3$ gives the weight W in pounds of a $\frac{3}{4}$-in. galvanized pipe with protective thread caps. The variable x represents the length of the pipe in feet.

a) Graph the equation giving the linear relationship between the length and weight of the pipe.

b) What is the slope of the equation? What is the physical significance of the slope in the problem?

c) Find the weight of an 8-ft-long piece of pipe.

d) If a piece of pipe weighs 19.85 lb, how long is the pipe?

23. The equation $C = 450x + 1200$ gives the cost C in dollars when x French wine barrels are produced.

a) Graph the equation giving the linear relationship between the number of barrels produced and the cost of producing those barrels.

b) While you may draw the graph as a line, what should be noted about the graph?

c) What is the slope of the equation? What is the physical significance of the slope in the problem?

d) Find the cost of producing 100 barrels.

e) How many wine barrels may be completed if you are allowed to spend $100,000?

24. You purchase an SUV for $26,500. A year later the car is worth only $24,800. If the value of the car continues to depreciate at that rate,

a) Find the linear equation that determines the value of the car based on the number of years you own it.

b) When will the car be worth $500?

25. As a weather balloon rises in altitude from sea level to 6 mi above sea level, the temperature decreases at a fairly constant rate. If the temperature is 59°F at sea level and 55.5°F at 1000 ft,

a) Find the linear equation that relates the temperature t to the altitude a.

b) What is the temperature at an altitude of 24,000 ft?

26. When water turns to ice, its volume increases about 9%. For example, 100 ml of water has a volume of 109 ml after being frozen. Find the linear function that determines the volume of water after it has been frozen.

27. The foundation for a brick wall rises 10 in. above ground level. The bricks used for the wall are 8 in. high.

a) Find a function that determines the height of the wall in inches based on the number of layers of brick that have been laid.

b) Sketch a graph of the function.

c) Find the equation that would give the height of the wall in feet.

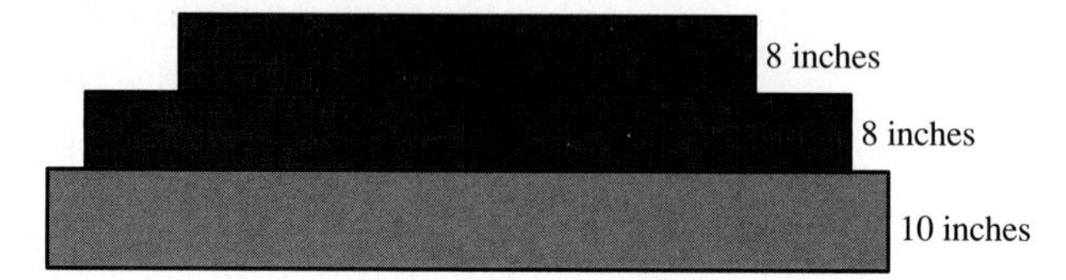

28. A fish tank is sitting on a stand 27 in. high. The tank is 36 in. tall and is being filled with water. The water is rising in the tank at a rate of 3 in./min.

a) Find a function that determines the distance the water level is above the floor at any given moment.

b) What is the domain of the function?

c) Sketch a graph of the function.

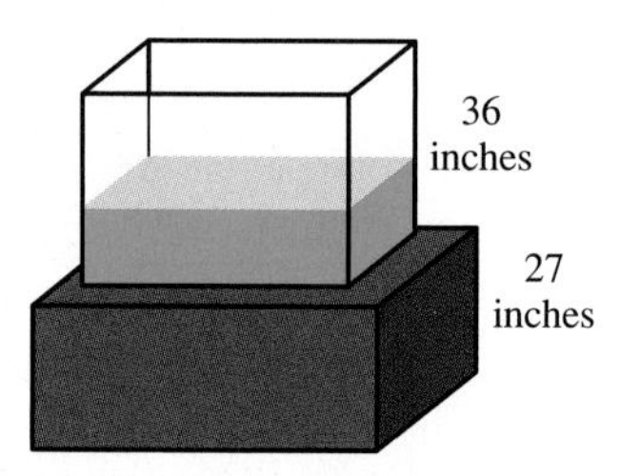

29. At higher altitudes, water boils at a lower temperature. At sea level, water boils at approximately 212°F. In Asheville, N.C., at 2000 ft, water boils at approximately 208°F.

a) Assuming that this relationship is linear, find an equation that relates altitude to the boiling temperature of water.

b) Find the boiling temperature of water at the top of Pike's Peak, Colorado, elevation 14,110 ft.

30. The Morgan Hill Car Rental Agency charges a one-time fixed charge of $35 for a Ford Taurus and a mileage charge of $0.20 per mile.

a) Write a linear equation to determine the cost (C) of driving m miles.

b) Use the equation to find the cost of driving from San Francisco to Los Angeles and back (806 miles).

c) How far can you drive for $200?

31. According to the tax rate schedules given in the 2001 Federal Income Tax booklet, a single person with a taxable income over $297,350 will owe a tax of $93,374 plus 39.1% of the amount of taxable income over $297,350.

a) Write a linear equation to determine the tax (T) for (x) amount of taxable income over $297,350.

b) Use the formula to determine the amount of tax owed by a single person making $45,000 more than $297,350.

c) According to the equation in part (a), a single person owing $120,000 in taxes has how much in taxable income?

32. For residential customers in Pleasanton, California, during June 2002, Pacific Gas and Electric Company determined the cost of electricity used according to the following: baseline usage of 379.5 kwh (kilowatt hours) cost $47.76, and each additional kilowatt hour up to 113.9 kwh costs $0.14321.

a) Write a linear equation to determine the cost (C) for electricity usage of (k) kwh above the baseline quantity to the additional 113.9 kwh.

b) If a home used 462 kwh in a month, what was the charge for electricity usage?

c) If the electricity costs for a month were $58.75, how many kilowatt hours above the baseline were used?

33. Is there a linear relationship between weight and height of men and women? The lists that follow show average weights of 20- to 24-year-old Americans by height.

Average Weight in Pounds of 20- to 24-Year-Old Americans

Men		Women	
Height	**Weight**	**Height**	**Weight**
5′7″	153	5′0″	112
5′9″	162	5′2″	120
5′11″	171	5′4″	128

a) Why does a linear equation fit these lists?

b) Find the equation that determines the weight for 20- to 24-year-old men and for 20- to 24-year-old women.

c) Check the equations out on various people. Do the equations work? When? For whom?

34. Windchill factor is a combination of the actual temperature and wind speed. The wind makes it feel colder than it really is. Below are the windchill Fahrenheit temperatures when the wind speed is 10 mph.

Actual Temperature	Windchill Temperature (at 10 mph)
40°F	28°F
30°F	16°F
20°F	3°F
10°F	−9°F
0°F	−22°F
−10°F	−34°F
−20°F	−46°F
−30°F	−58°F

a) Explain why a linear function does not exactly fit this chart.

b) Find a linear equation that relates windchill to the actual temperature by using the first two pairs of information.

c) How much error will that equation have in determining the windchill temperature when the actual temperature is −30°F?

d) Plot the given windchill data and the graph of the equation from part (b) on the same coordinate system.

e) How does the linear equation compare to the actual data?

35. In a football game, 56 points were scored. If the scoring consisted of touchdowns with the extra point (7 points each) and field goals (3 points each), in how many different ways could the 56 points be scored?

36. A skilled archer while shooting at a standard archery target got all her arrows in the center circle (9 points each) or in the next concentric circle (7 points each). What are all the possible ways in which she could have scored a total of 189 points?

section 6.2

Quadratic Models

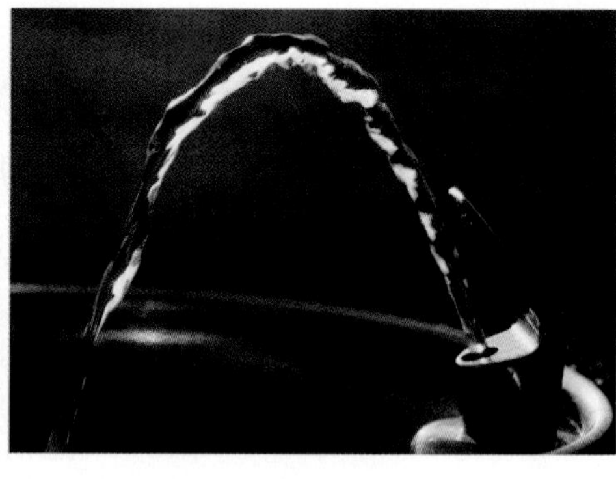

The fireworks display on Independence Day (courtesy of CORBIS) and the drinking fountain (Photographed by Philip James) can be algebraically described by parabolic functions.

The linear function studied in the previous section is only one of many functions that can be used as a mathematical model. There are situations in which the appropriate mathematical model is a quadratic equation with its parabolic graph. In this section, we will look at this model with a primary focus on sporting events. The path that is traced when athletes or objects travel through the air and return to the ground can be represented by a parabola. Quadratic functions can be used to model those paths. Before we can analyze such parabolas in sports, we will review the algebra of quadratic functions.

QUADRATIC FUNCTIONS

A **quadratic function** is a function whose graph is a parabola and has an equation of the form

$$y = ax^2 + bx + c, \qquad \text{where} \quad a \neq 0.$$

To graph a quadratic function, it is best to find the vertex (the maximum or minimum point) of the parabola it represents. The x coordinate of the vertex can be determined by the following equation.

$$x = \frac{-b}{2a} \quad \text{where} \quad \begin{cases} \text{the parabola opens upward if } a > 0 \\ \text{the parabola opens downward if } a < 0 \end{cases}$$

The y coordinate of the vertex can be found by substituting the value found for x into the quadratic function.

EXAMPLE 1

Find the vertex and graph of $y = x^2 - 6x + 4$.

Solution:

The x coordinate of the vertex is

$$x = \frac{-b}{2a} = \frac{-(-6)}{2(1)} = \frac{6}{2} = 3.$$

Substitute $x = 3$.

The y coordinate of the vertex is

$$y = x^2 - 6x + 4$$

$$y = (3)^2 - 6(3) + 4$$

$$y = 9 - 18 + 4$$

$$y = -5.$$

Thus, the vertex is $(3, -5)$. Since $a > 0$, the parabola opens upward, and the vertex is a minimum point.

To graph the parabola, calculate and plot points using x values to the left and right of the vertex.

x	y	
0	4	
1	−1	
2	−4	
3	−5	(vertex)
4	−4	
5	−1	
6	4	

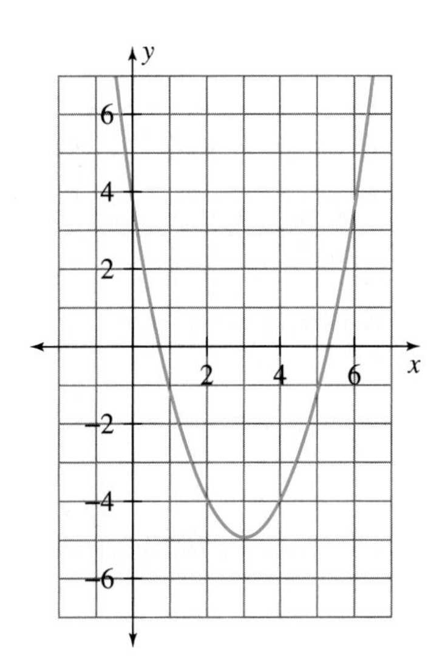

EXAMPLE 2

Graph $y = -2x^2 + 5x$.

Solution: Find the vertex.

$$x = \frac{-b}{2a} = \frac{-5}{2(-2)} = 1.25$$

$$y = -2(1.25)^2 + 5(1.25) = 3.125$$

Since $a < 0$, the parabola opens downward and the vertex (1.25, 3.125) is the maximum point of the parabola.

Calculate and plot points using x values to the left and right of the vertex.

x	y
−1	−7
0	0
1	3
1.25	3.125
2	2
3	−3

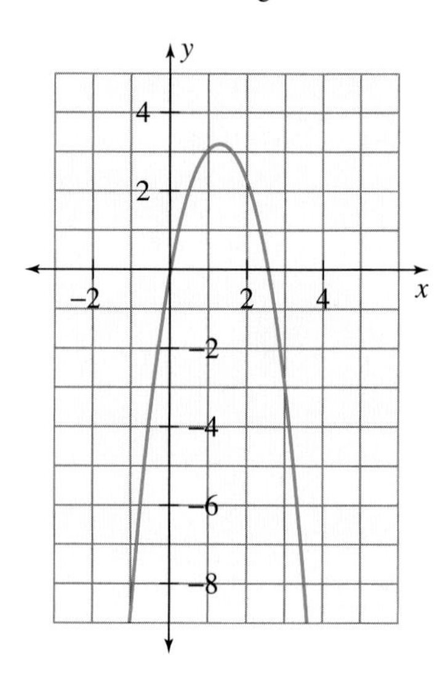

QUADRATIC FORMULA

If the y value of a quadratic function is zero, we have the quadratic equation

$$ax^2 + bx + c = 0.$$

We solve for x and find the x intercepts of the graph of the quadratic function by using a calculator and the quadratic formula.

Quadratic Formula

$$x = \frac{-b \pm \sqrt{b^2 - 4ac}}{2a}$$

where $a \neq 0$, and a, b, and c are the constants in the quadratic equation.

If the quantity under the square root sign is negative, the solutions are not real numbers. Since most calculators use only real numbers, your calculator might give you an error message when the square root of a negative number is attempted. The solutions to the quadratic equations in this text are real numbers. Therefore, if the error message occurs, please check your computation.

EXAMPLE 3

Solve for x: $x^2 - 5x + 7 = 4$.

Solution: First, get zero on one side of the equation by adding -4 to both sides of the equation.

$$x^2 - 5x + 7 = 4$$

$$x^2 - 5x + 3 = 0$$

$$x = \frac{-(-5) \pm \sqrt{(-5)^2 - 4(1)(3)}}{2(1)}$$ Use the quadratic formula with $a = 1, b = -5$, and $c = 3$.

$$x = \frac{5 \pm \sqrt{25 - 12}}{2} = \frac{5 \pm \sqrt{13}}{2} \approx 4.30 \text{ or } 0.70$$

We now return our attention to investigating parabolas in sports.

EXAMPLE 4

A competitor in the annual Calaveras County Frog Jumping Contest held in Angels Camp, California, was Rosie the Ribiter. Her three consecutive jumps totaled 21 ft $5\frac{3}{4}$ in. If her first jump was 7 ft (84 in.) in length and reached a height of 18 in. at its apex, find the quadratic function that models the parabolic path taken by Rosie on that first of three jumps.

Solution: Let

x = length of jump in inches

y = height of jump in inches

$(0, 0)$ = starting point of jump

$(84, 0)$ = landing point of jump

The apex of the jump occurs half way between $x = 0$ and $x = 84$, so the vertex of the parabola is $(42, 18)$. We will now use the general equation for a quadratic function.

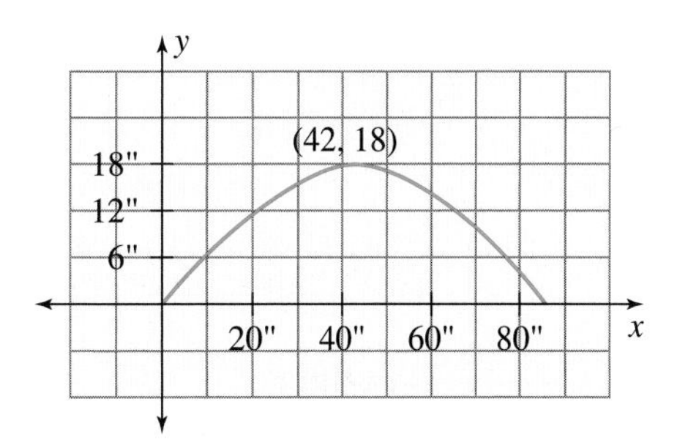

a) Since (0, 0) is on the graph,

$$y = ax^2 + bx + c$$
$$0 = a(0)^2 + b(0) + c$$
$$0 = c.$$

b) Since (84, 0) is on the graph and $c = 0$,

$$y = ax^2 + bx$$
$$0 = a(84)^2 + b(84)$$
$$0 = 7056a + 84b.$$

c) Since (42, 18) is also on the graph,

$$y = ax^2 + bx$$
$$18 = a(42)^2 + b(42)$$
$$18 = 1764a + 42b.$$

d) Solve the simultaneous system of equations obtained from parts (b) and (c).

$$\begin{aligned} 7056a + 84b = 0 \quad &\rightarrow \quad 7056a + 84b = 0 \\ [1764a + 42b = 18](-2) &\rightarrow \underline{-3528a - 84b = -36} \\ &\qquad 3528a = -36 \\ &\qquad a = \frac{-1}{98} \end{aligned}$$

Substituting that value for a, we get $b = \frac{6}{7}$.

Thus, the quadratic function that models the first jump of Rosie the Ribiter is

$$y = \frac{-1}{98}x^2 + \frac{6}{7}x.$$

This equation can be used to further analyze the jump. For example, how far from the takeoff was Rosie the Ribiter when she was 10 in. off the ground? By replacing y with 10 in., we can solve for the distance from the takeoff point.

$$y = \frac{-1}{98}x^2 + \frac{6}{7}x$$

$$10 = \frac{-1}{98}x^2 + \frac{6}{7}x$$

$$980 = -x^2 + 84x \qquad \text{multiplying both sides by 98}$$

$$x^2 - 84x + 980 = 0 \qquad \text{getting zero on one side}$$

$$x = \frac{84 \pm \sqrt{7056 - 3920}}{2} \qquad \text{using the quadratic formula}$$

$$x = 14 \text{ or } 70$$

Thus, Rosie was 10 in. off the ground at two times in her jump, 14 in. and 70 in. from her takeoff.

EXAMPLE 5

When a football is kicked for a 40-yard field goal, the parabolic path of the ball can be modeled by the quadratic function $y = -0.033x^2 + 1.42x$, where a coordinate system is established at the point where the ball is kicked. The x coordinate represents the distance in yards from the point on the ground where the ball was kicked, and the y coordinate represents the height of the kick in yards.

a) When the ball is 15 yd from the goal posts, what is its height?

b) The crossbar of the goal is 10 ft above the ground. How far above the crossbar did the ball cross the goal line?

c) How far behind the goal line did the ball hit the ground?

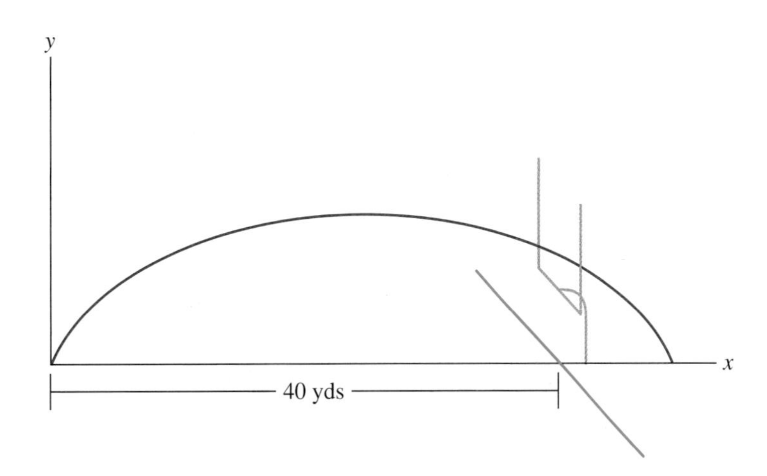

Solution:

a) When the ball is 15 yd from the goal posts, $x = 40 - 15 = 25$. Substituting $x = 25$ into the quadratic equation modeling the path of the football, we get

$$\begin{aligned} y &= -0.033x^2 + 1.42x \\ &= -0.033(25)^2 + 1.42(25) \\ &= -20.625 + 35.5 \\ &= 14.875 \text{ yd.} \end{aligned}$$

b) When the ball passes over the crossbar, $x = 40$. Substituting $x = 40$ into the equation we get

$$\begin{aligned} y &= -0.033x^2 + 1.42x \\ &= -0.033(40)^2 + 1.42(40) \\ &= -52.8 + 56.8 \\ &= 4 \text{ yd.} \end{aligned}$$

Since the crossbar is 10 ft high, the ball passes 2 ft above the crossbar.

c) When the ball hits the ground, $y = 0$.

Substituting $y = 0$ into the quadratic equation modeling the path of the football, we get

$$\begin{aligned} 0 &= -0.033x^2 + 1.42x \\ 0 &= (-0.033x + 1.42)x \end{aligned}$$

$$x = 0 \text{ or } x = \frac{-1.42}{-0.033} = 43.0 \text{ yd.}$$

Thus, the ball lands 3 yd beyond the plane of the crossbar.

EXAMPLE 6

A free throw in basketball follows a parabolic path. Suppose the ball leaves the player's hands with the center of the ball 7 ft above the ground and the ball reaches an apex of 13 ft halfway to the basket. The center of the basket is 10 ft above the floor and 14 ft from the free-throw line. Sketch a graph of the path the ball takes during a free throw. Find a quadratic function that models the free throw.

Solution: If we set up a rectangular coordinate system at the free-throw line, where the h = height of the free throw in feet and x = horizontal distance in feet, we can establish the position of the center of the ball at three points (x, h): start at (0, 7), apex at (7, 13), basket at (14, 10).

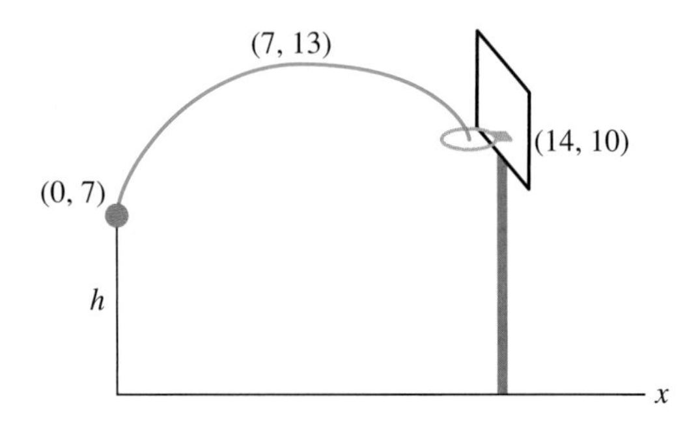

We must find the quadratic function $h = ax^2 + bx + c$ that passes through the points. Substituting the values for x and h from the three known points, we can find the values for a, b, c.

$$h = ax^2 + bx + c$$

At (0, 7), $7 = a(0)^2 + b(0) + c \Rightarrow c = 7$

At (7, 13), $13 = a(7)^2 + b(7) + c \Rightarrow 49a + 7b + 7 = 13$

At (14, 10), $10 = a(14)^2 + b(14) + c \Rightarrow 196a + 14b + 7 = 10$

Solving the system of equations for a and b, we get the following.

$$\begin{aligned} 49a + 7b = 6 &\xrightarrow{(-2)} -98a - 14b = -12 \\ 196a + 14b = 3 &\longrightarrow \underline{196a + 14b = 3} \\ &\qquad\qquad 98a = -9 \\ &\qquad\qquad a \approx -0.092 \end{aligned}$$

$$\begin{aligned} 49(-0.092) + 7b &= 6 \\ 7b &= 10.508 \\ b &\approx 1.5 \end{aligned}$$

Thus, the equation that models the flight of the free-throw is $h = -0.092x^2 + 1.5x + 7$.

EXAMPLE 7

The flight of an Olympic Games discus throw can be modeled with the quadratic function $y = -0.004x^2 + 0.8x + 5$, where x is the distance along the ground in feet and y is its height above the ground in feet. What is the distance of the throw modeled by the quadratic function?

Solution: To find the distance of the throw, we find the x value when the discus hits the ground. When the discus hits the ground, the y value equals zero. Setting $y = 0$ and solving using the quadratic formula, we get the following.

$$\begin{aligned} y &= -0.004x^2 + 0.8x + 5 \\ 0 &= -0.004x^2 + 0.8x + 5 \\ x &= \frac{-0.8 \pm \sqrt{(0.8)^2 - 4(-0.004)(5)}}{2(-0.004)} \\ &= \frac{-0.8 \pm \sqrt{0.64 + 0.08}}{-0.008} \\ &= \frac{-0.8 \pm 0.8485}{-0.008} \\ &= 206.06 \quad \text{or} \quad -6.06 \end{aligned}$$

Thus, the discus traveled 206.06 feet.

In the exercises that follow, you will encounter other situations that can be modeled with quadratic functions.

section 6.2

PROBLEMS ➤ Explain ➤ Apply ➤ Explore

➤ Explain

1. Describe how to find the vertex of a parabola.
2. What is the quadratic formula? What do the a, b, and c represent in the formula?
3. When an object follows the parabolic path $y = ax^2 + bx + c$, how do you determine when the object hits the ground?
4. When an object follows the parabolic path $y = ax^2 + bx + c$, how do you determine at what height the object begins its flight?

➤ Apply

In Problems 5–8, find the vertex, graph, and x intercepts of each parabola.

5. $y = -x^2 + 5$
6. $y = -x^2 + 4x$
7. $y = 0.3x^2 + 1.8x - 2$
8. $y = -x^2 + 6x - 5.5$
9. The path traveled by a golf ball hit with a 9-iron can be modeled with the quadratic function, $y = -0.042x^2 + 5x$, where x is the distance in yards from the point it was hit and y is the height of the golf ball in feet. Assuming that the ground is level,

   a) Find the maximum height reached by the ball.

   b) How far from where it was hit does the ball hit the ground?

10. The path that the feet of a typical college pole vaulter take to the top of the crossbar in a successful attempt can be described by the quadratic function $h = -\frac{1}{8}d^2 + 3d$, where $h =$ the height of his feet above the ground measured in feet, $d =$ the horizontal distance from his takeoff point to the bar measured in feet, and $0 \leq d \leq 12$.

a) Graph the path followed by the vaulter's feet.

b) How high did the vaulter's feet get on the vault?

11. In a ramp-to-ground jump, a motorcyclist left the ramp at an angle of 24° and a speed of 88 ft/s (60 mph). His jump can be modeled by the quadratic function $h = -0.0025d^2 + 0.45d + 15$, where d is the distance in feet along the ground from where the motorcycle left the ramp and h is the height of the motorcycle in feet.

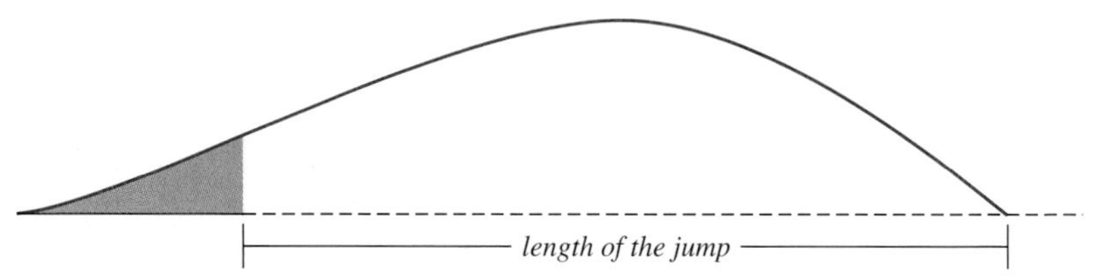

a) How far above the ground was the motorcycle when it left the ramp?

b) How long was the jump?

c) What is the maximum height reached by the motorcycle?

Explore

12. At a local frog jumping contest, Rivet's jump can be approximated by the equation $y = -\frac{1}{6}x^2 + 2x$, and Croak's jump can be approximated by $y = -\frac{1}{2}x^2 + 4x$, where $x =$ the length of the jump in feet and $y =$ the height of the jump in feet.

a) Which frog jumped higher? How high did it jump?

b) Which frog jumped farther? How far did it jump?

13. A punter kicks a football from a point 2 ft above the ground at the goal line. The low flying ball hits the ground 42 yd straight down the field. If 35 yd down the field, the ball has a height of 10 ft,

a) Find the quadratic function that expresses the relationship between the distance (d) along the ground in yards and the height (h) in feet of the football.

b) What is the maximum height reached by the football?

c) How far from the goal line is the ball 11 ft above the ground?

14. A diver does a swan dive from a platform 20 ft above the water. The path followed by the feet of the diver can be approximated by a parabola as shown in the figure.

a) Find the quadratic equation that gives a mathematical description of the dive.

b) How far from the point directly below the takeoff point will the diver's feet enter the water?

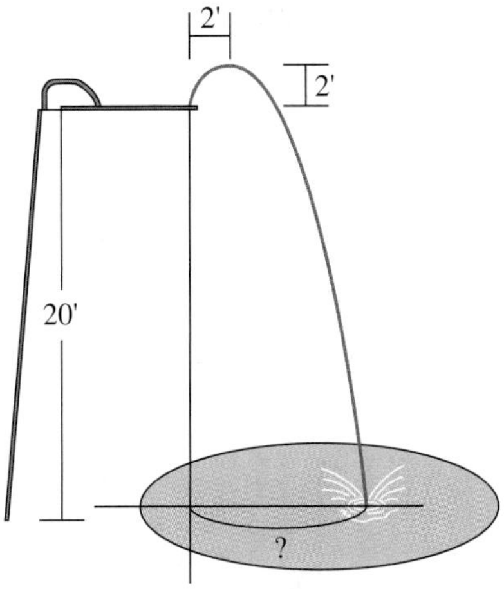

15. The approximate distance it takes to stop a car, based on the speed the car is traveling, is given in the following chart.

Miles per Hour	Stopping Distance (ft)
25	62
35	106
45	161
50	195
55	228
65	306

a) Find a quadratic function based on the stopping distance for 25 and 50 mph, and the fact that at 0 mph the stopping distance is 0 feet.

b) Use that equation to predict the stopping distances for 55 mph and 65 mph and compare them with the distances given in the chart.

c) How accurate is the quadratic function?

d) If the equation continues to be valid for higher speeds, how many feet would it take to stop a drag racer that reaches a speed of 230 mph?

16. To get maximum distance of an arrow shot from a bow, the arrow should be aimed at about a 45° angle with the horizontal. The equation of motion for such an arrow can be approximated by the following.

$$y = \frac{-32x^2}{v_0^2} + x \quad \text{where} \quad \begin{cases} x = \text{horizontal distance traveled in feet} \\ y = \text{height reached in feet} \end{cases}$$

Assume that an arrow leaves the bow with a speed of 192 ft/s.

a) Find the equation of motion for the arrow.

b) Find the height the arrow reaches.

c) Find the distance the arrow reaches.

d) How much higher would the arrow get if it is shot directly upward?

17. The numbers shown below, 1, 3, 6, 10, . . . , are called the triangular numbers. The first triangular number has a value of 1, the second has a value of 3, the third has a value of 6, and so forth. The relationship between the value of the triangular number V and its position in the sequence n is quadratic.

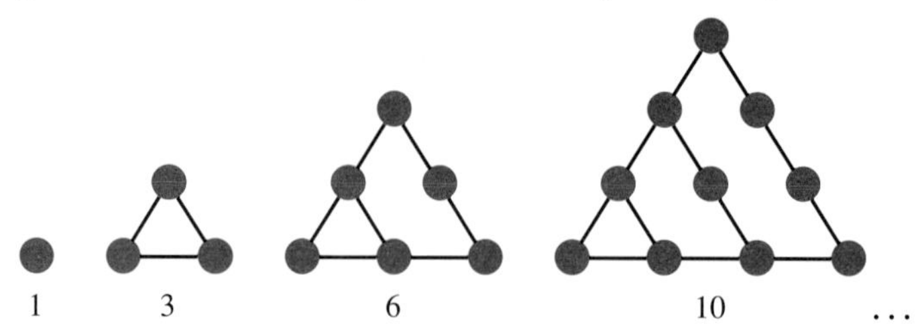

a) Find the equation that expresses V in terms of n.

b) Find the 100th triangular number.

18. A dog breeder has 260 ft of fencing material to make a kennel for the dogs along the side of a garage as shown in the figure. Let x represent the width of the kennel and y represent the length of the kennel.

a) Find the equation that represents y in terms of x,

b) Find the values for x and y that give the maximum area for the kennel. (*Hint:* Find the vertex of a parabola.)

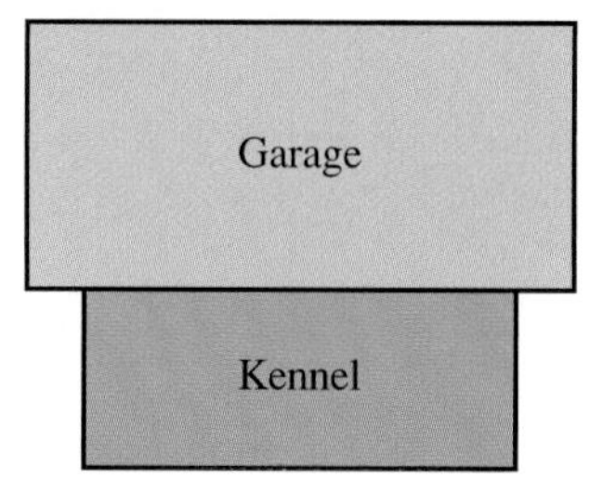

section 6.3

Exponential Models

In the earlier sections of this chapter, we studied two types of models: linear and quadratic. Linear equations are used to model situations that have a steady rate of increase or decrease, and quadratic equations are used to model situations displaying parabolic behavior. In this section, we will examine exponential functions.

EXPONENTIAL FUNCTIONS

An **exponential function** is a function that contains an exponent that is a variable and a base that is a constant, such as $y = 2^x$, $y = 1.025^x$, $y = e^x$, and $y = 10^x$. The base of an exponential function is greater than zero and not equal to one. The equation

$$y = b^x \quad \text{where} \quad b > 0, \quad b \neq 1, \quad x \text{ is a real number}$$

defines the basic exponential function. You must understand that there is a significant difference between a quadratic and an exponential function. You can see this difference by examining a chart of values and the graphs of the quadratic function $y = x^2$ and the exponential function $y = 2^x$.

$y = x^2$	
x	y
-3	9
-2	4
-1	1
0	0
1	1
2	4
3	9
4	16
5	25
6	36

$y = 2^x$	
x	y
-3	$\frac{1}{8}$
-2	$\frac{1}{4}$
-1	$\frac{1}{2}$
0	1
1	2
2	4
3	8
4	16
5	32
6	64

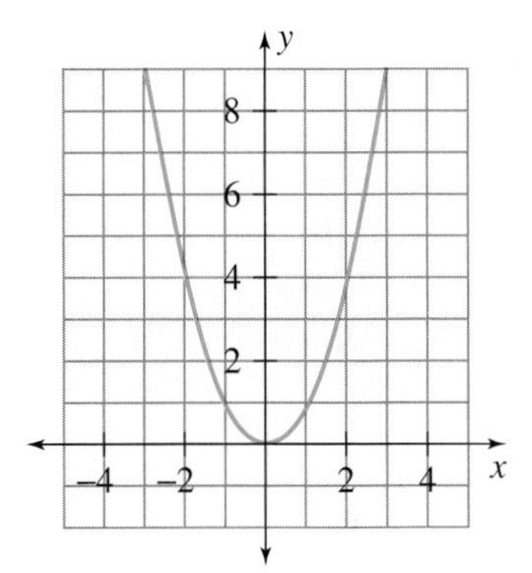

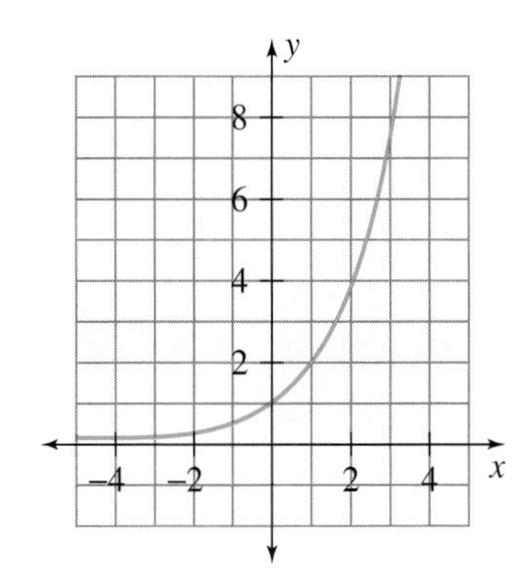

EXAMPLE 1

Graph the exponential function $y = 2^{-x} = \left(\frac{1}{2}\right)^x$.

Solution: Calculate a chart of values for (x, y) and plot the points.

$y = 2^{-x}$	
x	y
-3	8
-2	4
-1	2
0	1
1	$\frac{1}{2}$
2	$\frac{1}{4}$
3	$\frac{1}{8}$

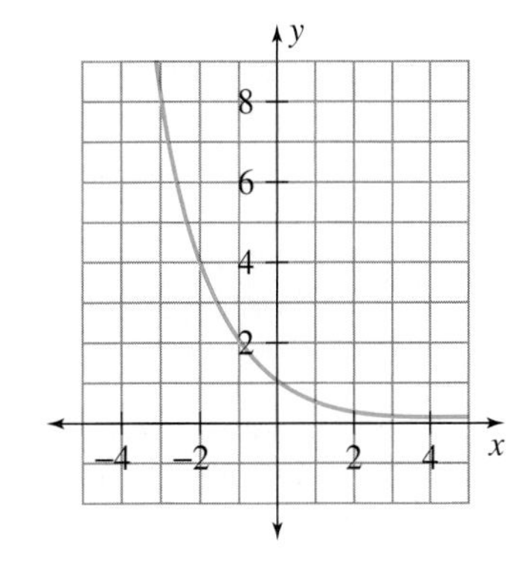

By examining exponential functions of the form $y = b^x$, where $b > 0$ and $b \neq 1$, we can summarize the following properties of an exponential function.

1. Its graph is a continuous curve that has the shape of a "banana" or "hockey stick."
2. Its graph passes through the point (0, 1).
3. Its graph approaches the x axis but never touches it.
4. If $b > 1$, then b^x increases as x increases.
5. If $0 < b < 1$, then b^x decreases as x increases.

In this section, we will consider exponential functions that include other constants along with the basic exponential form of b^x. No matter what constants are used, the graph of the exponential function retains the basic "banana" or "hockey stick" shape. The graph of an exponential function can be determined by calculating and plotting a sufficient number of points that satisfy the function.

EXAMPLE 2

Graph $y = 4 + 10^x$.

Solution: With the help of a calculator, determine and plot points that satisfy the exponential function. A sufficient number of points must be determined so that the exponential function can be graphed.

x	y
−3	4.001
−2	4.01
−1	4.1
0	5
1	14
2	104

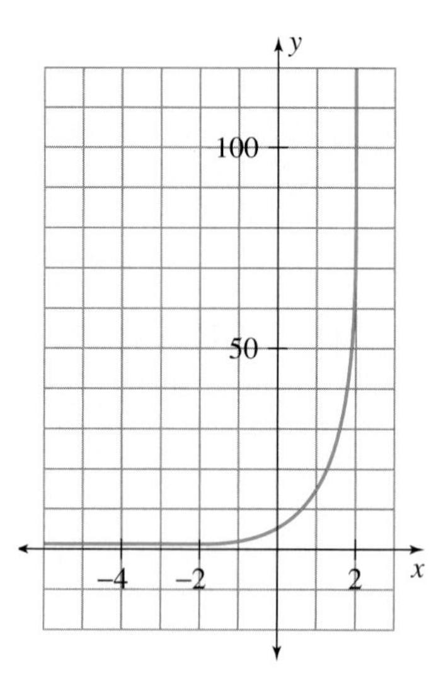

EXAMPLE 3

Graph $y = 50e^{0.3x}$.

Solution: With the help of a calculator, determine and plot points that satisfy the exponential function. A sufficient number of points must be determined so that the exponential function can be graphed.

x	y
−3	20.3
−2	27.4
−1	37.0
0	50
1	67.5
2	91.1
3	123.0

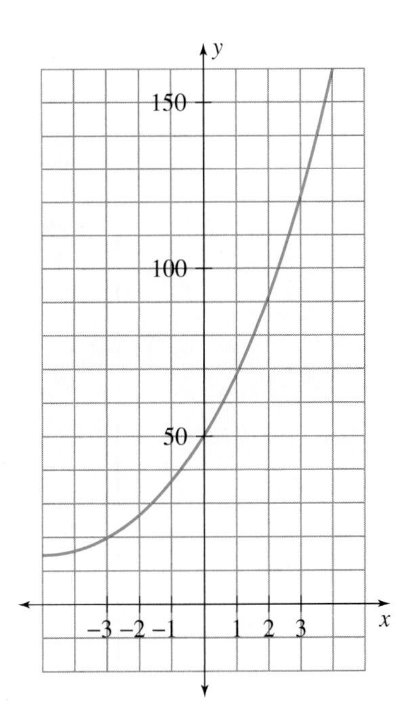

EXPONENTIAL MODELS

As we saw in the previous sections, mathematical models can be used to answer questions and to make predictions about observed events. Exponential functions can be used as mathematical models of real-life situations and scientific phenomena in which, as the values for one variable increase at a steady rate, the values for the other variable either

a) decrease rapidly and then decrease slowly, approaching a limiting value, or

b) increase slowly and then increase more and more rapidly.

An example of the latter can be seen in the common practice of determining the amount of money in an account after interest is compounded.

EXAMPLE 4 *Compound Interest*

Suppose you invest \$5000 in an account that earns 1% interest on the amount in the account each month. If you make no withdrawals from or deposits to the account, how much will you have in the account after ten years?

Solution: You will see in Chapter 9 that the amount in the account can be determined by an exponential function. If \$5000 is deposited in an account earning 1% interest a month, the amount (A) in the account after (n) months is

$$A = 5000(1.01)^n.$$

Thus, in ten years (120 months), we substitute $n = 120$ and get

$$A = 5000(1.01)^{120}$$
$$A = \$16{,}501.93.$$

To give you a better understanding of an exponential function, let's examine the graph of $A = 5000(1.01)^n$ for a 40-year period. By observing the distance between amounts on the vertical scale, you will notice that the amount in the account increases slowly at the start and increases more rapidly as time goes on. For example, the increase between the tenth and twentieth years is $37,960.83, while between the twentieth and thirtieth years, the increase is $125,285.44.

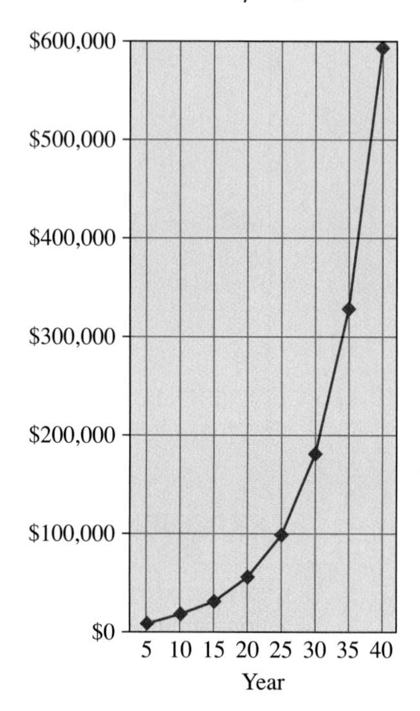

In the next example, we see that an exponential function can model a situation in which quantities increase by repeated multiplication.

EXAMPLE 5 *Exponential Growth*

Suppose you decide to form a new club that meets every Friday. At the first meeting, just you and a friend show up. The next week, each of you brings a new member to the meeting. Now the club has four members. The third Friday, each of the four members brings another person to the meeting, making a club of eight people. If every Friday each present member brings a new member to join the club, how many members will the club have after six months?

Solution:

Week	Members
1	2
2	4
3	8
4	16

To answer the question, we could continue the table until we reached the 26th week, but that would be a lot of work. Let's look at that problem a little more closely. We want to find a relationship between the week w the meeting is held and the number of members M at the meeting, assuming that every present member always brings a new member to the next meeting. By analyzing the data above, the number of members at any meeting is a power of 2 ($2 = 2^1$, $4 = 2^2$, $8 = 2^3$, $16 = 2^4$, etc.), the exponent being the week of the meeting. So the equation that determines M is

$$M = 2^w.$$

If $w = 26$, $M = 2^{26} = 67{,}108{,}864$.

As you can see, the growth in membership is astounding. By the 21st week, the increase in the number of members per week is over a million. In fact, by the 28th meeting every person in the United States would be a member of your club. This example gives us a good indication of what is meant by a quantity growing exponentially.

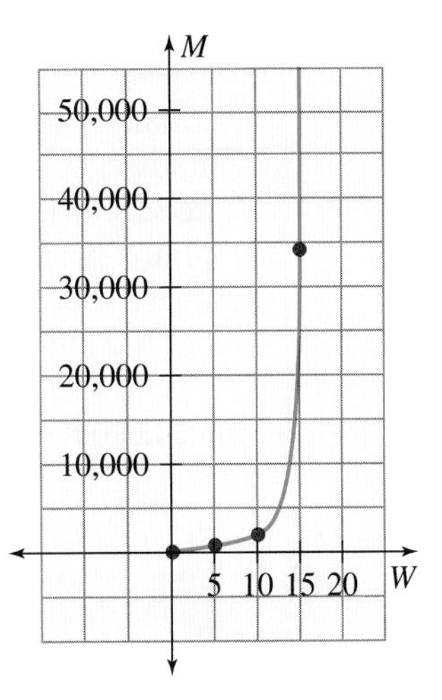

NOTE: The graph should be a discrete graph consisting of just the dots. The curve was added to make the exponential effect more evident.

EXAMPLE 6 *Exponential Growth of a Population*

Population increases when there are more births than deaths. Thomas Robert Malthus (1798) determined a model for predicting population based on the assumption that the rate of births B and the rate of deaths D remain constant and no other factors are considered. In this model the population P is given by the following exponential function.

$$P = P_0 e^{kt} \quad \text{where} \quad \begin{cases} P = \text{population at any time} \\ P_0 = \text{initial population} \\ k = \text{annual growth rate } (B - D) \end{cases}$$

In 1970, the population of the United States was 205,052,000, the birth rate was 18.4 per 1000 population, and the death rate was 9.5 per 1000 population. Use this information to estimate the number of people in the United States in 1991.

Solution: Using $P_0 = 205{,}052{,}000$, $t = 21$ years, and

$$k = \frac{18.4}{1000} - \frac{9.5}{1000} = 0.0089$$

with the formula $P = P_0e^{kt}$ gives

$$P = 205{,}052{,}000e^{(0.0089)(21)}$$
$$P \approx 247{,}191{,}566.$$

The actual population at the end of 1991 was 252,688,000. As you can see, the **Malthusian population model** did not give the exact 1991 population. One of the reasons for this is that both the birth rate and death rate have changed since 1970. The model did, however, give a reasonable approximation on the facts that were available in 1970.

EXAMPLE 7 *Atmospheric Pressure*

Atmospheric pressure is produced by the weight of air from the top of the atmosphere as it presses down upon the layers of air below it. At sea level, air pressure is about 14.7 pounds per square inch. As the distance from the earth's surface increases, the air pressure decreases. This phenomenon can be observed when a sealed bag of potato chips becomes puffed out like a balloon when taken into the mountains. The following exponential function relating air pressure (P) and altitude (a) can approximate the atmospheric pressure at altitudes up to 50,000 ft.

$$P = 14.7(10)^{-0.000018a} \quad \text{where} \quad \begin{cases} P = \text{pressure measured in lb/in.}^2 \\ a = \text{altitude measured in feet} \end{cases}$$

The graph of this function is as follows.

a	P
0	14.7
10,000	9.7
30,000	4.2

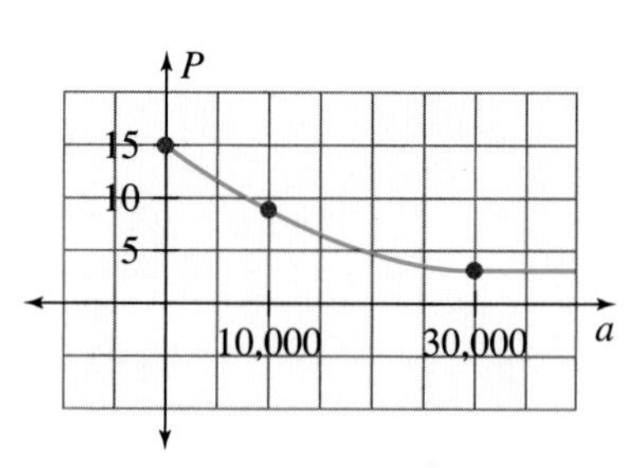

What is the air pressure on the top of the world's tallest mountain, Mount Everest in Nepal-Tibet, which has an altitude of 29,028 ft?

Solution:

$$P = 14.7(10)^{-0.000018a}$$
$$P = 14.7(10)^{-0.000018(29{,}028)}$$
$$P \approx 4.4 \text{ lb/in.}^2$$

EXAMPLE 8 *Exponential Data*

According to the CTIA Semi-Annual Wireless Survey, the number of U.S. cellular telephone subscribers from 1988 to 2000 in thousands of subscribers is as follows.

Year	1988	1990	1992	1994	1996	1998	2000
Subscribers	2,069	5,283	11,033	24,134	44,043	69,209	109,478

Find a mathematical model for the data. How many U.S. cellular telephone subscribers does this model predict in 2004?

Solution: The number of subscribers is increasing more rapidly as time progresses. The graph of the data suggests that an exponential function could be a good model to predict the number of U.S. cellular telephone subscribers.

U.S. Cellular Telephone Subscribers

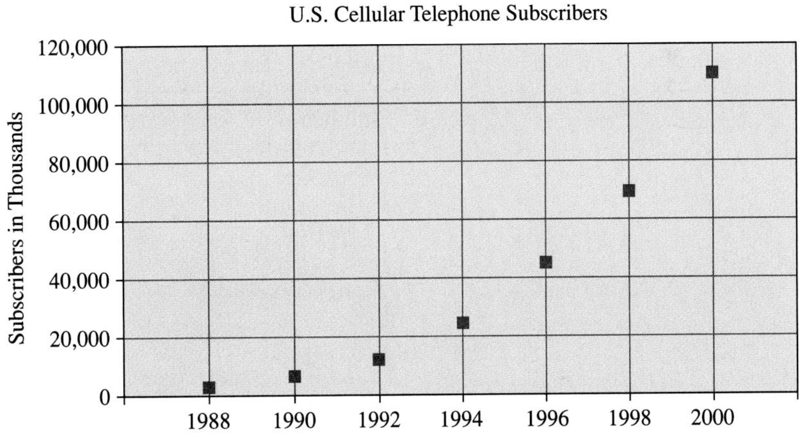

A standard exponential growth formula should work here.

$$A = Ce^{kt} \qquad \text{where} \quad \begin{cases} A = \text{amount} \\ t = \text{time} \end{cases}$$

To determine the exponential function that fits the data, we find the constants by using 1988 as the starting year ($t = 0$) and 2000 as the ending year ($t = 12$). We substitute the values $t = 0$, $A = 2069$ and $t = 12$, $A = 109{,}478$ to find C and k.

$$t = 0, \qquad A = 2069 \qquad \begin{cases} A = Ce^{kt} \\ 2069 = Ce^{k(0)} \\ 2069 = Ce^{0} \\ 2069 = C \end{cases}$$

$$\begin{aligned} A &= 2069e^{kt} \\ 109{,}478 &= 2069e^{k(12)} \\ 52.9135 &= e^{12k} \\ \ln(52.9135) &= \ln(e^{12k}) \\ 3.9687 &= 12k \ln e \\ 3.9687 &= 12k \\ 0.331 &\approx k \end{aligned}$$

Thus,

$$A = 2069e^{0.331t}.$$

To predict the number of subscribers in 2004, substitute $t = 16$ into the above equation.

$$A = 2069e^{0.331t}$$
$$A = 2069e^{0.331(16)}$$
$$A = 412{,}842$$

Since A is measured in thousands, there should be about 412,842,000 subscribers in 2004.

You might wonder about the accuracy of this exponential model. One way to check this is to compare its values to the data values and to graph both sets of values on the same coordinate system.

Year	1988	1990	1992	1994	1996	1998	2000
Data	2,069	5,283	11,033	24,134	44,043	69,209	109,478
Equation	2,069	4,011	7,776	15,075	29,226	56,660	109,844

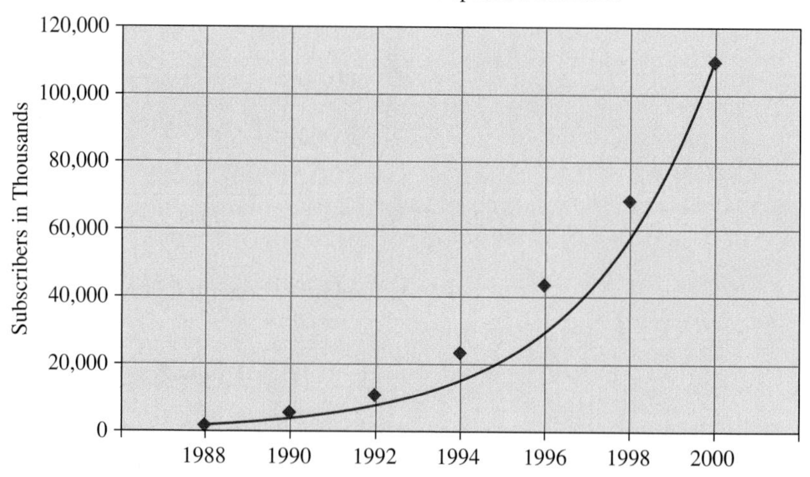

You can see that the graphs are similar but the exponential function underestimates the number of subscribers in some years. It is not a perfect model, but it does give us a way to make predictions.

This section has attempted to show you that exponential functions can be used as mathematical models for actual occurrences in the world. There are many other situations in which these functions are used, but many of them are beyond the scope of this book and require a greater knowledge of the areas in which they are used. The problems that follow, however, will let you experiment with other exponential functions used as mathematical models.

section 6.3

PROBLEMS ● Explain ● Apply ● Explore

Explain

1. Compare the graphs of an exponential function and a quadratic function. Which function seems more appropriate to describe the growth of a population? Explain.
2. In the exponential function $y = b^x$, where $b > 1$, what happens to the graph as
   a) x gets larger?
   b) x gets smaller?
3. In the exponential function $y = b^x$, where $0 < b < 1$, what happens to the graph as
   a) x gets larger?
   b) x gets smaller?
4. In the exponential function $y = b^x$, why does $b \neq 1$?
5. In the exponential function $y = b^x$, suppose $b = -4$. Determine the value of y if $x = -2, -1, 0, 1, 2$. What happens if $x = \frac{1}{2}$? Use these results to explain why the base of an exponential function is positive.
6. Explain why 10^x and e^x must be greater than zero for all real values of x.

Apply

In Problems 7–12, sketch a graph of each exponential function.

7. $y = 0.25e^x$
8. $y = 50e^{0.3x}$
9. $y = 500 + 10^x$
10. $y = 200 - 10^{-x}$
11. $y = 100(1.06)^{12x}$
12. $y = 200(1.015)^{4x}$
13. If the $5000 of Example 4 were deposited in a savings account that paid interest daily, the interest rate would be 0.000328767 ($0.12 \div 365$) each day and the amount A in the account after n days would be given by the exponential function, $A = 5000(1.000328767)^n$.
    a) Sketch a graph of that exponential function.
    b) Find the amount in the account after one year.
    c) Find the amount in the account after 20 years.
14. Suppose Parker Brothers determines that the profit P for a board game that is on the market for t years is given by the following equation.

$$P = 6000 + 20{,}000(3)^{-0.2t}$$

    a) Sketch the graph of this profit function.
    b) What is the profit after 25 years?
    c) What does the graph and the answer tell us about the profit for the board game?
    d) Is this exponential function a logical model for the profit from a board game? Give reasons for your answer.

15. Parker Brothers in Problem 14 is trying to stimulate sales of the board game through 30 days of television advertising in an area that has 250,000 viewers. The number of viewers V in thousands who are made aware of the board game after t days of advertising is expected to be $V = 250 - 250e^{-0.04t}$.

a) Sketch a graph of this exponential function for the 30 days.

b) How many viewers were made aware of the board game after one day?

c) How many viewers were made aware of the board game after two weeks?

d) How many viewers were made aware of the board game after 30 days?

16. If an amount of principal P is invested at an annual rate r expressed as a decimal and is compounded continuously, the amount A in the account at the end of t years is given by the exponential function, $A = Pe^{rt}$. If $5000 is invested at an annual rate of 9%,

a) Write the exponential function that determines the amount in the account at the end of any year.

b) Sketch a graph of the function.

c) Find the amount in the account after 20 years.

Explore

17. In an attempt to promote world peace, you decide to start a chain letter. You send a peace message to five friends asking each of them to send copies of the message to five of their friends by the end of the week. Suppose this process continues and every person sends the message on time to five new people. A chart of the number of people receiving messages each week would look like this.

Week	Number of People Receiving Messages
1	5
2	25
3	125
4	625

a) Graph the data using the number of weeks on the horizontal axis and the number of people on the vertical axis.

b) Find an exponential function relating the number of people p that receive the peace message to the number of weeks w.

c) How many people would receive the message during the 12th week?

18. Suppose you are gainfully employed and earn $200 a day. You are, however, offered a temporary job doing similar work for three weeks in which you will be paid $0.01 the first day, $0.02 the second day, $0.04 the third day, and so on. Your daily wage will continue to double for each of the 21 days. Would it be more profitable for you to take the temporary job or keep your regular job for the three weeks? Justify your answer mathematically.

19. The population of the Soviet Union as of January 1, 1987, was 282,000,000. It had a birth rate of 20.1 per 1000 population and a death rate of 9.8 per 1000 population. Using the Malthusian population model, determine an estimate of the population in the Soviet Union in 2003.

20. Answer the same questions as in Problem 19 but for the country of Ethiopia. In 1989, Ethiopia had a population of 48,898,000, a birth rate of 49.9 per 1000, and a death rate of 25.4 per 1000.

21. Using the exponential function that gives a model for atmospheric pressure, find the atmospheric pressure on the top of Mount McKinley in Alaska, altitude 20,320 ft.

22. Suppose you take up the game of golf. You keep a record of your average score (the average number of strokes it takes to complete a round of golf) for each month of playing golf. Explain why an exponential function might make a good mathematical model for your scores in this endeavor.

23. The number of compact discs shipped, according to the Recording Industry Association of America in Washington, D.C., is displayed in the following chart.

Year	1992	1994	1996	1998	2000
Compact Discs (in millions)	407.5	662.1	778.9	847.0	942.5

The data seem to indicate that the number shipped can be modeled with an increasing exponential function, $A = Ce^{kt}$. Using 1992 as $t = 0$ and 2000 as $t = 8$, determine an exponential function to model the data and use that function to predict the number of compact discs shipped in 2004.

24. Using another technique, a different exponential function that seems to fit the data of Problem 23 is $A = 1071.2 - 663.7e^{-0.205t}$. Using $t = 0, 2, 4, 6, 8$, compare that equation and the equation from Problem 23 to the data. Using $A = 1071.2 - 663.7e^{-0.205t}$, predict the number of CDs shipped in 2004. Which exponential model is better? Explain.

25. The number of cassettes shipped, according to the Recording Industry Association of America in Washington, D.C., is displayed in the following chart.

Year	1992	1994	1996	1998	2000
Cassettes (in millions)	366.4	345.4	225.4	158.5	76.0

The data seems to indicate that the number shipped can be modeled with a decreasing exponential function, $A = Ce^{kt}$. Using 1992 as $t = 0$ and 2000 as $t = 8$, determine an exponential function to model the data and use that function to predict the number of cassettes shipped in 2004.

26. Using other techniques, an exponential function that seems to fit the data of Problem 25 is $A = 2727.6 - 2361.2e^{0.0145t}$. Using $t = 0, 2, 4, 6, 8$, compare that equation and the equation from Problem 25 to the data. Using $A = 2727.6 - 2361.2e^{0.0145t}$, predict the number of cassettes shipped in 2004. Which exponential model is better? Explain.

27. A queen, wishing to reward a faithful maid, agreed to grant her one wish. The maid replied that she was a very humble woman and only wanted some corn as her reward. The maid requested that the corn be given to her in the following manner: Upon a chess board, place two kernels of corn on the first square, 4 on the second, 8 on the third, 16 on the fourth, and so on until the last (64th) square. At first the queen refused, saying that this was not a just reward for such a faithful maid. However, after the maid insisted, the queen ordered a servant to bring in a bag of corn and give the maid her desired reward. How much corn did the maid receive? If one pound of corn contains about 3500 kernels, how many pounds of corn did the maid receive?

28. Exponential functions are also used in other areas, such as in the decay of radioactive material (half-life) and bacterial growth. Do some research on one of these areas and on any other area which uses exponential functions as a mathematical model. Explain how the function is used and give some examples.

section 6.4

Logarithmic Models

LOGARITHMIC FUNCTIONS

You might recall from algebra that inverse functions are functions that have opposite effects. In arithmetic, multiplication and division are inverses of each other. For example, if you choose a number, say 7, and multiply it by 5, you get an answer of 35. If you then divide that answer by 5, you get back the 7 you started with. In algebra, equations such as $y = 2x$ and $y = \frac{1}{2}x$ are inverse functions. If you choose a number, say $x = 8$, and substitute $x = 8$ into the equation $y = 2x$, you get 16. Substituting $x = 16$ into the second equation, $y = \frac{1}{2}x$, gives you the answer 8, the number you started with.

Similarly, the inverse of the exponential function is the logarithmic function. For example, the inverse of the exponential function $y = 10^x$ is the function $y = \log x$. To see this inverse relationship, choose a number, say $x = 3$. Substituting $x = 3$ into the equation $y = 10^x$ gives us the answer $10^3 = 1000$. Substituting this result for x in the function $y = \log x$ gives us $\log 1000$.

Log $1000 = 3$ is equivalent to the statement $10^3 = 1000$. The exponential function gives the result of raising

CLOSE TO HOME JOHN McPHERSON

Knowing that their kids would inevitably snoop for their Christmas gifts, the Vurtmans planted decoy gifts where they would easily be found

10 to the third power as its answer, while the logarithmic function gives the exponent of 10 as its answer. A logarithm is simply an exponent. Although logarithms can be evaluated with different bases, in this chapter we will use only common logarithms and natural logarithms. **Common logarithms** use the base 10 and are denoted by the LOG (**log**) button on a calculator. **Natural logarithms** use base e ($e \approx 2.71828$) and are denoted by LN (**ln**) on a calculator. The logarithmic functions used in the text are

$$y = \log x, \text{ which is equivalent to } x = 10^y \text{ and}$$
$$y = \ln x, \text{ which is equivalent to } x = e^y.$$

Since 10 raised to any power or e raised to any power is always a positive quantity, x is always positive. We can, therefore, only take the log or ln of positive quantities.

As we did with exponential functions, logarithmic functions can be graphed by calculating and plotting a sufficient number of points.

EXAMPLE 1

Graph $y = \ln x$.

Solution: With the help of a calculator, determine and plot points that satisfy the equation. Remember to choose x values greater than 0, since it is not possible to compute the logarithm of a negative number or zero.

x	y
0.5	−0.69
1	0
5	1.61
10	2.30
20	3.00
30	3.40

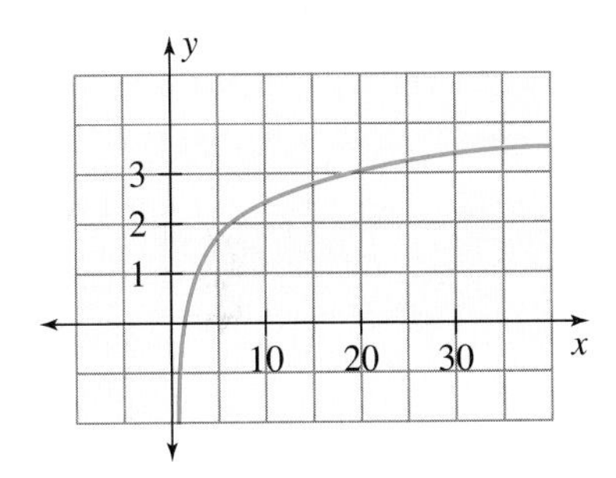

If we include other constants in a basic common or natural logarithmic function, the graph of the logarithmic function can be moved up or down or the rate at which it increases or decreases can be changed. The graphs of these functions have the same basic shape as the basic logarithmic function. They can be graphed by determining and plotting points with the help of a calculator. In this process, you must remember that you cannot take the logarithm of a negative number or zero.

EXAMPLE 2

Graph $y = 1.2 \log x$.

Solution: With the help of a calculator, determine and plot points that satisfy the logarithmic function. Since it is impossible to take logarithms of negative numbers or zero, use $x > 0$.

x	y
0.5	−0.36
1	0
5	0.84
10	1.20
20	1.56
30	1.77

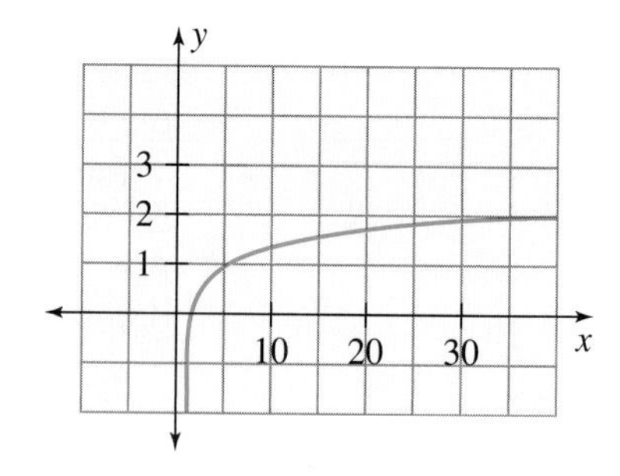

EXAMPLE 3

Graph $y = 4.7 + \ln(x - 2)$ where $x \geq 4$.

Solution: With the help of a calculator, determine and plot points that satisfy the logarithmic function using $x \geq 4$.

x	y
4	5.4
6	6.1
8	6.5
10	6.8
20	7.6

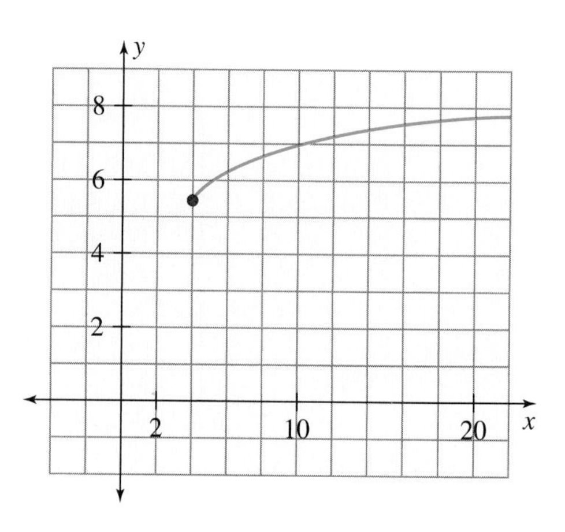

LOGARITHMIC MODELS

As we saw in the previous section, exponential functions can be used as mathematical models of situations in which a quantity experiences a period of gradual increase followed by a period of rapid increase. Logarithmic functions can also be used as mathematical models of real-life situations and scientific phenomena. Since logarithmic functions are inverses of exponential functions, they react in a manner that is opposite to the exponential function. The logarithmic functions we will be studying exhibit

a) rapid initial increase followed by a long period of gradual increase, or

b) rapid initial decrease followed by a long period of gradual decrease, but never approaching a limiting value.

EXAMPLE 4 *Height of Children*

A logarithmic function can be used to approximate the change in the height of a child as the child grows older. By age 2, most children have reached 50% of their adult or mature height. It takes approximately 16 years for the child to attain full adult height. A function that allows for a large initial change and then a gradual increase is a logarithmic function. In fact, the mature height of boys aged 0–16 can be approximated by the following function.

$$P = 29 + 48.8\log(A + 1) \qquad \text{where} \quad \begin{cases} P = \text{percentage of adult height} \\ A = \text{age in years} \end{cases}$$

The graph of this function is as follows.

A	*P*
0	29
5	67
10	80
15	88

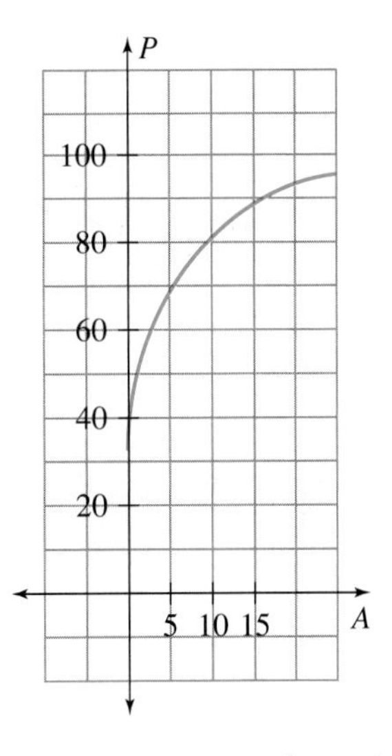

We can use that logarithmic model to answer questions about the growth of boys such as the following.

a) Approximately what percentage of his adult height is a boy at the age of 7?

b) If that seven-year-old boy is 52 inches tall, how tall can we expect him to be when he is an adult?

Solution:

a) Substituting $A = 7$ into the formula gives the following.

$$P = 29 + 48.8\log(A + 1)$$
$$P = 29 + 48.8\log(7 + 1) \approx 73$$

So a seven-year-old is 73% of his mature height.

b) From part (a), the 52-inch-tall boy is 73% of his mature height. So if $M =$ the mature height,

$$52 = 0.73M$$
$$M = 52 \div 0.73 \approx 71.2 \text{ in.}$$

Thus, we can expect the boy to become a little over 5′11″ tall.

EXAMPLE 5 *Newton's Law of Cooling*

From experiments on cooling bodies, Isaac Newton concluded that over moderate temperature ranges, the rate at which an object changes temperature is proportional to the difference between the temperature of the object and the temperature of the surrounding air.

If a hot cup of 200°F coffee is taken outdoors where the temperature is 35°F, it will begin to cool. If after 1 minute, the temperature of the coffee is 170°F, according to Newton's Law of Cooling, the time t it takes for the coffee to reach a temperature x is given by the following formula.

$$t = 25.5 - 5\ln(x - 35) \qquad \text{where} \quad \begin{cases} t = \text{time to reach temperature } x \\ x = \text{temperature of coffee} \\ \quad (35° < x \leq 170°) \end{cases}$$

Use this function to determine the following.

a) When is the temperature of the coffee 98.6°F?

b) When is the temperature of the coffee 40°F?

Solution:

a) Substituting $x = 98.6$°F into the equation gives

$$t = 25.5 - 5\ln(98.6 - 35) \approx 4.7 \text{ min.}$$

b) Substituting $x = 40$°F into the equation gives

$$t = 25.5 - 5\ln(40 - 35) \approx 17.5 \text{ min.}$$

EXAMPLE 6

According to Nielsen Media Research, the number of U.S. households in millions with cable television from 1981 to 2001 was as follows.

Year	1981	1983	1985	1987	1989	1991	1993	1995	1997	1999	2001
Households with Cable TV	23.2	34.1	39.9	45.0	52.6	55.8	58.8	63.0	65.9	67.6	69.0

Find a mathematical model for the data. How many U.S. households does this model predict will have cable TV in 2006?

Solution: The number of households has an initial rapid increase followed by years of gradual increase. The graph of the data suggests that a logarithmic function could be a good model to predict the number of U.S. households with cable television.

U.S. Households with Cable TV

Households in Millions

80
70
60
50
40
30
20
10
0

1985
1990
1995
2000

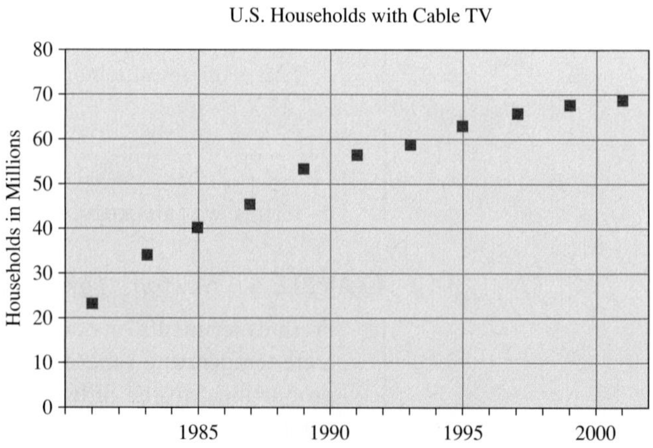

A standard logarithmic function should work here.

$$y = a + b \ln x \qquad \text{where} \quad \begin{cases} y = \text{number of households} \\ x = \text{year } (x = 1 \text{ for 1981, etc.}) \\ a, b = \text{constants} \end{cases}$$

NOTE: We start with an x value of 1, since log(0) does not exist.

To determine the logarithmic function that fits the data, we find the constants by using 1981 as the starting year ($x = 1$) and 2001 as the ending year ($x = 21$). We substitute the values $x = 1$, $y = 23.2$ and $x = 21$, $y = 69.0$ to find a and b.

$$x = 1, \quad y = 23.2 \quad \begin{cases} y = a + b \ln x \\ 23.2 = a + b \ln(1) \\ 23.2 = a + b(0) \\ 23.2 = a \end{cases}$$

$$x = 21, \quad y = 69.0 \quad \begin{cases} y = a + b \ln x \\ 69.0 = 23.2 + b \ln(21) \\ 69.0 = 23.2 + b(3.0445) \\ 45.8 = 3.0445b \\ 15.0 \approx b \end{cases}$$

Thus, $y = 23.2 + 15.0 \ln x$.

To predict the number of households in 2006, substitute $x = 26$ into the above equation.

$$y = 23.2 + 15.0 \ln x$$
$$y = 23.2 + 15.0 \ln(26)$$
$$y = 23.2 + 48.9$$
$$y \approx 72.1$$

Since y is measured in millions, there should be about 72,100,000 U.S. households with cable TV in 2006.

You might wonder about the accuracy of this logarithmic model. One way to check this is to compare its values to the data values and graph both sets of values on the same coordinate system.

Year	1981	1983	1985	1987	1989	1991	1993	1995	1997	1999	2001
Data	23.2	34.1	39.9	45.0	52.6	55.8	58.8	63.0	65.9	67.6	69.0
Equation	23.2	39.7	47.3	52.3	56.2	59.2	61.7	63.8	65.7	67.4	68.9

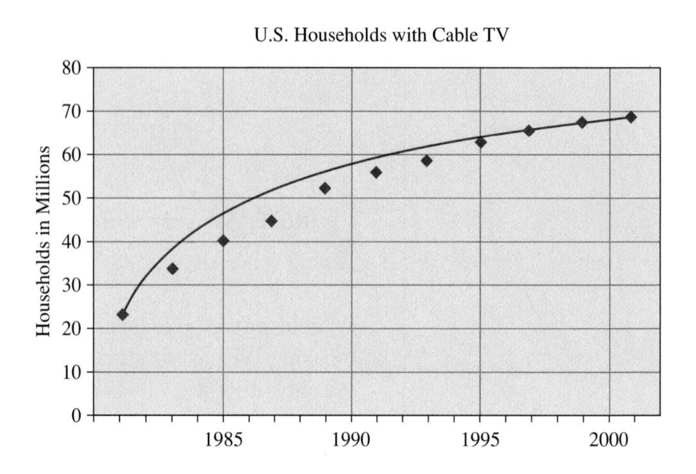

You can see that the graphs are similar but the logarithmic function overestimates the number of households in some years. It becomes more accurate at the end of the time period. For that reason, it should generate fairly good predictions for the years beyond 2001.

This section has attempted to show you that logarithmic functions can be used as mathematical models for actual occurrences in the world. There are many other situations in which these functions are used, but many of them are beyond the scope of this book and require knowledge of the areas in which they are used. The problems that follow will let you experiment with other logarithmic functions used as mathematical models.

PROBLEMS ◆ Explain ◆ Apply ◆ Explore

◆ Explain:

1. What are the basic characteristics of the function $y = \log x$?
2. Describe the general shape of the logarithmic models discussed in this section.
3. Suppose you take up the sport of weight lifting. You keep a record of the maximum number of pounds you can lift at the end of each week. Explain why a logarithmic function might make a good mathematical model for predicting the amount of weight you can lift from week to week.
4. You are advertising a new computer game. Consumer research is keeping track of how many people have heard of your new game (vertical axis of a graph) as the number of weeks of advertising continues (horizontal axis). Explain why a logarithmic function might make a good mathematical model for estimating the number of people who have heard of your game as the number of weeks of advertising increases.
5. What does your calculator do when you try to take the common logarithm of a negative number?
6. What does your calculator do when you try to take the natural logarithm of a negative number?

Apply:

In Problems 7–14, sketch a graph of each logarithmic function.

7. $y = \ln 2x$

8. $y = \log 5x$

9. $y = 4.2 \ln(x - 2.9)$

10. $y = 5.3 \log(x + 1)$

11. $y = 32 + 48.5 \log(x + 2)$

12. $y = 5 + 14.3 \ln(2x + 1.2)$

13. $y = \ln(3x - 1) - 15$

14. $y = -20 + \log(3x + 1)$

15. At age 5, a girl's height is approximately 62% of her full adult height. At age 15, she has reached about 98% of her adult height. The logarithmic function below gives an approximate percentage P of adult height a girl has reached at any age A from 5 to 15 years.

$$P = 62 + 35 \log(A - 4)$$

a) At age 10, what percentage of her height has a girl reached?

b) If the girl is 4′6″ at age 10, how tall can she expect to be as an adult?

16. A logarithmic model to approximate the percentage P of adult height a male has reached at any age A from 13 to 18 is

$$P = 16.7 \log(A - 12) + 87.$$

a) Sketch a graph this function.

b) What does the graph tell us about males that have reached the age of 18?

17. A roast, cooking for 2 hours, is taken out of the oven when the meat thermometer reads 140°F and is placed in a kitchen that is 68°F. After 6 minutes the thermometer reads 132°F. Newton's Law of Cooling states that the time t for the roast to get to a temperature x is given by

$$t = 217.9 - 50.94 \ln(x - 68) \qquad \text{where} \quad 68° < x \leq 140°.$$

a) In how many minutes will the internal temperature of the roast be 110°F?

b) In how many minutes will the internal temperature of the roast be 88°F?

c) Sketch a graph of this function.

18. If \$1000 is invested in an account that earns 1% interest each month, the number of months n for the account to grow to an amount A is given by the formula

$$n = -694.2 + 231.4 \log A$$

where $A \geq \$1000$ and no withdrawals or other deposits are made to the account.

a) Sketch a graph of this function.

b) How many years would it take for the account to grow to \$1 million?

Explore

19. If 1000 *E. coli* bacteria are placed in a culture, the time t in hours it takes for the bacteria to grow to an amount A can be approximated by the formula

$$t = -172.7 + 25 \ln A \qquad \text{where } A \geq 1000.$$

a) Sketch a graph of this function.

b) How many hours would it take for the culture to contain a million bacteria?

20. If \$500 is deposited into a savings account that pays 3.65% annual interest, compounded daily, the number of years (y) that it takes for \$500 to grow to an amount x is given by the equation $y = 27.4 \ln x - 170.3$.

a) How long would it take for your money to double?

b) How long would it take for you to have \$5000 in the account?

21. Radioactive elements such as uranium and plutonium actually decrease in mass over a period of time. The half-life of an element is the length of time required for the mass of an element to decay to one-half the original amount. Suppose a nuclear storage site has been contaminated by 100 kilograms of radioactive plutonium. The half-life of plutonium is 24,360 years. The equation relating the mass (m) of the plutonium to the elapsed time (t) is

$$t = \frac{24{,}360}{\ln 0.5} \ln \frac{m}{100}.$$

a) Find the time required for the 100 kilograms of plutonium to decay to 50 kilograms.

b) Find the time required for the 100 kilograms of plutonium to decay to 1 kilogram.

22. According to the U.S. Department of Commerce, the percent of U.S. households with a computer is as follows.

Year	1998	2000
Percent with Computers	42.1%	51.0%

Indications are that this percentage will increase at a slower and slower rate over the next ten years. Thus, it is an ideal situation for a logarithmic model.

a) Find a logarithmic function $y = a + b \ln x$ to model the data using $x = 1$ for 1998 and $y =$ the percent with computers.

b) Use the equation to predict the percent of households with computers in 2010.

c) In what year should 75% of U.S. households have computers?

23. According to the M Street Corporation in Littleton, New Hampshire, the number of U.S. radio stations is as follows.

Year	1994	1995	1996	1997	1998	1999
Radio Stations	9,778	9,889	9,991	10,207	10,292	10,444

a) Why does a logarithmic model seem appropriate?

b) Find a logarithmic function $y = a + b \ln x$ to model the data using $x = 1$ for 1994, $x = 6$ for 1999, and $y =$ the number of radio stations.

c) Use the equation to predict the number of radio stations in 2001. The actual numbers of radio stations in 2001 was 10,561. How accurate was the logarithmic model?

d) In what year should the number of radio stations reach 10,900?

CHAPTER 6 SUMMARY

Key Terms, Concepts, and Formulas

The important terms in this chapter are:	**Section**
Common logarithm: A logarithmic function that uses a base of 10, denoted by **log.**	6.4
Diophantine equation: An equation with integer coefficients and integer solutions.	6.1
Exponential model: A representation of phenomena using an exponential function.	6.3
Linear model: A representation of phenomena using a linear function of the form $y = mx + b$.	6.1
Logarithmic model: A representation of phenomena using a logarithmic function.	6.4
Malthusian population model: A method of predicting population growth based on constant birth and death rates.	6.3
Natural logarithm: A logarithm that uses a base of e, denoted by **ln.** ($e \approx 2.71828$.)	6.4
Quadratic function: A function of the form $y = ax^2 + bx + c$.	6.2

After completing this chapter, you should be able to:	
1. Find and graph linear functions that serve as mathematical models for situations in which the rate at which quantities change remains constant.	6.1
2. Set up and use Diophantine equations to model scoring in sports.	6.1
3. Find and graph quadratic functions whose parabolic shapes serve as models for given situations.	6.2
4. Find and graph exponential functions as models to analyze various situations.	6.3
5. Find and graph logarithmic functions as models to analyze various situations.	6.4

CHAPTER 6 REVIEW

Section 6.1

1. Graph the following linear functions.
   a) $y = -2x + 3$
   b) $P = 240t + 300$, where $1 \leq t \leq 10$
   c) $P = 240t + 300$, where t is a whole number, $1 \leq t \leq 10$
2. Your club is selling cans of soft drinks at a festival. The cost for the booth and ice is \$135.00, and you make \$0.77 on each can of soft drink you sell.
   a) Explain why a linear function could be used to model the profit at the booth.
   b) Write a linear equation to determine the profit (P) based on the number of cans of soft drinks (x) that you sell.
   c) If you sold 40 cases of soft drinks (24 cans in a case), how much profit would you have?
   d) How many cans of soft drinks must you sell to break even (have a profit of zero)?
   e) Sketch a graph of the profit equation.
3. At age 9, Krista had a stamp collection containing 102 stamps. When she entered high school at the age of 14, her collection had grown to 1567 stamps. Assume that Krista continues to collect stamps at that rate.
   a) Find a linear function that, based on her age, predicts the number of stamps she owns.
   b) Predict how many stamps she will have when she graduates from high school at age 18.
   c) Determine at what age her stamp collection would contain more than 10,000 stamps.
4. In a track and field meet, a team scored 68 points with only first place finishes (5 points each) and second place finishes (3 points each). Set up a Diophantine equation and determine in how many ways this could be accomplished.

Section 6.2

5. Find the vertex and graph the following quadratic functions.
   a) $y = -x^2 + 6x$
   b) $y = x^2 - 6x + 8$
   c) $y = -0.2x^2 + 4.8x - 1.2$
6. After a world-class shot-putter releases the shot put, the flight of the shot put can be modeled with a quadratic function.

$$y = -\frac{x^2}{64} + x + 7 \quad \text{where} \quad \begin{cases} y = \text{height above the ground in feet} \\ x = \text{horizontal distance in feet} \end{cases}$$

   a) How far above the ground does the shot put leave the competitor's hand?
   b) What is the maximum height reached by the shot put?
   c) Sketch a graph of the flight of the shot put.
   d) What is the distance of the toss (the x coordinate when the shot put hits the ground)?
7. A golfer hits a golf ball with a pitching wedge, and as it passes over the flag stick it is 3 yd above the ground. The ball hits the ground 100 yd from where it started just 2 yd past the flag stick.
   a) Set up a coordinate system for the flight of the golf ball and determine the coordinates of three known positions of the golf ball.
   b) Find the quadratic equation that models the flight of the golf ball.
   c) What is the maximum height reached by the golf ball?

Section 6.3

8. Graph the following exponential functions.
   a) $y = 0.25e^{1.5x}$
   b) $y = 15 - 4^{-x}$

9. The Quick Calculator Company's training department determines that after completing the basic training program, a new employee will be able to assemble (Q) calculators per day.

$Q = 90 - 30e^{-0.5t}$ where $t =$ the number of weeks worked after training

a) How many calculators can a new employee assemble per day after the basic training?

b) How many calculators can an employee with 1 week of experience, 5 weeks of experience, and 20 weeks of experience assemble per day?

c) Sketch a graph of the exponential function.

d) How many calculators is a very experienced worker expected to assemble per day?

10. In "Hobbes' Internet Timeline v5.6" (© 2002 Robert H. Zakon), a table of the growth of Internet sites is given. Here is some of the data from that table.

Year	1996	1997	1998	1999
No. of sites (in millions)	.6	1.7	3.6	9.6

Source: www.zakon.org/robert/internet/timeline.

a) Why does an exponential function seem appropriate to model the growth of the number of Internet sites?

b) Using the exponential function $A = Ce^{kt}$, where $A =$ the number of Internet sites in millions and $t = 0$ for 1996 and $t = 3$ for 1999, find an exponential model for the data.

c) Compare the results of the function with the actual data for 1996–1999. How does the model compare with the data?

d) Use the model to predict the number of Internet sites in 2000 and 2001. The actual number of sites was 25.7 million in 2000 and 36.3 million in 2001. How did the model's results compare with those data?

e) What can you say about the exponential model for the number of Internet sites?

Section 6.4

11. Graph the following logarithmic functions.

a) $y = 3 \ln x$

b) $y = 5.6 + 2.3 \log x$

12. Systolic blood pressure measures the amount of pressure in the blood vessels when the heart contracts and pushes blood through the circulatory system. A child's normal systolic pressure can be approximated by the logarithmic function $p = 18 + 19.4 \ln x$, where p is the pressure and x is the child's weight in pounds.

a) Find the systolic pressure of a 60-lb child, an 80-lb child, and a 100-lb child.

b) Sketch a graph of the systolic pressure for children.

13. According to the National Telecommunications and Information Administration of the U.S. Department of Commerce, the percentage of U.S. households with Internet access is as follows.

Year	1998	2000
Percent with Access	26.2	41.5

Indications are that this percentage will increase but at a slower and slower rate. Thus, a logarithmic model might be appropriate to predict the percentage of households with Internet access in future years.

a) Using the logarithmic function $y = a + b \ln x$, where $y =$ the percentage with Internet access and $x =$ the year ($x = 1$ for 1998), find the logarithmic model for the data.

b) Use the model to predict the percentage of households with Internet access in 2006.

c) Use the model to predict when 75% of the U.S. households will have Internet access.

CHAPTER 6 TEST

1. Sketch graphs of the following functions.
   a) $y = 5x - 4$
   b) $y = -x^2 + 6x - 1$
   c) $y = 0.5e^{2x}$
   d) $y = 3 + 1.5 \ln x$

2. Scientists use the Kelvin temperature scale where the lowest possible temperature (absolute zero) is zero degrees Kelvin (0°K). The linear function that relates the Kelvin scale to the Celsius scale is as follows.

$K = C + 273$ where K = temperature on Kelvin scale, C = temperature on Celsius scale

   a) Sketch a graph of this function.
   b) At what Celsius temperature is absolute zero?
   c) Since water boils at 100°C, at what Kelvin temperature does it boil?

3. If a person walks at 5 mph for an hour, the approximate number of calories burned per hour, based on the person's weight, is given in the following chart.

Weight in Pounds	Calories Burned per Hour
110	440
132	500
154	560
176	620
198	680

   a) Explain why a linear function would be an appropriate model for this data.
   b) Find the linear function that determines the calories c burned per hour based on the weight w of the person walking at 5 mph.
   c) If you weigh 160 lb and walk at 5 mph, how many calories does your body burn per hour?

4. The first (bottom) row of a huge stack of logs in a lumber yard has 247 logs. The second row has 245 logs, the third row has 243 logs, and so on. Assume that each successive row continues to contain exactly two fewer logs than the previous row.

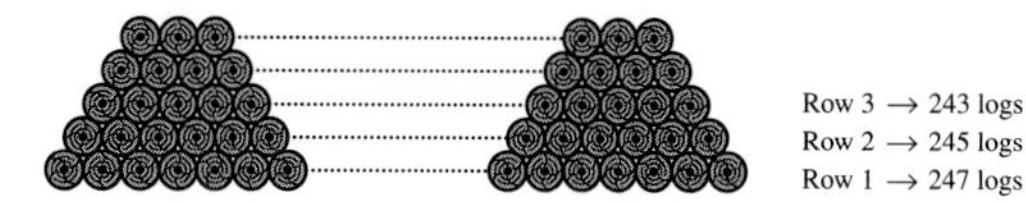

   a) Explain why a linear function could be used to predict the number of logs L in any row r.
   b) Find the linear function that determines the number of logs in any given row.
   c) Sketch a graph of the function found in part (b).
   d) Find how many logs are in the 50th row.
   e) Find how many rows of logs the stack contains.

5. A rock is thrown vertically upward at 88 ft/s (60 mph) from a sheer cliff in the Grand Canyon, 5000 ft above the Colorado River. A quadratic function that approximates the height of the rock at any given time is

$$h = -16t^2 + 88t + 5000.$$

   a) Sketch a graph of this quadratic function for $t \geq 0$.
   b) What is the maximum height the rock reaches?
   c) How long does it take for the rock to reach the Colorado River?

6. In the game of horseshoes, points are scored with horseshoes around the stake (3 points each), horseshoes closest to the stake (1 point each), and horseshoes leaning on the stake (2 points each). You scored 21 points, which included three horseshoes leaning on the stake. Set up a Diophantine equation and determine in how many ways you could have accomplished that winning score.

7. In the 1968 Olympics in Mexico City, Bob Beamon of the United States electrified the

track and field world with his 29′2.5″ leap in the long jump. Assume that the path of the jump can be approximated by a parabola where the highest point occurred at the middle of the jump and at that point his feet were 4′6″ off the ground.

a) Find the quadratic function that determines the height of this jump as a function of the length of the jump where distances are measured in inches.

b) How far past the takeoff board was Bob Beamon when his feet were 30 in. off the ground.

8. The ancient Greeks studied the pentagonal numbers as shown. The first pentagonal number is 1, the second one is 5, the third one is 12, the fourth one is 22, and so on. The algebraic function that relates the value of a pentagonal number to what term it is in the sequence of numbers is a quadratic function.

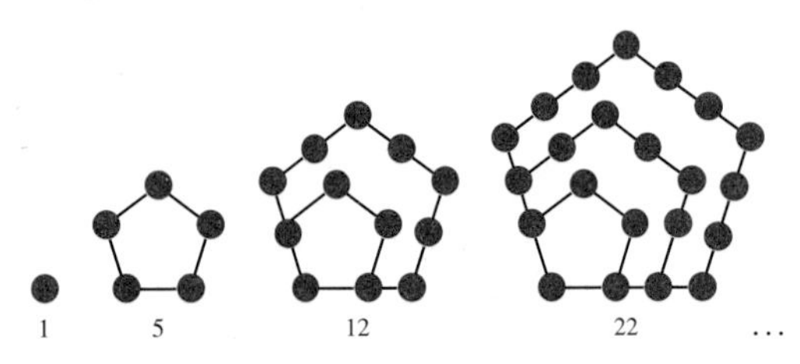

a) Find the quadratic function that determines the value of a pentagonal number P based on what term t it is in the sequence.

b) Sketch a graph of that function.

c) Determine the 100th pentagonal number and describe its geometric shape.

9. If the guaranteed rate of return on an investment of \$6000 is compounded annually at 10% per year, the exponential function that determines the amount A that investment is worth at time t is given by the following.

$$A = 6000(1.1)^t \quad \text{where} \begin{cases} A = \text{amount investment is worth} \\ t = \text{time in years} \end{cases}$$

a) Sketch a graph of this exponential function for $t \geq 0$.

b) Determine the value of the investment after 5 years, 10 years, and 30 years.

10. Suppose the total cost for manufacturing a certain toy truck is given by the equation

$$C = 500 + 400 \ln x \quad \text{where} \begin{cases} C = \text{total cost} \\ x = \text{number of trucks} \\ \text{and } 1 \leq x \leq 3000. \end{cases}$$

a) Sketch a graph of this cost function.

b) Determine the cost for manufacturing 1000 trucks and 2000 trucks.

c) Determine the cost for manufacturing each truck when 1000 trucks and 2000 trucks are produced. What happens to the cost per truck as the number of trucks manufactured increases?

11. According to the Recording Industry Association of America the number of music videos and music DVDs shipped in millions from 1998–2000 is shown in the following chart.

Year	1998	1999	2000
Music Videos (in millions)	27.2	19.8	18.2
Music DVDs (in millions)	0.5	2.5	3.3

The indication is that the number of videos shipped will experience a period of gradual decrease while the number of DVDs shipped will increase and more rapidly. Thus, a decreasing logarithmic function would probably predict the number of videos shipped, and an increasing exponential function would probably predict the number of DVDs shipped.

a) Find a logarithmic function of the form $y = a + b \ln x$ to model the data for number of music videos shipped and use that model to predict the number of videos shipped in 2004. (Use $x = 1$ for 1998.)

b) Find an exponential function of the form $y = Ce^{kx}$ to model the data for number of music DVDs shipped and use that model to predict the number of DVDs shipped in 2004. (Use $x = 0$ for 1998.)

12. The Parker Brothers game Monopoly™ has a game board with 40 spaces where game tokens can land, and a bank that has $15,140 in play money. Suppose you place $3 on the start (Go), $9 on the next space (Mediterranean Avenue), $27 on the next space (Community Chest), and so on. You keep tripling the amount placed on each space as you go around the board.

a) Find an exponential function that determines the amount A placed on each space S of the Monopoly board.

b) How much must you place on the ninth space (Vermont Avenue)? Will you have enough money to do that?

c) The Parker Brothers Company prints about $40 billion in play money each year. With a year's worth of play money would you be able to put the required amount on the 24th space (Illinois Avenue)?

d) How much play money would you need for the 40th space (Boardwalk)? How many years of play money production by Parker Brothers would you need to put the required amount on Boardwalk?

7

CHAPTER SEVEN

Geometry and Art

Kristin Shinoda of Ohlone College used tesselations to create *Warriors and Dragons*.

OVERVIEW

In this chapter, you will learn that there is more to geometry than the definitions, postulates, theorems, proofs, and measurement formulas you might have studied in previous geometry courses. The chapter starts with a quick look at Euclidean geometry (the geometry that is taught from grade school through high school) and then discusses the existence of non-Euclidean geometry. After that, the chapter leads you on a short journey through aspects of geometry that have an artistic flair, such as perspective, the golden ratio, polygons, starts, and tessellations. The chapter ends with a look at fractal geometry, a geometry that combines the use of algebraic equations and computers to produce some interesting visual effects.

7 A SHORT HISTORY OF GEOMETRY

The exact origin of geometry is not known, but its roots are believed to date back before recorded history. There is evidence that intuitive concepts of geometry are universal. For example, prehistoric men and women probably realized that the shortest distance between two points is the straight line joining the two points. A basic understanding of geometric shapes is evident in the tools, weapons, and shelters designed by prehistoric men and women. Further, their drawings and handicrafts show a concern for spatial relationships. Their pottery, baskets, and weaving display examples of symmetry and sequences of designs. This concern for spatial relationships probably originated in the wonder of the world around them. Perhaps they marveled at the concentric circles formed when a rock is thrown into a pool of water, or the network of hexagons found in a beehive, or the intricate patterns found in snowflakes. Though we can only speculate on their knowledge of geometry, prehistoric humans' interest in geometry seems to have originated from basic intuition, practical needs, and the aesthetics of order and design.

This stamp, issued by the Maldive Islands in 1988, shows Euclid of Alexandria at work on the principles of geometry.

The first recorded evidence of geometry can be found with the Babylonians and Egyptians (3000–300 B.C.). Cuneiform tablets, hieroglyphic papyri, inscriptions on temples and tombs, and construction feats show a variety of practical uses of geometric concepts. Since both civilizations were largely agricultural, much of their geometry was developed to parcel out land, determine areas and perimeters, and calculate volumes of their granaries. In Egypt, the periodic flooding of the Nile River made surveying the land for the reestablishment of boundary lines a necessity. With the use of ropes, Egyptian surveyors, called rope-stretchers, accurately redetermined agricultural plots in the Nile Valley after the annual flooding had subsided. The Egyptians also showed their skill at measurement in the construction of their pyramids. For example, the Great Pyramid of Gizeh has a square base with sides that are about 756 feet long. Amazingly, the difference in the lengths of these sides is less than two-thirds of an inch. Babylonian irrigation canals and their beautifully constructed temples, such as the Hanging Gardens of Babylon, also showed practical uses of geometric concepts. The Egyptians and Babylonians are credited with the formation of a geometry that consisted of practical uses of measurement techniques.

The Chinese were also early pioneers in the study of geometry, as is seen in *K'iu-ch'ang Suan-shu,* or Arithmetic in Nine Sections, which was probably prepared by Chóu-kung around 1100 B.C. However, it is believed that Ch'ang Ts'ang collected writings of the ancients and wrote the version of *K'iu-ch'ang Suan-shu* that has been passed down through history. In this book, the Chinese showed an understanding of determining areas and volumes of geometric objects, finding lengths of sides of figures, and using what is now called the Pythagorean Theorem some 500 years before the time of Pythagoras.

The Greeks took the empirical geometry of the Babylonians and Egyptians and began to show that geometric truths could be abstracted from the practical situations in which they arise. In fact, the word *geometry* comes from the Greek words *geo,* meaning "earth," and *met'ron,* meaning "measure." However, the Greeks took the study of geometry far beyond measurement. They developed a geometric system based on logic, in which geometric facts follow from generally accepted statements called axioms or postulates. The position of the Greeks on the method and importance of geometry is summarized in quotes from two Greek scholars. Anaxagoras (499–427 B.C.) stated, "Reason rules the world," and Plato (430–347 B.C.) stated, "God eternally geometrizes." Contributions by Greek philosophers and mathematicians such as Thales, Pythagoras, Socrates, Plato, and Aristotle culminated with the work of Euclid of Alexandria (c. 300 B.C.), *Elements,* a collection of 13 books of which the first six and the last three are devoted to geometry. In *The Elements,* Euclid logically developed and summarized the geometry that was known up to that time. This "Euclidean" model dominated the study of geometry for thousands of years and is still the basis of geometry courses taught at the secondary level.

Following Euclidean geometry, the major advancements in geometry settled into three distinct categories: analytic/algebraic geometry, projective/descriptive geometry, and non-Euclidean geometry.

Analytic/Algebraic Geometry

In about 1629, Pierre de Fermat began applying methods of algebra to geometric objects. His work included the determination of equations for lines, circles, ellipses, parabolas, and hyperbolas. However, his work was not published until after his death in 1679. In 1637, *La Geometrie,* a discourse by René Descartes on analytic geometry, was published. In this discourse, Descartes introduced the *xy* coordinate system and allowed for the graphic representation of geometric curves given by equations. Because the work of Descartes was published first, he is credited with founding analytic (algebraic) geometry. In the Descartes-Fermat scheme, points became pairs of numbers and curves became sets of points generated by algebraic equations. Geometry became "algebraized."

Projective/Descriptive Geometry

In the 15th and 16th centuries, a focus of geometry was to obtain the correct perspective in representing what one sees. Architects Filippo Brunelleschi (c. 1400) and Leon Alberti (c. 1435), along with artists Pietro Franceschi (c. 1490), Leonardo da Vinci (c. 1500), and Albrecht Dürer (c. 1525), worked on representing three-dimensional objects on a two-dimensional surface. This interest in projecting a space figure onto another surface led to the development of projective and descriptive geometry. The mathematics of these geometries led to an understanding of perspective, became the foundation of architectural and mechanical drawing, and initiated investigations of topology.

Non-Euclidean Geometry

The fifth postulate (assumption) in Euclid's *Elements* has been the source of much thought and mathematical investigation. It says,

> That, if a straight line falling on two straight lines makes the interior angles on the same side less than two right angles, the two straight lines, if produced indefinitely, meet on that side on which the angles are less than two right angles.

A simpler and more intuitive equivalent of the postulate was formulated by John Playfair in 1795.

Through a given point only one parallel can be drawn to a given straight line.

In error, mathematicians thought that this postulate could be shown to be true as a result of logical deductions from Euclid's other postulates. Attempts to do this led to what is known today as Lobachevskian geometry and Riemannian geometry. Lobachevskian geometry is based on the assumption that more than one parallel line can be drawn to a given line through an external point, while Riemannian geometry's assumption is that no parallel line can be drawn through the point. Some of the details of these geometries are discussed in Section 7.1.

CHAPTER 7 PROJECTS

Research Projects

1. One of the reasons for the creation of geometry was its use in the measurement of lengths, perimeters, areas, and volumes. Give at least five examples of how these practical applications of geometry are still used today.
2. The pyramids of Egypt are early examples of practical geometry at work. Discuss the geometry involved in the construction of the pyramids.
3. Who were the Pythagoreans? What did this society do and what were some of its beliefs?
4. What are optical illusions? Find examples of them. Artists such as M. C. Escher have used them to produce interesting results. Investigate optical illusions in art.

Math Projects

1. Constructions with a straightedge and a compass were a major concern of early geometric studies. What are four geometric operations that can be performed with such constructions? Show how these are done. What are the three famous construction problems that have been proven to be impossible?
2. Using the techniques given in this chapter, create your own work of "geometric art."
3. Using fractal-generating computer software, generate some fractals and clearly show its self-similarity properties. (*Note:* Computer software that generate fractals can be found on the Internet.)

Euclidean and Non-Euclidean Geometry

Bruce Cohen's *Interior with Brick Wall Out Doorway* uses familiar shapes from Euclidean geometry. (Courtesy of the artist.)

EUCLIDEAN GEOMETRY

The geometry that is studied in elementary and secondary schools is based on the system of geometry contained in *Elements,* a book written by the Greek mathematician Euclid in about 300 B.C. This work summarized much of the geometry known up to that time and presented the first systematic treatment of geometry. Starting with definitions, postulates, and common notions, Euclid proceeded to prove numerous propositions of a geometry now called **Euclidean geometry.** The intent of this section is not to cover all of Euclidean geometry but to make you aware of the development of this mathematical system and to summarize some of the basic components of a system of geometry.

Undefined Terms

One of the most important features of a mathematical system is that the terms used in the system must have clear definitions. When making a definition, you should use words that are simpler than the term being defined. However, if you continually try to define terms by using simpler words, the process could conceivably go on forever.

There comes a time when a definition must use a term whose meaning is assumed to be clear. Such terms are called **undefined terms.** They are used to begin the process of defining new terms. We will begin our study of Euclidean geometry by considering its undefined terms: point, line, and plane.

A point can be described as a location or position. It can be represented by a dot and is usually named by an uppercase letter. A line may be described as the set of points arranged along a straight path. It extends indefinitely in two opposite directions and can be named by either a single lowercase letter or by two points on the line with a double-headed arrow placed above the letters. A plane can be described as the set of all points that form a flat surface extending indefinitely in all directions. It can be represented by means of a parallelogram and named by an uppercase letter placed in one corner of the parallelogram.

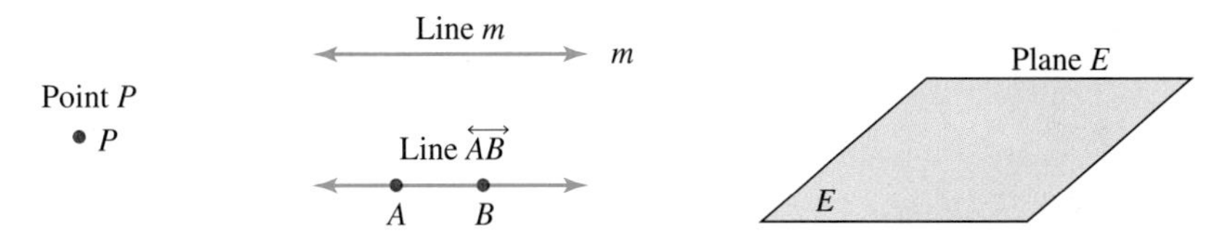

The preceding paragraph merely gives an intuitive description of a point, a line, and a plane. These are not considered definitions. Point, line, and plane are not defined in the Euclidean geometric system. However, as we proceed in the development of this geometric system, we will become more aware of the properties of these three basic geometric objects.

Definitions

With the use of the undefined terms, we can now give meanings to other terms and expressions of geometry. Here are some basic definitions:

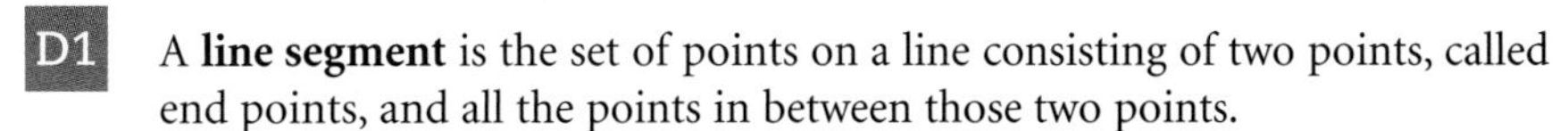

D1 A **line segment** is the set of points on a line consisting of two points, called end points, and all the points in between those two points.

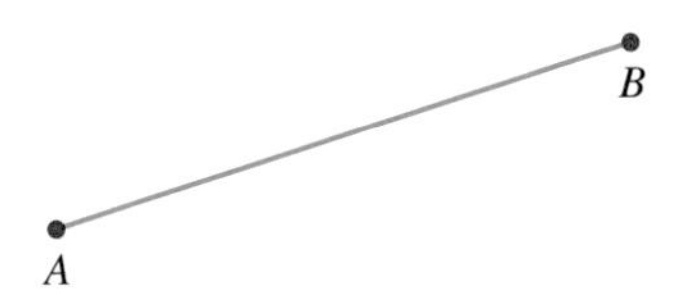

The line segment joining the points A and B is written $\overline{AB}$. The length of the line segment $\overline{AB}$ is written as AB.

D2 A **ray** is the figure formed by extending a line segment in only one direction.

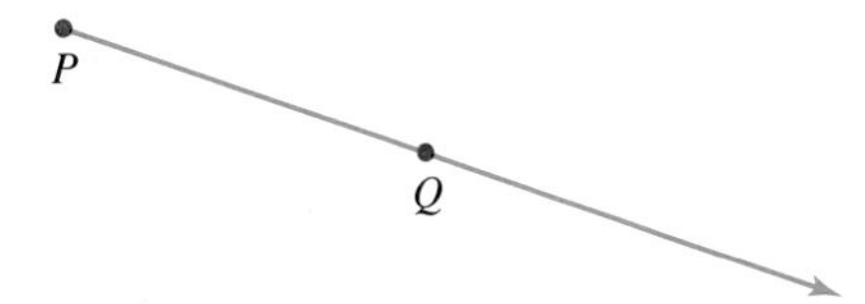

A ray with endpoint at point P extending through point Q is written $\overrightarrow{PQ}$.

D3 An **angle** is the figure formed by two rays or line segments with a common end point.

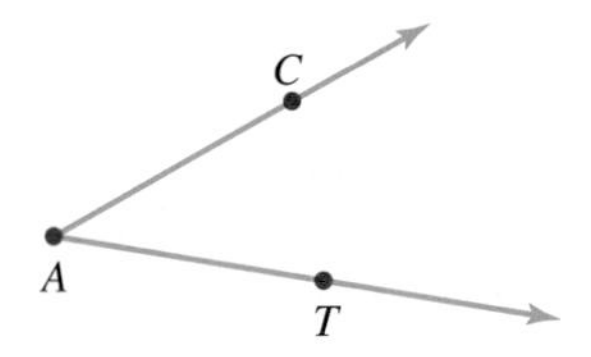

$\angle CAT$ has a vertex at point A which is the endpoint of $\overrightarrow{AC}$ and $\overrightarrow{AT}$.

D4 Types of angles: A **straight angle** is an angle that forms a line and measures 180°; a **right angle** is an angle that has a measure of 90°; an **acute angle** has a measure of between 0° and 90°; an **obtuse angle** has a measure of between 90° and 180°.

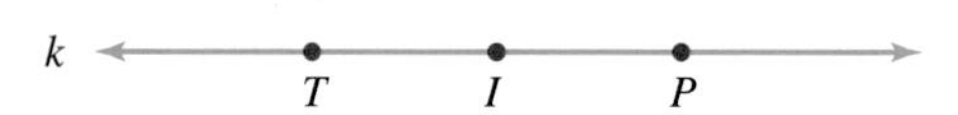

$\angle TIP$ is a straight angle

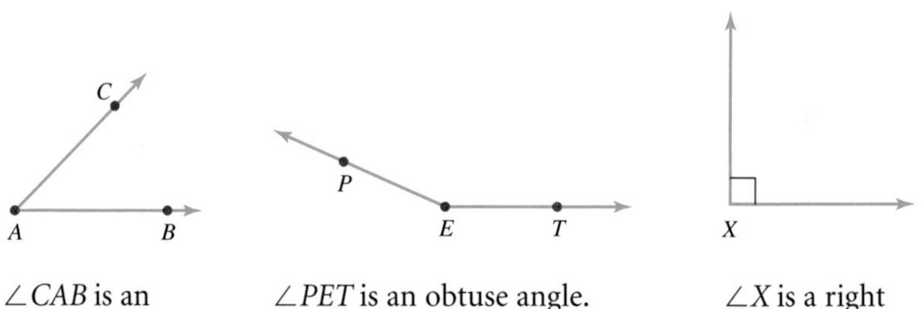

$\angle CAB$ is an acute angle.

$\angle PET$ is an obtuse angle.

$\angle X$ is a right angle.

D5 A **triangle** is a figure consisting of three line segments determined by three points that are not on the same line.

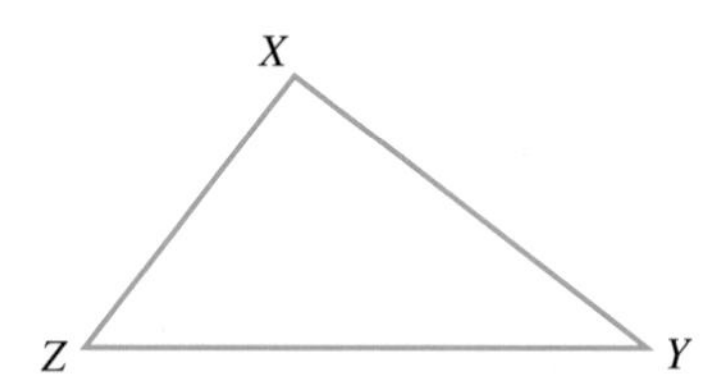

$\triangle XYZ$ has three vertices at the points X, Y, and Z. It has three sides, $\overline{XY}$, $\overline{YZ}$, and $\overline{ZX}$. It has three angles $\angle X$, $\angle Y$, and $\angle Z$.

D6 An **exterior angle** is an angle formed outside a triangle by one side of the triangle and the extension of another side of the triangle.

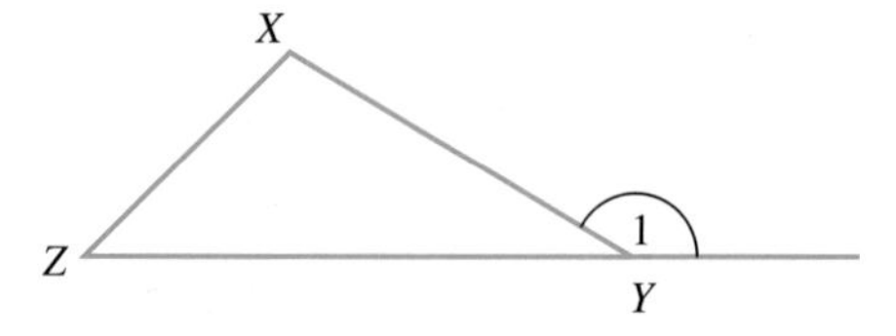

$\angle 1$ is an exterior angle for $\triangle XYZ$.

D7 Two lines, rays, or line segments are **perpendicular** ($\perp$) if they intersect and form a right angle. Two lines, rays, or line segments are **parallel** ($\|$) if they lie in the same plane and the lines that contain them do not intersect.

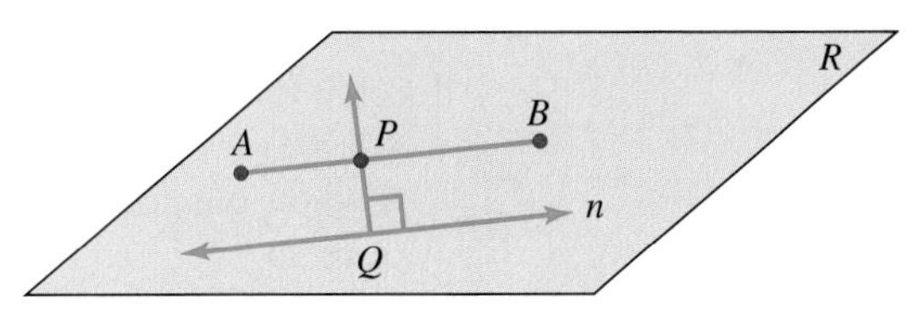

EXAMPLE 1

In plane R, segment $\overline{AB}$ is parallel to line n, written $\overline{AB} \| n$, and ray $\overrightarrow{QP}$ is perpendicular to line n, written $\overrightarrow{QP} \perp n$.

Axioms and Postulates

If we are to develop a system of geometry, each proposition must logically follow from other propositions. A system of geometry must start with some propositions that are accepted as true. We simply cannot deduce all statements. These assumptions are called **axioms** and **postulates.** Though these terms are used interchangeably, axioms refer to assumptions from arithmetic or algebra, whereas postulates refer to assumptions from geometry. Axioms and postulates are sometimes called self-evident truths. However, we shall see in our study of non-Euclidean geometry that this is not necessarily the case. We will start our study with axioms and postulates based on those of Euclid.

Axioms

 A quantity may be substituted for its equal in any expression.

EXAMPLE 2

If $\angle A + \angle B = \angle C$ and $\angle A = 90°$, then $90° + \angle B = \angle C$.

A2 If quantities are equal to the same quantity, then they are equal to each other.

EXAMPLE 3

If $A = C$ and $G = C$, then $A = G$.

A3 If equal quantities are added to or subtracted from equal quantities, the results are equal.

EXAMPLE 4

If $\angle A = \angle R$ and $\angle B = \angle K$, then

$$\angle A + \angle B = \angle R + \angle K \quad \text{and} \quad \angle A - \angle B = \angle R - \angle K.$$

A4 If equal quantities are multiplied by the same quantity or divided by the same nonzero quantity, the results are equal.

EXAMPLE 5

If $x = y$, then $5x = 5y$ and $x \div 7 = y \div 7$.

A5 A whole quantity is equal to the sum of its parts and is greater than any one of them.

EXAMPLE 6

$AD = AB + BC + CD$ and $AD > AB$, $AD > BC$, $AD > CD$, $AD > AC$, $AD > BD$.

Postulates

P1 Through two given points, one and only one line can be drawn.

EXAMPLE 7

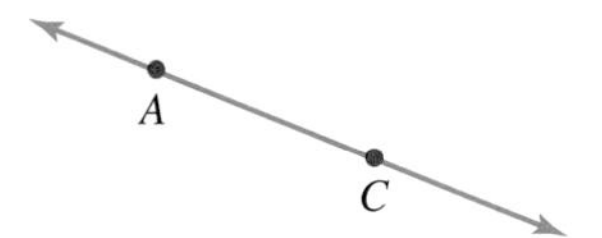

Through the points A and C only one line can be drawn.

P2 A line segment can be extended indefinitely in both directions.

EXAMPLE 8

The line segment $\overline{PQ}$ can be extended into line $\overleftrightarrow{PQ}$.

P3 If two points of a line lie in a plane, then the line through the two points lies in the plane.

EXAMPLE 9

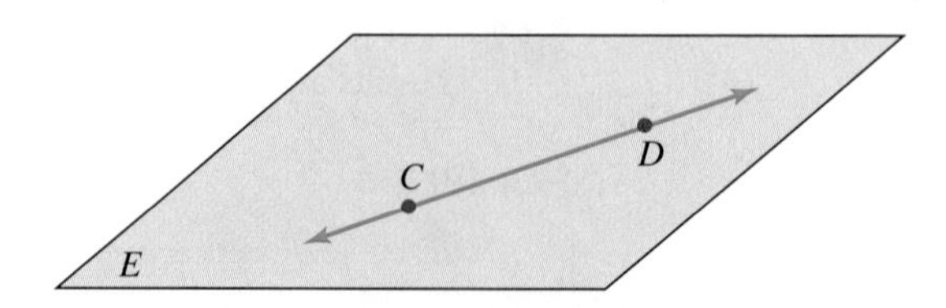

Points C and D lie in plane E. Therefore, the entire line passing through C and D lies in plane E.

P4 To every pair of points there corresponds a unique positive number called its distance.

EXAMPLE 10

The line segment $\overline{PQ}$ has a unique distance; $PQ = 4$ cm.

P5 **Parallel Postulate**
Through a given point, only one parallel can be drawn to a given line.

EXAMPLE 11

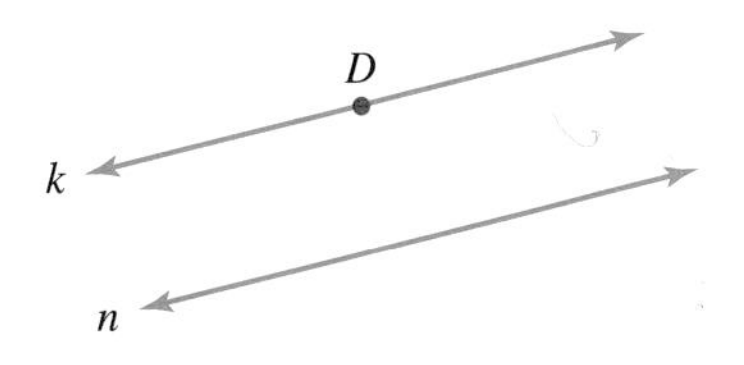

Through point D, line k is the only line that can be drawn parallel to line n.

P6 To every angle there corresponds a unique number between 0° and 180° called the measure of the angle.

EXAMPLE 12

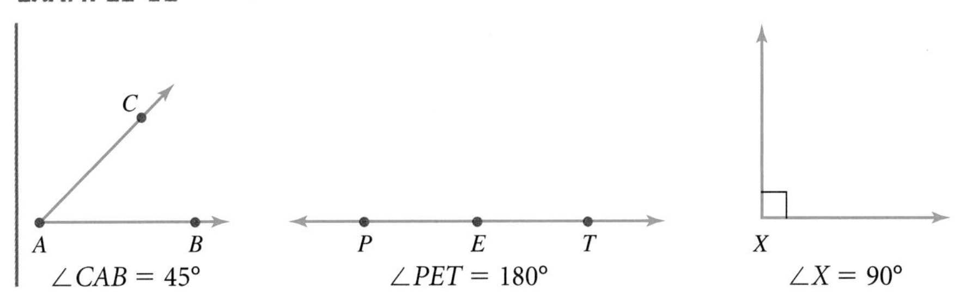

$\angle CAB = 45°$ $\angle PET = 180°$ $\angle X = 90°$

With the establishment of axioms, postulates, and definitions in this section, we can demonstrate the logical development of and prove some propositions of Euclidean geometry. Such proven propositions are called **theorems.** The **proof** of a theorem consists of a series of statements that logically show that the theorem is true based on axioms, postulates, definitions, and other proven theorems. The following theorems are examples of propositions from Euclidean geometry that can be logically deduced from the definitions, axioms, postulates, and theorems.

Theorems

If a line segment lies in a plane, then the line containing the segment lies in the plane.

T2 If two distinct lines intersect, they intersect in at most one point.

T3 If two distinct lines in the same plane are parallel to the same line, then they are parallel to each other.

T4 An exterior angle of a triangle is greater than either nonadjacent interior angle.

T5 **The Triangle-Sum Theorem:**
The sum of the measures of the angles of a triangle is 180.°

These propositions and many others constitute the propositions of Euclidean geometry. There are, however, other geometries.

NON-EUCLIDEAN GEOMETRY

Any system of geometry that uses a consistent set of postulates, with at least one postulate that is not logically equivalent to one of Euclid's postulates, is a **non-Euclidean geometry.** In particular, mathematicians wondered whether Euclid's parallel postulate was really a postulate. The statement of the postulate by Euclid seemed to lack the clarity and the "self-evident" character of his other postulates. In fact, Euclid did not use the parallel postulate until the proof of his 29th proposition. Mathematicians therefore attempted to derive the parallel postulate from Euclid's other postulates. These attempts, however, met with little success. It was not until the late 19th and early 20th centuries that mathematicians Eugenio Beltrami (1835–1900), Felix Klein (1849–1929), and Henri Poincaré (1854–1912) finally established that the parallel postulate could not be deduced from Euclid's other postulates. The attempts to prove the parallel postulate did, however, generate some interesting results. The work of the Jesuit priest Girolamo Saccheri (1667–1733) laid the groundwork for the creation of the two principal non-Euclidean geometries that bear the names of the mathematicians who spent their lifetimes studying them: Lobachevskian geometry and Riemannian geometry.

Russian mathematics professor Nicolai Lobachevsky (1793–1856), Hungarian army officer Janós Bolyai (1802–1860), and renowned German mathematician Carl Gauss (1777–1855) accepted the postulates of Euclid but replaced the parallel postulate with the following postulate:

Lobachevskian Parallel Postulate

Through a point that is not on a line, there is more than one line parallel to the given line.

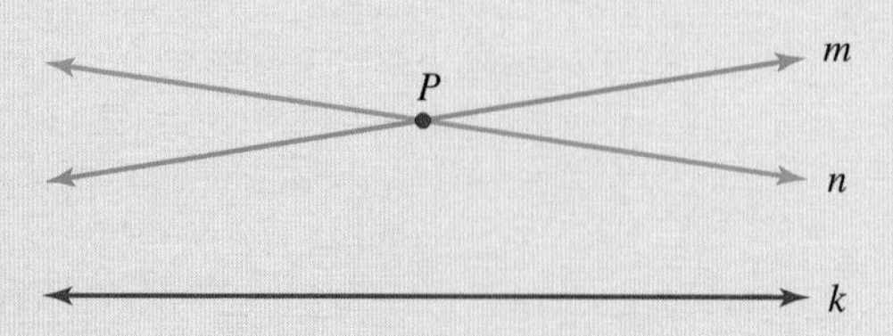

Through point P, lines m and n are parallel to line k ($m \parallel k$ and $n \parallel k$).

The German mathematician and student of Carl Gauss, Georg Riemann (1826–1866), replaced Euclid's parallel postulate with this:

Riemannian Parallel Postulate

Through a point that is not on a line, there is no line parallel to the given line.

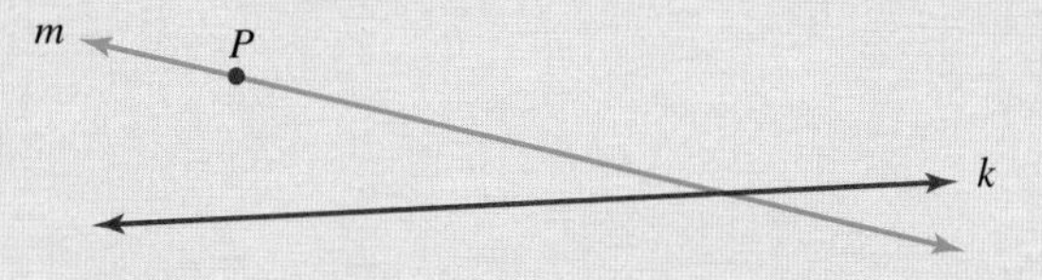

Any line m through point P will intersect line k.

These assumptions led to the creation of geometries that were logically consistent and contained no contradictions within themselves. However, the theorems deduced from these postulates contradicted some of the well-established theorems of Euclidean geometry. So controversial was the idea that one could deny a postulate of Euclid and arrive at conclusions that contradicted Euclid that even the renowned mathematician Carl Gauss was reluctant to publish his findings. Eventually, these non-Euclidean geometries were accepted by the mathematical community. However, the question of which geometry actually gives an accurate description of physical space remained. Euclidean geometry was so widely accepted as accurately describing physical space that these non-Euclidean geometries were considered as mere mental exercises. Mathematicians believed that the geometry of the physical world must be Euclidean. Although it is true that Euclidean geometry corresponds with our intuitive ideas about our surroundings, we would like to present some models and observations to suggest the existence of non-Euclidean geometries.

A Riemannian Model

Besides changing Euclid's parallel postulate, Georg Riemann questioned whether lines extended infinitely. He proposed that lines trace back on themselves, like circles. You could traverse a circle endlessly, but its length is still finite. Euclid's second postulate said that a line segment can be extended indefinitely in both directions. Riemann modified Euclid's second postulate to state that a line is endless but not necessarily infinite. Furthermore, do parallel lines never meet? Never? If one looks down a set of railroad tracks, it seems that the tracks get closer together. Maybe parallel lines in space eventually do meet.

If we accept the two assumptions that lines are endless but not infinite and that parallel lines do not exist, we form a different geometry, the geometry of Georg Riemann. Its postulates and theorems cannot be easily displayed on the flat plane of Euclidean geometry. They can, however, be visualized on the surface of a sphere. In Riemannian geometry, the Euclidean plane becomes a sphere, the Euclidean line becomes a great circle on the sphere, and the Euclidean point becomes a point on the sphere along with its antipodal point (point on the sphere farthest from the first point). For example, although we consider the North and South Pole to be different points, in Riemannian geometry the North and South Poles are considered together as one point. Such a model allows us to visualize the propositions of Riemannian geometry. The great circles of a sphere have a finite length. Since any two great circles intersect, there are no parallel lines in such a model.

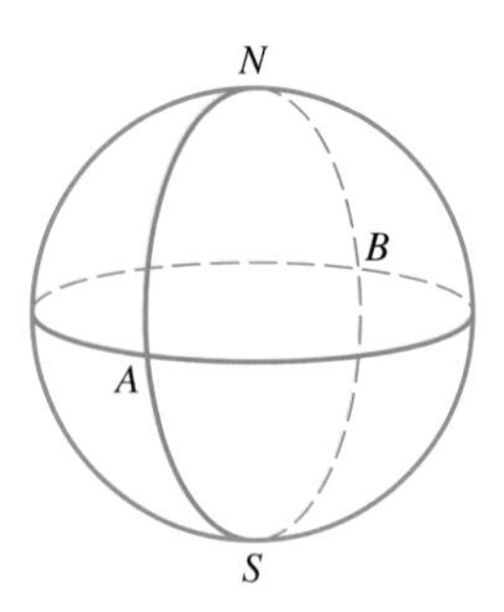

Consequences of the Riemannian Parallel Postulate

In Riemannian geometry, all theorems that are consequences of the parallel postulate are different from those in Euclidean geometry. One of the most significant results of the Riemannian parallel postulate is that the sum of the measures of the angles of a triangle is greater than 180° rather than equal to 180°, as found in Euclidean geometry. If we look at the triangle drawn on the spherical model for Riemannian geometry in the figure, we see that $\triangle ANB$ has two 90° angles, so the sum of the angles of $\triangle ANB$ is greater than 180°.

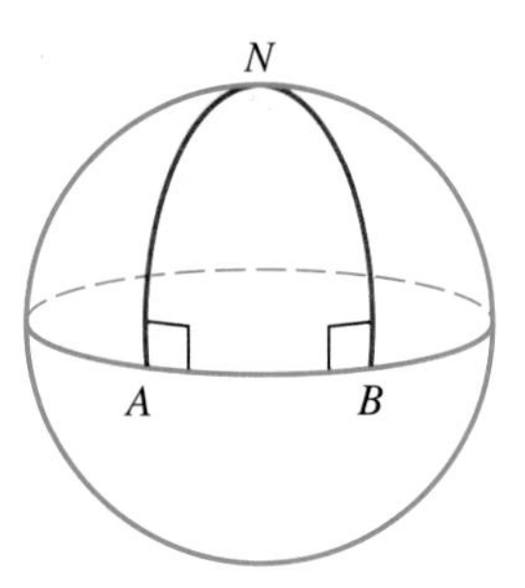

Riemannian Triangle-Sum Theorem

The sum of the measures of the angles of a triangle is greater than 180°.

A Lobachevskian Model

With Riemannian geometry, we introduced the possibility that lines are not necessarily straight. The same holds true in Lobachevskian geometry. In fact, scientists believe that space is curved. A ray of light traveling through space does not take a straight path. According to Einstein's theory of relativity, the path of a ray of light is affected by the gravitational field of large objects in space. The gravitational field causes the ray of light to bend. The lines in Lobachevskian geometry also bend, as can be seen in the model in the figure.

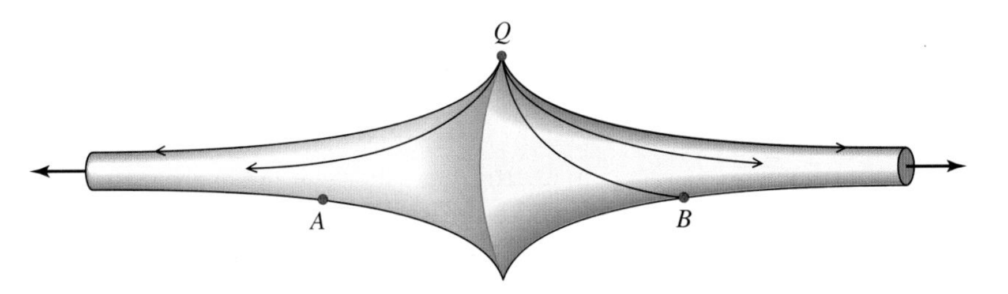

The model resembles two attached trumpets with the small ends extending indefinitely, a shape called a pseudosphere. The pseudosphere corresponds to the plane in Euclidean geometry and to the sphere in Riemannian geometry. Lobachevskian lines consist of two symmetric curves meeting at a point at the widest part of the pseudosphere. Through point *Q*, there is more than one "line" parallel to the "line" through *A* and *B*. Through point *Q* on the pseudosphere, you will also notice that "line" *QB* intersects "line" *AB* at point *B*.

Consequences of the Lobachevskian Parallel Postulate

In Lobachevskian geometry, all theorems that are consequences of the parallel postulate are different from those in Euclidean geometry. A notable result of the Lobachevskian parallel postulate is that the sum of the measures of the angles of a triangle is less than 180° rather than equal to 180° (as found in Euclidean geometry) or greater than 180° (as found in Riemannian geometry). If we look at an equiangular triangle drawn on the pseudosphere model, each angle is less than 60° because the sides of the triangle curve inward. The sum of the angles of the triangle is visibly less than 180°.

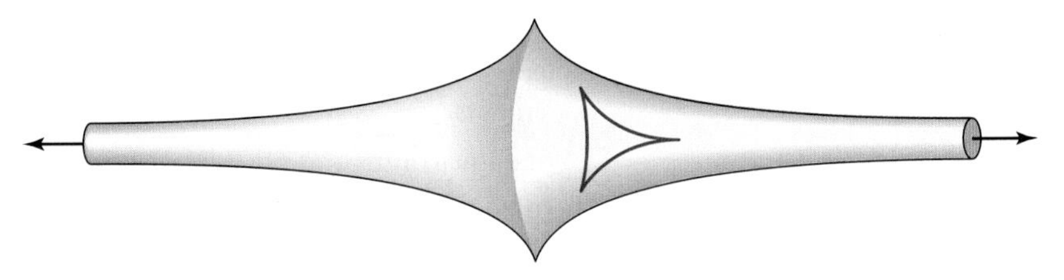

Lobachevskian Triangle-Sum Theorem

The sum of the measures of the angles of a triangle is less than 180°.

Is the World Euclidean, Riemannian, or Lobachevskian?

The conclusions of each geometry are true on the basis of the initial assumptions of each geometry, but is the world Euclidean, Riemannian, or Lobachevskian? We might never decide whether Euclidean, Riemannian, or Lobachevskian geometry adequately describes our world, since mathematics does not establish truths about the physical world. Geometry gives us a set of logical conclusions based on possible perceptions and assumptions about the physical universe. The mathematics of each geometry has an intrinsic beauty and logic. Instead of asking, "Is the world Euclidean, Riemannian, or Lobachevskian?" one might wonder, "Which geometry is the best to apply in a given situation?" Humankind's experience over thousands of years suggests that in working with measurement, travel, construction, and design, Euclidean geometry seems most useful. On the other hand, advancements in science over the last century suggest that in investigating outer space or analyzing the inner space of atoms, properties of non-Euclidean geometries may be appropriate models.

PROBLEMS ▶ Explain ▶ Apply ▶ Explore

▶ Explain

1. Explain why it is necessary to have undefined terms and postulates in the development of a deductive system of geometry.
2. What is the difference between a ray and a line segment?
3. What are the differences between straight, right, acute, and obtuse angles?
4. Given lines a and b, explain the difference between $a \parallel b$ and $a \perp b$. Make a sketch of each.
5. Even though $\overline{RS}$ does not intersect line m, explain why $\overline{RS}$ is not parallel to line m.

6. Explain why the following definition of parallel segments is faulty: "Two segments are parallel if they do not intersect." Give a good definition for parallel segments.
7. What are axioms and postulates? What are the differences and similarities?
8. Explain the difference between a postulate and a theorem in geometry.
9. Explain why the figure is not possible for lines m and n in Euclidean geometry.

10. Explain why the figure is not possible in Euclidean geometry for points A and B and the line through A and B in plane E.

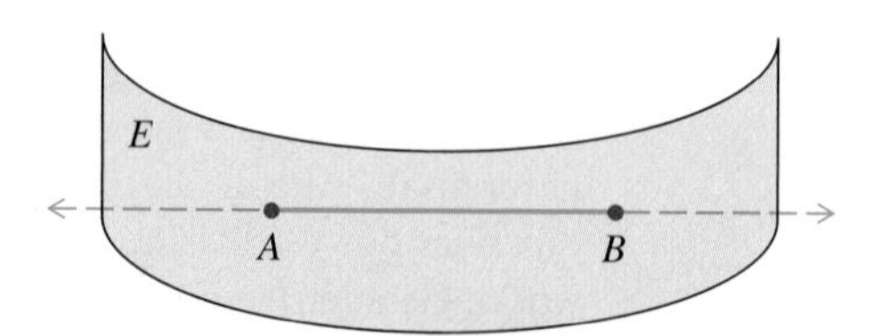

11. The transitive property of equality states that if $a = b$ and $b = c$, then $a = c$. Explain why this is equivalent to Axiom A2.

12. Describe the parallel postulates of Euclidean, Riemannian, and Lobachevskian geometries. Explain how they are the same and how they are different.
13. Contrast and compare the triangle-sum theorems of Euclidean, Riemannian, and Lobachevskian geometries.

Apply

In Problems 14–21, in the Euclidean plane, make a sketch that satisfies the conditions.

14. Obtuse $\angle FUN$, where $\overline{UF}$ and $\overline{UN}$ are line segments.
15. Acute $\angle RUN$, where points R, U, and N lie in plane M.
16. Line j and line k, where line r makes a right angle with line j and intersects line k.
17. Line j and line k, where line r is perpendicular to both line j and line k.
18. $\overline{FX} \perp \overrightarrow{FG}$, line $y \parallel \overline{FX}$.
19. Right angle $\angle HER$ with $\overline{HB} \perp \overline{ER}$.
20. Right angle $\angle HER$ with $\overline{BE}$ making an acute angle with $\overline{HE}$.
21. $\triangle RAD$ with an exterior angle at each vertex.

In Problems 22–26, use specific points, lines, planes, and angles to make a sketch of each theorem and state the conclusion of the theorem.

22. Theorem T1.
23. Theorem T2.
24. Theorem T3.
25. Theorem T4.
26. Theorem T5.

In Problems 27–30, in a Riemannian model, make a sketch that satisfies the conditions.

27. $\triangle ABC$ in which $\angle A$, $\angle B$, and $\angle C$ are right angles.
28. $\triangle REI$ with $RE = EI = RI$ and $\angle E$ is an obtuse angle.
29. Acute $\angle FAD$ and $\overline{HE} \perp \overline{AD}$.
30. $\overleftrightarrow{XY}$ and $\overline{QV} \perp \overline{XY}$.

In Problems 31–34, in a Lobachevskian model, make a sketch that satisfies the conditions.

31. Right $\triangle ABC$ with line m intersecting $\overline{AB}$.
32. Acute $\angle FAD$ and $\overrightarrow{HE} \perp \overline{AD}$.
33. Lines h, k, and n with $h \parallel k$, $n \parallel k$, and $h \parallel n$.
34. Lines h, k, and n with $h \parallel k$, $n \parallel k$, and h not parallel to n.

Explore

35. Euclid defined a point as "that which has no part." Do you think this is a good definition? Why or why not?
36. Euclid defined a line as "breadthless length." Do you think this is a good definition? Why or why not?
37. Using a standard dictionary, show how giving the definition of the word "dimension" can lead you in a circular path; that is, continuing to define the words of previous definitions could lead you back to the word you began with.

38. Show how Axiom A3 can be used to solve for x in the equation $5x - 3 = 4x + 9$.

39. Show how Axioms A3 and A4 can be used to solve for x in the equation $5 + 6x = 4x - 11$.

40. Draw three pairs of parallel lines. Draw a line that intersects each pair of parallel lines. Use a protractor to measure each set of corresponding angles as shown in the figure. Corresponding angles $\angle 1$ and $\angle 2$, are marked on the figure.

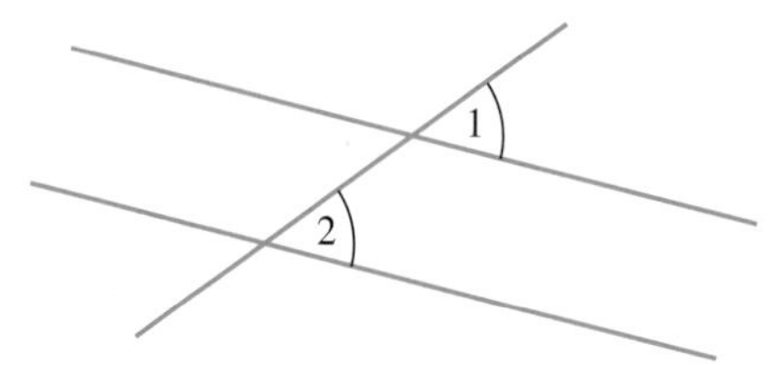

What seems to be true about these angles? Formulate a precise statement of your conjecture.

41. Draw three pairs of parallel lines. Draw a line that intersects each pair of parallel lines. Use a protractor to measure each set of alternate interior angles. Alternate interior angles, $\angle 1$ and $\angle 2$, are marked on the figure.

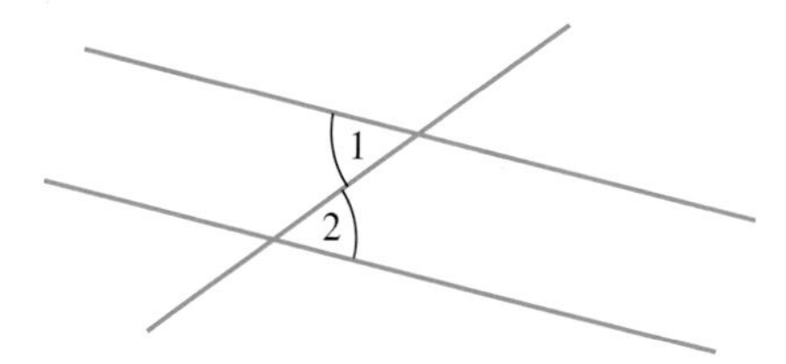

What seems to be true about these angles? Formulate a precise statement of your conjecture.

42. Draw three large triangles and an exterior angle for each triangle. Use a protractor to measure the exterior angle and the interior angles of each triangle. Using your results, make some conjectures about exterior angles and interior angles of a triangle.

43. What does the existence of non-Euclidean geometry tell us about mathematics and the truth about the physical world?

44. Make a sketch of a ray in the Riemannian plane. Explain why the ray is actually a line.

45. One of the theorems from Lobachevskian geometry states that parallel lines are not always the same distance apart. Use the Lobachevskian model to demonstrate this theorem.

In Problems 46–49, use the fact that a rectangle is a four-sided plane figure that has four right angles.

46. Draw a four-sided plane figure with three right angles in the Lobachevskian plane. What type of angle is the fourth angle?

47. Draw a four-sided plane figure with three right angles in the Riemannian plane. What type of angle is the fourth angle?

48. Explain why rectangles do not exist in Riemannian geometry.

49. Explain why rectangles do not exist in Lobachevskian geometry.

In Problems 50–52, use the Lobachevskian triangle-sum theorem to explain why each is possible.

50. In a right triangle, the sum of the two non-right angles is less than 90°.

51. Each angle of an equiangular triangle is less than 60°.

52. A triangle can have only one right angle or one obtuse angle.

In Problems 53–55, use the Riemannian triangle-sum theorem to explain why each is possible.

53. Each angle of an equiangular triangle is greater than 60°.

54. In a right triangle $\triangle ABC$, if $\angle A$ is a right angle, then $\angle B + \angle C > 90°$.

55. A triangle can have two right angles.

***56.** The five theorems of geometry stated in this section constitute a very small selection of the theorems of Euclidean geometry. Find five other theorems of Euclidean geometry. Make a labeled sketch of each theorem and state the conclusion of each theorem.

section 7.2

Perspective

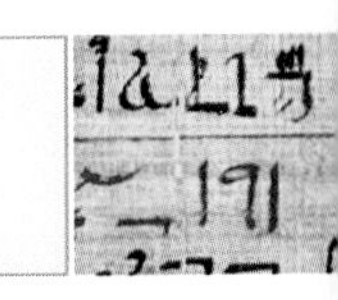

Have you ever looked at a beautiful scene and wondered how an artist can draw it on a piece of paper? The paper is flat (two-dimensional), and the visual scene is in real space (three-dimensional). How can the artist place lines and shapes on the paper to display this real space? This art of drawing objects in such a way as to give them depth and show their distance from the observer is called **perspective.** With a few simple techniques and geometric operations, the artist is able to give a drawing the illusion of three-dimensional space. When perspective is used, the depth and distance of the actual three-dimensional scene are conveyed on the two-dimensional surface. With the use of the techniques presented in this section, you too can draw the three-dimensional world on a piece of paper.

OVERLAPPING SHAPES

One of the easiest techniques for creating a sense of depth is using **overlapping shapes.** When shapes overlap each other in a drawing, an illusion of shallow space is created. When you see one shape placed in front of another, you are given the impression that the scene is not flat. You are led to conclude that the scene is three dimensional. Here are some examples of this technique.

Mary Cassatt was among many late 19th century artists who adopted this style of shallow perspective. In her etching, *The Letter* (1891), there are many subtle uses of overlapping shapes to create the illusion of depth. The desk that projects into the picture overlaps the woman's legs. The hands with the envelope overlap the body and face. The letter overlaps the top of the desk. The overlapping causes us to conclude that a three-dimensional scene appears on the two-dimensional aquatint etching.

Detail from *The Letter* by Mary Cassatt. (National Gallery of Art)

Diminishing Sizes

Since objects near to us appear larger than objects that are farther away, a second method of conveying a sense of depth is by systematically making objects smaller. This technique of **diminishing sizes** also creates the illusion of depth. Like the technique of overlapping shapes, it causes us to conclude that a three-dimensional scene appears on a two-dimensional surface. Here is an example of that technique.

The technique of overlapping shapes and diminishing the size can be combined to create an even greater illusion of depth. In the example that follows, the trees decrease in size and overlap each other.

In André Derain's *The Old Bridge* (1910), the illusion of space is created by the positioning of smaller and smaller dwellings, progressing up the hill away from the bridge.

The Old Bridge by André Derain. (National Gallery of Art)

ATMOSPHERIC PERSPECTIVE

A third technique that can serve to create the illusion of depth on a flat surface is a gradual lessening of clarity and visual strength. When we look into the distance, objects at a farther distance are often characterized by diminishing color and values of shading. The hills close by seem clear and colorful, with defined details and shadows, while the hills farther away appear softer and less detailed. As objects recede into the background, they become less distinct. Individual trees become a forest.

People with faces become anonymous forms in a crowd. It seems that the effect of the earth's atmosphere makes the objects less identifiable as they get farther away. Therefore, the technique that is used to produce this effect is called **atmospheric perspective.**

We can demonstrate this technique by shading the trees of the previous example. We can decrease the intensity of the shading and the crispness of the edges of the trees as we progress from the first to the last tree. A deeper, more illusionist effect can be seen in the figure that follows.

Chinese and Japanese ink and wash landscapes are famous for the use of this method of conveying space and distance. Leonardo da Vinci developed an oil painting technique that creates this effect by scrubbing across previously painted forms with an almost dry brush. In his *Mona Lisa* (1505), he created softness of edges, a veil of mystery, and a subtle depth in the background using this "sfumato" or "smoky effect" method.

Cattleya Orchid and Three Brazilian Hummingbirds by Martin Johnson Heade. (National Gallery of Art)

In Martin Johnson Heade's *Cattleya Orchid and Three Brazilian Hummingbirds* (1871), you will notice that the flower, birds, and trees at the front of the painting are very bright and clear. As you visually progress beyond the foreground, the objects become less defined and engulfed in an eerie mist.

ONE-POINT PERSPECTIVE

These methods of creating depth were not enough for the artists of the Renaissance who desired to draw and paint more realistically. They wanted their paintings to be a window through which one looks at the real world. They wanted to frame reality. To accomplish this, they needed to understand how to describe the environment of human-made structures, such as streets, buildings, bridges, fences, and walls. Their study of the classical past failed to reveal any logical solutions. However, in Florence, Italy (c. 1400), Filippo Brunelleschi discovered a "perspective" technique that solved the problem for the artists. Brunelleschi was an architect but also a lover of classical thought, philosophy, art, and mathematics. While his accomplishments in architecture were many, his greatest contribution was the discovery of linear, **one-point perspective.** This technique enabled artists to realistically represent three-dimensional space on a two-dimensional canvas by using the simple geometry of converging lines. When a drawing is created with a one-point perspective, objects seem to converge to a single fixed point. Objects are systematically shortened as they recede into the distance.

Creating One-Point Perspectives

To create a one-point perspective drawing, the artist must first focus on what is to be drawn, often by using a small rectangle as a frame or viewfinder. The artist establishes the window through which the scene will be drawn (see Figure 7.1).

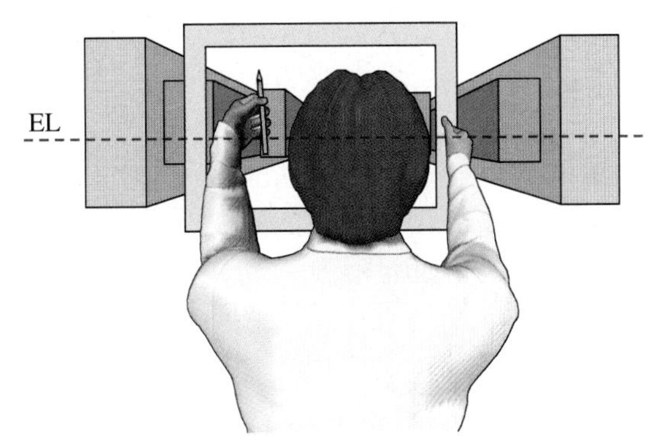

FIGURE 7.1

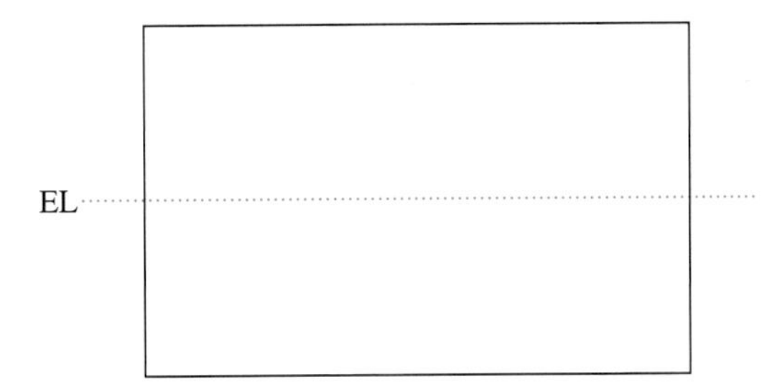

FIGURE 7.2

Once the limits of the scene have been determined, the following steps can be used to draw the scene in a one-point perspective.

1. Looking straight ahead, a line is placed across the frame of the picture to establish where everything should be located in terms of being above or below eye level. Such a line is called the **eye level line.** In Figure 7.2, the eye level line (EL) is placed near the center of the frame.
2. To create the illusion of space, a point is established on the eye level line to which all receding lines converge. This point is called the **vanishing point** (VP). For example, to draw a three-dimensional box, the front of the box can be given the illusion of receding to the vanishing point by connecting the corners of the box to the vanishing point as shown in Figure 7.3.
3. The box can be given a sense of depth by drawing the back of the box using segments that are parallel to the front of the box and have endpoints on the lines to the vanishing point as shown in Figure 7.4.
4. By darkening some of the lines and erasing all nonessential lines, the box can be seen as a solid object receding into space toward a fixed point as shown in Figure 7.5.
5. If the artist wants to add other objects to the drawing, the eye level line and the vanishing point remain the same. In Figure 7.6, the box becomes a part of a more complex scene.

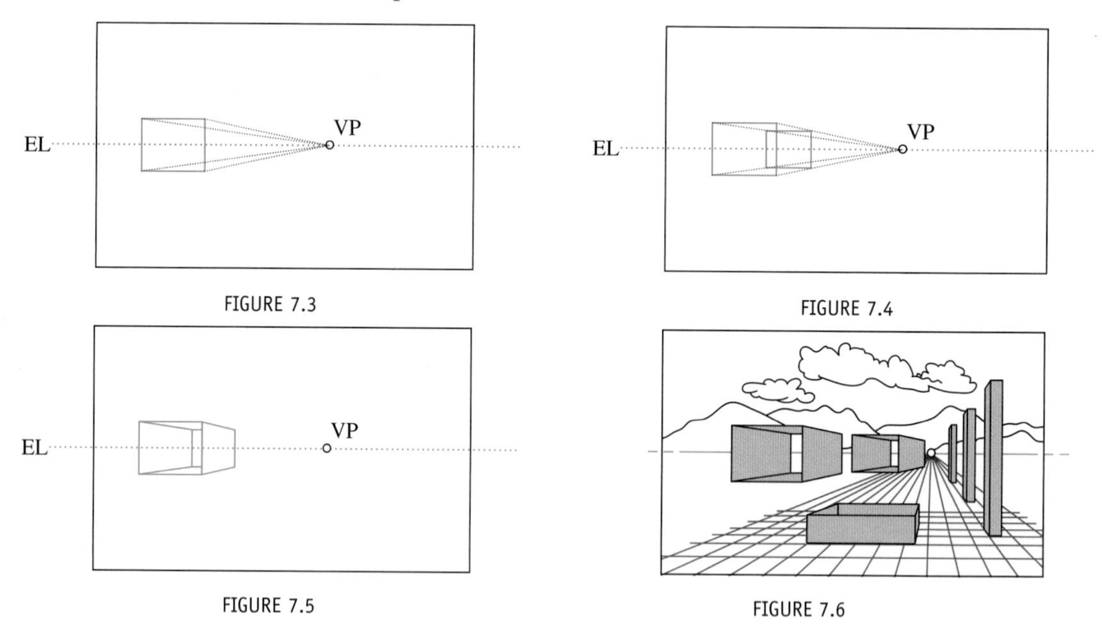

FIGURE 7.3

FIGURE 7.4

FIGURE 7.5

FIGURE 7.6

Varying Positions of Eye Level Line and Vanishing Point

In creating the illusion of space on a two-dimensional surface, there are an unlimited number of ways you can use the basic procedure for one-point perspectives. Changing the eye level line and vanishing point can create very different effects. The figures that follow show variations when the subjects being drawn are in different relative positions to the eye level line. The changes caused by moving the vanishing point are left as exercises at the end of the section.

1. Most of the subject is above the eye level line.

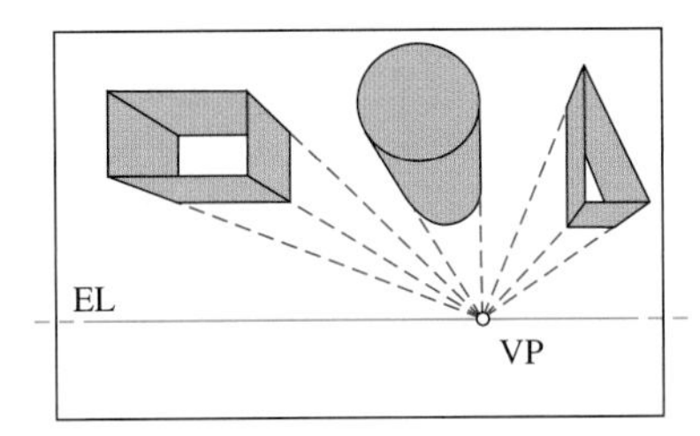

Notice that the eye level line is drawn low across the frame and the objects connected to the vanishing point appear above you.

2. Most of the subject is below the eye level line.

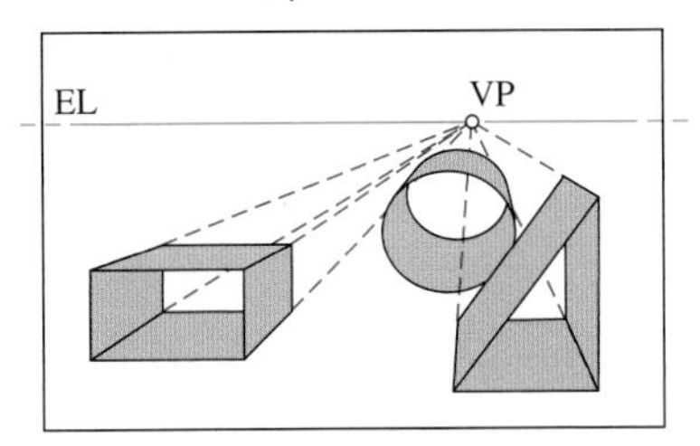

Notice that the eye level line is drawn high across the frame and the objects connected to the vanishing point appear below you.

3. The subject is distributed above, below, and on the eye level line.

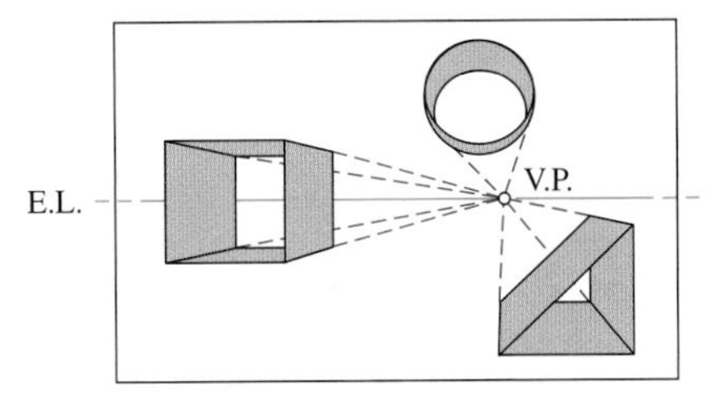

Notice that the eye level line is drawn horizontally near the center of the frame and the objects connected to the vanishing point appear above, below, and at eye level.

In Vincent van Gogh's *Flower Beds in Holland* (1883), you can see that he used a one-point perspective to give us a beautiful view of his homeland.

Detail from *Flower Beds in Holland* by Vincent van Gogh. (National Gallery of Art)

Equally Spaced Objects

If equally spaced vertical objects such as poles, posts, or trees are drawn in a one-point perspective, they not only get shorter, they also get closer together. A very efficient geometric method is used to determine the spacing between such upright objects in a one-point perspective. Here is how it is done.

1. The first upright ($\overline{AB}$) is drawn, and the endpoints are connected to the vanishing point on the eye level line (see Figure 7.7).
2. A second upright ($\overline{CD}$) is drawn at an arbitrary distance from the first upright.
3. The third upright is mathematically placed at E. Point E is positioned by finding where the line drawn from the top of the first upright (A) through the midpoint of the second upright (M) intersects the line to the vanishing point.
4. Other poles are systematically located and drawn between the lines to the vanishing point by using the third step.

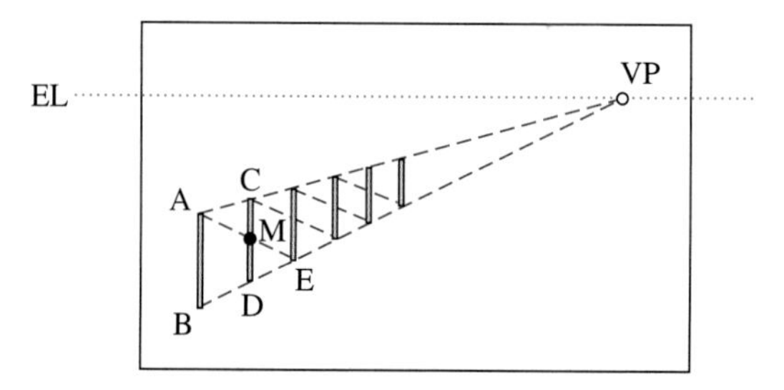

FIGURE 7.7

Proportions in a One-Point Perspective

In a one-point perspective, objects of the same size are drawn proportionally smaller as they get closer to the vanishing point. This proportion is determined once the first object has been drawn and its distance from the vanishing point has been established. If another object of the same size is positioned closer to the vanishing point, the ratio of its height to its distance from the vanishing point is the same as the ratio of the height to the distance of the first object. If we let h_1 be the height of the first object, d_1 be its distance from the vanishing point, h_2 be the height of the added object, and d_2 be its distance from the vanishing point, we get the following proportion for the objects in the drawing.

$$\frac{h_1}{d_1} = \frac{h_2}{d_2}$$

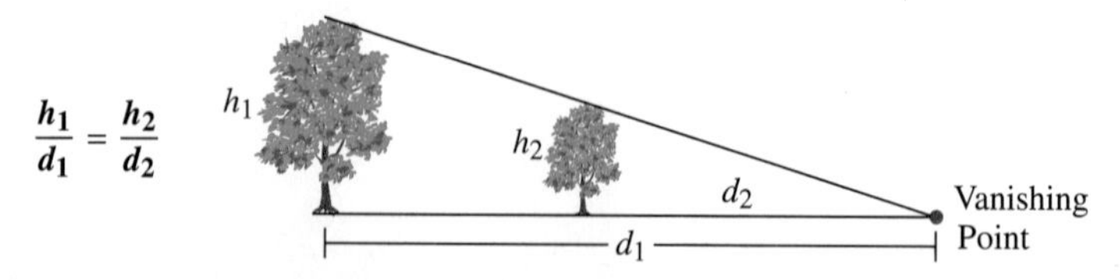

This proportion enables us to determine the heights or distances of other congruent objects that are positioned in a line of objects.

EXAMPLE 1

Suppose the first tree in the drawing that follows is 5 in. tall and 18 in. from the vanishing point.

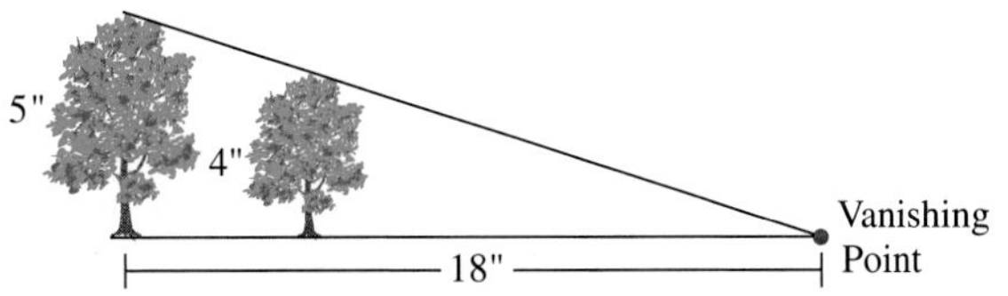

5"
4"
18"
Vanishing Point

a) How far from the vanishing point should you place a second tree that is to be 4 in. tall in the drawing?

b) What is the distance between the first tree and the second tree?

Solution:

a) Using the proportion for a line of objects of the same size in a one-point perspective, we get the following solution. Let $h_1 = 5$, $d_1 = 18$, $h_2 = 4$, and d_2 be unknown.

$$\frac{h_1}{d_1} = \frac{h_2}{d_2}$$

$$\frac{5}{18} = \frac{4}{d_2}$$

$$5d_2 = 18(4)$$

$$5d_2 = 72$$

$$d_2 = 14.4$$

Thus, the second tree is placed 14.4 in. from the vanishing point.

b) Since the first tree is positioned 18 in. from the vanishing point and the second tree is positioned 14.4 in. from the vanishing point, the distance between the trees on the drawing is $18 - 14.4 = 3.6$ in.

EXAMPLE 2

Suppose in the drawing of the previous example you want to draw another tree 3 in. from the vanishing point. How tall should you make the tree in the drawing?

Solution: We again use the proportion for a line of objects of the same size this time with $h_1 = 5$, $d_1 = 18$, $d_2 = 3$, and h_2 being unknown.

$$\frac{h_1}{d_1} = \frac{h_2}{d_2}$$

$$\frac{5}{18} = \frac{h_2}{3}$$

$$18h_2 = 5(3)$$

$$18h_2 = 15$$

$$d_2 = \frac{15}{18} = \frac{5}{6}$$

Thus, the tree should be $\frac{5}{6}$ in. tall.

MORE COMPLEX PERSPECTIVES

Artists have gone beyond the one-point perspective discovered by Brunelleschi in 1400. They have perfected two-point and three-point perspectives and have integrated different techniques into a single painting. It is beyond the scope of this book to show you how these more complex perspectives are created. Our objective is to demonstrate how simple constructions from geometry can be used to create artistic effects. However, we will give you some examples of these more complex techniques.

Two-Point Perspective

In *The Square at St. Mark's* by Canaletto (1735), you will notice that if you follow the receding lines to the right and left there are two vanishing points on the eye level line. You will also notice that both of these vanishing points are off the canvas.

The Square of St. Mark's by Canaletto. (National Gallery of Art)

Three-Point Perspective

In David McLaughlin's *Delphic Vision* (1993), in addition to vanishing points to the right and left, the view sweeps upward, high above the ground, to a third point. This three-point perspective along with the overlapping of shapes, diminishing sizes, and atmospheric perspective give this watercolor a stunning three-dimensional effect on McLaughlin's two-dimensional canvas.

Delphic Vision by David McLaughlin. (Courtesy of the artist)

If you experiment with the techniques described in this section, you can draw objects on a piece of paper that give an illusion of being three dimensional. The exercises that follow will give you practice is doing just that.

section 7.2

PROBLEMS ▸ Explain ▸ Apply ▸ Explore

▸ Explain

1. What is perspective?
2. What is the perspective technique of overlapping shapes? Sketch an example.
3. What is the perspective technique of diminishing sizes? Sketch an example.
4. What is the technique of atmospheric perspective? Sketch an example.
5. What is a one-point perspective? Sketch an example.
6. What are the eye level lines and vanishing points? Sketch an example.
7. What is the basic procedure for creating a one-point perspective?
8. In a one-point perspective, what procedure is used to position equally spaced objects of the same size?

▸ Apply

9. Use the following shape and make a sketch that gives a sense of depth with
   a) overlapping shapes.
   b) diminishing sizes.
   c) atmospheric perspective.
10. Use the following shape and make a sketch that gives a sense of depth with
   a) overlapping shapes.
   b) diminishing sizes.
   c) atmospheric perspective.

11. Create solid objects in a one-point perspective using the shapes given as the front of the object and the eye level line and vanishing point as shown.

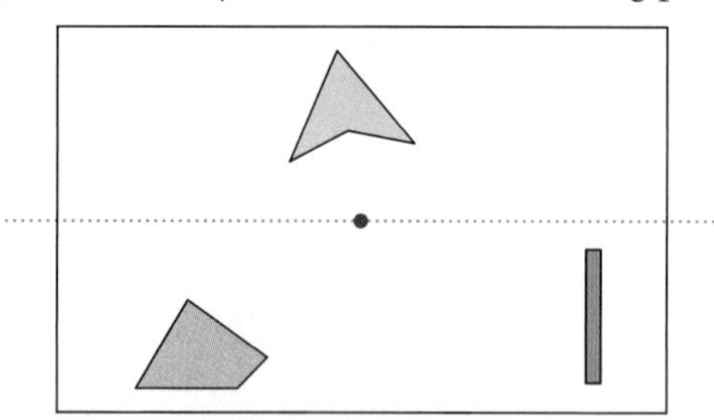

12. Create solid objects in a one-point perspective using the shapes given as the front of the object and the eye level line and vanishing point as shown.

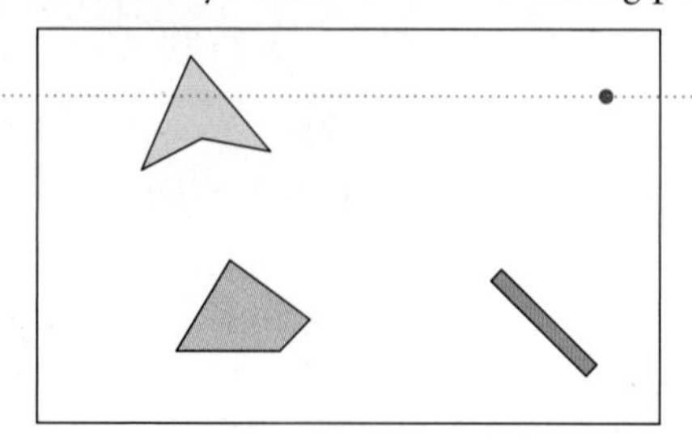

13. Create solid objects in a one-point perspective using the shapes given as the front of the object and the eye level line and vanishing point as shown.

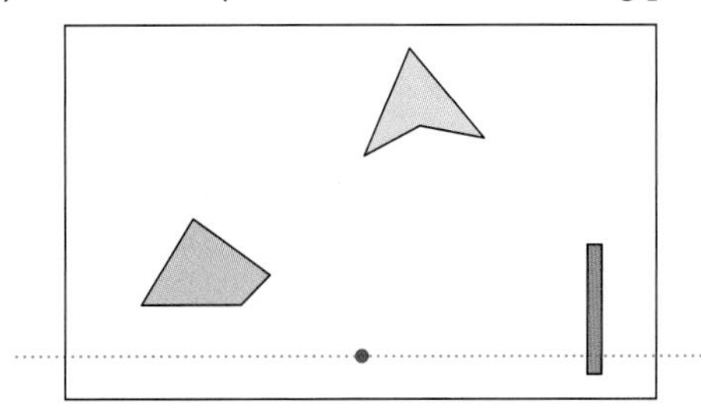

14. Draw a one-point perspective showing three billboards along a two-lane road that has a word on each billboard. The words on the billboards give the message, "EAT AT AL'S." The start of the road, the first billboard, the eye level line, and the vanishing point are shown in Figure 7.8

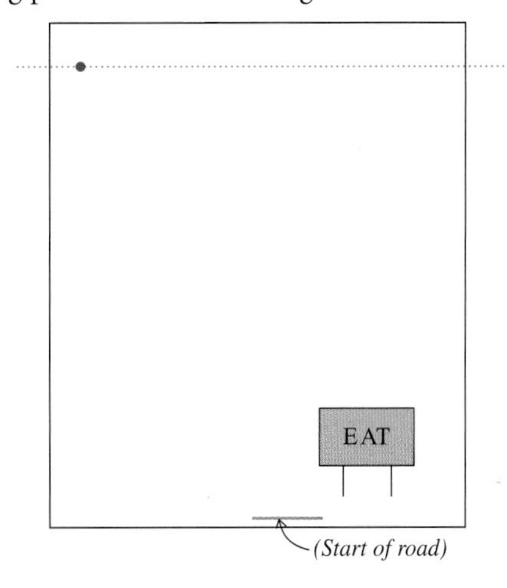

FIGURE 7.8

15. A scene showing a road through the desert is shown in Figure 7.9. There is a series of equally spaced telephone poles along the road. However, only one pole is shown in the picture. Using a one-point perspective, draw at least five more poles.

FIGURE 7.9

16. Put "LIFE" in perspective by using the picture frame, eye level line, and the vanishing point shown in Figure 7.10. Use a one-point perspective to show the word "LIFE" systematically receding toward the vanishing point.

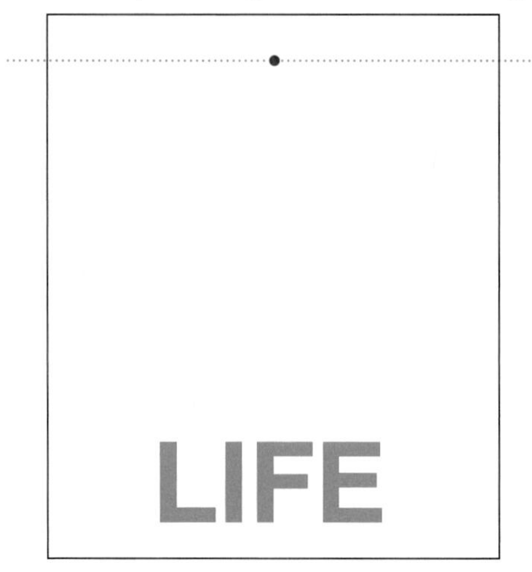

FIGURE 7.10

17. Put "FIT" in perspective by using the picture frame, eye level line, and the vanishing point shown in Figure 7.11. Use a one-point perspective to show the word "FIT" systematically receding toward the vanishing point.

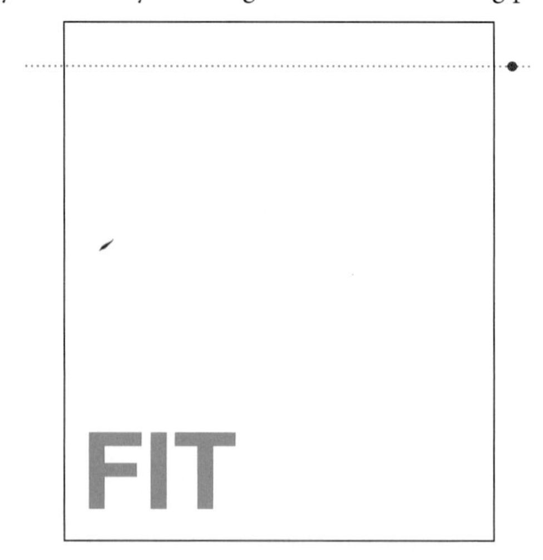

FIGURE 7.11

In Problems 18–23, a series of utility poles of the same size are a part of a one-point perspective drawing. The letters a, b, c, d, and e represent heights or distances in the drawing as shown in Figure 7.12. Use the method in Example 1 to determine the desired heights and distances in each problem.

18. $a = 3$ in., $e = 12$ in., $b = 2$ in., find d and c.

19. $a = 4$ ft, $e = 20$ ft, $b = 2.5$ ft, find d and c.

20. $a = 4$m, $e = 10$ m, $d = 2$ m, find b and c,

21. $a = 12.2$ cm, $e = 24.5$ cm, $d = 4.5$ cm, find b and c.

22. $b = 3.6$ cm, $d = 1.2$ cm, $c = 6$ cm, find e and a.

23. $a = 8$ in., $c = 5$ in., $e = 20$ in., find d and b.

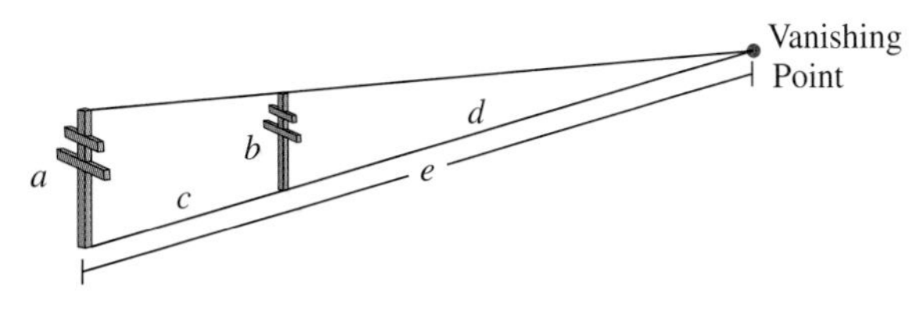

FIGURE 7.12

Explore

24. The wall along a stretch of a four-lane freeway has the pattern as shown in the figure that follows. Use a one-point perspective to give the wall an illusion of running along the freeway. Choose your own eye level line, vanishing point, and picture frame.

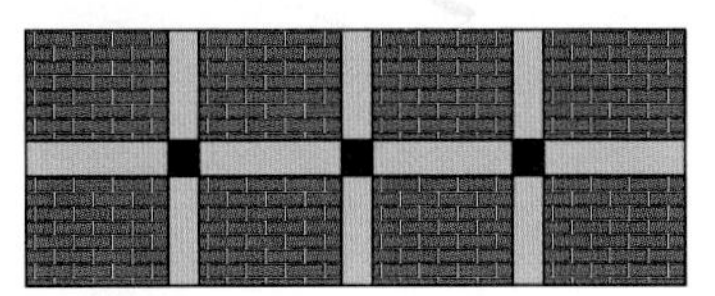

25. A wooden fence runs along a horse pasture. Use a one-point perspective to draw the fence, giving it an illusion of being three dimensional. Choose your own eye level line, vanishing point, and picture frame.

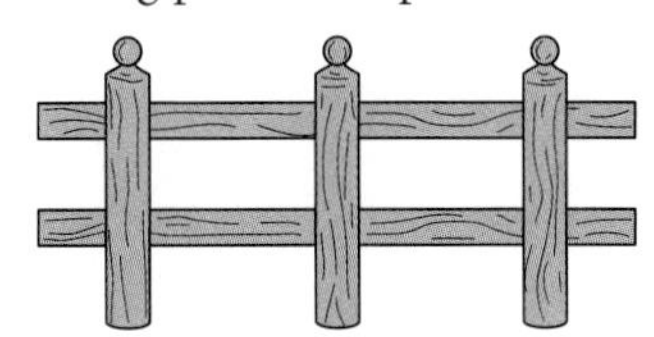

26. In the painting that follows, *The New Road* by Grant Wood (1939), what methods of perspective are evident? Explain.

The New Road by Grant Wood.
(National Gallery of Art)

27. In the painting that follows, *Quadrille at the Moulin Rouge* by Henri de Toulouse-Lautrec (1892), what methods of perspective are evident? Explain.

Detail from *Quadrille at the Moulin Rouge* by Henri de Toulouse-Lautrec. (National Gallery of Art).

28. In the painting that follows, *Cappriccio of Roman Ruins* by Marco Ricci (1720), what methods of perspective are evident? Explain?

Capriccio of Roman Ruins by Marco Ricci. (National Gallery of Art)

29. In the painting that follows, *Death and the Miser* by Hieronymus Bosch (1490), what methods of perspective are evident? Explain.

Detail from *Death and the Miser* by Hieronymus Bosch. (National Gallery of Art)

30. Examine some art of the ancient Egyptians. Do they make use of perspective? If so, which of the techniques discussed in this chapter are used?

31. Are any methods of perspective used in the art that follows?

Tomb Hieroglyphs at the Valley of the Queens, Luxor, Egypt. (Photograph by Wolfgang Kaehler)

32. An artist is using a one-point perspective to draw a line of Boeing 767 airliners parked along a runway. The first Boeing 767 in the line is to be 10 cm tall, and the last Boeing 767 is to be 1.5 cm tall. If the last Boeing 767 is drawn

8 cm from the vanishing point, how far from the last Boeing 767 should he draw the first Boeing 767 airliner?

33. An artist wants to have a fence running along a field in a one-point perspective drawing. She decides to use a 3-in. vertical post at the start of the fence and a vanishing point that is 16 in. away from the vertical post. If she wants the last vertical post positioned in the perspective to be $\frac{1}{2}$ in. tall, how far from the first pole should she place the last pole?

34. Examine some art of the ancient Chinese or Japanese. Do they make use of perspective? If so, which of the techniques discussed in the chapter are used?

35. Look down a set of railroad tracks or double center lines along a straight road. What do you notice? What perspective technique is suggested by this visual experience.

36. Look at paintings and drawings in an art book. Find examples of the perspective techniques discussed in this chapter. Make photocopies of the art and clearly indicate the perspective techniques used.

37. Examine art shown in other sections of this text. Find an example of these perspective techniques:

a) overlapping shapes.

b) diminishing sizes.

c) atmospheric perspective.

d) one-point perspective.

Golden Ratios and Rectangles

We continue our look at geometry and art by studying various geometric forms from a more aesthetic point of view. We look at the visual beauty contained in geometric objects and investigate some of the mathematics behind the creation of geometric designs.

Ancient architects, sculptors, and artists used a ratio of distances in their work that they deemed pleasing to the eye. Called the **Golden Ratio,** it is based on the division of a line segment into two parts such that the ratio of the longer piece to the shorter piece is the same as the ratio of the entire line segment to the longer piece. The Greek letter **phi,** ϕ (pronounced fī), was adopted in the early 20th century to represent this ratio because it is the first letter in the name of the Greek sculptor Phidias, who made extensive use of the Golden Ratio in his work.

The Golden Ratio

$$\phi = \frac{a}{b} = \frac{a+b}{a}$$

a b

The exact value of the Golden Ratio can be determined by performing some algebraic manipulations on the ratio $a/b = (a + b)/a$. In the derivation that follows, we use the quadratic formula. If you need a review of the quadratic formula, see Section 6.2.

$$\frac{a}{b} = \frac{a+b}{a} \qquad \text{Golden Ratio}$$

$$a^2 = b(a+b) \qquad \text{Cross multiplying}$$

$$a^2 - ba - b^2 = 0 \qquad \text{Subtracting } b(a+b) \text{ from both sides}$$

$$a = \frac{b \pm \sqrt{b^2 + 4b^2}}{2} \qquad \text{Solving for } a \text{ by the quadratic formula}$$

Therefore,

$$a = \frac{b + b\sqrt{5}}{2} \qquad \text{The length } a \text{ must be positive.}$$

Thus,

$$\frac{a}{b} = \frac{\dfrac{b + b\sqrt{5}}{2}}{b} = \frac{b + b\sqrt{5}}{2}\left(\frac{1}{b}\right)$$

$$\phi = \frac{1+\sqrt{5}}{2} = 1.618033988\ldots$$

The Golden Ratio is an irrational number because it contains $\sqrt{5}$. Its decimal representation is a nonterminating, nonrepeating decimal. However, because of errors inherent in measurement and for ease in computation, we will use 1.62 as the value of the Golden Ratio in this chapter.

A **Golden Rectangle** is a rectangle whose sides form the Golden Ratio. The Greeks believed that a rectangle having this ratio was more pleasing to the eye than was any other rectangle. In 1876, the psychologist Gustav Fechner analyzed the responses of many people about which rectangles they found most pleasing. From his research, he concluded that people prefer rectangular shapes that are approximately Golden Rectangles. Experiments carried out by Witmer (1894), Lalo (1905), and Thorndike (1917) arrived at similar conclusions.

The Golden Rectangle

$$\frac{l}{w} = \frac{l+w}{l} = \phi$$

EXAMPLE 1

By measuring the rectangles below, determine whether they approximate Golden Rectangles.

Solution: Measuring the rectangles, we find the dimensions of rectangle A to be 3 cm and 7 cm, whereas rectangle B has dimensions of 2.1 cm and 3.4 cm. If we take the ratio of the *longer* side to the *shorter* side in each rectangle, we get

Rectangle *A* $7/3 \approx 2.33$, and that is not close to the Golden Ratio of approximately 1.62.

Rectangle *B* $3.4/2.1 \approx 1.62$, which is approximately the same as the Golden Ratio.

Thus, rectangle A is not a Golden Rectangle, but rectangle B is very close to a Golden Rectangle.

EXAMPLE 2

Find a point on $\overline{AB}$ such that $\overline{AB}$ is divided into segments that form the Golden Ratio.

Solution: Let X be a point that divides $\overline{AB}$ into the Golden Ratio as shown.

According to the Golden Ratio, $a/b = (a + b)/a \approx 1.62$. Since the length of $\overline{AB}$ is 7 cm, $a + b = 7$, and we get

$$\frac{7}{a} \approx 1.62$$

$$1.62a \approx 7$$

$$a \approx 4.3$$

Thus, the point X is 4.3 cm from point A.

The Golden Ratio has an interesting history. Objects that contain the Golden Ratio have been studied and admired through the ages. In the rest of this section, we investigate some situations in which the Golden Ratio occurs.

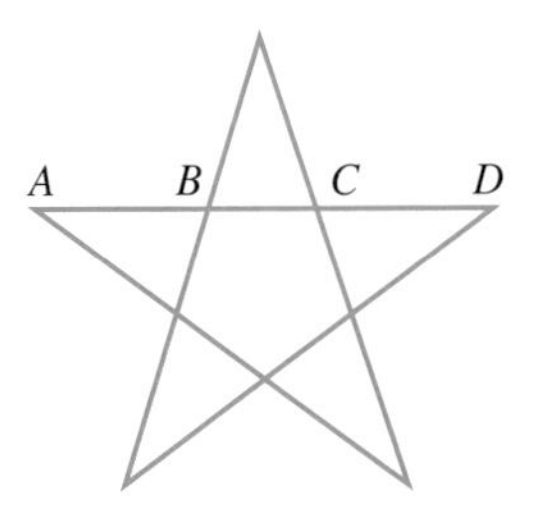

FIGURE 7.13

EXAMPLE 3

The pentagram, which dates back to ancient Babylonian and Greek cultures, was the mystic symbol and badge of the Society of Pythagoras. The pentagram contains the Golden Ratio many times. By actual measurement, find two occurrences of the Golden Ratio in the pentagram in Figure 7.13.

Solution: The ratios of AC to CD and BD to AB are approximately 1.62.

The Golden Ratio can also be found in the measurements of buildings of antiquity such as the Parthenon on the Acropolis in Athens. You can verify that the ratio of the length to the width of the overall dimensions of the Parthenon is approximately the Golden Ratio by measuring the scale drawing of the Parthenon in Figure 7.14.

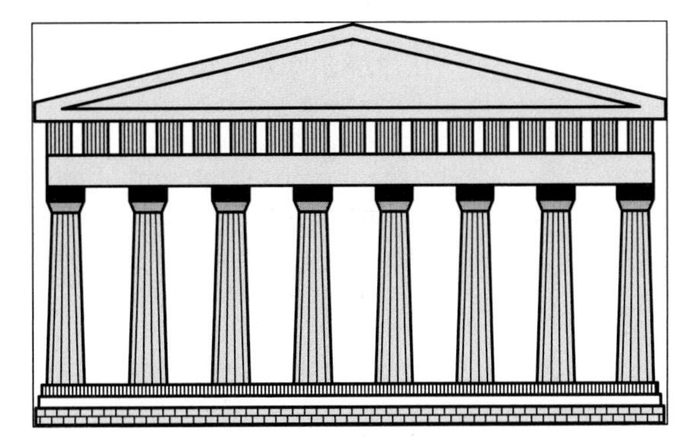

FIGURE 7.14
The Parthenon on the Acropolis in Athens

In *Geometry of Art and Life,* the author, Matila Ghyka, writes that the "perfect" human face can be viewed as sequences of Golden Ratios.

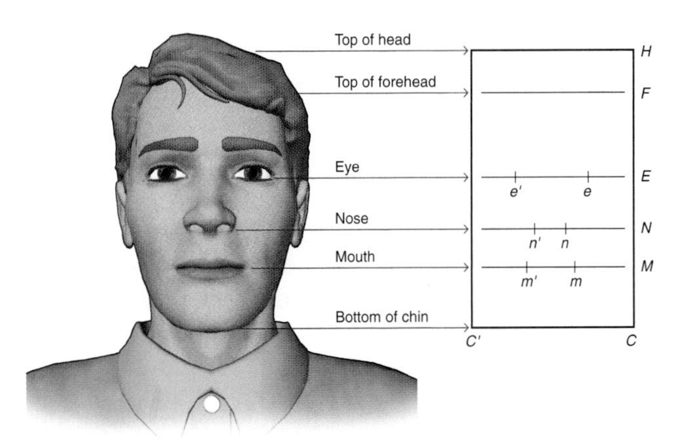

$$C'C = \text{Width of the face}$$
$$e'e = \text{Distance between outsides of eyes}$$
$$n'n = \text{Distance between outsides of nostrils}$$
$$m'm = \text{Distance between ends of mouth}$$

$$\frac{HC}{C'C} = \frac{HE}{FE} = \frac{EC}{NC} = \phi, \frac{FE}{EN} = \frac{EM}{EN} = \frac{NC}{MC} = \phi, \frac{C'C}{e'e} = \frac{e'e}{m'm} = \frac{m'm}{n'n} = \phi$$

EXAMPLE 4

Consider a "perfect" face that measures 2.5 in. between the outside edges of the eyes. How wide and how long is such a face?

Solution: In the figure, $C'C$ represents the width of the face, HC represents the height of the face, and $e'e$ represents the distance between the eyes.

The following Golden Ratios can be used.

$$\frac{C'C}{e'e} = \frac{C'C}{2.5} \approx 1.62 \qquad C'C \approx 2.5(1.62) \approx 4.05$$

$$\frac{HC}{C'C} = \frac{HC}{4.05} \approx 1.62 \qquad HC \approx 4.05(1.62) \approx 6.56$$

Thus, such a face has a width of about 4 in. and a height of about $6\frac{1}{2}$ in.

Leonardo Fibonacci (c. 1200) studied sequences of numbers in which each successive number is the sum of the two previous numbers: 1, 1, 2, 3, 5, 8, 13, 21, 34, 55 .... In 1877, Edward Lucas named that sequence of numbers the **Fibonacci numbers.** This sequence has the unique property that the ratio of successive terms gets close to the Golden Ratio.

$$\frac{1}{1} = 1 \qquad \frac{2}{1} = 2 \qquad \frac{3}{2} = 1.5$$

$$\frac{5}{3} \approx 1.6666667 \qquad \frac{8}{5} = 1.6 \qquad \frac{13}{8} = 1.625$$

$$\frac{21}{13} \approx 1.6153846 \qquad \frac{34}{21} \approx 1.6190476 \qquad \frac{55}{34} \approx 1.6176470\ldots$$

EXAMPLE 5

Find the ratio of the 15th and 16th Fibonacci numbers. How does this ratio compare to the Golden Ratio?

Solution: The 15th Fibonacci number is 610, and the 16th is 987; $987/610 \approx 1.6180328$. The first six digits of that ratio match those of the Golden Ratio.

The Golden Ratio was referred to by various names over the centuries. Luca Pacioli called it *divina proportione* (divine proportion) in 1509. Johann Kepler called it *sectio divina* (divine section) in 1610. The term "Golden Ratio" or "Golden Section" came into use around 1840. The Golden Ratio has stimulated mathematical interest for centuries and is still of interest to designers, botanists, sculptors, composers, and artists. The work of artists Georges Seurat, Piet Mondrian, and Juan Gris and some musical compositions by Bela Bartok utilize the Golden Ratio.

We end this section with a beautiful mathematical curve. It is known as the logarithmic spiral and is created by curves formed within Golden Rectangles. The spiral has no ending point. It grows outward or inward indefinitely, but its shape remains unchanged. In nature, this spiral can be seen in the successive chambers of the nautilus seashell.

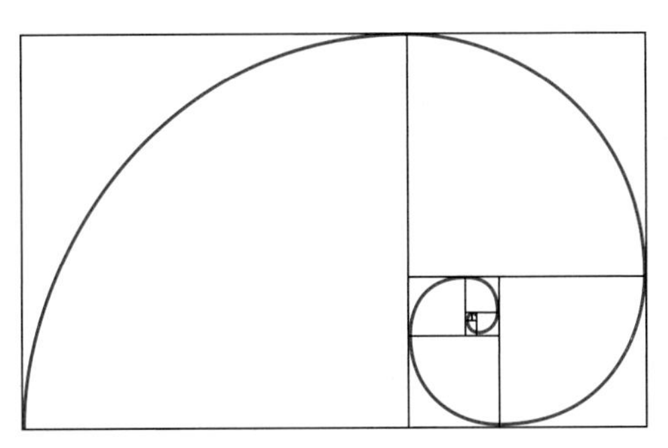

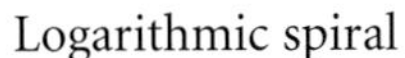
Logarithmic spiral

Nautilus seashell

section 7.3

PROBLEMS  Explain  Apply  Explore

Explain

1. What is the Golden Ratio?
2. Why were the Greeks fascinated with the Golden Ratio?
3. What is the importance of the Golden Ratio in art?
4. How do you find the point that divides a line segment into the Golden Ratio?
5. What is a Golden Rectangle?
6. If you knew the measurement of the longer side of a Golden Rectangle, how would you find the length of the shorter side?
7. If you knew the measurement of the shorter side of a Golden Rectangle, how would you find the length of the longer side?
8. What are the Fibonacci numbers and what relationship do they have to the Golden Ratio?

Apply

In Problems 9–12, find a point that approximately divides each segment into the Golden Ratio.

9. ______________________

10. ____________________________

11. __________________________________

12. ____

In Problems 13–16, determine which of the following rectangles are approximately Golden Rectangles.

13.

14.

15.

16.

In Problems 17–23, one dimension of a Golden Rectangle is given. Find the two possible values for the other dimension of the Golden Rectangle.

17. 23 ft

18. 4.5 mi

19. 56.8 m

20. 45.5 cm

21. 9 cm

22. $10\frac{3}{4}$ in.

23. 36 in.

Explore

If you were to construct a "Golden Cross," you could make the ratio of the height (h) to the width (w) of the cross equal to the Golden Ratio and the ratio of the bottom section of the cross (b) and the top section of the cross (t) equal to the Golden Ratio. Find the width, top section, and bottom section of the cross if the height of the cross is as given in Problems 24–26.

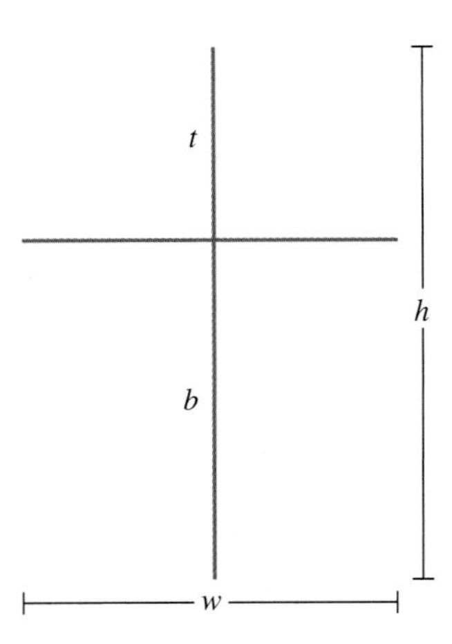

24. $h = 72$ in.

25. $h = 16$ cm

26. $h = 20$ ft

27. Study the ratios within the human face shown in Example 4. By analyzing at least two faces or pictures of faces, what can you conclude about the ratios or the average of the ratios?

28. The architect Le Corbusier (c. 1940) developed the modular system of harmonious proportions. In this system, he established that in the visually "perfect" human form:

    a) The ratio of distance from the bottom of the neck to the naval to the distance from the top of the head to the bottom of the neck is the Golden Ratio.

    b) The ratio of the distance from the naval to the knee to the distance from the knee to the bottom of the foot is the Golden Ratio.

    Measure people or pictures of people to determine if and when these ratios are golden. Do the averages of the ratios you found approximate the Golden Ratio?

We can define a "Golden Box" as one whose height, width, and length satisfy the Golden Ratio. We are assuming here that the height is the shortest dimension and the length is the longest dimension of the box.

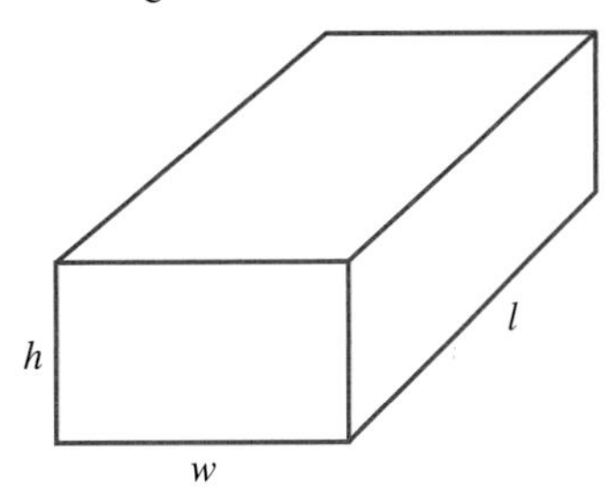

$$\frac{w}{h} = \frac{l}{w} \approx 1.62$$

In Problems 29–34, one dimension of a "Golden Box" is given. Find the dimensions that are not given.

29. $h = 5$ in.

30. $l = 5$ in.

31. $w = 9$ cm

32. $w = 5$ in.

33. $h = 9$ cm

34. $l = 9$ cm

35. Construct a "Golden Box" that has a height of 3 in.

36. Measure various boxes. Do they form "Golden Boxes," as defined? Find the average of the heights, widths, and lengths of the boxes. Is the ratio of these averages approximately the Golden Ratio?

37. Explain why three consecutive Fibonacci numbers can be used to give approximate dimensions of "Golden Boxes."

38. By finding the ratio of the length to the width of ten common rectangular-shaped objects, such as boxes, cards, cushions, doors, appliances, and windows, determine which objects approximate Golden Rectangles and which objects do not approximate Golden Rectangles. Find the average of the ten ratios. How does your result compare to that of psychologist Gustav Fechner, who concluded that the average of the ratios of the sides of common rectangles was approximately the Golden Ratio? Since Golden Rectangles are supposed to be the most pleasing to the eye, why aren't all manufactured rectangular objects Golden Rectangles?

39. A "Golden Can" could be defined as a can that has its height greater than its diameter, where the ratio of its height to its diameter is equal to the Golden Ratio. Measure ten cans of different sizes. Which ones approximate the Golden Ratio, and which do not? Find the average of the ten ratios. How does this average compare to the Golden Ratio? Since the Golden Ratio is supposed to be the most pleasing to the eye, why aren't all cans "Golden Cans"?

40. Are Golden Rectangles really more pleasing to the eye? To help answer this question, do the following:

- **a)** On separate cards, cut out rectangles that are 1 cm by 5 cm, 2 cm by 5 cm, 3 cm by 5 cm, 4 cm by 5 cm, and 5 cm by 5 cm.
- **b)** Place the cards in a random order and ask ten people to select the rectangle that they find the most pleasing or attractive.
- **c)** Tabulate your results and calculate the percentage of the people choosing each rectangle.
- **d)** Use your results to answer the question posed at the beginning of the problem.
  (*Note:* The 3 by 5 cm rectangle is approximately a Golden Rectangle.)

41. Consider a parallelogram whose sides form the Golden Ratio (i.e., 3 cm by 5 cm). The shape of the parallelogram is determined by the angle between two adjacent sides, as shown below.

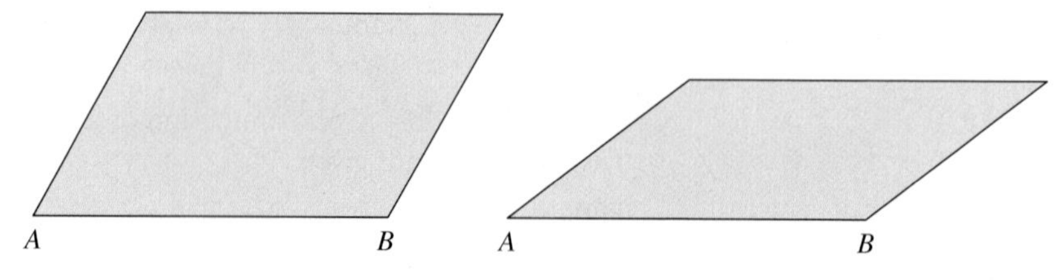

To make the parallelogram even more "golden," let us also make the ratio of the measures of $\angle A$ to $\angle B$ approximately the Golden Ratio. Such a parallelogram would have $\angle A \approx 69°$ and $\angle B \approx 111°$. Is such a "Golden Parallelogram" more pleasing to the eye than other parallelograms? Construct an experiment to answer that question. (See Problem 40.)

42. In 1876, psychologist Gustav Fechner did experiments with ellipses that had the ratio of the length of the major axis to the length of the minor axis approximately equal to the Golden Ratio. Again, he discovered that people found that ellipses close to "Golden Ellipses" were more pleasing to the eye. Use the ellipses shown to test Fechner's findings in present society. (*Note:* The first ellipse is the "Golden Ellipse.")

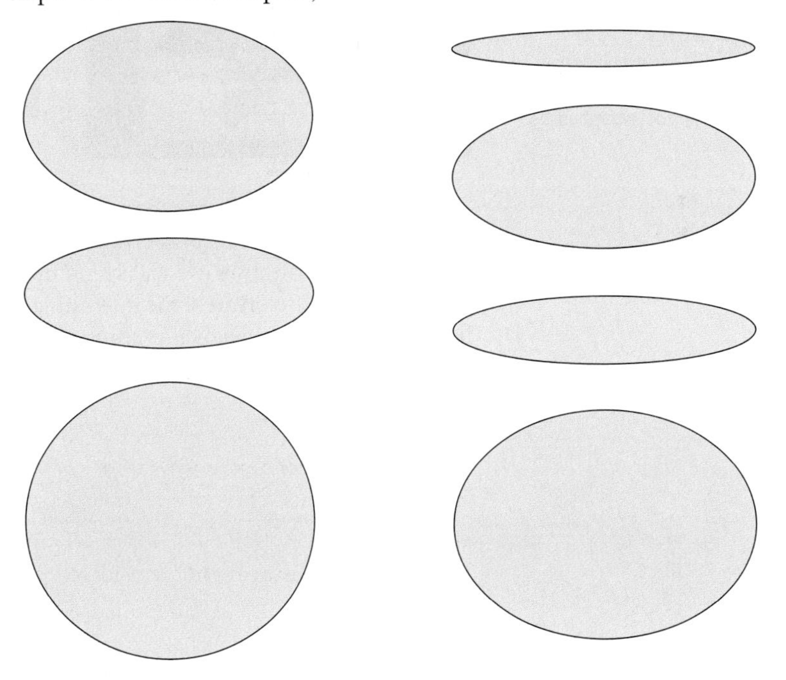

43. In Problem 41, find the exact value for $\angle A$ so that the ratio of $\angle A$ to $\angle B$ is the Golden Ratio. (*Hint:* Use the facts that $\angle A + \angle B = 180°$ and that the ratio of $\angle B$ to $\angle A$ is $(1 + \sqrt{5})/2$.

44. The Golden Ratio ϕ is the only positive number that if decreased by 1 equals its reciprocal. Show algebraically that $\phi - 1 = \frac{1}{\phi}$.

45. In Example 5, we showed that the ratio of terms in the Fibonacci sequence approaches the Golden Ratio. Consider any sequence in which we choose any two starting numbers and generate the terms of the sequence by adding two consecutive numbers as in the Fibonacci sequence. For example, if we chose 5 and 2 as the starting numbers, the sequence would be 5, 2, 7, 9, 16, 25, 41, 66, 107, ... Examine the ratio of successive terms of any such sequence, as we did for the Fibonacci sequence in Example 5. What do you notice about successive ratios of terms?

section 7.4

Polygons and Stars

(*Left*) Star Quilt photographed by Roman Soumar shows an artist's use of tessellations. (*Right*) Ukrainian artists use polygons and stars in this basket of Easter Eggs (Pysanky) photographed by James A. Sugar.

The Golden Rectangle is just one type of a class of geometric objects called polygons. Polygons and their properties are carefully examined in most treatments of geometry. In this section, we will spend most of our efforts investigating how polygons can be used to create some interesting geometric designs. A **polygon** is a closed figure in a plane formed by line segments that intersect each other only at their end points. The following are examples of polygons.

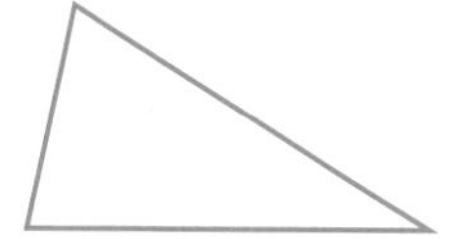

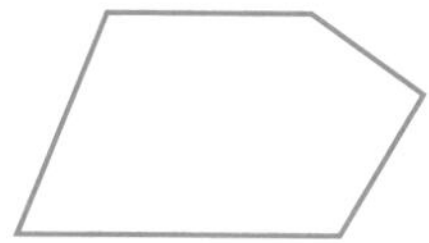

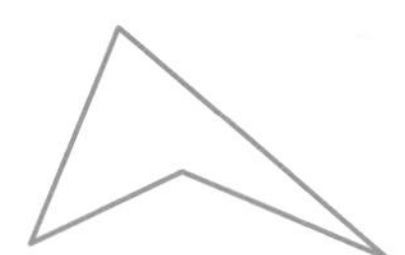

The following figures are not considered polygons.

(a)

(b)

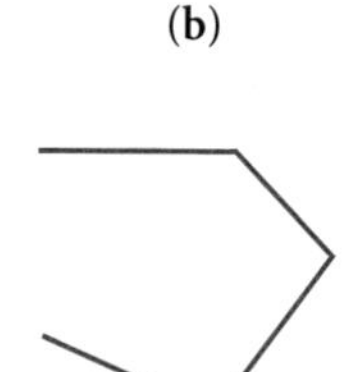

(c)

EXAMPLE 1

Explain why each of the figures shown in (a), (b), and (c) is not a polygon.

Solution: Each figure has a property that contradicts the definition of a polygon.

a) The sides intersect at a point other than at the end points.

b) It is not a closed figure.

c) The top is a curve, not a line segment.

The line segments that form a polygon are called its **sides,** and the points at which the sides meet are called **vertices** (plural of "**vertex**"). Polygons are named by the number of sides they contain. Specific names are given to polygons with 3 to 12 sides. Other polygons are usually referred to as n-gons, where n represents the number of sides of the polygon. For example, a polygon with 13 sides is called a 13-gon.

Polygons are classified as concave or convex. A polygon is **concave** if an extension of one of its sides enters the interior of the polygon; a polygon is **convex** if the extensions of its sides do not enter the interior of the polygon. Some polygons are classified as **regular polygons.** These polygons have sides of equal length and angles of equal measure. Some common objects with shapes that are regular polygons are pizza boxes, stop signs, and honeycombs.

The following is a list of the first ten polygons and their names. In column I, you will find regular polygons; in column II, convex polygons; and in column III, concave polygons.

Number of Sides	Name	I	II	III
3	Triangle		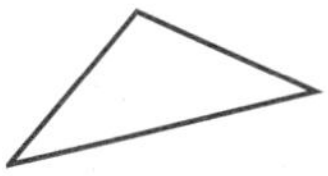	
4	Quadrilateral	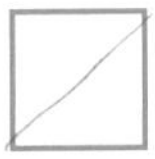		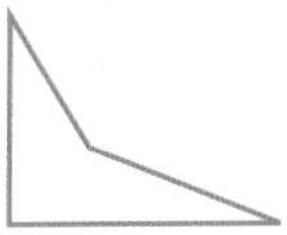
5	Pentagon			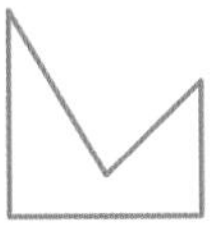
6	Hexagon		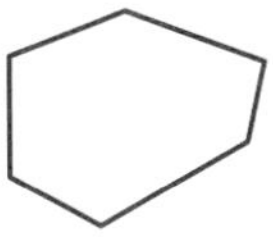	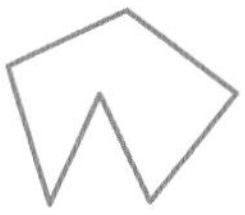
7	Heptagon		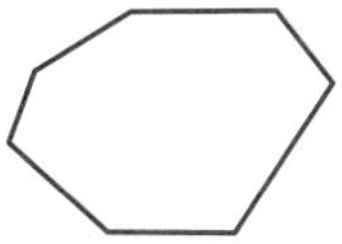	
8	Octagon		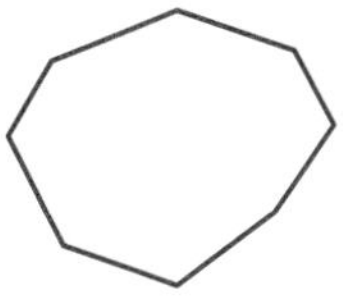	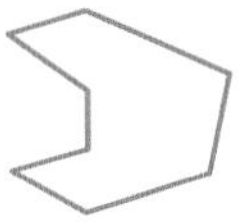
9	Nonagon		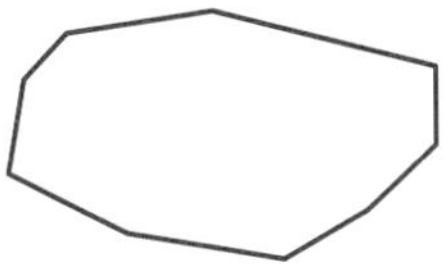	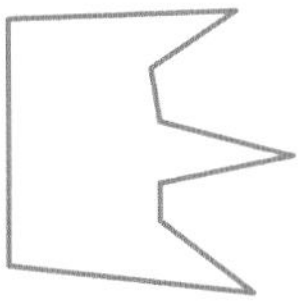
10	Decagon	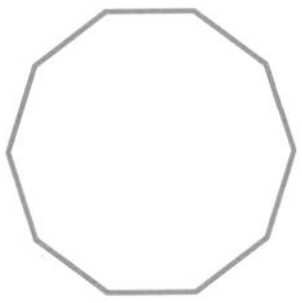	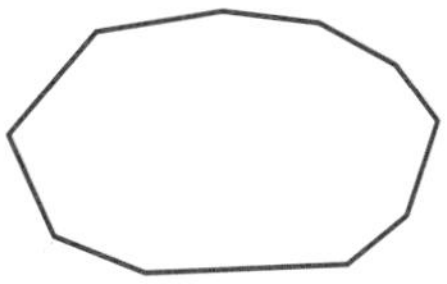	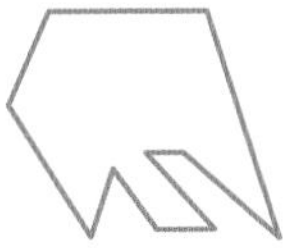

continued

Number of Sides	Name	I	II	III
11	Undecagon	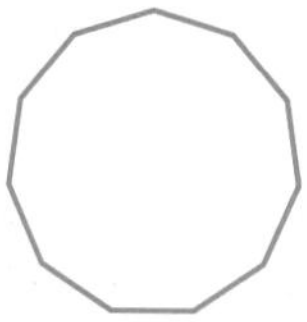	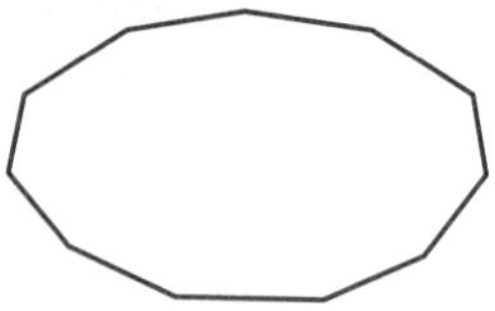	
12	Dodecagon	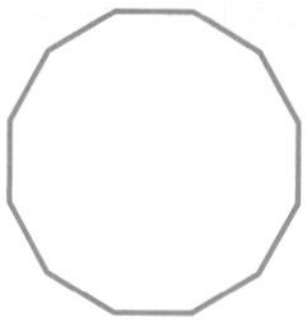	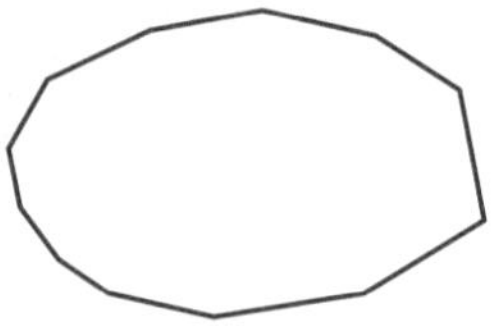	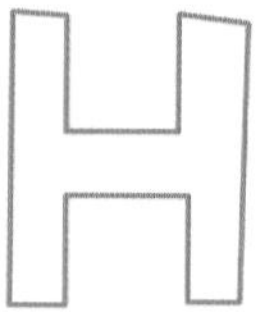

EXAMPLE 2

In each figure, (a) give the name of the polygon and classify it as concave or convex and (b) determine whether it is a regular polygon.

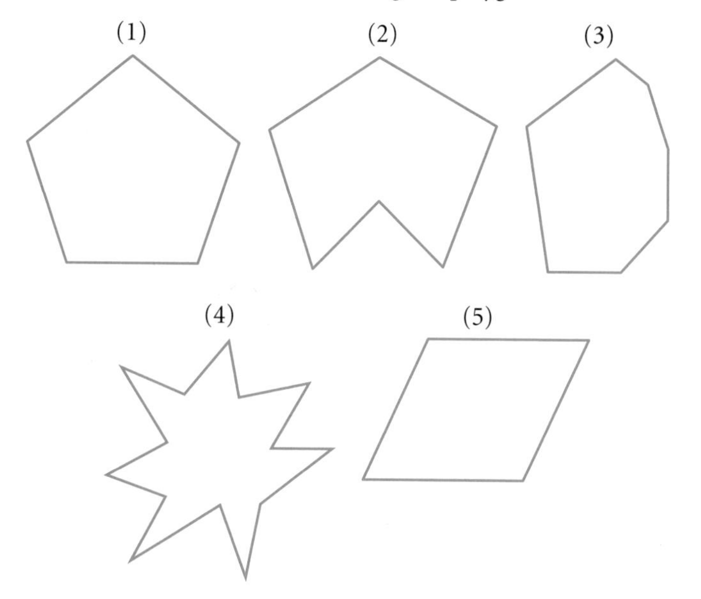

Solution:

a) **1.** Pentagon, convex

2. Hexagon, concave

3. Heptagon, convex

4. 14-gon, concave

5. Quadrilateral, convex

For example, Figure 2 is concave because an extension of the side $\overline{AB}$ enters the interior of the hexagon.

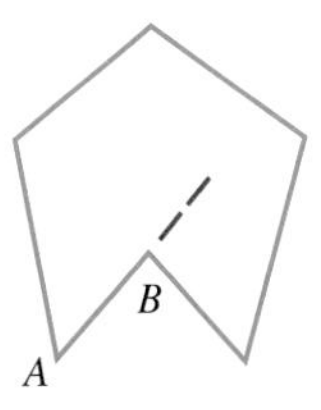

b) The only regular polygon is Figure 1. Figure 5 is not a regular polygon because its angles are not of equal measure.

Polygons and Their Angles

The angles of a polygon have a unique property. The sum of the angles for each type of polygon is always the same. For example, in Section 7.1, we learned in the triangle-sum theorem (T5) that the sum of the angles of a triangle is 180°. This means that in any triangle the sum of its three angles equals 180°. A proof of this theorem can be found in the original work, *Elements* by Euclid, and in most geometry textbooks. We will not present a proof for the triangle-sum theorem here, but we will give you a visual justification for the theorem. If you fold $\triangle ABC$ along the dotted lines as shown in the figure on the left, you will obtain the figure on the right.

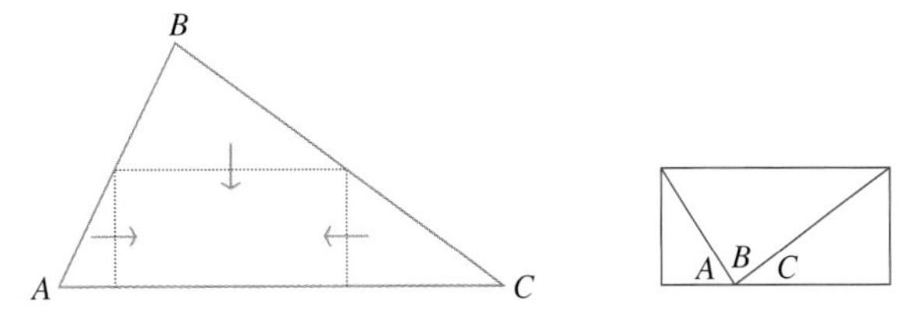

In the figure, $\angle A$, $\angle B$, and $\angle C$ form a straight angle. Since straight angles measure 180°, $\angle A + \angle B + \angle C = 180°$. The fact that the sum of the angles of a triangle is 180° will allow us to find the sum of the angles for any polygon. To determine this sum, we will cover a given convex polygon with nonoverlapping triangles originating at one vertex and use these triangles to determine the sum of the angles of the polygon.

Sides		Triangles	Sum of Angles
4	1 2	2	$180°(2) = 360°$
5	1 2 3	3	$180°(3) = 540°$
6	1 2 3 4	4	$180°(4) = 720°$

These examples show a relationship between the number of nonoverlapping triangles that cover a convex polygon and the number of sides. The number of triangles is 2 less than the number of sides. Thus, if the polygon has n sides, there are $n - 2$ triangles. Since the sum of the angles of each triangle is 180°, the sum of the angles (S) of the convex polygon is as follows.

Sum of the Angles of a Polygon

$$S = 180(n - 2)$$

(*Note:* this formula also works for concave polygons.)

If the polygon is a regular polygon, each of its angles has the same measure. Thus, to find the measure of each angle (A) of a regular polygon, we take the sum of its angles and divide by the number of angles or sides (n) in the polygon.

Each Angle of a Regular Polygon

$$A = \frac{180(n - 2)}{n}$$

EXAMPLE 3

Find the sum of the angles in a regular decagon and find the measure of each of its angles.

Solution: A decagon has 10 sides. Thus, we use $n = 10$ in the formulas for the sum of the angles of a polygon and each angle of a regular polygon.

$$\begin{aligned} S &= 180(n - 2) \\ &= 180(10 - 2) = 180(8) = 1440° \\ A &= \frac{180(n - 2)}{n} \\ &= \frac{180(10 - 2)}{10} = \frac{1440}{10} = 144° \end{aligned}$$

To complete our discussion of polygons, we will investigate methods for actually drawing them and creating designs based on them. Polygons have probably been admired and studied since prehistoric times. These shapes are present in almost every object made by man. The simple beauty and completeness of polygons, especially regular polygons, have stimulated many creative designs over the centuries. Designs using polygons can be found in religious symbols such as the Star of David; in logos such as the pentagon emblem on Chrysler cars; in modern quilts, flooring, and wallpaper; and on wheel covers of cars and trucks. We are confident that this excursion into geometry will enable even the "unartistic" to create some beautiful results.

DRAWING REGULAR POLYGONS

If you study the regular polygons shown earlier in this section, you will notice that the vertices are equally spaced and that, as the number of sides increases, the regular polygon appears to be more and more like a circle. This suggests a method for actually creating regular polygons.

For example, to draw a regular pentagon, find five equally spaced points on a circle and use those points as vertices of the regular pentagon. We will use the fact that if an arrow with its endpoint at the center of a circle rotates completely around the circle like the second hand of a clock, it passes through 360°. Therefore, to create five equally spaced points on the circle, we simply divide 5 into 360. If we take the answer 72, and mark off five angles at the center of the circle (central angles) of 72°, we find the desired points on the circle to use as vertices of the regular pentagon.

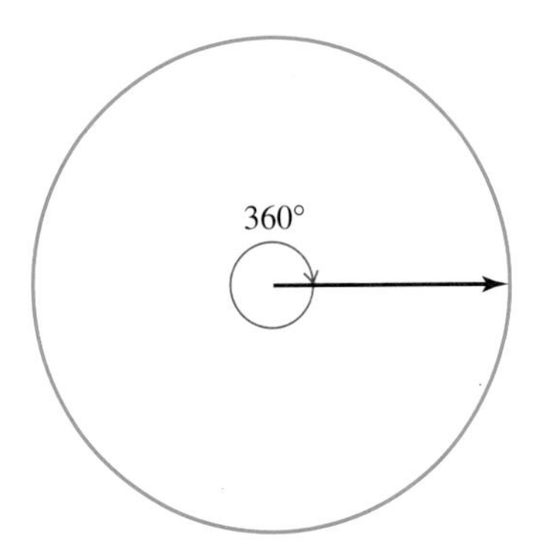

Let's actually draw a regular pentagon. With a compass, draw a large circle. Using a protractor, measure five 72° angles at the center of the circle. Extend the angles until they intersect the circle. Connect the five points marked on the circle to form the pentagon.

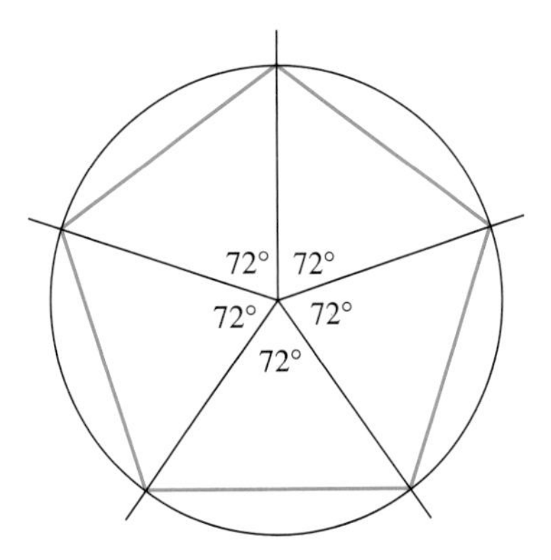

EXAMPLE 4

Draw a regular octagon.

Solution: Using the method described in the creation of the pentagon, we should mark off eight central angles of 45° each, since $360 \div 8 = 45$.

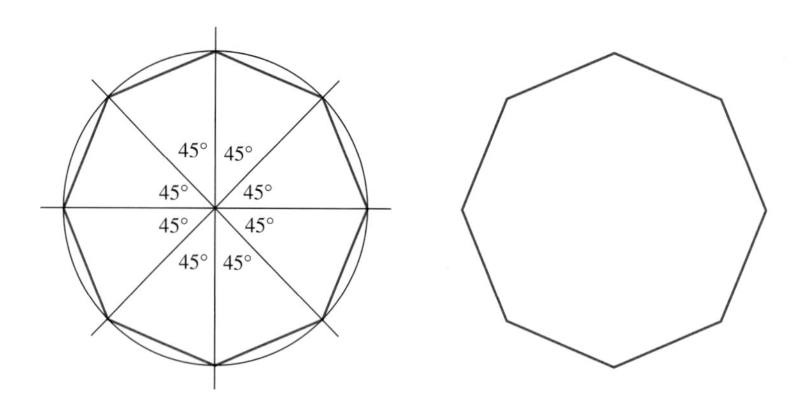

DRAWING REGULAR STARS

Finding equally spaced points on a circle can also be used to create some interesting geometric "stars." Consider seven equally spaced points on a circle obtained by marking off seven central angles of approximately 51.4° ($360 \div 7 \approx 51.429$). Instead of connecting each point to create a regular heptagon, connect every second point or every third point.

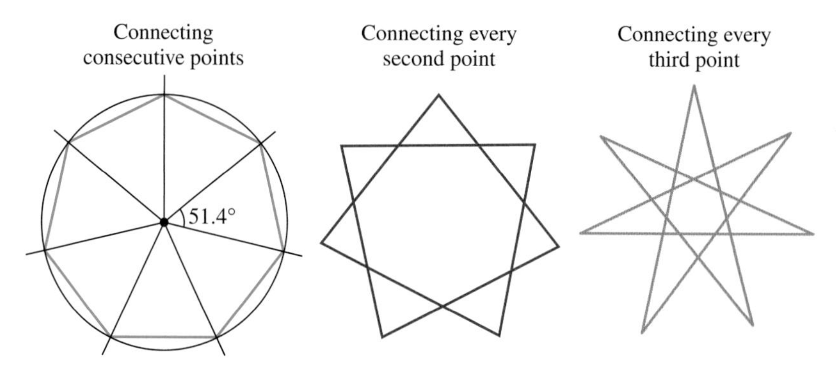

EXAMPLE 5

Draw a 12-point star by connecting every fifth point of a regular dodecagon.

Solution: Find 12 equally spaced points on a circle by marking off 12 central angles of 30° ($360 \div 12 = 30$).

Mark off each 30° central angle.

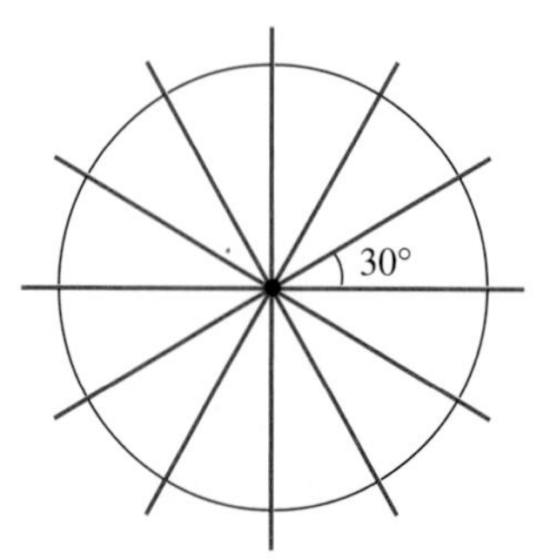

Connect every fifth point.

The process of locating equally spaced points on a circle can be used to create regular polygons and a variety of symmetrical stars. That technique combined with others such as those listed below can be used to create some interesting geometric designs.

1. Construct stars using points on a circle that are not equally spaced.
2. Create the illusion of curved lines by marking off equally spaced points on both sides of an angle and systematically connecting points on one side of the angle to points on the opposite side.

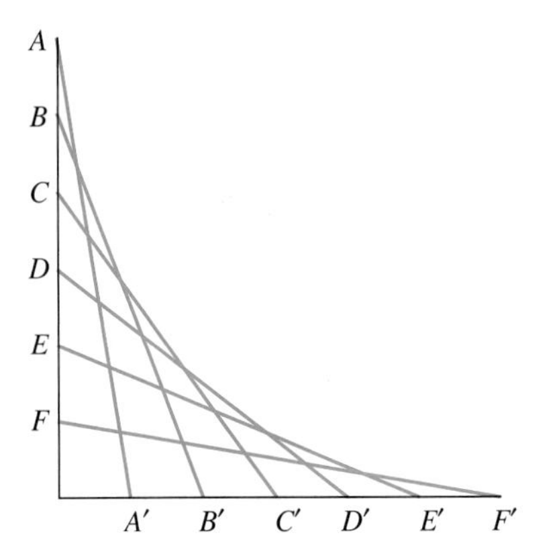

3. Use different colors to systematically shade various regions of the star.
4. Combine different polygons and stars along with various lines in the same design.

The results of experimenting with those and other techniques can be surprising. Don't be afraid to let your creative juices flow. Geometry can be beautiful. What follows is an example of a geometric design using some of those techniques.

section 7.4

PROBLEMS ▸ Explain ▸ Apply ▸ Explore

▸ Explain

1. What is a polygon?

In Problems 2–5, explain why each figure is not a polygon.

2.

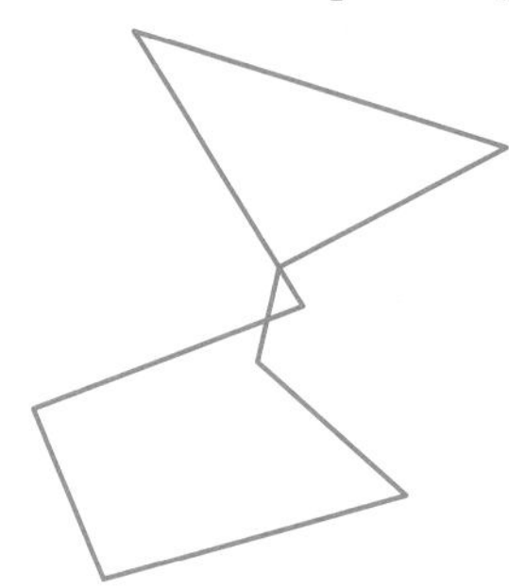

3.

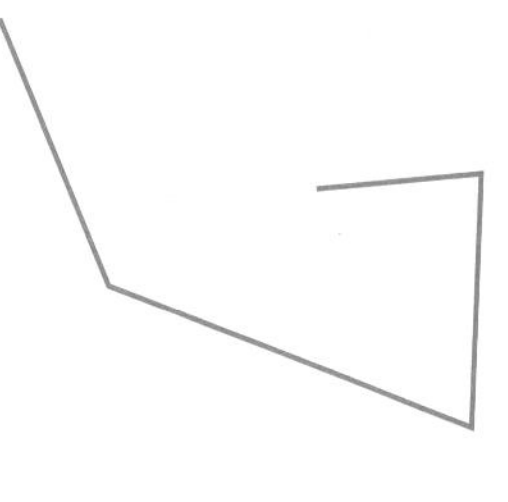

4.

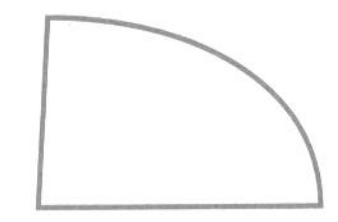

5.

6. What is a regular polygon?

In Problems 7–10, explain why each figure is not a regular polygon.

7.

8.

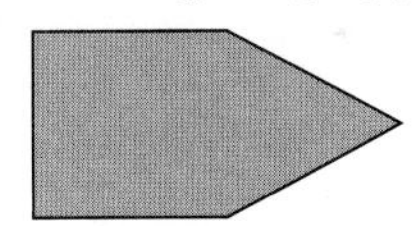

9.

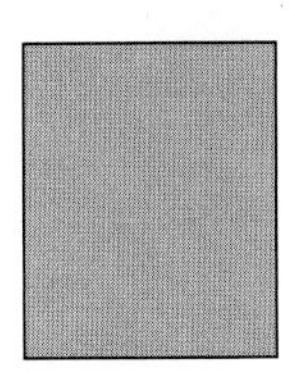

10.

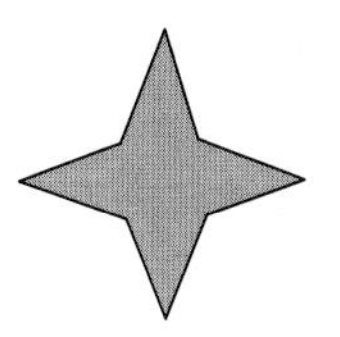

11. What are convex and concave polygons?

In Problems 12–15, give the name of each polygon and explain why it is concave or convex.

12.

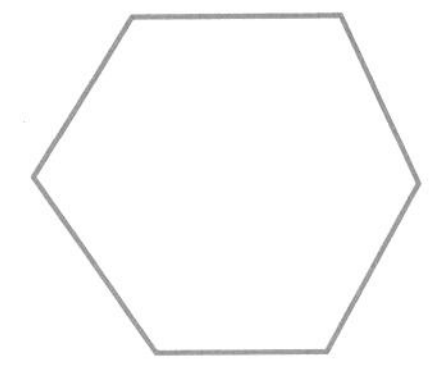

13.

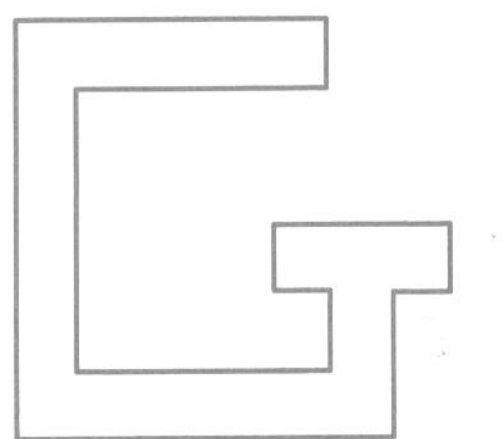

14.

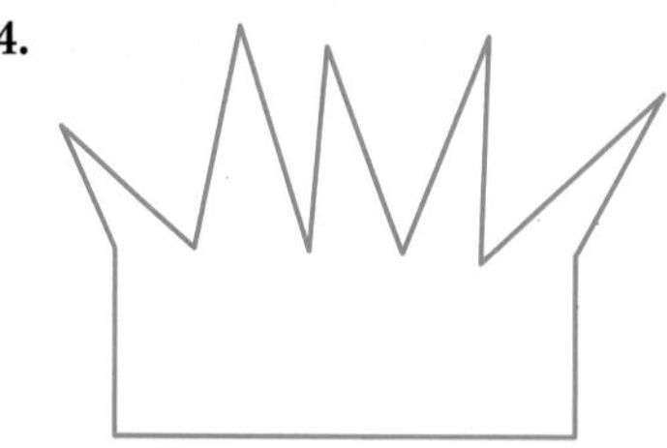

15.

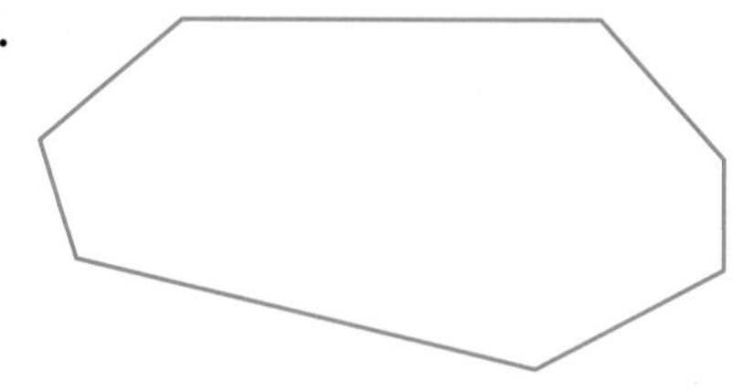

16. What steps should you follow to draw a regular polygon of n sides?

17. What steps should you follow to draw a regular star of n points?

18. How do you find the sum of the angles of a polygon?

19. How do you find the measure of each angle of a regular polygon?

Apply

In Problems 20–26, make a sketch of each polygon and find the sum of the angles of each polygon.

20. **a)** Concave quadrilateral
b) Convex quadrilateral

21. **a)** Concave pentagon
b) Convex pentagon

22. **a)** Concave hexagon
b) Convex hexagon

23. **a)** Concave 13-gon
b) Convex 13-gon

24. **a)** Concave 14-gon
b) Convex 14-gon

25. **a)** Concave decagon
b) Convex decagon

26. **a)** Concave dodecagon
b) Convex dodecagon

27. Complete the following chart to determine the central angle needed to find equally spaced points of a circle.

Number of Points	Central Angle	Number of Points	Central Angle
5	______	13	______
6	______	14	______
7	______	15	______
8	______	16	______
9	______	18	______
10	______	20	______
11	______	30	______
12	______	36	______

In Problems 28–35, draw the regular polygon indicated and find the measure of each angle of the polygon.

28. Regular hexagon

29. Regular octagon

30. Regular nonagon

31. Regular undecagon

32. Regular 18-gon

33. Regular 20-gon

34. Regular 15-gon

35. Regular 14-gon

Explore

In Problems 36–42, draw the regular star indicated. Color different regions in the star to give it an "artistic" effect.

36. A 10-point star connecting every second point

37. A 10-point star connecting every third point

38. A 12-point star connecting every fifth point

39. A 15-point star connecting every sixth point

40. A 20-point star connecting every fifth point

41. A 20-point star connecting every ninth point

42. A 30-point star connecting every tenth point

43. The designs of many wheel covers on cars and trucks use the techniques discussed in this section. Examine and sketch the designs of three different wheel covers, noting the make and model of the vehicle.

44. Designs for wheel covers on cars and trucks can be created by using the techniques discussed in this section. Use these techniques to design your own wheel cover.

45. Create an original geometric design using a combination of any techniques presented in this section.

46. Find a proof of the triangle-sum theorem in Euclid's *Elements* or in a geometry textbook. What theorems are used in the proof?

section 7.5

Tessellations

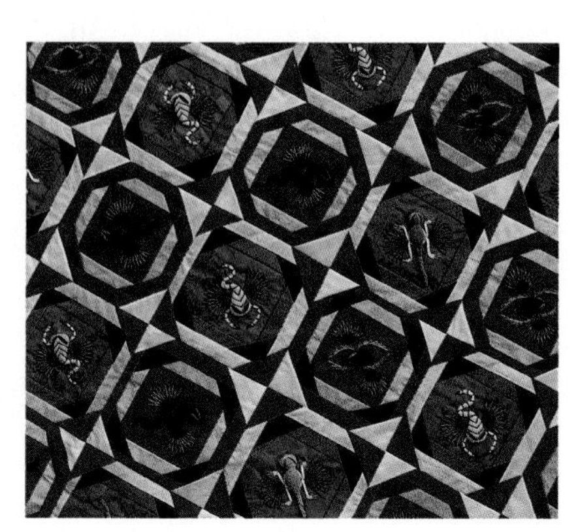

(*Left*) Anthony Cooper's photograph, *Fissures in Sun-Baked Clay,* shows a natural tessellation of a pond at summer's end. (*Right*) Quilt with scorpion pattern shows an artist's use of tesselations. (Courtesy of Corbis)

The study of polygons in the previous section enabled us to create various geometric designs based on placing equally spaced points on a circle. In this section, we investigate combining polygons into geometric patterns called tessellations. The word *tessellation* comes from the Latin word *tessella,* which is a small square tile used in Roman mosaics. A **tessellation** is a pattern of one or more congruent shapes that covers an area in a plane without overlapping or leaving any gaps. A tessellation actually covers the entire plane if its basic pattern is continued in all directions. However, we will be concerned with using tessellations only to cover a small region in the plane. Simple tessellations or tilings are very common. They can be seen, for example, on walls, on clothing, in beehives, or in works of art.

Tessellating squares, shower wall

Tessellating equilateral triangles, designer belt

Tessellating regular hexagons, honeycomb

FIGURE 7.15

Tessellations were used on window lattices in China; on painted ceilings in Egypt; on the mosques in Granada, Spain; on textile and basket patterns of the native peoples of Peru, India, Ghana, America, Mexico, and other countries; and in the decorative arts of the ancient Greeks, Romans, Arabs, Japanese, Persians, and Celts. Tessellations can be observed in nature in the cells of an onion skin, in the design of a spider's web, and in the arrangement of seeds in a sunflower. Tessellations can be seen in our modern society as designs on wallpaper, linoleum, parquet flooring, ceramic tiles, patchwork quilts, crocheted placemats, and lace tablecloths. Tessellations have been made popular by the work of Dutch artist M.C. Escher. Inspired by the ornamental art of the Moors, he created intriguing designs that truly integrate mathematics and art. Tessellations can also be seen in the work of modern kinetic and optical artists such as Bridget Riley and Victor Vasarely and in the intricate tilings of Heinz Voderberg and Roger Penrose. In this section, we introduce some of the principles and methods of creating basic tessellations.

TESSELLATING WITH A REGULAR POLYGON

The only regular polygons that tessellate the plane are the equilateral triangle, the square, and the regular hexagon, as shown in Figure 7.15. Any of the other regular polygons would overlap at a vertex and therefore not tessellate the plane.

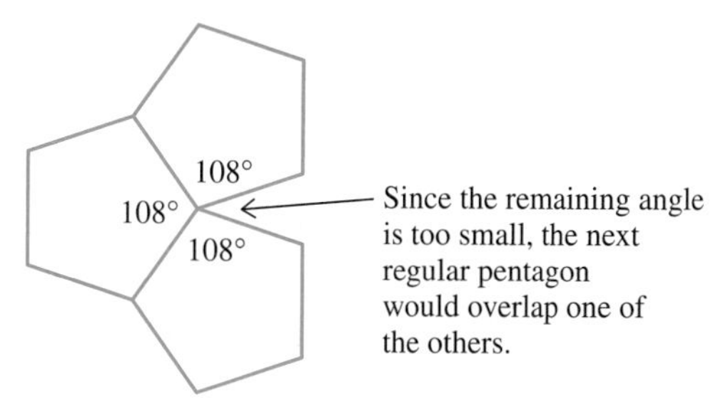

For example, if regular pentagons were placed around one vertex, as shown in the figure, overlapping would occur by the fourth regular pentagon. A similar situation occurs for any other regular polygon with the number of sides other than 3, 4, or 6. Table 7.1 will be helpful in verifying that fact. Its proof is a starred exercise in this section's problem set.

TABLE 7.1 Regular Polygons

Number of Sides	Measure of Each Angle	Number of Sides	Measure of Each Angle
3	60°	11	$147\frac{3}{11}$°
4	90°	12	150°
5	108°	15	156°
6	120°	18	160°
7	$128\frac{4}{7}$°	20	162°
8	135°	24	165°
9	140°	36	170°
10	144°	42	$171\frac{3}{7}$°

EXAMPLE 1

Show that it is not possible to tessellate the plane with regular octagons.

Solution:

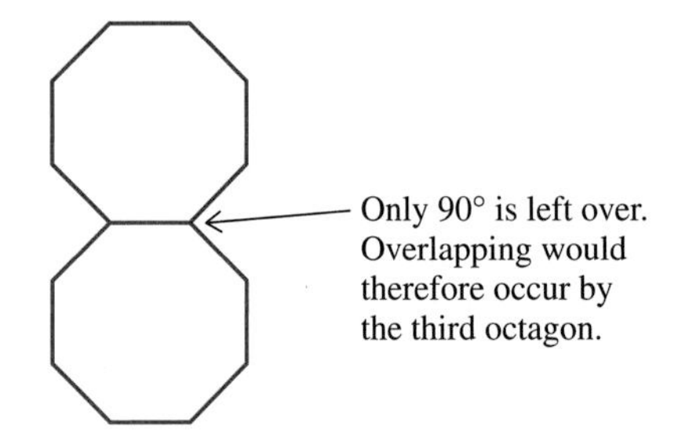

Since each angle of an octagon is 135°, two octagons placed at the same vertex would have an angle sum of 270°. That would leave 90° at the vertex. Thus, overlapping would occur if another octagon is placed at that vertex.

TESSELLATING WITH A TRIANGLE OR QUADRILATERAL

We have seen that only three regular polygons tile the plane. If we allow the use of nonregular polygons, any triangle or quadrilateral can tessellate the plane. The basis for this is the fact that to tessellate the plane, there must be no overlapping of polygons and no gaps left by the polygons. Overlapping occurs when the sum of the angles placed around a vertex is greater than 360°, and gaps occur when the sum is less than 360°. Since the sum of the angles of a triangle is 180° and the sum of the angles of a quadrilateral is 360°, triangles and quadrilaterals can be placed around a vertex so that the sum of the angles is exactly 360°. In the case of the triangle, we need to use each angle of the triangle twice. In the case of the quadrilateral, we need to use each angle only once around a vertex.

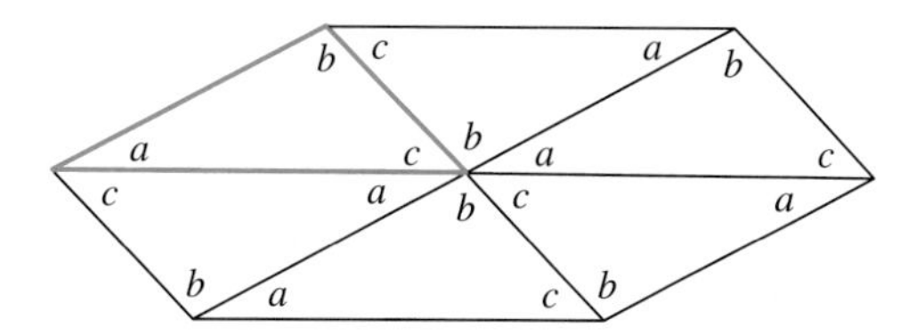

Around the vertex at the center of the figure the three angles of the triangle appear twice. Since $a + b + c = 180°$, the sum of the six angles is exactly 360°, and no overlapping occurs.

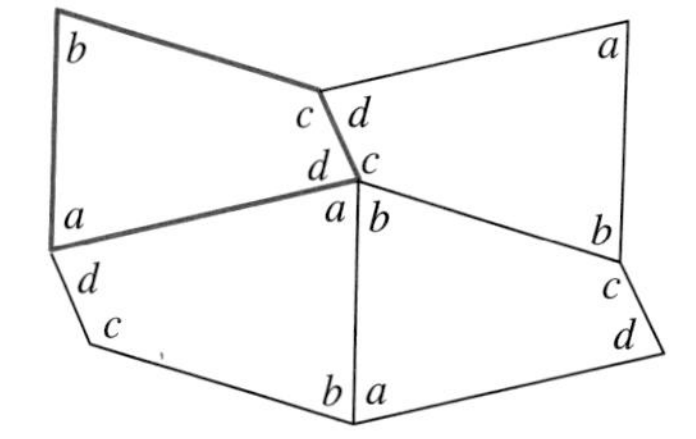

Around the vertex at the center of the figure the four angles of the quadrilateral appear. Since $a + b + c + d = 360°$, no overlapping occurs.

By actually tessellating the plane with different triangles and quadrilaterals, you will be convinced that any triangle or quadrilateral tessellates the plane.

EXAMPLE 2

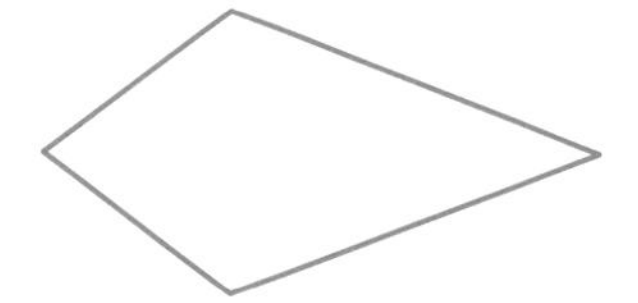

Arrange the angles of the kite-shaped quadrilateral in the figure so that four of them can be placed around a vertex without overlapping each other.

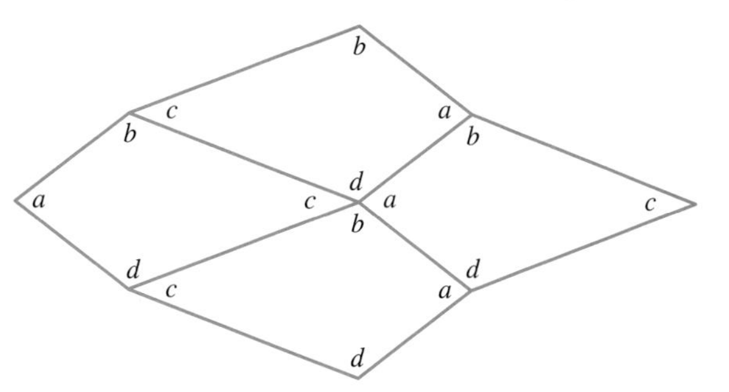

Solution: By sliding and turning over the quadrilateral, we can arrange four of them around one vertex as shown.

You are probably wondering how one actually tessellates an area with a triangle or quadrilateral. This process, as you might have guessed, requires some mathematical operations that move geometric figures: translations and reflections.

Translations	**Reflections**
The triangle is translated a distance that is equal to each of its sides in the direction of each side.	**The triangle is reflected across one of its sides, across a hrozintal line, and across a vertical line.**

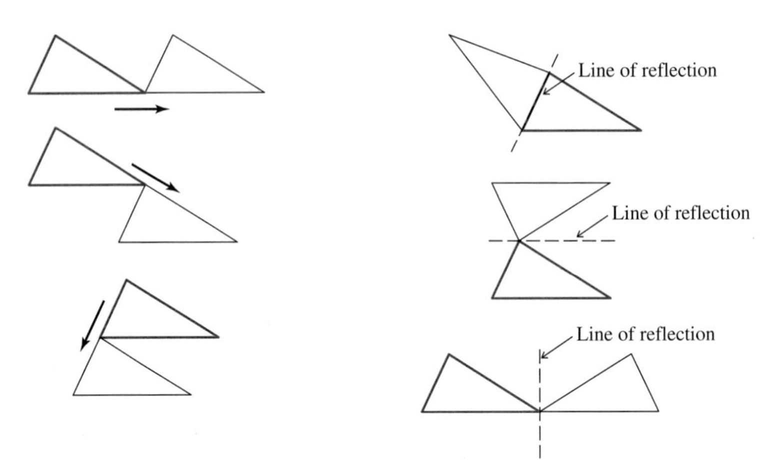

A **translation** slides an object a certain distance in a certain direction. A **reflection** gives the mirror image of an object across a certain line called the line of reflection. A reflection has the effect of turning an object over and placing it on the opposite side of its line of reflection.

Tessellations with a triangle can be formed by using translations or a combination of reflections and translations. Let us examine both methods.

Tessellating with a Triangle Using Translations

To tessellate an area with a triangle, we can translate the triangle in the direction of two sides of the triangle a distance that is equal to the sides of the triangle. The tessellation can be created by repeatedly using two translations.

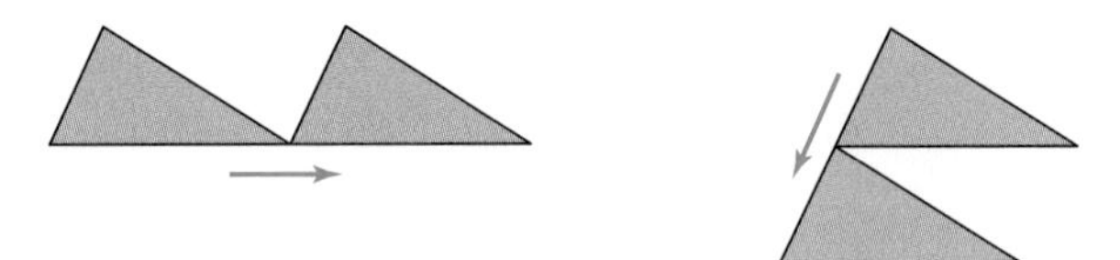

The details of actually creating the tessellation are as follows:

1. Create a primary line of triangles by repeatedly translating the original triangle in the direction of one of the sides of the triangle.

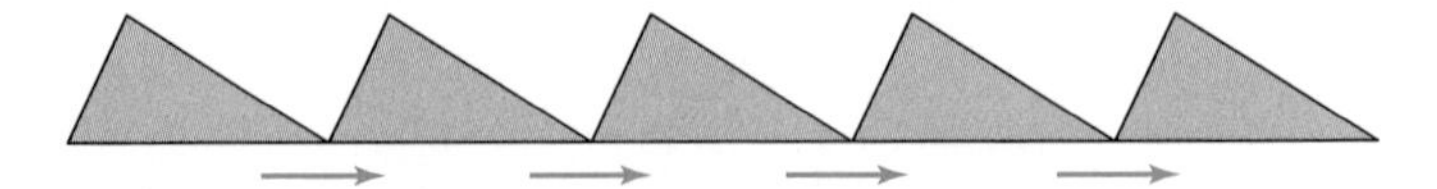

2. Using each triangle in that primary line, repeatedly translate in the direction of another side of the triangle.

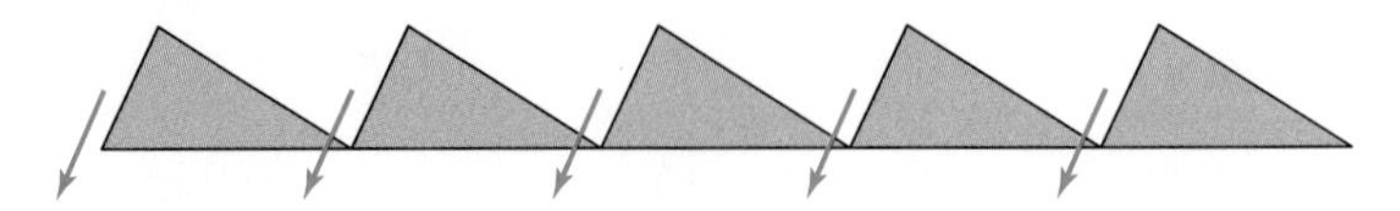

3. Connect any missing lines, and the tessellation will be complete. You will notice that in the process, blank spaces created half of the triangles.

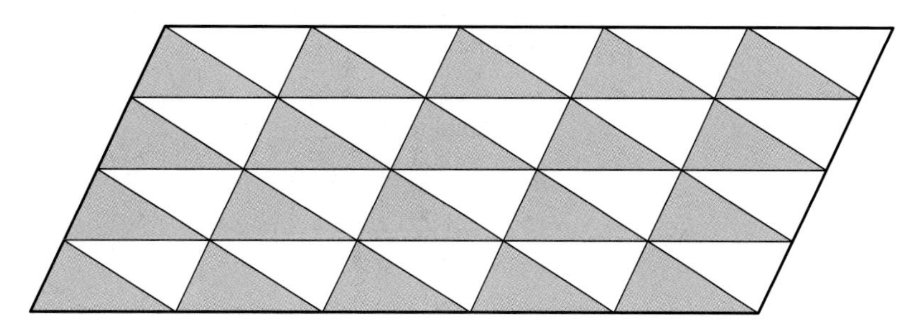

EXAMPLE 3

Using the method shown in the previous example, create a tessellation with the triangle in the figure.

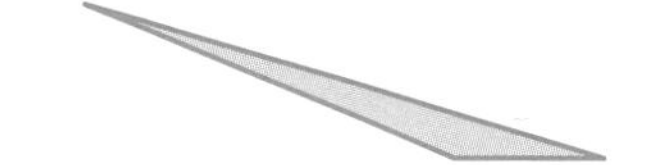

Solution: We can tessellate with the triangle by translating the triangle in the two directions shown.

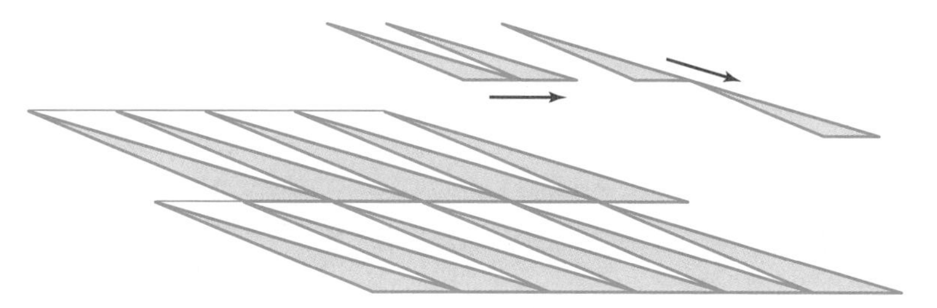

Tessellating with a Triangle Using Translations and a Reflection

A second method of tessellating an area using a triangle involves a reflection and translations. The steps for creating such a tessellation from $\triangle ABC$ are as follows:

1. Reflect the triangle $\triangle ABC$ across line AC, forming quadrilateral $ABCB'$.

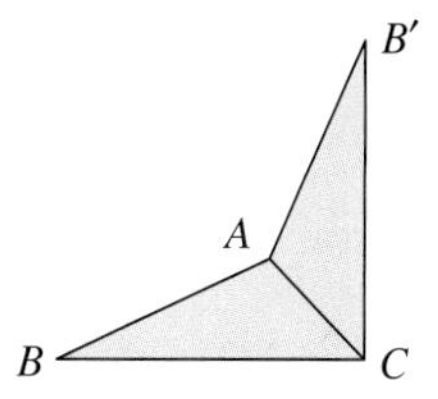

2. Use translations of quadrilateral $ABCB'$ in the direction of its diagonals $\overline{AC}$ and $\overline{BB'}$ as shown.

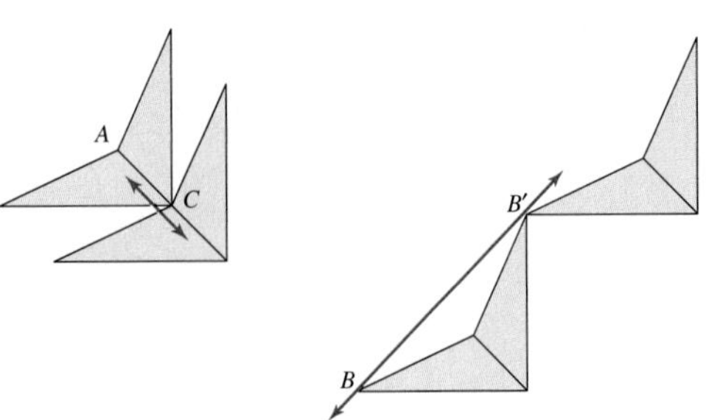

3. Create a primary line of quadrilaterals using the first translation, and then translate each of those quadrilaterals using the second translation. The blank spaces will form quadrilaterals tessellating in a direction opposite of that of ray AC.

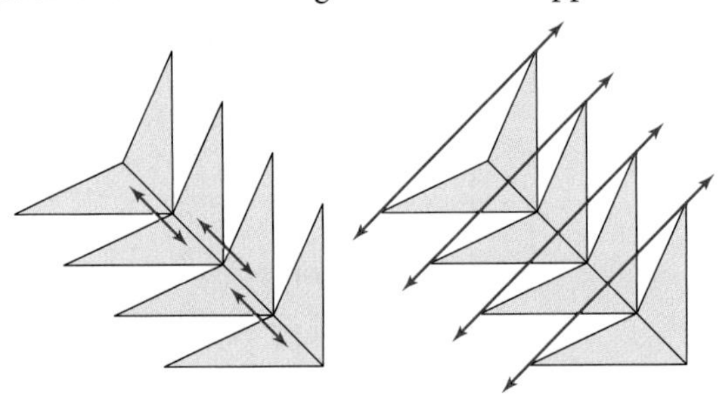

The result of reflecting the original triangle and then translating the resulting quadrilateral will produce the following tessellation.

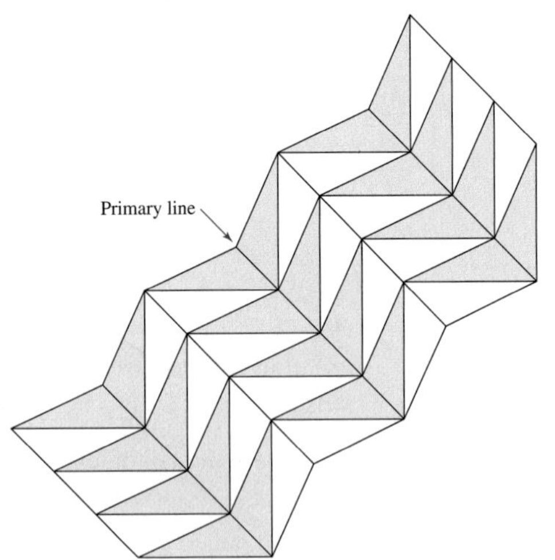

This method will also make it possible to tessellate the plane with any quadrilateral. All you need to do is translate the quadrilateral in the direction of its diagonals.

EXAMPLE 4

Tessellate an area with the quadrilateral shown.

Solution: To tessellate the plane with the quadrilateral, repeatedly translate the quadrilateral in the direction of both diagonals as shown.

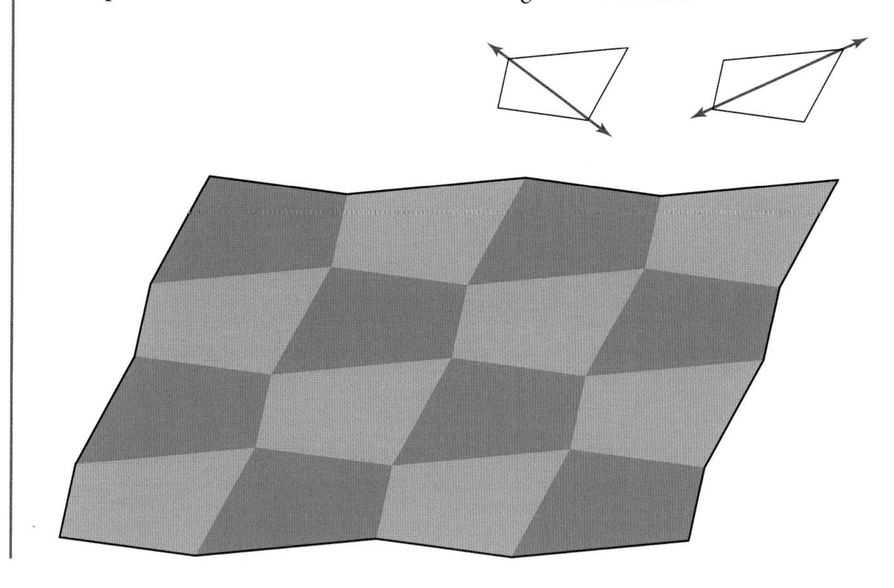

TESSELLATING WITH OTHER POLYGONS

It is possible to tessellate the plane with certain pentagons and hexagons, but it is impossible to tessellate the plane with single polygons that have more than six sides. Some examples of tessellating pentagons and hexagons follow.

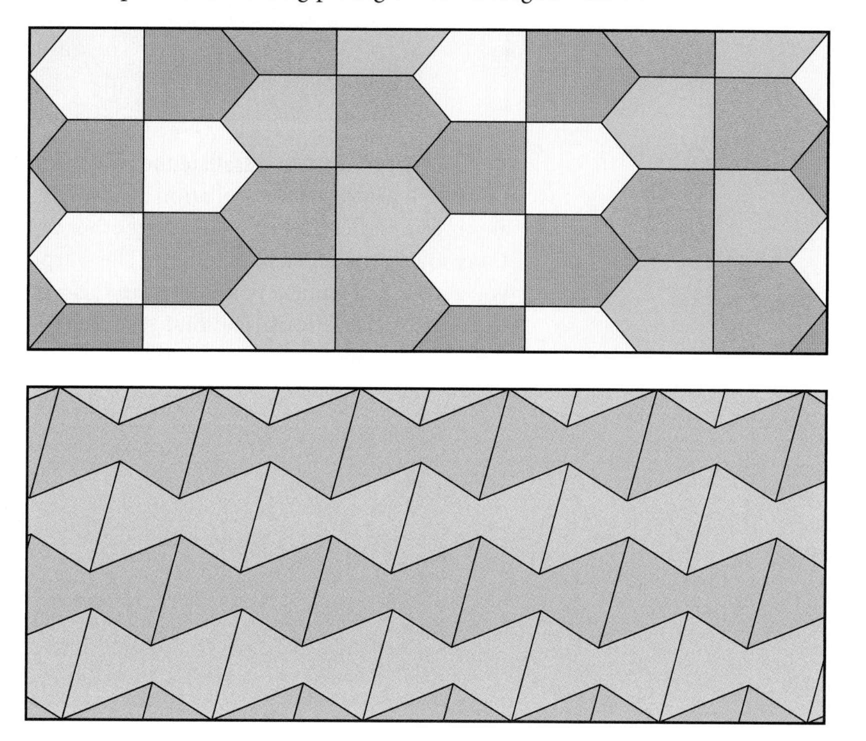

TESSELLATING WITH MORE THAN ONE POLYGON

The plane can also be tessellated with more than one polygon. There are a multitude of ways to tessellate an area by using nonregular polygons, but there are only 21 possible tessellations using combinations of regular polygons. Remember that for polygons to tessellate the plane, the sum of the angles at any vertex must be exactly 360°. Table 7.1, which lists the size of each angle of a regular polygon, will be helpful in determining which combination of regular polygons can be used to tessellate the plane. The tessellations that follow are examples of using more than one polygon to tile the plane.

EXAMPLE 5

Explain why it is possible to tessellate a plane by using two regular octagons and one square at each vertex.

Solution: The measure of an angle of a regular octagon is 135°, and the measure of an angle of a square is 90°. The sum of the angles at one vertex of two octagons and a square is 360° (135 + 135 + 90 = 360). With such a sum, there will be no overlapping of polygons.

The next tessellation has two regular octagons and one square at each vertex.

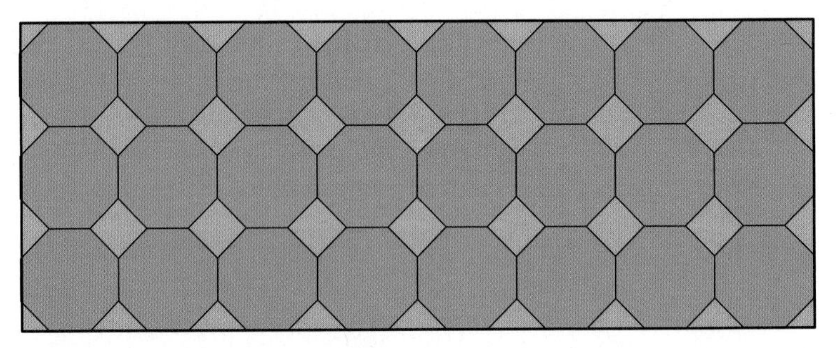

The list of ways to tessellate the plane with polygons is endless. This section has presented some basic tessellation principles. There is much more to be learned about techniques used in generating other types of tessellations and using tessellations to create Escher-like designs. The purpose of this section was simply to give you a taste of the many possibilities in this area of mathematics. We end this section with three tessellations. The first two incorporate stars into a tessellation, and the third shows how simple modifications of a basic tessellation (regular hexagons) can produce some interesting results.

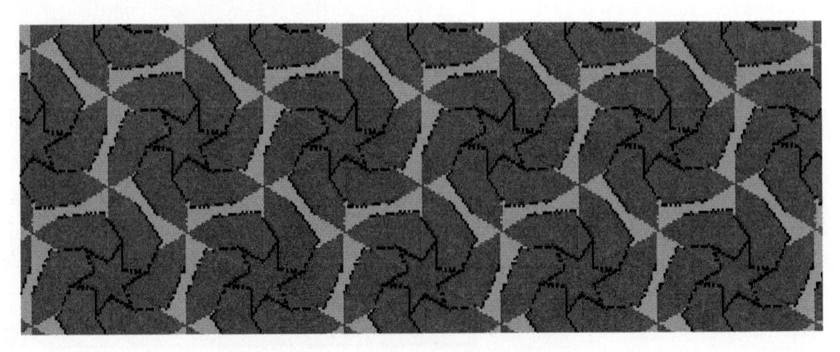

section 7.5

PROBLEMS ◐ Explain ◐ Apply ◐ Explore

◐ Explain

1. What is a tessellation?
2. What is the origin of the word *tessellation*?
3. Which three regular polygons can tessellate the plane?
4. Explain what must happen at each vertex for polygons to tessellate an area.
5. What is a translation?
6. What is a reflection?
7. If a given triangle is to tessellate an area with only translations, what steps need to be taken?
8. If a given triangle is to tessellate an area with only reflections and translations, which steps need to be taken?
9. How can a region be tessellated with a quadrilateral?
10. Give at least three examples of tessellations in your home.

◐ Apply

11. Show why it is impossible to tessellate the plane with regular heptagons.
12. Show why it is impossible to tessellate the plane with regular nonagons.
13. Show why it is impossible to tessellate the plane with regular decagons.

In Problems 14–17, tessellate a region with the triangle shown, using only translations.

14.

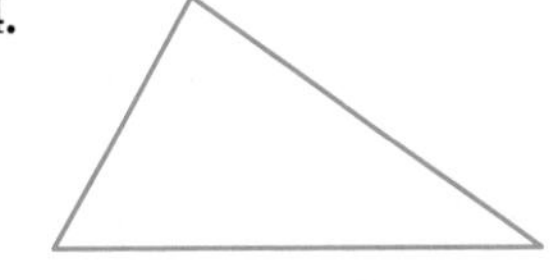

15.

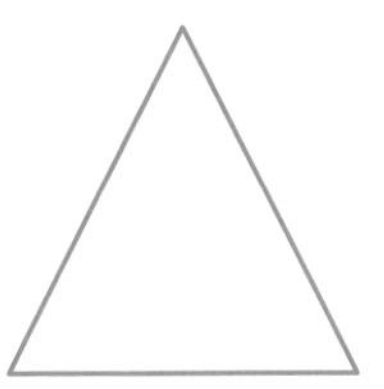

16.

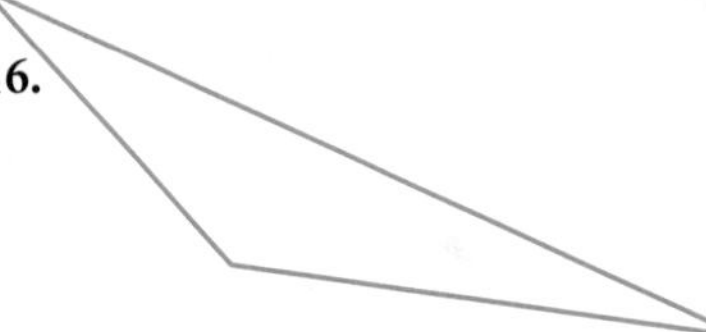

17.

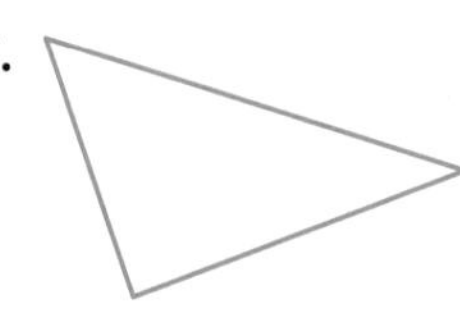

In Problems 18–21, tessellate a region with the triangle shown, using a reflection and translations.

18.

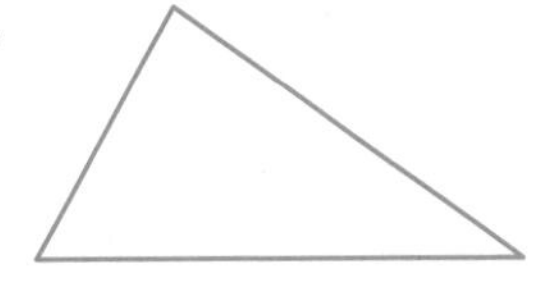

19.

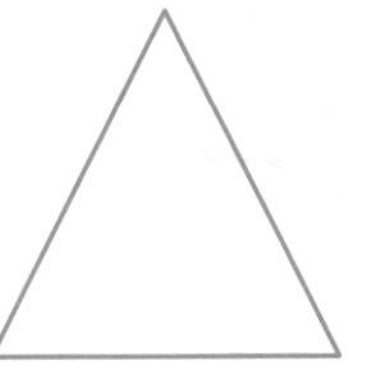

20.

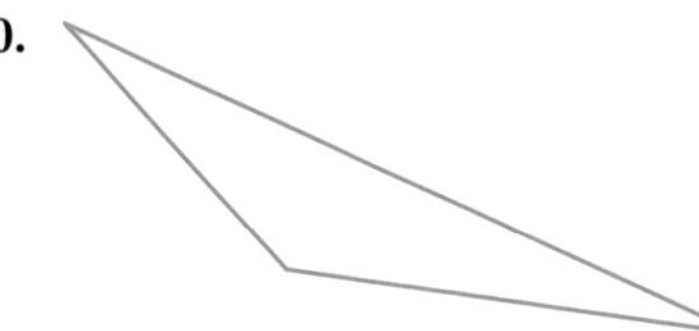

21.

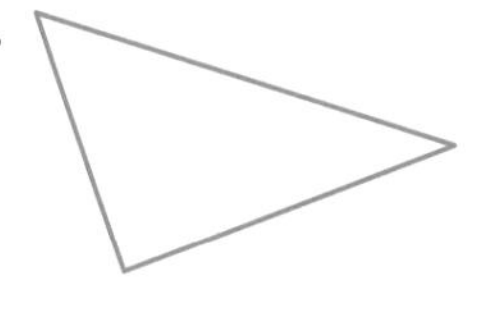

In Problems 22–25, tessellate a region with the quadrilateral shown.

22.

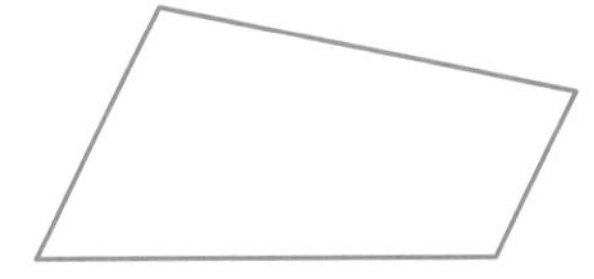

23.

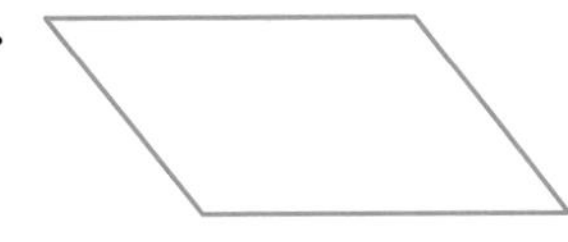

24.

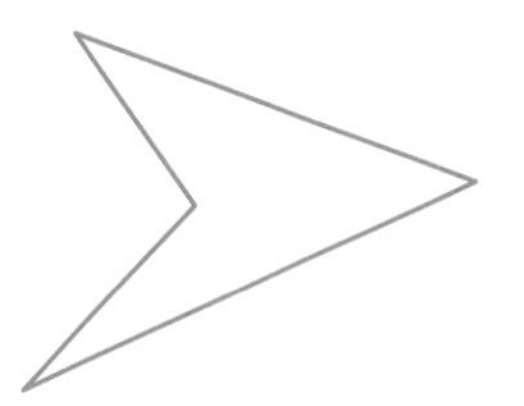

25.

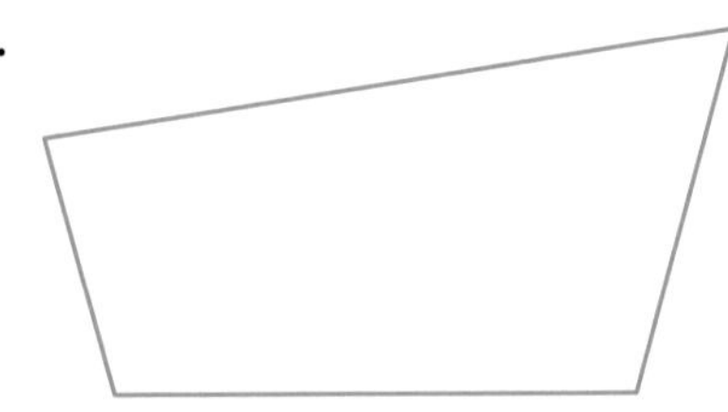

Explore

26. Explain why it is impossible to tessellate a plane with the following regular polygons located at one vertex: one regular hexagon, one regular pentagon, and one square.

27. Explain why it is impossible to tessellate a plane with the following regular polygons located at one vertex: two dodecagons and one octagon.

28. Tessellate a region with two regular hexagons and two equilateral triangles at each vertex.

29. Tessellate a region with one regular hexagon, two squares, and one equilateral triangle at each vertex.

30. Tessellate a region with one regular dodecagon, one regular hexagon, and one square at each vertex.

31. There are 21 possible combinations of regular polygons that can be placed around a vertex to tessellate the plane. Find at least ten of these combinations. (*Hint:* The sum of the angles around a vertex must be 360°.)

32. Draw sketches of the polygons for the combinations found in Problem 31.

33. Using the principles and techniques of the last two sections, create your own tessellation.

34. If you examine things you see each day, you will see many examples of natural and human-made tessellations. Find at least five examples of tessellations and make sketches of them.

35. If a box is packed with objects of the same size and shape so as to leave no gaps or empty spaces, a three-dimensional tessellation is formed. Sometimes it is not possible to pack a box so that a perfect tessellation is created. To save space and packaging costs, the objective for a manufacturer is to minimize the area of the gaps or empty spaces. Suppose a company that manufactures wooden miter boxes for the arts and crafts market is currently packing 12 boxes in a sturdy container. The miter box, the dimensions of its front face, and the top of a packed container are shown in the figure. This, however, is not the most efficient way to pack the container. More miter boxes can be put in the container if it is packed more efficiently.

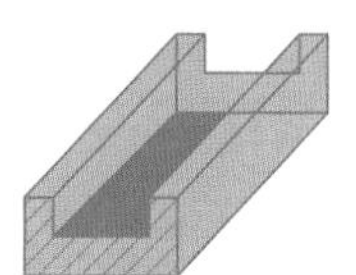

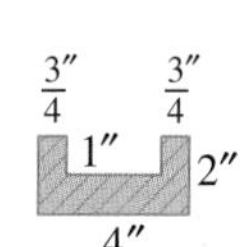

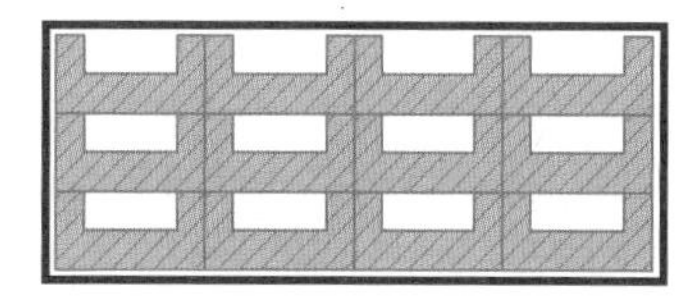

a) Find a way to pack 14 miter boxes in the container.

b) Make a sketch of the top surface of the container.

36. Suppose the manufacturing company in Problem 35 also makes L-shaped wooden shelves that have the same length as the miter boxes and are packed in the same size container as the miter boxes. The inside dimensions of the top of the packing container are 16.1 in. by 6.1 in. The L-shaped shelf and the dimensions of its front face are shown in the figure.

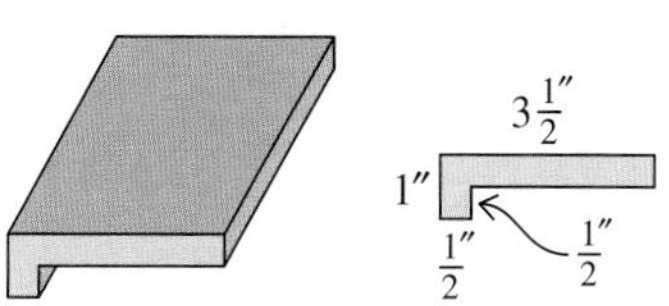

a) Find the maximum number of L-shaped shelves that can be packed in the container.

b) Make a sketch of the top surface of the container.

c) How much empty space is on the top surface of the packed carton?

37. Using the formula for each angle (A) of a regular polygon with n sides, show that the only regular polygons that tessellate the plane are the equilateral triangle, square, and regular hexagon. (*Note:* The expression $\frac{360}{A}$ must be an integer.)

section 7.6

Fractals

(*Left*) A photo of ferns in Mt. Rainier, Washington. (Courtesy of Craig Tuttle)
(*Right*) Fractal design of a fern. (Courtesy of Pat O'Hara)

Since the time of the ancient civilizations, geometry has been used to measure and describe the world in which we live. We have come to believe that much of the world can be described by the basic shapes of geometry. The planets and stars are in the shape of spheres; curved paths such as roads can be measured with straight line segments; a parabola can be used to describe the path of a thrown object; the orbits of the planets are elliptical.

As science has progressed, new discoveries have demonstrated that many of our early ideas about using simple geometric shapes to describe natural phenomena are only partially true. The earth is not really a sphere. It bulges along the equator. In addition, although the surface of a sphere is smooth, the surface of the earth is dimpled with craters, canyons, and oceans and has rolling hills and towering mountains jutting from its surface. The simple geometrical models of our predecessors do not accurately describe a world that has since been more carefully observed.

In addition to reexamining old models, scientists are taking a look at phenomena that were previously considered too complex to be described by mathematics. Why are ferns constructed the way they are? Is there a pattern to the branching designs of oak trees? In an oil spill, what determines the depth to which the oil penetrates the surface? How can the static of a radio transmission be described mathematically? Is there a way to describe the fluctuations of cotton prices over the last 100 years? These phenomena are too complex for the simple geometric models of earlier times. As with the imperfect surface of the earth, a new geometry is needed to accurately describe these details.

Through the use of computers in the 1970s and 1980s, we have seen the arrival of a new area of mathematics called **fractal geometry.** This geometry can model the complex situations mentioned above. The word *fractal,* coined by the mathematician Benoit Mandelbrot, comes from the same Latin root as does the word *fraction.*

Mandelbrot uses this term because simple one-, two-, and three-dimension figures such as lines, squares, and spheres cannot describe complex phenomena. To do so requires objects that have fractional dimension. In this section, we give a short history of fractals, describe what is meant by a fractional dimension, describe how fractals can be drawn, and show some examples of one of the most famous fractal sets, the Mandelbrot set.

HISTORY

The 19th century was a time when new ideas extended mathematics outside the realm of the observed physical world. Mathematicians such as Georg Cantor, Giuseppe Peano, and David Hilbert used ideas that were at the frontiers of mathematical thought. One such idea was to have a "curve" (a path with length but no thickness) fill a two-dimensional area. One such curve, called Hilbert's curve, has its first four generations as shown here.

In each drawing, the boldface curve wraps around the previous generation of the curve, shown by the thinner line. There are two things to notice. First, the basic shape remains the same in each drawing. As the pictures become more detailed, the basic shape is simply drawn with a finer scale. Second, the amount of white space in each drawing is being reduced. Hilbert's remarkable curve eventually passes through all the points in the space, "filling" the entire space.

A question arising from the concept of space-filling curves led to a discovery that became a critical part of Mandelbrot's fractional dimensions. If a curve is a one-dimensional object having length but no thickness and a square is a two-dimensional object having both length and width, what is the dimension of a space-filling curve? Since it is a curve, it should be a one-dimensional object. However, it fills the two-dimensional space of a square. In 1919, two mathematicians, Felix Hausdorff and A. S. Besicovitch, published a work that answered this question. A space-filling curve has a dimension that is between 1 and 2, the dimension approaching 2 as the complexity of the curve increases.

A second major factor that influenced Mandelbrot in the development of fractals was the work on iteration of functions* and complex numbers** by the French mathematicians Pierre Fatou and Gaston Julia. Their work, along with the availability of the computer, enabled Mandelbrot to make many of his discoveries.

CONSTRUCTING SNOWFLAKES, CARPETS, AND OTHER FRACTALS

To construct a fractal, we start with a geometric figure and divide it into smaller versions of itself. We then replace some of the smaller versions, as specified by the rule that generates the fractal. First let's look at a few interesting figures from fractal geometry.

The von Koch Snowflake

The von Koch snowflake is constructed by starting with the line segments forming an equilateral triangle, dividing each of the line segments into three equal sections, and replacing the middle section of each segment by two additional sections. Shown here are the first three iterations of this process. If we start with a line segment, the second figure is constructed from four line segments whose lengths are one-third the length of the original line segment. As we repeat this process, notice that all four of the segments of the second diagram have been used to form the third diagram.

Since the initial step in the construction of the von Koch snowflake is to start with an equilateral triangle, we must perform this process on each of the original sides. Doing this on each side of the triangle creates a six-pointed star.

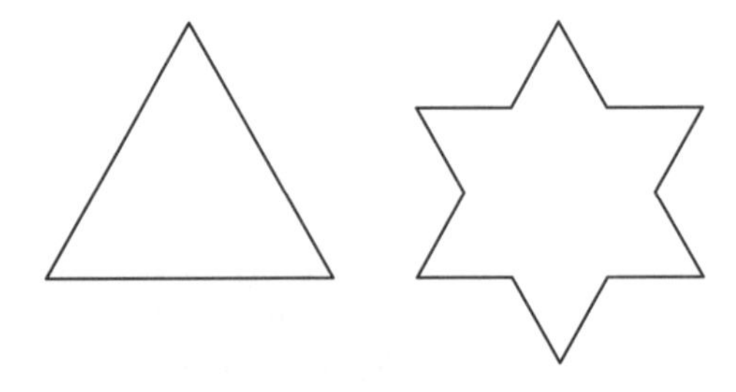

*When a function is iterated, it is used repeatedly. For example, suppose we are using the function $f(x) = \sqrt{x}$. If we pick a number, let's say 10, we want to know what happens when the number is put into the function, and that result is in turn put into the function and the process is continued for some time. With our example, $\sqrt{10} \approx 3.1623$, $\sqrt{3.1623} \approx 1.7783$, $\sqrt{1.7783} \approx 1.3335$, and so on.

**Complex numbers are of the form $a + bi$, where a and b are real numbers and $i = \sqrt{-1}$.

If we repeat this process for each of the 12 segments in the six-pointed star, we obtain a star with 18 points. This, in turn, gives us a 66-point star.

NOTE: Be sure to notice that the new points are constructed on each segment of the previous diagram.

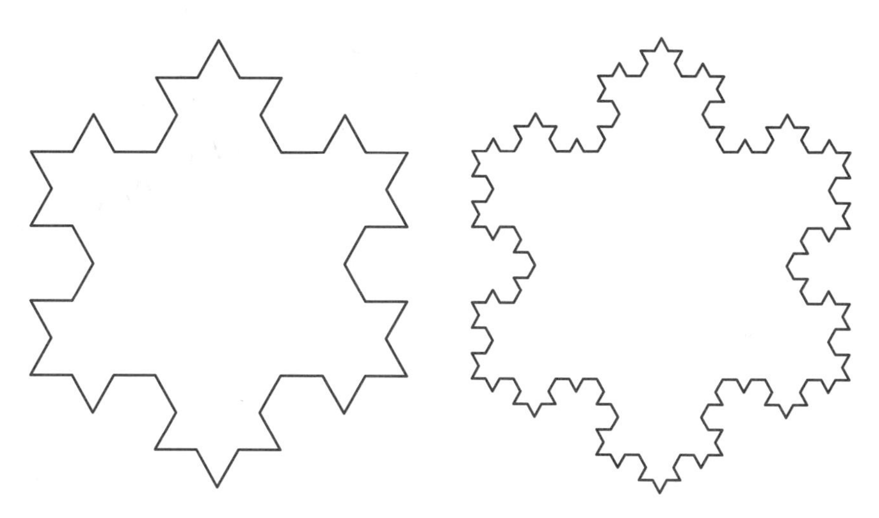

If we continue this process, we arrive at the next two iterations of the von Koch snowflake.

EXAMPLE 1

A fractal whose first two iterations are shown in Figures 7.16 and 7.17 is constructed by dividing each line segment into five equal segments and replacing the middle segment by three new segments, one-fifth the length of the original segment. Draw the next iteration of the fractal.

Solution: To construct the next iteration of the fractal, we need to understand the basic construction process. This consists of dividing each line segment into five equal-length sections and constructing a new three-sided figure to replace the middle of the five sections. This must be done for each line segment in the figure. Therefore, starting with one side of Figure 7.17, we have the next iteration given in Figure 7.18.

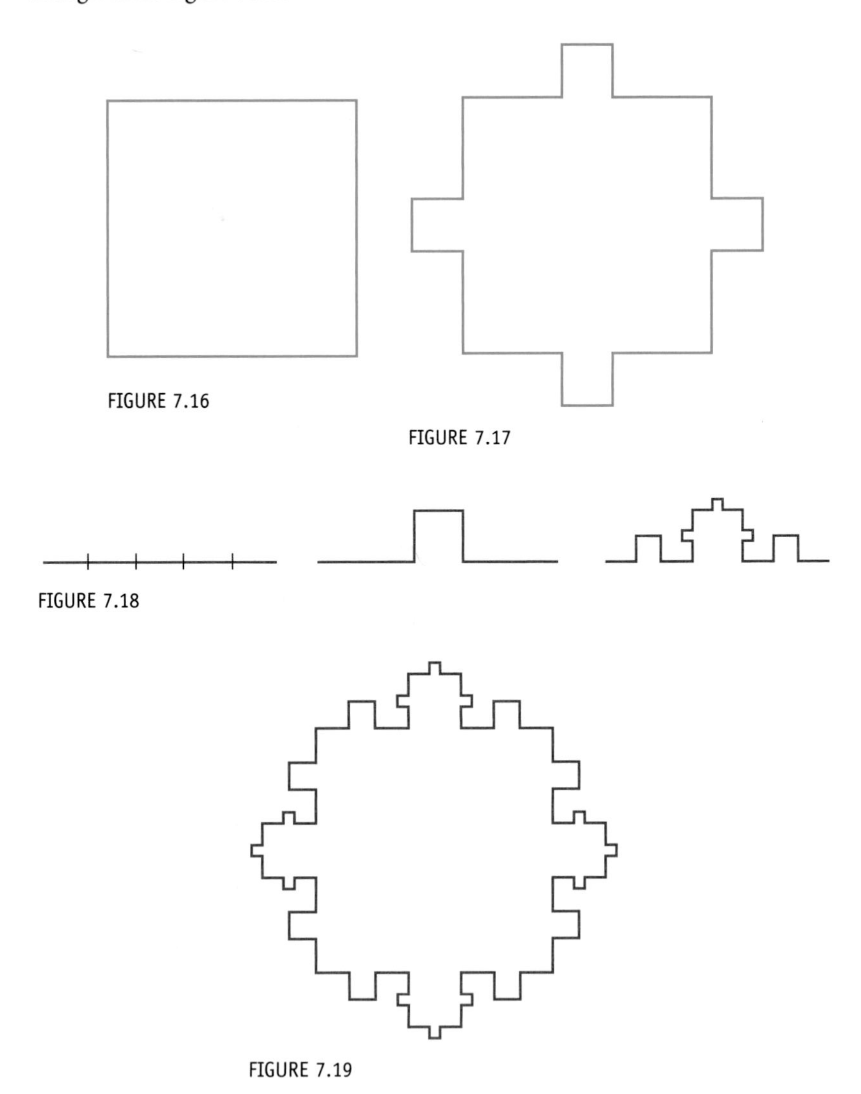

FIGURE 7.16

FIGURE 7.17

FIGURE 7.18

FIGURE 7.19

By applying this process to each side of Figure 7.18, we have the third generation of the fractal shown in Figure 7.19.

The Sierpinski Carpet

With the von Koch snowflake and the fractal shown in Example 1, the construction of the fractal was done by working with line segments. The following fractal is created by working with a two-dimensional object, a square. As shown in Figure 7.20, the basic step of the construction is to start with a square, divide it into nine smaller squares of equal size, and remove the center square.

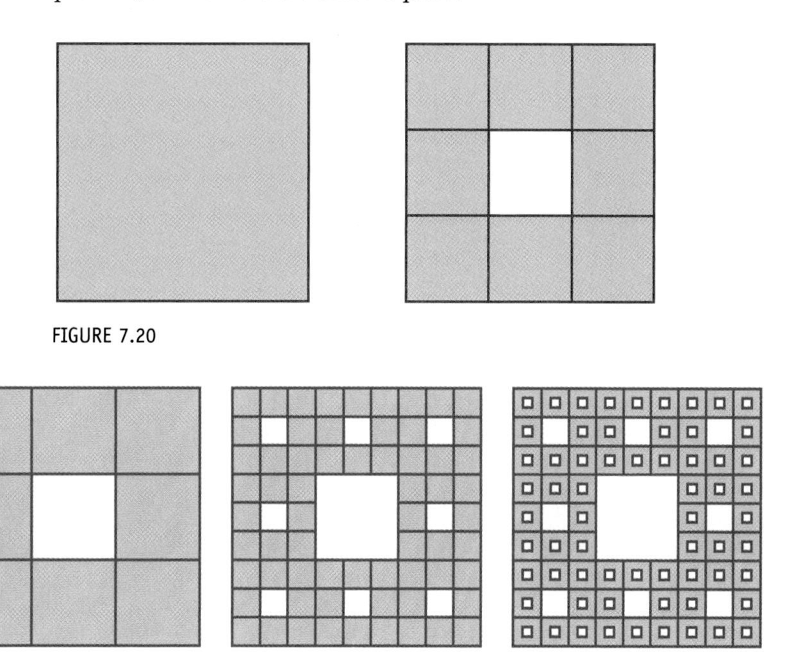

FIGURE 7.20

FIGURE 7.21

The Sierpinski carpet is shown in Figure 7.21. Starting with a square, it uses the process just described to create a plane region with progressively smaller holes.

DIMENSION OF A FRACTAL

The formula for the dimension of a fractal is as follows.

The Dimension of a Fractal

$$d = \frac{\log N}{\log (1/r)} \qquad \text{where}$$

r = ratio of the length of the new object to the length of the original object

N = the number of new objects

Using this formula, we can determine the dimension of fractals.

EXAMPLE 2

Find the dimension of the von Koch snowflake.

Solution: To determine the dimension of the von Koch snowflake, we examine the basic construction process, the change that occurs in one step of the construction.

Since each line segment is divided into three equal sections, $r = \frac{1}{3}$. Because the construction step involves replacing the middle of the three sections by two new sections, the new figure consists of four segments (each of which has a length that is one-third the length of the original segment); therefore, we have $N = 4$. Using the dimension formula gives the dimension of the von Koch snowflake as

$$d = \frac{\log N}{\log (1/r)} = \frac{\log 4}{\log\left(\frac{1}{1/3}\right)} = \frac{\log 4}{\log 3} \approx 1.26.$$

EXAMPLE 3

Determine the dimension of the fractal generated in Example 1.

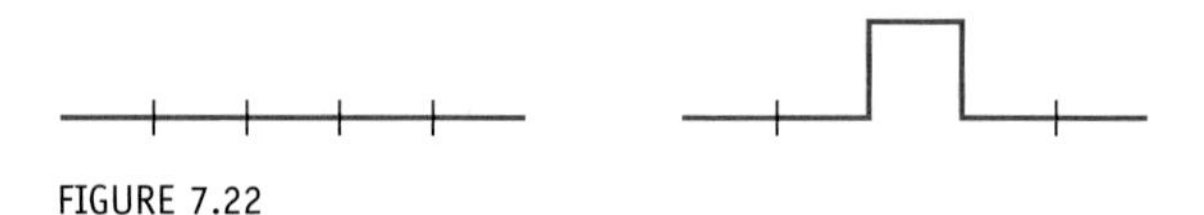

FIGURE 7.22

Solution: To determine the dimension of the fractal, we again examine the basic construction process. Referring to Figure 7.22 and to the directions for creating the fractal, we see that the fractal is created by dividing each line segment into five equal-length subsections. Each of the original line segments is replaced by the new shape. This shape is created from seven of the small subsections. Therefore, using the dimension formula with $N = 7$ and $r = \frac{1}{5}$, we have

$$d = \frac{\log N}{\log (1/r)} = \frac{\log 7}{\log\left(\frac{1}{1/5}\right)} = \frac{\log 7}{\log 5} \approx 1.21.$$

EXAMPLE 4

Find the dimension of the Sierpinski carpet.

Solution: Since we are replacing one square by eight smaller ones, $N = 8$. Since the length of the side of a square is one-third the length of the side of the previous square, $r = \frac{1}{3}$. Using the dimension formula gives

$$d = \frac{\log N}{\log (1/r)} = \frac{\log 8}{\log\left(\frac{1}{1/3}\right)} = \frac{\log 8}{\log 3} \approx 1.89.$$

What Do We Mean by the *Dimension* of an Object?

To understand the concept of a fractional dimension, we first need to examine what we mean when we say that a figure is two dimensional. Let's look at a square with a side of length 1. The area of the square is given by $A = 1^2 = 1$. Dividing each side of the square by 5 gives 25 miniature versions of the original square.

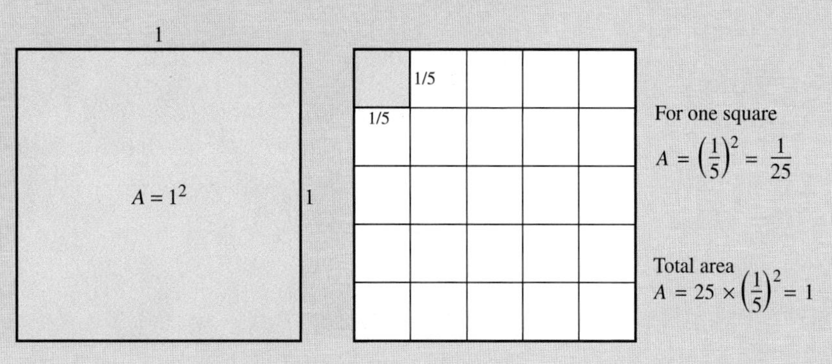

Each of the 25 new squares has sides of length $\frac{1}{5}$ and an area of $\left(\frac{1}{5}\right)^2$. This gives a total area for the 25 squares of $A = 25 \times \left(\frac{1}{5}\right)^2 = 1$.

In general, if we let r be the ratio of the length of the new object to the length of the original object, N be the number of new objects, and d be the dimension of the figure, we can say $Nr^d = 1$. For the two-dimensional squares, we have used $N = 25$, $r = \frac{1}{5}$, and $d = 2$.

To determine a formula for the dimension of an object, we solve for d in this formula.

$$Nr^d = 1$$

We start by dividing both sides by r^d and using the rules of exponents.

$$N = \frac{1}{r^d}$$

$$N = \left(\frac{1}{r}\right)^d$$

Next, we take the common log of both sides and apply the rules of logarithms.

$$\log N = \log \left(\frac{1}{r}\right)^d$$

$$\log N = d \log \left(\frac{1}{r}\right)$$

Finally, dividing by $\log \frac{1}{r}$ gives the desired equation.

$$\frac{\log N}{\log (1/r)} = d$$

The Mandelbrot Set

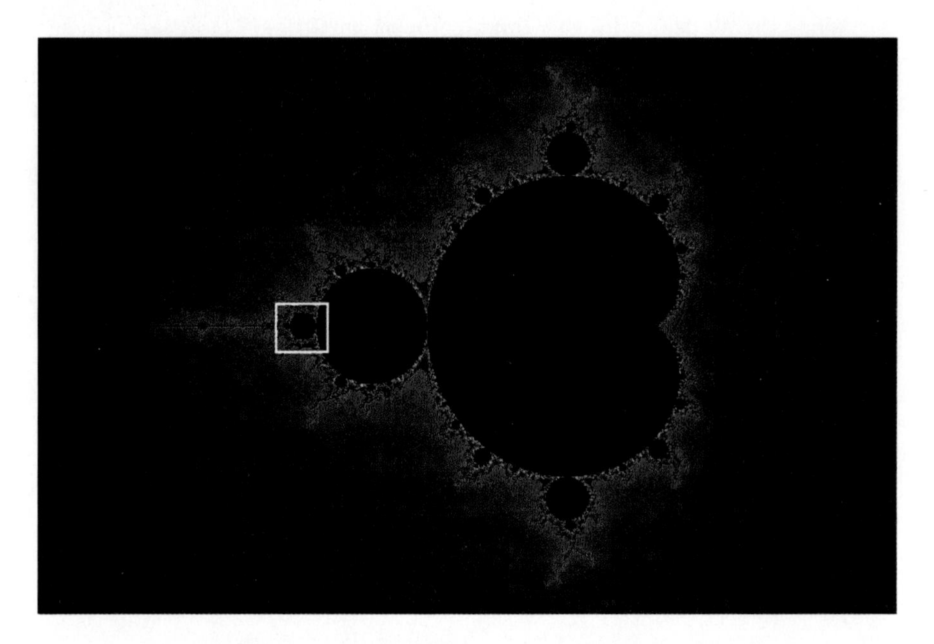

This diagram is called the Mandelbrot set. It is more complex than the fractals discussed earlier, but it still demonstrates the same repetitive patterns exhibited by the other fractals. Notice that the general shape of a circle with a little nob sticking off one side is repeated, at smaller scales, throughout the diagram. If we take the portion of the diagram located in the box and magnify it, we obtain the same structure. This process can be continued indefinitely, each new iteration showing similar characteristics.

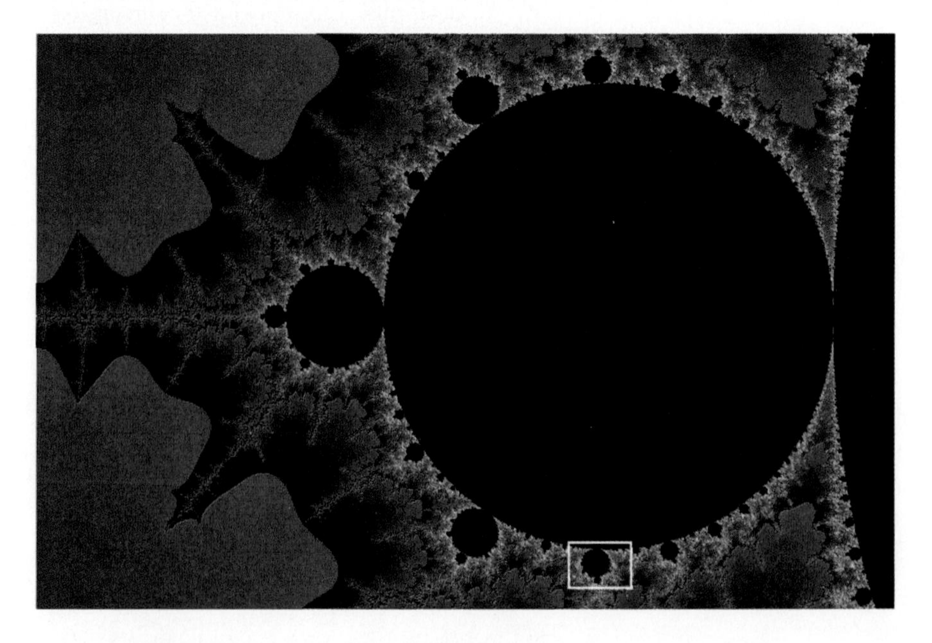

Examining the small nob at the bottom center of this diagram again gives a new, smaller version of the same diagram.

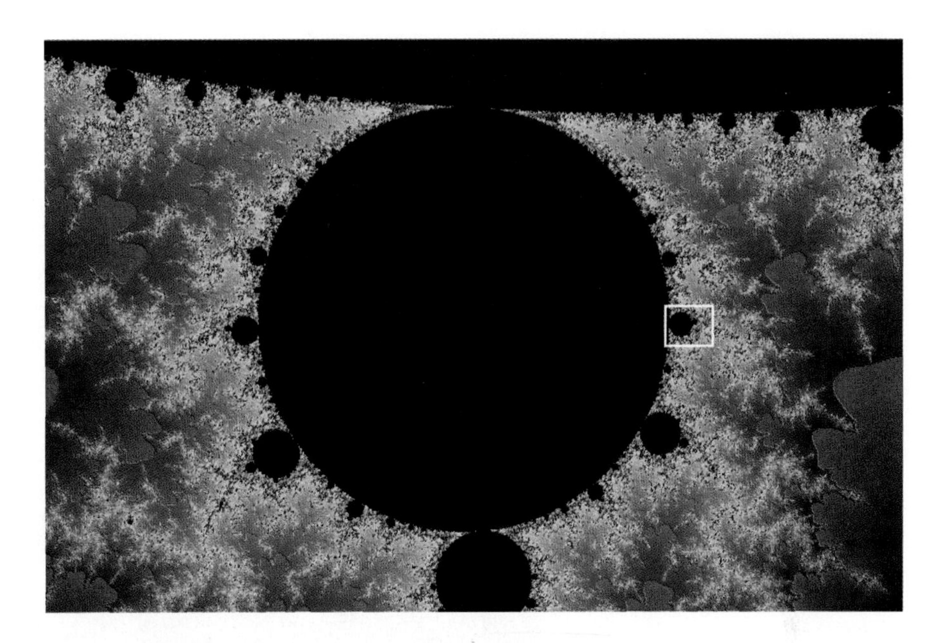

If we again zoom in on another nob that was only a dot on the original diagram, we get the following:

This set is generated by a very short set of instructions (see page 413), but it provides insights into many physical phenomena. The idea of a portion of a set being similar to itself (called self-similarity) can be seen in the mapping of coastlines. For example, the coastline of California has many small bays and peninsulas. These are usually measured on a scale of thousands of feet. If we examine a small portion of the coastline more closely, the smaller section is also seen to have inlets and peninsulas, this time measured on a scale of hundreds of feet. If this process is continued, we can examine a section of coastline only a fraction of an inch long. In this very small section, we can again see inlets in the form of grains of sand. As we examine the coastline on increasingly more detailed levels, we continue to see the same degree of intricacy.

This phenomenon of self-similarity can be seen in many other situations. Cotton prices since 1900 have fluctuated owing to the vagaries of supply and demand. Yet when Mandelbrot and the Harvard economist Hendrik Houthakker closely examined the fluctuations, they found that a decade-by-decade pattern of prices showed the same type of fluctuation as yearly prices or monthly prices. Noise in telephone lines has certain, apparently random, patterns when the signals are examined on an hourly basis. When the same signal is examined at the level of seconds or minutes, the same patterns, at smaller scales, reappear. Like the von Koch snowflake and the Sierpinski carpet, the same patterns reappear as the situation is examined in finer detail.

The intricate fractals that follow were created with a computer and a fractal-generating equation.

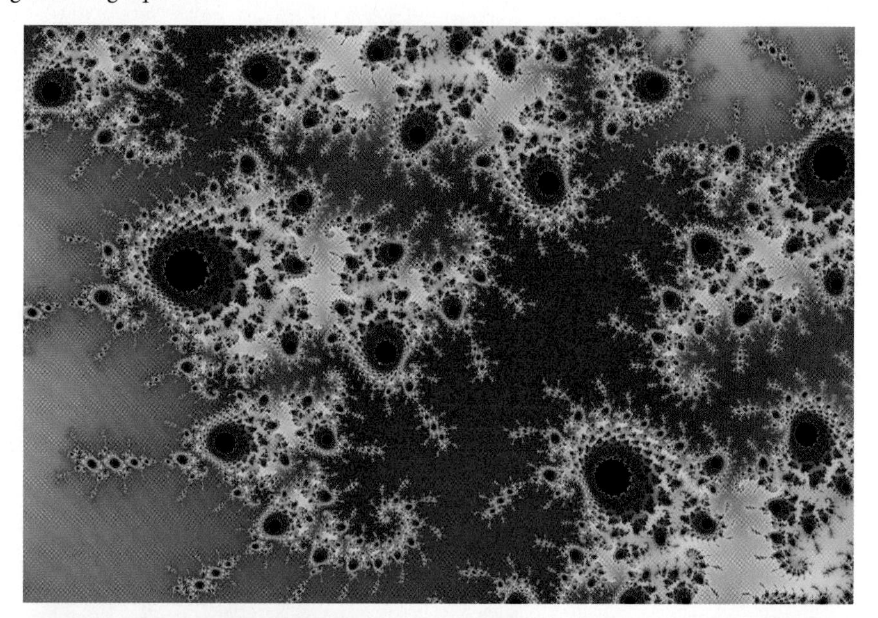

The Mathematics Behind the Mandelbrot Set

In algebra, you might have learned that complex numbers are numbers of the form $a + bi$, where $i = \sqrt{-1}$. You might also have learned that the magnitude of a complex number $z = a + bi$ is given by $\sqrt{a^2 + b^2}$.

The Mandelbrot set is generated by using the equation $z_n = (z_{n-1})^2 + C$, where z_n and C are complex numbers. A complex number z_0 is chosen and substituted into the equation for the z value on the right side of the equation. For each complex number chosen, the equation is used repeatedly until the magnitude of z exceeds 2 or until it is determined that the magnitude of z will probably never exceed 2. If the final result exceeds 2, the starting point z_0 is not in the Mandelbrot set. If the final result does not exceed 2, the starting point is in the Mandelbrot set.

For example, let's look at $z = 0.75 + 0i$ and assume that the constant $C = 0$. Since the first z value is 0.75, we write $z_1 = 0.75$. This gives the following sequence of values.

$$\begin{aligned} z_2 &= (0.75)^2 = 0.5625 \\ z_3 &= (0.5625)^2 = 0.3164 \\ z_4 &= (0.3164)^2 = 0.1001 \\ z_5 &= (0.1001)^2 = 0.0100, \ldots \end{aligned}$$

It is apparent that this sequence of values will approach zero and hence is less than 2. Therefore, the starting point $z_1 = 0.75 + 0i$ is in the Mandelbrot set.

If we start with a point such as $z_1 = 0.75 + 1.3i$, we arrive at the following sequences of values.

$$\begin{aligned} z_2 &= (0.75 + 1.3i)^2 = -1.1275 + 1.95i \\ z_3 &= (-1.1275 + 1.95i)^2 = -2.5312 - 4.3973i \\ z_4 &= (-2.5312 - 4.3973i)^2 = -12.9293 + 22.2609i \end{aligned}$$

Since these numbers are growing quickly, we can see that the magnitude will exceed the limit of 2. Hence, the point $z = 0.75 + 1.3i$ is not in the Mandelbrot set. Although these calculations are tedious, the availability of computers has made the computations and the drawing of the Mandelbrot set and other fractals a reality.

PROBLEMS ● Explain ● Apply ● Explore

● Explain

1. What is a fractal and what properties does a fractal possess?
2. What is the dimension of a fractal? What formula is used to detemine the dimension of a fractal? Explain what each variable in the formula represents.
3. What does it mean for a fractal to have a dimension of 1.6?
4. What does it mean for a fractal to have a dimension of 2.6?
5. What does self-similarity refer to when discussing fractals?
6. What is meant by a space-filling curve?
7. Examine a leaf of a fern. Explain why the parts of a leaf have properties of fractals.
8. Explain why cirrus clouds can be more accurately depicted by fractals than by standard geometric shapes.

● Apply

9. Create a fractal by starting with a square, dividing each line segment into three equal lengths, and replacing the middle third of each side with three line segments whose lengths are one-third the length of the original segment. The first iteration is shown here. Repeat this process and draw the next iteration of this fractal.

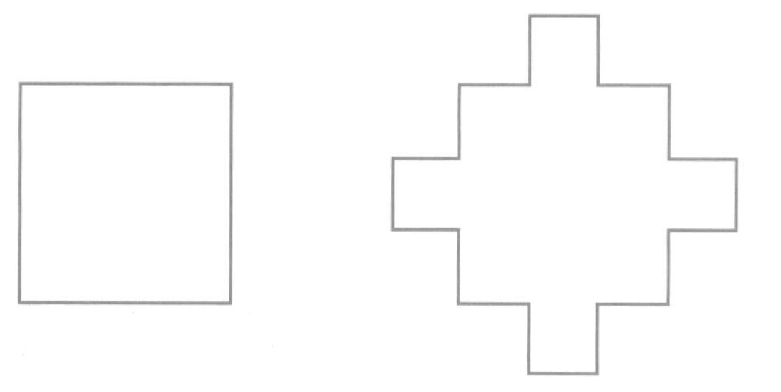

10. Create a fractal by starting with a square, dividing each line segment into five equal sections, and replacing the second and fourth sections of each segment with three line segments whose lengths are equal to the section they are replacing. The first iteration is shown here. Repeat this process and draw the next iteration of this fractal.

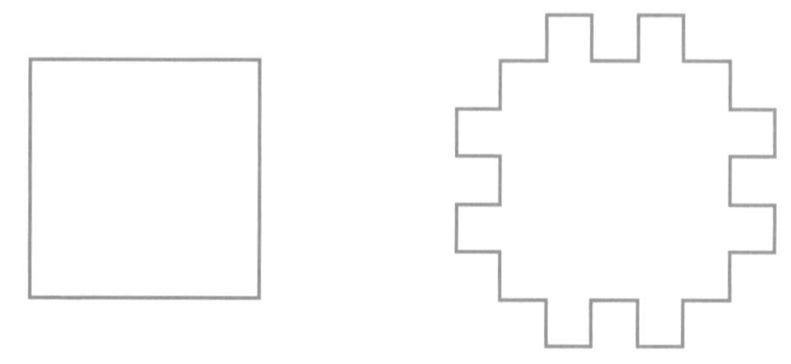

11. Create a fractal by starting with a line segment, dividing the segment into four equal lengths, and replacing both the second and third sections by two segments whose lengths are one-fourth the length of the original segment. The first iteration is shown here. Repeat this process and draw the next iteration of the fractal.

12. Create a fractal by starting with a line segment, dividing the segment into five equal lengths, and replacing both the second and fourth sections by two segments whose lengths are one-fifth the length of the original segment. The first iteration is shown here. Repeat this process and draw the next iteration of the fractal.

13. A tree-like fractal can be created by starting with a line segment, dividing it into three equal segments, and replacing the top two segments with four equal segments in a Y-shaped pattern as shown. In the next iteration this same construction is performed on each of the five segments of the Y-shaped figure. Draw the next iteration of this fractal.

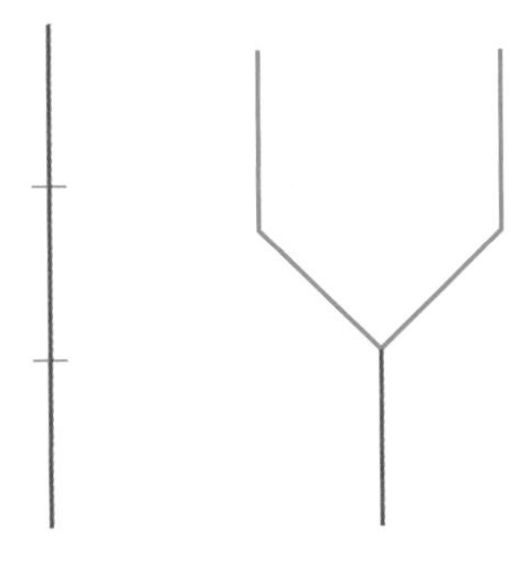

14. A tree-like fractal can be created by starting with a line segment, dividing it into four equal segments, and replacing the top three segments with four equal segments in a Y-shaped pattern as shown. In the next iteration this same construction is performed on each of the five sgements of the Y-shaped figure. Draw the next iteration of this fractal.

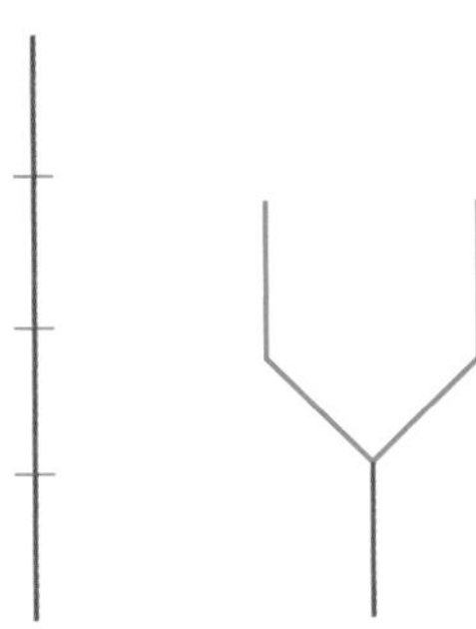

Explore

15. Create your own fractal by starting with a line segment, dividing it into equal parts, and systematically replacing some of the parts by other segments. Repeat this process three times.

16. Create your own fractal by starting with an object and altering it by adding smaller version(s) of the original object. Repeat this process three times.

17. Find the dimension of the fractal in Problem 9.

18. Find the dimension of the fractal in Problem 10.

19. Find the dimension of the fractal in Problem 11.

20. Find the dimension of the fractal in Problem 12.

21. Find the dimension of the fractal in Problem 13.

22. Find the dimension of the fractal in Problem 14.

23. A fractal is created by the following process. A line segment of length 1 is drawn. It is divided into five equal sections. The second and fourth of these sections are then removed. This process is continued indefinitely. What is the dimension of this fractal?

24. A fractal is created by the following process. Start with a regular pentagon. Cut each side into five equal-length sections and replace the middle section with another regular pentagon. This process is continued indefinitely. What is the dimension of this fractal?

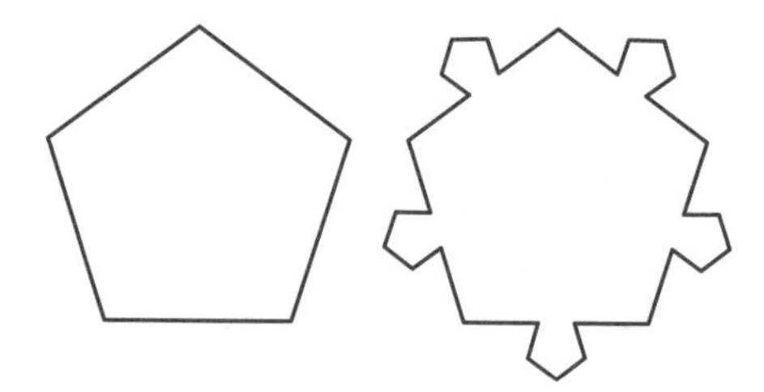

***25.** Investigate some of the other areas of study to which fractals are being applied.

***26.** The repetitive fractal patterns generated by computers can have a stunning visual effect. Find examples that show this "artistic" effect of fractals.

***27.** What is a Sierpinski gasket? Draw the first three iterations of the gasket.

***28.** What is a Cantor dust? In what area of study is it being applied?

CHAPTER 7 SUMMARY

Key Terms, Concepts, and Formulas

The important terms in this chapter are:	Section
Acute angle: An angle that has a measure between 0° and 90°.	7.1
Angle: The figure formed by two rays or line segments with a common end point.	7.1
Atmospheric perspective: A technique of creating an illusion of depth on a two-dimensional surface by a gradual lessening of clarity and visual strength.	7.2
Axiom: A proposition that is assumed to be true. Historically, axioms referred to propositions of algebra or arithmetic.	7.1
Concave polygon: A polygon in which an extension of at least one of its sides enters the interior of the polygon.	7.4
Convex polygon: A polygon in which extensions of its sides do not enter the interior of the polygon.	7.4
Diminishing sizes: A technique of creating an illusion of depth on a two-dimensional surface by systematically making objects smaller.	7.2
Euclidean geometry: A phrase used to describe a system of geometry based on *Elements* written by the Greek mathematician Euclid.	7.1
Exterior angle: An angle formed outside a triangle by one side of the triangle and an extension of another side of the triangle.	7.1
Eye level line: A horizontal line across a picture in a one-point perspective that shows the observer's focus when looking straight ahead.	7.2
Fibonacci numbers: The sequence of numbers 1, 1, 2, 3, 5, 8, 13, 21, 34, 55, . . . , where each successive number is the sum of the two previous numbers.	7.3
Fractal: An object with a fractional dimension.	7.6
Golden Ratio (ϕ): The ratio between distances a and b such that $\phi = \frac{a}{b} = \frac{a+b}{a} \approx 1.62.$	7.3
Golden Rectangle: A rectangle whose sides form the Golden Ratio.	7.3
Line segment: The set of points on a line consisting of two endpoints and all the points on the line between those two points.	7.1
Non-Euclidean geometry: A phrase used to describe a system of geometry that assumes a consistent set of postulates with at least one that is not logically equivalent to Euclid's postulates.	7.1
Obtuse angle: An angle that has a measure between 90° and 180°.	7.1
One-point perspective: A technique of creating an illusion of depth on a two-dimensional surface by using the geometry of converging lines.	7.2

Overlapping shapes: A technique of creating an illusion of depth on a two-dimensional surface by placing shapes in front of one another. 7.2

Parallel: Two lines, rays, or line segments are parallel (||) if they lie in the same plane and the lines that contain them do not intersect. 7.1

Parallel postulate: A postulate of geometry that establishes the number of lines that are parallel to a given line through an external point. 7.1

Perpendicular: Two lines, or line segments are perpendicular ($\perp$) if they intersect and form a right angle. 7.1

Perspective: The art of drawing objects on a two-dimensional surface so as to give an illusion of depth and show distance from the observer. 7.2

Polygon: A closed figure in a plane formed by the line segments that intersect each other only at their endpoints. 7.4

Postulate: A proposition that is assumed to be true. Historically, postulates referred to propositions of geometry. 7.1

Proof: A series of statements that logically show a conclusion follows from the hypothesis. 7.1

Ray: The figure formed by extending a line segment in only one direction. 7.1

Reflection: Movement of a shape in a plane by finding its mirror image across a specified line. 7.5

Regular polygon: A polygon that has sides of equal length and angles of equal measure. 7.4

Right angle: An angle that has a measure of 90°. 7.1

Sides: Line segments that form a polygon. 7.4

Straight angle: An angle that forms a line and has a measure of 180°. 7.1

Tessellation: A pattern of one or more congruent shapes that covers an area in a plane without overlapping or leaving any gaps. 7.5

Theorem: A proven proposition. 7.1

Translation: Movement of a shape in a plane by sliding it in a certain direction for a specified distance. 7.5

Triangle: A figure consisting of three line segments, determined by three points not on the same line. 7.1

Triangle-sum theorem: A theorem of geometry that states the sum of the measures of the angles of a triangle. 7.1

Undefined terms: Terms that have meanings assumed to be clear. Point, line, and plane are undefined terms in geometry. 7.1

Vanishing point: A fixed point on the eye level line in a one-point perspective to which receding lines coverage. 7.2

Vertex (pl. **vertices**)**:** The intersection point of sides of a polygon. 7.4

After completing this chapter, you should be able to:

		Section
1.	Explain and use undefined terms, definitions, axioms, postulates, and theorems in a system of Euclidean geometry.	7.1
2.	State the differences between Euclidean, Lobachevskian, and Riemannian geometries, including the parallel postulate, model, and triangle-sum theorem of each system.	7.1
3.	Show depth on a two-dimensional surface using overlapping shapes, diminishing sizes, atmospheric perspective, or a one-point perspective.	7.2
4.	Find a point on a line segment that divides it into the Golden Ratio.	7.3
5.	Determine whether a rectangle is approximately a Golden Rectangle and explain the significance of Golden Rectangles in art, architecture, and other areas.	7.3
6.	Design and construct regular polygons and stars.	7.4
7.	Determine the sum of the angles of a polygon and each angle of a regular polygon.	7.4
8.	Create tessellations with a triangle or quadrilateral using translations and reflections.	7.5
9.	Recognize basic properties of fractals.	7.6
10.	Given a geometric figure and a method of construction, draw the next iteration of a fractal and determine the dimension of the fractal.	7.6

CHAPTER 7 REVIEW

Section 7.1

1. Explain why it is necessary to have undefined terms and postulates in a geometric system.
2. Show how Axioms A3 and A4 can be used to solve for x in the equation $6x + 5 = 2x - 3$.
3. Draw $\angle BIG$ with $\overline{GA} \perp \overrightarrow{IG}$ on models for the plane in Euclidean, Lobachevskian, and Riemannian geometries.
4. Draw lines, h, j, and k intersecting at point P on models for the Euclidean plane, the Riemannian plane, and the Lobachevskian plane.
5. Explain why squares do not exist in Lobachevskian and Riemannian geometries.

Section 7.2

6. What techniques did Giovanni Paolo Panini use in his painting *The Interior of the Pantheon, Rome,* found on page 480, to make the inside of the building seem immense?
7. Examine the Egyptian wall painting from the tomb of Menna at the start of Section 1.1. Why does it have such a "flat" visual impression?

8. Use a one-point perspective to draw a barbed wire fence along a country road.
9. An artist's one-point perspective shows a series of crosses in a cemetery. In the picture, a cross that is 6 inches tall is 10 inches from the vanishing point. How tall should the artist make a similar cross if she plans to place the cross 3 inches from the vanishing point.

Section 7.3

10. What importance did the Greeks place on the Golden Ratio?
11. A stage has a length of 22 feet. Where would you place the podium so that its position divides the stage into a Golden Ratio? Why would you want to do this?
12. One side of a rectangular window is 54 inches. To make the window a Golden Rectangle, what are two possible measurements for the other dimension of the window.
13. A kite is a quadrilateral with two distinct pairs of equal adjacent sides as shown below. Create a definition for a "Golden Kite." Draw some of these "Golden Kites." (*Hint:* There are two different ratios of segments that could be used to make the "Golden Kite.")

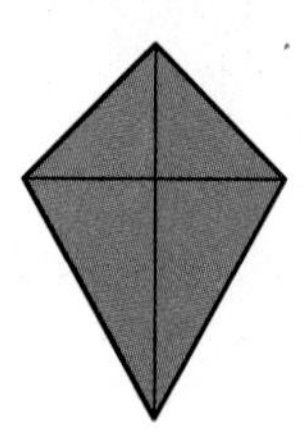

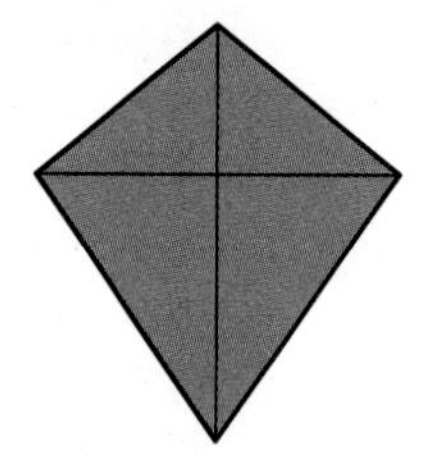

Section 7.4

14. Draw a convex dodecagon, a concave dodecagon, and a regular dodecagon.
15. Find the sum of the angles for the polygons in Problem 14.
16. Find the measure of each angle of the regular polygon in Problem 14.
17. Using the regular polygon in Problem 14, create a star by connecting every fourth point.

Section 7.5

18. Explain why a regular dodecagon cannot tessellate a region.
19. Tessellate a region with a regular hexagon, squares, and equilateral triangles.
20. Tessellate a region with the triangle

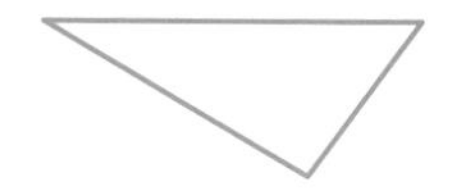

    a) Using translations.
    b) Using a reflection and translations.
21. Tessellate a region with the following figures.
    a)

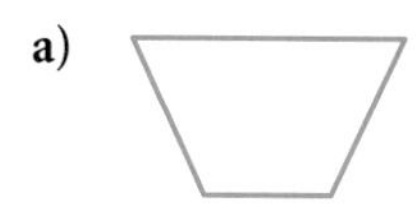

    b)

Section 7.6

22. Create a fractal by starting with a line segment, dividing it into three equal lengths, and replacing the middle segment with five sides of a regular hexagon. The first iteration of the fractal is shown below. Repeat this process and sketch the next iteration of the fractal.

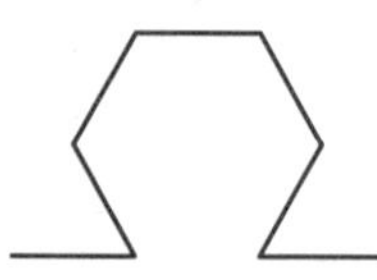

23. Find the dimension of the fractal in Problem 22.
24. The dimension of a fractal is $d \approx 1.62$. Explain what that means? Why might the objects in this dimension be "special"?

CHAPTER 7 TEST

1. Explain the difference between
   a) Undefined terms and definitions.
   b) Postulates, axioms, and theorems.
2. Draw $\triangle TRY$ with $\angle T$ an obtuse angle on the model for the plane in Euclidean, Lobachevskian, and Riemannian geometries.
3. In light of the non-Euclidean geometries, explain the statement "Mathematics does not give truth about the real world."
4. Using the given shape and each of the following techniques, make a sketch that gives a sense of depth.
   a) Overlapping shapes
   b) Diminishing sizes
   c) Atmospheric perspective
5. Use a one-point perspective with the given shapes, line at eye level, and vanishing point to make solid objects of varying lengths.

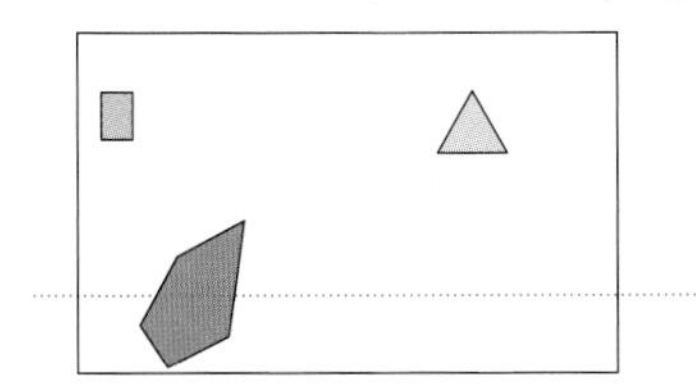

6. Find a point that approximately divides the segment into a Golden Ratio.

7. If the width of a rectangle is to be 8.5 in., how long should its length be to make it a Golden Rectangle?
8. Construct the following regular polygons:
   a) dodecagon b) 18-gon
9. Construct stars from the regular polygons in Problem 8 by connecting the following points:
   a) Every third point of the polygon
   b) Every fifth point of the polygon
10. Tessellate a region with the following triangles:
    a) Using the only translations.

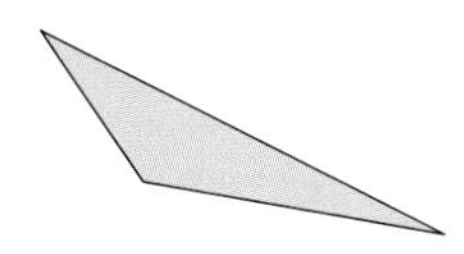

    b) Using reflections and translations.

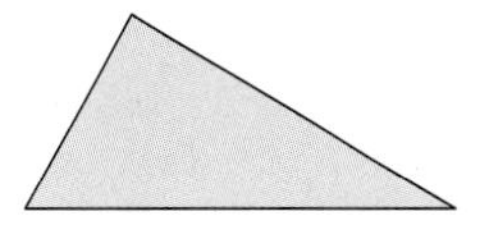

11. Tessellate a region with the following figures:
    a)
    b)

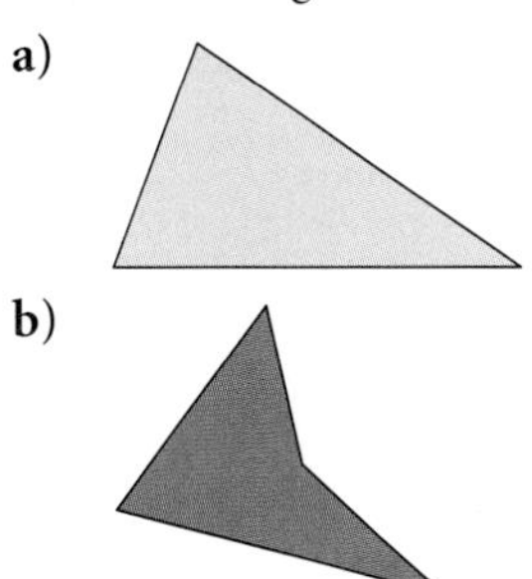

12. Explain why it is possible to tessellate a plane with a square, a regular hexagon, and a regular dodecagon placed at each vertex, but it is impossible to tessellate a plane with an equilateral triangle, a regular pentagon, and a regular heptagon placed at each vertex.
13. Use the following procedure to construct a fractal. Start with a rectangle whose length is twice its width. Divide each line segment into four sections of equal length. Replace the center two sections with three sides of a rectangle, as shown in the diagram. The length of the longer of the three added segments should equal one-half the length of the original segment. The length of the other two added segments should be one-fourth the length of the original segment. The first iteration is shown. Repeat this process and draw the next iteration of the fractal.

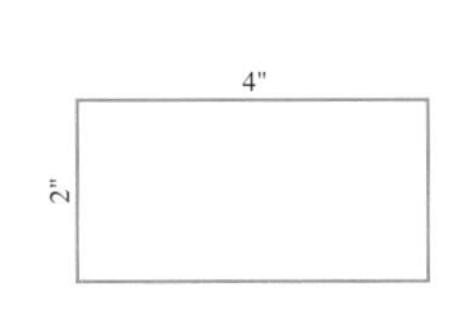

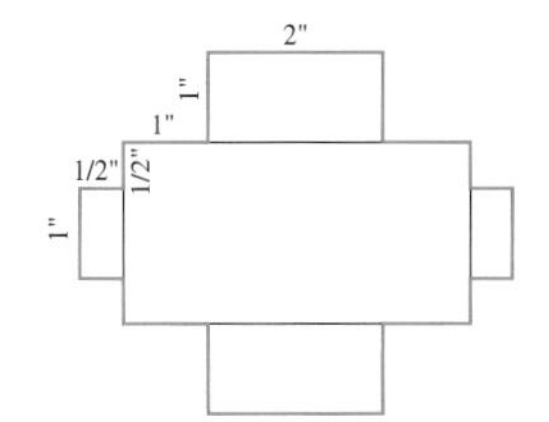

14. Determine the dimension of the fractal discussed in Problem 13. Remember that if the length of the original line segment is 4 in., the length of the next iteration of that segment is formed by six sections, for a total of 6 in.

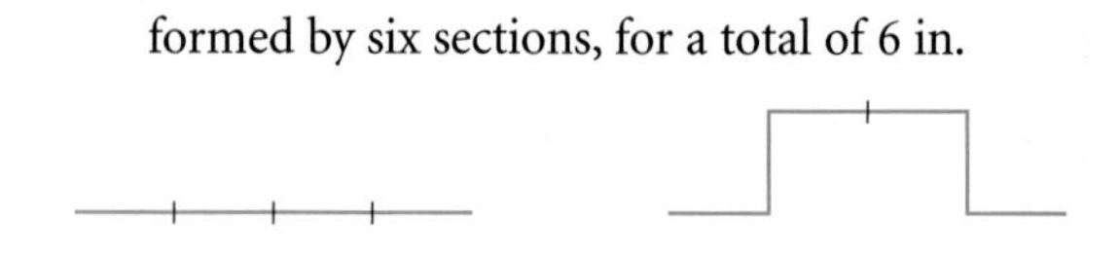

CHAPTER EIGHT

Trigonometry: A Door to the Unmeasurable

Trigonometry is used in many aspects of sailing: establishing location by the stars, plotting courses, and positioning sails for maximum speed. Even the size of the triangular sails, which can be seen in Raoul Dufy's *Regatta at Cowes,* is a factor in maximizing speed. (Natioal Gallery of Art)

OVERVIEW

In this chapter, you will learn how to find unknown sides and angles of triangles using trigonometry. With the use of trigonometric formulas and a scientific calculator, you will be able to determine distances that you may not be able to measure directly. For example, you will be able to find the height of a building, the width of a canyon, the altitude of a hot air balloon, or the distance a cannonball travels. You will see that mathematics can be used to solve some very practical problems. You will see that a knowledge of trigonometry opens a door to determining unmeasurable distances.

8

A SHORT HISTORY OF TRIGONOMETRY

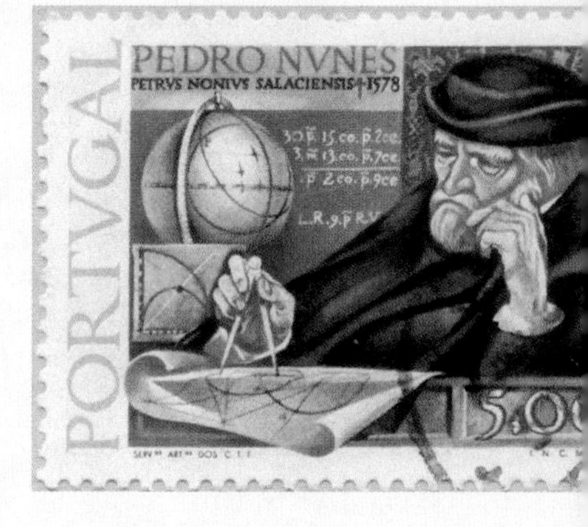

On the left, the 1983 San Marino stamp displays the Greek mathematician Pythagoras along with the geometrical representation of the Pythagorean Theorem. On the right, in the Portuguese-issued stamp commemorating the 500th anniversary of his birth, Pedro Nunes of Portugal shows the instruments and diagrams he used in navigation in the 1500s.

As the understanding of numbers, algebra, and geometry developed, ancient scholars turned their attention to a quantitative study of the sun, the moon, the planets, and the stars. This interest in astronomy led to the formation of what is known today as trigonometry. The word *trigonometry* comes from the Greek words *tri'gonon* (triangle) and *met'ron* (measure). It began as the study of the relationship between arcs and chords of a circle, as in a bow and a bowstring, progressed to the study of triangles on the surface of a sphere, and is now commonly known as the study of the relationship between the angles and sides of a triangle.

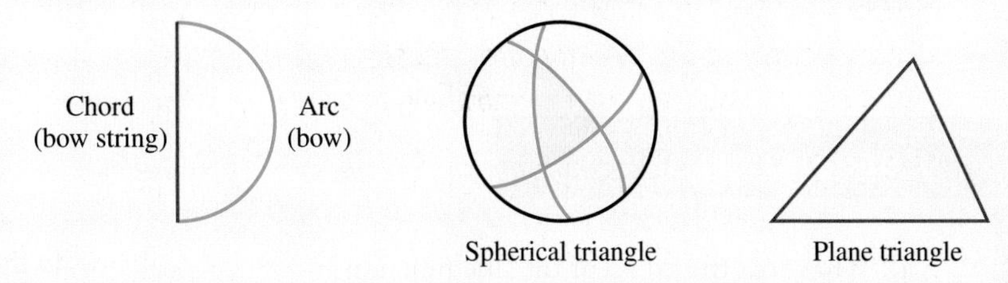

There is evidence that astronomy (requiring trigonometry) was studied by the ancient Babylonians and Chinese. Ancient Egyptians investigated the ratios of sides of a triangle as recorded in the Ahmes papyrus (1550 B.C.). However, the first significant contributions to trigonometry were made by the Greeks. Motivated by their interest in astronomy, they systematically studied the relationship between arcs and chords of a circle. The Greek mathematician and astronomer Hipparchus of Nicea (c. 140 B.C.) wrote the first systematic study of trigonometry. He developed methods of studying spherical triangles, is given credit for dividing the circle into 360 degrees, and is considered the founder of trigonometry.

During the Dark Ages in Europe, the development of trigonometry moved to the East. Hindu astronomers displayed their knowledge of trigonometry in the book *Sûrya Siddhanta* (c. 400) and in the works of Varâhamihira (c. 505) and Aryabhata (c. 510). The Arabs continued the development of trigonometry by building on the work of the Hindus. Contributions during the ninth and the tenth centuries by mathematicians Al-Battânî, Nasîr ed-dîn al-Tûsî, and others are the main reasons this period is called the golden age of Arabian mathematics.

As Europe emerged from the Dark Ages and moved into the Renaissance, European contributions to the development of trigonometry again appeared. Beginning in 1220 with Leonardo Fibonacci's summary of Greek trigonometry in *Practica Geometriae* and continuing through 1600 with Bartholomaus Pitiscus's work, the first text to have the title *Trigonometry,* many European mathematicians contributed to the development of trigonometry. By 1658, the names of the trigonometric functions had taken their present forms, and by 1750, trigonometry had become more than merely a tool for astronomers and surveyors. It was used by mathematicians John Bernoulli, Abraham de Moivre, Leonhard Euler, Isaac Newton, and Gottfried Leibniz in analyzing mathematical concepts, studying complex numbers, modeling periodic phenomena, and studying mathematical physics.

In the 21th century, trigonometry is still important. It is used in the study of electricity, light, sound, radio, television, and other areas in which values occur repeatedly. Trigonometry is also used in finding distances that cannot be measured directly, an important technique that is used by surveyors, astronomers, navigators, engineers, architects, and others. In this chapter, you will learn more about trigonometry and see how it is used to solve problems in some of these areas.

CHAPTER 8 PROJECTS

Research Projects

1. What does the graph of the sine function $y = \sin x$ look like in a rectangular coordinate system? What are the amplitude and period of that sine curve? Examine some phenomena that exhibit characteristics of sine curves, such as sound and electrical waves, biorhythm charts, and seismograph readings.
2. This chapter deals with the study of triangles on a plane surface. This is not the only trigonometry that has been investigated. The study of triangles on a sphere is a part of spherical trigonometry. What are some properties of spherical triangles? What are some of the formulas used in spherical trigonometry? How do these formulas differ from those of plane trigonometry? Why is spherical trigonometry of considerable importance in navigation and astronomy?
3. The study of the motion of planets and stars led to the development of trigonometry. However, the theories of the astronomers themselves are quite interesting. Compare and contrast the astronomy of Ptolemy (A.D. 85–165), Nicolaus Copernicus (1473–1543), Tycho Brahe (1546–1601), and Johannes Kepler (1571–1630). Make diagrams of the movements of the earth, sun, moon, and planets according to each astronomer.

Math Projects

1. Use the trigonometric methods described in this chapter to find the height of a building or mountain in your area. Explain in detail how you did this and compare your results with the actual height of the building or mountain as published in the building specifications or on a map. If there is a difference in the results, explain what might have caused the difference.
2. Section 8.6 presents equations to determine the distance a shot put travels based on the height, angle, and velocity at which it is tossed. Suppose the height and velocity were the same (e.g., $h = 7$ ft and $v = 44$ ft/sec) but you changed the launch angle. The object of this project is to determine what angle would generate the longest toss. Some coaching manuals advise a launch angle of 45°. Is that really the best? Analyze the distance generated by angles below and above 45°. Graph the flight obtained from each angle and determine which angle produces the longest distance traveled by a shot put.
3. To find the square root of a number, we simply press some keys on a calculator and get the answer. How could you find a square root without a calculator? The objective of this project is to actually find some square roots by researching and using the following non-calculator techniques.
   - **a)** A geometric technique in which lengths of the hypotenuse of right triangles are the square roots of whole numbers (a root spiral).
   - **b)** A paper-and-pencil method similar to the long division technique.

section 8.1

Right Triangles, Sine, Cosine, Tangent

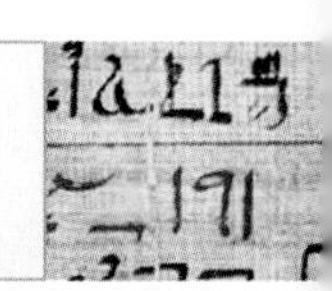

Before beginning our study of trigonometry, we will review some facts about right triangles. A **right triangle** is a triangle with a 90° angle. The two other angles are acute angles (between 0° and 90°). The sides of the right triangle that form the 90° angle are the **legs** of the right triangle, while the other side, the one opposite the 90° angle, is the **hypotenuse** of the right triangle. The hypotenuse is always the longest side of a right triangle. In $\triangle ABC$,

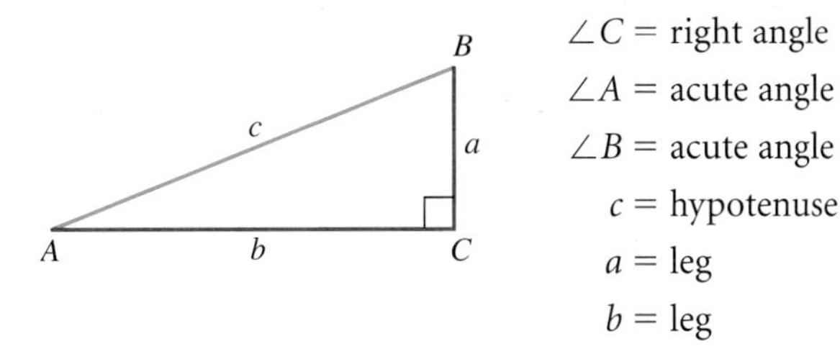

$\angle C =$ right angle
$\angle A =$ acute angle
$\angle B =$ acute angle
$c =$ hypotenuse
$a =$ leg
$b =$ leg

Notice the following:

1. Vertices are labeled with uppercase letters, and sides are labeled with lowercase letters.
2. A side of the triangle has the same letter as the angle opposite (across) from it. Side a is opposite $\angle A$, side b is opposite $\angle B$, and side c is opposite $\angle C$.
3. A leg of a right triangle can be referred to by its orientation with respect to an acute angle of the triangle. The leg across from an acute angle is called the *opposite side*, and the leg that is a part of the angle is called the *adjacent side*. With reference to $\angle A$, a is the opposite side and b is the adjacent side. With reference to $\angle B$, b is the opposite side and a is the adjacent side. In either case, c is the hypotenuse.

THE PYTHAGOREAN THEOREM

The Greek mathematician Pythagoras (c. 582–501 B.C.) is given credit for the basic theorem concerning the sides of a right triangle: the **Pythagorean Theorem.** Geometrically, it states that the sum of the squares on the legs of a right triangle is equal to the square on the hypotenuse. In algebraic terms, if the lengths of the legs of a right triangle are a and b and the length of the hypotenuse is c, then $a^2 + b^2 = c^2$.

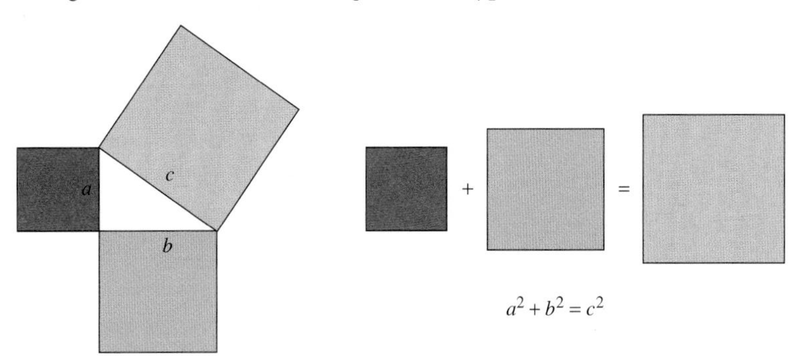

This theorem can be used to find any side of a right triangle when two sides are known.

EXAMPLE 1

If the legs of a right triangle are 5 in. and 12 in., find the length of the hypotenuse.

Solution: In the Pythagorean Theorem, let $a = 5$, $b = 12$, and solve for c.

$$\begin{aligned} a^2 + b^2 &= c^2 \\ 5^2 + 12^2 &= c^2 \\ 25 + 144 &= c^2 \\ 169 &= c^2 \\ \pm\sqrt{169} &= c \\ 13 &= c \end{aligned}$$

The hypotenuse is 13 in. (*Note:* The positive square root is used since a distance cannot be negative.)

EXAMPLE 2

In the right triangle below, $a = 6.5$ cm, $r = 3.3$ cm, find the length of d.

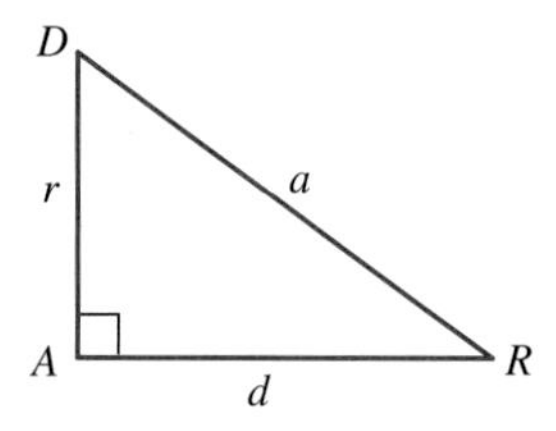

Solution: Since a is the hypotenuse (opposite the 90° angle), r and d are legs. Applying the Pythagorean Theorem, we get the following.

$$\begin{aligned} r^2 + d^2 &= a^2 \\ 3.3^2 + d^2 &= 6.5^2 \\ 10.89 + d^2 &= 42.25 \\ d^2 &= 31.36 \\ d &= \sqrt{31.36} = 5.6 \text{ cm} \end{aligned}$$

THE TRIG FUNCTIONS

The right triangle is the basic building block of trigonometry. The fundamental trigonometric (trig) functions, **sine** (sin), **cosine** (cos), and **tangent** (tan) are simply ratios of the lengths of the sides of right triangles. In the definitions of the trig functions that follow, whenever the label of a side of the triangle is used, we are referring to the length of that side.

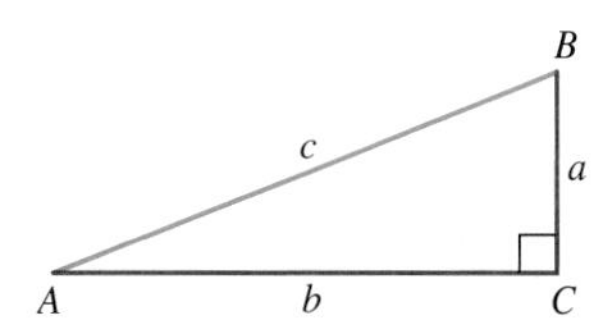

$$\sin A = \frac{\text{side opposite } \angle A}{\text{hypotenuse}} = \frac{a}{c} \qquad \sin B = \frac{\text{side opposite } \angle B}{\text{hypotenuse}} = \frac{b}{c}$$

$$\cos A = \frac{\text{side adjacent to } \angle A}{\text{hypotenuse}} = \frac{b}{c} \qquad \cos B = \frac{\text{side adjacent to } \angle B}{\text{hypotenuse}} = \frac{a}{c}$$

$$\tan A = \frac{\text{side opposite } \angle A}{\text{side adjacent to } \angle A} = \frac{a}{b} \qquad \tan B = \frac{\text{side opposite } \angle B}{\text{side adjacent to } \angle B} = \frac{b}{a}$$

A chant-line acronym formed from the first letters of the definitions can be used to help remember them.

"soh cah toa"

$$\text{sin—}\frac{\textbf{o}\text{pposite}}{\textbf{h}\text{ypotenuse}} \quad \text{cos—}\frac{\textbf{a}\text{djacent}}{\textbf{h}\text{ypotenuse}} \quad \text{tan—}\frac{\textbf{o}\text{pposite}}{\textbf{a}\text{djacent}}$$

The three other trig functions—cosecant, secant, and cotangent—are the reciprocals of the sine, cosine, and tangent. They will not be covered in this text.

EXAMPLE 3

In the right triangles below, find sin A, cos A, tan A, sin B, cos B, and tan B.

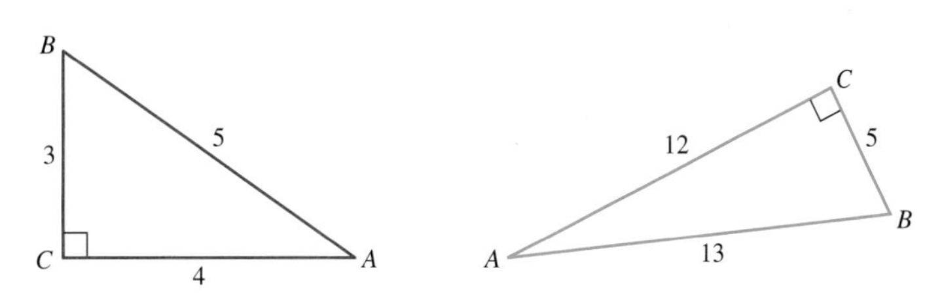

Solution:

First triangle:

$$\sin A = \frac{opp}{hyp} = \frac{3}{5} \qquad \sin B = \frac{opp}{hyp} = \frac{4}{5}$$

$$\cos A = \frac{adj}{hyp} = \frac{4}{5} \qquad \cos B = \frac{adj}{hyp} = \frac{3}{5}$$

$$\tan A = \frac{opp}{adj} = \frac{3}{4} \qquad \tan B = \frac{opp}{adj} = \frac{4}{3}$$

Second triangle:

$$\sin A = \frac{opp}{hyp} = \frac{5}{13} \qquad \sin B = \frac{opp}{hyp} = \frac{12}{13}$$

$$\cos A = \frac{adj}{hyp} = \frac{12}{13} \qquad \cos B = \frac{adj}{hyp} = \frac{5}{13}$$

$$\tan A = \frac{opp}{adj} = \frac{5}{12} \qquad \tan B = \frac{opp}{adj} = \frac{12}{5}$$

USING A CALCULATOR TO FIND TRIGONOMETRIC FUNCTIONS

A scientific calculator can be used to find the ratio of sides for any given angle measure. If the calculator is set in *degree* mode, it will give you the sine ratio of an angle measured in degrees with the SIN key. Check your calculator instruction booklet or ask your instructor to learn how to do this. For example, many TI calculators will give you the sin 30° by pressing the following keys:

The display will give you the answer, 0.5. Thus, $\sin 30° = 0.5$. This means that in *any* right triangle with a 30° angle, the ratio of the side opposite the 30° angle to the hypotenuse of the right triangle is 0.5 or 1/2. The size of the triangle does not matter. All right triangles with the same acute angles are similar triangles with proportional sides. The trig functions have the same values for a given angle no matter what the size of the triangle.

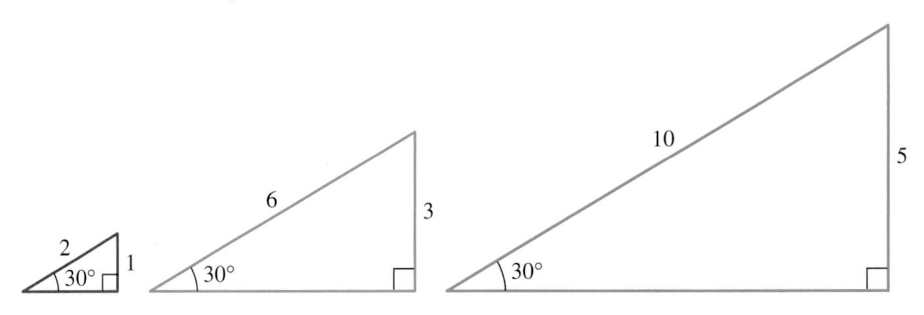

EXAMPLE 4

Use a calculator to find the following, rounded off to four decimal places.

a) sin 42° **b)** cos 67° **c)** tan 54.6° **d)** sin 5.79° **e)** tan 3.14°

Solution:

a) 0.6691 **b)** 0.3907 **c)** 1.4071 **d)** 0.1009 **e)** 0.0549

You must keep in mind that when you press the SIN, COS, or TAN key for a given angle, the calculator is giving you a ratio of sides of a right triangle in decimal form.

The calculator can also be used to determine an angle of a right triangle when the ratio of two sides is known. For example, if the ratio of the length of the opposite side to the hypotenuse is 6/10, $\sin A = 0.6$. If the $\sin A = 0.6$, a calculator will give you the measure of $\angle A$ by using the inverse sin key. Check your calculator instruction booklet or ask your instructor to learn how to do this. For example, many TI calculators will do this by pressing the following keys:

The display will give you an answer, $36.8698976458 \approx 36.9°$. This means that if an angle of a right triangle has the ratio of its opposite side to its hypotenuse equaling 6/10, then the angle is about 36.9°.

EXAMPLE 5

Find an angle that satisfies each of the following trig functions. Round off answers to the nearest tenth of a degree.

a) $\cos B = 0.2588$ **b)** $\tan A = 3.29$ **c)** $\sin X = 2/9$

Solution:

a) $B \approx 75.0°$ **b)** $A \approx 73.1°$ **c)** $X \approx 12.8°$

The exercises that follow will test your understanding of the Pythagorean Theorem and the basic definitions of the trig functions: sine, cosine, tangent.

section 8.1

PROBLEMS ○ Explain ○ Apply ○ Explore

○ Explain

1. What is a right triangle?
2. What are the legs of a right triangle? What is the hypotenuse of a right triangle?
3. What is the Pythagorean Theorem?
4. Can you use the Pythagorean Theorem to find side c in the triangle below? Why or why not?

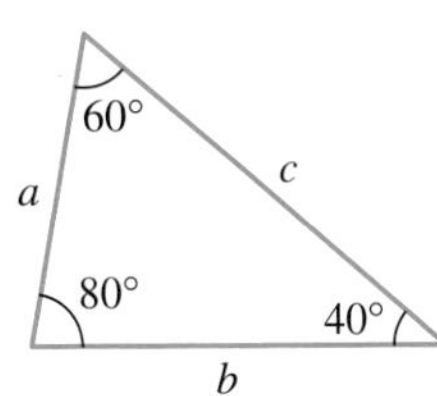

5. Give the right triangle definitions of the sine, cosine, and tangent functions.
6. In a right triangle, if $\sin R = \frac{p}{t}$, what do R, p, and t represent?
7. In a right triangle, if $\cos M = \frac{k}{f}$, what do M, k, and f represent?
8. In a right triangle, if $\tan H = \frac{r}{x}$, what do H, r, and x represent?
9. What is soh cah toa?

○ Apply

In the right triangles below, determine which sides are legs and which side is the hypotenuse.

10.

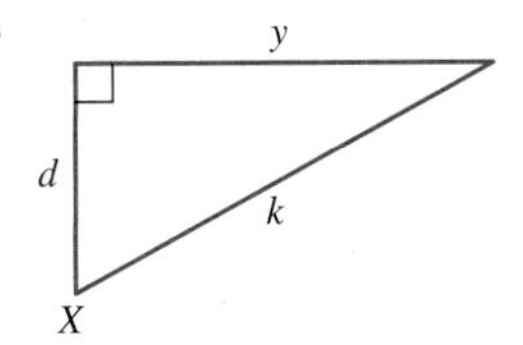

11.

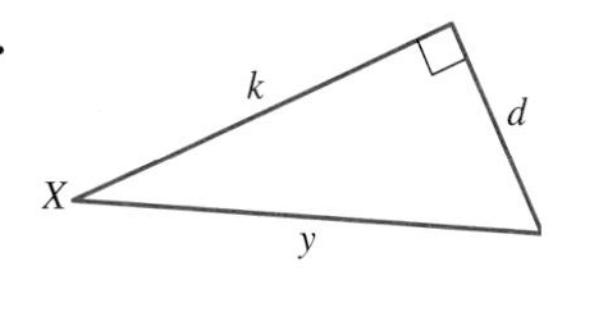

12.

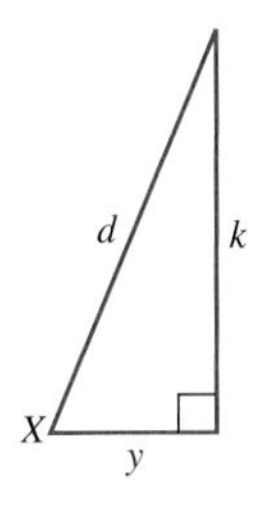

13. For each of the right triangles given in Problems 10–12, find the sides that are opposite from and adjacent to $\angle X$.

In Problems 14–18, find the length of the side that is not given.

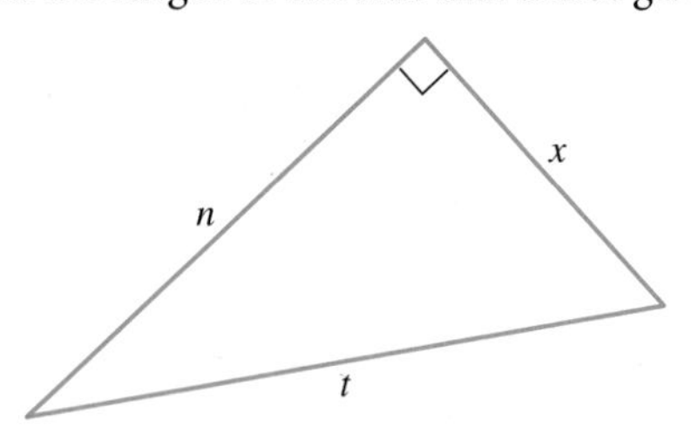

14. $x = 7, n = 24$

15. $x = 4.3, n = 9.7$

16. $n = 9, t = 41$

17. $n = 3.24, t = 67.24$

18. $x = 54.1, t = 76$

In Problems 19–21, the triangles have right angles at *C*. Find sin *A*, cos *A*, tan *A*, sin *B*, cos *B*, and tan *B*.

19.

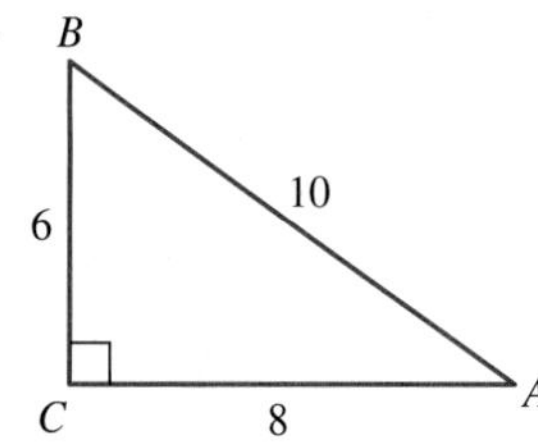

20.

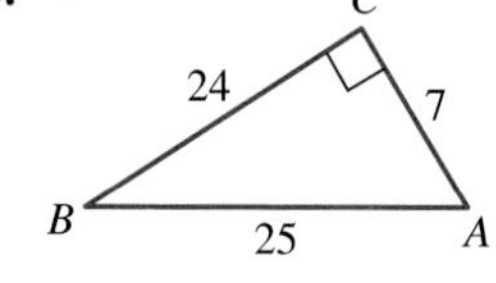

21.

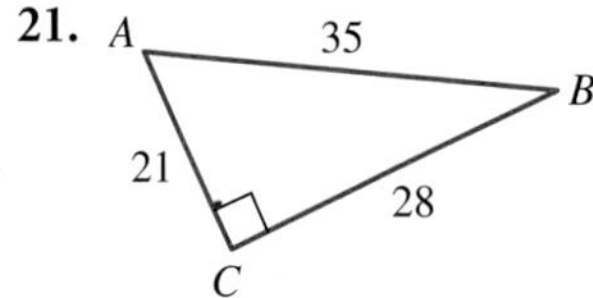

In Problems 22–27, use a calculator to find the value of each trigonometric function. Round-off answers to four decimal places.

22. $\sin 59°$

23. $\sin 12.6°$

24. $\cos 82°$

25. $\cos 5.9°$

26. $\tan 33°$

27. $\tan 76.3°$

In Problems 28–33, use a calculator to find the angle for each trigonometric function. Round-off answers to the nearest tenth.

28. $\sin A = 0.8974$

29. $\cos Y = 0.123$

30. $\tan B = 6.548$

31. $\sin F = 5/12$

32. $\cos K = 4/11$

33. $\tan J = 24/7$

Explore

A **Pythagorean triple** is a set of three natural numbers that satisfies the Pythagorean Theorem. In Problems 34–39, verify that each set of numbers is a Pythagorean triple.

34. 3, 4, 5

35. 36, 48, 60

36. 20, 21, 29

37. 33, 56, 65

38. 12, 35, 37

39. 5, 12, 13

40. Is 17, 27, 47 a Pythagorean triple? Why or why not?

In Problems 41–49, give answers rounded to the nearest tenth.

41. Find the acute angles of the right triangle with sides measuring: 3, 4, 5.

42. Find the acute angles of the right triangle with sides measuring: 36, 48, 60.

43. Find the acute angles of the right triangle with sides measuring: 20, 21, 29.

44. Find the acute angles of the right triangle with sides measuring: 33, 56, 65.

45. Find the acute angles of the right triangle with sides measuring: 12, 35, 37.

46. Find the acute angles of the right triangle with sides measuring: 5, 12, 13.

In Problems 47–49, use the box shown below.

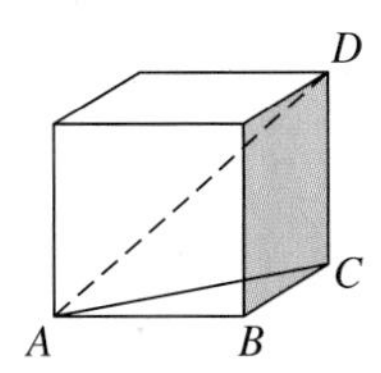

47. If $AB = 8$, $BC = 6$, and $CD = 7$, find the length of $\overline{AC}$ and $\overline{AD}$ and the measure of $\angle DAC$.

48. If $AB = 7.5$, $BC = 7.5$, and $CD = 7.5$, find the length of $\overline{AC}$ and $\overline{AD}$ and the measure of $\angle DAC$.

49. If $AB = 3.9$, $BC = 6.5$, and $CD = 5.2$, find the length of $\overline{AC}$ and $\overline{AD}$ and the measure of $\angle DAC$.

section 8.2

Solving Right Triangles

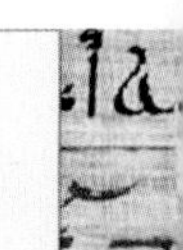

The Lighthouse at Honfleur by Georges Seurat. For the navigator at sea, a lighthouse serves as a permanent reference point for charting a course. (National Gallery of Art)

The trigonometric functions can be used to help **solve triangles,** that is, to find unknown sides and angles of the triangles. A triangle is solved when the lengths of all three sides and the measures of all three angles are known. Typically, you know three parts of the triangle (sides or angles) and must find the values of the other three parts. This can be done by using the three trig functions (sin, cos, tan), the Pythagorean Theorem, the fact that the sum of the angles of a triangle equals 180°, some algebra, and a scientific calculator. To establish some uniformity in the answers, we will use the following round-off rules in the rest of this chapter.

Round-Off Rules

The final answers to angles and sides of triangles that have more than one decimal digit will be rounded off to the nearest tenth. The final answers will be computed by using intermediate results that have been rounded off to four decimal places (to the right of the decimal point).

EXAMPLE 1

In the right triangle below, find $\angle B$, a, and b.

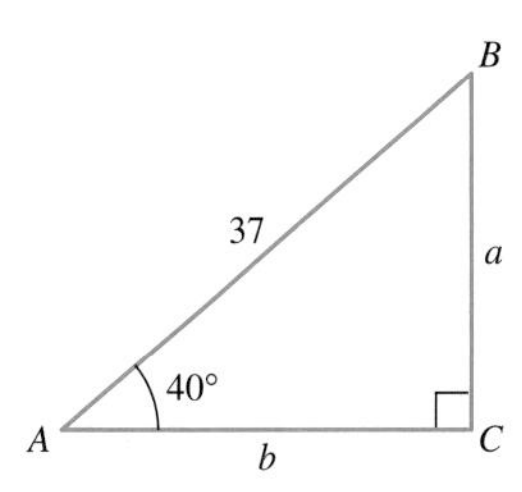

Solution: Since the sum of the angles of a triangle is 180°, we get the following:

$$\begin{aligned} A + B + C &= 180^\circ \\ 40^\circ + B + 90^\circ &= 180^\circ \\ 130^\circ + B &= 180^\circ \\ B &= 50^\circ \end{aligned}$$

Since we know the hypotenuse and side a is the side opposite the 40° angle, we use sin 40° to find a.

$$\begin{aligned} \sin 40^\circ &= \frac{a}{37} \\ 0.6428 &= \frac{a}{37} \\ 37(0.6428) &= a \\ 23.8 &= a \end{aligned}$$

Since we know the hypotenuse and side b is the side adjacent to the 40° angle, we use cos 40° to find b.

$$\begin{aligned} \cos 40^\circ &= \frac{b}{37} \\ 0.7660 &= \frac{b}{37} \\ 37(0.7660) &= b \\ 28.3 &= b \end{aligned}$$

Note: In the above example, we would not use the tangent to find a and b, since the ratio would have two unknowns, $\tan 40^\circ = \frac{a}{b}$, and we cannot solve for two unknowns in one equation.

EXAMPLE 2

In the right triangle below, find $\angle X$, $\angle Y$, and z.

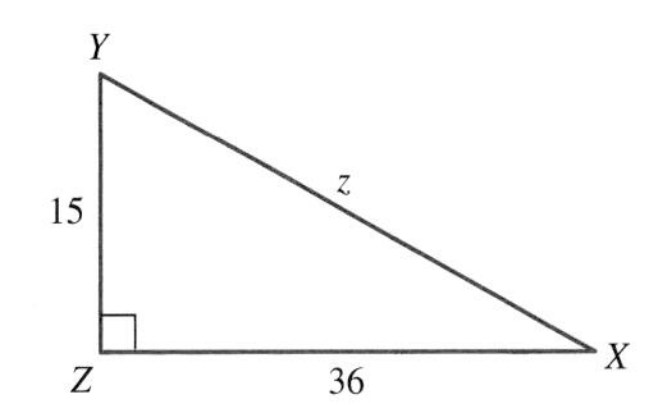

Solution: To find $\angle X$, we can use the tangent function since we know the sides opposite and adjacent to $\angle X$. That would leave only one unknown in the equation, X.

$$\tan X = \frac{15}{36}$$

$$\tan X = 0.4167$$

$$X = 22.6°$$

Since we know that the sum of the measures of the angles of a triangle is 180°, we can find the measure of $\angle Y$.

$$X + Y + Z = 180°$$

$$22.6° + Y + 90° = 180°$$

$$Y = 67.4°$$

Finally, using the Pythagorean Theorem, we can determine the value of z.

$$x^2 + y^2 = z^2$$

$$15^2 + 36^2 = z^2$$

$$225 + 1296 = z^2$$

$$1521 = z^2$$

$$39 = z$$

EXAMPLE 3

If an acute angle of a right triangle is 78° and the side adjacent to that angle is 48 cm, how long is the hypotenuse of the right triangle?

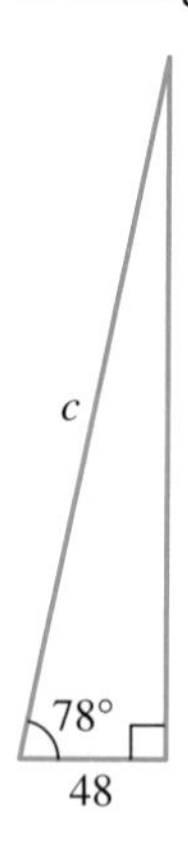

Solution: Let $c =$ the length of the hypotenuse.

$$\cos 78° = \frac{48}{c}$$

$$0.2079 = \frac{48}{c}$$

$$0.2079c = 48$$

$$c = 230.9 \text{ cm}$$

EXAMPLE 4

In $\triangle FUN$, $\angle F$ is the right angle, $u = 26.3$, and $N = 19.2°$. Find the parts of the triangle that are not given.

Solution: First draw $\triangle FUN$, placing the given facts on the triangle and labeling unknown sides. We must find n, f, and $\angle U$.

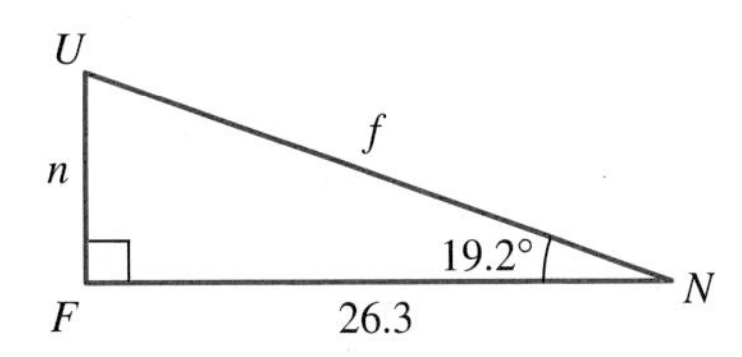

$\angle U$ can be found by using the fact that the sum of the angles of a triangle is 180°.

$$F + U + N = 180°$$
$$90° + U + 19.2° = 180°$$
$$109.2° + U = 180°$$
$$U = 70.8°$$

To find n, we can use the tangent function.

$$\tan 19.2° = \frac{n}{26.3}$$
$$0.3482 = \frac{n}{26.3}$$
$$0.3482(26.3) = n$$
$$9.2 = n$$

To find f, we can use the cosine function.

$$\cos 19.2° = \frac{26.3}{f}$$
$$0.9444 = \frac{26.3}{f}$$
$$0.9444f = 26.3$$
$$f = 27.8$$

EXAMPLE 5

In the figure below, find the length of side h.

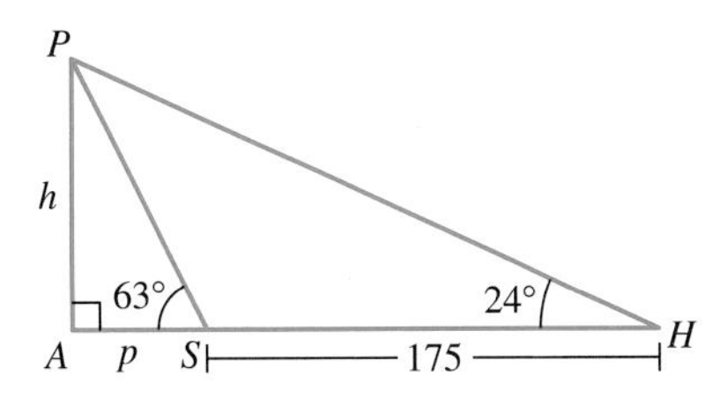

Solution: Since there is no right triangle with only one unknown side, we start by separating the figure into two right triangles: $\triangle SAP$ and $\triangle HAP$.

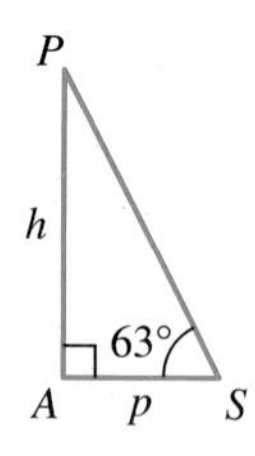

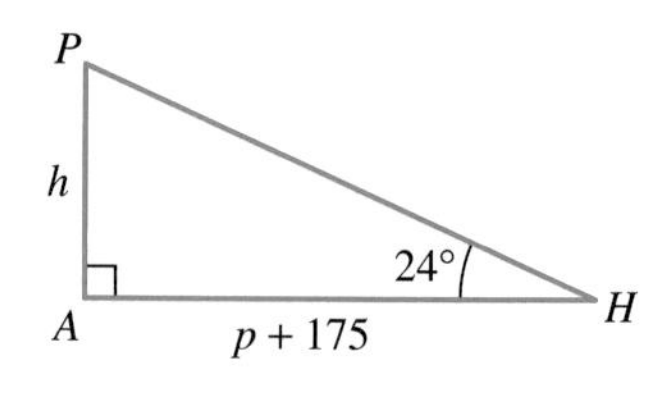

Using right $\triangle SAP$:

$$\tan 63° = \frac{h}{p}$$
$$1.9626 = \frac{h}{p}$$
$$1.9626p = h$$
$$p = \frac{h}{1.9626}$$

Using right $\triangle HAP$:

$$\tan 24° = \frac{h}{p + 175}$$
$$0.4452 = \frac{h}{p + 175}$$
$$0.4452(p + 175) = h$$
$$0.4452p + 77.91 = h$$
$$0.4452p = h - 77.91$$
$$p = \frac{h - 77.91}{0.4452}$$

Since we have two expressions for p, we can set those equal to each other and solve for h.

$$\frac{h}{1.9626} = \frac{h - 77.91}{0.4452}$$
$$0.4452h = 1.9626(h - 77.91)$$
$$0.4452h = 1.9626h - 152.9062$$
$$-1.5174h = -152.9062$$
$$h = 100.8$$

This gives 100.8 ft for the length of side h.

In the problems that follow, you will solve right triangles by using the techniques shown in this section. Hopefully, things will go smoothly. However, depending on how you round off intermediate results, your answers might differ from those in the Answer Section. If you find that this is the case, round off intermediate results to four decimal places as stated in the round-off rules at the beginning of this section.

section 8.2

PROBLEMS ▶ Explain ▶ Apply ▶ Explore

▶ Explain

1. What does "solving a triangle" indicate you should do?
2. What is the round-off rule for calculating angles of a triangle?
3. What is the round-off rule for calculating sides of a triangle?
4. What is the round-off rule for intermediate results in solving a triangle?

5. If the two legs of a right triangle are known, how do you find the acute angles of the triangle?
6. If one leg and the hypotenuse of a right triangle are known, how do you find the acute angles of the triangle?
7. If all three sides of a right triangle are known, how do you find the acute angles of the triangle?

Apply

Solve the triangles in Problems 8–15.

8. 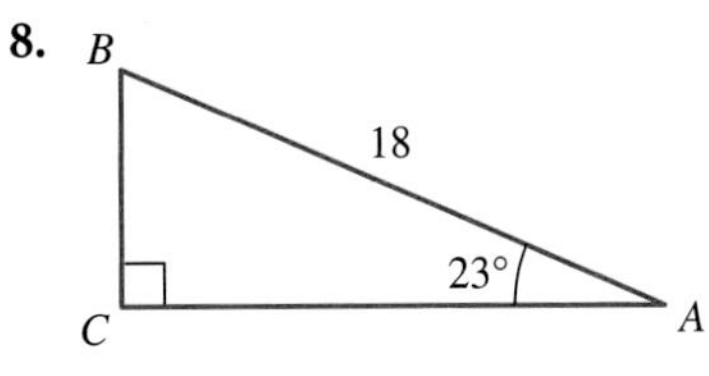

9. 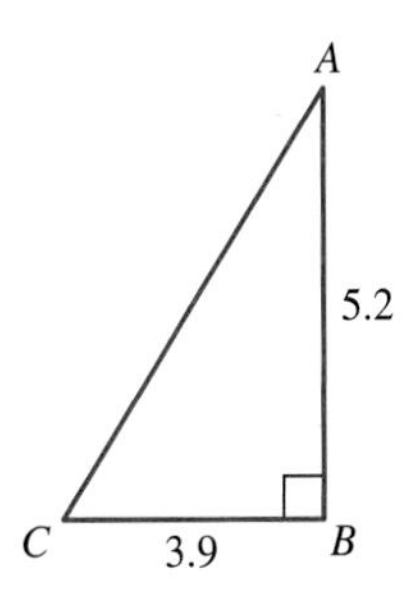

10. 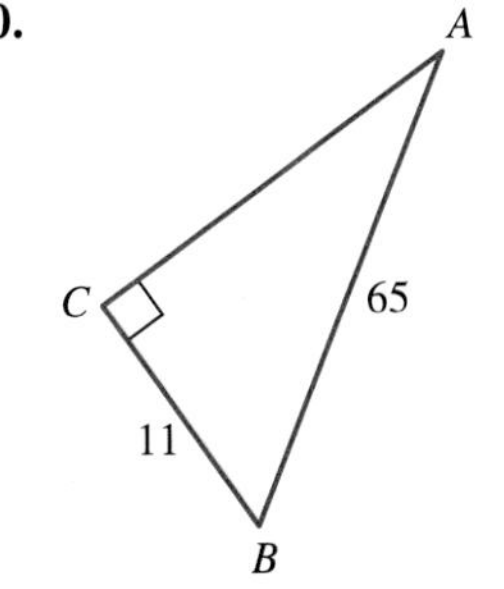

11. 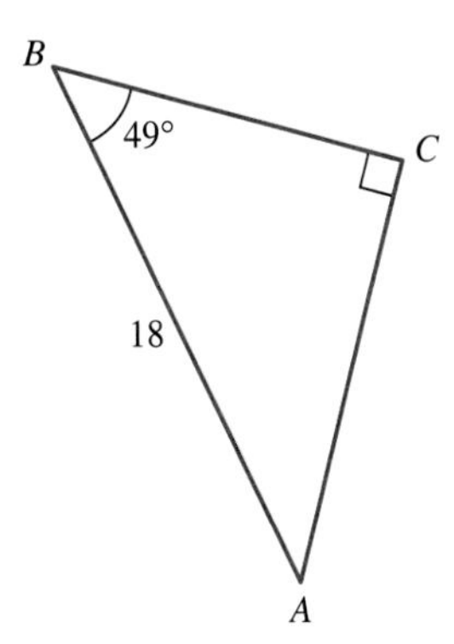

12. 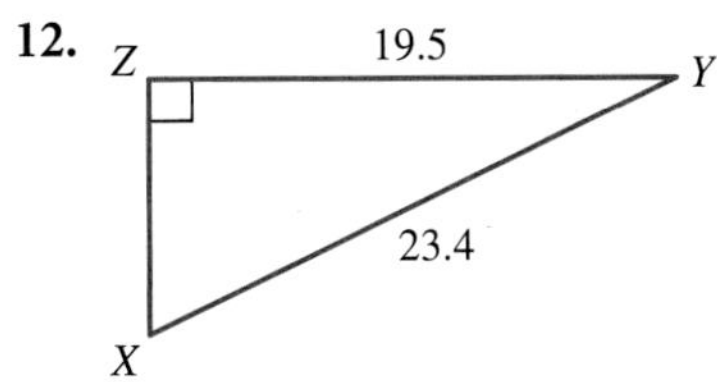

13. 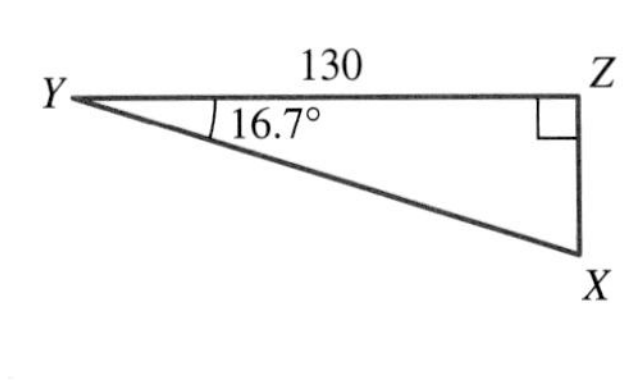

14. 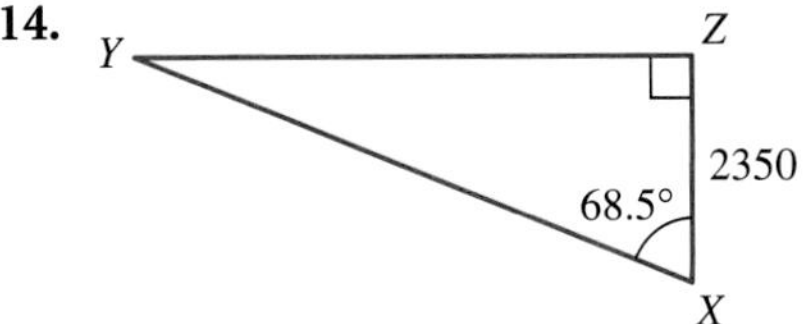

15. 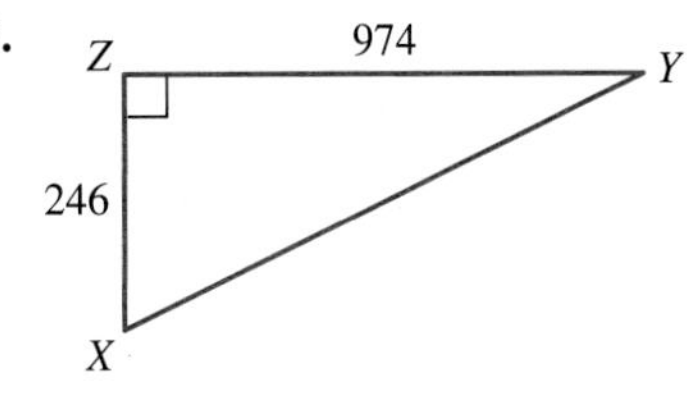

Explore

In Problems 16–21, find the sides and angles that are not given in $\triangle TIP$, where $I = 90°$.

16. $t = 9, p = 40$

17. $t = 102, i = 2602$

18. $P = 72°, p = 56.5$

19. $P = 15.4°, p = 208$

20. $T = 17.5°, p = 112$

21. $T = 23°, p = 44.6$

In Problems 22–25, determine the value of h.

22.

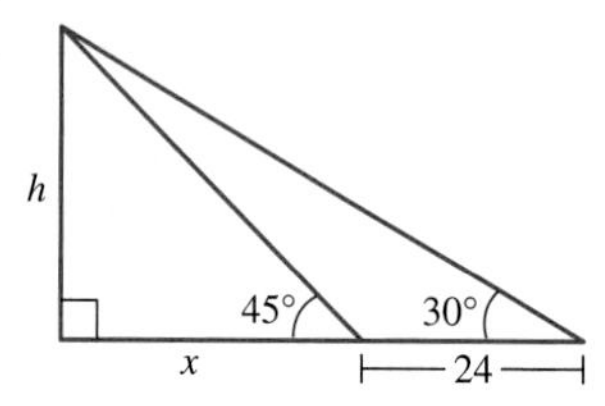

23.

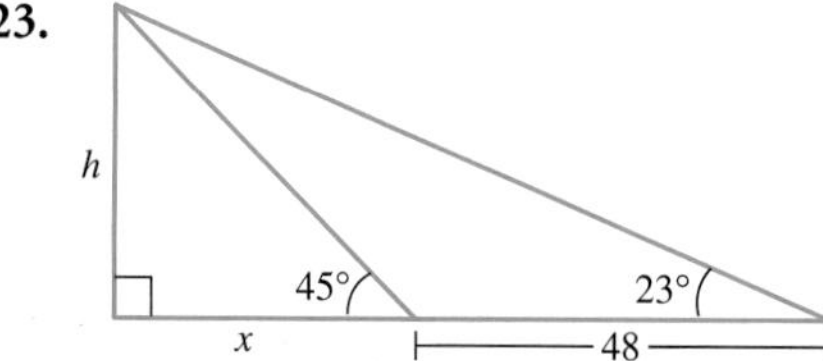

24.

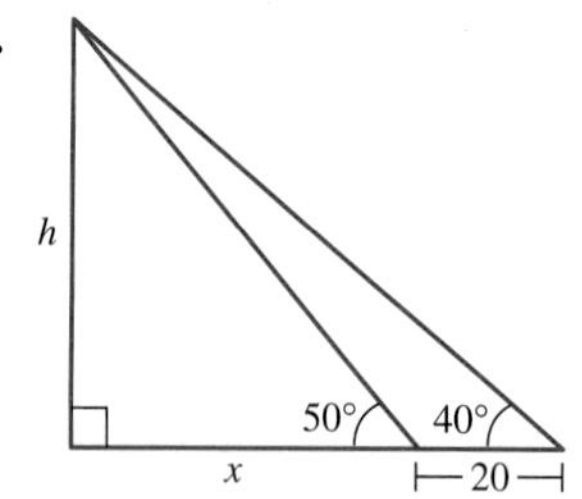

25.

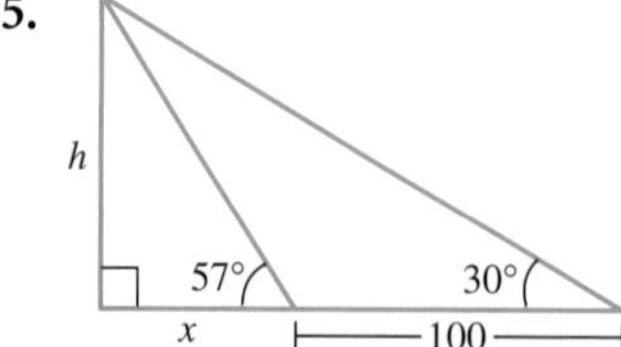

In Problems 26 and 27, determine the value of x.

26.

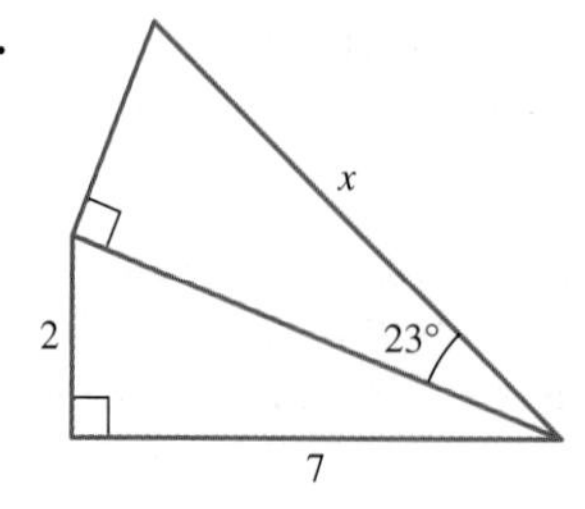

27.

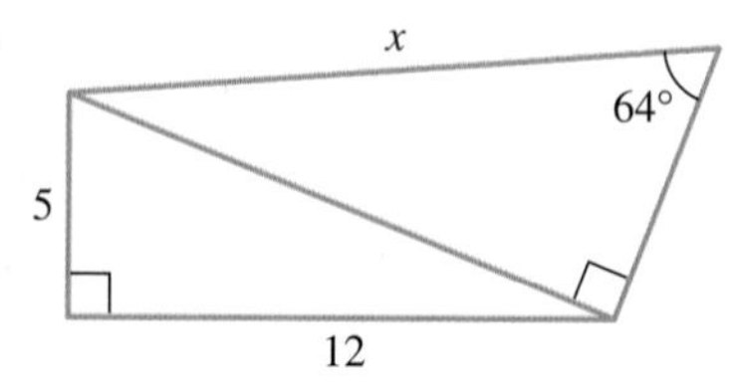

section 8.3

Right Triangle Applications

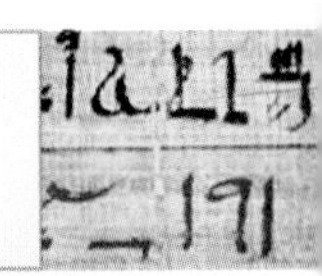

Eight Bells by Winslow Homer shows one of the devices used by navigators in determining angles. (National Gallery of Art)

The trigonometric functions studied in the previous section can be used to solve many different kinds of practical problems. In solving these problems, you will find it helpful to do the following:

1. Sketch the situation and the right triangle involved.
2. Write the known angles or sides on the triangle.
3. Use letters to represent the unknown angles or sides.
4. Use trig functions to help solve the triangle, remembering to have only one unknown quantity in a trigonometric equation.

EXAMPLE 1

The instruction booklet for a 50-ft fire ladder states that, for safety reasons, when the ladder is leaning against a vertical surface, the angle the ladder makes with the horizontal ground must be from 60° to 75°. Under those constraints, what are the minimum and maximum distances that the ladder will reach on the wall?

Solution: The minimum angle and maximum angle are shown in the right triangles. If we let $h =$ the height the ladder reaches up the wall, using the sine function, we can solve for h in each case.

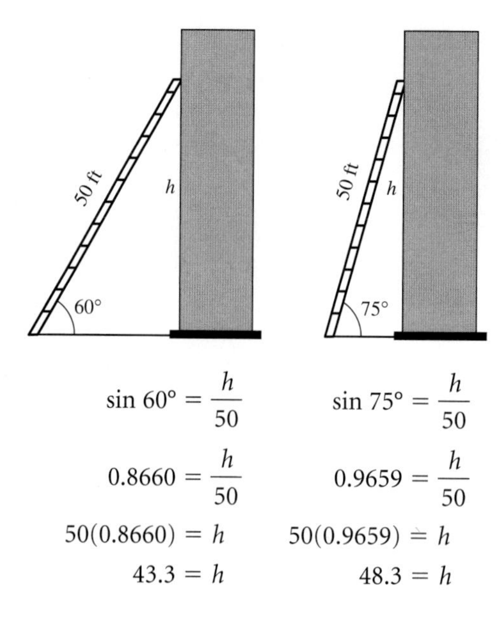

$$\sin 60° = \frac{h}{50} \qquad \sin 75° = \frac{h}{50}$$

$$0.8660 = \frac{h}{50} \qquad 0.9659 = \frac{h}{50}$$

$$50(0.8660) = h \qquad 50(0.9659) = h$$

$$43.3 = h \qquad 48.3 = h$$

Thus, the ladder will safely reach heights between 43.3 and 48.3 ft on the wall.

EXAMPLE 2

A small airplane takes off from an airport at an angle of 32.3° with level ground. Three-fourths of a mile (3960 ft) from the airport is a 1500-ft peak in the flight path of the plane. If the plane continues that angle of ascent, find (a) its altitude when it is above the peak and (b) how far it will be above the peak.

Solution:

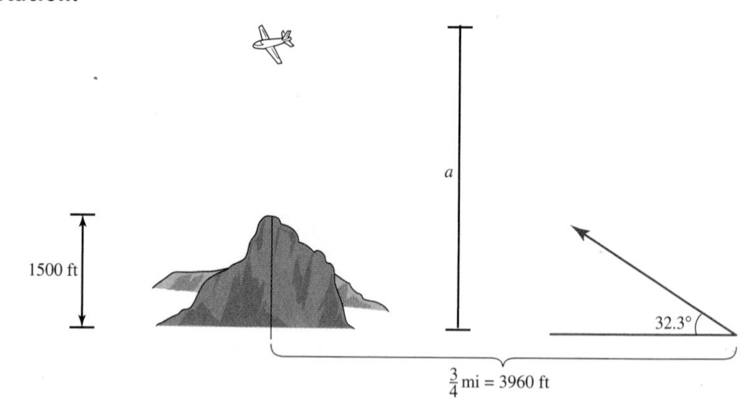

Let a = the altitude of the plane.

$$\tan 32.3° = \frac{a}{3960}$$

$$0.6322 = \frac{a}{3960}$$

$$3960(0.6322) = a$$

$$2503.5 = a$$

Thus, (a) the altitude of the plane is 2503.5 ft, and (b) it is 2503.5 ft − 1500 ft = 1003.5 ft above the peak.

EXAMPLE 3

The swimming area at Shadow Cliffs Lake is roped off with floating markers as shown in the figure. If a section of the beach is 350 yd long and the ropes make a 90° angle at the platform, how long is the swim from one corner on the beach along the floating markers to the middle of the stationary platform and then to the other corner on the beach?

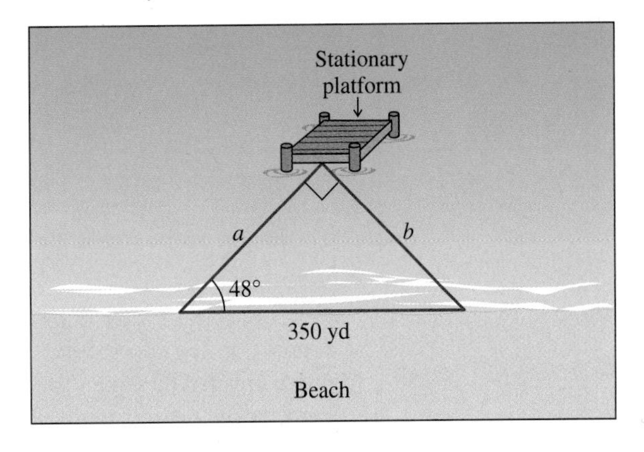

Solution: Let a and b = the legs of the right triangle. We need to find $a + b$.

$$\cos 48° = \frac{a}{350} \qquad \sin 48° = \frac{b}{350}$$

$$0.6691 = \frac{a}{350} \qquad 0.7431 = \frac{b}{350}$$

$$350(0.6691) = a \qquad 350(0.7431) = b$$

$$234.2 = a \qquad 260.1 = b$$

Thus, the distance is $234.2 + 260.1 = 494.3$ yd.

ANGLES OF ELEVATION AND DEPRESSION

Many problems with right triangles involve the angle made with an imaginary horizontal line. An angle between such a horizontal line and the line of sight to an object that is above the horizontal is called the **angle of elevation,** and the angle made between such a horizontal line and the line of sight to an object that is below the horizontal is called the **angle of depression.** Instruments such as transits and sextants can be used to measure such angles.

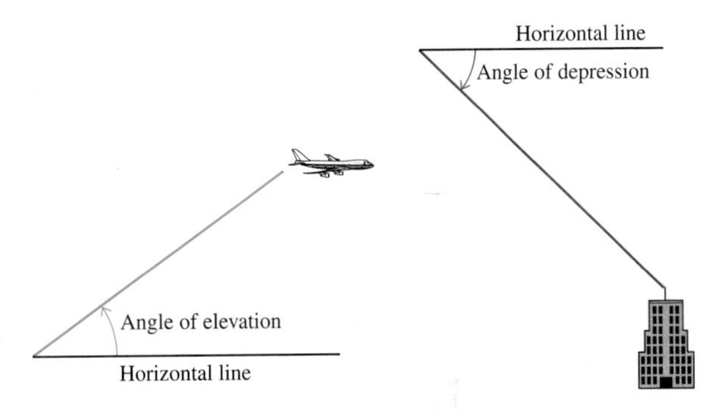

EXAMPLE 4

On level ground, at a point 75 ft from the base of a tree, the angle of elevation to the top of the tree is 70°. Find the height of the tree.

Solution: Let h = the height of the tree in feet.

$$\tan 70° = \frac{h}{75}$$

$$2.7475 = \frac{h}{75}$$

$$75(2.7475) = h$$

$$206.1 = h$$

The height of the tree is 206.1 ft.

EXAMPLE 5

A point on the edge of the Grand Canyon in Arizona is 4600 ft above the Colorado River. The angle of depression to the middle of the canyon floor is 16°. Find, to the nearest tenth of a mile, the horizontal distance to a point directly above the middle of the canyon floor.

Solution: Let x = the horizontal distance to the middle of the Grand Canyon floor from one side.

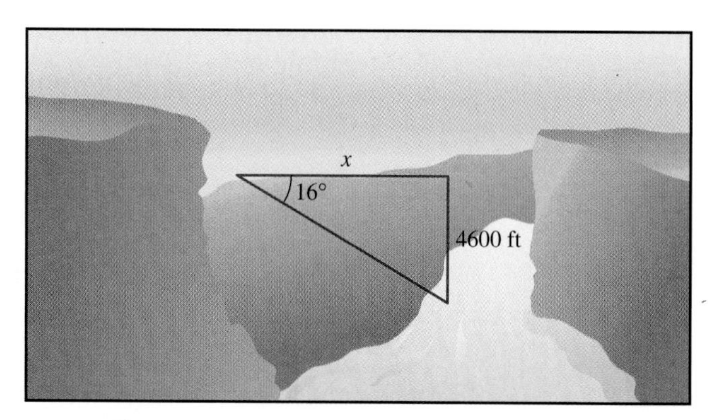

$$\tan 16° = \frac{4600}{x}$$

$$0.2867 = \frac{4600}{x}$$

$$0.2867x = 4600$$

$$x = \frac{4600}{0.2867}$$

$$x = 16{,}044.6 \text{ ft} \approx 3.0 \text{ mi}$$

EXAMPLE 6

The sign on the side of a straight uphill stretch of highway reads, "9% GRADE NEXT 5 MI."

a) Find the angle of elevation of the highway.

b) Determine the change in altitude in feet after driving 5 mi.

Solution: A 9% grade indicates that the slope of the road is 9/100, which means the highway rises 9 ft vertically for every 100-ft change in the horizontal direction.

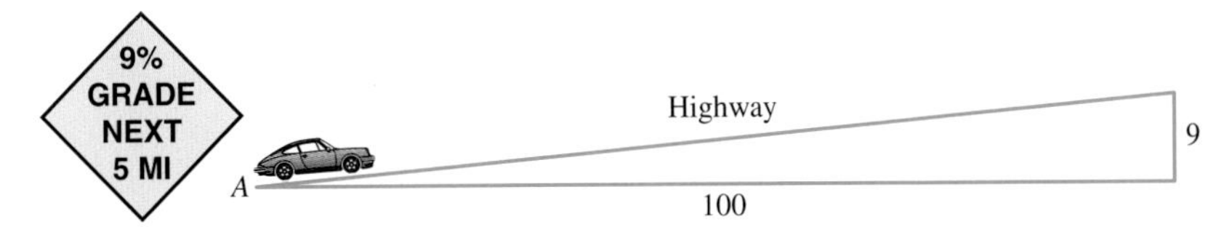

a) To find the angle of elevation, find $\angle A$.

$$\tan A = \frac{9}{100} = 0.09$$

$$A = 5.1°$$

b) Using the result from part (a), we get the following.

$$5 \text{ mi} = 5(5280) = 26{,}400 \text{ ft}$$

A triangle with vertex A, angle 5.1°, hypotenuse 26,400 ft, and vertical side a.

$$\sin 5.1° = \frac{a}{26{,}400}$$

$$0.0889 = \frac{a}{26{,}400}$$

$$0.0889(26{,}400) = a$$

$$a = 2347.0$$

Thus, the angle of elevation of the highway is 5.1°, and after 5 mi of driving, there is a change in altitude of 2347.0 ft.

EXAMPLE 7

To find the height of Mission Peak near the Ohlone College campus, a student went to the football field with a transit and found that the angle of elevation to the top of the peak was 9.8°. At 100 yd (300 ft) from that point and in line with the first measurement, the angle of elevation was 9.6°. How far above the football field is the top of Mission Peak?

Solution: The following figure shows Mission Peak along with the two measurements of the angles of elevation. We need to find the value of a to find the height of Mission Peak above the football field. Both right triangles $\triangle BCA$ and $\triangle BCD$, will be used in solving for a.

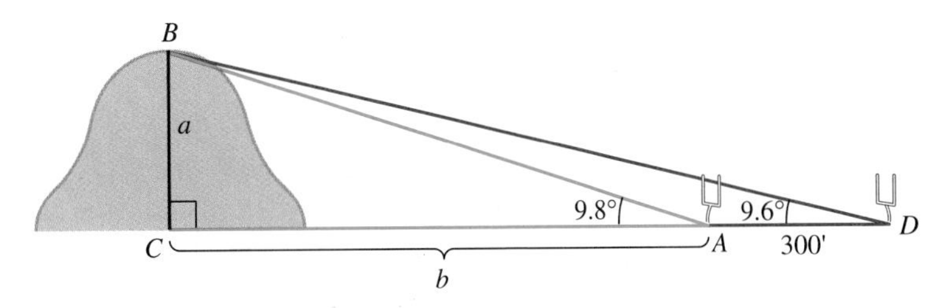

In $\triangle BCA$,

$$\tan 9.8° = \frac{a}{b}$$

$$0.1727 = \frac{a}{b}$$

$$0.1727b = a$$

$$b = \frac{a}{0.1727}$$

In $\triangle BCD$,

$$\tan 9.6° = \frac{a}{b + 300}$$

$$0.1691 = \frac{a}{b + 300}$$

$$0.1691(b + 300) = a$$

$$0.1691b + 50.73 = a$$

$$0.1691b = a - 50.73$$

$$b = \frac{a - 50.73}{0.1691}$$

Since we have found two different expressions for b, we can set them equal to each other and solve for a.

$$\frac{a}{0.1727} = \frac{a - 50.73}{0.1691}$$

$$0.1691a = 0.1727(a - 50.73)$$

$$0.1691a = 0.1727a - 8.7611$$

$$-0.0036a = -8.7611$$

$$a = 2433.6$$

Therefore, Mission Peak is approximately 2433.6 ft above the football field.

section 8.3

PROBLEMS ▶ Explain ▶ Apply ▶ Explore

▶ Explain

1. What four steps would be helpful in solving practical problems involving right triangles?
2. What is an angle of elevation?
3. What is an angle of depression?
4. Why can you not use the sine, cosine, or tangent functions on a triangle that has angles of 66°, 34°, and 80°?
5. Explain why the title of this chapter, "Trigonometry: A Door to the Unmeasurable," is appropriate in this section.

▶ Apply

In Problems 6–10, solve using the round-off rules established in Section 8.2.

6. An Eagle Scout taking trigonometry finds the distance across a river by doing the following:
   a) He stands on the bank of the river and chooses a boulder directly across the river as a marker.
   b) He paces off 25 yd along the bank.
   c) From that point, he approximates that the angle back to the rock is 60°.

d) Being well prepared, he takes out his calculator, uses trig, and determines that the river is 43.3 yd wide.

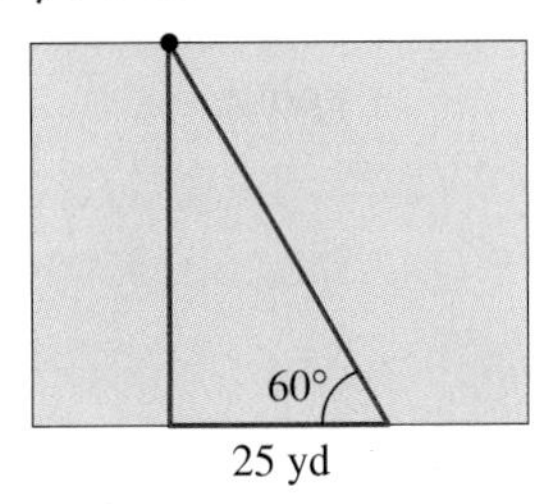

Show that the Eagle Scout was correct in his approximation.

7. A sign on a straight stretch of freeway reads, "10% GRADE altitude 3000 ft." If the next altitude sign you see states that the altitude is 4350 ft, how many miles have you driven since seeing the 10% grade sign?

8. Because of road construction, a 36-mi section of a highway is closed. The detour makes a 55° angle with the highway as shown. How many additional miles does one travel using the detour?

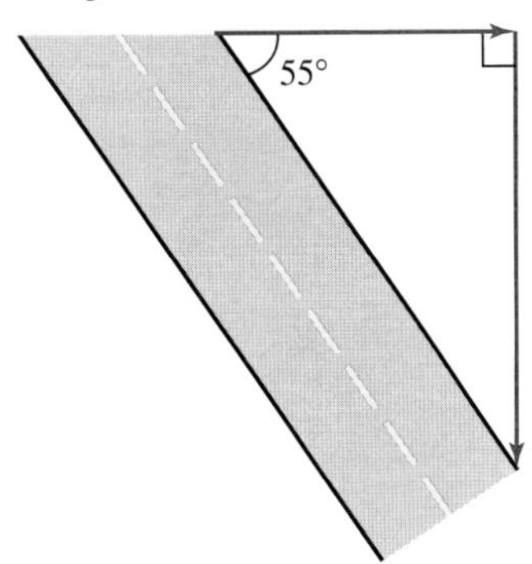

9. To find the distance across a lake, a tree on the opposite side of the lake was taken as a marker, and 60 m was walked off, as shown in the figure. If the angle to the tree was 75°, how far is it across the lake?

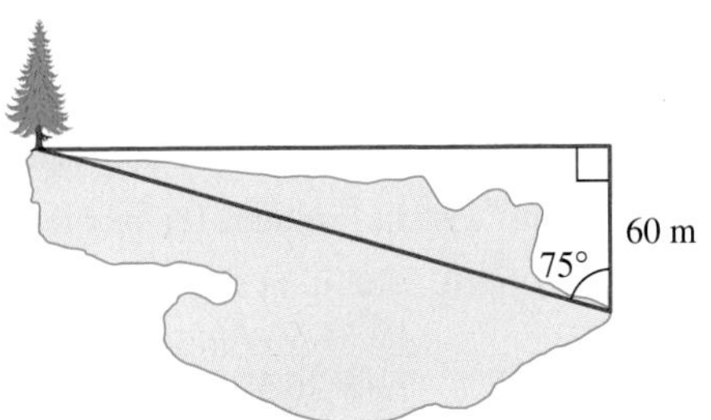

10. After lifting off from its launch site, a hot air balloon floats along over an open field with the top of its basket 250 ft above ground level. If the angle of depression to its takeoff point is 2°, how far is the balloon from its takeoff point?

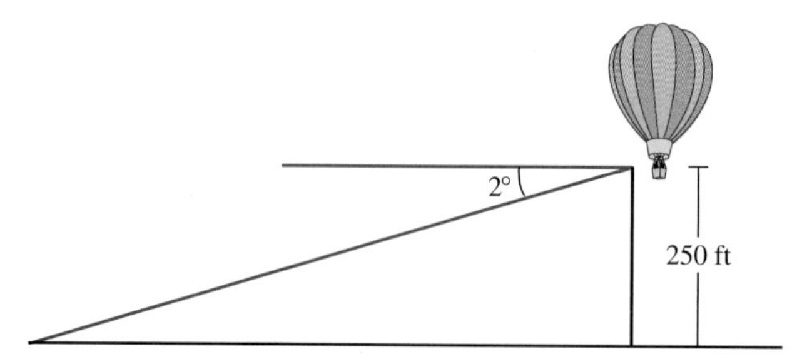

Explore

In Problems 11–23, make a sketch of the situation and solve using the round-off rules established in Section 8.2.

11. A guy wire, which supports a vertical circus tent pole, makes a 76.5° angle with level ground. If the guy wire is secured to the ground 12 ft away from the bottom of the pole, how tall is the pole and how long is the guy wire?

12. The pitch of a roof is 5/12. That means the roof rises 5 ft for each 12-ft change in the horizontal direction. What is the angle of elevation of the roof? If the actual length of the roof is 32 ft, how much does the roof rise in that distance?

13. How far up a vertical wall will a 32-ft ladder reach if it makes a 67° angle with the level ground?

14. Radar indicates that the distance to an approaching airplane is 8.6 mi and the angle of elevation to the plane is 26°. At what altitude is the plane flying?

15. From a point at eye level (5 ft off the ground) the angle of elevation to the top of a radio transmitting tower is 70°. If the person measuring this angle is 40 ft from the tower, what is the height of the tower?

16. The angle of depression from the roof of a 310-ft office building to the bottom of a statue in the Civic Center is 8.2°. If the ground between the building and the statue is level, find the distance from the building to the statue.

17. The angle of depression from the top of the largest Khufu Pyramid (height 482 ft) in Egypt to the bottom of a nearby marker measures 12.7°. If the ground is level, how far is the top of the pyramid from the bottom of the marker?

18. A sea-to-air guided missile shot from a submarine leaves the water at an angle of elevation of 18.6° traveling at 480 ft/s. If the missile maintains a constant angle of ascent and the same speed, how far above sea level will it be after 30 s?

19. In Problem 18, how long will it take the missile to reach an altitude of 10,000 ft?

20. A ship at sea measures the angle of elevation to the top of a cliff on shore to be 14.8°. After traveling a half mile closer to the cliff, the angle of elevation is now 26.5°. How many feet is the top of the cliff above sea level?

21. A tennis player hits an overhead smash 35 ft from the net. If the ball is hit 10.5 ft off the ground and just clears the net at a height of 3 ft, how far from the net and at what angle will the ball strike the ground?

22. In *Tactics for Trout*, the author, Dave Hughes, states that a trout has a 97° angle of vision in any direction. Furthermore, he states that because of the reflection of light, a trout cannot see objects above the water surface within a 10° angle. The implication is that a trout has both a region within its range of vision and a region not in its range of vision.

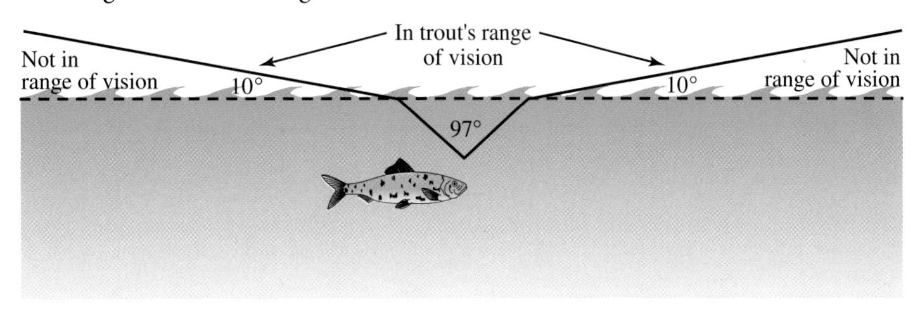

A knowledge of the range of vision of a trout can help fly fisherman determine how close he can get to a trout without being seen. If a trout is at a depth of 2 ft, how close can a 6-ft person get to the trout without being seen by the trout? (*Hint:* The angle at the trout's eye is 48.5°, and you need to find the length of *d*.)

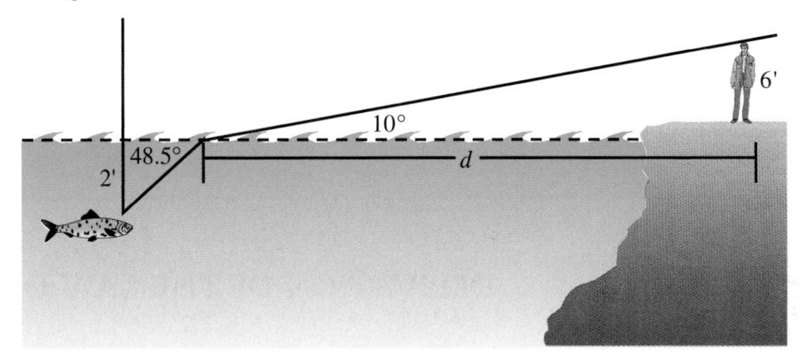

23. In Problem 22, if a trout is at a depth of 1 ft, how close can a 5-ft person get to the trout without being seen by the trout?

section 8.4

The Laws of Sines and Cosines

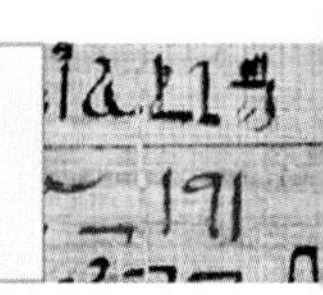

The three trigonometric functions that were studied in the previous sections were defined and used in reference to right triangles. The next two laws of trigonometry, the Law of Sines and the Law of Cosines, will allow us to solve triangles that are not right triangles. A triangle with each angle measuring less than 90° is called an **acute triangle,** and a triangle with one angle greater than 90° is called an obtuse triangle. The Laws of Sines and Cosines can be used to solve acute and obtuse triangles. However, in this survey of trigonometry, we will apply them only to acute triangles that have a unique solution.

THE LAW OF SINES

The **Law of Sines** gives a relationship between the sides and angles of a triangle. It states that the ratio of the length of the side of any triangle to the sine of the angle opposite that side is the same for all three sides of the triangle. That is, for any $\triangle ABC$,

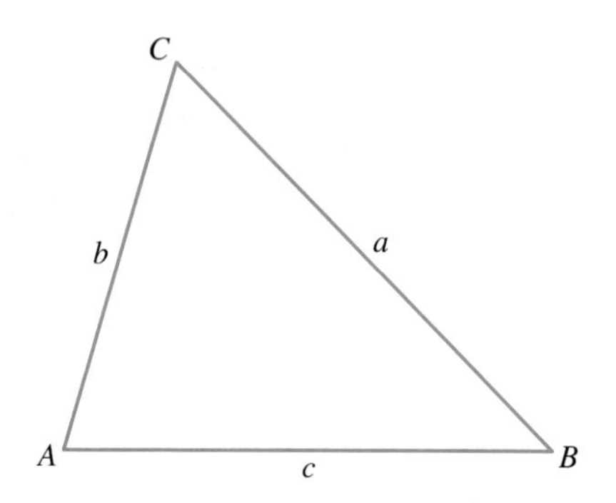

Law of Sines

$$\frac{a}{\sin A} = \frac{b}{\sin B} = \frac{c}{\sin C}.$$

DERIVATION OF THE LAW OF SINES

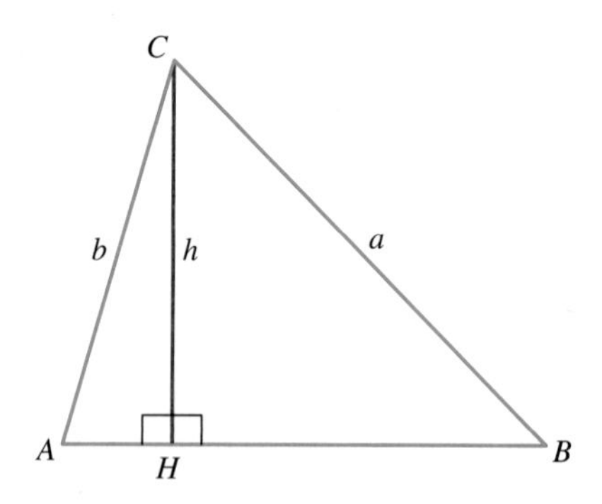

Let h be the length of the segment from C forming right angles with $\overline{AB}$ at H.

In $\triangle ACH$,	In $\triangle CHB$,
$\sin A = \dfrac{h}{b}$	$\sin B = \dfrac{h}{a}$
$b \sin A = h$	$a \sin B = h$

Since we have found two expressions for h, they must be equal. Setting them equal to each other, we get the following.

$$a \sin B = b \sin A$$

$$\frac{a \sin B}{(\sin A)(\sin B)} = \frac{b \sin A}{(\sin A)(\sin B)} \qquad \text{[dividing both sides by } (\sin A)(\sin B)\text{]}$$

$$\frac{a}{\sin A} = \frac{b}{\sin B}$$

If we use the same process with a segment from B making right angles with $\overline{AC}$, we can show that the third ratio, $c/\sin C$, is equal to the two ratios given above.

The Law of Sines can be used to solve any triangle in which two angles and one side are known. In each example, you will notice that the ratio of the sine of a known angle and its opposite side will be set equal to a ratio that has either an unknown angle or an unknown side.

EXAMPLE 1

In $\triangle ABC$, $A = 40°$, $B = 60°$, and $b = 9$. Find the measure of $\angle C$ and the lengths of a and c.

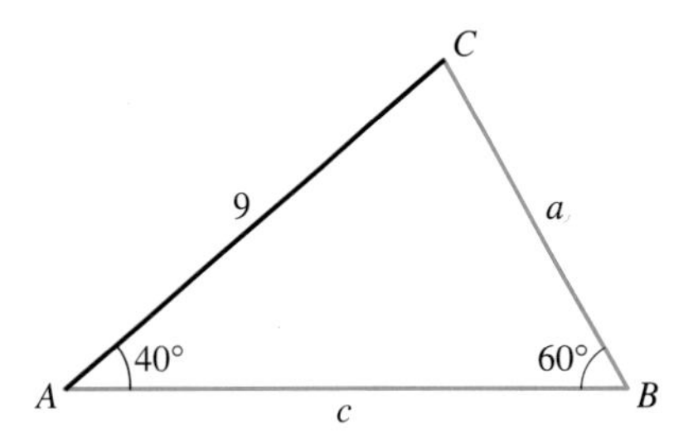

Solution: $C = 80°$, since the sum of the angles of a triangle is $180°$ and $180° - 40° - 60° = 80°$.

To find sides a and c, we use the Law of Sines.

$$\frac{a}{\sin A} = \frac{b}{\sin B} \qquad \frac{c}{\sin C} = \frac{b}{\sin B}$$

$$\frac{a}{\sin 40°} = \frac{9}{\sin 60°} \qquad \frac{c}{\sin 80°} = \frac{9}{\sin 60°}$$

$$\frac{a}{0.6428} = \frac{9}{0.8660} \qquad \frac{c}{0.9848} = \frac{9}{0.8660}$$

$$0.8660a = 9(0.6428) \qquad 0.8660c = 9(0.9848)$$

$$0.8660a = 5.7852 \qquad 0.8660c = 8.8632$$

$$a = 6.7 \qquad c = 10.2$$

NOTE: The Law of Sines should be used to solve any triangle in which two angles and one side are known. This situation is referred to as angle-angle-side (AAS) or angle-side-angle (ASA). The Law of Sines can also be used when two sides and an angle opposite one of the sides is known (SSA). Since this case may lead to more than one solution, it will not be discussed in this survey of trigonometry.

EXAMPLE 2

In $\triangle XYZ$, $Y = 86.2°$, $Z = 21.5°$, and $y = 110$. Find the measure of $\angle X$ and the lengths of sides x and z.

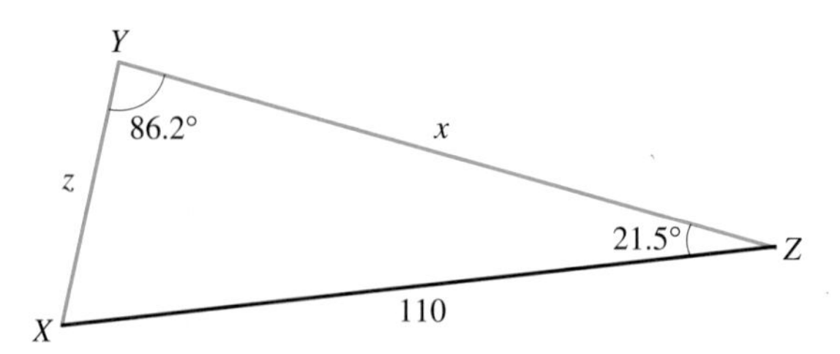

Solution: $X = 180° - 86.2° - 21.5° = 72.3°$. To find sides x and z, we use the Law of Sines.

$$\frac{x}{\sin X} = \frac{y}{\sin Y} \qquad \frac{z}{\sin Z} = \frac{y}{\sin Y}$$

$$\frac{x}{\sin 72.3°} = \frac{110}{\sin 86.2°} \qquad \frac{z}{\sin 21.5°} = \frac{110}{\sin 86.2°}$$

$$\frac{x}{0.9527} = \frac{110}{0.9978} \qquad \frac{z}{0.3665} = \frac{110}{0.9978}$$

$$0.9978x = 110(0.9527) \qquad 0.9978z = 110(0.3665)$$

$$0.9978x = 104.797 \qquad 0.9978z = 40.315$$

$$x = 105.0 \qquad z = 40.4$$

THE LAW OF COSINES

As was seen in the previous example, the Law of Sines gives a very efficient means for solving acute triangles in which two angles and one side are known. However, it is not possible to use the Law of Sines to solve a triangle in which the three sides (SSS) or one angle and the two sides that form the angle (SAS) are known. The law of trigonometry that enables us to solve triangles with those conditions is called the **Law of Cosines.** It states that for any $\triangle ABC$,

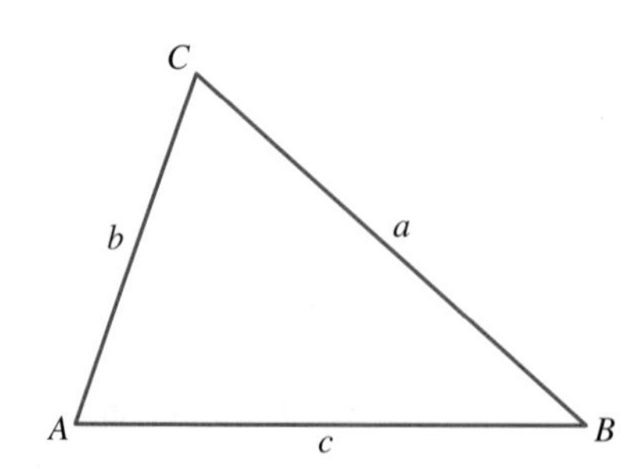

Law of Cosines

$$a^2 = b^2 + c^2 - 2bc \cos A$$
$$b^2 = a^2 + c^2 - 2ac \cos B$$
$$c^2 = a^2 + b^2 - 2ab \cos C$$

DERIVATION OF THE LAW OF COSINES

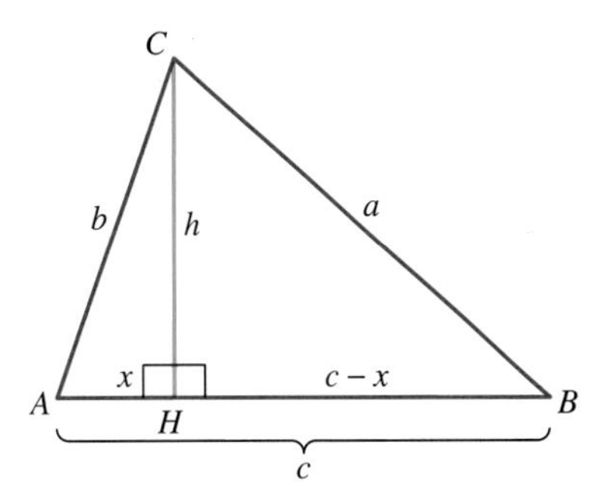

Let h be the length of a segment from C forming right angles with $\overline{AB}$ at H. Using the Pythagorean Theorem in right $\triangle CHB$,

$$a^2 = (c - x)^2 + h^2$$
$$a^2 = c^2 - 2cx + x^2 + h^2.$$

In right $\triangle CHA$, we have $x^2 + h^2 = b^2$

Substituting, we get

$$a^2 = c^2 - 2cx + b^2$$
$$a^2 = b^2 + c^2 - 2cx.$$

Also in $\triangle CHA$, $\cos A = \dfrac{x}{b}$, so $b \cos A = x$.

$$\therefore a^2 = b^2 + c^2 - 2cb \cos A$$
$$a^2 = b^2 + c^2 - 2bc \cos A$$

If we use the same process, drawing segments from A or B that make right angles with the sides opposite those angles, we can derive the other forms of the Law of Cosines.

Notice that the variable of the triangle used on the left of the equal sign corresponds to the angle opposite that side on the right of the equal sign. That is, in $a^2 = b^2 + c^2 - 2bc \cos A$, a^2 on the left corresponds to $\cos A$ on the right side of the equation. Recognizing that this is true for all three forms of the Law of Cosines makes the task of remembering the Law of Cosines a little easier.

EXAMPLE 3

In $\triangle ABC$, $A = 64°$, $b = 18$, and $c = 39$. Use the Law of Cosines to find (a) side a and (b) B.

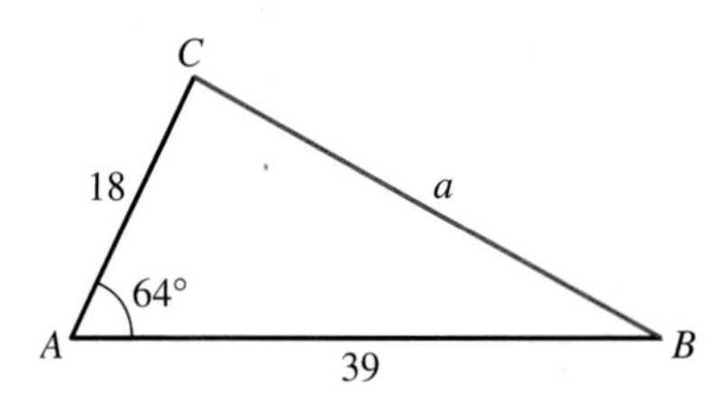

Solution:

a) Since two sides and the included angle are known (SAS), this triangle can be solved by the Law of Cosines.

$$a^2 = b^2 + c^2 - 2bc \cos A$$
$$a^2 = 18^2 + 39^2 - 2(18)(39) \cos 64°$$
$$a^2 = 324 + 1521 - 1404(0.4384)$$
$$a^2 = 1229.4864$$
$$a = \sqrt{1229.4864} = 35.1$$

b) Since we now know a, b, and c, we can find B from the Law of Cosines, where B is the only unknown quantity.

$$b^2 = a^2 + c^2 - 2ac \cos B$$
$$18^2 = 35.1^2 + 39^2 - 2(35.1)(39) \cos B$$
$$324 = 1232.01 + 1521 - 2737.8 \cos B$$
$$-2429.01 = -2737.8 \cos B$$
$$0.8872 = \cos B$$
$$27.5° = B$$

NOTE: The Law of Cosines can be used to solve any triangle in which a side, an angle, and a side (SAS) or three sides (SSS) are known.

EXAMPLE 4

In $\triangle USA$, $u = 38$, $s = 42$, and $a = 29$. Find the measures of $\angle U$, $\angle S$, and $\angle A$.

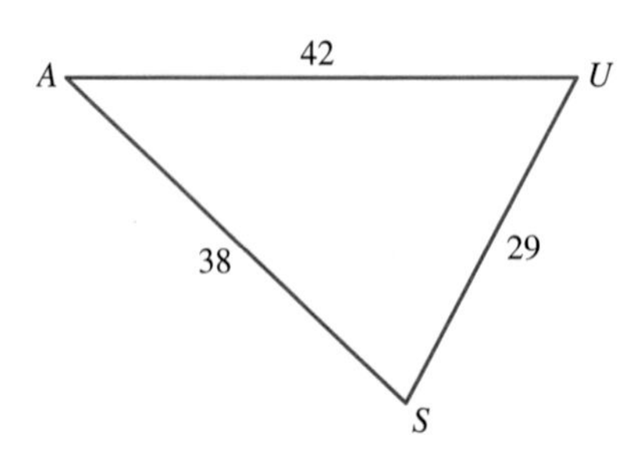

Solution: Since the three sides of the triangle are known, we first find U by writing the Law of Cosines, using u, s, a, and U.

$$\begin{aligned} u^2 &= s^2 + a^2 - 2sa \cos U \\ 38^2 &= 42^2 + 29^2 - 2(42)(29) \cos U \\ 1444 &= 1764 + 841 - 2436 \cos U \\ -1161 &= -2436 \cos U \\ 0.4766 &= \cos U \\ 61.5° &= U \end{aligned}$$

We can now find S by using the Law of Cosines again. However, it is easier to use the Law of Sines with u and $\sin U$ forming the known ratio and S being the only unknown quantity.

$$\begin{aligned} \frac{u}{\sin U} &= \frac{s}{\sin S} \\ \frac{38}{\sin 61.5°} &= \frac{42}{\sin S} \\ \frac{38}{0.8788} &= \frac{42}{\sin S} \\ 38 \sin S &= 42(0.8788) \\ \sin S &= 0.9713 \\ S &= 76.2° \end{aligned}$$

Since we now know two of the angles of $\triangle USA$, we can find A by using the fact that the sum of the angles of a triangle equals 180°.

$$A = 180° - 61.5° - 76.2° = 42.3°$$

section 8.4

PROBLEMS ○ Explain ○ Apply ○ Explore

○ Explain

1. What is the Law of Sines?
2. What is the Law of Cosines?
3. Explain when it is necessary to use the Law of Sines or the Law of Cosines.
4. In this section, what should be known about an acute triangle to use the Law of Sines?
5. In this section, what should be known about an acute triangle to use the Law of Cosines?

Apply

In Problems 6–14, use the Law of Sines to find the designated side in $\triangle SUN$.

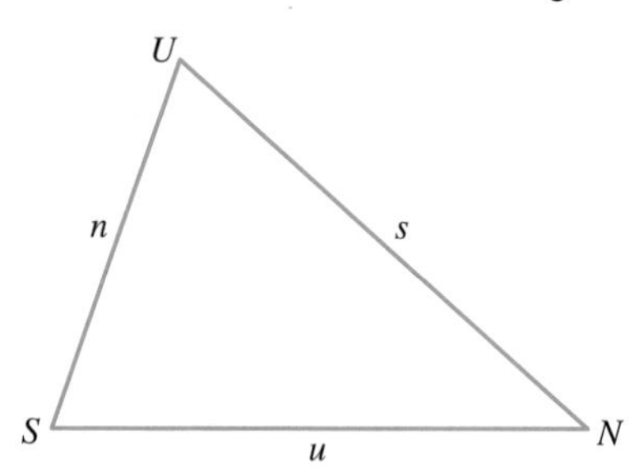

6. $S = 36°$, $U = 70°$, $s = 7$; find u.
7. $S = 56.5°$, $U = 62.9°$, $s = 25$; find n.
8. $S = 9°$, $U = 87°$, $n = 34.6$; find s.
9. $S = 47°$, $U = 60°$, $n = 450$; find u.
10. $S = 28.7°$, $N = 71.5°$, $u = 56$; find n.
11. $S = 66.6°$, $N = 55.5°$, $s = 44.4$; find u.
12. $S = 57°$, $N = 57°$, $n = 7.5$; find s.
13. $N = 12.5°$, $U = 80.25°$, $u = 1240$; find n.
14. $N = 75°$, $U = 30°$, $s = 1.23$; find u.

In Problems 15–20, use the Law of Cosines to find the designated angle or side in $\triangle DOG$.

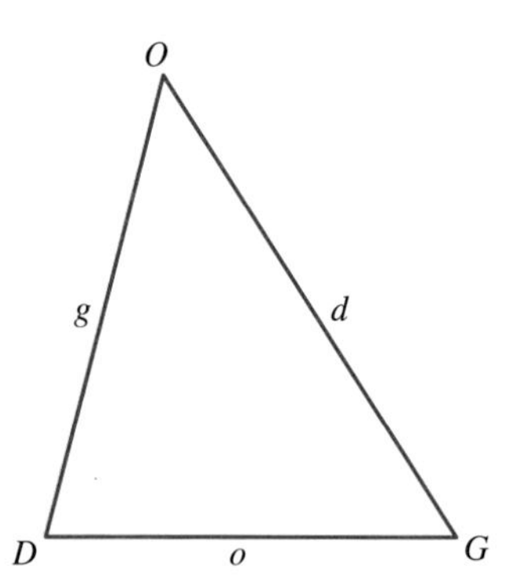

15. $D = 73°$, $o = 12$, $g = 34$; find d.
16. $O = 40.6°$, $g = 23$, $d = 45.5$; find o.
17. $G = 37.25°$, $d = 5.5$, $o = 6.75$; find g.
18. $d = 15$, $o = 16$, $g = 17$; find D.
19. $d = 9.4$, $o = 8.3$, $g = 5.8$; find O.
20. $d = 189$, $o = 213$, $g = 220$; find G.

Explore

21. Solve $\triangle CAT$ if $A = 56.7°$, $T = 44.7°$, and $c = 32.1$

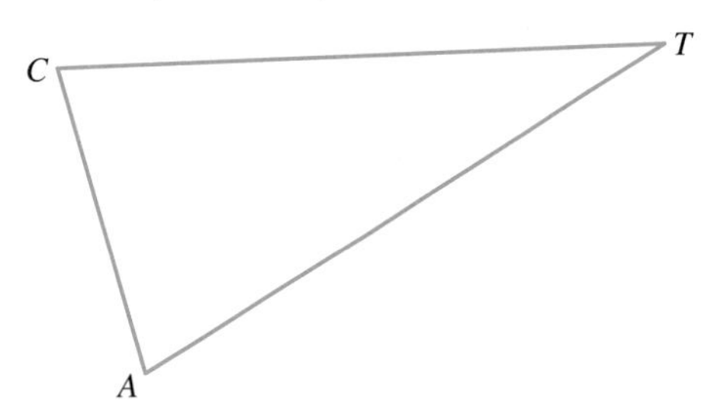

22. Solve $\triangle KEY$ if $Y = 55°$, $E = 55°$, and $k = 100$.

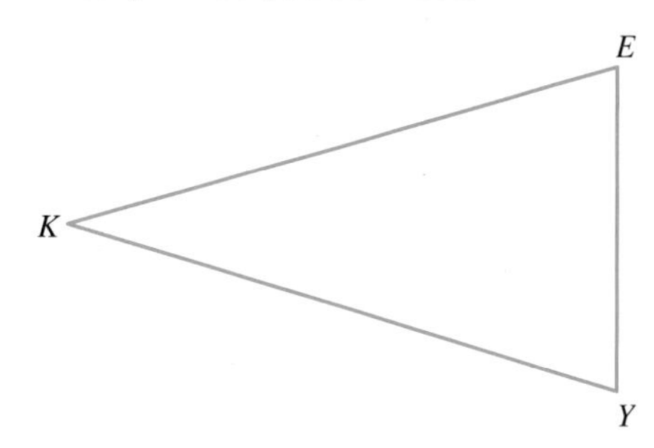

23. Solve $\triangle CUP$ if $c = 19.8$, $p = 20.6$, and $u = 14.4$.

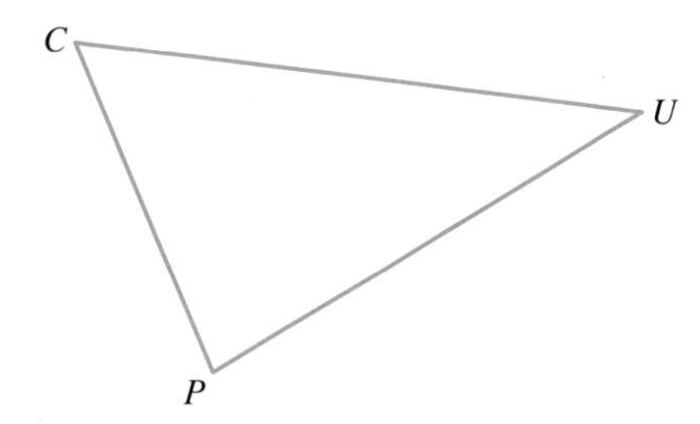

24. Solve $\triangle CAR$ if $c = 106$, $a = 155$, and $r = 127$.

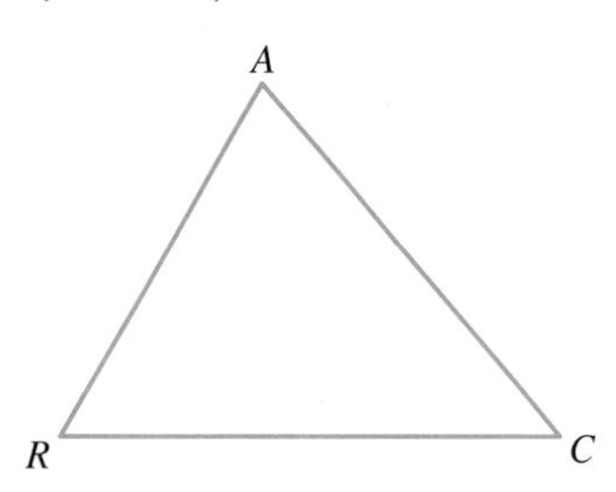

25. In $\triangle GET$, $E = 67.5°$, $g = 23.5$, and $t = 45.3$. Find the measures of the parts of the triangle that are not given.

26. In $\triangle MET$, $t = 345.5$, $E = 32.2°$, and $M = 58.5°$. Find the measures of the parts of the triangle that are not given.

27. In $\triangle PET$, $p = 20$, $e = 23$, and $t = 28$. Find the measures of the parts of the triangle that are not given.

28. In $\triangle JET$, $j = 78.8$, $E = 52.2°$, and $J = 45°$. Find the measures of the parts of the triangle that are not given.

29. Verify that the Law of Sines applies to right triangles by determining the unknown sides and angles of $\triangle RAD$, where $A = 90°$, $R = 65°$, and $AR = 20$, using:

a) The Law of Sines.

b) Right triangle trig.

30. Verify that the Law of Cosines applies to right triangles by determining the angles of a 3-4-5 triangle using:

a) The Law of Cosines.

b) Right triangle trig.

***31.** Determine why it is impossible to have a triangle $\triangle ABC$ with sides $a = 10$, $b = 12$, and $c = 23$.

section 8.5

Acute Triangle Applications

Surveyors make extensive use of trigonometry. (Courtesy of Keith Dannemiller/CORBIS)

In Section 8.3, we saw how the trigonometry of right triangles can be applied to many different kinds of problems. In Example 7 of that section, we showed how to solve a problem using two right triangles and a great deal of algebra. In this section, we will see how the Laws of Sines and Cosines can simplify this process.

EXAMPLE 1

Two fire-lookout stations are 15 mi apart, with station B directly east of station A. Both stations spot a fire on a mountain to the north. The line of sight to the fire from station A makes a 37.3° angle with a line running between the two stations (the east-west line), and the line of sight to the fire from station B makes a 54.2° angle with the east-west line. How far is the fire from station A?

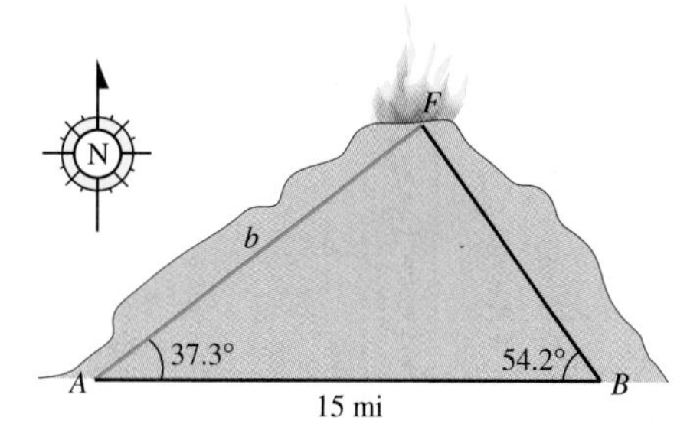

Solution: The figure shows the information given in the problem. Since $A = 37.3°$ and $B = 54.2°$, $F = 180° - 37.3° - 54.2° = 88.5°$. Since we now know $\angle F$ and the side opposite that angle, we can apply the Law of Sines to find side b.

Let b = distance to the fire from A.

$$\frac{f}{\sin F} = \frac{b}{\sin B}$$

$$\frac{15}{\sin 88.5°} = \frac{b}{\sin 54.2°}$$

$$\frac{15}{0.9997} = \frac{b}{0.8111}$$

$$0.9997b = 15(0.8111)$$

$$b = 12.2$$

The distance to the fire from station A is approximately 12.2 mi.

EXAMPLE 2

The bottom of a hot air balloon is tethered at the top of a small hill with 200 ft of rope. Because of the wind blowing from the west, the angle of elevation to the balloon is 67.2°. If the hill makes an angle of 9° with the horizontal and the rope to the balloon is taut, how far from a point 56 ft down the hill is the bottom of the balloon?

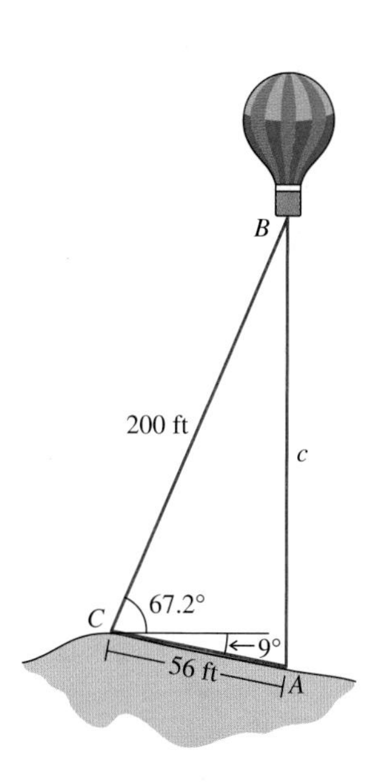

Solution: The hot air balloon along with the given information is shown in the figure. The angle at C is 76.2° (67.2° + 9°). In $\triangle BCA$, a side, an angle, and a side are known (SAS). This allows us to use the Law of Cosines to find the height of the balloon above the ground. Let c represent the height of the bottom of the balloon above the ground.

$$c^2 = a^2 + b^2 - 2ab \cos 76.2°$$

$$c^2 = 200^2 + 56^2 - 2(200)(56)(0.2385)$$

$$c^2 = 40{,}000 + 3136 - 5342.4$$

$$c^2 = 37{,}793.6$$

$$c = 194.4 \text{ ft}$$

EXAMPLE 3

A mining company digs a 750 yd horizontal mine shaft into a hill with an incline of 17°. How far up the hill should an air shaft, making an 87° angle with the hill, be drilled so that it will meet the mine shaft? How long is the air shaft?

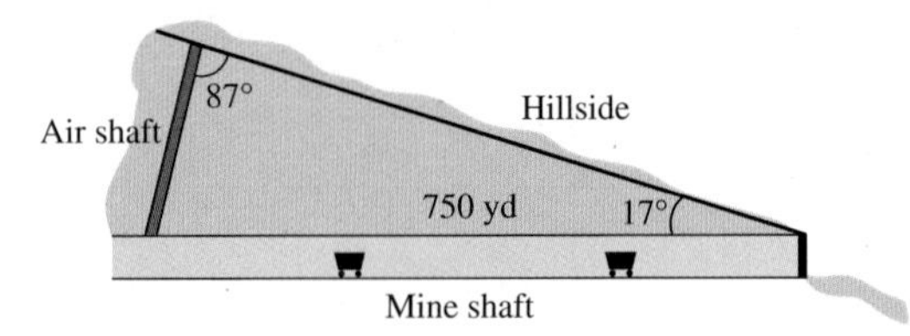

Solution: The angle where the air shaft meets the mine shaft is 76° (180° − 87° − 17°). Since two angles and a side are known (AAS), the Law of Sines can be used to find the distance up the hill where the air shaft is to be drilled and the length of the air shaft.

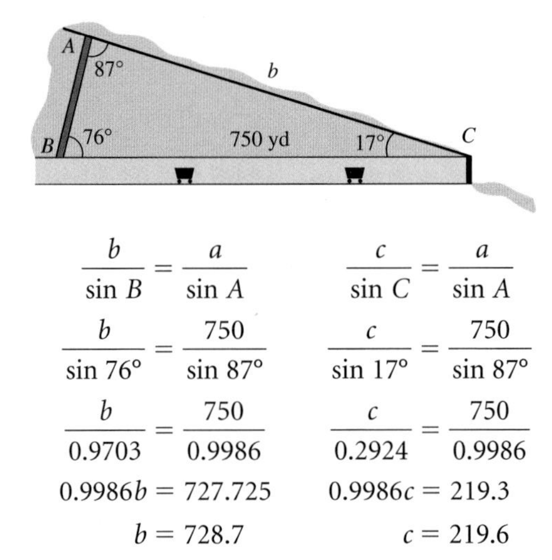

$$\frac{b}{\sin B} = \frac{a}{\sin A} \qquad \frac{c}{\sin C} = \frac{a}{\sin A}$$

$$\frac{b}{\sin 76°} = \frac{750}{\sin 87°} \qquad \frac{c}{\sin 17°} = \frac{750}{\sin 87°}$$

$$\frac{b}{0.9703} = \frac{750}{0.9986} \qquad \frac{c}{0.2924} = \frac{750}{0.9986}$$

$$0.9986b = 727.725 \qquad 0.9986c = 219.3$$

$$b = 728.7 \qquad c = 219.6$$

Therefore, the shaft should be drilled 728.7 yd up the hill and should be 219.6 yd long.

NAVIGATION PROBLEMS

In navigation problems, one of the common ways to give the course of a plane or ship is in terms of a **bearing.** A bearing is an acute angle measured from a north-south line toward either the east or the west. In this system, either N (north) or S (south) is written, followed by an acute angle, then E (east) or W (west). The following examples show how this system is used to give a direction.

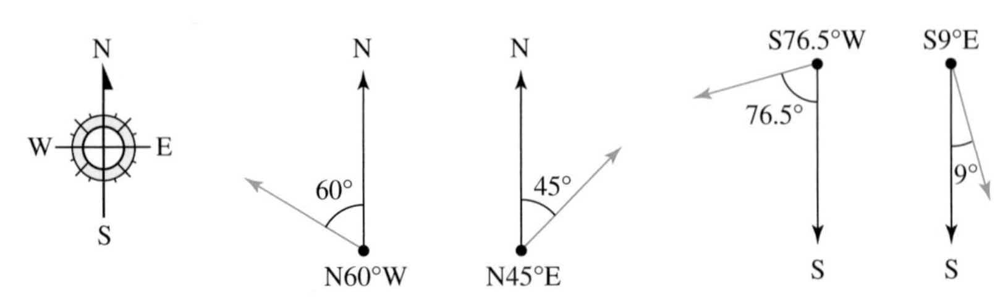

EXAMPLE 4

An airplane leaves an airport and flies directly north at 450 mph, while a second airplane leaves the airport at the same time flying at 360 mph on a bearing of N70°E. How far apart are the airplanes after 2.5 hr?

Solution: Traveling at 450 mph, after 2.5 hr the plane flying directly north travels 1125 mi (2.5 × 450) and the plane flying N70°E at 360 mph travels 900 mi (2.5 × 360). Placing that information on the diagram shows that we have the SAS situation, so we will use the Law of Cosines.

Let $a =$ the distance between the planes after 2.5 hr.

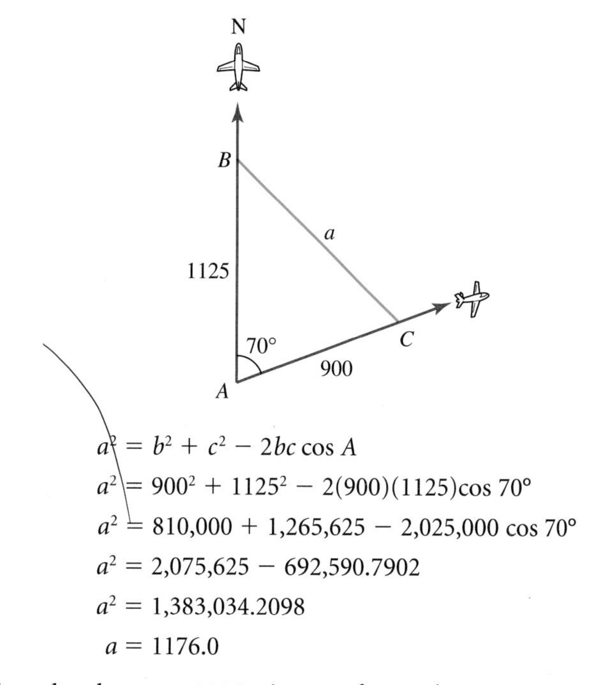

$$a^2 = b^2 + c^2 - 2bc \cos A$$
$$a^2 = 900^2 + 1125^2 - 2(900)(1125)\cos 70°$$
$$a^2 = 810{,}000 + 1{,}265{,}625 - 2{,}025{,}000 \cos 70°$$
$$a^2 = 2{,}075{,}625 - 692{,}590.7902$$
$$a^2 = 1{,}383{,}034.2098$$
$$a = 1176.0$$

Therefore, the planes are 1176 mi apart after 2.5 hr.

EXAMPLE 5

A sport-fishing boat leaves Bob's Pier heading directly east. After traveling for 50 mi, the captain hears a fishing report on the radio, which entices him to turn the boat and proceed on a bearing of S42°W for 27 mi. How far is the boat from Bob's Pier, and what bearing should the boat have originally taken to arrive at the fishing spot?

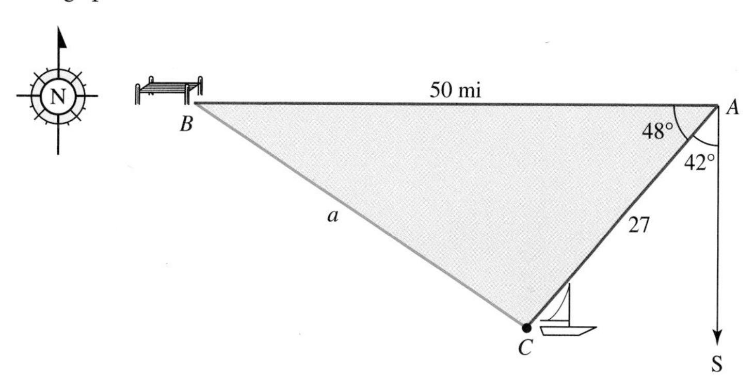

Solution: The figure shows the course taken by the fishing boat (starting at point B). Since a line going east forms a 90° angle with one going south, $\angle BAC = 90° - 42° = 48°$, and since we have an SAS situation, we can use the Law of Cosines to find the distance a from Bob's Pier.

$$a^2 = b^2 + c^2 - 2bc \cos 48°$$
$$a^2 = 27^2 + 50^2 - 2(27)(50)(0.6691)$$
$$a^2 = 729 + 2500 - 1806.57$$
$$a^2 = 1422.43$$
$$a = 37.7 \text{ miles}$$

Since we know the three sides of the triangle (SSS), we will use the Law of Cosines to find the bearing from Bob's Pier to point C.

$$b^2 = a^2 + c^2 - 2ac \cos B$$
$$27^2 = 37.7^2 + 50^2 - 2(37.7)(50) \cos B$$
$$729 = 1421.29 + 2500 - 3770 \cos B$$
$$0.8468 = \cos B$$
$$32.1° = B$$

Since bearing is measured from the north-south line, we need to subtract 32.1° from 90° to find the angle made by the line from Bob's Pier to the fishing spot. Thus, the bearing angle is $90° - 32.1° = 57.9°$, and the actual bearing is S57.9°W.

In solving acute triangles, you will find it helpful to do the following:

1. Sketch the situation and the triangle involved.
2. Write the known angles and sides on the triangle.
3. Use letters to represent the unknown angles and sides.
4. Find the unknown angle by subtracting the sum of the two angles from 180° if two angles are known.
5. For triangles in which two angles and a side are known, use the Law of Sines.
6. For triangles in which a side, an angle, and a side (SAS) or three sides (SSS) are known, use the Law of Cosines.

The trigonometric methods shown in this chapter give you powerful tools for solving problems that contain triangles. The problems that follow will show you more applications that involve both right and acute triangles.

section 8.5

PROBLEMS ● Explain ● Apply ● Explore

● Explain

1. What six steps are helpful in solving application problems involving acute triangles?
2. Explain how a bearing is measured in navigation problems.
3. Explain what a bearing of N22°E indicates. Make a sketch of that bearing.
4. Explain what a bearing of S75°W indicates. Make a sketch of that bearing.
5. Explain why the title of this chapter, "Trigonometry: A Door to the Unmeasurable," is appropriate in this section.

● Apply

6. A tall fir tree growing on the side of a hill makes a 70° angle with the hill. From a point 60 ft up the hill, the angle of elevation to the top of the tree is 65° and the angle of depression to the bottom of the tree is 20°. How tall is the fir tree?

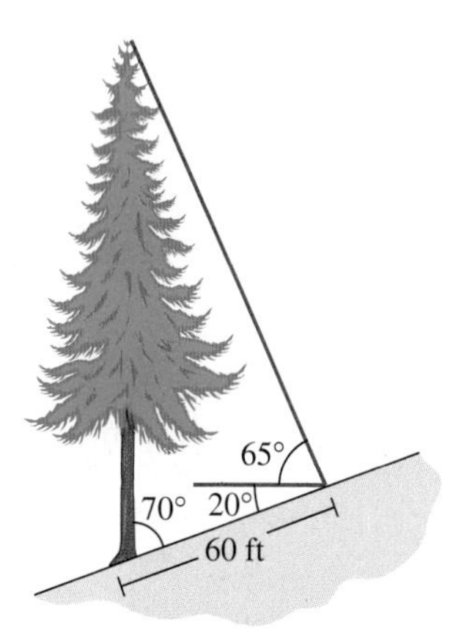

7. A large helium balloon advertising a sale at a local auto dealer is tethered by a rope at a west end of a level car lot. A wind blowing from the west causes the balloon to have an angle of elevation of 73°. If from the other end of the car lot, 500 ft away, the angle of elevation to the balloon is 52°, how far above the car lot is the balloon?

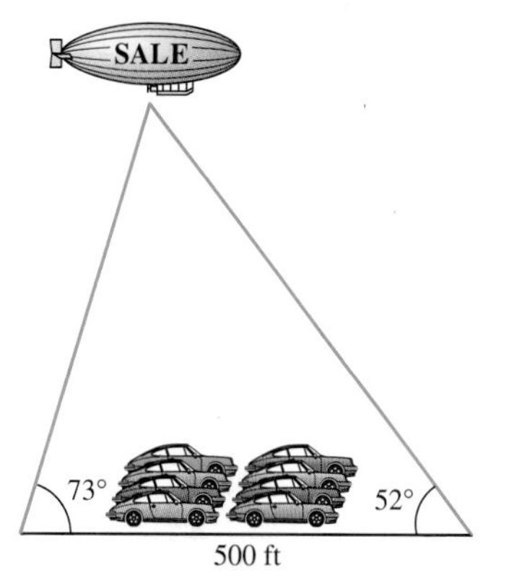

8. To find the distance between a fire-lookout station and a cabin on the other side of a canyon, a ranger uses a transit to measure angles at two different points as shown in the figure. Using the information shown, determine the distance from the fire-lookout station to the cabin.

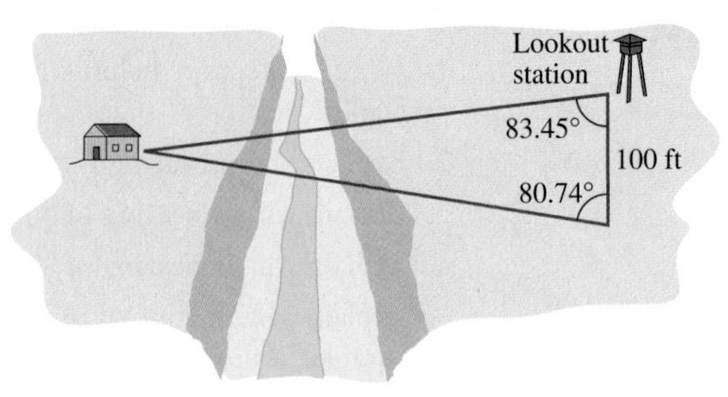

9. The walls of a trench form a 65° angle. To get out of the trench, highway workers place the bottom of a 25-ft plank on one side of the trench and lean the top of the plank on the other side. If the bottom of the plank makes an 80° angle with the wall of the trench, how far up the other side does the top of the plank reach?

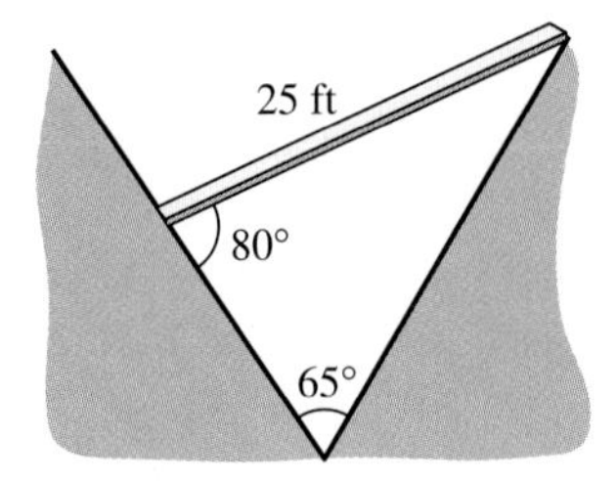

10. To find the distance across Mallard Cove at Bass Lake, a Girl Scout troop made the measurements in the diagram. What is the distance across Mallard Cove?

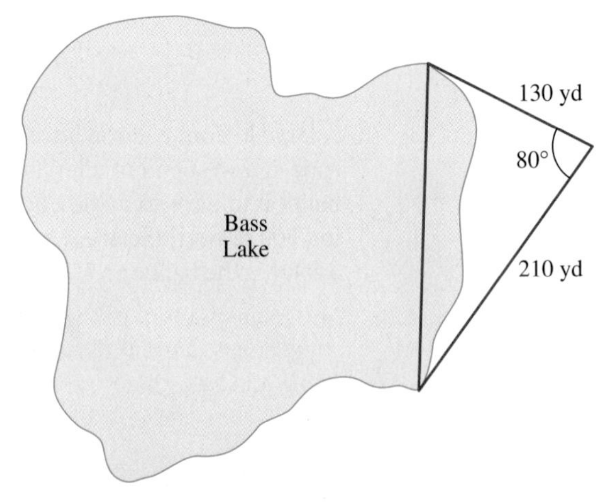

11. An A-frame mountain cabin is 30 ft wide. If the roof of the cabin makes a 63° angle with the base of the cabin, what is the length of the roof from ground level to the peak of the roof?

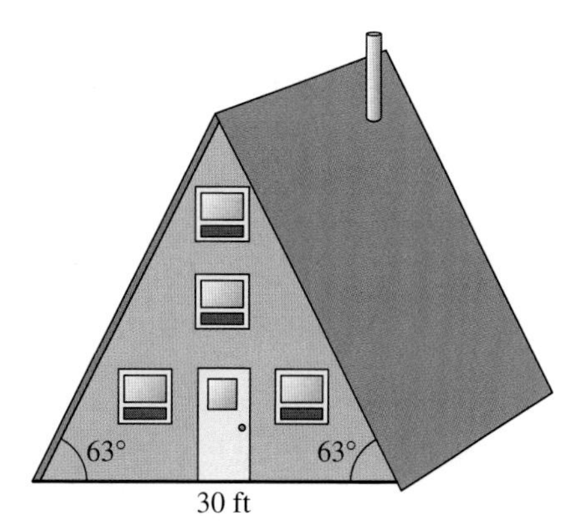

Explore

12. A pilot planned to fly a small plane from Kodiak, Alaska to Tikchik Lakes, a distance of 300 mi. After flying for 2 hr at 150 mph, the pilot realized that she had been flying on a course that was off by 2°. How far from Tikchik Lakes was the airplane at this point? At what angle from the present course must the pilot turn to arrive at Tikchik Lakes?
13. Two cross-country skiers are at the bottom of a ravine. The sides of the ravine form a 72° angle. The first skier goes straight up one side of the ravine while the second one goes straight up the other side. After 20 min, the first skier is 1600 m up one side of the ravine while the second skier is only 1000 m up the other side. How far apart are the two skiers?
14. A ship leaves Miami and cruises for 74 mi on a bearing of S60°E. Another ship leaves at the same time and sails for 56 mi on a bearing of S20°E. How far apart are the two ships?
15. An airplane leaves Denver at 1:00 P.M. on a bearing of N20°E flying at 280 mph. At 1:30 P.M., a second airplane leaves on a bearing of N15°W flying at 375 mph. If the two airplanes continue flying on those courses, how far apart will they be at 3:30 P.M.?
16. How long is an escalator if it makes a 32° angle with the floor and carries people a vertical distance of 20 ft between floors?
17. From a point on the floor, the angle of elevation to the top of a door is 48°, while the angle of elevation to the ceiling above the door is 59°. If the door is 8 ft tall, what is the vertical distance from the floor to the ceiling?
18. The pitch of a roof can be determined by taking the tangent function of the angle of elevation of the roof. Find the pitch of both sides of the roof in the figure and determine the distance between the ground and the peak of the roof.

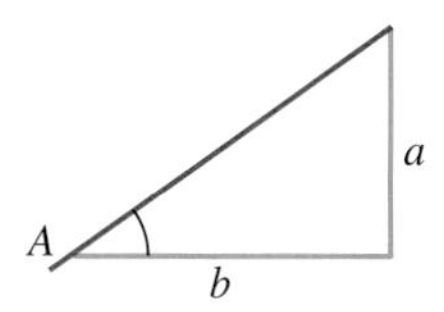

$$\text{Pitch} = \frac{a}{b} = \tan A$$

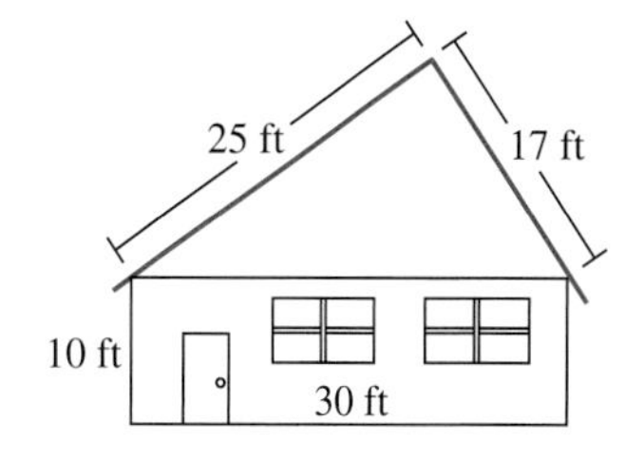

19. A triangular-shaped piece of property has sides that are 136 ft, 125 ft, and 178 ft. What are each of the angles formed by the property lines of this triangular lot?

20. A parallelogram-shaped lot has sides that are 120 m and 232 m. If the angle between the two sides is 58°, how long is the diagonal of the parallelogram that is opposite that angle?

21. When a kite is flying high in the sky, you will notice that the kite string does not make a straight line to the kite. Realizing this, find a method for determining the height above the ground for a kite that is flying in the air if (a) the ground is level and (b) the ground is not level.

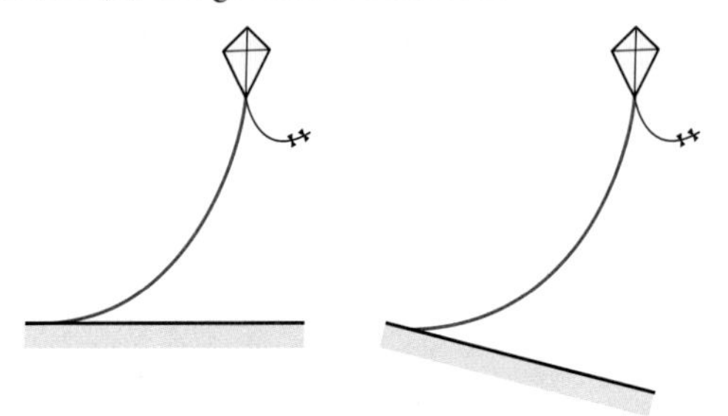

22. A rhombus is a parallelogram with equal sides, as shown. If $\angle A$ is an acute angle, find a formula for the length of the diagonal of the rhombus that is opposite $\angle A$. That is, if $d =$ the length of the diagonal, represent d in terms of x and A.

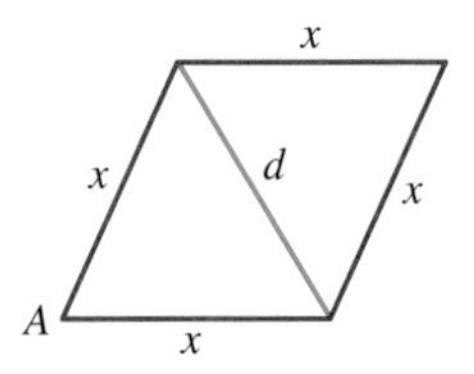

23. A hot air balloon is hovering over a lake. From one side of the lake the angle of elevation to the bottom of the balloon is 54°, and from the other side of the lake, 1660 ft away, the angle of elevation is 69°. How far is the bottom of the hot air balloon above the lake?

24. Explain how a person in a ship that is traveling parallel to the coast could use a lighthouse, the speed of the ship, a device for measuring angles, and a knowledge of trigonometry to determine the distance from the ship to the lighthouse.

section 8.6

The Motion of a Projectile

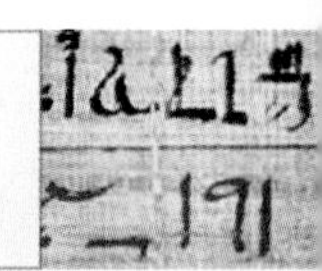

Robbie Knievel jumping the fountain at Caesar's Palace in Las Vegas, Nevada. (Courtesy of Betmann/CORBIS)

In general, it is difficult to analyze the motion of an object that is hurled or shot forward, such as a motorcycle, cannonball, shot put, golf ball, or bullet. However, if we assume that gravity is the only force acting on the object once it is launched, we can use trigonometry to help develop a mathematical model to analyze its motion. Let us consider a ramp to ground jump by a motorcycle jumper. If the motorcycle leaves the top of the ramp 15 feet above the ground at 60 miles per hour (88 feet per second) with an angle of elevation of 24°, how long is the jump? The flight of the motorcycle would follow a path as shown.

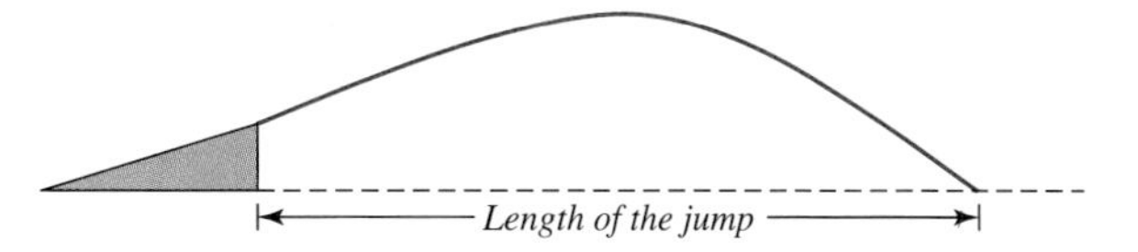

If gravity is not considered and the motorcycle travels in a straight line at a constant speed, we can place the motorcycle in a vertical xy coordinate system at ground level. The distance traveled by the motorcycle in t seconds would equal $88t$ (distance = rate × time). This can be modeled by using the triangle shown below.

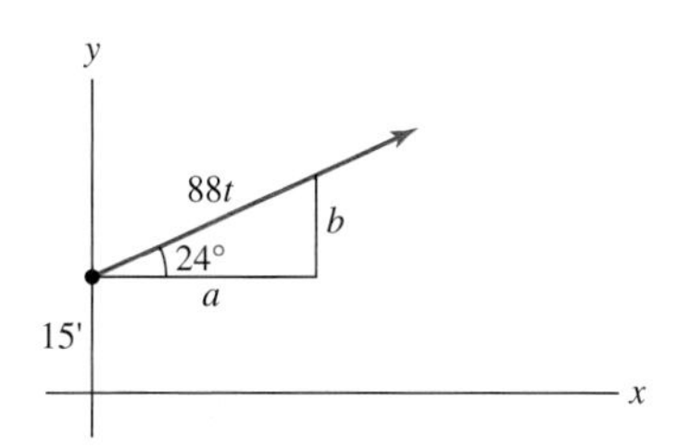

Using the trig functions of sine and cosine, we get the following results.

$$\sin 24° = \frac{b}{88t} \quad \text{or} \quad b = (88t)\sin 24°$$

$$\cos 24° = \frac{a}{88t} \quad \text{or} \quad a = (88t)\cos 24°$$

Thus, the x and y coordinates for such a projectile are:

$$x = a = (88\cos 24°)t$$
$$y = b + 15 = (88\sin 24°)t + 15.$$

Since gravity pulls the motorcycle toward the earth, it affects the y coordinate of the motorcycle. Physicists have determined that at time t, the effect of gravity (g) in this situation is $-1/2gt^2$. Using the acceleration due to gravity, $g = 32$ ft/sec^2, we get the equations that model the motion of the motorcycle.

$$x = (88\cos 24°)t$$
$$y = (88\sin 24°)t + 15 - 16t^2$$

A set of equations in which x and y are defined as functions of another variable are called **parametric equations.** If we substitute decimal values for cos 24° and sin 24°, we get parametric equations that model the motion of the motorcycle.

$$x = (88 \times 0.9135)t = 80.3880t$$
$$y = (88 \times 0.4067)t + 15 - 16t^2 = -16t^2 + 35.7896t + 15$$

EXAMPLE 1

What is the position of the motorcycle after 2 seconds?

Solution: Since we determined the equations of motion to be

$$x = 80.3880t$$
$$y = -16t^2 + 35.7896t + 15.$$

Let $t = 2$ and get

$$x = 80.3880(2) = 160.8$$
$$y = -16(2)^2 + 35.7896(2) + 15 = 22.6.$$

The motorcycle travels 160.8 ft in the horizontal direction and is 22.6 ft above the ground.

EXAMPLE 2

How long will it take for the motorcycle to hit the ground? What is the length of the jump?

Solution: When the motorcycle hits the ground, its y coordinate must be zero. Setting $y = 0$ and using the quadratic formula gives

$$-16t^2 + 35.7896t + 15 = 0$$

$$t = \frac{-35.7896 \pm \sqrt{35.7896^2 - 4(-16)(15)}}{2(-16)}$$

$$= 2.6 \text{ sec.}$$

Substituting $t = 2.6$ into the equation for x will give the length of the jump.

$$x = 80.3880t = 80.3880(2.6) = 209.0 \text{ ft}$$

If we consider any projectile where gravity is the only force acting on it once it has been launched, the following set of equations models the **motion of the projectile.**

Motion of a Projectile

$$x = (v \cos A)t \qquad y = -16t^2 + (v \sin A)t + h \qquad \text{where} \quad \begin{cases} t = \text{time in seconds} \\ A = \text{angle of elevation} \\ v = \text{launch velocity in } {}^{\text{ft}}/_{\text{sec}} \\ h = \text{height at launch in feet} \end{cases}$$

By assuming that gravity is the only force acting on the projectile once it is launched, we are assuming that the projectile is traveling in a vacuum where other factors such as air resistance, spin, drag, and friction are not considered. However, the equations are reasonably accurate in the real world if the projectile is fairly heavy and travels close to the earth at a relatively low speed. Therefore, the following examples focus on the flight of motorcycles, cannonballs, shot puts, and hammer throws rather than those of golf balls, bullets, and arrows.

EXAMPLE 3

A cannonball is shot with an angle of elevation of 60° from a point 10 ft above the ground and a launch velocity of 200 feet per second.

a) What are its equations of motion?

b) What is its position after 6 seconds?

c) How long will it take to hit the ground?

d) How far along the ground from its original position will it hit the ground?

Solution:

a) Using $v = 200$, $h = 10$, and $A = 60°$, we get the equations of motion.

$$x = (200 \cos 60°)t = 100t$$

$$y = -16t^2 + (200 \sin 60°)t + 10 = -16t^2 + 173.2051t + 10$$

b) When $t = 6$, we get

$$x = 100(6) = 600$$

$$y = -16(6)^2 + 173.2051(6) + 10 = 473.2.$$

The cannonball travels 600 ft in the horizontal direction and is 473.2 ft above the ground.

c) When the cannonball hits the ground, its y coordinate must be zero. Setting $y = 0$ and using the quadratic formula, we find t.

$$-16t^2 + 173.2051t + 10 = 0$$

$$t = \frac{-173.2051 \pm \sqrt{173.2051^2 - 4(-16)(10)}}{2(-16)}$$

$$= 10.9 \text{ sec}$$

d) Substituting $t = 10.9$ into the equation for x will tell us how far from its original position it will hit the ground.

$$x = 100t = 100(10.9) = 1090 \text{ ft}$$

A graph of the motion of the cannonball can be shown in a rectangular coordinate system by plotting its (x, y) points at values for time t. The chart that follows shows the x and y coordinates of the cannonball using $x = 100t$ and $y = -16t^2 + 173.2t + 10$ for various values of t.

t	x	y
0	0	10
2	200	292.4
4	400	446.8
6	600	473.2
8	800	371.6
10	1000	142

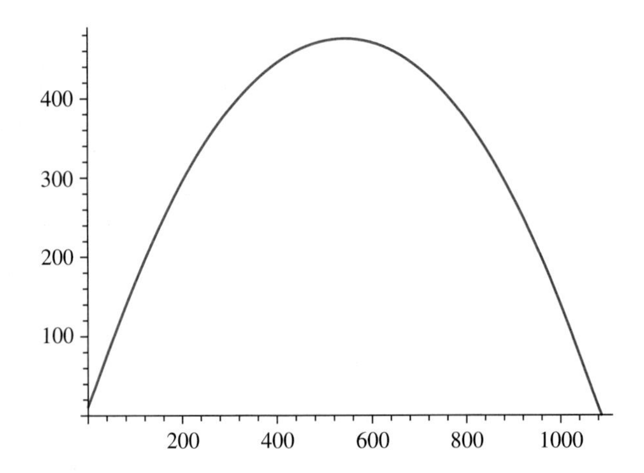

EXAMPLE 4

An Olympic hammer thrower threw the hammer 230 ft. If the hammer was released 3 ft above the ground at a 40° angle of elevation,

a) How long was the hammer in the air?

b) What was the initial velocity of the hammer?

Solution

a) In the formulas for the motion of a projectile, $h = 3$ and $A = 40°$.

$$x = (v \cos 40°)t \approx 0.766vt$$

$$y = -16t^2 + (v \sin 40°)t + 3 \approx -16t^2 + 0.643vt + 3$$

Since the hammer hits the ground at 230 ft, we set $x = 230$ and $y = 0$. We then solve the system of equations for both v and t.

$$0.766vt = 230 \qquad \text{or} \qquad v = \frac{230}{0.766t}$$

$$-16t^2 + 0.643vt + 3 = 0$$
$$-16t^2 + 0.643\left(\frac{230}{0.766t}\right)t + 3 = 0$$
$$-16t^2 + 193.068 + 3 = 0$$
$$-16t^2 = -196.068$$
$$t^2 = 12.2542$$
$$t = 3.5$$

b) Substituting $t = 3.5$ into $v = \dfrac{230}{0.766t}$ gives

$$v = \frac{230}{0.766(3.5)} \approx 85.8.$$

The hammer was in the air for 3.5 sec and was thrown at a velocity of 85.8 ft/sec.

The equations developed in this section will enable you to solve problems involving the motion of projectiles. In some of the problems, the launch velocity will be given in miles per hour (mph). To work with the equations of motion as given, however, you must change miles per hour to feet per second. This can be done by using the following conversion factor.

$$1 \text{ mph} = \frac{22}{15} \text{ ft/sec}$$

For example, 90 mph $= 90\left(\dfrac{22}{15}\right) = 132$ ft/sec.

section 8.6

PROBLEMS ○ Explain ○ Apply ○ Explore

○ Explain

1. What is a projectile and what assumption was made in deriving the equations of motion of a projectile?
2. What real-world factors are not taken into consideration in the equations of motion of a projectile?
3. In the equations of motion of a projectile, the letters x, y, t, A, v, and h are used. What does each letter represent and in what unit is each measured?
4. Why does the section concentrate on the flight of motorcycles, cannonballs, shot puts, and hammer throws rather than the flight of golf balls, bullets, and arrows?
5. What are parametric equations? Why are the equations for the motion of a projectile parametric equations?
6. What happens to the flight of a projectile as the launch angle approaches 90°?

Apply

7. The equations for a projectile are $x = 140(\cos 75°)t$ and $y = -16t^2 + 140(\sin 75°)t + 58$. What is the initial position of the projectile and what is its position after 5 seconds?
8. The equations for a projectile are $x = 320(\cos 15°)t$ and $y = -16t^2 + 320(\sin 15°)t + 200$. What is the initial position of the projectile and what is its position after 6 seconds?
9. The equations for a projectile are $x = 140(\cos 75°)t$ and $y = -16t^2 + 140(\sin 75°)t + 58$. Make a table of values and sketch the graph of the flight of the projectile.
10. The equations for a projectile are $x = 320(\cos 15°)t$ and $y = -16t^2 + 320(\sin 15°)t + 200$. Make a table of values and sketch the graph of the flight of the projectile.
11. A cannonball is shot from a point 15 ft above level ground with an initial velocity of 250 ft/sec. If the launch angle is 36°, how far from the launch point does the ball hit the ground? Make a table of values and sketch a graph of the flight of the cannonball.
12. Suppose the cannonball in Problem 11 is shot from an initial height of 25 ft. How far from the launch point does the ball hit the ground? Make a table of values and sketch a graph of the flight of the cannonball.
13. A high school athlete throws a 12-lb shot put with a launch angle of 39°, initial height of 6 ft, and velocity of 40 ft/sec. What is the length of the throw?
14. If the athlete in Problem 13 makes the same toss but with an angle of 42°, what is the length of the throw?
15. In ancient times, catapults were used to toss rocks in military encounters. If a catapult releases a 25-lb rock from ground level at 60 mph at a 45° angle, how long is the rock in the air and how far from its original position does the rock land?
16. If the catapult in Problem 15 releases the rock at a 47° angle, how long is the rock in the air and how far from its original position does the rock land?
17. If Barry Bonds hits a baseball at a 36° angle from 3 ft above the ground with a velocity of 70 mph, how long is the ball in the air and how far from home plate will the ball land?
18. If the velocity in Problem 17 is 75 mph, how long is the ball in the air and far how from home plate will the ball land?

Explore

19. Suppose baseball pitcher Randy Johnson threw a ball from home plate at 90 mph. If he released the ball 8.5 ft above the ground at 35° angle of elevation, could the ball reach the center field fence 421 ft away?
20. If the ball in Problem 19 was thrown at 80 mph, could the ball reach the center field fence?
21. The world record for the shot put, 75 ft 10.25 in., was set on May 20, 1990, by Randy Barnes of the United States. If the toss was made at a 42° angle from

7.5 ft above the ground, how long was the shot put in the air? At what speed (in mph) did he release the shot?

22. The world record for the women's hammer throw, 249′3″, was set on May 13, 1999, by Mihaela Melinte of Romania. If the toss was made at a 38° angle from 3 ft above the ground, how long was the hammer in the air? At what speed (in mph) did she release the hammer?

23. A baseball is hit from a cliff at a point 600 ft above a lake. If the ball is hit at a 20° angle of elevation with an initial velocity of 50 mph, how far will the ball travel in the horizontal direction before it hits the water?

24. Suppose that, because of physical and equipment limitations, the motorcycle jumper at the beginning of the section (24° at 60 mph) must decrease the launch angle of the jump 2° for each 2-mph increase in launch velocity. Examine the actual launch angles used by motorcycle jumpers, 14° to 24°, and the corresponding velocity to determine the angle and velocity that would produce the longest jump.

CHAPTER 8 SUMMARY

Key Terms, Concepts, and Formulas

The important terms in this chapter are:	**Section**
Acute triangle: A triangle with each angle measuring less than 90°.	8.4
Angle of depression: An angle made between a horizontal line and a line to an object that is below the horizontal line.	8.3
Angle of elevation: An angle made between a horizontal line and a line to an object that is above the horizontal line.	8.3
Bearing: A measure of direction to a north-south line.	8.5
Cosine: In a right triangle, the ratio of the side adjacent to an angle to the hypotenuse.	8.1
Hypotenuse: The side opposite the 90° angle in a right triangle.	8.1
Law of Cosines: The relationship between the sides and angles of a triangle which states that for any $\triangle ABC$,	8.4

$$a^2 = b^2 + c^2 - 2bc \cos A \text{ or}$$
$$b^2 = a^2 + c^2 - 2ac \cos B \text{ or}$$
$$c^2 = a^2 + b^2 - 2ab \cos C.$$

Law of Sines: The relationship among the sides and angles of a triangle which states that for any $\triangle ABC$, 8.4

$$\frac{a}{\sin A} = \frac{b}{\sin B} = \frac{c}{\sin C}.$$

Legs: The sides of a right triangle that form the 90° angle. 8.1

Motion of a projectile: The path of a projectile where gravity is the only force acting on it once it has been launched. Its equations are as follows. 8.6

$$x = (v \cos A)t \qquad \text{where} \quad \begin{cases} t = \text{time in seconds} \\ A = \text{angle of elevation} \\ v = \text{launch velocity in ft/sec} \\ h = \text{height at launch in feet} \end{cases}$$
$$y = -16t^2 + (v \sin A)t + h$$

Parametric equations: A set of equations in which x and y are defined as functions of another variable. 8.6

Pythagorean Theorem: The relationship of the sides of a right triangle that states that for any right triangle, where a and b are the legs and c is the hypotenuse, 8.1

$$a^2 + b^2 = c^2.$$

Pythagorean triple: A set of three natural numbers that satisfy the Pythagorean Theorem. 8.1

Right triangle: A triangle with a 90° angle. 8.1

Sine: In a right triangle, the ratio of the side opposite an angle to the hypotenuse. 8.1

Soh cah toa: An acronym for remembering **s**in **o**pposite/**h**ypotenuse, **c**os **a**djacent/**h**ypotenuse, **t**an **o**pposite/**a**djacent. 8.1

Solving triangles: The process of finding unknown angles and sides of a triangle. 8.2

Tangent: In a right triangle, the ratio of the side opposite an angle to the side adjacent to an angle. 8.1

After completing this chapter, you should be able to:	**Section**
1. Apply the Pythagorean Theorem to right triangles.	8.1
2. Define the sine, cosine, and tangent trigonometric functions in reference to right triangles.	8.1
3. Solve right triangles using the sine, cosine, and tangent functions.	8.2
4. Use trigonometric functions to solve various application problems involving right triangles.	8.3
5. Use the Law of Sines and the Law of Cosines to solve acute triangles.	8.4
6. Use the Law of Sines and Law of Cosines to solve various application problems involving acute triangles.	8.5
7. Use a set of parametric equations to analyze the motion of projectile where gravity is the only force acting on the projectile once it has been launched.	8.6

CHAPTER 8 REVIEW

Section 8.1

1. If the legs of a right triangle are 28 in. and 45 in., how long is its hypotenuse?
2. A 26-in. TV has a square screen with a 26-in. diagonal. How long is each side of the TV screen?
3. Use the triangle below to find an expression for the sin X, cos X, and tan T.

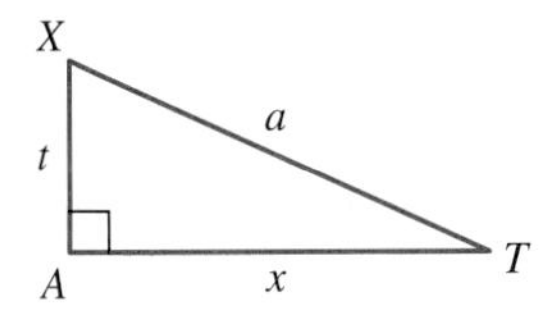

4. Find the value of each of the following.
   a) sin 67° b) tan 44.3°
5. Find the measure of the angle (N) in each of the following.
   a) $\sin N = 0.2345$
   b) $\tan N = 36$.

Section 8.2

In Problems 6 and 7, solve the right triangles.

6. X
   21
   A 28 T

7. S
   36.9
   23°
   U A

8. In $\triangle CUP$, $\angle C$ is a right angle, $U = 74°$, and $c = 102.5$. Solve the triangle.
9. What are the angles of a triangle whose sides are the Pythagorean triple 20, 21, 29?

Section 8.3

10. Explain the difference between an angle of elevation and an angle of depression.
11. From a point 50 m from the bottom of a radio tower, the angle of elevation to the top of the tower is 67°. Find the height of the tower.
12. To determine the altitude of an approaching airplane, Debra and Charles measured the angle of elevation to the plane at the same time from locations that were 2000 ft apart, as shown in the figure. What is the altitude of the airplane?

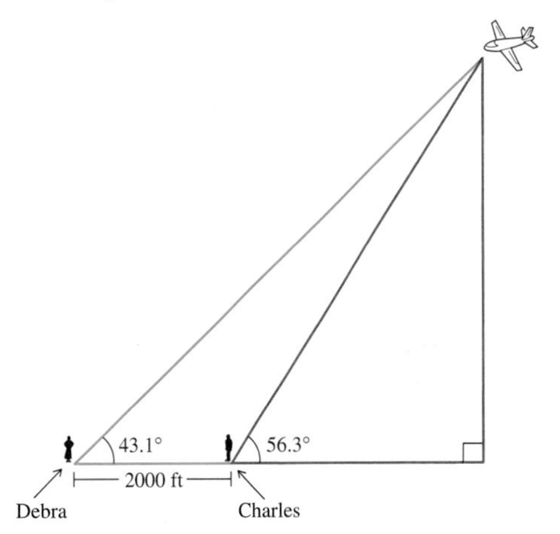

Section 8.4

In Problems 13 and 14, find the parts of the triangles that are not given.

13. A
    6.4
    35°
    62°
    S T

14.

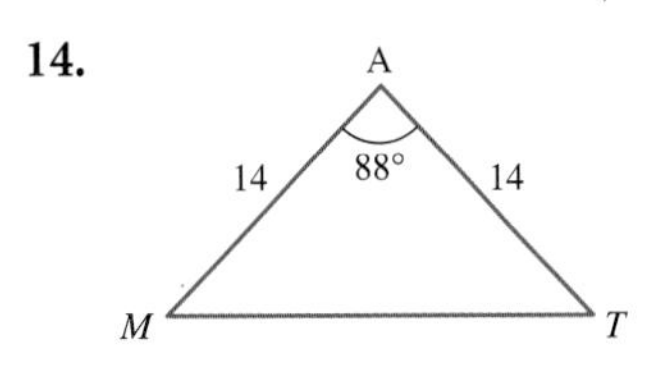

15. Solve $\triangle CAP$, $C = 25.3°$, $A = 74°$, and $p = 102.5$.

16. Solve $\triangle NAP$, $n = 89$, $a = 50.8$, and $p = 97.6$.

Section 8.5

17. To chart the movement of a polar bear, scientists attached a radio transmitter to its neck. Two tracking stations are monitoring the radio signals from the bear. Station B is 10 miles directly east of station A. On Monday, station A measured the direction to the bear at N43°E, and station B measured it at N30°W. Three days later the directions to the bear from the two tracking stations were N24°E and N30°W, respectively. How much farther from station B was the polar bear after those three days?

18. With the use of modern electronic equipment, the distance for various field events in a track meet can be measured without the use of measuring tapes. For example, in the discus throw, after a competitor has made a fair throw from the discus ring to point D, an electronic transmitter placed at D sends a signal to a device in the official's booth above the track. The device then determines the angles at D and at B. Since the distance from the device to the center of the discus ring is a known distance, the length of the throw can be found using trigonometry.

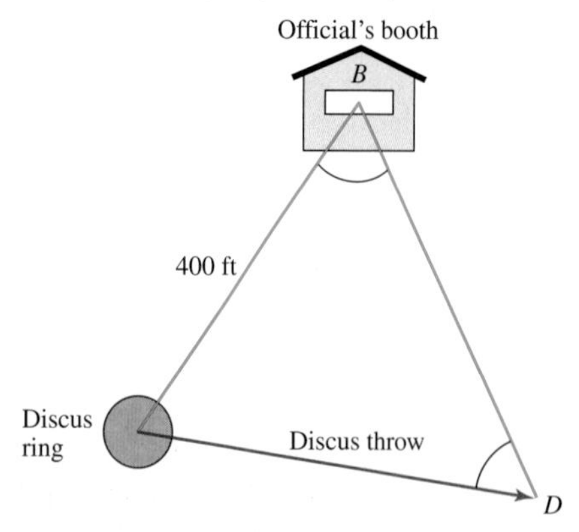

At an invitational track meet, the distance from the receiving device to the center of the discus ring is 400 feet and the radius of the discus ring is 4 feet. If the angle at D is 50.1° and the angle at B is 24.6°, determine the length of a discus throw. Since a discus throw is measured from the outer edge of the discus ring, you need to subtract 4 ft to account for the radius of the ring.

19. Two lookout stations, which are 25 miles apart along the coast on a north-south line, spot an approaching yacht. One lookout station measures the direction to the yacht at N33°E, while the other station measures the direction to the yacht at S62°E. How far is the yacht from each lookout station? How far is the yacht from the coast?

20. To determine the distance to an oil platform in the Pacific Ocean from two points on a beach, a surveyor measures the angles to the platform from the beach. The angle made with the shoreline from one point is 83° and from the other point is 78.6°. If it is 950 yards between the points, what are the distances to the oil platform from each point?

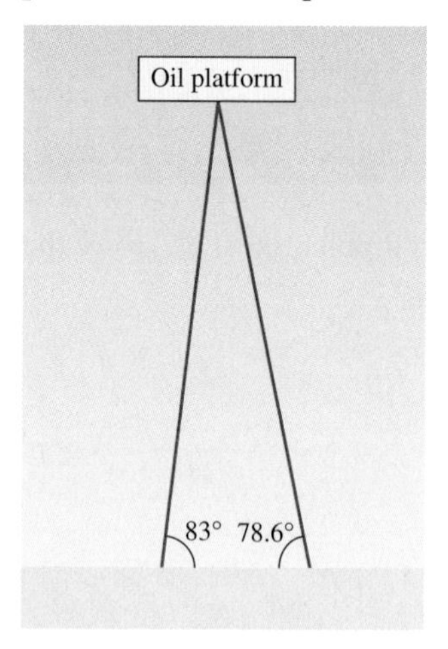

Section 8.6

21. The parametric equations for the flight of a projectile are given by $x = 88(\cos 55°)t$ and $y = -16t^2 + 88(\sin 55°)t + 10$. At what position does the projectile start? Where is the projectile after 4 seconds? Sketch a graph the flight of the projectile.

22. How much farther will a rock thrown at a velocity of 60 mph than at one of 50 mph travel if both are thrown from a point 20 ft above ground with an angle of elevation of 32°?

23. The best shot put by an American woman occurred on June 25, 1988, when Ramona Page tossed the shot put 66.2 ft. If the shot put was released at a point 6 ft above the ground with an angle of elevation of 42°, at what velocity did she release the shot put?

CHAPTER 8 TEST

Solve the triangles in Problems 1–3.

1.

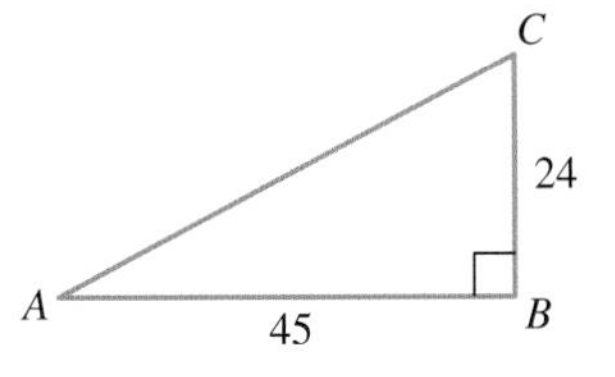

2.

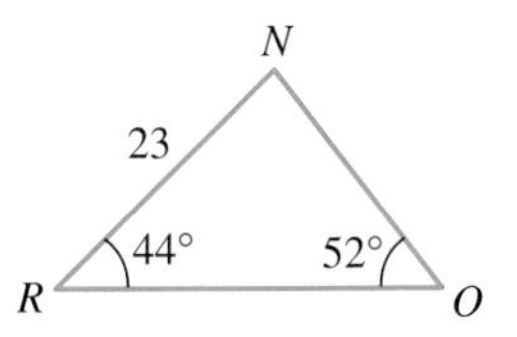

3.

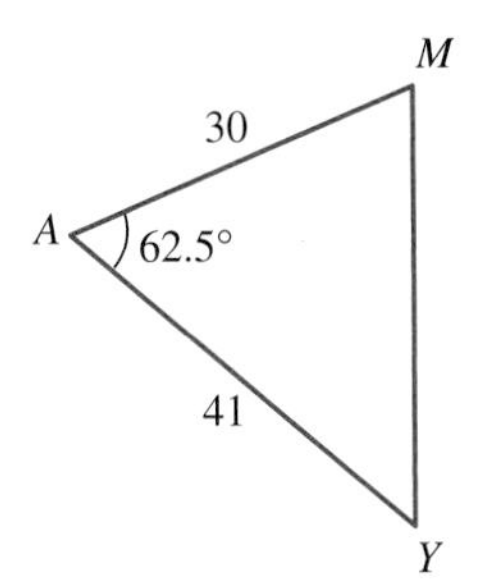

4. In $\triangle CAR$, $\angle C$ is a right angle, $R = 54°$, and $c = 77.8$. Find the length of side a.

5. In $\triangle CAN$, $c = 110$, $a = 73$, and $n = 89$. Find the measure of $\angle C$.

6. What angle does a 30-ft plank make with level ground if it is placed so that it reaches 20 ft up a vertical wall?

7. The Leaning Tower of Pisa in Italy is 177 feet tall and leans at an angle of about 84.5°.

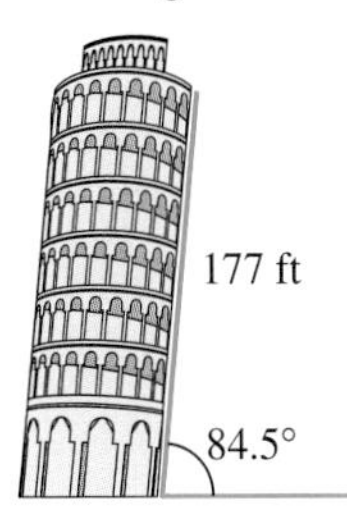

   a) If a ball is dropped from the edge of the tower and strikes the ground at a 90° angle, how far from the base of the tower does the ball hit the ground and how far does the ball fall?

   b) If, because of the wind, a ball dropped from the tower takes a straight path to the ground and strikes the ground at an 80° angle of elevation to the top of the tower, how far from the base of the tower does it hit the ground and how far did the ball fall?

8. Two airplanes leave an airport at 8:00 A.M. on different runways. One flies on a bearing of N64.5°W at 315 mph, and the other flies on a bearing of S27°W at 295 mph. How far apart will the airplanes be at 10:00 A.M.?

9. Three islands are located in the South Pacific. Island *A* is located 252 miles directly west of Island *B*. Island *C* is located to the north of Island *A* and Island *B*. If the distance from Island *A* to Island *C* is 195 miles and the distance from Island *B* to Island *C* is 287 miles, on what bearing should one navigate to go from Island *A* to Island *C* and from Island *B* to Island *C*?

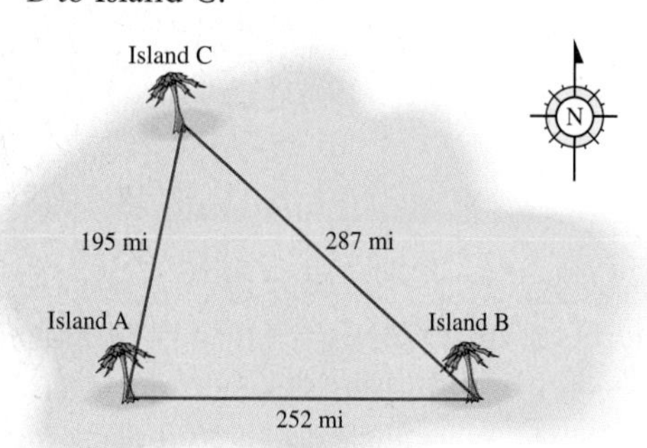

10. If you throw a rock with a velocity of 60 mph from a cliff 90 ft above a lake with an angle of elevation of 35°, how far into the lake will the rock splash?

11. On June 7, 1987, Natalya Lisovskaya of Russia established a world record of the woman's shot put of 74.25 ft. If the shot put was released at a point 6 ft above the ground with an angle of elevation of 40°, at what velocity did she release the shot put?

CHAPTER NINE

Finance Matters

Financial transactions, as seen in Masaccio's *The Rendering of the Tribute Money*, have always been a formal and highly regulated exchange. (Art Resources)

OVERVIEW

Handling financial dealings is one of the responsibilities of an adult in today's society. In this chapter, you will learn how to determine the amount of interest earned in various types of accounts, using both simple and compound interest. You will learn how to calculate the value of an annuity and the amount of a loan payment and to see the benefits of paying off a loan early. If you ever plan to buy a car or save for retirement, this chapter will pique your interest.

A SHORT STORY OF INTEREST

Interest has been charged for the use of money since Babylonian times (2000–500 B.C.). Ancient cuneiform tablets show that the Sumerians used both simple and compound interest. A tablet from 1700 B.C. contains a problem concerning how long it would take an amount of money to double if interest was compounded annually at a certain rate. Interest rates have varied greatly throughout history. Babylonian interest rates were 20%, and rates in Cicero's Rome (c. 60–43 B.C.) reached 48%. By the time of Justinian (A.D. 483–565), rates were limited by law to 6%, but the Indian mathematician Bhāskara (c. 1150) mentions interest rates as high as 60%. In 1304, interest rates in Nuremberg were an astounding 220%.

Historically, lending money for a fee has been opposed. The Greek philosophers Plato and Aristotle wrote against the taking of interest. In fact, Aristotle believed that money is, by nature, "barren" and that the birth of money from money was "unnatural"; therefore, he condemned the taking of interest.

Men such as Martin Luther and Thomas Aquinas argued that the Scriptures specifically forbade the taking of interest. The Roman Catholic Church was officially against usury, the charging of interest, until the 1830s, and penalties for disobeying Church law included being denied a Christian burial. Public feeling about usury was so strong that in the Middle Ages, it was believed that a prolonged rainstorm was caused by the burial of an Italian moneylender in consecrated ground. In an attempt to stop the downpour, the body was disinterred from the grave and thrown into the Po River. In spite of the strong feelings against charging interest, lending still occurred. Powerful banks were founded in Venice in the late 1100s and in Genoa and Barcelona in the early 1400s. The success of these areas in commerce and the arts can be partially attributed to the availability of money from interest-charging banks.

Banks were also started by Jewish families. Since Jews were not associated with the Catholic Church, there was no conflict with charging interest. However, the stereotype that Jews charged unreasonably high interest rates during this time is not necessarily correct. Christian banks operated under a much higher risk in the lending of money because of possible censure by the Church. Because of the risk involved, they often charged higher rates.

One method of avoiding conflict with the Church over the charging of interest was by calling it something else. A lender would agree to lend a sum of money at no charge if the money was returned within a specified time. If not, the borrower paid the lender an additional fee. The fee was computed by the lender (hence being as high as he chose) and was the difference between the lender's current financial standing and what his standing would have been had the money been repaid on time. The word *interest* comes from the Latin *Id quod inter est,* or "that which is in between."

Since the availability of money is necessary in a modern economy, negative attitudes toward interest have diminished. However, if the interest rate is extremely high, it is still called usury and is still viewed with disfavor.

Detail from *The Interior of the Pantheon, Rome* by Giovanni Paolo Panini illustrates the classic architectural style that we associate with financial institutions. (National Gallery of Art)

CHAPTER 9 PROJECTS

Research Projects

1. Contact a loan company and obtain all the necessary information so that you can explain the loan application process, the required fees, and how the calculations are made. Also, include all of the forms used for the application process.
2. Obtain information regarding Private Mortgage Insurance (PMI). Under what circumstances will a lender require PMI? How much does PMI cost? How can you eliminate PMI from your mortgage?
3. Obtain information about the purpose and use of escrow accounts. Find out how, when, and why escrow accounts are used. Discuss the use of escrow accounts both during the loan application process and during the lifetime of the loan.

Math Projects

1. Use the Internet to find the cost of the car of your dreams. Include the cost of all the options you desire. Do further research to find the terms and rates for financing such a car. Perform the calculations to determine the monthly payments to purchase the car. Make sure you include sales tax and license fees, if these apply in your state, and add 3% to the cost for dealer preparation. How much do you actually pay for the car? How much interest do you pay?
2. It is July 1, 2004, and you have just had a new baby and are developing a plan to finance her college education. Research indicates that you can expect 2% growth on long term investments, after accounting for the effect of inflation. This means that you can do the calculations using today's prices. Suppose you deposit $5000 into an account, plan on contributing $200 each month through June 1, 2026, and anticipate the following expenses:
   - **a)** July 1, 2022: Give your daughter $5000 to help her buy a car.
   - **b)** Aug. 1, 2022: Give your daughter $1200 for books and tuition at the local community college.
   - **c)** Jan. 1, 2023: Give your daughter $1200 for books and tuition at the local community college.
   - **d)** July 1, 2023: Give your daughter $1200 for books and tuition at the local community college.
   - **e)** Jan 1. 2024: Give your daughter $1200 for books and tuition at the local community college.
   - **f)** Every month, July 1, 2024, through May 1, 2026: Give your daughter $2500 for university related expenses.

   On June 1, 2026, you close the account and give your daughter the balance. How much do you give her?
3. An amortization schedule is a table listing the amount of principal and interest that are included in each loan payment and the balance of the loan that remains after each payment. Such tables are often included with the annual mortgage statement sent by mortgage companies to their customers. If you have a $1200 loan at 13.2%, compounded quarterly, repaid in 1 year with quarterly payments of $325.15, the amortization table would look as follows.

Payment	Interest Repaid	Principal Repaid	Loan Balance
			$1,200.00
$325.15	$39.60	$285.55	$ 914.45
$325.15	$30.18	$294.97	$ 619.48
$325.15	$20.44	$304.71	$ 314.77
$325.16	$10.39	$314.77	$ —

a) Write an amortization schedule for an 18-month car loan of $18,000. The loan has a 9.84% annual interest, compounded monthly.

b) Suppose a person decides to add an extra $50 each month. Recalculate the amortization schedule.

section 9.1

Percents

The banker of the 1500s, as portrayed by Jan Gossaert in *Portrait of a Merchant,* performed hours of quill-and-paper computation and bookkeeping that would take the personal computer using basic accounting software only a few minutes. (National Gallery of Art)

Percents are frequently used in business transactions. For example, the sales tax rate might be 7.5%, the interest rate on a 30-year loan might be 9.16%, a department store might be offering a 25% discount on all items, or inflation might be increasing at a rate of 1.3% per year. The ability to work with percents is essential to an investigation of the mathematics of finance.

WHAT IS A PERCENT?

The word **percent** means "out of one hundred." A score of 93% on a test means that if the test was worth 100 points you scored 93 points. Percents give us a standard way of comparing data. For example, to compare three quizzes in which your scores were 22 out of 25 points, 27 out of 30 points, and 43 out of 50 points, we represent each score as an amount out of 100.

$$22 \text{ out of } 25 = \frac{22}{25} = \frac{88}{100} = 0.88 = 88\%$$

$$27 \text{ out of } 30 = \frac{27}{30} = \frac{90}{100} = 0.90 = 90\%$$

$$43 \text{ out of } 50 = \frac{43}{50} = \frac{86}{100} = 0.86 = 86\%$$

Since each score is represented as an amount out of 100, we can see that a score of 27 out of 30 was the highest mark of the three quizzes. The three quiz scores, when written as an amount out of 100, can be represented by a fraction, a decimal, or a percent.

Since a percent means out of 100, to change a percent to a decimal, you drop the percent sign and divide by 100. This can be quickly accomplished by moving the decimal point two places to the left and dropping the percent sign.

$$79\% = \frac{79}{100} \text{ or } 79\% = \underset{\Leftarrow\Leftarrow}{7\ 9}.\% = 0.79$$

$$5.3\% = \frac{5.3}{100} \text{ or } 5.3\% = \underset{\Leftarrow\Leftarrow}{\ \ 5}.3\% = 0.053$$

To change a decimal to a percent, use the reverse process. Instead of moving the decimal point two places to the left, move the decimal point two places to the right. Instead of dropping the percent sign, attach the percent sign.

$$0.34 = 0.\underset{\Rightarrow\Rightarrow}{3\ 4} = 34\%$$

$$0.045 = 0.\underset{\Rightarrow\Rightarrow}{0\ 4}\ 5 = 4.5\%$$

To change a fraction to a percent, first change the fraction to a decimal by dividing the denominator into the numerator and then change the decimal to a percent.

$$\frac{4}{5} = 4 \div 5 = 0.8 = 0.\underset{\Rightarrow\Rightarrow}{8\ \ } = 80\%$$

$$\frac{11}{16} = 11 \div 16 = 0.6875 = 0.\underset{\Rightarrow\Rightarrow}{6\ 8}\ 75 = 68.75\%$$

EXAMPLE 1

Change 33.65% to a decimal.

Solution: To change 33.65% to a decimal, move the decimal point two places to the left and drop the percent sign.

$$33.65\% = \underset{\Leftarrow\Leftarrow}{3\ 3}.65\% = 0.3365$$

EXAMPLE 2

Change $\frac{73}{121}$ into a percent rounded off to the nearest tenth of a percent.

Solution: To change $\frac{73}{121}$ into a percent, divide 73 by 121, move the decimal point two places to the right, and round off the result to the nearest tenth.

$$\frac{73}{121} = 73 \div 121 \approx 0.6033 = 0.\underset{\Rightarrow\Rightarrow}{6\ 0}\ 33 \approx 60.3\%$$

FINDING THE PERCENT OF A NUMBER

Percents are often given as a rate that is used to find a portion of a number. A sales tax rate of 7.5% means that the tax is 7.5% of the price. A 25% discount means that the amount of the discount is 25% of the price. When a percent is used as a rate in this context, you must find the percent of a number. A percent of a number is found by multiplying the percent and the number. You must remember to convert the percent to a decimal before performing the multiplication.

EXAMPLE 3

Find 7.5% of 360.

Solution: 7.5% is changed to a decimal and multiplied by 360.

$$7.5\% \text{ of } 360 = 0.075 \times 360 = 27$$

EXAMPLE 4

If the sales tax rate is 8%, how much sales tax is charged on a $99.49 coat and how much is actually paid for the coat?

Solution: The sales tax is 8% of the cost of the coat.

$$\begin{aligned} 8\% \text{ of } \$99.49 &= 0.08 \times \$99.49 \\ &= \$7.9592 \\ &\approx \$7.96 \end{aligned}$$

Note: Dollar amounts should be rounded off to the nearest cent. The sales tax is added to the cost of the coat to get the actual amount paid.

$$\text{Cost including sales tax} = \$99.49 + \$7.96 = \$107.45$$

EXAMPLE 5

A television set retails for $759.95. During a sale, a store gives a 25% discount. What is the sale price?

Solution: The discount is 25% of the $759.95 retail price.

$$\text{Discount} = 0.25 \times \$759.95 \approx \$189.99$$

The discount is subtracted from the price of the TV to get the final price.

$$\text{Sale price} = \$759.95 - \$189.99 = \$569.96$$

EXAMPLE 6

The amount paid for gasoline at a service station was \$17.99. If this amount included 28.5% in various taxes, what is the cost of the gasoline without tax?

Solution: The \$17.99 amount is the sum of the cost of the gasoline and the 28.5% tax. Let $x =$ the cost of the gasoline without the tax. The tax is 28.5% of x or $0.285x$. Since the total cost of the gasoline is \$17.99, we have the following equation.

$$x + 0.285x = \$17.99$$
$$1.285x = 17.99$$
$$\frac{1.285x}{1.285} = \frac{17.99}{1.285}$$
$$x = \$14.00$$

MARKUPS AND MARKDOWNS

Percents are also used as a basis for **marking up** (increasing) or **marking down** (decreasing) the price of an item. For example, a retailer might mark up the wholesale cost of an item 40% before it is sold in the store or mark down the retail price of an item 20% during a sale.

EXAMPLE 7

To make a profit, a college bookstore marks up the wholesale cost of books 40% to obtain the retail price for books sold in the bookstore.

a) If an algebra book has a wholesale cost of \$24.00, what is the retail price?

b) If the retail price marked in a biology book is \$49.49, what is the wholesale cost of the book?

Solution:

a) The amount of markup is obtained by multiplying the percent of the markup by the cost. The amount of the markup is then added to the wholesale cost.

$$\text{Retail price} = \text{Cost} + \text{Markup}$$
$$= \$24.00 + 0.40(24.00)$$
$$= \$24.00 + \$9.60$$
$$= \$33.60$$

b) Let $x =$ the wholesale cost of the book.

$$\text{Retail price} = \text{Cost} + \text{Markup}$$
$$\$49.49 = x + 0.40x$$
$$\$49.49 = 1.4x$$
$$\frac{\$49.49}{1.4} = \frac{1.4x}{1.4}$$
$$\$35.35 = x$$

EXAMPLE 8

The retail price of a dress was \$78.00. However, when the dress was placed on the sale rack, it was marked down 20%. Then, at a Moonlight Sale, the dress was marked down an additional 30%.

a) What was the cost of the dress after the final 30% markdown?

b) Is the final sale price equal to a 50% discount on the original price of the dress?

Solution:

a) The amount of the markdown is obtained by multiplying the percent of the markdown by the price. The amount of the markdown is then subtracted from the price. To find the final sale price in this problem, find the price after a 20% markdown. Then mark down that price by an additional 30%.

$$\begin{aligned}\text{Price after the 20\% markdown} &= \text{Price} - \text{Markdown}\\ &= \$78.00 - 0.20(\$78.00)\\ &= \$78.00 - \$15.60\\ &= \$62.40\\ \text{Price after the 30\% markdown} &= \$62.40 - 0.30(\$62.40)\\ &= \$62.40 - \$18.72\\ &= \$43.68\end{aligned}$$

b) To answer this question, compare the price after a 50% discount and the \$43.68 sale price obtained previously.

$$\begin{aligned}\text{Price after the 50\% markdown} &= \text{Price} - \text{Markdown}\\ &= \$78.00 - 0.50(\$78.00)\\ &= \$78.00 - \$39.00\\ &= \$39.00\end{aligned}$$

The sale price after a 50% discount is less than the price after successive 20% and 30% markdowns.

PERCENT INCREASE AND DECREASE

Percents are also used as a standard way to describe changes in prices, costs, profits, and other quantities. You might read that the cost of living increased 1.2% or the sales for a company decreased 12.6%. How are such percents determined? They are calculated by taking the amount of the increase or decrease (A), dividing it by the starting or base amount (B), and changing the resulting decimal into a percent by moving the decimal point two places to the right.

$$\textbf{Percent increase or decrease} = \frac{A}{B} \quad \text{where} \quad \begin{cases} A = \text{amount of increase or decrease} \\ B = \text{base (starting) amount} \end{cases}$$

EXAMPLE 9

If the price of a dozen eggs increased from \$1.25 to \$1.40, what was the percent increase in the price of eggs?

Solution: The amount of the increase is \$1.40 − \$1.25 = \$0.15. The base (starting) amount is \$1.25.

$$\text{Percent increase} = \frac{A}{B}$$

$$= \frac{0.15}{1.25} = 0.12 = 12\%$$

EXAMPLE 10

In 2002, the sales for a company were \$2,475,000; in 2003, the sales dropped to \$1,950,000. What was the percent decrease in sales?

Solution: The amount of the decrease was \$2,475,000 − \$1,950,000 = \$525,000. The base (starting) amount is \$2,475,000.

$$\text{Percent decrease} = \frac{A}{B}$$

$$= \frac{525{,}000}{2{,}475{,}000} \approx 0.2121 \approx 21.2\%$$

section 9.1

PROBLEMS ▸ Explain ▸ Apply ▸ Explore

▸ Explain

1. What is a percent?
2. How do you change a decimal to a percent?
3. How do you change a percent to a decimal?
4. How do you change a fraction to a percent?
5. If the sales tax rate in your county is 5.2%, how is the actual amount you must pay for a taxable item determined?
6. If a store is giving a 30% discount on all items, how is the discount price for each item determined?
7. What is the difference between a markup and a markdown? How is each calculated?
8. In determining the percent increase in sales for a company, what information is needed and how is the percent increase calculated?
9. In determining the percent decrease in enrollment at a college, what information is needed and how is the percent decrease determined?

Apply

10. Change the following percents to decimals.

a) 9% b) 0.7% c) 17.5% d) 92.16% e) 120%

11. Change the following percents to decimals.

a) 5% b) 6.7% c) 9.125% d) 234% e) 0.03%

12. Change the following decimals to percents.

a) 0.02 b) 0.123 c) 0.6075 d) 1.34 e) 0.003

13. Change the following decimals to percents.

a) 0.35 b) 0.06 c) 0.0025 d) 2.015 e) 0.75

14. Change the following fractions to exact percents.

a) 3/4 b) 7/8 c) 31/50 d) 21/32 e) 39/40

15. Change the following fractions to percents accurate to the nearest tenth of a percent.

a) 5/7 b) 2/3 c) 16/29 d) 34/995 e) 5/64

16. If the sales tax rate is 4.5%, find the amount of tax on items that are priced at

a) $34.00 b) $125.95 c) $675.79

17. If the sales tax rate is 7%, find the amount of tax on items that are priced at

a) $34.00 b) $125.95 c) $675.79

18. Find the original price of an item that is marked $74.00 if that marked price represents a discount of

a) 20% b) 30% c) 45%

19. Find the original price of an item that is marked $17.50 if that marked price represents a discount of

a) 15% b) 40% c) 75%

20. A sporting goods store marks up the wholesale price of fishing lures 60%. Find the retail price for a lure that wholesales for

a) $1.79 b) $2.99 c) $3.50

21. A toy store marks up the wholesale price of toys 35%. Find the retail price of a toy that wholesales for

a) $20.00 b) $45.60 c) $109.99

22. A jewelry store marks up the cost of a ring 40%. Find the cost of a ring that retails for

a) $79.95 b) $285.99 c) $1599.00

23. A hardware store marks up the cost of paint 25%. Find the cost of a gallon of paint that retails for

a) $17.00 b) $25.50 c) $29.99

24. Find the percent increase in price.

a) $5.00 to $5.75 b) $32.50 to $34.58 c) $25,000 to $27,000

25. Find the percent decrease in price.

a) $9.00 to $7.79 b) $432 to $216.50 c) $4999 to $3000

Explore

26. If the inflation rate is predicted to be 2.3% a year, how much will $175.00 in groceries cost a year later?

27. A compact disc player now sells for $379.99. If the cost for consumer goods decreased 0.3% in the past month, what was the cost of the compact disc player one month ago?

28. If a radio that costs $100.00 is marked up 35% and later the marked-up price is marked down 35%, how much below cost is the radio being sold for?

29. The manufacturer's selling price on a pair of shoes is $55.00.

a) If a shoe outlet marks up that price 40%, what is the retail price of the pair of shoes?

b) If a shoe outlet marks down that retail price by 40%, how much of a loss is the shoe outlet taking on the pair of shoes?

30. According to the IRS Tax Rate Schedule for the year 2001, if you are single with taxable income over $65,550 but not over $136,750, your tax is $14,645 plus 30.5% of the amount over $65,550. If your taxable income is $70,000, how much tax do you owe?

31. According to the IRS Tax Rate Schedule for the year 2001, if you are married and filing a joint return with taxable income over $45,200 but not over $109,250, your tax is $6780 plus 27.5% of the amount over $45,200. If your taxable income is $105,000, how much tax do you owe?

32. Which stock showed the greater percent increase in selling price:

a) A share of Netco stock that sold for $54.50 on Monday and $57.25 on Friday or

b) A share of Trapco stock that sold for $20.75 on Monday and $22.25 on Friday?

33. Which stock showed the greater percent decrease in selling price:

a) A share of Mitek stock that sold for $37 3/8 on 11/8/02 and $29 1/2 on 11/8/03 or

b) A share of Yotek stock that sold for $8 5/8 on 11/8/02 and $6 1/4 on 11/8/03?

34. A bank made $234,567,000 in loans during the second quarter of the year. If 62% of that amount was made for home loans and 85% of the amount of home loans was made for single-family dwellings, how much money was loaned for single-family dwellings?

35. To encourage large orders, a manufacturer offers a series of discounts. The discounts for orders of amount (A) are as follows:

If $\$1000 \le A \le \$5000$, the discount is 8.5%.

If $\$5000 < A \le \$15{,}000$, the discount is 8.5% on the first $5000 and 9.5% on the next $10,000.

If $A > \$15{,}000$, then the amount of the discount is 8.5% of the first $5000, 9.5% of the next $10,000, and 10.5% of the amount over $15,000.

What is the discount given on an order of

a) $4200? **b)** $9890? **c)** $19,750?

Simple Interest

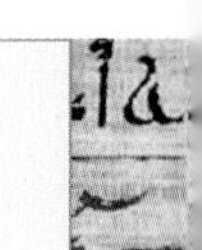

Interest is the fee charged for the use of money. If we deposit money in a bank, the bank may use the money to provide loans for other customers. In return for the use of the money, the bank will pay a certain percentage of the amount invested. In a similar manner, if we borrow money from a bank, we will be required to pay interest to the bank in return for the privilege of using the money.

One way to calculate interest is to use **simple interest.** Simple interest indicates that the interest is earned only at the end of the specified time and is earned only on the amount deposited.

The formula used to calculate simple interest is given by the following.

> Simple Interest
>
> $$I = Prt \quad \text{where} \quad \begin{cases} I = \text{Interest} \\ P = \text{Principal or amount deposited} \\ r = \text{Interest rate} \\ \end{cases}$$

EXAMPLE 1

If you deposit \$1500 in a bank for three years at an annual rate of 9%, find the amount of simple interest you will earn.

Solution: Since $I = Prt$, $I = (1500)(0.09)(3) = \$405$.

EXAMPLE 2

If you deposit \$1500 in the bank for three years and the bank is paying simple interest of 1.5% each month, find the amount of interest earned.

Solution: Because the time and the interest rate are not given in the same units of time, you cannot merely substitute the numbers into the formula as we did in Example 1.

Since the rate is per month, change the time of 3 years into months by multiplying by 12. This gives

$$t = 3 \times 12 = 36 \text{ months.}$$

We now have the interest rate per month and the time in months, so we can substitute $P = \$1500$, $r = 0.015$, and $t = 36$ months into $I = Prt$, giving

$$I = (1500)(0.015)(36) = \$810.$$

EXAMPLE 3

If you earned \$500 on a \$12,000 investment that earned simple interest for 18 months, what is the annual interest rate?

Solution: To find the annual interest rate, first convert the time into years. This gives

$$t = 18 \div 12 = 1.5 \text{ years.}$$

Now substituting $P = \$12{,}000$, $I = \$500$, and $t = 1.5$ into the formula $I = Prt$ gives

$$500 = (12{,}000)(1.5)r$$
$$500 = 18{,}000r$$
$$r = \frac{500}{18{,}000} \approx 0.0278 = 2.78\%.$$

If interest is left in an account along with the principal, the amount in the account is the total of the principal and the interest. Therefore, the amount (A) is given by

$$A = P + I.$$

Substituting $I = Prt$ gives $A = P + Prt$. Factoring P out of the right-hand side gives $A = P(1 + rt)$.

Amount in a Simple Interest Account

$$A = P(1 + rt) \quad \text{where} \quad \begin{cases} A = \text{Amount in the account (including interest)} \\ P = \text{Principal} \\ r = \text{Interest rate} \end{cases}$$

EXAMPLE 4

If you deposit \$1500 in a bank for three years and the bank pays 9% simple interest per year, find the amount in the account after the interest has been added to the account.

Solution: Substituting $P = \$1500$, $r = 0.09$, and $t = 3$ into $A = P(1 + rt)$ gives

$$A = 1500[1 + 0.09(3)] = 1500(1.27) = \$1905.$$

Simple interest is sometimes used to calculate the interest charged on credit cards. The following example shows how this works.

EXAMPLE 5

On a typical credit card statement, you will see a table such as the one below.

Current Billing Period: 30 days	Average Daily Balances	Daily Periodic Rates	Annual Percentage Rates	Periodic Finance Charges
Purchases	\$2,315.66	0.04655%	16.99%	\$32.34
Cash Advances	\$500.00	0.05477%	19.99%	\$ 8.22

How are the periodic finance charges determined?

Solution: In the statement, the average daily balance is the principal (P), daily periodic rates is (r), and the current billing period is the time (t). Since the rates on purchases and cash advances are different, the interest on each is calculated separately.

Using $I = Prt$, we get the following.

Purchases:	Cash advances:
$I = P \times r \times t$	$I = P \times r \times t$
$= 2315.66 \times 0.0004655 \times 30$	$= 500 \times 0.0005477 \times 30$
$= 32.34$	$= 8.22$

The final example in this section considers the case in which you earn interest on any interest that was added to your account in earlier years.

EXAMPLE 6

Suppose a bank pays Bill 6% simple interest each year on the amount in the account for the entire year. He deposits \$1000 on January 1. If Bill lets the interest accumulate in the account, how much is in the account after 1 year? If the simple interest for the second year is determined from the amount in the account after the first year's interest is added, how much is in the account after two years? Repeat this process to determine the amount in the account after three years.

Solution: Using the formula $A = P(1 + rt)$, with $P = \$1000$, $r = 0.06$, and $t = 1$ year, we have

$$A = 1000(1 + 0.06) = \$1060.00.$$

Thus, during the second year, \$1060 is the principal in the account. The amount in the account at the end of the second year is

$$A = 1060(1 + 0.06) = \$1123.60.$$

Similarly, during the third year, \$1123.60 is the principal in the account, so the amount at the end of the third year is

$$A = 1123.60(1 + 0.06) = \$1191.02.$$

As you can see from Example 6, using this method to calculate the amount in the account after 30 years would be a very tedious process. The amount in the account during each intermediate year must be calculated before you arrive at the final amount. In the next section, we will find an easier way to do this.

PROBLEMS ○ Explain ○ Apply ○ Explore

○ Explain

1. Explain what is meant by *principal.*
2. Explain how simple interest is calculated.
3. If you know the simple interest rate, the amount of interest earned, and the principal, explain how you can determine the length of time of the investment.

4. If you know the total amount in an account earning simple interest, the principal, and the time that the principal has been in the account, explain how you can determine the interest rate.
5. If the interest rate is given as a monthly rate and time is given in years, what step(s) must you take to determine the simple interest?
6. If the interest is given as an annual rate and time is given in months, what step(s) must you take to determine the simple interest?

Apply

In Problems 7–22, use the simple interest formula $I = Prt$ and $A = P(1 + rt)$ and the given information to find the indicated value.

7. $P = \$2000$ 5% annually $t = 4$ years Find I.
8. $P = \$3000$ 4% annually $t = 6$ years Find I.
9. $P = \$2500$ 0.75% monthly $t = 3$ months Find I.
10. $P = \$3500$ 0.5% monthly $t = 6$ months Find I.
11. $P = \$3000$ 4% annually $t = 6$ years Find A.
12. $P = \$2000$ 5% annually $t = 4$ years Find A.
13. $P = \$35{,}000$ 4% annually $t = 4$ months Find A.
14. $P = \$25{,}000$ 5% annually $t = 3$ months Find A.
15. $P = \$1000$ 0.0247% daily $t = 1$ year Find A.
16. $P = \$1000$ 0.0329% daily $t = 1$ year Find A.
17. $A = \$3100$ 4% annually $t = 6$ years Find P.
18. $A = \$6000$ 5% annually $t = 4$ years Find P.
19. $P = \$3000$ 4% annually $A = \$3300$ Find t.
20. $P = \$2000$ 5% annually $A = \$2350$ Find t.
21. $P = \$2000$ $A = \$3000$ $t = 2$ years Find r (annually).
22. $P = \$2500$ $A = \$2600$ $t = 3$ months Find r (monthly).

Explore

23. Nina deposits \$3400 into a savings account earning simple interest at 6.3% annually. She intends to leave the money in the bank for eight months. How much money, including both principal and interest, can she withdraw at the end of this time?
24. Dylan deposits \$4700 into a savings account earning simple interest at 4.6% annually. He intends to leave the money in the bank for six months. How much money, including both principal and interest, can he withdraw at the end of this time?
25. You win \$4700 in a charity drawing and decide to deposit it into a savings account earning simple interest at 5.51% annually. You intend to leave the money in the bank until the account is worth \$5000. How long must you wait?
26. Ed purchased 500 shares of stock for \$23.63 per share. After holding the stock for 18 months, he sells the stock for \$26.37 per share. Assuming that there are

no commissions on either the purchase or sale of the stock, what was the annual simple interest rate earned on the investment?

27. Angelina purchased 2000 shares of stock for $87.88 per share. After holding the stock for 18 months, she sells the stock for $93.12 per share. Assuming that there are no commissions on either the purchase or sale of the stock, what was the annual simple interest rate earned on the investment?

28. Frances earns $6350 for completing a special project at work and decides to deposit it into a savings account. What annual simple interest rate is needed to have $7000 in one year?

29. The Radoviches need $20,000 as the down payment for a house. If they currently have $18,500 in a bank account, what annual simple interest rate must the bank pay the Radoviches so that the account will have the total down payment after one year?

30. In 1993, Salvador purchased a 10-year bond with a face value of $25,000. The purchase price was $14,285.71. If the bond is redeemed for its face value in 2003, what is the simple interest rate on the bond?

31. Phuong Lan purchased a tax-free bond with a face value of $50,000. The purchase price was $31,250. If the bond is redeemed for its face value in ten years, what is the simple interest rate on the bond?

32. Stanley Chang is investing $13,157.89 in bonds that can be redeemed for $35,000. If the annual simple interest rate is 8.3%, how long must the bonds be held?

33. **So You Want to Be a Millionaire:** You have purchased bonds for $200,000 and will redeem them for $1,000,000. If the annual simple interest rate earned by the bonds is 8.5%, how long must you hold the bonds?

34. Suppose you have a savings account that earns 5% interest annually and you let the interest accumulate in your account at the end of each year. If the account initially has a balance of $10,000, use the procedure given in Example 6 to find how much is in the account at the end of each of the next three years.

35. Suppose you have a savings account that earns 2.5% interest annually and you let the interest accumulate in your account at the end of each year. If the account initially has a balance of $1000, use the procedure given in Example 6 to find how much is in the account at the end of each of the next three years.

36. On your credit card statement, you see the table given below. Find the periodic finance charge on your purchases.

Current Billing Period: 30 days	Average Daily Balances	Daily Periodic Rates	Annual Percentage Rates	Periodic Finance Charges
Purchases	$715.66	0.02712%	9.99%	
Cash Advances	$920.00	0.03833%	13.99%	

37. On a typical credit card statement, you see the table given above. Find the periodic finance charge on your cash advances.

Compound Interest

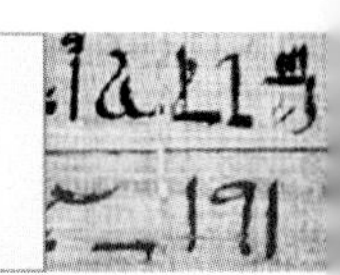

In Example 6 of Section 9.2, we discussed what would happen if the interest was added to the account at the end of each year. Each successive year's earnings included interest on the principal interest left in the account from the year before. This method of calculation is called **compound interest.**

The problem turned out to be difficult because the amount in the account had to be computed for each year to get the final result. If the interest was to be added at the end of each month, over 100 calculations would be necessary for a nine-year time period. Fortunately, there is a better way to do this.

Let us examine what is done when the amount in the account is computed. We will use the formula $A = P(1 + rt)$ with $t = 1$, giving $A = P(1 + r)$. The amount in the account at the beginning of the year is multiplied by $(1 + r)$ to give the amount in the account at the end of the year. This final amount is used as the principal for the next year.

Continuing this process, we have the following table.

Year Number	Amount at the Beginning of the Year	Amount at the End of the Year
1	P	$P(1 + r)$
2	$P(1 + r)$	$[P(1 + r)](1 + r) = P(1 + r)^2$
3	$P(1 + r)^2$	$[P1 + r)^2](1 + r) = P(1 + r)^3$
4	$P(1 + r)^3$	$[P(1 + r)^3](1 + r) = P(1 + r)^4$
5	$P(1 + r)^4$	$[P(1 + r)^4](1 + r) = P(1 + r)^5$

Notice that the exponent is the same as the year. This pattern gives a formula to obtain the amount in the account at the end of any year.

$$A = P(1 + r)^t \qquad \text{where} \quad \begin{cases} r = \text{annual interest rate} \\ t = \text{time in years} \end{cases}$$

Now try Example 6 of Section 9.2 again.

EXAMPLE 1

Suppose a bank pays Bill 6% interest, compounded annually. He deposits $1000 on January 1. If he lets the interest accumulate in the account, how much does he have after one year? After two years? After three years?

Solution: By the formula $A = P(1 + r)^t$, after one year Bill has

$$A = 1000(1 + 0.06) = \$1060.00.$$

For the second year, Bill has

$$A = 1000(1 + 0.06)^2 = 1000(1.12360) = \$1123.60.$$

For the third year, Bill has

$$A = 1000(1 + 0.06)^3 = 1000(1.19102) = \$1191.02.$$

Note that the results of the calculations are the same as in Example 6 of Section 9.2. More important, we can immediately calculate the results for any year, without computing the amount from any previous year.

EXAMPLE 2

Carol is depositing \$1500 into an account earning 9%, compounded annually. How much money will be in the account after 25 years?

Solution: Use the formula $A = P(1 + r)^t$, with $P = \$1500$, $r = 0.09$, and $t = 25$.

$$A = 1500(1 + 0.09)^{25}$$
$$= 1500(8.623081) = \$12{,}934.62$$

Now suppose that the interest is added to the account at the end of every month rather than at the end of the year. If the annual interest rate is 6%, the monthly interest rate is $6\% \div 12 = 0.06 \div 12 = 0.005$. Over the course of one year, interest is compounded (added to the account) 12 times. Thus, the expression for the amount in the account at the end of the year is

$$A = P\left(1 + \frac{0.06}{12}\right)^{12(1)} = P(1.005)^{12}.$$

This suggests the compound interest formula for any rate of compounding.

Compound Interest Formula

$A = P(1 + r)^n$ where

- A = amount in the account after n time periods
- P = present value of the account (the amount deposited)
- r = periodic interest rate
  = (annual interest rate ÷ number of periods per year)
- n = number of time periods
  = (number of years × number of periods per year)

A **period** is the time interval between successive additions of interest to an account. For example, if interest is compounded monthly, the number of periods per year is 12. If the interest is compounded daily, the number of periods per year is 365.

EXAMPLE 3

Carol is depositing \$1500 into an account earning 9%, compounded monthly. How much money will be in the account after 25 years?

Solution: Use the formula $A = P(1 + r)^n$, $P = \$1500$, $r = 0.09 \div 12 = 0.0075$, and $n = 25 \times 12 = 300$.

$$A = 1500(1 + 0.0075)^{300}$$
$$= 1500(9.408415) = \$14{,}112.62$$

Compare this result with that of Example 2. Notice that by compounding more frequently, the amount of interest earned has increased. What do you think will happen if interest is compounded daily?

EXAMPLE 4

If \$1500 is deposited into an account earning 9%, compounded daily, how much money will be in the account after 25 years?

Solution: Use $r = 0.09 \div 365 = 0.0002466$ and $n = 25 \times 365 = 9125$.

$$\begin{aligned} A &= 1500(1 + 0.0002466)^{9125} \\ &= 1500(9.4872386) = \$14{,}230.86 \end{aligned}$$

(Note: If more digits are used in the value for r, $A = \$14{,}227.66$.)

Starting with the compound interest formula, we can solve for the amount in the account (A), or the **present value** (P). The next problem involves solving for P, the present value of the account.

EXAMPLE 5

How much money must be deposited into an account that earns 6%, compounded monthly, so that \$20,000 can be withdrawn in seven years?

Solution: Use $r = 0.06 \div 12 = 0.005$ and $n = 7 \times 12 = 84$.

$$\begin{aligned} 20{,}000 &= P(1 + 0.005)^{84} \\ 20{,}000 &= P(1.520370) \\ P &= 20{,}000 \div 1.520370 = \$13{,}154.69 \end{aligned}$$

HOW LONG WILL IT TAKE?

Most people don't have \$13,154.69 available to deposit into an account. Suppose we have only half this amount, \$6577.35. If this money is deposited into the account used in Example 5, will it take twice as long to reach the desired \$20,000?

Solving this type of problem involves solving for the exponent in an equation and requires using logarithms. Let's look at a problem with a known solution to review how this is done.

EXAMPLE 6

Solve $5^n = 25$.

Solution: We know that the answer to this question is $n = 2$. The algebra required to reach this answer is as follows.

Start with the original equation: $5^n = 25$

Take the natural logarithm of both sides: $\ln(5^n) = \ln(25)$

Move the exponent to the front: $n \ln(5) = \ln(25)$

Use a calculator to do the arithmetic: $n = \dfrac{\ln(25)}{\ln(5)} \approx \dfrac{3.218876}{1.609438} = 2$

Now that we have the necessary math tools, let's solve our problem.

EXAMPLE 7

If \$6577.35 is deposited into an account that earns 6%, compounded monthly, how long will it take for the account to be worth \$20,000?

Solution: Using the formula for compound interest, $A = P(1 + r)^n$, with $r = 0.06 \div 12 = 0.005$, $P = \$6577.35$, and $A = \$20{,}000$, we get the following.

$$20{,}000 = 6577.35(1 + 0.005)^n$$
$$\frac{20{,}000}{6577.35} = 1.005^n$$
$$3.040738 = 1.005^n$$
$$\ln(3.040738) = \ln(1.005^n)$$
$$\ln(3.040738) = n \ln(1.005)$$
$$n = \frac{\ln(3.040738)}{\ln(1.005)} = 223$$

Now we need to either think or worry. If $n = 223$ is measured in years, we will never live to see this money. The good news is that since the interest in the account was compounded monthly, n is measured in months. Therefore, we will have the \$20,000 in 223 months or $223 \div 12 = 18.58$ years, or 18 years, 7 months. Notice that by reducing the initial deposit by half, we must more than double the number of years to reach the same account balance.

SOLVING FOR THE INTEREST RATE

In some instances, you might want to determine the interest rate. Doing so requires the use of one of the rules of exponents. As we have done before, let's first look at an example with a known solution and then proceed to a problem using real-life numbers.

EXAMPLE 8

Solve $x^3 = 8$.

Solution: The algebra technique used to solve for x is to use another exponent, one that will eliminate the existing exponent, 3.

	$x^3 = 8$
Raise both sides to the 1/3 power:	$(x^3)^{1/3} = 8^{1/3}$
On the left side, the exponents multiply to 1:	$x^1 = 8^{1/3}$
Complete the calculation:	$x = 2$

EXAMPLE 9

If \$6577.35 is deposited into an account, with interest compounded monthly, what annual rate is needed to accumulate \$20,000 in the account after seven years?

Solution: Use $n = 7 \times 12 = 84$, $P = \$6577.35$, and $A = \$20{,}000$.

$$20{,}000 = 6577.35(1 + r)^{84}$$
$$\frac{20{,}000}{6577.35} = (1 + r)^{84}$$
$$3.040738 = (1 + r)^{84}$$
$$3.040738^{1/84} = \left[(1 + r)^{84}\right]^{1/84}$$
$$1.013327 = 1 + r$$
$$r = 0.0133$$

Since the interest is compounded monthly, we have an annual rate of $0.0133 \times 12 = 0.1596$, or 15.96%.

CREDIT CARDS AND COMPOUND INTEREST

EXAMPLE 10

As seen in Example 5 of Section 9.2, on a typical credit card statement, you will see a table such as the one below.

Current Billing Period: 30 days	Average Daily Balances	Daily Periodic Rates	Annual Percentage Rates	Periodic Finance Charges
Purchases	$2,315.66	0.04655%	16.99%	$32.34
Cash Advances	$500.00	0.05477%	19.99%	$8.22

We showed how finance charges are calculated using simple interest. However, the actual method involves the compound interest formula and requires knowing the beginning balance on the account and the dates of any purchases and payments.

In this example, at the beginning of the period, June 4, the balance was $2469.10. On June 14, a $400 payment was recorded, and on June 17, a $172.15 purchase was recorded. Calculate the periodic finance charge.

Solution: *From June 4 through June 14*: In the first ten-day period, you are charged interest, compounded daily, on the balance of $2469.10. Using the compound interest formula with $P = 2469.10$, $r = 0.0004655$, and $n = 10$, the balance is

$$A = 2469.10(1 + 0.0004655)^{10} = \$2480.62$$

and the interest charged can be calculated by the change in the balance,

$$I = 2480.62 - 2469.10 = \$11.52.$$

From June 15 through June 17: In the following three-day period, you are charged interest, compounded daily, on the balance of $2480.62 less your $400 payment. Using the compound interest formula with $P = 2480.62 - 400 = \$2080.62$, $r = 0.0004655$, and $n = 3$, the balance is

$$A = 2080.62(1 + 0.0004655)^3 = \$2083.52$$

and the interest charged can be calculated by the change in the balance,

$$I = 2083.52 - 2080.62 = \$2.90.$$

From June 18 through July 3: In the following 17-day period, you are charged interest, compounded daily, on the balance of $2083.52 plus your purchase of $172.15. Using the compound interest formula with $P = 2083.52 + 172.15 = \$2255.67$, $r = 0.0004655$, and $n = 17$, the balance is

$$A = 2255.67(1 + 0.0004655)^{17} = \$2273.59$$

and the interest charged can be calculated by the change in the balance,

$$I = 2273.59 - 2255.67 = \$17.92.$$

Thus, the total finance charge for the period is $\$11.52 + \$2.90 + \$17.92 = \$32.34$.

EFFECTIVE INTEREST RATE

When a bank advertises its interest rates, it often gives two rates of interest: the nominal interest rate and the effective interest rate. The **nominal interest rate** is the interest rate that we have used throughout this section, namely, the annual rate. The **effective interest rate** is the simple interest rate which, over a one-year period, provides the same amount of interest as the nominal rate. In terms of algebra, this is written as follows.

$$\text{Simple Interest} = \text{Compound Interest}$$
$$P\left[1 + (r_{eff})t\right] = P(1 + r)^{nt} \qquad \text{Cancel the } P\text{'s and let } t = 1.$$
$$1 + r_{eff} = (1 + r)^{n}$$
$$r_{eff} = (1 + r)^{n} - 1$$

Thus, the effective rate may be found with the following formula.

Effective Interest Rate

$$r_{eff} = (1 + r)^{n} - 1 \qquad \text{where} \quad \begin{cases} r = \text{periodic interest rate} \\ n = \text{number of periods per year} \end{cases}$$

Effective interest rates can be used to compare accounts with different nominal interest rates. An account with a greater effective rate earns more interest.

EXAMPLE 11

A bank offers two savings accounts. One account has 6.3% interest, compounded semiannually, while the other account has 6.24% interest, compounded monthly. Determine the effective rate for each account and determine which account will earn more interest.

Solution: For the 6.3% account, $r = 0.063 \div 2 = 0.0315$, and $n = 2$. The effective interest rate is

$$r_{eff} = (1 + 0.0315)^{2} - 1 = 0.06399 = 6.399\%.$$

For the 6.24% account, $r = 0.0624 \div 12 = 0.0052$, and $n = 12$. The effective interest rate is

$$r_{eff} = (1 + 0.0052)^{12} - 1 = 0.06422 = 6.422\%.$$

Since the effective rate for the second account is greater, the second account will earn more interest.

section 9.3

PROBLEMS • Explain • Apply • Explore

Explain

1. What information is needed to determine the amount in an account earning compound interest?
2. What information is needed to determine the present value of an account earning compound interest?
3. Why does an account earning compound interest accumulate money faster than an account earning simple interest?
4. When compound interest is computed, why is the annual interest rate divided by the number of periods per year?
5. When compound interest is computed, why is the number of years multiplied by the number of periods per year?
6. How do simple and compound interest differ?
7. What is an "effective" interest rate?
8. When solving for which variable in the compound interest formula will you need to use logarithms?

Apply

In Problems 9–24, use the compound interest formula $A = P(1 + r)^n$ and the given information to determine the value of the specified variable. In all cases, the interest rates are given as annual rates.

9. $P = \$2000$	6% compounded monthly	$t = 4$ years	Find A.	
10. $P = \$3000$	9% compounded monthly	$t = 6$ years	Find A.	
11. $P = \$6000$	5% compounded daily	$t = 3$ years	Find A.	
12. $P = \$5000$	10% compounded daily	$t = 4$ years	Find A.	
13. $A = \$6000$	6% compounded monthly	$t = 2$ years	Find P.	
14. $A = \$5000$	12% compounded monthly	$t = 3$ years	Find P.	
15. $A = \$7500$	6% compounded quarterly	$t = 5$ years	Find P.	
16. $A = \$5000$	7.2% compounded quarterly	$t = 10$ years	Find P.	
17. $A = \$5000$	9% compounded monthly	$P = \$3500$	Find t.	
18. $A = \$6000$	6% compounded monthly	$P = \$2000$	Find t.	
19. $A = \$15{,}000$	7.8% compounded monthly	$P = \$5000$	Find t.	
20. $A = \$5500$	4% compounded monthly	$P = \$1000$	Find t.	
21. $A = \$5000$	$P = \$3500$	$t = 2$ years	compounded annually	Find r.
22. $A = \$6000$	$P = \$2000$	$t = 9$ years	compounded annually	Find r.
23. $A = \$75{,}000$	$P = \$25{,}000$	$t = 15$ years	compounded monthly	Find r.
24. $A = \$60{,}000$	$P = \$20{,}000$	$t = 12$ years	compounded monthly	Find r.

In Problems 25–28, determine the effective interest rate.

25. Determine r_{eff} if the nominal rate is 6% compounded monthly.
26. Determine r_{eff} if the nominal rate is 9% compounded monthly.

27. Determine r_{eff} if the nominal rate is 13% compounded weekly.
28. Determine r_{eff} if the nominal rate is 5.2% compounded weekly.

Explore

29. Which is the better investment for a gift of $1000, an 8% account compounded annually or a 7.8% account compounded daily? Each investment is for one year.
30. Which is the better investment for a gift of $3000, an 8.5% account compounded annually or an 8.3% account compounded daily? Each investment is for one year.
31. Antonio deposits $15,000 into a certificate of deposit that guarantees 6.6% annual interest rate, compounded quarterly. How much will be in the account at the end of five years?
32. The Lee family has decided to invest $12,000 in an account that pays 7.3% interest, compounded daily. What is the value of the account after 12 years?
33. The Park family has $10,000 invested in a money market fund that pays 5.475% interest, compounded daily. The account has a three-year term. For tax purposes, the Parks must know the amount of interest earned during each year. Determine the amount of interest earned in each of the three years.
34. A relative has decided to establish a bank account for your newborn daughter that will pay for some of her future college expenses. It is intended that the amount be worth $10,000 eighteen years from now. Assuming that the account will earn 7.6%, compounded quarterly, how much money should be deposited into the account.
35. A business has decided to place a certain amount of current profits into a bank account earning 9%, compounded monthly, for a period of five years. At the end of that time, the business will use the money to purchase $15,000 worth of new equipment. How much should the business deposit into the account?
36. After moving to the United States from Eastern Europe, the Marvineviches invested their savings of $4000 into an account earning 7.2% interest, compounded monthly. How much interest was earned by the account during each of the first three months?
37. **So You Want to Be a Millionaire:** Suppose you invest $10,000 in an account that earns 9% compounded monthly. After eight years, you withdraw the entire amount and deposit it into an account that earns 10% compounded quarterly. How long do you have to leave it in the bank so that you have $1,000,000?
38. An interitance from your great aunt in Des Moines of $7000 is deposited into an account that earns 7.5% compounded monthly. After eight years, you withdraw the entire amount and deposit it into an account that earns 10% compounded quarterly. This second account requires you to keep the money in the account for seven years. How much money have you accumulated at the end of those seven years?
39. Uncle Bill and Aunt Marilyn plan on buying a vacation home. They have $20,000 to invest and want to make a down payment of $40,000 on the home.

If the best investment currently available is a 12% account, compounded daily, how long do they have to wait until they have enough money to make the down payment.

40. You, being the wise parent, decide to invest $25,000 for your newborn child's college education. Setting your sights high, you aim for an education at Vine Covered University (VCU). Estimating the future costs at VCU, you arrive at a figure of $150,000. If you can get a 9% annual rate, compounded monthly, how old will your child be when you have the funds to send her to VCU?

41. After extensive negotiations, the Indians of Manhattan Island agree with the banks of Amsterdam to invest $24 at 5% interest, compounded daily. After keeping this investment for 400 years, the Indians decide to cash in the account. What is the account balance?

42. Determine the value of a $100 deposit at the end of one year if the account earns 12% interest, compounded

a) monthly.

b) weekly.

c) daily.

d) every second.

43. Your credit card had a beginning balance on June 1 of $3510.24. On June 14 (13 days later), a $700 payment was recorded, and on June 20 (6 days later), a $120.15 purchase was recorded. If the daily periodic rate is 0.02164%, calculate the finance charge for the month?

44. Your credit card had a beginning balance on June 1 of $4259.64. On June 11 (10 days later), a $180 payment was recorded, and on June 28 (17 days later), a $520.67 purchase was recorded. If the daily periodic rate is 0.030822%, calculate the finance charge for the month.

45. Your credit card had a beginning balance on April 1 of $2153.64. On April 12 (11 days later), a $900 payment was recorded, and on April 20 (8 days later), a $120.15 purchase was recorded. If the daily periodic rate is 0.02164%, calculate the finance charge for the month.

46. Your credit card had a beginning balance on April 1 of $4259.64. On April 9 (8 days later), a $1000 payment was recorded, and on April 20 (11 days later), a $320.92 purchase was recorded. If the daily periodic rate is 0.030822%, calculate the finance charge for the month.

Annuities

As you were doing the problems in Section 9.3, the situations might have seemed a little beyond your current financial status. For many, the idea of depositing \$10,000 is not a realistic situation. Most people are more likely to save a little money every month or every week than to make one large deposit. An account in which money is deposited at the end of each period is called an ordinary **annuity.**

Suppose you deposit \$100 into an account at the end of every month for four months and interest is compounded monthly at an annual rate of 12%. How much is in the account at the end of four months?

To solve this problem, consider the compound interest earned by each deposit using the formula discussed in the previous section, $A = P(1 + r)^n$.

The first deposit is in the bank for three months, so its value will be $\$100(1 + 0.01)^3$.

The second deposit is in the bank for two months, so its value will be $\$100(1 + 0.01)^2$.

The third deposit is in the bank for one month, so its value will be $\$100(1 + 0.01)^1$.

The fourth deposit has a value of \$100 (since it has not earned any interest).

Adding these four terms gives

$$100 + 100(1.01) + 100(1.01)^2 + 100(1.01)^3 = \$406.04.$$

This method certainly solved the problem. However, if these monthly deposits continued for 30 years, this would mean 360 deposits. This process would soon become very tiresome. There is a formula that is used to determine the sum of all the payments in an annuity without doing all the individual computations.

Annuity Formula

$$S = PMT\left[\frac{(1 + r)^n - 1}{r}\right] \quad \text{where} \quad \begin{cases} S = \text{amount of the annuity} \\ PMT = \text{amount of each deposit} \\ r = \text{periodic interest rate} \\ n = \text{the number of deposits} \end{cases}$$

We will be using PMT (which stands for payment) as a reminder that an annuity is different from an account with one deposit. For an annuity, you make deposits into the account every period.

EXAMPLE 1

Suppose you deposit \$100 every month for 20 years into an account earning 6%, compounded monthly.

a) What is the value of the account after 20 years?

b) What were your total deposits?

c) How much interest was earned?

Solution:

a) Using $PMT = 100$, $r = 0.06 \div 12 = 0.005$, and $n = 20 \times 12 = 240$ gives

$$S = 100\left[\frac{(1 + 0.005)^{240} - 1}{0.005}\right] = 100\left[\frac{3.310204 - 1}{0.005}\right] = \$46{,}204.09.$$

b) The total deposits were 240 payments of \$100 each, which gives

$$240 \times 100 = \$24{,}000.00.$$

c) The amount of interest earned is the difference between the amount in the annuity and the amount deposited. Therefore, the interest is

$$\$46{,}204.09 - \$24{,}000.00 = \$22{,}204.09.$$

EXAMPLE 2

Suppose that when Hector retires, he has an annuity worth \$120,000. The annuity was created by using equal monthly payments over a period of ten years. If the account had an interest rate of 8.1%, compounded monthly, how much were the monthly payments?

Solution: Using $S = 120{,}000$, $r = 0.081 \div 12 = 0.00674$, and $n = 10 \times 12 = 120$ gives the following.

$$120{,}000 = PMT\left[\frac{(1 + 0.00675)^{120} - 1}{0.00675}\right]$$

$$120{,}000 = PMT\left[\frac{2.241799 - 1}{0.00675}\right]$$

$$120{,}000 = PMT[183.970163]$$

$$\frac{120{,}000}{183.970163} = PMT$$

$$PMT = \$652.28$$

EXAMPLE 3

Suppose we deposit \$100 every month into an account earning 6%, compounded monthly. How long must we continue making deposits so that the account is worth \$100,000?

Solution: The number of payments n is to be determined. Using the annuity formula

$$S = PMT\left[\frac{(1 + r)^n - 1}{r}\right]$$

along with $PMT = 100$, $r = 0.06 \div 12 = 0.005$, and $S = 100{,}000$, we have

$$100{,}000 = 100\left[\frac{(1 + 0.005)^n - 1}{0.005}\right].$$

Dividing both sides by 100 gives

$$1000 = \left[\frac{(1 + 0.005)^n - 1}{0.005}\right].$$

Completing the algebra gives the following.

$$5 = 1.005^n - 1$$
$$6 = 1.005^n$$
$$\ln 6 = \ln(1.005^n)$$
$$\ln 6 = n \ln 1.005$$
$$n = \frac{\ln 6}{\ln 1.005} = 359.2 \approx 360 \text{ months, or 30 years}$$

There is an important thing to notice here. Even though the calculation for n is 359.2, the answer is 360 months, not 359 months. The reason is that if we made only 359 payments, the value of the account would be slightly less than \$100,000. By making 360 payments, the account is slightly over \$100,000.

PROBLEMS ➤ Explain ➤ Apply ➤ Explore

➤ Explain

1. What is an ordinary annuity?
2. What is the difference between an ordinary annuity and the type of account discussed in Section 9.3?
3. How can you determine the total amount of interest earned by an ordinary annuity?
4. How can you determine the total amount you deposit into an ordinary annuity?

➤ Apply

$$S = PMT\left[\frac{(1+r)^n - 1}{r}\right] \quad \text{where} \quad \begin{cases} S = \text{amount in the annuity} \\ PMT = \text{amount of each deposit} \\ r = \text{periodic interest rate} \\ n = \text{number of deposits} \end{cases}$$

Use the annuity formula and the given information to solve the following problems. In all cases, interest rates are given as annual rates.

5. $PMT = \$200$ 6% compounded monthly $t = 5$ years Find S.
6. $PMT = \$100$ 4% compounded quarterly $t = 2$ years Find S.
7. $PMT = \$150$ 6% compounded quarterly $t = 25$ years Find S.
8. $PMT = \$50$ 4% compounded monthly $t = 30$ years Find S.
9. $S = \$20{,}000$ 4% compounded quarterly $t = 10$ years Find PMT.
10. $S = \$35{,}000$ 12% compounded monthly $t = 15$ years Find PMT.
11. $S = \$200{,}000$ 6% compounded monthly $t = 20$ years Find PMT.
12. $S = \$435{,}000$ 9% compounded quarterly $t = 30$ years Find PMT.

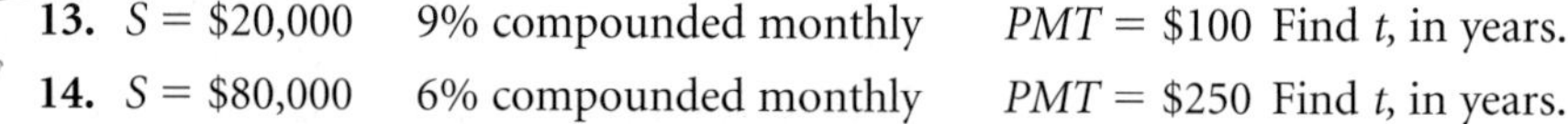

13. $S = \$20{,}000$ 9% compounded monthly $PMT = \$100$ Find t, in years.
14. $S = \$80{,}000$ 6% compounded monthly $PMT = \$250$ Find t, in years.
15. $S = \$20{,}000$ 7.2% compounded monthly $PMT = \$50$ Find t, in years.
16. $S = \$80{,}000$ 8% compounded monthly $PMT = \$250$ Find t, in years.

Explore

17. Sanjay has decided to invest $500 each quarter into a retirement account that has annual earnings of 9.3%, compounded quarterly. If Sanjay continues his investments for 25 years, how much money will he have in the retirement account?

18. Antonio will be retiring in 15 years. With his children grown and finished with college, he can save money toward his retirement. He decides to deposit $800 per month into a mutual fund that is earning 6.72%, compounded monthly. How much will Antonio have in the account when he retires?

19. George listened when his banker told him to start saving money in an Individual Retirement Account (IRA). Beginning when he was 22, George deposited $100 every month into an account earning 9% compounded monthly.

    a) How much will be in the account when George retires at age 70?

    b) How much of this money did George deposit?

    c) How much of this money is interest?

    George's brother Skippy decided to spend the first ten years buying himself toys. He reasoned that he could accumulate more money than George if he deposited $200 every month starting at age 32. Skippy also plans to retire at age 70 and to use the same 9% account as George.

    d) How much money will be in Skippy's account?

    e) How much of this money did Skippy deposit?

    f) How much of this money is interest?

20. After hearing all the advertisements about becoming wealthy when you retire, you decide to contribute to a mutual fund that averages a 13% return per year.

    a) If you contribute $2500 each year for the next 25 years, how much will be in the account?

    b) How much of this money did you deposit?

    c) How much of this money is interest?

21. When Jill was first hired by SemiTechCorp as a design engineer, she had sufficient income to deposit $400 each month into an IRA paying 9% interest, compounded monthly. The monthly deposits lasted for ten years.

    a) How much was in the account at the end of ten years?

    b) Owing to family responsibilities and investments in real estate, Jill was not able to continue with these IRA contributions. Instead, she deposited the entire IRA account into a 25-year certificate of deposit earning 11% compounded quarterly. What was the value of the account when it matured?

22. When Lisa was first hired by the Environmental Protection Agency as a research scientist, she had sufficient income to deposit $600 each quarter into an IRA paying 10% interest, compounded quarterly. The quarterly deposits lasted for 12 years.

    a) How much was in the account at the end of 12 years?

b) Because of her parents' nursing home costs, Lisa was not able to continue these deposits. Instead, she deposited the entire IRA account into a 30-year certificate of deposit earning 12% compounded monthly. What was the value of the account when it matured?

23. In hopes of driving a luxurious Belchfire 8088, Bert has been setting aside $150 each month into an account earning 6%, compounded monthly. How long will it take to accumulate the necessary $58,000 purchase price of the car?

24. Grandpa has decided to set up a college fund for his newborn grandson. How much should he deposit every month into an account paying 7.5% interest compounded monthly so that the account will be worth $30,000 by the time his grandson is 18?

25. **So You Want to Be a Millionaire:** You have decided to make monthly deposits into an account earning 6.3% interest, compounded monthly. If you want to have $1,000,000 in 20 years, how much should you deposit each month?

26. Anticipating the need for $150,000 as a college fund for their children, the Cleavers deposit $300 every month into an account earning 7.8% compounded monthly. How long will it take to accumulate the college fund?

27. Akiko is making monthly deposits into an annuity that will be worth $200,000 in 30 years. The annuity earns 7.8% annual interest, compounded monthly.

a) What are the monthly payments into the annuity?

b) What is the value of the annuity after 15 years?

28. Juanita is making quarterly deposits into an annuity that will be worth $175,000 in 35 years. The annuity earns 8.4% annual interest, compounded quarterly.

a) What are the quarterly payments into the annuity?

b) What is the value of the annuity after 25 years?

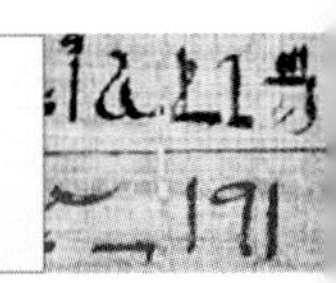

section 9.5 Loans

(Courtesy of Images.com/CORBIS)

Borrowing money from a bank is the most common way of buying a house or a car. In this section, we look at the mathematics of fixed-rate **loans,** ones in which the interest rates do not change. Although we will not show the derivation, the formula for loans is as follows.

Loan Formula

$$L = PMT\left[\frac{1-(1+r)^{-n}}{r}\right] \quad \text{where} \quad \begin{cases} L = \text{amount of the loan} \\ PMT = \text{amount of each deposit} \\ r = \text{periodic interest rate} \\ n = \text{the number of payments} \end{cases}$$

This formula can be used in several ways. First, if the amount of the payments, the length of time remaining on the loan, and the interest rates are known, the formula can be used to determine the amount of the loan.

EXAMPLE 1

Maria has a car loan with monthly payments of \$295.61. The loan is for three years and has an annual interest rate of 12%. What is the amount of her car loan?

Solution: Use $PMT = 295.61$, $r = 0.12 \div 12 = 0.01$, and $n = 3 \times 12 = 36$, to calculate her loan.

$$L = 295.61\left[\frac{1-(1+0.01)^{-36}}{0.01}\right]$$

$$L = 295.61\left[\frac{1-0.698925}{0.01}\right]$$

$$L = 295.61[30.107505]$$

$$L = \$8900.08$$

Another use for the loan formula is to determine the loan payments if you know the amount of money being borrowed.

EXAMPLE 2

Suppose Bill receives a loan from his bank for \$9000 and he makes monthly payments for four years. If the annual interest rate is 6%, compounded monthly, what are the monthly payments?

Solution: We know $L = 9000$, $r = 0.06 \div 12 = 0.005$, and $n = 4 \times 12 = 48$ and must determine the payments, *PMT.* Substituting these values, we get

$$9000 = PMT\left[\frac{1-(1+0.005)^{-48}}{0.005}\right]$$

$$9000 = PMT\left[\frac{1-0.787098}{0.005}\right]$$

$$9000 = PMT[42.580318]$$

$$PMT = \$211.37.$$

We can use the loan formula to find out the total amount paid on a loan, the total interest paid on a loan, and the balance of a loan at any point during the life of the loan.

EXAMPLE 3

When buying a \$180,000 home in Milpitas, Theresa made a down payment of \$40,000 and took out a loan for the remaining \$140,000. The loan has a 30-year term with monthly payments and an annual rate of 10.8%.

a) What is the monthly payment?

b) What is the total of the payments over the 30 years?

c) How much interest will be paid on the loan?

d) What is the balance of the loan after ten years?

Solution:

a) Use $L = 140{,}000$, $r = 0.108 \div 12 = 0.009$, and $n = 30 \times 12 = 360$.

$$140{,}000 = PMT\left[\frac{1-(1+0.009)^{-360}}{0.009}\right]$$

$$140{,}000 = PMT\left[\frac{1-0.039736}{0.009}\right]$$

$$140{,}000 = PMT[106.696041]$$

$$PMT = \frac{140{,}000}{106.696041} = \$1312.14$$

b) 360 payments of \$1312.14 give $360 \times 1312.14 = \$472{,}370.40$.

c) The interest is the difference between the total payments and the value of the loan.

$$\$472{,}370.40 - \$140{,}000 = \$332{,}370.40$$

d) To determine the balance of the loan after ten years, we calculate the amount of a loan (L) if payments of \$1312.14 are made for the remaining 20 years of the loan at the same rate ($r = 0.009$).

$$L = 1312.14\left[\frac{1 - (1 + 0.009)^{-240}}{0.009}\right]$$

$$L = 1312.14\left[\frac{1 - 0.116445}{0.009}\right]$$

$$L = 1312.14[98.172800]$$

$$L = \$128{,}816.46$$

As we saw in Example 3, interest on the home loan exceeded \$300,000. Therefore, when paying off a loan, people often decide to accelerate the payments to lower the amount of interest paid. When you **accelerate the payments,** the actual loan payments are greater than what is required by the terms of the loan. In the next example, we look at the effects of accelerating the payments of a loan.

EXAMPLE 4

Theresa took out a loan for \$140,000. The loan had a 30-year term with monthly payments and an annual interest rate of 10.8%. The loan payments are \$1,312.14. Instead, Theresa decides to pay \$1500 each month.

a) How long did it take to pay off the loan?

b) What was the total amount paid on the loan?

c) What was the total interest paid?

d) How much does Theresa save by accelerating her loan payments?

Solution:

a) The question asks us to determine n. Using the loan formula,

$$L = PMT\left[\frac{1 - (1 + r)^{-n}}{r}\right]$$

along with $PMT = 1500$, $L = 140{,}000$, and $r = 0.108 \div 12 = 0.009$, we have

$$140{,}000 = 1500\left[\frac{1 - 1.009^{-n}}{0.009}\right].$$

Dividing by 1500 gives

$$93.3333 = \left[\frac{1 - 1.009^{-n}}{0.009}\right]$$

Completing the algebra we get the following.

$$0.84 = 1 - 1.009^{-n}$$
$$1.009^{-n} = 1 - 0.84$$
$$1.009^{-n} = 0.16$$
$$\ln(1.009^{-n}) = \ln(0.16)$$
$$-n\ln(1.009) = \ln(0.16)$$
$$n = \frac{\ln(0.16)}{-\ln(1.009)} = 204.54$$

Therefore, Theresa has to make 205 monthly payments (17 years and 1 month). Notice that this cuts almost 13 years off the duration of the loan!

b) The total amount paid on the loan is given by 205 payments of $1500 each.

$$205 \times 1500 = \$307{,}500$$

c) The total interest paid is the difference between the value of the loan and the total payments. Therefore, the total interest paid is

$$\$307{,}500 - \$140{,}000 = \$167{,}500.$$

d) If Theresa had made the required payments of $1312.14, she would have paid a total of

$$\$1312.14 \times 360 = \$472{,}370.40.$$

Accelerating her payments resulted in a savings of

$$\$472{,}370.40 - \$307{,}500 = \$164{,}870.40.$$

REFINANCING

When interest rates drop, people who have mortgages at a high interest rate will often want to refinance their loans. **Refinancing** a loan is the process of paying off an existing loan and replacing it with a new loan. In addition to refinancing to get a lower interest rate, some people also refinance their mortgages so that they can draw on the value of their home to help meet other expenses, such as paying for an extensive remodeling job or to help pay for their children's college expenses.

To refinance, the balance of the existing mortgage must be known. While this number is often available from the mortgage company, being able to calculate the number yourself will help you determine whether refinancing is a feasible idea in your specific situation. The next example shows how you can determine the balance of your existing loan.

EXAMPLE 5

Anita and Carlos originally purchased their home with a $20,000 down payment and a $180,000 loan. The loan has monthly payments for 30 years and has an annual interest rate of 9.6%, compounded monthly. After making payments for seven years, they are considering refinancing their mortgage at a lower rate of 6.6%, compounded monthly.

a) What are the payments on the original loan?

b) What is the balance of the original loan after seven years?

c) If they refinance the remaining balance at 6.6%, compounded monthly, for 30 years, what will the new loan payments be?

d) How much money will they save over the life of the loan by refinancing?

Solution:

a) To determine the original loan payments, we use the loan formula with $L = 180{,}000$, $r = 0.096 \div 12 = 0.008$ and $n = 30 \times 12 = 360$.

$$180{,}000 = PMT\left[\frac{1 - (1.008)^{-360}}{0.008}\right]$$

$$180{,}000 = PMT[117.902287]$$

$$PMT = \$1526.69$$

b) To determine the loan balance after making seven years of payments, we can use the loan formula again. This time, the amount of the payments is known, \$1526.69, and the length of time remaining on the loan is 23 years. Thus, using $PMT = \$1526.69$, $r = 0.008$, and $n = 23 \times 12 = 276$, we get

$$L = 1526.69\left[\frac{1 - (1.008)^{-276}}{0.008}\right]$$

$$L = 1526.69[111.138726]$$

$$L = \$169{,}674.38.$$

This means that after making payments for seven years, the balance of the loan has decreased from \$180,000 to \$169,674.38.

c) To determine the new loan payments, we use $L = 169{,}674.38$, $r = \dfrac{0.066}{12} = 0.0055$, and $n = 30 \times 12 = 360$.

$$169{,}674.38 = PMT\left[\frac{1 - (1.0055)^{-360}}{0.0055}\right]$$

$$169{,}674.38 = PMT[156.578125]$$

$$PMT = \$1083.64$$

d) Instead of making payments of \$1526.69 for another 23 years, they will make payments of \$1083.64 for 30 years.

$$\begin{aligned}\text{Total savings} &= 1526.69 \times (23 \times 12) - 1083.64 \times (30 \times 12)\\ &= 1526.69 \times 276 - 1083.64 \times 360\\ &= 421{,}366 - 390{,}110 = \$31{,}256\end{aligned}$$

Thus, the total savings will be \$31,256.

BUYING A HOUSE

The loan formula that we have been using is the standard method of calculating the loan payments on a mortgage. However, the process of buying a house with the help of a mortgage is a more complicated situation, especially if you are a first-time buyer just meeting the minimum requirements established by the loan company. To further develop the scenario of buying a house, we first introduce you to some additional financial terms that you may encounter. Following that, we see how these details affect your finances.

One of the major benefits of buying a house with the help of a mortgage is that, often, the entire amount of interest paid on the mortgage is deductible from your federal income tax. In our example, we assume this to be true. Calculating this figure accurately is beyond the scope of this course; however, we estimate your tax savings for the first year of a new mortgage as 20% of your monthly payments.

The first expense that you often encounter when obtaining a mortgage is a fee called **points.** Points are a one-time fee charged by the lender and are calculated as a percentage of your loan amount. For a \$150,000 loan with 2 points, it would cost you $0.02 \times \$150{,}000 = \$3000$. Typically, if you are willing to pay more points, the lender will give you a lower interest rate on your loan. Points can be included in the loan or paid in cash. In our examples and the exercises, we assume that the points are paid in cash and are in addition to the down payment.

The lender also charges the borrower several other fees to cover the administrative costs of processing the loan application. These might be called notary fees, document preparation fees, courier services, or loan origination fees. Finally, certain loans require additional payments into an escrow account and for Private Mortgage Insurance (PMI). For the purpose of our examples and the exercises, we ignore these costs.

EXAMPLE 6

You are currently renting an apartment for \$1500 per month in Sacramento, California, and have decided to buy a \$250,000 house. You buy the house using a \$30,000 down payment and a \$220,000 loan. The loan is for 30 years and has an annual interest rate of 8.25%. The lender is charging 2 points for the loan. Your federal tax savings in the first year can be estimated as 20% of your monthly payment.

a) What is your monthly mortgage payment?

b) What is the balance of your mortgage after one year?

c) How much did your mortgage decrease?

d) What were the points paid on the loan?

e) What are your monthly tax savings?

f) How does the cost of renting for one year compare to the first year costs of buying a home? Assume that you are paying the points in this first year.

Solution:

a) Using $n = 30 \times 12 = 360$, $r = 0.0825 \div 12 = 0.006875$, and $L = 220{,}000$, we have

$$220{,}000 = PMT\left[\frac{1 - (1.006875)^{-360}}{0.006875}\right], \text{ so } PMT = \$1652.79.$$

b) Since there are 29 years left on the mortgage, use $n = 29 \times 12 = 348$, $r = 0.006875$, and PMT $= \$1652.79$, giving

$$L = 1652.79\left[\frac{1 - (1.006875)^{-348}}{0.006875}\right] = \$218{,}251.88.$$

c) The mortgage decreased by $\$220{,}000 - \$218{,}251.88 = \$1748.12$.

d) The points paid on the loan are given by:

$$\text{Points} = 0.02 \times 220{,}000 = \$4400.$$

e) The monthly tax savings is given by:

$$\text{Tax Savings} = 0.20 \times 1652.79 = \$330.56.$$

f) Your costs for renting for one year are $12 \times 1500 = \$18{,}000$. In the first year, the purchase costs are given by the following.

$$\begin{aligned}\text{Cost} &= \text{Monthly Payments} + \text{Points} - \text{Tax Savings} - \text{Decrease in Mortgage}\\ &= (12 \times 1652.79) + 4400 - (12 \times 330.56) - 1748.12 = \$18{,}518.64\end{aligned}$$

In the first year, buying the house is more expensive than renting. However, the points are a one-time cost and you are investing money in a piece of property that you expect to appreciate in value. This opportunity is lost if you rent. Thus, your annual cost of ownership will be less than the cost of renting.

PROBLEMS ▶ Explain ▶ Apply ▶ Explore

section 9.5

▶ Explain

1. In a fixed-rate loan, the payments remain the same each month. Does the amount of interest paid each month remain the same? Explain.
2. In a fixed-rate loan, the payments remain the same each month. Does the amount of principal repaid each month remain the same? Explain.
3. What does it mean to *accelerate* the payments on a loan?
4. When solving for which variable in the loan formula, will you need to use logarithms?
5. When comparing the cost of renting versus buying a place to live, what factors should be considered?
6. What are points?

▶ Apply

$$L = PMT\left[\frac{1-(1+r)^{-n}}{r}\right] \quad \text{where} \quad \begin{cases} L = \text{amount of the loan} \\ PMT = \text{amount of each deposit} \\ r = \text{periodic interest rate} \\ n = \text{number of payments} \end{cases}$$

In Problems 7–18, use the loan payment formula and the given information to find the indicated value. In all cases, interest rates are given as annual rates.

7. $PMT = \$200$ 6% compounded monthly $t = 5$ years Find L.
8. $PMT = \$100$ 4% compounded quarterly $t = 2$ years Find L.
9. $PMT = \$250$ 8% compounded quarterly $t = 5$ years Find L.
10. $PMT = \$100$ 5.4% compounded monthly $t = 3$ years Find L.
11. $L = \$20{,}000$ 4% compounded quarterly $t = 10$ years Find PMT.
12. $L = \$35{,}000$ 12% compounded monthly $t = 15$ years Find PMT.
13. $L = \$120{,}000$ 8% compounded quarterly $t = 20$ years Find PMT.
14. $L = \$235{,}000$ 9.3% compounded monthly $t = 30$ years Find PMT.

15. $L = \$20{,}000$ 9% compounded monthly $PMT = \$200$ Find t, in years.
16. $L = \$80{,}000$ 6% compounded quarterly $PMT = \$2000$ Find t, in years.
17. $L = \$180{,}000$ 8.4% compounded quarterly $PMT = \$5000$ Find t, in years.
18. $L = \$150{,}000$ 6.3% compounded monthly $PMT = \$1200$ Find t, in years.

Explore

19. Bill and Judy are buying a seaside cottage in Bolinas. The mortgage will be \$109,000, to be repaid monthly in 15 years. Find the monthly payments if the interest rate is 10.5%, compounded monthly.
20. The Hopkins family is purchasing a Wave Cruiser yacht costing \$75,000. If the down payment is \$60,000 and the loan is \$15,000, find the monthly payments on their five-year, 15% loan (compounded monthly).
21. You have decided to purchase a Toyota, using your savings and an \$8000 loan. If the loan is at 13.2%, compounded monthly, and has monthly payments for four years, find the
    - **a)** monthly payment.
    - **b)** total paid over four years.
    - **c)** total interest paid.
22. Olivia's Visa™ card has a balance of \$4250.00. She plans to pay it off in three years, using equal monthly payments. The interest rate is 20.4%, compounded monthly. Assuming that no additional charges are made to the account, find the
    - **a)** monthly payment.
    - **b)** total paid over three years.
    - **c)** total interest paid.
23. Isaac plans to accelerate the payments on his \$5000 car loan. The original loan had an interest rate of 13.5% compounded monthly for four years, and Isaac plans on paying \$250 per month.
    - **a)** How much were the original loan payments?
    - **b)** How long will it take Isaac to pay off the loan with his \$250 payments?
    - **c)** How much will the accelerated payments save Isaac over the life of the loan?
24. Ron and Dianne plan to accelerate the payments on their \$200,000 home loan. The original loan had an interest rate of 10.5% compounded monthly for 30 years.
    - **a)** How much were the original loan payments?
    - **b)** How long will it take to pay off the loan if they pay \$100.00 extra each month?
    - **c)** How much will the accelerated payments save over the life of the loan?
25. Phil and Angelika are refinancing a mortgage. The existing loan is a 30-year mortgage for \$175,000 at 9.6%, compounded monthly. They have made payments on the loan for nine years. Find the remaining balance on the loan.

26. A corporation has a mortgage on its main facility. The original value of the mortgage was \$1,200,000. The loan is a 30-year, fixed-rate loan at 8.4%, compounded monthly. Find the balance of the loan after six years.

27. Mike and Melissa have a choice of two loans. The first loan has a 7.23% annual rate, compounded monthly, for 15 years. The other loan is a 30-year loan at 7.5%, compounded monthly. Either loan will be for \$150,000.

a) Calculate the payment for each loan.

b) Calculate the total paid on each loan.

c) Which loan is a better choice? Explain.

28. Alan and Karen have a choice of two loans. The first loan has a 8.23% annual rate, compounded monthly, for 15 years. The other loan is a 30-year loan at 8.06%, compounded monthly. Either loan will be for \$290,000.

a) Calculate the payment for each loan.

b) Calculate the total paid on each loan.

c) Which loan is a better choice? Explain.

29. So You Want to Buy a House: You are currently renting an apartment in Westwood for \$1650 per month. You have the opportunity of buying a condo for \$180,000.

a) Suppose you have \$15,000 available for the down payment. The rest will be financed using a 30-year mortgage at 8.1%, compounded monthly. The loan will cost you 2 points. Compare the first-year cost of renting versus buying including the estimated tax savings of 20% of the monthly payments.

b) You have been told that if you can make a down payment of \$36,000, you can get a 7.95% loan with 1.5 points. Compare the first-year cost of renting versus buying including the estimated tax savings of 20% of the monthly payments. Be sure to discuss any advantages and disadvantages of the larger down payment.

CHAPTER 9 SUMMARY

Key Terms, Concepts, and Formulas

The important terms in this chapter are:	**Section**
Accelerated payments: Loan payments that are greater than the payments required by the terms of the loan.	9.5
Annuity: A type of savings account into which money is deposited on a regular basis.	9.4
Compound interest: The interest paid on both the principal and the accumulated interest.	9.3
Effective interest rate: The simple interest rate that earns an amount equivalent to the amount earned by the compound or nominal interest rate.	9.3
Interest: The fee that is charged for the use of money.	9.2
Loan: An amount of money that is borrowed.	9.5
Markdown: The amount by which a quantity is decreased, frequently given as a percent of the quantity.	9.1
Markup: The amount by which a quantity is increased, frequently given as a percent of the quantity.	9.1
Nominal interest rate: The compound interest rate.	9.3
Percent: Amount out of one hundred.	9.1
Period: The interval of time between successive additions of interest to an account.	9.3
Points: A fee that is sometimes required when obtaining a mortgage. It is a percentage of the mortgage.	9.5
Present value: The beginning value in an account; the amount deposited.	9.2
Refinancing: The process of paying of an existing loan and replacing it with a new loan.	9.5
Simple interest: Interest paid on only the principal of an account and added only once during the time period.	9.2

After completing this chapter, you should be able to:	**Section**
1. Find the percent of a number, determine markups and markdowns, and calculate percent increases and decreases.	9.1
2. Calculate interest, interest rates, annuities, and loan payments using the following formulas.	9.2–9.5

Simple interest: $A = P(1 + rt)$, $I = Prt$

Compound interest: $A = P(1 + r)^n$

Annuities: $S = PMT\left[\frac{(1 + r)^n - 1}{r}\right]$

Loans: $L = PMT\left[\frac{1 - (1 + r)^{-n}}{r}\right]$

3. Determine what type of situation (simple interest, compound interest, annuity, or loan) is being described in a problem and apply the appropriate formula(s). (9.2–9.5)
4. Compare and analyze the costs of renting versus buying a house. 9.5

CHAPTER 9 REVIEW

Section 9.1

1. a) Convert 31/32 to a percent.
   b) Convert 31.25% to a decimal.
2. A lawn mower retails for $279.99. This price reflects the dealer's cost plus a 20% markup. What is the dealer's cost?
3. According to the IRS Tax Rate Schedule for the year 2001, if you are single with taxable income over $65,550 but not over $136,750, your tax is $14,645.00 plus 30.5% of the amount over $65,550. If your taxable income is $120,000, how much tax do you owe?
4. Bill's average golf score dropped from 93 to 87 over a two-year period. Jill's average dropped from 88 to 83 over the same period. What was the percentage decrease in each of their scores? Who had the greater percentage decrease?

Section 9.2

5. a) Use the simple interest formula to find A if $P = \$2000$, $t = 3$ years, and the interest rate is 8.4%.
   b) Use the simple interest formula to find P if $A = \$2000$, $t = 3$ years, and the interest rate is 8.4%.
6. Alice invests $10,000 in an account that earns 8% simple interest. What is the value of the account after three years?
7. Bill paid $9300 for a bond. He sold it two years later for $10,000. What simple rate of interest did he earn?
8. Lorenz loaned a friend $500 for three months. The annual simple interest rate was 9%. How much money was Lorenz repaid by his friend?

Section 9.3

9. a) Use the compound interest formula to find A if $P = \$2000$, $t = 3$ years, and the interest rate is 8.4%, compounded monthly.
   b) Use the compound interest formula to find P if $A = \$2000$, $t = 3$ years, and the interest rate is 8.4%, compounded monthly.
10. How long will it take $5,000 to grow to $12,000 in an account that earns 7.3% interest compounded daily?
11. At what interest rate will it take $3,000 to grow to $24,000 in an account that earns interest compounded monthly for 20 years?
12. Find the effective interest rate of an account earning 6%, compounded monthly.

13. How much money should you deposit into a bank account earning 6% interest, compounded quarterly, so that you will have \$1500 in three years?

14. You deposit \$1000 into a savings account earning 7.2% interest, compounded monthly. What is the value of the account after 12 years?

15. Your credit card had a beginning balance on November 1 of \$2500.64. On November 12 (11 days later), a \$2200 payment was recorded, and on November 28 (16 days later), a \$621.67 purchase was recorded. If the daily periodic rate is 0.02164%, calculate the finance charge for the month.

Section 9.4

16. Use the annuity formula to find S if $PMT =$ \$200, $t = 3$ years, and the interest rate is 8.4%, compounded monthly?

17. Use the annuity formula to find PMT if $S =$ \$9000, $t = 3$ years, and the interest rate is 8.4%, compounded monthly.

18. How long will it take to have \$200,000 in an annuity if you deposit \$500 a month at an annual interest rate of 7%, compounded monthly?

19. Joe wants to create an annuity that will ensure a comfortable retirement. He estimates that he will need \$500,000 to meet his retirement plans. How much must Joe deposit each month for 32 years into an account earning 7.65% to meet his goal?

20. **So You Want to Be a Millionaire:** Stephanie plans on saving \$400 per month until she has a total of \$1,000,000. If her money is growing at an annual rate of 9.3%, compounded monthly, how long will she have to wait? What will be the total of her deposits?

Section 9.5

21. Use the loan formula to find L if $PMT =$ \$200, $t = 3$ years, and the interest rate is 8.4%, compounded monthly.

22. Use the loan formula to find PMT if $L =$ \$9000, $t = 3$ years, and the interest rate is 8.4%, compounded monthly.

23. How long will it take to pay off a \$200,000 loan if you pay \$2000 a month at an annual interest rate of 7%, compounded monthly?

24. Phil has a \$150,000 mortgage at 7.2% compounded monthly for 30 years. How long will it take him to pay off the mortgage if he increases his monthly payment by \$150?

25. The purchase of a \$15,000 automobile is to be accomplished with a 20% down payment. The remainder will be financed at 11.4%, compounded monthly, for three years.
    a) What are the monthly payments?
    b) What is the total amount paid for the car?
    c) How much interest was paid on the loan?

26. Cesar bought a car three years ago with a loan of \$10,000. The loan was for five years, with monthly payments of \$207.58. If the annual interest rate was 9%, compounded monthly, how much does Cesar owe on the loan now?

27. Monica is currently renting an apartment in Santa Cruz for \$1500 per month. She has the opportunity of buying a condo for \$150,000. She has \$10,000 available for the down payment. The rest will be financed by using a 30-year mortgage at 8.4%, compounded monthly. The loan will cost her 2 points. If the points are paid in cash when she receives the loan, compare the first year cost of buying versus renting including the 20% tax savings.

CHAPTER 9 TEST

1. A shirt with a retail price of $24.95 is placed on a clearance rack. Items on the rack are marked 25% off. If the sales tax is 6.5%, how much should you actually pay for the shirt after the discount is given and sales tax is included.
2. Determine the final price of a $2500 piano
   a) after two successive 15% markdowns.
   b) after a 30% markdown.
3. Your salary went from $2250 per month to $3000 per month, what is the percent increase in your salary?
4. A manufacturer's representative receives a commission that is 2.75% of her sales during the year. If she has a commission of $45,900 for the year, what were her total sales for the year?
5. A certain investment has earned simple interest for five years. The initial investment was $10,804.85. The account is now worth $14,700. What is the annual interest rate?
6. Suppose $1000 is deposited in an account earning 10% interest. Find the value of the account at the end of one year if the interest is compounded monthly.
7. Grandma wants to create an account for her newborn grandchild's college fund. She wants the account to be worth $25,000 in 18 years. If the account earns 8.1%, compounded monthly for the next 18 years, how much should she deposit?
8. After depositing $400 per quarter for ten years into an account earning 8% compounded quarterly, Amanda is forced to stop making deposits.
   a) How much is in the account at the end of the ten years?
   b) How much interest has the account earned?
   c) If the account continues to earn the same interest rate for another 25 years and all interest is accumulated in the account, how much will the account be worth 35 years from when Amanda made her initial deposit?
9. You have won a lottery worth $1,000,000 after taxes. If you deposit it into an account earning 6% annual interest, compounded quarterly, how long will it be until the account is worth $5,000,000?
10. You have won a lottery that pays $30,000 after taxes each year, for the rest of your life. If you deposit the $30,000 each year into an account earning 6% annual interest, compounded annually, how many years will it take for the account to be worth $1,000,000?
11. Alicia has been making monthly mortgage payments of $926.21 for 9 years. If the interest rate is 9.7%, compounded monthly, and the original loan was for 30 years, find the current balance of the loan.
12. What is the effective interest rate of an account earning 5.9%, compounded quarterly?
13. Phil has a $150,000 mortgage, at 7.2% compounded monthly for 30 years. How long will it take him to pay off the mortgage if he pays $2,500 each month?
14. Jon is currently renting an apartment in Berkeley for $1800 per month. He has the opportunity of buying a condo for $210,000. He has $30,000 available for the down payment. The rest will be financed by using a 30-year mortgage at 8.1%, compounded monthly. The loan will cost him 2 points paid in cash when receiving the loan. Compare the first year cost of buying versus renting, including the estimated tax savings of 20% of the monthly payments.
15. Your credit card has a beginning balance on June 1 of $7159.68. On June 11, a $1850 payment was recorded, and on June 28 a $320.67 purchase was recorded. If the daily periodic rate is 0.030822%, calculate the finance charge for the month.

CHAPTER TEN

Math from Other Vistas

This interpretation of Van Gogh's *Starry Night* was created as a Calculus project utilizing equations and Maple™. (Courtesy of Ohlone College Student, Gloria Guy)

OVERVIEW

In the nine previous chapters, we investigated the math that could follow the arithmetic, algebra, and geometry studied by the liberal arts student. That does not imply that there is no more mathematics to study. The study of math is limitless. In this chapter, we will introduce other areas of mathematics that would complement the previous studies of the liberal arts student.

In Sections 10.1 and 10.2, we will examine what is considered the culmination of arithmetic, algebra, geometry, and trigonometry: calculus. In Section 10.3, we will look at an ancient triangular array of numbers that has some interesting applications. In Section 10.4, we will look at some of the methods used in determining the winner of an election. In Section 10.5, we will investigate the way a representative body of leaders can be selected in a democratic system. Finally, in Section 10.6, we will examine a technique to maximize profits for a business.

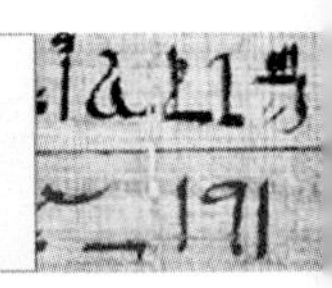

Differential Calculus

Sir Isaac Newton and Gottfried von Leibniz are considered the co-founders of calculus.

Mathematics is one of the subjects that is considered an essential part of the training of a well-educated person. For many, calculus is the culmination of such mathematical thinking. Calculus integrates arithmetic, algebra, geometry, and trigonometry into a study of two mathematical processes called *differentiation* and *integration.* In differential calculus, we investigate the slope (steepness) of a curve; in integral calculus, we examine areas of regions. These investigations began with ancient Greek mathematicians and were synthesized in the 1700s with the work of the co-founders of calculus, the physicist Sir Isaac Newton and the mathematician Gottfried Leibniz. Sections 10.1 and 10.2 introduce you to some of the findings of Newton and Leibniz. We begin this journey with a discussion of finding the slope of a curve.

DERIVATIVES

Let's look at the graph of the equation $y = 3x^2 + 1$ and draw a line that intersects the graph in two points: at point P, where $x = 0$ and at point Q, where $x = h$. The letter h is used to represent any value except zero. Now, find the slope of the line through points P and Q.

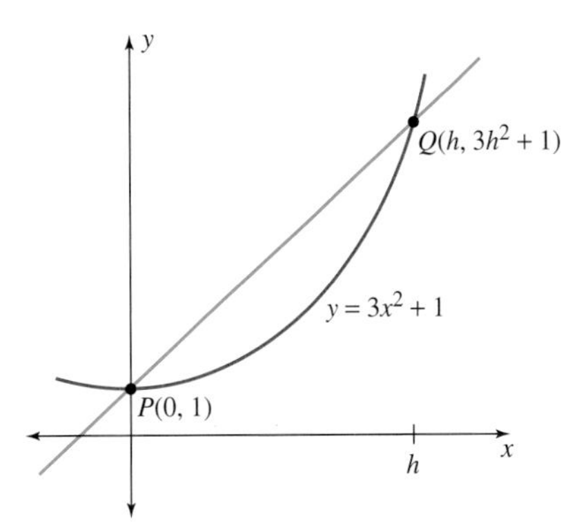

Since the slope of a line is given by

$$m = \frac{y_2 - y_1}{x_2 - x_1},$$

to find the slope of the line segment PQ, first find the y coordinates of the points P and Q. This can be done by substituting the x values into the equation $y = 3x^2 + 1$. For $x = 0$, $y = 1$, and $x = h$ gives $y = 3h^2 + 1$.

Using the points (0, 1) and $(h, 3h^2 + 1)$, we can find the slope of the line through the points P and Q.

$$m = \frac{y_2 - y_1}{x_2 - x_1} = \frac{3h^2 + 1 - 1}{h - 0} = \frac{3h^2}{h} = 3h \qquad \text{(provided that } h \neq 0\text{)}$$

We now know that the slope of the line through P and Q is given by $m = 3h$. This means that the slope of the line will depend on the value of h.

Examine the following series of diagrams that show point Q moving along the curve toward P. What happens to the slope of the line PQ?

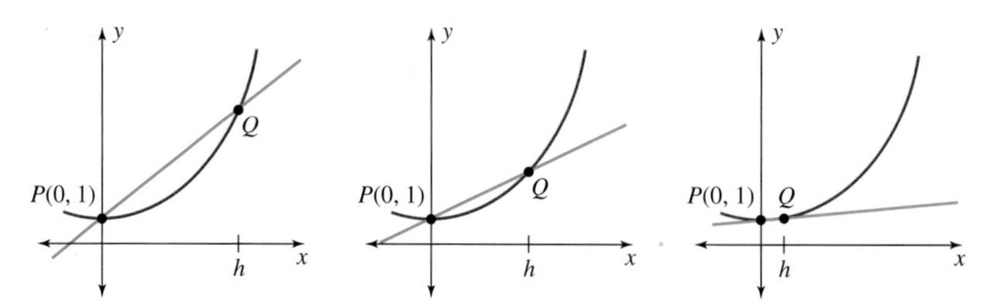

If you determined that the slope was getting close to zero as point Q moved closer to point P, you were correct. As h takes on smaller values, the distance between the points decreases and the slope of the line segment PQ approaches 0. This can also be seen by looking again at the slope formula and letting h get close to 0. Since $m = 3h$, if h approaches zero, then the slope of the line PQ approaches zero. In addition, as h gets close to 0, the line will intersect the graph at two points, P and Q, that are very close together. When, in a region very near to point P, the line intersects the curve only at the point P, the line is called the tangent line to the curve at the point P.*

The following pictures give three examples of tangent lines.

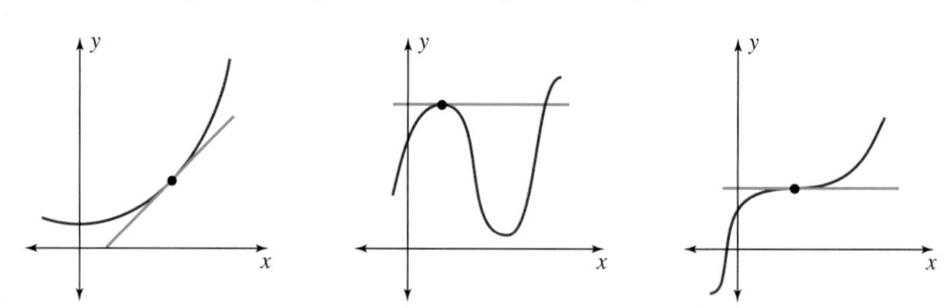

The following pictures give examples of lines that are not tangent to the curve.

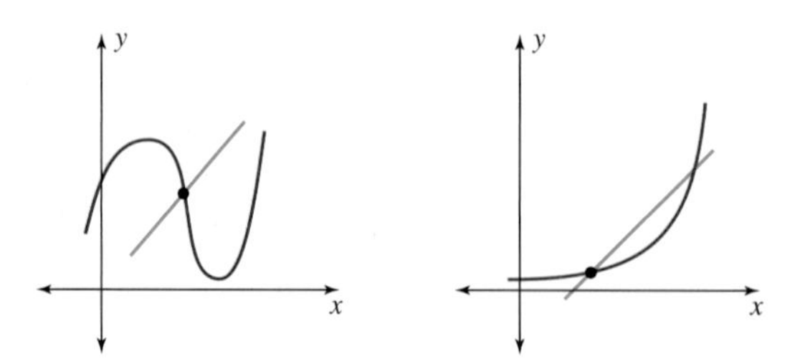

*This statement is true only for functions whose graphs do not include any sharp points or jumps in the graph. Functions whose graphs contain sharp points or jumps will not be discussed here.

Sharp point   Jump

We are discussing the slope of the tangent line because the slope of the curve at point P is defined as the slope of the tangent line at point P. The slope of the curve is such an important topic that mathematicians have a specific name for it. The slope of a curve, or function, at a point P is called the derivative of the function at the point P, and the process for finding a derivative is called differentiation.

EXAMPLE 1

Find the slope of the tangent line to the curve $y = x^2 + 4x$ at the point (2, 12).

Solution: We start by sketching a graph of the function. To find the slope we use two points P and Q. Point P is the point (2, 12). We choose point Q a small distance from point P, so we set the x coordinate of Q as $2 + h$. To find the y coordinate of Q, substitute $2 + h$ for x in the function $y = x^2 + 4x$. This gives

$$\begin{aligned} y &= (2 + h)^2 + 4(2 + h) \\ &= 4 + 4h + h^2 + 8 + 4h \\ &= 12 + 8h + h^2 \end{aligned}$$

Thus, point Q is given by $(2 + h, 12 + 8h + h^2)$.

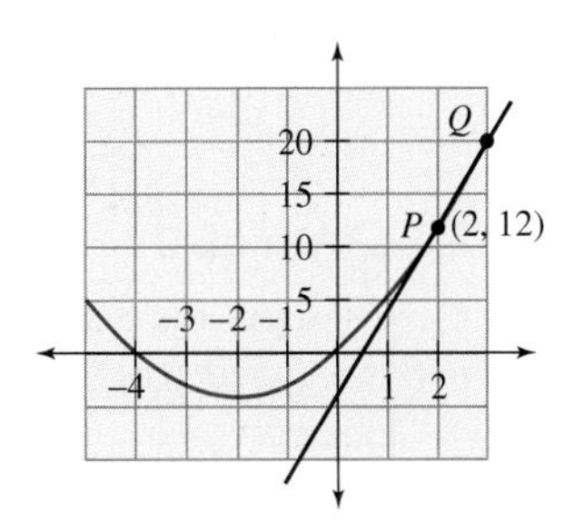

We can now use points $P(2, 12)$ and $Q(2 + h, 12 + 8h + h^2)$ to find the slope of the line through points P and Q.

$$m = \frac{(12 + 8h + h^2) - 12}{(2 + h) - 2} = \frac{8h + h^2}{h} = 8 + h \qquad \text{(provided } h \neq 0\text{)}$$

To find the slope of the tangent line to $y = x^2 + 4x$ at $x = 2$, let the points P and Q get close together. In other words, let h get close to 0. As h takes on values close to 0, the formula $m = 8 + h$ becomes $m = 8$. This implies that at $x = 2$, the slope of the tangent line to the curve is 8.

In Example 1, we found the slope of the tangent line at a particular point; we found at the point (2, 12) that the slope of the tangent line to $y = x^2 + 4x$ is 8. We also stated that the slope of the tangent line at a point is the same as the slope of the curve at that point. However, as we can see from the diagram, the slope of the curve changes, depending on the point being examined.

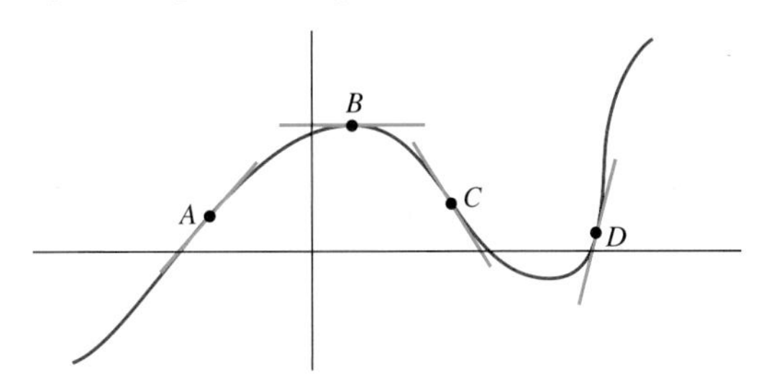

At points A and D, the slope of the tangent line is positive, the curve being steeper at point D than at A. At point C, the slope is negative, while the slope of the tangent line is approximately zero at point B. Since a slope represents the steepness of a line and since the steepness of a curve depends on the point being considered, the slope of a curve is not a constant. Instead, the slope of a curve is changing and can be found by a formula.

Shortcuts for Finding Derivatives

While it is possible to continue finding derivatives this way, it is tedious. However, mathematicians have developed many shortcuts for this process. The three shortcuts we will use are the following. In these shortcuts, you will see the expression $\frac{dy}{dx}$, a commonly used notation for the derivative.

1. If $f(x) = k$, where k is a constant, $\frac{dy}{dx} = 0$.

2. If $f(x) = kx^n$, then $\frac{dy}{dx} = knx^{n-1}$ $(n \neq 0)$.

3. The derivative of $f(x) \pm g(x)$ equals the sum (difference) of the derivatives of $f(x)$ and $g(x)$.

EXAMPLE 2

Find the derivative of the following functions.

a) $f(x) = 3$

b) $f(x) = 8x^5$

c) $f(x) = 4x^3 - 6x^2 + 5x + 1$

Solution:

a) For $f(x) = 3$, we use rule 1 to get $\frac{dy}{dx} = 0$.

b) For $f(x) = 8x^5$, we use rule 2 to get $\frac{dy}{dx} = 8 \cdot 5x^4 = 40x^4$.

c) For $f(x) = 4x^3 - 6x^2 + 5x + 1$, we use rule 2 for the derivatives of the first three terms and rule 1 for the derivative of the fourth term. From rule 2, the derivative of $4x^3$ is $4 \cdot 3x^2 = 12x^2$, the derivative of $6x^2$ is $6 \cdot 2x^1 = 12x$, and the derivative of $5x$ is $5 \cdot 1x^0 = 5$. By rule 1, the derivative of 1 is zero. Therefore, by rule 3, if $f(x) = 4x^3 - 6x^2 + 5x + 1$, $\frac{dy}{dx} = 12x^2 - 12x + 5$.

Applications of the Derivative

The first application of the derivative that we will consider is how the derivative can aid in the graphing of functions. In algebra, the graphing of parabolas is facilitated by finding the location of the vertex of the parabola. The vertex gives the location of the maximum or minimum point of the parabola. We now discuss how to use derivatives to determine the maximum or minimum points of any curve.

As can be seen in the diagram on page 531, a curve may have several maximum and minimum points. If we use the common English usage of the word maximum, there can only be one maximum or highest point for a graph. In mathematics, this is called an absolute maximum.

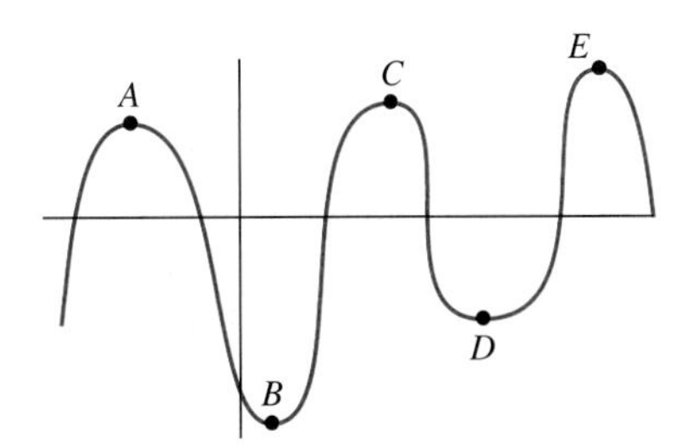

To indicate that we are concerned with all the points that are at the top of a curve, the term relative maximum is used. A relative maximum is a point that is higher than all the points very close by. Similarly, a relative minimum is a point that is lower than all the points very close by. The graph shows three relative maximums, located at points A, C, and E. There are also two relative minimums, located at B and D.

If we redraw the diagram, including the tangent lines at the relative maximum and relative minimum points, we will be able to see the role that the derivative plays in determining relative maximum and relative minimum points. Notice that each of the five tangent lines in the drawing is horizontal. Since the slope of a horizontal line is zero, we can determine the relative maximum and relative minimum points of a curve by setting the derivative equal to zero.

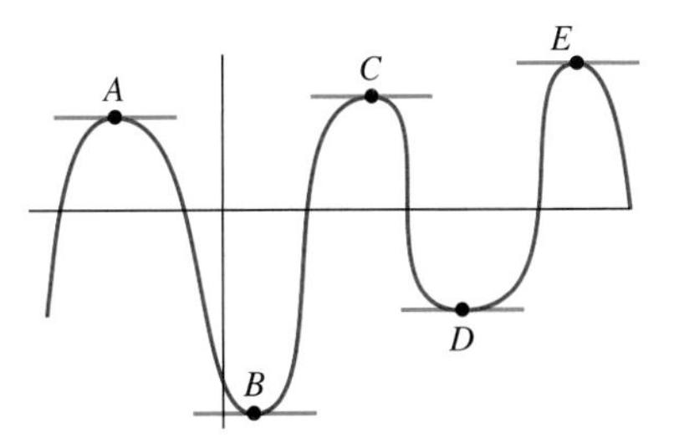

EXAMPLE 3

Find the relative maximum and minimum points of the curve $y = 6x^3 - 8x + 1$.

Solution: To find the relative maximum and minimum points, we first sketch the graph for integer values of x. From the graph, we can see that there is a relative maximum point near $x = -1$ and a relative minimum point near $x = 1$. However, we cannot tell from the graph exactly where the relative maximum and relative minimum points are located.

x	y
−2	−31
−1	3
0	1
1	−1
2	33

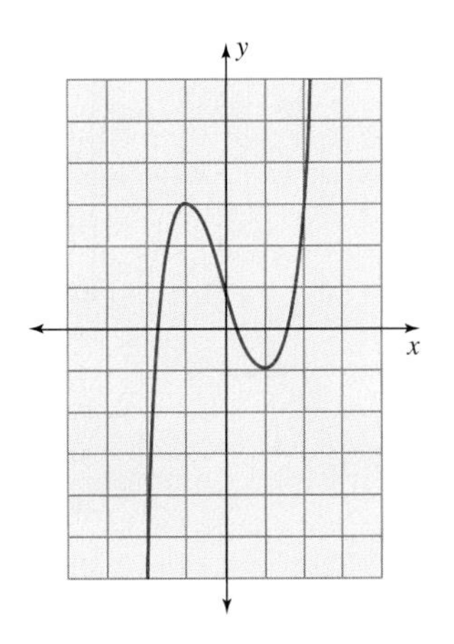

The relative maximum and relative minimum points occur where the slope of the tangent line is equal to zero. Therefore, we find the locations of the relative maximum and relative minimum points by setting the derivative equal to 0 and comparing the resulting values of x with the information from the graph.

From our rules for derivatives, we know that if $y = 6x^3 - 8x + 1$, $dy/dx = 18x^2 - 8$. We determine the location of the relative maximum and minimum points by setting the derivative equal to 0.

$$18x^2 - 8 = 0$$
$$18x^2 = 8$$
$$x^2 = 4/9$$
$$x = \pm 2/3$$

Since we have two solutions for x, we look at the graph to determine which point is the maximum and which is the minimum. From the graph, we find that the relative maximum point was located near $x = -1$. Therefore, we know that the curve reaches its relative maximum at $x = -2/3$.

To determine the y value at the relative maximum point, we substitute the value $x = -2/3$ into the equation $y = 6x^3 - 8x + 1$. This gives

$$y = 6\left(\frac{-2}{3}\right)^3 - 8\left(\frac{-2}{3}\right) + 1 = \frac{41}{9} \approx 4.56.$$

In the same way, we find that the relative minimum occurs at $x = 2/3$. This gives a y value of

$$y = 6\left(\frac{2}{3}\right)^3 - 8\left(\frac{2}{3}\right) + 1 = \frac{-23}{9} \approx -2.56.$$

Using this new information, we can redraw the graph including the exact locations of the relative maximum and relative minimum points.

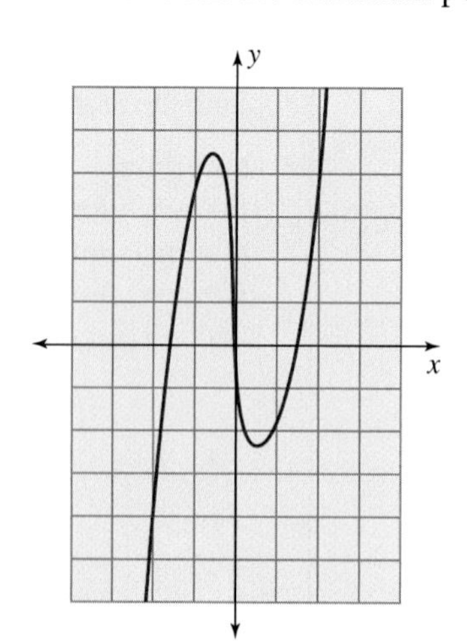

Calculus is a very powerful mathematical tool and has many applications. The standard calculus course requires two or three semesters to learn all its techniques and uses. Because of this, we cannot possibly study the many areas in which calculus plays a role. We do however, present some typical applications.

EXAMPLE 4

Find the dimensions of the rectangular region with maximum area that can be fenced with 120 ft of wire mesh if the fencing forms the perimeter of the region and one interior fence divides the area into two equal sections. (We know that there will be a maximum, since there is a given amount of available fencing.)

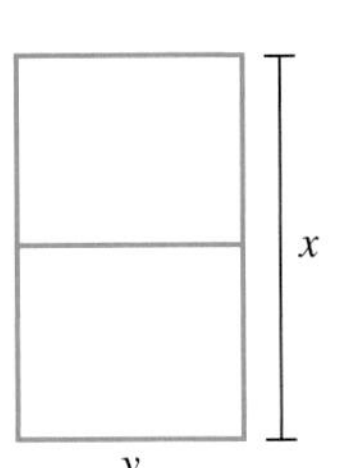

Solution: We start by drawing the rectangle and writing equations that model the situation. The total length of fencing is given by

$$2x + 3y = 120.$$

We also know that the total area of the rectangle is

$$A = xy.$$

Since we want to find the maximum area, we express the area as a function of one variable. Solving the first equation for y, we have

$$\begin{aligned} 2x + 3y &= 120 \\ 3y &= 120 - 2x \\ y &= \frac{120 - 2x}{3} = 40 - \frac{2}{3}x. \end{aligned}$$

Substituting y into the equation for area gives

$$A = x\left(40 - \frac{2}{3}x\right) = 40x - \frac{2}{3}x^2.$$

We now find the derivative and use it to determine the maximum value of the area.

$$\frac{dA}{dx} = 40 - \frac{2}{3}(2x) = 40 - \frac{4}{3}x$$

Setting the derivative equal to zero and solving for x gives the following.

$$\begin{aligned} 40 - \frac{4}{3}x &= 0 \\ 40 &= \frac{4}{3}x \\ 40\left(\frac{3}{4}\right) &= x \\ 30 &= x \end{aligned}$$

Since $y = 40 - (2/3)x$, substituting $x = 30$ gives $y = 20$. This means that the maximum area for the region fenced with the 120 ft of wire mesh is $30 \times 20 = 600$ sq ft.

EXAMPLE 5

A closed rectangular box is to be constructed from 2400 sq cm of cardboard. If the box has a square base of side x and height y, find the values of x and y that form the box of maximum volume. (We know there will be a maximum, since there is a given amount of available cardboard.)

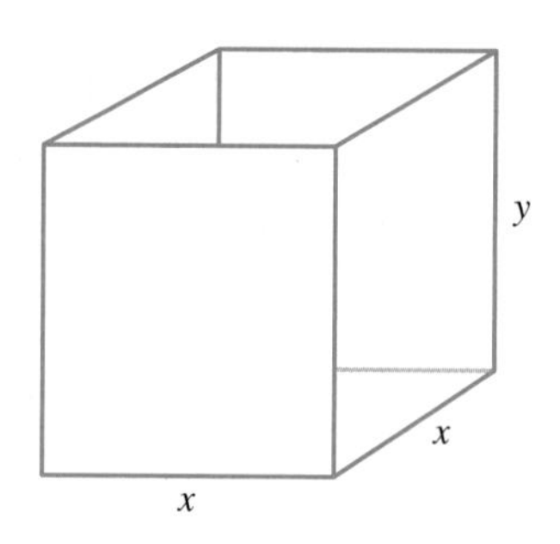

Solution: Since the box has a square base, the area of the top and the bottom of the box is given by x^2. The four sides of the box each have area xy, so the total surface area (S.A.) of the box is

$$\text{S.A.} = 2x^2 + 4xy.$$

Since the total amount of cardboard available is 2400 sq cm, we have the equation

$$2400 = 2x^2 + 4xy.$$

The problem asks us to find the maximum volume of the box. Since the volume of a box equals length × width × height, we have

$$V = x^2y.$$

Since this expression has two variables, we use the surface area equation to solve for y in terms of x, and substitute for y in the volume equation.

$$2400 = 2x^2 + 4xy$$
$$2400 - 2x^2 = 4xy$$
$$\frac{2400 - 2x^2}{4x} = y$$

Substituting this value of y into the expression for the volume gives the following.

$$V = x^2y$$
$$V = x^2\left(\frac{2400 - 2x^2}{4x}\right)$$
$$V = 600x - \frac{x^3}{2}$$

To determine the maximum, find the derivative, set the derivative equal to 0, and solve for x.

$$\frac{dV}{dx} = 600 - \frac{3x^2}{2}$$
$$0 = 600 - \frac{3x^2}{2}$$
$$\frac{3x^2}{2} = 600$$
$$x^2 = 400$$
$$x = 20$$

(Since x represents a dimension of the box, x must be positive.)

Substituting $x = 20$ into $y = \dfrac{2400 - 2x^2}{4x}$ gives $y = 20$. Therefore, the box of maximum volume for a given surface area is a cube. In this case, the cube has dimensions $20 \times 20 \times 20$.

In summary, this section has introduced two topics. The first topic was a study of some of the shortcuts used to find a derivative. The second topic was how the derivative can be used to determine the relative maximum or relative minimum value of a function. The problem set that follows will give you some practice in working with both topics.

section 10.1

PROBLEMS ▶ Explain ▶ Apply ▶ Explore

▶ Explain

1. What is a tangent line to a curve?
2. What is meant by the slope of the curve at a point?
3. Draw a curve where $\dfrac{dy}{dx}$ is positive at $x = 3$. Draw the tangent line to the curve at $x = 3$.
4. Draw a curve where $\dfrac{dy}{dx} = 0$ at $x = 3$. Draw the tangent line to the curve at $x = 3$.
5. If $f(x)$ is a constant function, what is its derivative and what does that mean?
6. What steps should you take in order to determine the relative minimum or maximum points for a curve?

▶ Apply

7. The graph represents the function $y = f(x)$. Determine which point(s) satisfy the following conditions.

   a) $\dfrac{dy}{dx} = 0$

   b) $\dfrac{dy}{dx}$ is positive

   c) $\dfrac{dy}{dx}$ is negative

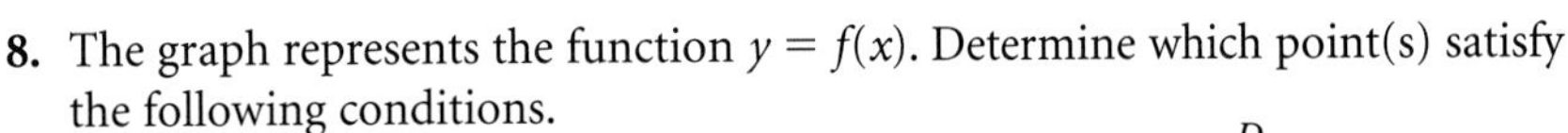

8. The graph represents the function $y = f(x)$. Determine which point(s) satisfy the following conditions.

   a) $\dfrac{dy}{dx} = 0$

   b) $\dfrac{dy}{dx}$ is positive

   c) $\dfrac{dy}{dx}$ is negative

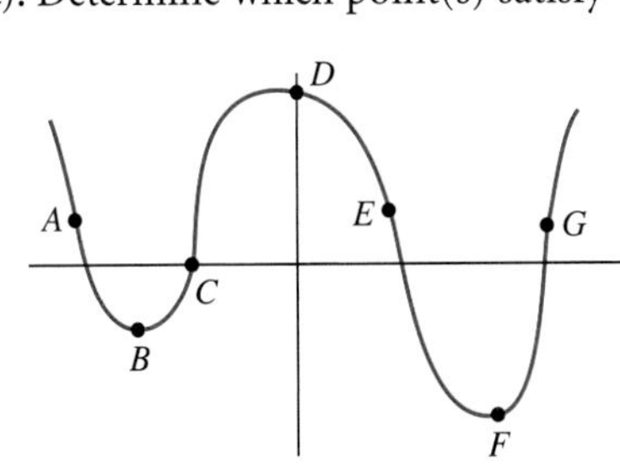

9. On the graph, the value of the derivative at $x = 1$ is 0.5, and the derivative at $x = 4$ is 2. What can you say about the value of the derivative at $x = 2$?

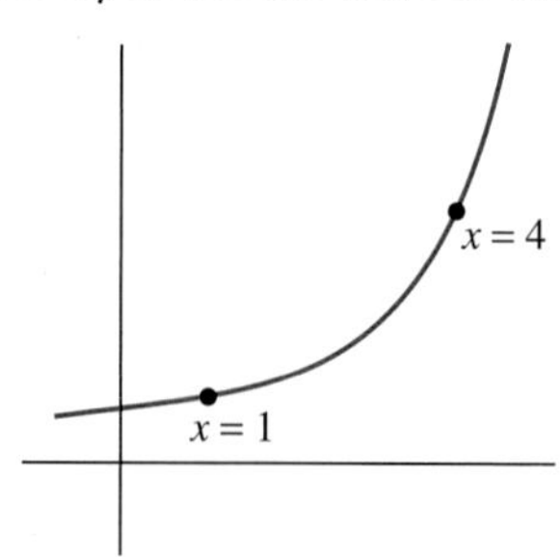

10. On the graph, the value of the derivative at $x = 1$ is -3, and the value of the derivative at $x = 4$ is -1. What can you say about the value of the derivative at $x = 3$?

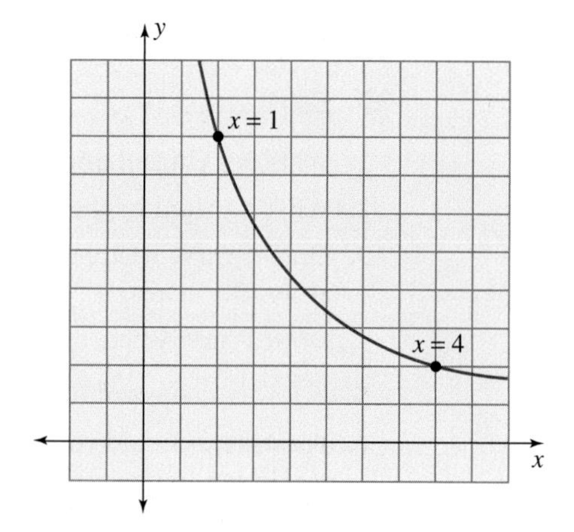

Find the derivatives of the functions in Problems 11–24.

11. $f(x) = 2x + 5$

12. $f(x) = x^2 + 5x$

13. $y = -x^2 + 9$

14. $f(x) = x^3 - 6x$

15. $y = 3x^3 - 5x - 7$

16. $f(x) = -2x^3 + 8x^2$

17. $f(x) = x^4 + x^3 + x^2 + x + 1$

18. $f(x) = -4x^4 + 3x^2 - 16$

19. $y = 4x^5$

20. $y = 3x^8$

21. $y = 4x^5 - 5x^2$

22. $y = 2x^5 - 3x^2$

23. $f(x) = x^5 - 3x^2 + 7x + 5$

25. a) Sketch the graph of the line $y = 3x + 7$. Does this graph have any relative maximum or relative minimum points?

b) Find the derivative of $y = 3x + 7$. Can the derivative equal 0? If so, find the value(s) of x where $dy/dx = 0$. If not, explain.

26. a) Sketch the graph of $y = x^2 + 2x$. Does this graph have any relative maximum or relative minimum points?

b) Find the derivative of $y = x^2 + 2x$. Find the point where $dy/dx = 0$.

c) Does the value found in part (b) give a relative maximum or a relative minimum point?

27. **a)** Sketch the graph of $y = -x^2 + 4x$. Does this graph have any relative maximum or relative minimum points?

b) Find the derivative of $y = -x^2 + 4x$. Find the value(s) of x where $dy/dx = 0$.

c) Does the value found in part (b) give a relative maximum or a relative minimum point?

28. **a)** Sketch the graph of $y = -x^2 + 6x$. Does this graph have any relative maximum or relative minimum points?

b) Find the derivative of $y = -x^2 + 6x$. Find the value(s) of x where $dy/dx = 0$.

c) Does the value found in part (b) give a relative maximum or a relative minimum point?

29. **a)** Plot points to sketch the graph of $y = 2x^3 - 6x$.

b) Find the derivative of $y = 2x^3 - 6x$.

c) Find the value(s) of x where $dy/dx = 0$.

d) Determine which value gives a relative maximum and which gives a relative minimum.

30. **a)** Plot points to sketch the graph of $y = 2x^3 - 6x + 12$.

b) Find the derivative of $y = 2x^3 - 6x + 12$.

c) Find all the points where $dy/dx = 0$.

d) Determine which value gives a relative maximum and which gives a relative minimum.

Explore

31. The path traveled by a frog is given by $h = \frac{-1}{98}x^2 + \frac{6}{7}x$, where x is the horizontal distance traveled and h is the height in inches at any point along the path.

a) Determine the x value that gives the maximum height.

b) Determine the maximum height.

32. A ball is thrown vertically in the air with an initial speed of 139.33 ft/s from an initial height of 9 ft. The equation describing the height of the ball at any time is $h = -16t^2 + 139.33t + 9$, where t is the time in seconds and h is the height in feet.

a) Determine the time when the ball reaches its maximum height.

b) Determine the maximum height.

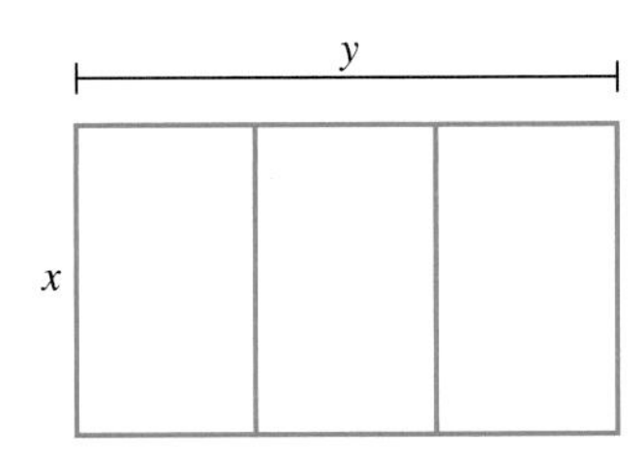

33. Use calculus to solve the following problem. A chicken farmer wants to build a fenced area for her free-roaming chickens. Because she wants to separate the different breeds, she plans to have three adjacent pens, as shown in the diagram. If 1200 ft of fencing is available, what is the largest total area that can be fenced off for the chickens?
(*Hint:* Area $= xy$ and total fencing $= 4x + 2y$.)

34. Use calculus to solve the following problem. An orchid fancier is building a fiberglass enclosure next to his house, as shown in the diagram. Since the house will be used for one side of the enclosure, only three sides will need to

be enclosed by the fiberglass. The roof of the greenhouse is to be built of some other material. If 60 ft of the fiberglass walls are available, what is the maximum rectangular area that can be enclosed?
(*Hint:* Area $= xy$ and total length of walls $= 2x + y$.)

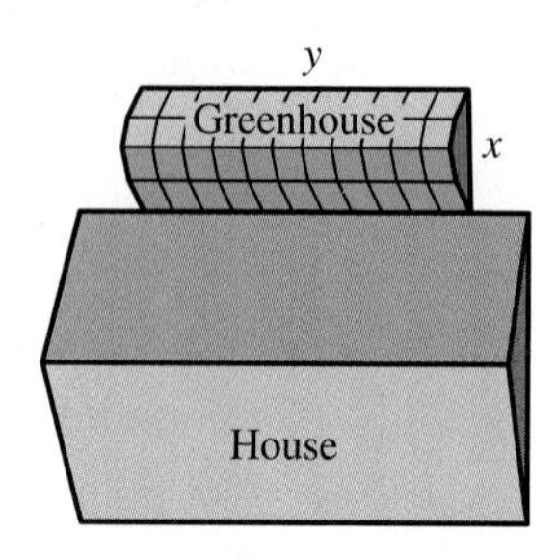

35. Use calculus to solve the following problem. A contractor wants to build a covered box that is twice as long as it is wide. The height is to be determined by the amount of wood available. If the contractor has 48 sq ft of plywood available, find the dimensions of the box with maximum volume.
(*Hint:* Volume $= 2x^2h$ and total plywood $= 2(2x^2) + 2(2xh) + 2(xh)$.)

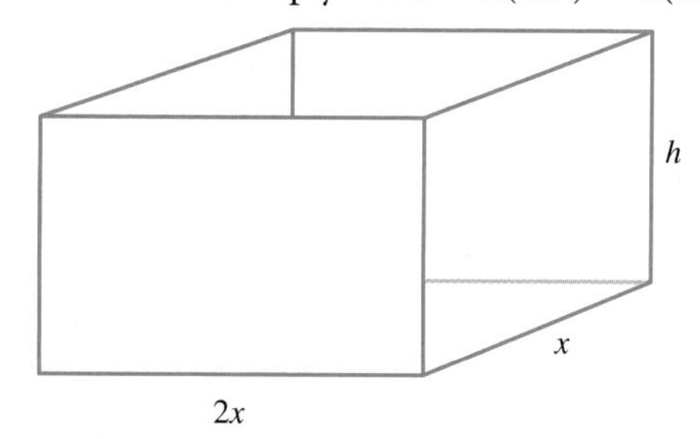

36. A rectangle is drawn with sides along the x and y axes and one vertex on the line $y = -2x + 16$ as shown in the figure below. What is the area of the largest rectangle that can be drawn satisfying these conditions?
(*Hint:* The area of the rectangle is $A = xy$.)

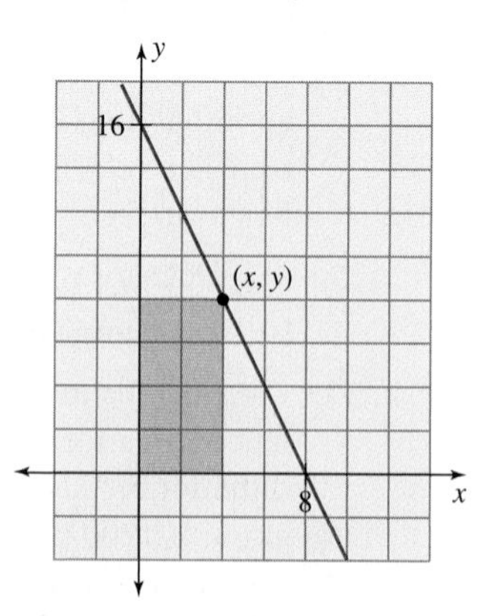

37. A manufacturer of aerobics shoes determines that the average cost for producing a pair of shoes is given by

$C = 80{,}000x^{-1} + 0.0002x + 8.2$ where x is the number of shoes manufactured.

a) Find the production level that minimizes the average cost.

b) What is the minimum average cost?

38. A wood sculptor who creates custom headboards determines that the profit from selling the headboards can be approximated by

$P = 6x^2 - 0.02x^3$ where x is the number of headboards sold.

a) Find the number of headboards that must be sold to maximize the profit.

b) What is the maximum profit?

section 10.2

Integral Calculus

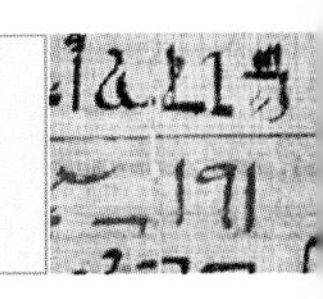

Calculus can be used to analyze the curves, volumes, and areas of structures such as the *Interior of Saint Peter's, Rome* by Giovanni Paolo Panini. (National Gallery of Art)

The next topic we consider is integration. Integration is a method for finding the area of a region by filling the region with shapes of known area. Integration has many uses in calculus; however, we will restrict our investigation to finding the area bounded by a curve in a plane. Many methods, such as the Eudoxian method of exhaustion and Cavalieri's indivisibles, have been used to find areas. We will use the method of rectangles, developed by Gilles Persone de Roberval (1602–1675) of France. It is essentially the method used to introduce integration to a modern calculus class.

The problem is to find the area of the region bounded by the lines $y = 2x$, $y = 0$, and $x = 4$. Using the formula for the area of a triangle, the area is $(1/2)(4)(8) = 16$.

However, we will use a method involving rectangles so that we can arrive at a process for finding any area.

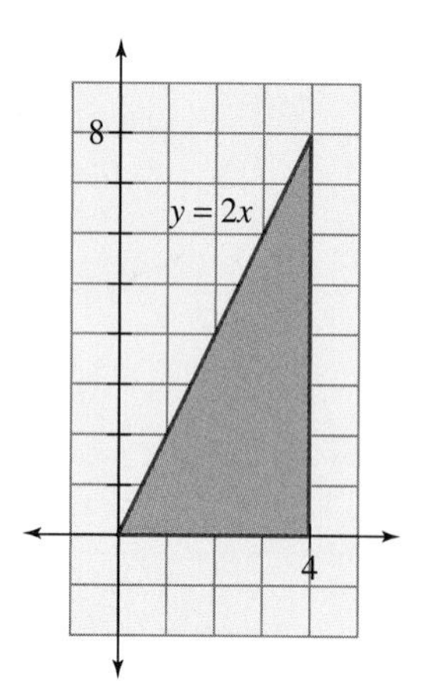

The first step in this method is to divide the region into a number of rectangles. The more rectangles we use, the greater the accuracy of our result. In the first diagram there are four rectangles inside the triangle (one of them has a height of 0), whereas the second diagram has the rectangles extending outside of the triangle. From here, we can anticipate that the area of the lower rectangles will be less than the area of the region, and the area of the upper rectangles will exceed the area of the region.

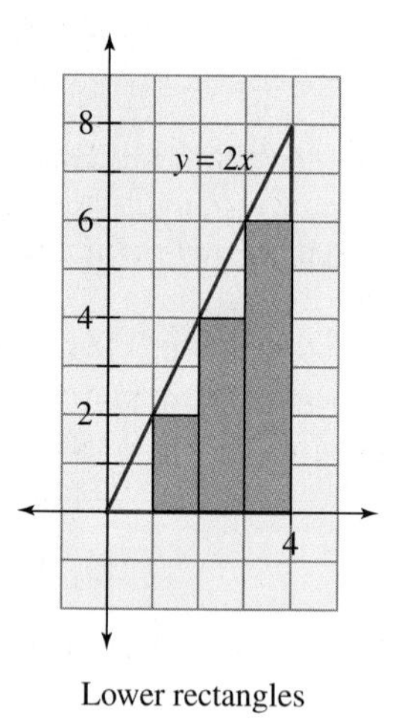

Lower rectangles

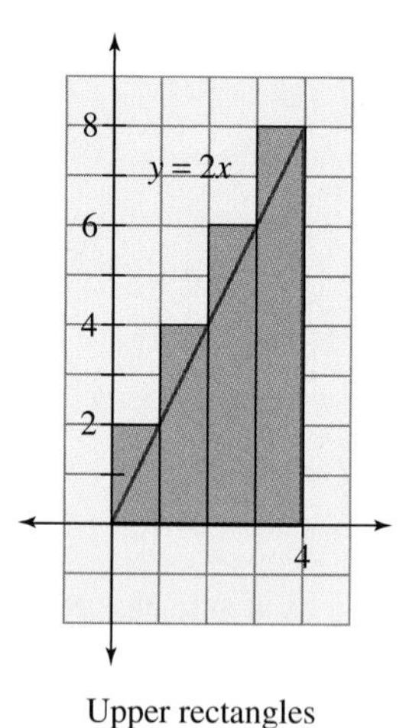

Upper rectangles

An improvement on this method is to use rectangles that approximate the region even more closely. Instead of using rectangles that are obviously smaller or larger than the desired area, we use rectangles that intersect the curve at the midpoint of their top edges, as shown in the following diagram. The upper left corner of each rectangle is outside the region and therefore will provide an overestimate of the area. However, each rectangle has some space above it, underestimating the area of the region. The result should be a value that gives a more accurate approximation of the actual area of the region.

x	y
0.5	1
1.5	3
2.5	5
3.5	7

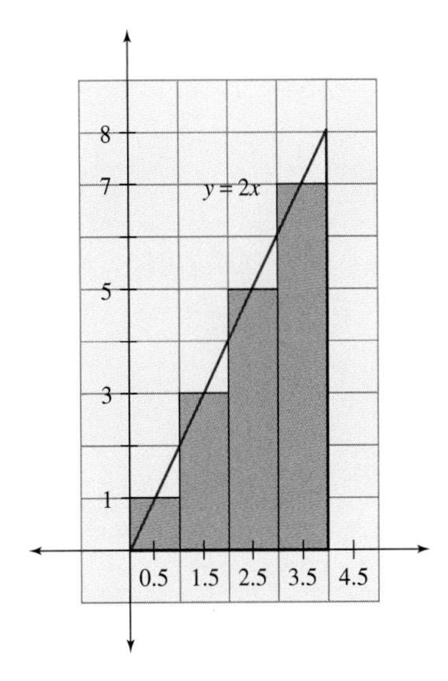

To estimate the area of the region, find the area of the four rectangles. Each rectangle has width 1. We need only find the heights of the rectangles. To do this, we use the fact that the line $y = 2x$ crosses the top of each rectangle at its midpoint. Therefore, the x values are 0.5, 1.5, 2.5, and 3.5. Thus, the heights of the rectangles can be found by substituting these x values into the equation $y = 2x$ as shown in the table. The area of the four rectangles is

$$A = \text{width} \times \text{height of the four rectangles}$$
$$= 1 \times 1 + 1 \times 3 + 1 \times 5 + 1 \times 7 = 16.$$

Through the use of four rectangles, intersecting the line $y = 2x$ at the midpoints of their upper edges, we were able to find the correct area. Although we will usually be able to find only the approximate area, the rectangle method appears to work and can be applied to other regions.

In summary, the rectangle method can be used to estimate the area bounded by a curve by dividing the desired area into rectangles with the midpoint of the top edge of each rectangle intersecting the curve. The area of the region is estimated by calculating the sum of the areas of the rectangles.

EXAMPLE 1

Approximate the area of the region bounded by the x axis, the line $x = 4$, and the curve $y = x^2/4$.

a) Use the rectangle method with four rectangles.

b) Use the rectangle method with eight rectangles.

Solution:

a) Our first step will be to sketch the desired region. As before, the region has been divided into four rectangles. The midpoints of the rectangles are 0.5, 1.5, 2.5, and 3.5. Substituting each of these x values into the equation $y = x^2/4$ gives the following y values:

x	y
0.5	0.0625
1.5	0.5625
2.5	1.5625
3.5	3.0625

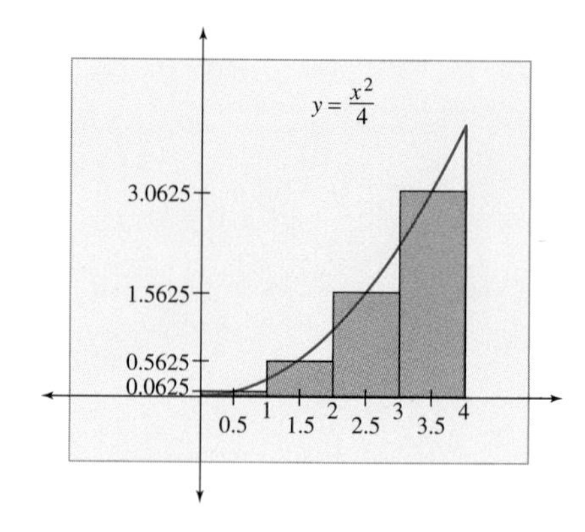

Since the width of each of the rectangles is 1, the sum of the areas of the rectangles is as follows.

$$\begin{aligned} A &= \text{width} \times \text{height of the four rectangles} \\ &= 1 \times 0.0625 + 1 \times 0.5625 + 1 \times 1.5625 + 1 \times 3.0625 = 5.25 \end{aligned}$$

b) We now repeat this process with eight rectangles. Since we want to create eight rectangles in the interval 0 to 4, the width of each rectangle is 1/2, since $(4 - 0)/8 = 1/2$. We can now find the midpoint of each rectangle and its corresponding height, the y value. Since the width of the first rectangle is 1/2 or 0.5, the midpoint of the first rectangle is 0.25. Adding 0.5 gives the midpoint of the second rectangle as 0.75. Repeating this process gives the midpoints of each of the eight rectangles. Using the width of 0.5 and the heights as given by the y values, we have the following.

x	y
0.25	0.0156
0.75	0.1406
1.25	0.3906
1.75	0.7656
2.25	1.2656
2.75	1.8906
3.25	2.6406
3.75	3.5156

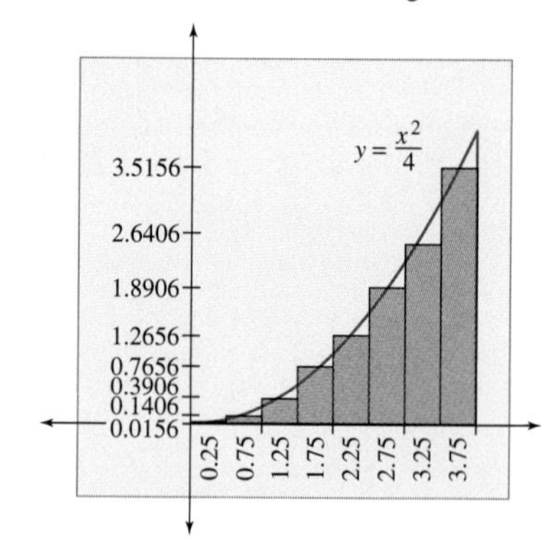

$$\begin{aligned} A &= \text{width} \times \text{height of the four rectangles} \\ &= 0.5 \times 0.0156 + 0.5 \times 0.1406 + 0.5 \times 0.3906 \\ &\quad + 0.5 \times 0.7656 + 0.5 \times 1.2656 + 0.5 \times 1.8906 \\ &\quad + 0.5 \times 2.6406 + 0.5 \times 3.5156 \\ &= 5.3124 \end{aligned}$$

ANTIDERIVATIVES AND DEFINITE INTEGRALS

While the process of determining an area by dividing it into small pieces has been used for 2000 years, it requires a large number of calculations. One of the great discoveries in mathematics was that determining such an area could be greatly simplified through the use of calculus.

In the preceding section, we introduced the process of finding a derivative. An antiderivative of a function $f(x)$ is some other function $F(x)$ such that the derivative of $F(x)$ is $f(x)$. In other words, it is the reverse of the derivative operation. For example, the derivative of x^2 is $2x$. Therefore, an antiderivative of $2x$ is x^2.

One of the most important discoveries of Newton and Leibniz is that calculating an area can be accomplished by using an antiderivative. In particular, if we start with a function $2x$, the area will involve the function x^2. Finding an area by reversing the derivative operation is finding a definite integral and uses the notation $\int_a^b f(x)\,dx$. It is read "the integral from a to b of $f(x)\ dx$."

The process for computing a definite integral, known as the Fundamental Theorem of Calculus, involves two steps.

Step 1: Find a function $F(x)$ that has a derivative of $f(x)$. $F(x)$ is called an antiderivative of $f(x)$.

Step 2: Calculate $F(b) - F(a)$.

EXAMPLE 2

Calculate the definite integral $\int_0^4 2x\,dx$.

Solution:

Step 1: In this example, $f(x) = 2x$. Since the derivative of x^2 is $2x$, an antiderivative is $F(x) = x^2$.

Step 2: The value of $\int_0^4 2x\,dx = F(4) - F(0) = 4^2 - 0^2 = 16$.

Notice that this is the same value as we found for the area of the triangle in the beginning of this section.

Shortcuts for Finding Antiderivatives

In the preceding section, we used shortcuts to find the derivative. Since finding an integral is the reverse of finding a derivative, we have similar shortcuts for finding antiderivatives. Here are two of them.

1. An antiderivative of a constant k is given by kx.
2. An antiderivative of kx^n is given by $\frac{kx^{n+1}}{n+1}$, where $n \neq -1$.

EXAMPLE 3

Find an antiderivative of each of the following:

a) x^3

b) $6x^2$

c) 2

d) $x^3 + 6x^2 - 2$

Solution:

a) Using shortcut 2 with $n = 3$, an antiderivative of x^3 is

$$\frac{x^{3+1}}{3+1} = \frac{x^4}{4}.$$

b) Using shortcut 2 with $n = 2$ and $k = 6$, an antiderivative of $6x^2$ is

$$\frac{6x^{2+1}}{2+1} = \frac{6x^3}{3} = 2x^3.$$

c) Using shortcut 1, an antiderivative of 2 is $2x$.

d) By putting these results together, an antiderivative of $x^3 + 6x^2 - 2$ is

$$\frac{x^4}{4} + 2x^3 - 2x.$$

Using a Definite Integral to Find an Area

In Example 1, we used rectangles and found that the area under the curve $f(x) = \frac{x^2}{4}$ on the interval $0 \le x \le 4$ was approximately 5.31. We now will use a definite integral to find the exact value of the area.

EXAMPLE 4

Use a definite integral to find the area under $f(x) = \frac{x^2}{4}$ on the interval $0 \le x \le 4$.

Solution: The area is given by the definite integral

$$\int_0^4 \frac{x^2}{4} dx = \int_0^4 \frac{1}{4} x^2 dx.$$

Step 1: Using shortcut 2 with $k = \frac{1}{4}$ and $n = 2$, we get an antiderivative

$$F(x) = \frac{1}{4}\left(\frac{x^{2+1}}{2+1}\right) = \frac{1}{4}\left(\frac{x^3}{3}\right) = \frac{x^3}{12}.$$

Step 2: The area is given by

$$F(4) - F(0) = \frac{4^3}{12} - \frac{0^3}{12} = \frac{64}{12} = \frac{16}{3} = 5\frac{1}{3} \approx 5.333\ldots$$

The Fundamental Theorem of Calculus gives the exact area of the region. Notice that while the answer from Example 1 is not exact, it is still important in real-world situations in which the region is not determined by a mathematical function. Consider the following example, in which you are finding the area of an irregularly shaped lawn.

EXAMPLE 5

A lawn fertilizer company suggests that Weed-O-Burn be applied at a rate of 2 pounds per 1000 square feet. Because of the potency of the fertilizer, Chauncy the gardener wants to accurately determine the area of the irregularly shaped lawn, shown in the figure. His only tool is a 100-ft tape measure. How can Chauncy determine the area? What is the approximate area?

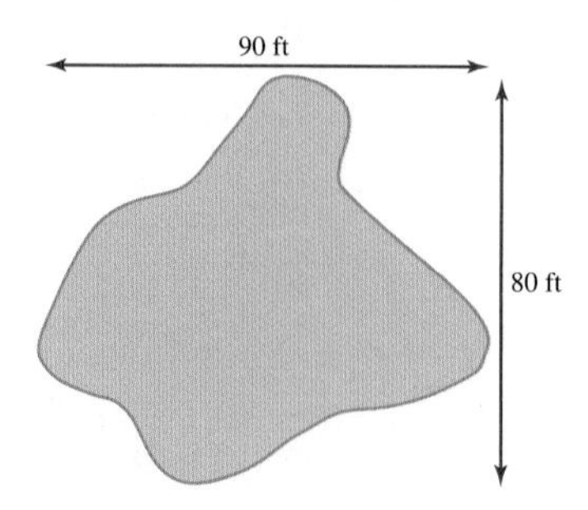

Solution: To determine the area, Chauncy decides to divide the lawn into 10 ft-wide strips and measure the length of each strip. This will give him a set of rectangles whose combined areas will approximate the area of the lawn. There will be eight rectangles ($80 \div 10$), each being 10 ft wide. Chauncy measured the length of each rectangle. His results are given in the figure. Using the formula for the area of a rectangle, we can now find the approximate area of the lawn.

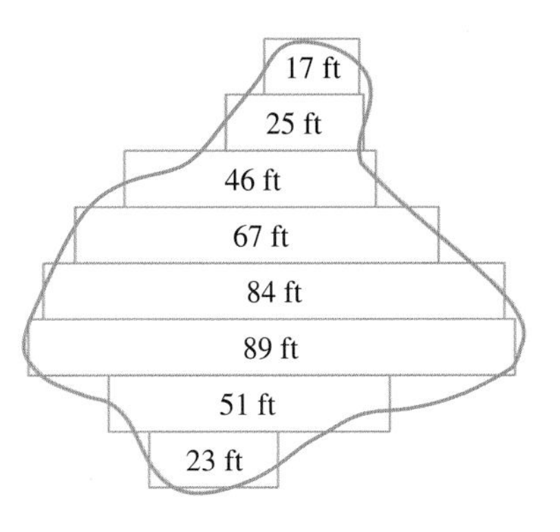

$$\begin{aligned} A &= \text{width} \times \text{length} \\ &= 10 \times 17 + 10 \times 25 + 10 \times 46 + 10 \times 67 + 10 \times 84 \\ &\quad + 10 \times 89 + 10 \times 51 + 10 \times 23 \\ &= 4020 \text{ sq ft of lawn} \end{aligned}$$

This section concludes our short discussion of calculus. In Sections 10.1 and 10.2, we showed that differentiation gives the slope of a curve and how, with the help of some shortcuts, the derivative can be applied to certain mathematical problems. We also showed that integration gives a method for determining the areas of regions bounded by curves. We also introduced shortcuts for doing integration.

We have presented a very brief overview of calculus. We have not attempted to give a complete explanation of the power and applicability of calculus. However, it is our hope that you have seen some mathematics that is intriguing enough for you to pursue additional studies in this area.

section 10.2

PROBLEMS ○ Explain ○ Apply ○ Explore

○ Explain

1. What is integration?
2. When calculating an area, what are "upper" rectangles? Explain what happens to the approximation of the area under a curve as the number of upper rectangles increases.
3. When calculating an area, what are "lower" rectangles? Explain what happens to the approximation of the area under a curve as the number of lower rectangles increases.
4. When calculating an area, what are "midpoint" rectangles? Explain what happens to the approximation of the area under a curve as the number of midpoint rectangles increases.

Apply

In Problems 5–8, sketch the given area and use the midpoint rectangle method to approximate the given area.

5. The area bounded by the lines $y = -2x + 24$, the x axis, and the y axis, using three rectangles.
6. The area bounded by the lines $y = -3x + 24$, the x axis, and the y axis, using four rectangles.
7. The area bounded by the lines $y = 2x + 4$, the x axis, the y axis, and the line $x = 8$, using four rectangles.
8. The area bounded by the lines $y = 3x + 12$, the x axis, the y axis, and the line $x = 8$, using three rectangles.

In Problems 9–12, sketch the given area and use the Fundamental Theorem of Calculus to find the given area.

9. The area bounded by the x axis, the line $x = 5$, and the curve $y = x^2$.
10. The area bounded by the x axis, the y axis, the line $x = 5$, and the curve $y = x^2 - 2x + 3$.
11. The area bounded by the x axis and the curve $y = 36 - x^2$.
12. The area bounded by the x axis and the curve $y = 12x - x^2$.

Explore

13. In the sketch of a small pond, the measurements are 25 ft apart and start 12.5 ft from the ends of the pond. Use the measurements to estimate the area of the pond.

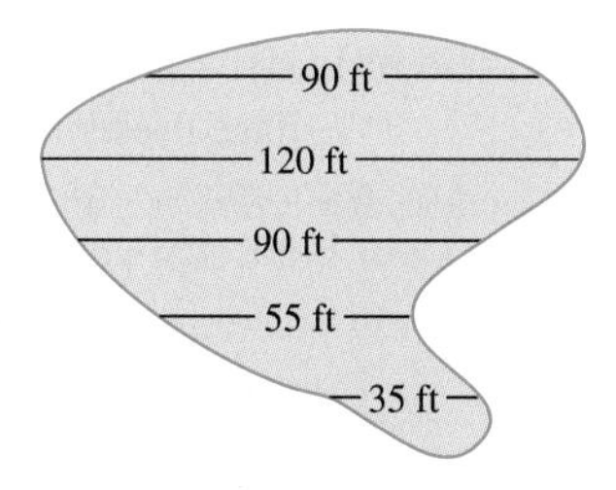

14. The following figure is drawn to scale. Trace the figure onto your paper. Use a metric ruler and the rectangle method with six horizontal rectangles to find the area of the figure in square centimeters.

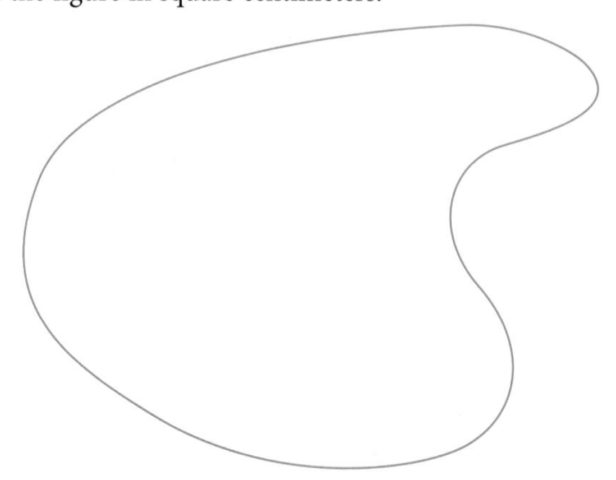

15. Consider the region bounded by the x axis, the line $x = 1$, the line $x = 4$, and the curve $y = \dfrac{1}{x}$.

a) Draw a sketch of the region.

b) Approximate the area of that region using three rectangles.

16. The equation $y = \sqrt{9 - x^2}$ gives the graph of the semicircle with a radius of 3 that is shown below.

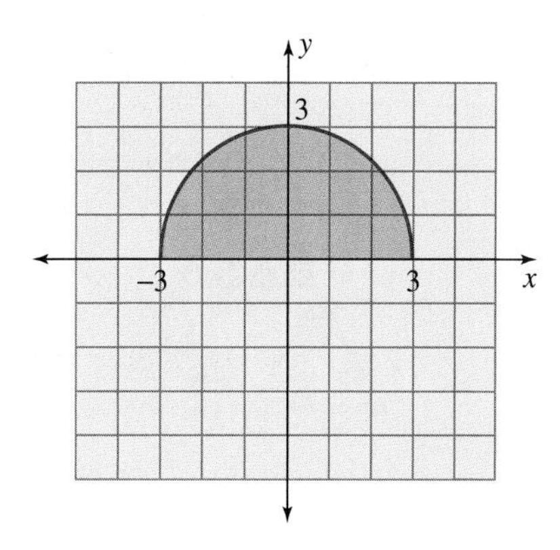

a) Approximate the area of the semicircle using six rectangles.

b) How does that approximation compare to the actual area of a semicircle with radius $r = 3$ given by the formula $A = \frac{1}{2}\pi r^2$.

17. The area of the ellipse, $\dfrac{x^2}{a^2} + \dfrac{y^2}{b^2} = 1$, is given by $A = \pi ab$. Consider an ellipse given the equation $\dfrac{x^2}{64} + \dfrac{y^2}{25} = 1$.

a) Find the area of the ellipse by using the formula. Distances should be measured in centimeters. Give the answer accurate to two decimal places.

b) Find the area of the ellipse using 8 "midpoint" rectangles with a width of 2 centimeters. The lengths of the rectangles will be determined by the equation. Round off your answer to two decimal places.

c) How accurate is the area determined by the rectangle method?

18. The area of a circle with radius r is given by $A = \pi r^2$. Consider a circle with a radius of 10 centimeters.

a) Find the area of the circle by using the formula. Give the answer accurate to two decimal places.

b) Draw the circle on an xy coordinate system with the center of the circle at the origin. Find the area of the circle by using ten "midpoint" rectangles with a width of 2 centimeters. The lengths of the rectangles will be determined by the equation $x^2 + y^2 = 100$. Round off your answer to two decimal places.

c) How accurate is the area determined by the rectangle method?

The Pascal–Yang Hui Triangle

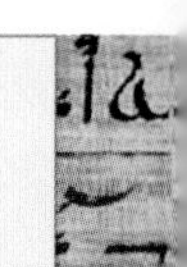

Using modular arithmetic, the Pascal–Yang Hui Triangle is displayed in color. Designed using *Mathematica* by Bob Bradshaw.

In your study of algebra, you might have encountered the triangular array of numbers shown below.

Row	
0	**1**
1	**1 1**
2	**1 2 1**
3	**1 3 3 1**
4	**1 4 6 4 1**
5	**1 5 10 10 5 1**
6	**1 6 15 20 15 6 1**
7	**1 7 21 35 35 21 7 1**
8	**1 8 28 56 70 56 28 8 1**
9	**1 9 36 84 126 126 84 36 9 1**
10	**1 10 45 120 210 252 210 120 45 10 1**

If you examine the triangle, you will see that each row starts and ends with the number 1 and the other entries in each successive row are obtained by adding the two numbers above it.

$$\begin{array}{ccccccccccc} & 1 & & 4 & & 6 & & 4 & & 1 & \\ 1 & & 5 & & 10 & & 10 & & 5 & & 1 \end{array}$$

EXAMPLE 1

Find row 11 of the triangle.

Solution: Row 11 starts and ends with 1. The other entries are sums of the two numbers above it as shown in the following diagram.

$$\begin{array}{cccccccccccc} 1 & 10 & 45 & 120 & 210 & 252 & 210 & 120 & 45 & 10 & 1 & \\ 1 & 11 & 55 & 165 & 330 & 462 & 462 & 330 & 165 & 55 & 11 & 1 \end{array}$$

That triangle is commonly called Pascal's Triangle after the French mathematician Blaise Pascal, who in 1653 published many of its patterns and uses. However, there is evidence that Chinese mathematicians described this triangular pattern some 500 years before Pascal. In China, it is called the Yang Hui Triangle. For this reason, we will refer to the triangle as the Pascal–Yang Hui Triangle. In this section, we will look at some of the uses of the Pascal–Yang Hui Triangle and examine some generalizations of the triangle.

BINOMIAL EXPANSION

The numbers in the triangle can be used in raising a binomial to a power. For example, to find $(1 + x)^4$, you could multiply $(1 + x)(1 + x)(1 + x)(1 + x)$ and get the following result.

$$\begin{aligned}(1 + x)^4 &= (1 + x)(1 + x)(1 + x)(1 + x)\\ &= (1 + 2x + x^2)(1 + 2x + x^2)\\ &= 1 + 2x + x^2 + 2x + 4x^2 + 2x^3 + x^2 + 2x^3 + x^4\\ &= 1 + 4x + 6x^2 + 4x^3 + 1x^4\end{aligned}$$

It took a lot of algebra to get the answer. Furthermore, using that straightforward expansion technique would make finding $(1 + x)^{10}$ a very difficult problem. However, the Pascal–Yang Hui Triangle can simplify the expansion of a binomial. If you look at the fourth row of the triangle, you will see the numbers 1, 4, 6, 4, 1. These numbers are the coefficients in the expansion of $(1 + x)^4$, and the exponents of x are the whole numbers from 0 to 4. Similarly, other rows of the triangle can be used to find the coefficients of expanding $(1 + x)$ to any whole number exponent.

EXAMPLE 2

Find the expansion of $(1 + x)^{10}$.

Solution: Row 10 of the triangle will give you the coefficients of the expansion, and the exponents of x will start at zero and increase by one in each successive term.

$$\begin{aligned}(1 + x)^{10} = 1 &+ 10x + 45x^2 + 120x^3 + 210x^4 + 252x^5 + 210x^6 + 120x^7 + 45x^8\\ &+ 10x^9 + 1x^{10}\end{aligned}$$

POLYNOMIAL EXPANSION

The Pascal–Yang Hui Triangle makes the expansion of a binomial a much easier problem, but what about the expansion of other polynomials? For example, what is the expansion of $(1 + x + x^2)^4$? We could again use a straightforward expansion technique, but that would really be a lot of work. If we systematically examine raising $(1 + x + x^2)$ to a power, we will see the emergence of a Pascal–Yang Hui-like triangle.

Expansion		Coefficients
$(1 + x + x^2)^0 = 1$	$\rightarrow$	1
$(1 + x + x^2)^1 = 1 + x + x^2$	$\rightarrow$	1 1 1
$(1 + x + x^2)^2 = 1 + 2x + 3x^2 + 2x^3 + x^4$	$\rightarrow$	1 2 3 2 1
$(1 + x + x^2)^3 = 1 + 3x + 6x^2 + 7x^3 + 6x^4 + 3x^5 + x^6$	$\rightarrow$	1 3 6 7 6 3 1

If you examine the triangle formed by the coefficients, you will see that each row starts and ends with the number one and the other entries in each sucessive row are obtained by adding the three numbers above it. We will refer to this triangle as *Triangle-3*. The coefficients of $(1 + x + x^2)^4$ can be obtained from the fourth row of Triangle-3. The braces in the chart that follows show how this is done.

Triangle-3

```
                1
             1  1  1
          1  2  3  2  1
       1  3  6  7  6  3  1
    1  4 10 16 19 16 10  4  1
```

The exponents of x will follow the pattern of the previous problem: Start with zero and increase by one in each successive term.

$$(1 + x + x^2)^4 = 1 + 4x + 10x^2 + 16x^3 + 19x^4 + 16x^5 + 10x^6 + 4x^7 + 1x^8$$

EXAMPLE 3

Find $(1 + x + x^2)^5$.

Solution: The coefficients of the expansion will be the fifth row of Triangle-3, and the exponents of x will successively be the whole numbers from 0 to 10. The braces below show how the fifth row is generated from the fourth row of Triangle-3.

```
       1   4  10   16  19  16   10   4   1
   1   5  15  30   45  51  45   30  15   5   1
```

$$(1 + x + x^2)^5 = 1 + 5x + 15x^2 + 30x^3 + 45x^4 + 51x^5 + 45x^6 + 30x^7 + 15x^8 + 5x^9 + 1x^{10}$$

Similar techniques can be used to raise higher-degree polynomials to a power. In the exercises at the end of the section, you will be asked to expand polynomials with four, five, and six terms by using Triangle-4, Triangle-5, and Triangle-6.

COMBINATIONS

The Pascal–Yang Hui Triangle can also be used in counting problems that were introduced in Chapter 3. The entries of the triangle give the number of groupings of r items that can be made from a set of n items without regard for the order of the items. These are called combinations and are noted by $C_{n,r}$. The formula used to determine the number of combinations is

$$C_{n,r} = \frac{n!}{r!(n - r)!}.$$

For example, if three people were selected from a group of seven people, the number of ways to do this is the number of combinations,

$$C_{7,3} = \frac{7!}{3!(7 - 3)!} = 35.$$

The Pascal–Yang Hui Triangle gives the number of combinations generated by the combination formula. For example, the seventh row of the triangle, 1, 7, 21, 35, 35, 21, 7, 1 are the answers to $C_{7,0} = 1$, $C_{7,1} = 7$, $C_{7,2} = 21$, $C_{7,3} = 35$, $C_{7,4} = 35$, $C_{7,5} = 21$, $C_{7,6} = 7$, $C_{7,7} = 1$.

EXAMPLE 4

Find the number of ways in which four items can be selected from a group of ten different items ($C_{10,4}$) using the Pascal–Yang Hui Triangle.

Solution: The tenth row of the triangle is 1, 10, 45, 120, 210, 252, 210, 120, 45, 10, 1. The first entry is $C_{10,0} = 1$, and the fifth entry is $C_{10,4} = 210$. There are 210 ways that a group of four people can be selected from a group of ten people.

"DICE" PROBLEMS

We will now show you how Pascal–Yang Hui Triangles can help solve counting problems involving dice. We will start by considering "two-sided dice"—coins with one dot placed on one side and two dots on the other side.

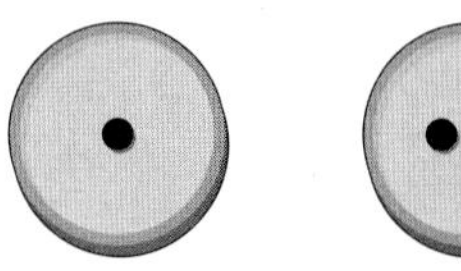

If a number of these "two-sided dice" are tossed, we will determine the number of ways in which the upward facing dots would have a certain sum. By systematically examining all possible outcomes when one, two, three, and four of the "two-sided dice" are tossed, we will observe an interesting pattern.

1. If one "two-sided die" is tossed, what is the number of ways that the coin lands on 1 or 2? By listing the outcomes, we get the following.

Sum	Outcome List	Number of Ways
1	● (1)	1
2	●● (2)	1

2. If two "two-sided dice" are tossed, what is the number of ways that the sum of the dots is 2, 3, or 4? By listing all possible outcomes, we get the following.

Sum	Outcome List	Number of Ways
2	(1, 1)	1
3	(1, 2), (2, 1)	2
4	(2, 2)	1

Note: (1, 2) means that one coin has one dot facing up and the other has two dots facing up.

3. If three "two-sided dice" are tossed, what is the number of ways that the sum of the dots is 3, 4, 5, or 6? By listing all possible outcomes, we get the following.

Sum	Outcome List	Number of Ways
3	(1, 1, 1)	1
4	(1, 2, 1), (2, 1, 1), (1, 1, 2)	3
5	(2, 2, 1), (1, 2, 2), (2, 1, 2)	3
6	(2, 2, 2)	1

Note: (1, 2, 2) means that one coin has one dot facing up and the others have two dots facing up.

4. If four "two-sided dice" are tossed, what is the number of ways that the sum of the dots is 4, 5, 6, 7, or 8? By listing all possible outcomes, we get the following.

Sum	Outcome List	Number of Ways
4	(1, 1, 1, 1)	1
5	(1, 2, 1, 1), (2, 1, 1, 1), (1, 1, 2, 1), (1, 1, 1, 2)	4
6	(2, 2, 1, 1), (1, 1, 2, 2), (2, 1, 1, 2), (1, 2, 1, 2), (1, 2, 2, 1), (2, 1, 2, 1)	6
7	(1, 2, 2, 2), (2, 1, 2, 2), (2, 2, 1, 2), (2, 2, 2, 1)	4
8	(2, 2, 2, 2)	1

If more "two-sided dice" are tossed, we could find the number of ways various sums can occur by listing all possible outcomes. This, however, would be quite tedious. Let's examine the results that we have so far. By organizing the total columns horizontally, we get the following.

Coins	Number of Ways for Each Sum								
1				1		1			
2			1		2		1		
3		1		3		3		1	
4	1		4		6		4		1

Notice that the Pascal–Yang Hui Triangle has appeared. By continuing the observed patterns, we can find the number of ways to get the sums of more "two-sided dice."

EXAMPLE 5

If eight "two-sided dice" are tossed, what is the number of ways that the sum of the upward-facing dots is 13?

Solution: The eighth row of the Pascal–Yang Hui Triangle will give us the number of ways that each sum can occur. With eight "two-sided dice" the smallest sum is 8, and each successive entry in that row will give the number of ways in which larger sums can occur.

Sum	8	9	10	11	12	13	14	15	16
	↕	↕	↕	↕	↕	↕	↕	↕	↕
No. of ways	1	8	28	56	70	56	28	8	1

Thus, the number of ways to have a sum of 13 with eight "two-sided dice" is 56.

"FOUR-SIDED DICE"

A "four-sided die" can be created by considering a wheel divided into four equal sectors with one, two, three, and four placed in the sectors. When one of these wheels is spun, a marker determines on which number the wheel stopped.

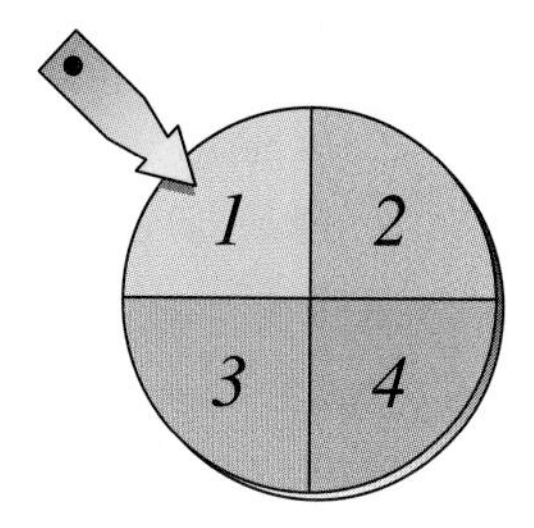

EXAMPLE 6

If five of these "four-sided dice" are spun, find the number of ways to get a sum of 12 from the numbers indicated by the markers.

Solution: Just as a systematic examination of possible outcomes for "two-sided dice" resulted in the Pascal–Yang Hui Triangle, such an analysis of "four-sided dice" would result in Triangle-4. That is, a triangular array of numbers with the first row having four ones and successive rows that start and end with one and all other entries are obtained by adding the four numbers above it. The braces below show how this is done. Notice that when there are fewer than four entries above a number, the missing entries are treated as zero.

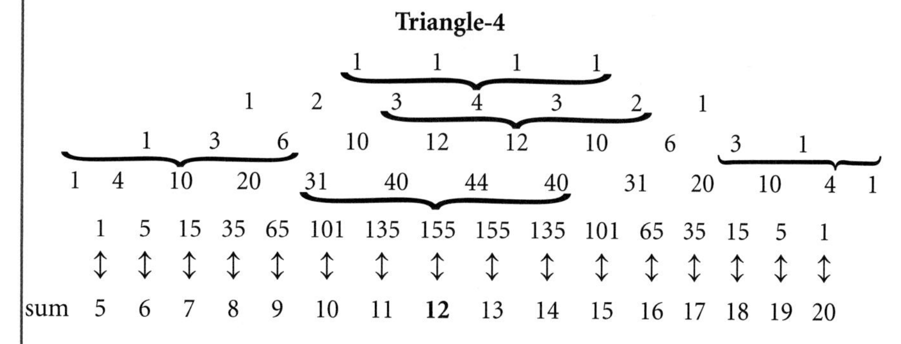

Since five of these dice are spun, in the fifth row, the first entry indicates the number of ways to get a sum of 5. The second entry indicates the number of ways to get a sum of 6; the third entry, a sum of 7; and so on. Thus, the number of ways to get a sum of 12 with five "four-sided dice" is 155.

The procedure developed for "two-sided dice" and "four-sided dice" can be generalized to dice with any number of faces or wheels with any number of equal sectors. Thus, by using Triangle-6, you will be able to solve a problem such as "If five standard dice are tossed, what is the number of ways that the sum of the dice is 16?" The solution to that problem is left to you as an exercise.

PROBLEMS ⊙ Explain ⊙ Apply ⊙ Explore

Explain

1. Why is what is commonly known as *Pascal's Triangle* called the *Pascal–Yang Hui Triangle* in this text?
2. What is meant by Triangle-3?
3. How do you get a row from the Pascal–Yang Hui Triangle from a previous row?
4. How do you get a row from the Triangle-3 from a previous row?
5. What does row 6 of the Pascal–Yang Hui Triangle give you when expanding a binomial?
6. What does the second row of Triangle-3 give, when expanding the trinomial $1 + x + x^2$?
7. What does the third row of Triangle-4 give, when expanding the polynomial $1 + x + x^2 + x^3$?
8. How do you get the number of combinations of five items from a set of nine different items ($C_{9,5}$) using the Pascal–Yang Hui Triangle?
9. What does row 6 of the Pascal–Yang Hui Triangle give you when working with combinations?
10. What does the third row of Triangle-4 give, when spinning "four-sided dice"?
11. What does the fourth row of Triangle 3 give, when spinning "three-sided dice"?

Apply

12. Expand $(1 + x)^6$.
13. Expand $(1 + x)^7$.
14. Expand $(1 + x + x^2)^4$.
15. Expand $(1 + x + x^2)^6$.
16. Expand $(1 + x + x^2 + x^3)^3$.
17. Expand $(1 + x + x^2 + x^3)^4$.
18. Find the number of ways to get a sum of 13 with nine "two-sided dice."
19. Find the number of ways to get a sum of 17 with nine "two-sided dice."
20. Find the number of ways to get a sum of 11 with five "three-sided dice."
21. Find the number of ways to get a sum of 13 with five "three-sided dice."
22. Find the number of ways to get a sum of 7 with four "four-sided dice."
23. Find the number of ways to get a sum of 12 with four "four-sided dice."
24. Find the number of ways to get a sum of 8 with six "four-sided dice."

Explore

25. The first two rows of Triangle-5 are shown.

$$\begin{matrix} & & 1 & 1 & 1 & 1 & 1 & & \\ 1 & 2 & 3 & 4 & 5 & 4 & 3 & 2 & 1 \end{matrix}$$

a) Find the next three rows.

b) Use Triangle-5 to expand $(1 + x + x^2 + x^3 + x^4)^3$.

c) Use Triangle-5 to expand $(1 + x + x^2 + x^3 + x^4)^4$.

d) If four "five-sided dice" are spun, in how many ways can you get a sum of 13?

e) If five "five-sided dice" are spun, in how many ways can you get a sum of 13?

26. The first two rows of Triangle-6 are shown.

$$\begin{array}{ccccccccccc} & & & 1 & 1 & 1 & 1 & 1 & 1 & & \\ 1 & 2 & 3 & 4 & 5 & 6 & 5 & 4 & 3 & 2 & 1 \end{array}$$

a) Find the next three rows.

b) Use Triangle-6 to expand $(1 + x + x^2 + x^3 + x^4 + x^5)^3$.

c) Use Triangle-6 to expand $(1 + x + x^2 + x^3 + x^4 + x^5)^4$.

d) If four standard dice are tossed, in how many ways can you get a sum of 16?

e) If five standard dice are tossed, in how many ways can you get a sum of 16?

section 10.4

Voting Systems

Politicians, like those shown in *The Senate* by William Gropper, use apportionment and voting patterns to set up their campaign strategy—where should they concentrate their campaigning for the biggest "vote-getting" impact. (The Museum of Modern Art, New York)

It's Friday evening, and it has been a long week of school and work. You and four friends are going to go to the movies to relax, but you are not able to agree on which movie to see. Vicky wants to see *Disney Extravaganza.* Wanda and Xavier want to see the latest remake of *Pride and Prejudice.* You and Zelma want to see *The Terminated.* Also available is *Gone with the Wind.* Since there is no clear choice and you are all enrolled in the same math class, the five of you to decide to try to work out the decision mathematically. The first step seems clear: Write out a table that shows each person's preferences. In this preference table, each person ranks the movies from 1 (first choice) to 4 (last choice.)

	Vicky	Wanda	Xavier	You	Zelma
Pride	3	1	1	4	4
Terminated	4	4	3	1	1
Disney	1	2	4	3	3
Gone	2	3	2	2	2

For example, the table shows that Wanda's first choice is *Pride and Prejudice,* her second choice is *Disney Extravaganza,* her third choice is *Gone with the Wind,* and her least favorite is *The Terminated.*

The term voting system is used to describe the many different ways in which this information can be analyzed. We will look at four of the most frequently used methods. They all have advantages and disadvantages in their use.

1. Plurality method
2. Plurality with elimination method
3. Borda count method
4. Pairwise comparison method

PLURALITY METHOD

The plurality method is the one commonly used in elections: Each voter gets to make one choice, and the candidate with the most votes wins. When using the plurality method, second, third, and subsequent choices do not count. This is a major disadvantage of the plurality method.

In our example, for the purposes of the plurality method, the information in the preference table reduces to

	No. of First-Place Votes
Pride	2
Terminated	2
Disney	1
Gone	0

In our example, the plurality method would not produce a decision, since two movies, *Pride and Prejudice* and *The Terminated,* are tied with two votes each.

PLURALITY WITH ELIMINATION

Plurality with elimination is the same as the plurality method if there is a winner. If there is no clear winner, the choice with the fewest first place votes is eliminated. If there is a tie for least-favorite choice, all of those choices are eliminated. A major disadvantage of this method is the need to conduct another election each time the election fails to produce a winner.

In a situation such as ours, in which the voters have ranked all their choices, there is no need to conduct another election. We simply need to eliminate the least favorite choice from the preference table.

Returning to our preference table, we have the following.

	Vicky	Wanda	Xavier	You	Zelma
Pride	3	1	1	4	4
Terminated	4	4	3	1	1
Disney	1	2	4	3	3
Gone	2	3	2	2	2

Since *Gone with the Wind* did not receive any first-place choices, it will be removed from the preference table. This also has the effect of changing other votes. For example, in Vicky's case, *Pride and Prejudice* moves up from number 3 to number 2, and *The Terminated* moves from number 4 to number 3. Making these changes in the preference table gives the following.

	Vicky	Wanda	Xavier	You	Zelma
Pride	2	1	1	3	3
Terminated	3	3	2	1	1
Disney	1	2	3	2	2

This leaves the number of first-place votes the same as it was before.

	No. of First-Place Votes
Pride	2
Terminated	2
Disney	1

Since this does not result in any changes to the number of first-place votes, we repeat the process, this time eliminating *Disney Extravaganza.*

	Vicky	Wanda	Xavier	You	Zelma
Pride	1	1	1	2	2
Terminated	2	2	2	1	1

Since *Pride and Prejudice* now has more first place votes than *The Terminated,* it is the winning movie.

BORDA COUNT METHOD

The Borda count method requires each voter to rank all the candidates, and each vote for a candidate has a weighting based on its rank. In our example, we assign a first-place vote a weight of 4 points, a second place vote a weight of 3 points, and so on. The candidate with the most total points is declared the winner.

Returning to our preference table, we have the following.

	Vicky	Wanda	Xavier	You	Zelma
Pride	3	1	1	4	4
Terminated	4	4	3	1	1
Disney	1	2	4	3	3
Gone	2	3	2	2	2

Thus, we can assign a point total for each movie. *Pride and Prejudice* has two first-place votes worth 4 points each, 0 second-place votes, 1 third-place vote worth 2 points, and 2 fourth-place votes worth 1 point each. This gives *Pride and Prejudice* a total of $(2 \times 4) + (0 \times 3) + (1 \times 2) + (2 \times 1) = 12$ points.

Summarizing the point totals for each movie, we have the following.

	points
Pride	(2 × 4) + (0 × 3) + (0 × 2) + (2 × 1) = 10
Terminated	(2 × 4) + (0 × 3) + (1 × 2) + (2 × 1) = 12
Disney	(1 × 4) + (1 × 3) + (2 × 2) + (1 × 1) = 9
Gone	(0 × 4) + (4 × 3) + (2 × 1) + (0 × 1) = 14

According the Borda count method, *Gone with the Wind* is the winning movie.

The above example shows what many feel is a major disadvantage of the Borda Count method. Even though not a single person made *Gone with the Wind* their first choice, it ends up with the highest ranking.

PAIRWISE COMPARISON METHOD

The pairwise comparison method compares each pair of candidates. The preferred candidate in each pair receives one point. The candidate with the most points is declared the winner.

In our example, we need to make six comparisons:

Pride and Prejudice versus *The Terminated*

Pride and Prejudice versus *Disney Extravaganza*

Pride and Prejudice versus *Gone with the Wind*

The Terminated versus *Disney Extravaganza*

The Terminated versus *Gone with the Wind*

Disney Extravaganza versus *Gone with the Wind*

To compare *Pride and Prejudice* versus *The Terminated,* return to the preference table and examine only those two movies.

	Vicky	Wanda	Xavier	You	Zelma
Pride	3	1	1	4	4
Terminated	4	4	3	1	1

Notice that three people (Vicky, Wanda, and Xavier) prefer *Pride and Prejudice,* while only two people (you and Zelda) prefer *The Terminated.* Therefore, *Pride and Prejudice* gets one point.

Comparing *Pride and Prejudice* versus *Disney Extravaganza* uses the following table.

	Vicky	Wanda	Xavier	You	Zelma
Pride	3	1	1	4	4
Disney	1	2	4	3	3

Disney Extravaganza gets a point, since it is preferred 3 to 2 over *Pride and Prejudice.*

Summarizing this information for all six comparisons, we have the following.

Comparison	Winner
Pride v. *Terminated*	*Pride*
Pride v. *Disney*	*Disney*
Pride v. *Gone*	*Gone*
Terminated v. *Disney*	*Terminated*
Terminated v. *Gone*	*Gone*
Disney v. *Gone*	*Gone*

Since *Gone with the Wind* won three of the six comparisons, it is declared the winning movie.

It's now too late to go to the movies, but you have learned something. The seemingly simple concept of voting can lead to unexpected complexities. In calculating the votes four different ways, there was no clear winner. Let's see how these four voting methods work when applied on a larger scale.

EXAMPLE 1

A town of 10,000 voters is conducting an election for mayor. The candidates are Alice Anderson (A), Bill Benson (B), and Carlita Cordova (C). Each voter lists all the candidates in order of preference, with the results shown in the following table. Apply each of the four voting methods to these results and see whether there is a clear winner.

	Number of People					
	2000	1500	2500	1000	1500	1500
1st place	*A*	*A*	*B*	*B*	*C*	*C*
2nd place	*B*	*C*	*A*	*C*	*A*	*B*
3rd place	*C*	*B*	*C*	*A*	*B*	*A*

Solution: The preference table shows the six possible orderings of the candidates and the number of people who voted for each order. For example, 2000 people choose the candidates in the order Alice, Bill, and Carlita.

Plurality Method: Using the plurality method, we are looking only at the first-place votes. Therefore, Alice and Bill are tied for first place, since each has 3500 first-place votes, compared to 3000 for Carlita.

Plurality with Elimination: Since there was no clear winner but Carlita is definitely in third place, the plurality with elimination method states that we should eliminate Carlita from the voting.

	Number of People					
	2000	1500	2500	1000	1500	1500
1st place	*A*	*A*	*B*	*B*	*A*	*B*
2nd place	*B*	*B*	*A*	*A*	*B*	*A*

This gives Alice 2000 + 1500 + 1500 = 5000 votes and Bill 2500 + 1000 + 1500 = 5000 votes, so the election remains a tie.

Borda Count Method: Using the Borda count method involves the same steps, but we also need to account for the numbers of voters. Returning to our preference table, we have the following.

	Number of People					
	2000	1500	2500	1000	1500	1500
1st place	*A*	*A*	*B*	*B*	*C*	*C*
2nd place	*B*	*C*	*A*	*C*	*A*	*B*
3rd place	*C*	*B*	*C*	*A*	*B*	*A*

If we use 3 points for a first-place vote, 2 points for a second-place vote, and 1 point for a third-place vote, Alice has 3500 first-place votes at 3 points each, 4000 second-place votes at 2 points each, and 2500 third-place votes at one point each. Repeating this calculation for each candidate gives the following.

Candidate	Points
Alice	$(3500 \times 3) + (4000 \times 2) + (2500 \times 1) = 21{,}000$
Bill	$(3500 \times 3) + (3500 \times 2) + (3000 \times 1) = 20{,}500$
Carlita	$(3000 \times 3) + (2500 \times 2) + (4500 \times 1) = 18{,}500$

Therefore, using the Borda count method, Alice is the winner.

Pairwise Comparison Method: The pairwise comparison method requires three comparisons:

Alice versus Bill

Alice versus Carlita

Bill versus Carlita

To compare Alice versus Bill, return to the preference table and ignore any information about Carlita. This gives the following.

	Number of People					
	2000	1500	2500	1000	1500	1500
1st place	*A*	*A*	*B*	*B*		
2nd place	*B*		*A*		*A*	*B*
3rd place		*B*		*A*	*B*	*A*

This results in a tie, with Alice and Bill each winning 5000 of the comparisons. Comparing Alice versus Carlita, we have the following table.

	Number of People					
	2000	1500	2500	1000	1500	1500
1st place	*A*	*A*			*C*	*C*
2nd place		*C*	*A*	*C*	*A*	
3rd place	*C*		*C*	*A*		*A*

This shows Alice having the advantage, 6000 votes to 4000 votes.

Finally, comparing Bill to Carlita gives the following.

	Number of People					
	2000	1500	2500	1000	1500	1500
1st place			*B*	*B*	*C*	*C*
2nd place	*B*	*C*		*C*		*B*
3rd place	*C*	*B*	*C*		*B*	

This shows Bill winning by 5500 to 4500 votes.

Summarizing this information, Alice and Bill are tied when compared to each other, and both beat Carlita. Again, there is no clear winner.

In conclusion, Alice can claim to be the winner if the Borda count method is used, but all other methods shows that she is tied with Bill. This is a very close vote, and Alice should not make claims about being the clear choice of the voters.

WHICH METHOD IS BEST?

As we went through these four methods, we pointed out a major disadvantage to each. There are several other voting methods that are not discussed in this section, and they also have advantages and disadvantages. Political scientists searched for a "perfect" method until 1951, when Kenneth Arrow, a researcher for the RAND Corporation showed that no perfect method exists. Instead, votes should be counted using a method that is agreed to before the election, and then the results should be accepted.

PROBLEMS ➲ Explain ➲ Apply ➲ Explore

➲ Explain

1. What is the main disadvantage of the plurality method?
2. What is the main disadvantage of the plurality with elimination method?
3. What is the main disadvantage of the Borda count method?
4. In the following table, it shows the preferences for different breeds of dogs in a vote taken by 18 people. Use the pairwise comparison method on the data to determine the most popular dog. Next, remove brittanies and cockers from the table and do the pairwise comparison again. Use the results to state a possible disadvantage of the pairwise comparison method.

	Number of People			
	8	4	5	1
Afghans	1	2	4	2
Brittanies	2	4	2	3
Cockers	3	3	1	4
Dachshunds	4	1	3	1

Apply

5. Six 12-year-old girls were asked to rank four children's books. The results are shown below. Apply each of the four voting methods to these results and see whether there is a clear favorite.

	Ann	Brenda	Cathy	Dawn	Elle	Francesca	Gloria
The Time Machine	1	2	1	2	4	4	2
The Mysterious Island	4	4	3	1	1	3	3
The Secret Garden	2	3	4	3	3	2	1
Wind in the Willows	3	1	2	4	2	1	4

6. Five 12-year-old boys were asked to rank four children's books. The results are shown below. Apply each of the four voting methods to these results and see whether there is a clear favorite.

	Gary	Harry	Jerry	Larry	Perry
Tom Swift and His Giant Robot	2	2	1	3	4
20,000 Leagues Under the Sea	3	4	3	1	1
Harry Potter, vol. 1	1	1	4	4	3
The Jungle Book	4	3	2	2	2

7. A college is conducting the elections for student body president. Three candidates are running for president, and 4000 students cast a ballot. Each voter lists all the candidates in order of preference, with the results shown in the following table. For example, 1200 voters chose candidate *A* as their first choice, *B* as their second choice, and candidate *C* as their third choice. Apply each of the four voting methods to these results and see whether there is a clear winner.

	Number of Students				
	1200	200	1000	1500	100
1st place	*A*	*A*	*B*	*C*	*B*
2nd place	*B*	*C*	*A*	*B*	*C*
3rd place	*C*	*B*	*C*	*A*	*A*

8. A college is conducting the elections for student body president. Three candidates are running for president, and 4000 students cast a ballot. Each voter lists all the candidates in order of preference, with the results shown in the following table. For example, 200 voters chose candidate *A* as their first choice, *B* as their second choice, and candidate *C* as their third choice. Apply each of the four voting methods to these results and see whether there is a clear winner.

	Number of Students				
	200	1200	1000	300	1200
1st place	*A*	*A*	*B*	*B*	*C*
2nd place	*B*	*C*	*A*	*C*	*A*
3rd place	*C*	*B*	*C*	*A*	*B*

Explore

9. A deteriorating economy is forcing the state legislature to make difficult choices while trying to create a balanced budget. Each of the 40 legislators lists all the options in order of preference, with the results shown in the following table. For example, 13 of the 40 legislators decided that they would prefer to first increase sales taxes, and then follow that by decreasing school funding, decreasing transportation funding, decreasing social service funding, and finally increasing income taxes.

	Number of Legislators				
	13	10	8	7	2
Increase sales taxes	1	5	2	3	5
Increase income taxes	5	1	3	2	4
Decrease school funding	2	2	5	4	5
Decrease transportation funding	3	4	1	5	2
Decrease social service funding	4	3	2	1	1

The legislators agreed to use the Borda count method with the following weighting: first choice: 5 points; second choice: 4 points; and so on. According to this voting method, in what order should these options be implemented?

10. Referring to the preference table in Problem 9, suppose the legislature decides that it wants to make everyone's first choice be of a greater weight. Therefore, the legislators agreed to use the Borda count method with the following weighting: first choice: 7 points; second choice: 4 points; third choice: 3 points; fourth choice: 2 points; last choice: 1 point. According to this voting method, in what order should these options be implemented?

11. Referring to the preference table in Problem 9, suppose the legislature decides that it wants to make everyone's first and second choices be of a greater weight. Therefore, the legislators agreed to use the Borda count method with the following weighting: first choice: 6 points; second choice: 6 points; third choice: 3 points; fourth choice: 2 points; last choice: 1 point. According to this voting method, what is the preferred order for the legislature as a group?

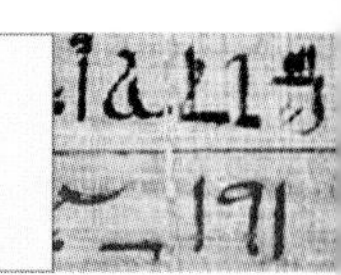

section 10.5 Apportionment

U.S. Capital building, Washington, D.C. (Courtesy of Richard Nowitz/CORBIS)

In this section, we will continue our math vistas by investigating the way a representative body of leaders can be selected in a democratic system by using different methods of apportionment. Apportionment is simply the process by which a given amount of a resource can be divided into parts. The section will present this topic by determining how the five counties of a state can divide its 50 legislative seats. We will determine the number of representatives for each of the five counties using different methods. Our investigations will show that logical methods of analyzing the data can produce different results.

The following table gives the populations of the different counties.

County	Population
Brooke	48,859
Hopkins	161,135
Isaac	87,194
Haley	596,270
Wallace	110,006
Total	1,003,464

The first assumption is that each seat in the legislature should provide representation for approximately the same number of people. Thus, since there are 50 seats, one way to determine the apportionment is to determine the percentage of the state's population that is in each region and multiply that percentage by 50. For example, for Brooke County, we have

$$\frac{48{,}859}{1{,}003{,}464} \times 50 \approx 2.43.$$

The results of the calculations for each region are presented in the Quotas column of Table 10.5.1. Since it is not possible to have a decimal number of seats apportioned to a county, we determine the number of seats by rounding off each quota to the nearest whole number.

TABLE 10.5.1 Number of Seats Using Rounding Method

County	Population	Quotas	No. Seats
Brooke	48,859	2.43	2
Hopkins	161,135	8.03	8
Isaac	87,194	4.35	4
Haley	596,270	29.71	30
Wallace	110,006	5.48	5
Totals	1,003,464	50.00	49

Notice that although the total number of seats should be 50, only 49 of the seats have been allocated. While rounding off the quotas seems to be a reasonable method, it fails to allocate all of the available seats. When the government of the United States was being established in the late 1780s, similar difficulties were encountered in setting up the apportionment for the seats in the House of Representatives. In search of a solution, several methods of apportionment were devised. We will look at three of these methods: those devised by Alexander Hamilton, Daniel Webster, and Thomas Jefferson.

The Hamilton method assigns an initial number of seats by taking the integer part of each quota. This may give a total number of seats that is fewer than the actual number of seats. In our example, the initial total number of seats is 48 (see Table 10.5.2). Since this allows for an additional two seats, the Hamilton method assigns those seats to the counties with the greatest decimal parts in their quotas. Since Haley (decimal part 0.71) and Wallace (0.48) have the greatest decimal parts in their quotas, they receive the additional seats.

TABLE 10.5.2 Number of Seats Using Hamilton's Method

County	Population	Quotas	Initial No. of Seats	No. Seats
Brooke	48,859	2.43	2	2
Hopkins	161,135	8.03	8	8
Isaac	87,194	4.35	4	4
Haley	596,270	29.71	29	30
Wallace	110,006	5.48	5	6
Total	1,003,464	50.00	48	50

DIVISOR METHODS

The apportionment methods of Jefferson and Webster are called divisor methods. Divisor methods apportion a resource by dividing the number in each group by a divisor and then rounding off the result to a nearby integer.

Jefferson's method works as follows. Since we have a total of 1,003,464 people in the state and 50 seats, each seat should represent $\frac{1{,}003{,}464}{50} \approx 20{,}069$ voters. This number 20,069 is our divisor. The number of representatives for each county is calculated by dividing the county population by the divisor and rounding down to the nearest whole number. This is shown in Table 10.5.3.

TABLE 10.5.3 Number of Seats Using Jefferson's Method

County	Population	Quotas	No. of Seats
Brooke	48,859	2.43	2
Hopkins	161,135	8.03	8
Isaac	87,194	4.35	4
Haley	596,270	29.71	29
Wallace	110,006	5.48	5
Totals	1,003,464	50.00	48

Since the total number of seats used, 48, is fewer than the 50 seats that are available, the divisor is decreased until the number of seats apportioned has increased to 50. The first divisor that achieves this goal is 19,234 instead of the initial 20,069. (See Table 10.5.4.)

TABLE 10.5.4 Number of Seats Using Jefferson's Method

County	Population	Pop./Divisor (19,234)	No. of Seats
Brooke	48,859	2.54	2
Hopkins	161,135	8.38	8
Isaac	87,194	4.53	4
Haley	596,270	31.00	31
Wallace	110,006	5.72	5
Totals	1,003,464		50

Webster's method is another divisor method. With Webster's method, you proceed as with Jefferson's method but round off to the nearest whole number rather than always rounding down. An example is given in Table 10.5.5.

TABLE 10.5.5 Number of Seats Using Webster's Method

		Divisor = 20,069		Divisor = 20,001	
County	Population	Quotas	No. of Seats	Pop./Divisor	No. of Seats
Brooke	48,859	2.43	2	2.44	2
Hopkins	161,135	8.03	8	8.06	8
Isaac	87,194	4.35	4	4.36	4
Haley	596,270	29.71	30	29.81	30
Wallace	110,006	5.48	5	5.50	6
Totals	1,003,464	50.00	49		50

WHY SO MANY METHODS?

We have introduced three methods that provide a clear way of determining the apportionment of a legislature if the number of available seats is known. There are other methods, such as the quota method, the Hill-Huntington method, and Lowndes' method. The Hill-Huntington method is currently being used to apportion the U.S. House of Representatives, but a discussion of it is beyond the scope of this course.

You might wonder why there are so many methods and which method is the best to use. The reason that there are many methods is that all of the methods have biases, some obvious and some hidden. As these problems were discovered, the various methods fell out of favor. For example, in the apportionment of the U.S. House of Representatives, Jefferson's method was used until 1842. Hamilton's method was used until 1901, when it was replaced by Webster's method. This was replaced, in turn, by the current method, Hill-Huntington, in 1941. In the remainder of this section, we will examine some of the problems associated with some of these methods.

The Bias in Jefferson's Method

Because the apportionment of the legislature requires some sort of rounding off of the quotas, one of the guidelines in setting up an apportionment procedure is given by the quota rule. The quota rule states that the number of seats apportioned to a group should be a whole number within one of the quota. For example, if a county's quota is 1.95, it should receive an apportionment of either 1 or 2.

Jefferson's method tends to favor states or counties with large populations. If we reexamine the apportionment of the state's legislature using Jefferson's method with respect to the quota rule, we see that Haley county has a quota of 29.71 (see Table 10.5.3), but it receives an apportionment of 31 seats, thus violating the quota rule (Table 10.5.4).

Two Paradoxes with Hamilton's Method

A problem that arose with Hamilton's method is now known as the Alabama paradox. In 1880, Hamilton's method was used to apportion the U.S. House of Representatives. In the Alabama paradox, the House was apportioned according to Hamilton's method. Using the same method with the same population figures, the size of the legislature was increased and the House reapportioned. The paradox arose because Alabama's apportionment decreased even though the total number of seats had increased and there was no change in the population. We can examine this paradox in the following example.

EXAMPLE 1

Consider the populations of the following three counties in a state with a 60-seat legislature. Using Hamilton's method, we get the following.

County	Population	Quotas	Initial No. of Seats	No. Seats
Anderson	3,010	30.21	30	30
Black Hills	2,710	27.20	27	27
Coldwater	258	2.59	2	3
Totals	5,978	60.00	59	60

What happens to the apportionment if the number of seats is increased to 61?

Solution: If the number of seats increases, the quota for each county must be recalculated. For Anderson County, we find

$$\frac{3010}{5978} \times 61 \approx 30.71$$

Repeating this procedure for the other two counties and computing the corresponding number of seats in the legislature, we have the following.

County	Population	Quotas	Initial No. of Seats	No. Seats
Anderson	3,010	30.71	30	31
Black Hills	2,710	27.65	27	28
Coldwater	258	2.63	2	2
Totals	5,978	61.00	59	61

Notice that though there was no change in the population of any of the three counties, Coldwater County lost one of its three seats.

A second paradox that affects Hamilton's method is called the population paradox. The population paradox occurs when reapportionment takes place and a region with a greater rate of increase in population loses seats to another region with a slower growth rate.

EXAMPLE 2

The following table gives the population figures for five counties, with populations given in thousands, and the associated apportionment of 50 seats as calculated by Hamilton's method.

Seats Computed Using Hamilton's Method

County	Population (in thousands)	Quotas	Initial No. of Seats	No. Seats
Anderson	1,732	9.60	9	10
Black Hills	1,503	8.33	8	8
Coldwater	784	4.34	4	4
Shasta	2,049	11.35	11	11
Tehachapi	2,957	16.38	16	17
Totals	9,025	50.00	48	50

Suppose the population of Anderson County increases by 5% while the population of Tehachapi County increases by 0.3%. The populations of the other three counties remain the same. Use Hamilton's method to calculate the new apportionment.

Solution: Since the population of Anderson County changed by 5%, the population increased by $1{,}732{,}000 \times 0.05 = 87{,}000$ or 87 thousand people. Therefore, the new population is $1732 + 87 = 1819$ thousand people. Similarly, multiplying the population of Tehachapi County by 0.003 and adding this amount to the original population results in a new population of 2966 thousand people.

Using Hamilton's method on these figures, we arrive at the following table.

Seats Computed Using Hamilton's Method

County	Population	Quotas	Initial No. of Seats	No. Seats
Anderson	1,819	9.97	9	10
Black Hills	1,503	8.24	8	8
Coldwater	784	4.30	4	5
Shasta	2,049	11.23	11	11
Tehachapi	2,966	16.26	16	16
Totals	9,121	50.00	48	50

Notice what has happened. Coldwater County's apportionment has increased from four seats to five seats despite having no growth in its population. In addition, Tehachapi County has suffered a decrease in the number of its representatives despite having a small increase in its population.

After seeing how Hamilton's method is subject to both the Alabama paradox and the population paradox and that Jefferson's method is biased for regions with larger populations, you might wonder whether there are similar difficulties with Webster's method or any other method. The answer is a definitive "Yes." In 1980, Michel L. Balinski and H. Peyton Young stated and proved what is known as the Balinski and Young impossibility theorem. The theorem states that it is mathematically impossible for any apportionment method to be flawless. If a method satisfies the quota rule, it will occasionally produce paradoxes. On the other hand, if the method does not produce paradoxes, in some situations it will violate the quota rule.

As a result of the impossibility theorem, you can see that an analysis of any apportionment method will result in the discovery of a flaw. Currently, the apportionment of the U.S. House of Representatives is accomplished by using the Hill-Huntington method. However, because of the impossibility theorem, the debate continues as to which method is best. Currently, mathematicians are debating the merits of Webster's method versus those of the Hill-Huntington method. Webster's method is considered by some to be the best method. It is theoretically possible that it will violate the quota rule. However, if it had been applied in every apportionment of the House of Representatives since 1790, no such violation would have occurred. Other mathematicians believe that the Hill-Huntington method is best because it minimizes the relative differences in size between districts. Will the method change? If it does, the U.S. Constitution says that it would require an act of Congress.

From the discussion in this section, you have observed how different analyses of the same set of data can lead to different conclusions and decisions. This points out the importance of understanding how statistics are computed and interpreted. Without this underlying understanding, you will not be able to fully use the power of statistics.

section 10.5

PROBLEMS ● Explain ● Apply ● Explore

● Explain

1. What is meant by "apportionment"?
2. Describe Hamilton's method.
3. Describe Jefferson's method.
4. Describe Webster's method.
5. What is the Alabama paradox?
6. What is the population paradox?
7. What is the quota rule?
8. What is the impossibility theorem?

● Apply

9. Suppose the state described at the beginning of the section has an Assembly of 100 members. Use the population statistics given in the text and Hamilton's method to determine the apportionment of the assembly.
10. Suppose the state described at the beginning of the section has an Assembly of 100 members. Use the population statistics given in the text and Webster's method to determine the apportionment of the assembly.
11. Suppose the state described at the beginning of the section has an Assembly of 100 members. Use the population statistics given in the text and Jefferson's method to determine the apportionment of the assembly.

A small state has three counties with the populations given in the following table. The State Assembly has 41 seats.

County	Population
Carlisle	110,993
New Port	441,946
Sonoma	113,229
Total	666,168

12. Use Hamilton's method to determine the apportionment of the Assembly.

13. Use Webster's method to determine the apportionment of the Assembly.

14. Use Jefferson's method to determine the apportionment of the Assembly.

Explore

In Problems 15–17, use the following information.

A school has the resources to offer 20 sections of math courses each semester. It is expected that there will be a total of 600 students wanting to take math courses.

Course	Number of Students
College Algebra	120
Calculus	80
Math for Business	235
Liberal Arts Math	165

15. Use Hamilton's method to determine the distribution of the number of sections of each course.

16. Use Webster's method to determine the distribution of the number of sections of each course.

17. Use Jefferson's method to determine the distribution of the number of sections of each course.

In Problems 18–20, use the following information.

A city has 45 garbage trucks that are in use five days a week. This provides for 225 different routes. The city is divided into six areas because of geographical and transportation limitations. The populations of these six areas are given in the chart.

District	Population
Bayview	25,456
Downtown	32,723
East End	27,568
Greenhaven	16,475
Ingleside	8,696
Riverview	11,458
Total	122,376

18. Use Hamilton's method to determine the number of garbage trucks in each district.

19. Use Webster's method to determine the number of garbage trucks in each district.

20. Use Jefferson's method to determine the number of garbage trucks in each disrict.

In Problems 21–23, use the following information.

A state has six congressional districts. The number of people that voted in the last congressional election in each district is given in the chart.

District	Voters
1	90,000
2	166,000
3	102,000
4	131,000
5	147,000
6	156,000

Suppose that 800 voting sites are to be used in an election.

21. Use Hamilton's method to allocate the number of voting sites in each district.

22. Use Webster's method to allocate the number of voting sites in each district.

23. Use Jefferson's method to allocate the number of voting sites in each district.

In Problems 24–26, use the following information.

A state has five congressional districts. The number of people that voted in the last congressional election in each district is given in the chart.

District	Voters
1	167,000
2	116,000
3	237,000
4	231,000
5	215,000

Suppose that a candidate plans to spend 150 days campaigning for the next election.

24. Use Hamilton's method to allocate the number of days that should be spent campaigning in each district.

25. Use Webster's method to allocate the number of days that should be spent campaigning in each district.

26. Use Jefferson's method to allocate the number of days that should be spent campaigning in each district.

Linear Programming

Another way to find a minimum is to use linear programming. (Copyright ©1999 by Thaves. Distributed from www.thecomics.com)

Our final math vista involves the objective of most businesses: to make money. In this section, we will present some mathematics that will enable a business to do exactly that. Suppose you are running a business. A goal for your business is to keep profits at a maximum as you continue to manufacture, package, and distribute your product. A typical situation for a company producing CD players might be as follows.

Lovelace Electronics manufactures two types of CD players: a standard model and a deluxe model. The profit from selling a single standard model is \$25, whereas the profit from selling a deluxe model is \$28. Although the profit would increase if you sold more CD players, production is restricted by the size of your assembly and testing facilities. Each CD player must be assembled and tested before being shipped. The assembly line is available for 24 worker-hours each shift. The testing line is available for 18 worker-hours each shift. The standard model requires one hour to assemble and one hour to test. The deluxe model requires two hours to assemble and one hour to test. How many of each type of CD player should be manufactured during each shift if the goal is to maximize the profit?

When tackling a problem containing so much information, it is important to determine the variables. Also, it is often helpful to organize the information using a table. Reading the last line of the problem, we find that the problem asks us to determine the number of each type of CD player to be manufactured so as to maximize the profit. Since there are two types of CD players being considered and we are also concerned about profit, it makes sense to have three variables. We will use the variables x, y, and P to represent the following:

$x =$ the number of standard model CD players

$y =$ the number of deluxe model CD players

$P =$ the profit

Organizing the information from the problem in the following table, we have the following.

	Standard Model	Deluxe Model	Time Available
Profit	\$25	\$28	
Assembly time	1 hr	2 hr	24 hr
Testing time	1 hr	1 hr	18 hr

Since profit will increase for each additional CD player manufactured, we have the following profit equation.

$$P = 25x + 28y$$

If we keep producing more of each model of CD player, the profit will continue to increase. However, the assembly and testing facilities are resources with limitations. As we determine the solution to the problem, we cannot exceed the amount of time available. However, if it will increase the profit, we can use less than the available amount of time. Therefore, the relationships for the time required are as follows.

$$\text{Assembly time:} \quad 1x + 2y \leq 24$$
$$\text{Testing time:} \quad 1x + 1y \leq 18$$

Notice that we use a "less than or equal" sign to allow for less than total usage of a resource.

A final consideration in the setup of the problem is that since x and y represent the number of CD players produced, x and y should be nonnegative values. Therefore, there are two more inequalities

$$x \geq 0 \qquad \text{and} \qquad y \geq 0$$

We have now completed the setup of our problem. We have changed the original word problem into a set of four inequalities and one equation. The inequalities are called constraints, and the equation is called the objective function. The goal is to find the values of x and y that maximize the value of the objective function while staying within the boundaries defined by the constraints.

$$\text{Maximize} \quad P = 25x + 28y \qquad \textbf{objective function}$$

$$\text{subject to} \quad \left.\begin{aligned} x + 2y &\leq 24 \\ x + y &\leq 18 \\ x &\geq 0 \\ y &\geq 0 \end{aligned}\right\} \quad \textbf{constraints}$$

Linear programming is a mathematical technique that is used to solve problems in which there is a stated objective function that must be either maximized or minimized and a series of constraints that limit the values of the variables. In these problems, both the objective function and the constraints must be linear. That is, none of the variables may have exponents or be multiplied by any other variables. There are many methods used in linear programming. In this book, we will limit the discussion to one called the graphing method. The graphing method relies on graphing the constraints and then determining the maximum or minimum value of the objective function.

To solve our problem using the graphing method, we first graph the four inequalities and shade the appropriate region. Notice that the graph is restricted to the first quadrant with its x and y axes, since the problem uses only nonnegative values for x and y.

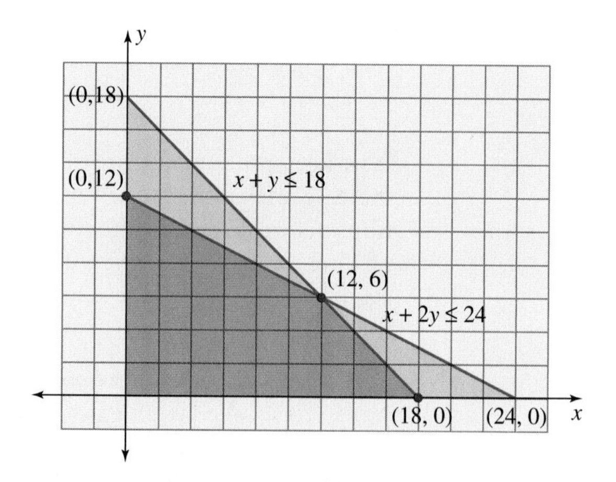

The region that is satisfied by all four constraints is shaded and is called the **feasible region**. The points within the region and on the boundary of the region all satisfy the constraints. Any point that is outside the region does not satisfy at least one of the constraints.

To determine the maximum value of the profit P, we use the Fundamental Theorem of Linear Programming.

The Fundamental Theorem of Linear Programming

The maximum or minimum value of the objective function will occur at a corner point of the feasible region.

To determine the maximum value of P, we substitute the values of the corner points into the objective function $P = 25x + 28y$ and determine its maximum value. Substituting the corner points (0, 0), (18, 0), (12, 6), and (0, 12) into the profit function, we get the following.

x	y	P
0	0	0
18	0	450
12	6	468
0	12	336

From this table, we can see that the maximum profit will be $468 and will occur when 12 standard model CD players and 6 deluxe models are produced per shift.

EXAMPLE 1

Suppose that the profit from the standard model CD players was reduced to $15 each and the profit from the deluxe model was increased to $32 each. Find the maximum profit.

Solution: The new profit function is $P = 15x + 32y$. Since the constraints are the same, we can use the feasible region and the corner points found above. Evaluating the new profit equation gives us the following.

x	y	P
0	0	0
18	0	270
12	6	372
0	12	384

Thus, the maximum profit is \$384 and occurs when 12 deluxe model CD players are produced per shift.

Notice that if the decision regarding production of the CD players is made solely on the basis of profit, the company would decide to produce none of the standard models. However, consumer demands will often influence the decision-making process, causing the company to continue producing a limited number of the less profitable models.

EXAMPLE 2

Sketch the region bounded by the following constraints.

$$2x + 3y \geq 18$$
$$8x + 2y \geq 32$$
$$x \geq 0$$
$$y \geq 0$$

Find the minimum value of the cost function $C = 20x + 16y$.

Solution: Sketching the graph, we have

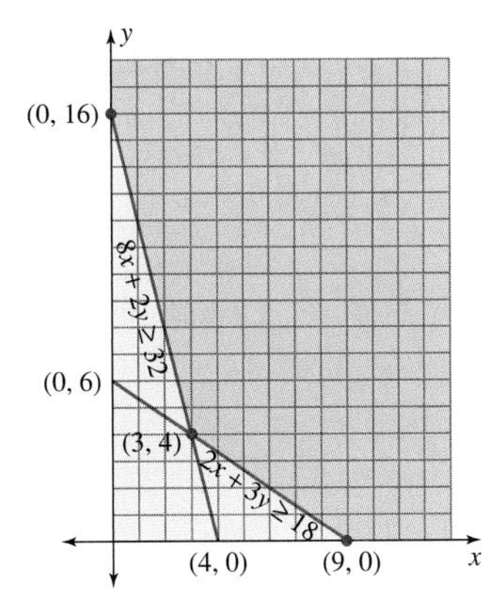

In this problem, the feasible region is not bounded on the top and the right side. However, using the corner points (9, 0), (3, 4), and (0, 16), we still find the solution by substituting the values of the corner points into the cost function, $C = 20x + 16y$.

x	y	C
9	0	180
3	4	124
0	16	256

Thus, we can see that the minimum cost is \$124 and occurs when $x = 3$ and $y = 4$.

EXAMPLE 3

Find both the maximum and minimum values of the function

$$R = 30x + 25y$$

given the following constraints.

$$\begin{aligned} 5x + y &\geq 14 \\ -5x + 6y &\leq 49 \\ 7x + 2y &\leq 77 \\ 4x + 9y &\geq 44 \\ x &\geq 0 \\ y &\geq 0 \end{aligned}$$

Solution: Sketching the graph, we have the following.

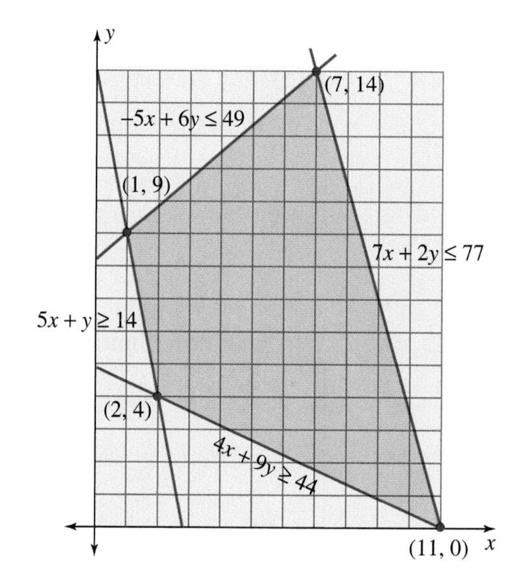

The corner points of the feasible region are (1, 9), (7, 14), (11, 0), and (2, 4). Evaluating $R = 30x + 25y$ at these points, we get the following.

x	y	C
1	9	255
7	14	560
11	0	330
2	4	160

Thus, the minimum value of R is 160 and occurs when $x = 2$ and $y = 4$. The maximum value of R is 560 and occurs when $x = 7$ and $y = 14$.

EXAMPLE 4

A veterinarian is creating a special diet for a sick dog. The diet will be a combination of two food mixes, A and B. Mix A contains 0.25 g of protein, 0.15 g of fiber, and 0.03 g of fat per gram of mix. Mix B contains 0.10 g of protein, 0.18 g of fiber, and 0.02 g of fat per gram of mix. The vet wants to create a diet that contains at least 25 g of protein and 27 g of fiber. How many grams of each type of food should be used to create a diet that contains the required amount of protein and fiber and at the same time contains the minimum amount of fat?

Solution: The first step in solving this problem is to arrange the information in a table.

	Protein	Fiber	Fat
Mix A	0.25	0.15	0.03
Mix B	0.10	0.18	0.02
Goal	≥ 25 g	≥ 27 g	minimum amount

Since the question asks for how many grams of mix A and mix B, we let the variables x and y represent the desired amounts, namely,

$$x = \text{number of grams of mix A}$$
$$y = \text{number of grams of mix B}$$

To begin, since we cannot use negative amounts of either mix A or mix B, we have

$$x \geq 0 \qquad \text{and} \qquad y \geq 0.$$

Since there is 0.25 gram of protein in each gram of mix A, there must be $0.25x$ grams of protein in x grams of mix A. Similarly, there must be $0.10y$ grams of protein in y grams of mix B. Thus, the total amount of protein from x grams of mix A and y grams of mix B is given by

$$0.25x + 0.10y.$$

Since the veterinarian wants a diet containing at least 25 g of protein, we obtain the constraint

$$0.25x + 0.10y \geq 25.$$

In a similar way, for the required amount of fiber, we obtain the constraint

$$0.15x + 0.18y \geq 27.$$

Since the vet wants to minimize the amount of fat, we have as an objective function

$$F = 0.03x + 0.02y.$$

Combining this information, we can state the linear programming problem.

$$\begin{array}{ll} \text{Minimize} & F = 0.03x + 0.02y \\ \text{subject to the constraints} & 0.25x + 0.10y \geq 25 \\ & 0.15x + 0.18y \geq 27 \\ & x \geq 0 \\ & y \geq 0 \end{array}$$

We graph the constraint $0.25x + 0.10y \geq 25$ by plotting the points $(100, 0)$ and $(0, 250)$, drawing the line through the points, and shading the region above the line. Similarly, we use the points $(180, 0)$ and $(0, 150)$ to graph the constraint $0.15x + 0.18y \geq 27$.

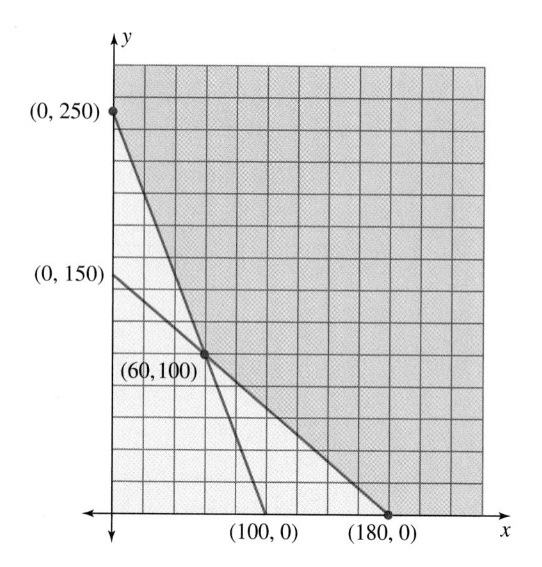

Evaluating the objective function $F = 0.03x + 0.02y$ at each of the corner points (0, 250), (60, 100), and (180, 0) gives the following.

x	y	F
0	250	5.0
60	100	3.8
180	0	5.4

Thus, we minimize the fat while still maintaining the required amounts of protein and fiber if we use 60 g of mix A and 100 g of mix B.

PROBLEMS ➤ Explain ➤ Apply ➤ Explore

➤ Explain

1. Explain what is meant by the feasible region.
2. Explain what is meant by the objective function.
3. Explain how the maximum value of a linear programming problem can be determined.
4. Explain how the minimum value of a linear programming problem can be determined.
5. Explain what is meant by a constraint.
6. Explain if each of the points A through E in the figure can be used as a possible maximum or minimum point for an objective function.

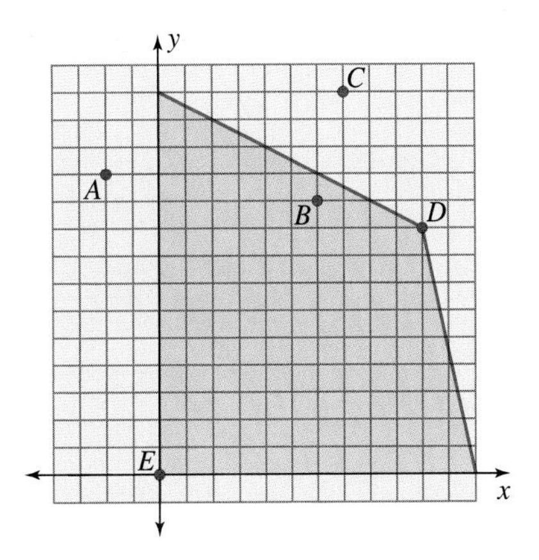

Apply

In Problems 7–14, find the desired maximum and minimum values over the desired region.

7. Maximize $P = 3x + 5y$

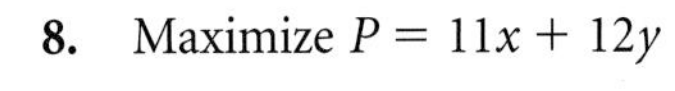

8. Maximize $P = 11x + 12y$

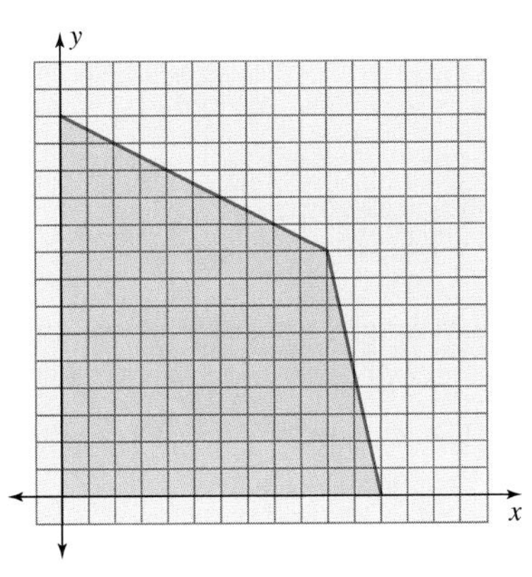

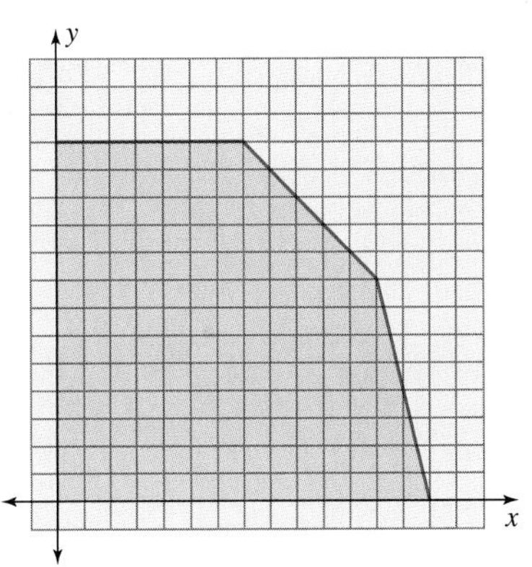

9. Minimize $P = 12x + 3y$

10. Minimize $P = 6x + 8y$

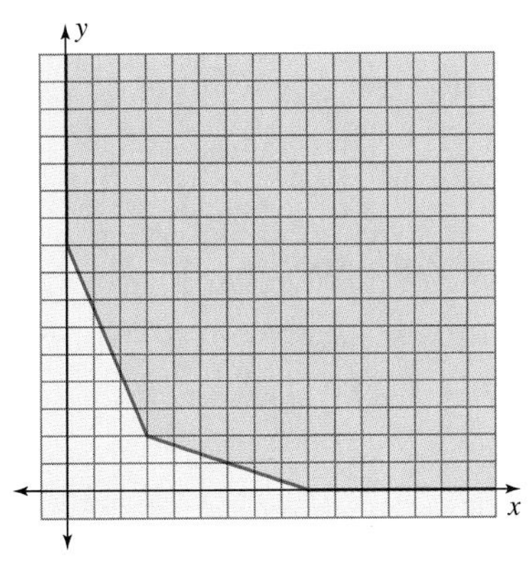

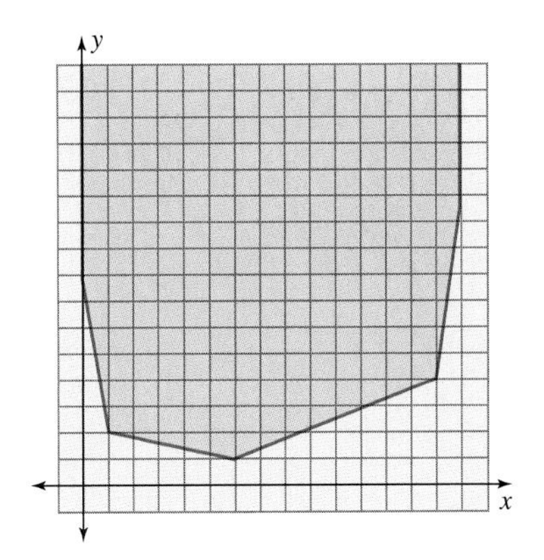

11. Maximize $P = 12x + 10y$

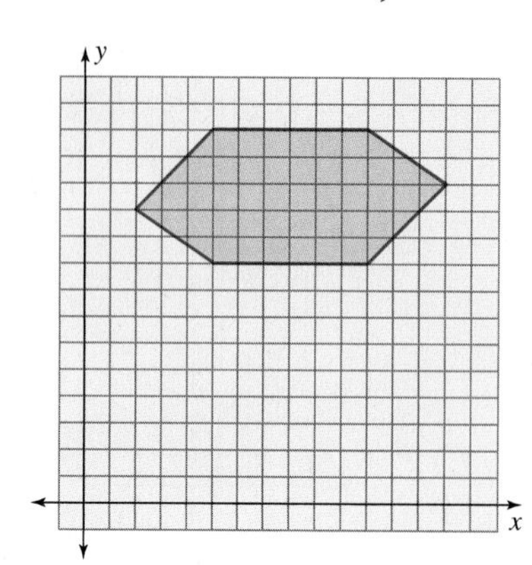

12. Minimize $P = 7x + 11y$

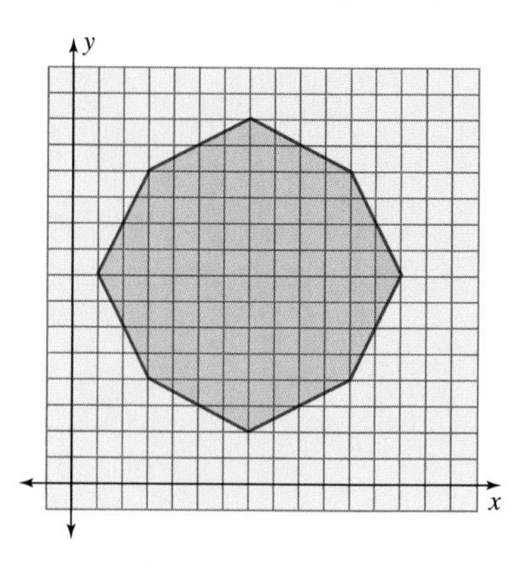

13. Maximize and minimize
$P = 2x + 7y$

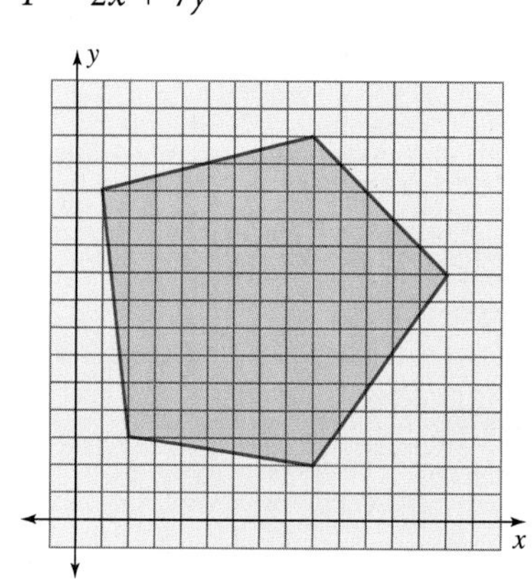

14. Maximize and minimize
$P = 7x + 8y$

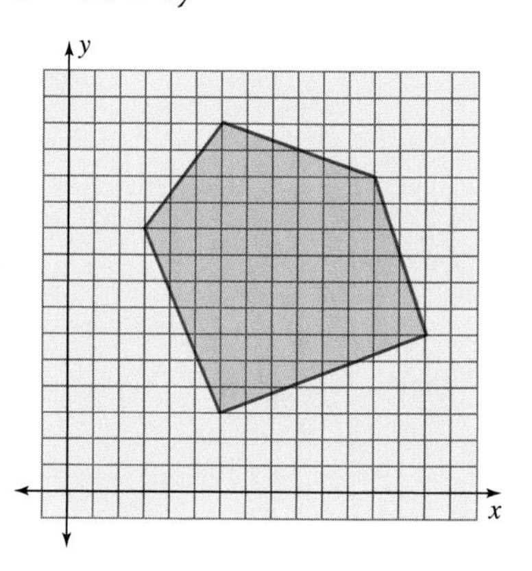

In Problems 15–24, solve the linear programming problem by sketching the region bounded by the constraints.

15. Maximize $P = 3x + 5y$

subject to the constraints $\begin{cases} 4x + y \leq 28 \\ 2x + 3y \leq 24 \\ x \geq 0 \\ y \geq 0 \end{cases}$

16. Maximize $P = 10x + 3y$

subject to the constraints $\begin{cases} x + 3y \leq 12 \\ 3x + 4y \leq 21 \\ x \geq 0 \\ y \geq 0 \end{cases}$

17. Maximize $P = 2x + 11y$

subject to the constraints $\begin{cases} 3x + 5y \leq 40 \\ 5x + 2y \leq 35 \\ x \geq 0 \\ y \geq 0 \end{cases}$

18. Maximize $P = 6x + 7y$

subject to the constraints $\begin{cases} x + 6y \leq 48 \\ 7x + 2y \leq 56 \\ x \geq 0 \\ y \geq 0 \end{cases}$

19. Minimize $C = 7x + 2y$

subject to the constraints $\begin{cases} 4x + y \geq 28 \\ 2x + 3y \geq 24 \\ x \geq 0 \\ y \geq 0 \end{cases}$

20. Minimize $C = x + 2y$

subject to the constraints $\begin{cases} x + 3y \geq 12 \\ 3x + 4y \geq 21 \\ x \geq 0 \\ y \geq 0 \end{cases}$

21. Minimize $C = 11x + 8y$

subject to the constraints $\begin{cases} 3x + 5y \geq 40 \\ 5x + 2y \geq 35 \\ x \geq 0 \\ y \geq 0 \end{cases}$

22. Minimize $C = 2x + 8y$

subject to the constraints $\begin{cases} x + 6y \geq 48 \\ 7x + 2y \geq 56 \\ x \geq 0 \\ y \geq 0 \end{cases}$

23. Find the maximum and the minimum of $P = 2x + 4y$

subject to the constraints $\begin{cases} x + y \leq 16 \\ 5x + y \leq 40 \\ -3x + 2y \leq 12 \\ 3x + 4y \geq 24 \end{cases}$

24. Find the maximum and the minimum of $P = 4x + 2y$

subject to the constraints $\begin{cases} x + y \leq 16 \\ 5x + y \leq 40 \\ -3x + 2y \leq 12 \\ 3x + 4y \geq 24 \end{cases}$

Explore

25. A veterinarian is creating a special diet for a sick cat. The diet will be a combination of two mixes, A and B. Mix A contains 0.25 g of protein, 0.15 g of fiber, and 0.03 g of fat per gram of mix. Mix B contains 0.10 g of protein, 0.18 g of fiber, and 0.03 g of fat per gram of mix. The vet wants to create a diet that contains at least 25 g of protein and 27 g of fiber. How many grams of each type of food should be used to create a diet that contains the required amount of protein and fiber and at the same time contains the minimum amount of fat? What is the minimum amount of fat?

26. Use the information from Problem 25 except that mix A contains 0.04 g of fat per gram of mix and mix B contains 0.1 g of fat per gram of mix. How many grams of each type of food should be used to create a diet that contains the required amount of protein and fiber and at the same time contains the minimum amount of fat? What is the minimum amount of fat?

27. An orchid-specialist wants to mix two types of fertilizers. One type of fertilizer contains 3% nitrogen, 2% phosphorus, and 1% potassium and costs $9 per pound. The other type of fertilizer contains 6% nitrogen, 1% phosphorus, and 1% potassium and costs $6 per pound. To properly feed all the orchids in the greenhouse, the specialist needs 27 lb of nitrogen, 12 lb of phosphorus, and 8 lb of potassium. Find the number of pounds of each fertilizer needed to minimize the total cost of the fertilizer while meeting the other requirements. What is the minimum cost?

28. Use the information in the preceding problem except that the first type of fertilizer costs $5/lb and the second type costs $7/lb. Find the number of pounds of each fertilizer needed to minimize the total cost of the fertilizer while meeting the other requirements. What is the minimum cost?

29. Suppose Huffy® produces two types of bicycles, mountain bikes and touring bikes. Suppose a mountain bike requires 2 hr to manufacture and 2 hr to assemble while a touring bike requires 2 hr to manufacture and 3 hr to assemble. Suppose a mountain bike provides Huffy with $70 of profit while a touring bike provides the company with $100 of profit. If Huffy has 40 hr/wk of manufacturing time and 42 hr/wk of assembly time, how many bikes of each type should they produce in order to maximize the profit? What is the maximum profit?

30. In the preceding problem, how could the profit be maximized if all conditions were the same except that the profit was $95 for a mountain bike and $80 for a touring bike? What is the maximum profit?

APPENDIX

NORMAL TABLE

This table gives the area under the standard normal curve, between $z = 0$ and the given value of z.

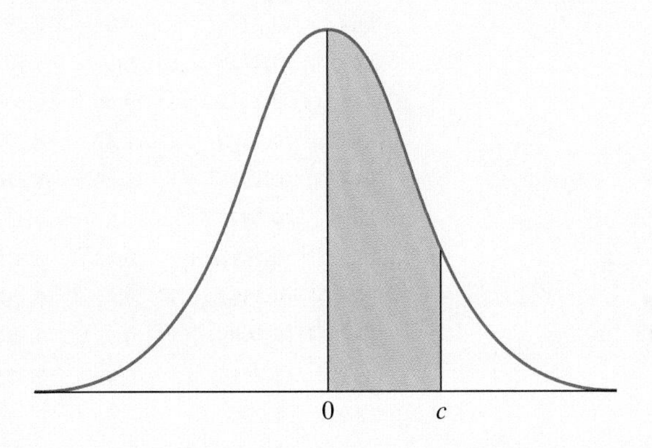

The values in the table give the area under the standard normal curve between $z = 0$ and $z = c$ where $z \geq 0$.

z	Area	z	Area	z	Area	z	Area
0.00	0.0000	0.07	0.0279	0.14	0.0557	0.21	0.0832
0.01	0.0040	0.08	0.0319	0.15	0.0596	0.22	0.0871
0.02	0.0078	0.09	0.0359	0.16	0.0636	0.23	0.0910
0.03	0.0120	0.10	0.0398	0.17	0.0675	0.24	0.0948
0.04	0.0160	0.11	0.0438	0.18	0.0714	0.25	0.0987
0.05	0.0199	0.12	0.0478	0.19	0.0753	0.26	0.1026
0.06	0.0239	0.13	0.0517	0.20	0.0793	0.27	0.1064

z	Area	z	Area	z	Area	z	Area
0.28	0.1102	0.68	0.2517	1.08	0.3599	1.48	0.4306
0.29	0.1141	0.69	0.2549	1.09	0.3621	1.49	0.4319
0.30	0.1179	0.70	0.2580	1.10	0.3643	1.50	0.4332
0.31	0.1217	0.71	0.2611	1.11	0.3665	1.51	0.4345
0.32	0.1255	0.72	0.2642	1.12	0.3686	1.52	0.4357
0.33	0.1293	0.73	0.2673	1.13	0.3708	1.53	0.4370
0.34	0.1331	0.74	0.2704	1.14	0.3729	1.54	0.4382
0.35	0.1368	0.75	0.2734	1.15	0.3749	1.55	0.4394
0.36	0.1406	0.76	0.2764	1.16	0.3770	1.56	0.4406
0.37	0.1443	0.77	0.2794	1.17	0.3790	1.57	0.4418
0.38	0.1480	0.78	0.2823	1.18	0.3810	1.58	0.4429
0.39	0.1517	0.79	0.2852	1.19	0.3830	1.59	0.4441
0.40	0.1554	0.80	0.2881	1.20	0.3849	1.60	0.4452
0.41	0.1591	0.81	0.2910	1.21	0.3869	1.61	0.4463
0.42	0.1628	0.82	0.2939	1.22	0.3888	1.62	0.4474
0.43	0.1664	0.83	0.2967	1.23	0.3907	1.63	0.4484
0.44	0.1700	0.84	0.2995	1.24	0.3925	1.64	0.4495
0.45	0.1736	0.85	0.3023	1.25	0.3944	1.65	0.4505
0.46	0.1772	0.86	0.3051	1.26	0.3962	1.66	0.4515
0.47	0.1808	0.87	0.3079	1.27	0.3980	1.67	0.4525
0.48	0.1844	0.88	0.3106	1.28	0.3997	1.68	0.4535
0.49	0.1879	0.89	0.3133	1.29	0.4015	1.69	0.4545
0.50	0.1915	0.90	0.3159	1.30	0.4032	1.70	0.4554
0.51	0.1950	0.91	0.3186	1.31	0.4049	1.71	0.4564
0.52	0.1985	0.92	0.3212	1.32	0.4066	1.72	0.4573
0.53	0.2019	0.93	0.3238	1.33	0.4082	1.73	0.4582
0.54	0.2054	0.94	0.3262	1.34	0.4099	1.74	0.4591
0.55	0.2088	0.95	0.3289	1.35	0.4115	1.75	0.4599
0.56	0.2123	0.96	0.3315	1.36	0.4131	1.76	0.4608
0.57	0.2157	0.97	0.3340	1.37	0.4147	1.77	0.4616
0.58	0.2190	0.98	0.3365	1.38	0.4162	1.78	0.4625
0.59	0.2224	0.99	0.3389	1.39	0.4177	1.79	0.4633
0.60	0.2257	1.00	0.3413	1.40	0.4192	1.80	0.4641
0.61	0.2291	1.01	0.3438	1.41	0.4207	1.81	0.4649
0.62	0.2323	1.02	0.3461	1.42	0.4222	1.82	0.4656
0.63	0.2357	1.03	0.3485	1.43	0.4236	1.83	0.4664
0.64	0.2389	1.04	0.3508	1.44	0.4251	1.84	0.4671
0.65	0.2422	1.05	0.3531	1.45	0.4265	1.85	0.4678
0.66	0.2454	1.06	0.3554	1.46	0.4279	1.86	0.4686
0.67	0.2486	1.07	0.3577	1.47	0.4292	1.87	0.4693

z	Area	z	Area	z	Area	z	Area
1.88	0.4699	2.28	0.4887	2.68	0.4963	3.08	0.49897
1.89	0.4706	2.29	0.4890	2.69	0.4964	3.09	0.49900
1.90	0.4713	2.30	0.4893	2.70	0.4965	3.10	0.49903
1.91	0.4719	2.31	0.4896	2.71	0.4966	3.11	0.49907
1.92	0.4726	2.32	0.4898	2.72	0.4967	3.12	0.49970
1.93	0.4732	2.33	0.4901	2.73	0.4968	3.13	0.49913
1.94	0.4738	2.34	0.4904	2.74	0.4969	3.14	0.49916
1.95	0.4744	2.35	0.4906	2.75	0.4970	3.15	0.49918
1.96	0.4750	2.36	0.4909	2.76	0.4971	3.16	0.49921
1.97	0.4756	2.37	0.4911	2.77	0.4972	3.17	0.49924
1.98	0.4761	2.38	0.4913	2.78	0.4973	3.18	0.49926
1.99	0.4767	2.39	0.4916	2.79	0.4974	3.19	0.49929
2.00	0.4772	2.40	0.4918	2.80	0.4974	3.20	0.49931
2.01	0.4778	2.41	0.4920	2.81	0.4975	3.21	0.49934
2.02	0.4783	2.42	0.4922	2.82	0.4976	3.22	0.49936
2.03	0.4788	2.43	0.4925	2.83	0.4977	3.23	0.49938
2.04	0.4793	2.44	0.4927	2.84	0.4977	3.24	0.49940
2.05	0.4798	2.45	0.4929	2.85	0.4978	3.25	0.49942
2.06	0.4803	2.46	0.4931	2.86	0.4979	3.26	0.49944
2.07	0.4808	2.47	0.4932	2.87	0.4979	3.27	0.49946
2.08	0.4812	2.48	0.4934	2.88	0.4980	3.28	0.49948
2.09	0.4817	2.49	0.4936	2.89	0.4981	3.29	0.49950
2.10	0.4821	2.50	0.4938	2.90	0.4981	3.30	0.49952
2.11	0.4826	2.51	0.4940	2.91	0.4982	3.31	0.49953
2.12	0.4830	2.52	0.4941	2.92	0.4982	3.32	0.49955
2.13	0.4834	2.53	0.4943	2.93	0.4983	3.33	0.49957
2.14	0.4838	2.54	0.4945	2.94	0.4984	3.34	0.49958
2.15	0.4842	2.55	0.4946	2.95	0.4984	3.35	0.49960
2.16	0.4846	2.56	0.4948	2.96	0.4985	3.36	0.49961
2.17	0.4850	2.57	0.4949	2.97	0.4985	3.37	0.49962
2.18	0.4854	2.58	0.4951	2.98	0.4986	3.38	0.49964
2.19	0.4857	2.59	0.4952	2.99	0.4986	3.39	0.49965
2.20	0.4861	2.60	0.4953	3.00	0.4987	3.40	0.49966
2.21	0.4864	2.61	0.4955	3.01	0.49869	3.41	0.49968
2.22	0.4868	2.62	0.4956	3.02	0.49874	3.42	0.49969
2.23	0.4871	2.63	0.4957	3.03	0.49878	3.43	0.49970
2.24	0.4875	2.64	0.4959	3.04	0.49881	3.44	0.49971
2.25	0.4878	2.65	0.4960	3.05	0.49886	3.45	0.49972
2.26	0.4881	2.66	0.4961	3.06	0.49889	3.46	0.49973
2.27	0.4884	2.67	0.4962	3.07	0.49893	3.47	0.49974

z	Area	*z*	Area	*z*	Area	*z*	Area
3.48	0.49975	3.62	0.49985	3.75	0.49991	3.88	0.49995
3.49	0.49976	3.63	0.49986	3.76	0.49992	3.89	0.49995
3.50	0.49977	3.64	0.49986	3.77	0.49992	3.90	0.49995
3.51	0.49978	3.65	0.49987	3.78	0.49999	3.91	0.49995
3.52	0.49978	3.66	0.49987	3.79	0.49993	3.92	0.49996
3.53	0.49979	3.67	0.49988	3.80	0.49993	3.93	0.49996
3.54	0.49980	3.68	0.49988	3.81	0.49993	3.94	0.49996
3.55	0.49981	3.69	0.49989	3.82	0.49993	3.95	0.49996
3.56	0.49982	3.70	0.49989	3.83	0.49994	3.96	0.49996
3.57	0.49982	3.71	0.49990	3.84	0.49994	3.97	0.49996
3.58	0.49983	3.72	0.49990	3.85	0.49994	3.98	0.49997
3.59	0.49986	3.73	0.49990	3.86	0.49994	3.99	0.49997
3.60	0.49984	3.74	0.49991	3.87	0.49995	4.00	0.49997
3.61	0.49985						

SELECTED ANSWERS

CHAPTER 1

Section 1.1

11. 121,000

13. 1,010,101

15. 1106

17. 734

19. 42,457

21. 7502

23. 37

25. 25,062

27. 11

29. 363

31. 57,068,640

33. **a)** X HHHH ΔΔΔΔ

b) 一
千
四
百
四
十

c) <<YYYYY

35. **a)** $\overline{\text{X}}$MMCCCXV

b) 一
万
二
千
三
百
一
十
五

c) $\overset{\alpha}{\text{M}}$ ͵βτιε

d) YYY <<YYYYYY <YYYYYY

e) ·
····
____
···
____
____

37. 1-800-YEA-MATH is the same as 1-800-932-6284. Answers will vary depending on the system of numeration chosen.

39. Additive systems are similar to the Attic Greek system. Answers will vary depending on the methods and symbols chosen. One possible solution is shown.

a) $75 = 8 \times 9 + 3 =$ ←←←←←←←← →→→

b) $366 = 4 \times 81 + 4 \times 9 + 6 =$
↓↓↓↓ ←←←← →→→→→→

c) $5280 = 7 \times 729 + 2 \times 81 + 1 \times 9 + 6 =$
↑↑↑↑↑↑↑ ↓↓ ← →→→→→→

d) $1000 = 1 \times 729 + 3 \times 81 + 3 \times 9 + 1 =$
↑ ↓↓↓ ←←← →

41. Answers will vary depending on the methods and symbols chosen.

43. **a)** In Egyptian, a carry takes place each time there are ten of the same symbol. When this happens, an additional symbol of the next higher value will replace the group of ten symbols.

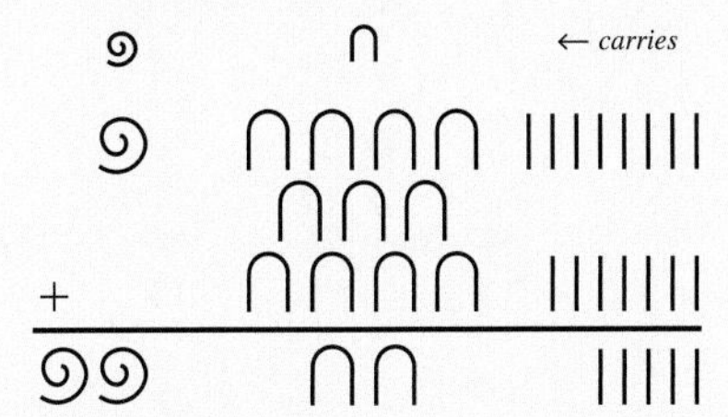

b) In Roman, a carry takes place each time there are more than four of the symbols I, X, C, M or more than one of the symbols V, L, or D. When this hap pens, an additional symbol of the next higher value will replace the value that has been exceeded.

	C	L	X	V	← *carries*
	C	L	XX	V	III
+			**XL**	**V**	**II**
	CC		**XX**	**V**	

c) In Chinese, a carry takes place each time there are more than nine in a particular place value. When this happens, one will be added to the next higher place value.

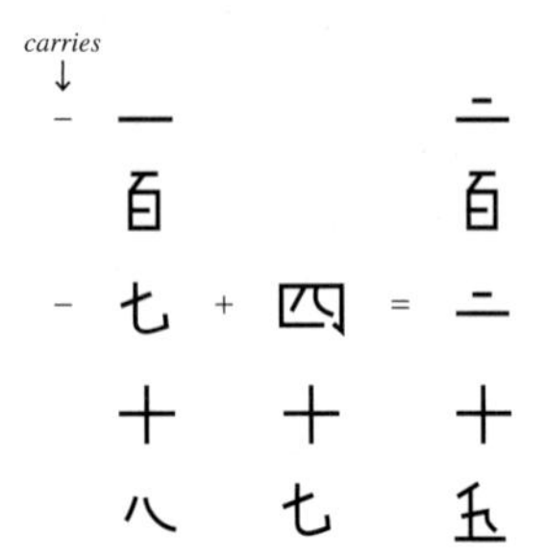

Section 1.2

7. $1 \times 10^2 + 3 \times 10^1 + 9 \times 10^0$

9. $4 \times 10^2 + 3 \times 10^1 + 7 \times 10^0 + 1 \times 10^{-1} + 5 \times 10^{-2}$

11. $3 \times 10^{-1} + 1 \times 10^{-2} + 4 \times 10^{-3}$

13. $5 \times 10^5 + 4 \times 10^4 + 3 \times 10^3 + 8 \times 10^2 + 6 \times 10^1 + 7 \times 10^0$

15. $5 \times 10^1 + 3 \times 10^0 + 1 \times 10^{-1} + 7 \times 10^{-2} + 1 \times 10^{-3}$

17. $6 \times 10^{-1} + 2 \times 10^{-2} + 1 \times 10^{-3} + 9 \times 10^{-4} + 3 \times 10^{-5}$

19. tens, hundredths, tens, ten thousands, tenths, hundredths, thousands, hundredths, ones, ones, hundred thousandths, thousandths

21. 7/20

23. 163/3000

25. 1/5

27. 3661/21,600

29. 1/3

31. 3/4

33. 1/63

35. 49/501

37. 743 1/5

39. 101 1/34

41. ◎ ◎ or ◎ ◎
|| ∩|| |||| ||

43. ◎ ◎
||| |||||

45. ◎ ◎
||| ||

47. . . . . + S = S . . . .

49. XII × III S. . . = XLV

51. $\rho\kappa \times \gamma\ \gamma\delta'\delta' = \upsilon\nu$

53. Answers will vary.

55. Answers will vary.

Section 1.3

7. 77

9. 5.625

11. 2952

13. 1132.25

15. 136

17. 15

19. 3101_5

21. 1736_8

23. 2160_7

25. $10B0_{12}$

27. 81_{16}

29. $313\ 126_9$

31. $11\ 011\ 110_2$

33. 2431_5

35. 556_8

37. $7D0_{16}$

39. $224\ 000\ 000_5$

41. $3\ 641\ 100_8$

43. $F4240_{16}$

45. 254_9

47. 3033_4

49. 150_7

51. 8

53. a) $45 = 1200_3$
$100 = 10201_3$
$200 = 21102_3$
$45 = 50_9$
$100 = 121_9$
$200 = 242_9$

b) As the base increases, the number of digits decreases.

c) Taken in pairs, the digits in the base 3 representation give the base 9 representation. For example, writing 21102_3 as 2_3 11_3 02_3, we have 2, 4, and 2. These are the digits in the base 9 representation of 200.

55. Answers will vary.

57. When using a base greater than 10, the telephone number will contain the same number of digits if you consider each digit separately. If you consider the digits as three and four digit numbers rather than separate digits, the new number will contain the same or fewer digits.

Section 1.4

9. a) 25_8 b) 15_{16}

11. a) 144_8 b) 64_{16}

13. a) 5656_8 b) BAE_{16}

15. $1\,100\ 101_2$

17. $100\ 101_2$

19. $101\ 011_2$

21. $1\ 001\ 001_2$

23. $1\ 100\ 011\ 110_2$

25. $100\ 000\ 100_2$

27. $110\ 001\ \ 110\ \ 001_2$

29. To be or not?

31. 0100 1000, 0100 1111, 0101 0000, 0100 0101

33. 0100 1101, 0110 1111, 0110 1101

35. Divide the binary number in groups of five digits. Using the place value for five binary digits, 16, 8, 4, 2, 1, find the value of each group of five digits. Remember to use A for 10, B for 11, C for 12, . . . , W for 32.

Section 1.5

13. a) $\sqrt{100}$, 19

b) 0, $\sqrt{100}$, 19

c) $-7, -\sqrt{64}, 0, \sqrt{100}, 19$

d) $-14.785, -7, -\sqrt{64}, -\frac{5}{16}, 0, 5\frac{7}{8}, 9.76555\ldots, \sqrt{100}, 19$

e) $\sqrt[5]{19}, \sqrt{8}, \pi$

f) $-14.785, -7, -\sqrt{64}, -\frac{5}{16}, 0, \sqrt[5]{19}, \sqrt{8}, \pi, 5\frac{7}{8}, 9.76555\ldots, \sqrt{100}, 19$

g) $-\sqrt{-25}, \sqrt{-8}$

h) π

15.

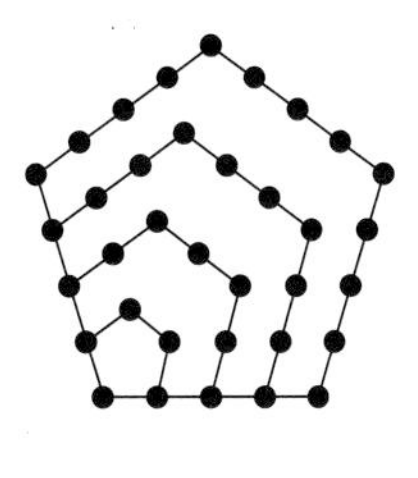

The fifth pentagonal number is 35.

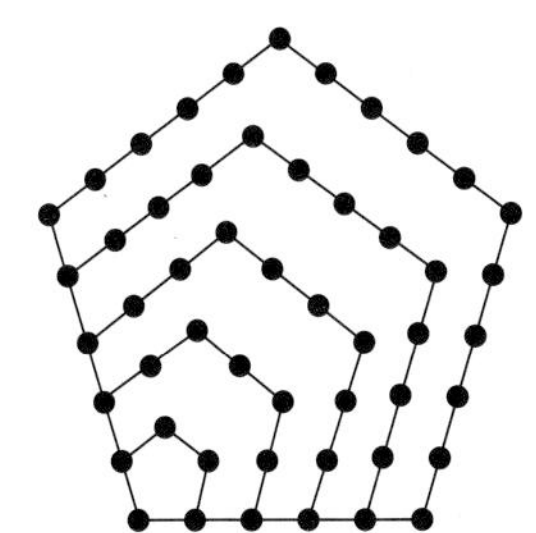

The sixth pentagonal number is 51.

17. a) prime, deficient

b) composite, deficient—proper factors are 1, 7, 11

c) composite, deficient—proper factors are 1, 5, 29

d) composite, abundant—proper factors are 1, 2, 4, 7, 14, 28, 71, 142, 284, 497, 994

e) composite, perfect—proper factors are 1, 2, 4, 8, 16, 32, 64, 127, 254, 508, 1016, 2032, 4064

f) composite, abundant—proper factors are 1, 2, 3, 4, 5, 6, 8, 9, 10, 12, 15, 18, 20, 24, 25, 30, 36, 40, 45, 50, 60, 72, 75, 90, 100, 120, 125, 150, 180, 200, 225, 250, 300, 360, 375, 450, 500, 600, 750, 900, 1000, 1125, 1500, 1800, 2250, 3000, 4500

19. Pattern: add the odd numbers (3, 5, 7, 9, 11, . . .) to proceed from term to term.

1, 4, 9, 16, 25, 36, 49, 64, 81, 100, 121, 144

21. The number 5 is a prime number. However, all other numbers ending in 5 are divisible by 5, so they are not prime.

23. 1.414 is a rational number and is approximately equal to $\sqrt{2}$ but $\sqrt{2}$ is an irrational number. The correct relationship is $\sqrt{2} > 1.414$.

CHAPTER 1 REVIEW

Review Section 1.1

1. a) 123

b) 𓍢∩∩|||

c) CXXIII

d) 一
百
二
十
三

e) ρκγ

f) ΗΔΔ|||

g) 123 = 2(60) + 3 = ∨∨ ∨∨∨

h) ·—
···

2. a) 14 b) 5614 c) 8945 d) 77,896
e) 25,560 f) 10,403

Review Section 1.2

3. a) $2 \times 10^1 + 5 \times 10^0 + 4 \times 10^{-1} + 7 \times 10^{-2}$

b) $1 \times 10^2 + 2 \times 10^1 + 0 \times 10^0 + 0 \times 10^{-1} + 4 \times 10^{-2} + 5 \times 10^{-3}$

4. a) 0.75

b) S . . .

c) ◎ ◎
|||| ||

d) ⌐⌐ <<<<∨∨∨∨∨

e) γδ′δ′

5. Answers will vary depending on what is selected.

Review Section 1.3

6. **a)** 194

b) 10 7/16 = 10.4375

c) 2140

d) 194 1/64 = 194.015625

7. **a)** $123 = 1\ 111\ 011_2$

b) $123 = 443_5$

c) $123 = 173_8$

d) $123 = A3_{12}$

e) $123 = 7B_{16}$

Review Section 1.4

8. Disks, tapes, and CDs store data using two different states—low/high voltage pulses, forward/backward magnetic fields, and dark/light marks. These two states are associated with the 0's and 1's of the binary numbers system.

9. 0100 0010, 0100 0100, 0110 1111, 0010 0000, 0110 1001, 0111 0100, 0010 0001, 0100 0010

10. 563_8, 173_{16}

11. $1\ 000\ 011_2$, 103_8, 43_{16}

12. **a)** $111\ 011\ 010_2$

b) $11\ 100\ 111_2$

Review Section 1.5

13. **a)** Both can be represented as ratios of integers.

$$6.55 = 6\frac{55}{100} = 6\frac{11}{20} = \frac{131}{20}$$

$$6.555\ldots = 6\frac{5}{9} = \frac{59}{9}$$

b) 3.14 can be represented as a ratio of integers so it is a rational number.

$$3.14 = 3\frac{14}{100} = \frac{314}{100} = \frac{157}{50}$$

π can not be expressed as a ratio integers so it is irrational.

c) $-\sqrt{9}$ is the real number -3. $\sqrt{-9}$ is an imaginary number.

14. **a)** Natural numbers: 2, $\sqrt[3]{125}$

b) Whole numbers: 0, 2, $\sqrt[3]{125}$

c) Integers: $-7, -\sqrt{16}, 0, 2, \sqrt[3]{125}$

d) Rational numbers: $-7.6, -7, -5\frac{2}{3}, -\sqrt{16}, 0, 2, 8.33\ldots, 10\frac{3}{5}, \sqrt[3]{125}$

e) Irrational numbers: $\sqrt[3]{127}$

f) Real numbers: all numbers except $\sqrt{-16}$

g) Imaginary numbers: $\sqrt{-16}$

h) Noninteger rational numbers: $-7.6, -5\frac{2}{3}, 8.33\ldots, 10\frac{3}{5}$

15. 400—composite, abundant

461—prime, deficient

496—composite, perfect

512—composite, deficient

16. By substituting $n = 0$ to $n = 9$, we get 41, 43, 47, 53, 61, 71, 83, 97, 113. If n is a multiple of 41, then all three terms of the polynomial will have 41 as a factor. Thus, the polynomial will have a factor of 41 and will not result in a prime number.

17. Sixth triangular number: 21
Sixth square number: 36
Sixth pentagonal number: 51

CHAPTER 1 TEST

1. 𓆼𓆼|||

2. MMIII

3. 二
千
三

4. $'\beta\gamma$

5. XXIII

6. <<<YYY <<YYY

7. —
═
•••

8. $11\ 111\ 010\ 011_2$

9. $31\ 003_5$

10. $7D3_{16}$

11. 10.666. . .

12. XS. .

13. ∩ |||||| 𓆼 || 𓆼

14. < ǂǂ <<<<

15. $\iota\beta\gamma'\gamma'$

16. 12 231.5

17. 753_8, $1EB_{16}$

18. $1\ 010\ 100_2$

19. $101\ 111\ 001_2$

20. a) $\sqrt{-400}$

b) $-5, \sqrt[3]{-27}, 0, 13$

c) $-\sqrt{40}, \sqrt[4]{19}, \pi$

d) 13

e) All numbers except those listed in (c) and $\sqrt{-400}$

f) All numbers except $\sqrt{-400}$

g) 0, 13

CHAPTER 2

Section 2.1

19. It is not a statement because it is neither true nor false.

21. It is not a statement because it is neither true nor false.

23. Statement

25. It is not a statement because it is neither true nor false.

27. My car is not in the shop.

29. I don't hate sitting around doing nothing.

31. That is not an example of an exponential equation.

33. Some fish cannot live under water.

35. All numbers are prime numbers.

37. No trees are always green.

39. Some of the numbers are positive.

41. *Converse:* If the phone is in use, then you get a busy signal.

Inverse: If you do not get a busy signal, then the phone is not in use.

Contrapositive: If the phone is not in use, then you do not get a busy signal.

43. *Converse:* If the point is 16″ from the center of the circle, then it is on the circle.

Inverse: If it is not a point on the circle, then it is not 16″ from the center of the circle.

Contrapositive: If the point is not 16″ from the center of the circle, then it is not on the circle.

45. *Converse:* If I am listening, then G. H. Mutton is speaking.

Inverse: If G. H. Mutton is not speaking, then I am not listening.

Contrapositive: If I am not listening, then G. H. Mutton is not speaking.

47. *Converse:* If the figure is not a hexagon, then it has five sides.

Inverse: If the figure does not have five sides, then it is a hexagon.

Contrapositive: If a figure is a hexagon, then it does not have five sides.

In problems 49–52 and 54–59, there are many other correct answers.

49. If x is an integer, then x is a real number.

51. If the product of two real numbers is zero, then at least one of them is zero.

53. Not possible

55. If x is a number, then x is a house.

57. If x is a whole number, then x is a prime number.

59. If x is a prime number, then x is a composite number.

61. A dog is a domesticated animal related to the wolf, fox, and jackal.

63. A book is a number of pieces of paper with printing or writing fastened together on one edge.

65. Answers will vary.

67. The contrapositive of the inverse of $A \rightarrow B$ is the converse of $A \rightarrow B$, namely $B \rightarrow A$.

Section 2.2

9. It is not correct. It contains a converse error. This would be correct:

When it is midnight, I am asleep.

It is midnight.

Therefore, I am asleep.

11. It is not correct. It contains an inverse error. This would be correct:

If you are a farmer in Polt County, then you grow corn.

Farmer Ron does not grow corn.

Therefore, Farmer Ron is not a farmer in Polt County.

13. It is a correct hypothetical syllogism.

15. If a whole number greater than 2 is even, then it is divisible by 2.

If a whole number is divisible by 2, then it is not a prime number.

Therefore, if a whole number greater than 2 is even, it is not a prime number.

17. Anyone that treats you with kindness is a nice person. My teacher, Mrs. Santos, is always very kind to me when I am sick.

19. If you are serious about school, you will have less time to watch TV.

21. One of many possible answers:

A: x is a whole number.

B: x is an integer.

C: x is a rational number.

D: x is an irrational number.

E: x equals $\sqrt{2}$

23. In a triangle, the longest side is opposite the largest angle and the shortest side is opposite the smallest angle.

25. The new quadrilaterals are parallelograms.

27. The angle formed by the line segments connecting the endpoints of the diameter to a point on a semicircle is a right angle.

29. A possible inductive argument:

The plane ride is long and cramped.
The hotels in Honolulu are big and impersonal.
The beaches in Honolulu are noisy and crowded.
The restaurants in Honolulu are often expensive.
Therefore, you should not go to Honolulu, Hawaii.

31. **a)** This is an inductive argument.

b) This is a deductive argument.

33. Some numbers are rational.

35. If the car is full of gasoline, we can see Vernal Falls.

37. By using the contrapositive of the third statement, we get, $P \rightarrow \sim S$.

39. One possible example is

P = The object is a square.
Q = The object is a rectangle with four congruent sides.
R = The object is a triangle.
S = The object has three sides.

Conclusion: If an object is a square, then it does not have three sides.

Section 2.3

7. M = it is midnight, A = asleep

$M \rightarrow A$
A
$\therefore M$

9. F = farmer in Polt County, C = grows corn

$F \rightarrow C$
$\sim C$
$\therefore \sim F$

11. S = scalene, W = two equal sides, H = three equal sides

$\sim S \rightarrow (W \vee H)$
$\sim H$
$\therefore W$

13. W = winning golfer, G = good hand-eye coordination, P = positive attitude

$W \rightarrow (G \wedge P)$
$\therefore (G \wedge P) \rightarrow W$

15. B = voted for Bush, N = voted for Nader, G = voted for Gore

$(B \vee N) \rightarrow \sim G$

17.

A	C	$A \rightarrow C$	$\sim C$	$(A \rightarrow C) \vee \sim C$
T	T	T	F	T
T	F	F	T	T
F	T	T	F	T
F	F	T	T	T

The statement is always true.

19.

A	$\sim A$	C	$\sim A \rightarrow C$	$\sim C$	$(\sim A \rightarrow C) \rightarrow \sim C$
T	F	T	T	F	F
T	F	F	T	T	T
F	T	T	T	F	F
F	T	F	F	T	T

The statement is true whenever C is false.

21.

A	B	$\sim B$	$A \wedge \sim B$	$\sim(A \wedge \sim B)$	$\sim A$	$\sim A \vee B$
T	T	F	F	**T**	F	**T**
T	F	T	T	**F**	F	**F**
F	T	F	F	**T**	T	**T**
F	F	T	F	**T**	T	**T**

Since both statements have the same truth values, the statements are equivalent.

23.

A	B	$A \vee B$	$\sim B$	$A \vee \sim B$	$(A \vee B) \wedge (A \vee \sim B)$
T	T	T	F	T	**T**
T	F	T	T	T	**T**
F	T	T	F	F	**F**
F	F	F	T	T	**F**

Since the statements have the same truth values, the statements are equivalent.

25.

P	Q	$P \rightarrow Q$	$(P \rightarrow Q) \wedge P$	$(P \rightarrow Q) \wedge P) \rightarrow Q$
T	T	T	T	T
T	F	F	F	T
F	T	T	F	T
F	F	T	F	T

Since the last column is always true, the argument is correct.

27.

A	$\sim A$	Q	$A \to Q$	$\sim A \wedge (A \to Q)$	$(\sim A \wedge (A \to Q)) \to A$
T	F	T	T	F	T
T	F	F	F	F	T
F	T	T	T	T	F
F	T	F	T	T	F

Since the last column is not always true, the argument is not correct.

29. E = Elections become TV popularity contests
G = good looking, smooth talking candidates get elected
$E \to G$
$\therefore \sim E \to \sim G$

E	G	$E \to G$	$\sim E$	$\sim G$	$\sim E \to \sim G$	$(E \to G) \to (\sim E \to \sim G)$
T	T	T	F	F	T	T
T	F	F	F	T	T	T
F	T	T	T	F	F	F
F	F	T	T	T	T	T

Since the last column is not always true, the argument is not correct.

31. M = High school graduates have poor math skills
W = High school graduates have poor writing skills
J = Less able to get a job in the computer industry
$M \to J$
$W \to J$
$\therefore M \to W$

M	W	J	$M \to J$	$W \to J$	$(M \to J) \wedge (W \to J)$	$M \to W$	$((M \to J) \wedge (W \to J)) \to (M \to W)$
T	T	T	T	T	T	T	T
T	T	F	F	F	F	T	T
T	F	T	T	T	T	F	F
T	F	F	F	T	F	F	T
F	T	T	T	T	T	T	T
F	T	F	T	F	F	T	T
F	F	T	T	T	T	T	T
F	F	F	T	T	T	T	T

Since the last column is not always true, the argument is not correct.

33. Constructing a truth table shows that the argument is incorrect if R is true, N is false and S is false.

Section 2.4

9. This flowchart describes the process which helps decide why a lawn mower will not start. The choices are limited to the lawn mower being out of gas, the spark plug wire being disconnected, the blade being engaged, or some unknown reason.

11. **a)** Stand.

b) Stand if the dealer shows a card $2 \leq x < 7$, otherwise draw another card and then reevaluate your hand.

c) Draw another card and then reevaluate your hand.

d) Draw another card and then reevaluate your hand.

e) Stand.

f) To show how to win the game, add the following to the existing flowchart.

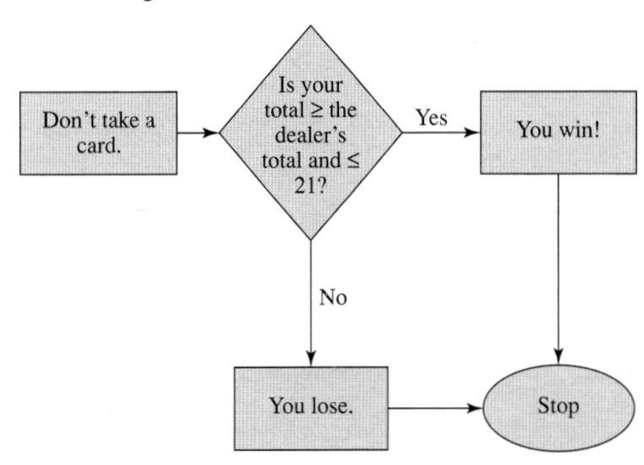

13.

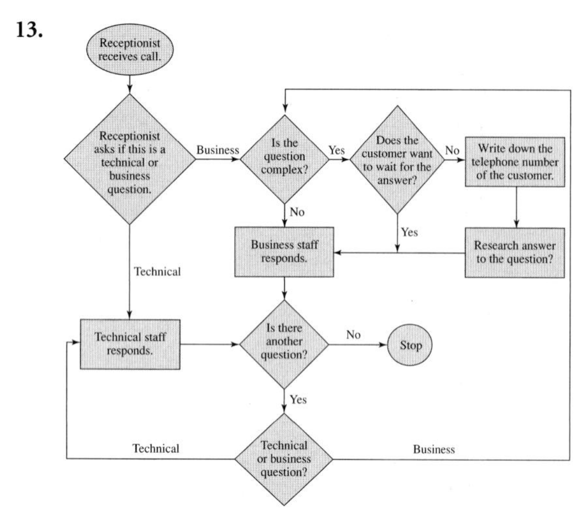

15.

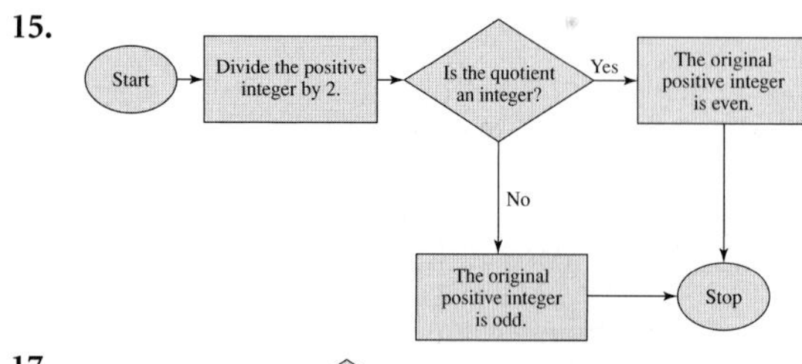

17.

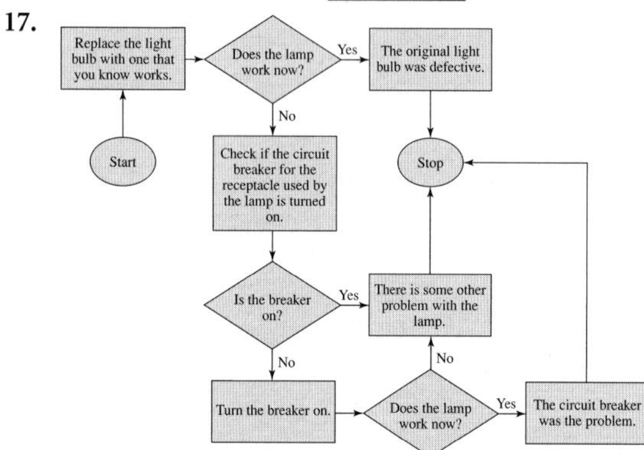

19.

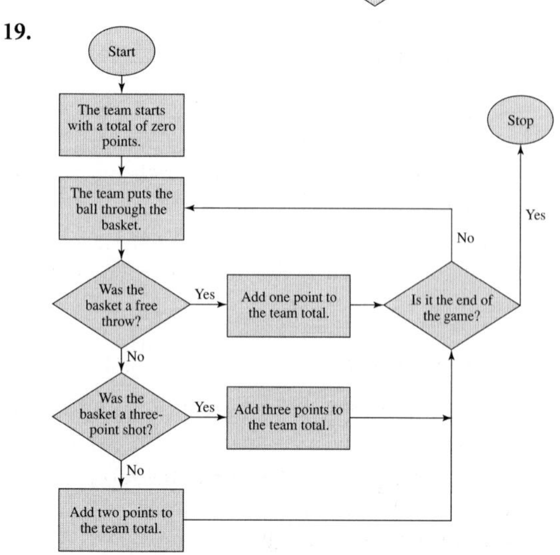

21.

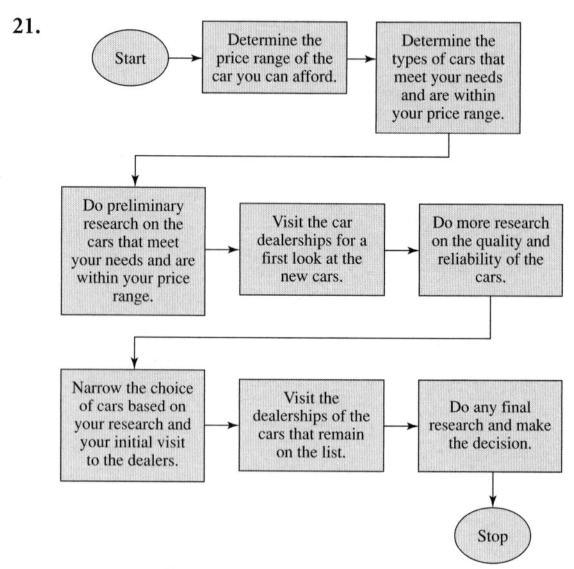

23. Answers will vary.

25.

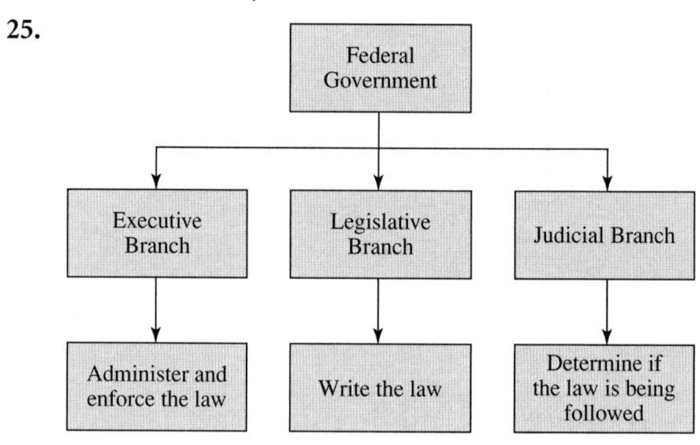

Section 2.5

9. Sum of numbers equals 18, 20, 22, and 24.

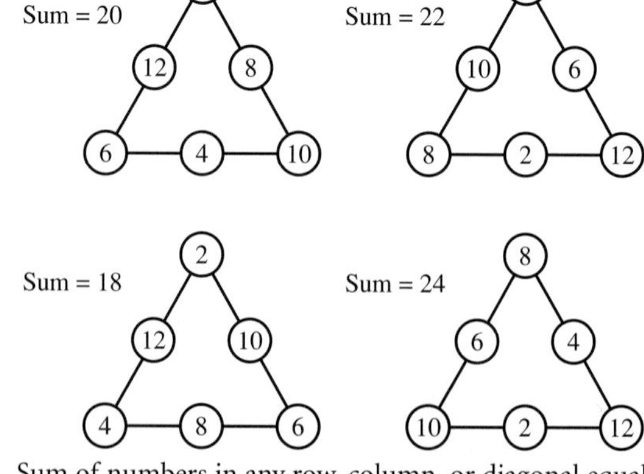

11. Sum of numbers in any row, column, or diagonal equals 27. Replace the number in each square of the 3×3 magic square with twice the number decreased by 1. If n = number in the magic square, replace it with $2n - 1$.

15	5	7
1	9	17
11	13	3

13.

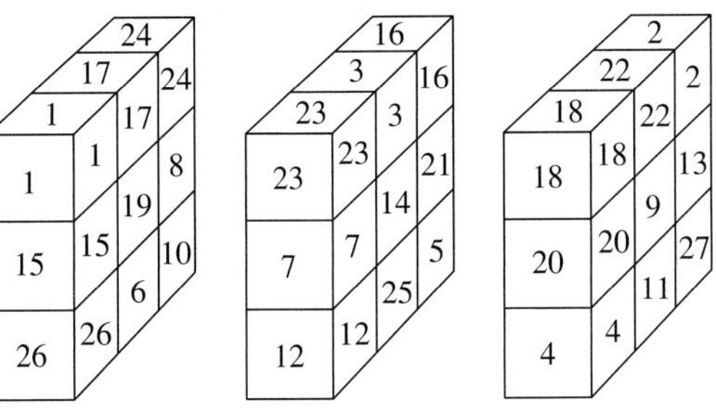

15. One of the possible solutions:

```
 ABC          132
- DE   ⇒    - 40
  GC           92
```

```
        GA               12
ABC )ADEC   ⇒   230 )2760
     ABC              230
      FEC              460
      FEC              460
```

17. Possible sum of legs are 25, 23, 27, 24, and 26.

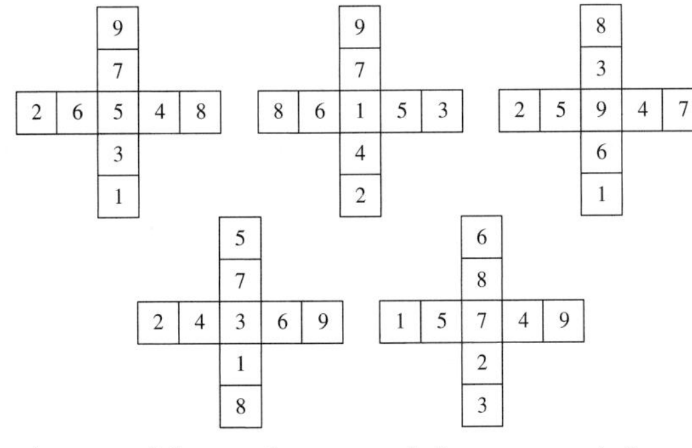

19. The sum of the numbers on each line segment is 8.

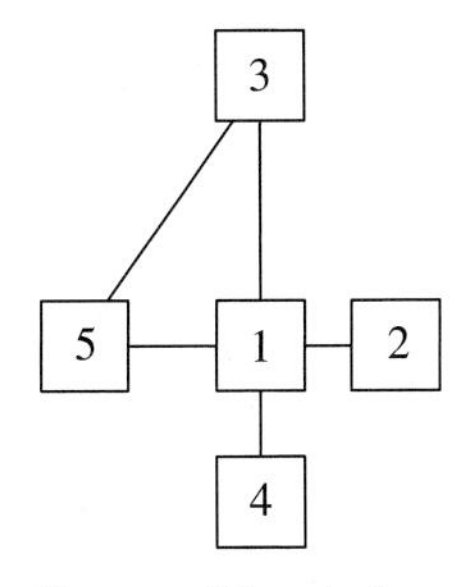

21. One possible solution:

```
  GOLF         7430
+ RAIN   ⇒   + 6591
 NOFUN        14021
```

23. One possible solution:

```
  FIVE         3475
- FOUR   ⇒   - 3260
   ONE          215
```

25. One possible solution:

```
   YOU          504
 ×   R   ⇒   ×    2
  COOL         1008
```

27. One possible solution:

```
      BIG            476
    × MAC          × 501
      BIG   ⇒        476
     AAA            000
   YESA            2380
   YESBIG         238476
```

29. One possible solution:

```
  SEND         9567
+ MORE   ⇒   + 1086
 MONEY        10653
```

31.

36	÷	6	×	1	**6**
÷		−		+	
9	−	3	×	5	**−6**
−		×		−	
8	−	4	×	2	**0**
−4		**−6**		**4**	

1 2 3 4 5 6 8 9

33.

5	÷	$\sqrt{9}$	×	6	10
+		+		−	
1	−	$\sqrt{4}$	−	7	−8
−		×		×	
8	+	3	+	2	13
−2		9		−8	

1 2 3 4 5 6 7 8 9

CHAPTER 2 REVIEW

Review Section 2.1

1. Some of my relatives do not live in Argentina.
2. If you fixed it, it was broken.
3. Math is the study of the relationship between quantities using numbers and symbols.
4. A statement in logic cannot be true or false at the same time. Questions, commands, and vague sentences are not considered statements.

Review Section 2.2

5. The argument has the converse error. The correct argument is:

   If there is a drought, you don't water the lawn.
   You water the lawn.
   Therefore, there is no drought.

6. By substituting $n = 0, 1, 2, 3, 4, 5, 6, 7, 8, 9, \ldots$, we get 41, 43, 47, 53, 61, 71, 83, 97, 113, which are all prime numbers. You may, therefore, make the conclusion that this formula always produces a prime number. However, when $n = 41$ the answer is 1681 (41×41) which is composite.

7. *Inductive:* I had a friend who drank too many beers and got in an accident. He wrecked his car and nearly killed himself. This might happen to you, so you should not drive.

   Deductive: If you drink excessively, your reactions and coordination are impaired.
   If your reactions and coordination are impaired, you may get in an accident.
   If you get in an accident, you could hurt yourself, your car, and others.
   If you could hurt yourself, your car, and others, you should not drive.
   Therefore, if you drink excessively, you should not drive.

8. Rearranging the statements and using the contrapositive for the first given statement, gives:

   If there are no reflections in the scene, you don't use a polarizing filter.
   If you don't use a polarizing filter, you are not photographing water.
   Therefore, if there are no reflections in the scene, you are not photographing water.

Review Section 2.3

9. D = Dog is a Brittany
   H = Hunt birds

   $D \to H$
   D
   $\therefore H$

10. The statement is true if B is true.

A	B	$\sim A$	$\sim B$	$\sim A \to \sim B$	B	$(\sim A \to \sim B) \to B$
T	T	F	F	T	T	T
T	F	F	T	T	F	F
F	T	T	F	F	T	T
F	F	T	T	T	F	F

11. Yes, the statements are equivalent.

A	B	$A \wedge B$	$\sim(A \wedge B)$	$\sim A$	$\sim B$	$\sim A \vee \sim B$
T	T	T	**F**	F	F	**F**
T	F	F	**T**	F	T	**T**
F	T	F	**T**	T	F	**T**
F	F	F	**T**	T	T	**T**

12. The argument is not correct. The last column of the truth table does not contain all true values.

   Let R = Election reforms
   B = Candidates buy elections

R	B	$\sim R$	$\sim R \to B$	$\sim B$	$R \to \sim B$	$(\sim R \to B) \to (R \to \sim B)$
T	T	F	T	F	F	F
T	F	F	T	T	T	T
F	T	T	T	F	T	T
F	F	T	F	T	T	T

Review Section 2.4

13.

Start
Is printer working?
Yes
Stop
No
Are printer and computer plugged in and switches turned on?
No
Plug in printer and computer; turn on switches.
Yes
Is printer cable connected to the computer?
No
Connect printer cable to the computer.
Yes
Does the printer have paper?
No
Put paper and ink into the printer.
Yes
Get help! (i.e., your 12-year-old brother).

If your printer is not working, you need to check and correct possible problems: Are the computer and printer turned on? Are the computer and printer plugged in? Is the printer cable connected to the computer? Does the printer have paper? If all these are checked and corrected, get some help from an expert. The flowchart simply organizes the process and gives a logical order to solving the problem.

Review Section 2.5

14.

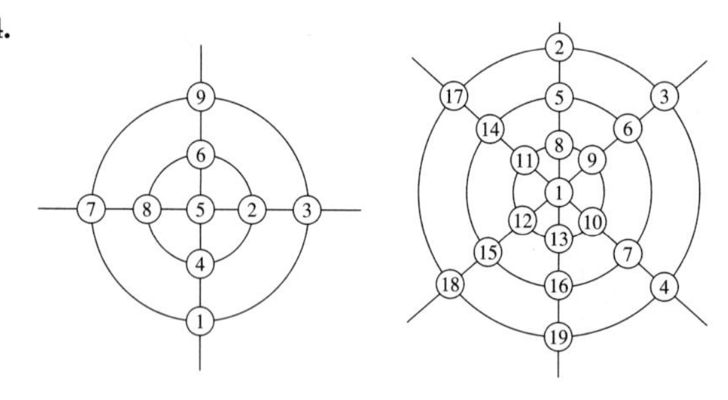

15.

```
  T E E          4 5 5
  H E E          7 5 5
+ H E E   ⇒    + 7 5 5
  J O K E        1 9 6 5
```

16.

```
  F U N          1 2 5
  × N      ⇒     × 5
  S U N          6 2 5
```

17.

36	÷	4	×	9	**81**
÷		+		×	
2	+	7	−	8	**1**
÷		−		÷	
6	×	1	÷	3	**2**
3		**10**		**24**	

1 2 3 4 5 6 7 8 9

CHAPTER 2 TEST

1. a) *Hypothesis:* "If it is after midnight." Conclusion: "I am in bed."

b) If I am in bed, it is after midnight.

c) If it is not after midnight, then I am not in bed.

d) If I am not in bed, then it is not after midnight.

2. Three components of a definition and the additional properties needed for a good definition can be found in Section 2.1 under definitions.

a) Not biconditional. Perpendicular lines are lines that intersect at right angles.

b) The statement uses the same root word in its definition and thus, it uses a word in the definition that may not be already understood.
Microscopic: that which can only be seen through a high powered magnifying device.

c) Not biconditional.
Natural numbers are whole numbers that are greater than zero.

3. One of many possible answers: If x is a natural number, then it is a real number.

4. The difference between inductive and deductive arguments is explained in Section 2.2. There are many possible arguments. Remember that an inductive argument is based on observing results, analyzing experiences, citing authorities, or presenting statistics. A deductive argument follows from accepted facts, assumptions, rules, or laws.

5. a) Using the rule for contrapositives: $Z \to \sim Y$ is equivalent to $Y \to \sim Z$. Thus, the argument is:

$$X \to Y$$
$$Y \to \sim Z$$
$$\sim Z \to P$$
$$\therefore X \to P$$

b) Using the rule for contrapositives: "If the geometry is Euclidean, then parallel lines exist." is equivalent to "If there are no parallel lines, then the geometry is non-Euclidean."
Thus the argument is:

If the geometry is Riemannian, then there are no parallel lines.
If there are no parallel lines, then the geometry is non-Euclidean.
If the geometry is non-Euclidean, then at least one of Euclid's postulates is changed.
∴ If the geometry is Riemannian, then at least one of Euclid's postulates is changed.

6. a) Let H = headache, G = grumpy, S = silent, $H \to (G \vee S)$.

b) Let G = good weather, B = play baseball, P = have a picnic, $\sim G \to (\sim B \wedge \sim P)$.

c) Let M = study math, G = good job, H = harder to advance, $(\sim M \wedge G) \to H$.

7.

A	B	$A \vee B$	$\sim B$	$A \vee \sim B$	$(A \vee B) \wedge (A \vee \sim B)$
T	T	T	F	T	**T**
T	F	T	T	T	**T**
F	T	T	F	F	**F**
F	F	F	T	T	**F**

Since the two statements have the same truth values, the statements are equivalent.

8. Let P = parrot, W = cracks walnuts, A = my animal. The argument is $(P \to W) \wedge (A \to \sim \mathbf{W})$

$$A \to \sim \mathbf{P}$$

P	W	A	$P \to W$	$\sim W$	$A \to \sim W$	$(P \to W) \wedge (A \to \sim W)$	$\sim P$	$A \to \sim P$	$((P \to W) \wedge (A \to \sim W)) \to (A \to \sim P)$
T	T	T	T	F	F	F	F	F	T
T	T	F	T	F	T	T	F	T	T
T	F	T	F	T	T	F	F	F	T
T	F	F	F	T	T	F	F	T	T
F	T	T	T	F	F	F	T	T	T
F	T	F	T	F	T	T	T	T	T
F	F	T	T	T	T	T	T	T	T
F	F	F	T	T	T	T	T	T	T

Since the last column is always true, the argument is correct.

9.

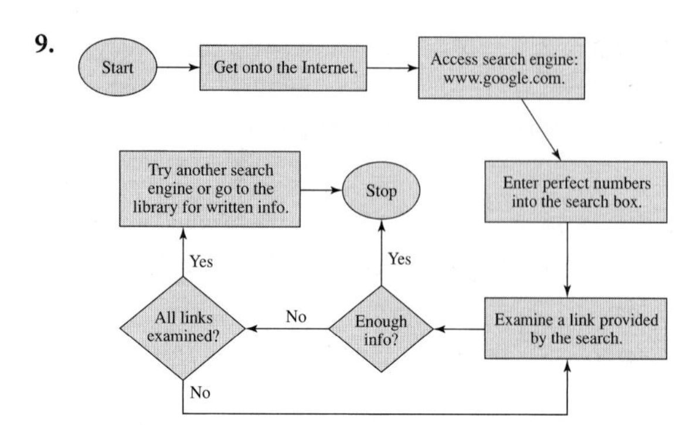

10.

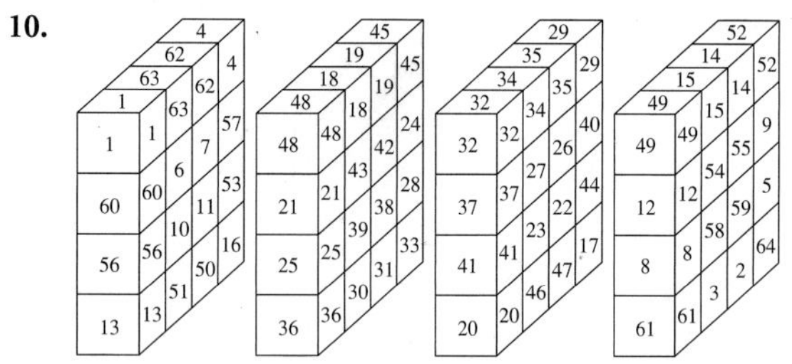

11.

$$\begin{array}{r} \text{YEA} \\ \times \quad 4 \\ \hline \text{MATH} \end{array} \Rightarrow \begin{array}{r} 968 \\ \times \quad 4 \\ \hline 3872 \end{array}$$

12.

72	÷	8	×	1	**9**
÷		−		×	
9	−	3	×	4	**−3**
−		−		+	
2	×	5	−	0	**10**
6		**0**		**4**	

0 1 2 3 4 5 8 9

CHAPTER 3

Section 3.1

11. a) $\{x|x$ is an even whole number $\leq 10\}$

    b) $\{0, 2, 4, 6, 8, 10\}$

13. True

15. One possible true statement is T $\not\subset$ F.

17. One possible true statement is $5 \in$ I.

19. One possible true statement is $5 \in$ I.

21. True

23. One possible true statement is F $\subset$ I.

25. $5k$

27. k^2

29. $A = \{$Michigan, Ohio, Kentucky, Illinois$\}$

31. $N = \{\ \}$

33. $n(P) = 12$

35. $n(B) = 30$

37. $3k \leftrightarrow 3^{k-1}$

Section 3.2

9. a) {apples, bananas, peaches, tomatoes, beans, peas, sprouts}

   b) {tomatoes}

11. 48 — 13. 15

15. $n(A \cap B) = -5$ — 17. 76

19. 93 — 21. 51

23. 7 — 29. 68

31. Empty set — 33. 125

Section 3.3

7. 30 — 9. 62

11. 1215 — 13. 2500

15. 1285 — 17. 53

19. 267 — 21. 254

23. a) *IV* — b) *II*

    c) *III* — d) *V*

    e) *I* — f) *VII*

    g) *VI* — h) *VIII*

25. a) 74 — b) 68

    c) 81 — d) 17

27. The shaded region is outside of both A and B.

Section 3.4

9. Permutations or Basic Counting Law

11. Basic Counting Law

13. Combinations

15. Basic Counting Law

17. Permutations or Basic Counting Law

19. Permutations or Basic Counting Law

21. Combinations

23. 16 — 25. 10^9

27. 3 — 29. 210

31. 1024 — 33. 2,598,960

35. a) 1024 — b) 10,240

37. 347,373,600 — 39. 190

41. 30 — 43. 175,760,000

45. 180 — 47. 38,760

CHAPTER 3 REVIEW

Review Section 3.1

1. a) $\{x|x$ is a whole number less than 20 that is divisible by 3$\}$

   b) $\{0, 3, 6, 9, 12, 15, 18\}$

3. The cardinality of a set is the number of elements in a set.

4. $1 + 11k$

Review Section 3.2

6. $T \cap S = \{12, 24, 26, 48\}$

7. 93 **8.** 51

Review Section 3.3

9. 5 **10.** 19 **11.** 27 **12.** 52

Review Section 3.4

14. 72 **15.** 190 **16.** 6 **17.** 3360

CHAPTER 3 TEST

3. a) 2380 **b)** 57,120

4. $3 + 4k$ **5.** 170

6. a) Use $5 \in I$ **b)** Correct
c) Use $F \subset I$ **d)** Correct

7. 72

8. a) 1056 **b)** 12,672 **c)** 164,736

10. 77 **11.** 52 **12.** 4

CHAPTER 4

Section 4.1

13. a) 1/52 **b)** 1/13 **c)** 10/13 **d)** 1/4
e) 1/2 **f)** 2/13

15. 195 **17.** 0.85 **19.** 0.40

21. a) 1/52 **b)** 1/13 **c)** 3/13 **d)** 1/4
e) 2/13 **f)** 19/259 **g)** 9/129

23. 0.99994

25. a) 8/25 **b)** 1/75 **c)** 4/75 **d)** 12/37
e) 24/73 **f)** 1/55

Section 4.2

5. a) 1/216 **b)** 1/72 **c)** 1/36

7. a) $0.000001539 \approx \frac{1}{649{,}740}$ **b)** $0.00001847 \approx \frac{1}{54{,}145}$

9. a) $0.002777 \approx \frac{1}{360}$ **b)** $0.03610 \approx \frac{1}{27.7}$

11. 1/56

13. a) $0.2272 \approx \frac{1}{4.4}$ **b)** $0.0006449 \approx \frac{1}{1551}$
c) $0.01209 \approx \frac{1}{82.7}$

15. a) 1/13,983,816 **b)** $0.00001845 \approx \frac{1}{54{,}201}$
c) $0.0009686 \approx \frac{1}{1032}$

17. 8/125 **19.** 1/190 **21.** 5/14 **23.** 1/10

25. a) 1/10 **b)** 1/5 **c)** 1

27. 1/907,200 **29.** $0.0000008227 \approx \frac{1}{1{,}215{,}450}$

Section 4.3

7. 0.60

9. 1:3

11. 10/11; 1/11

13. a) 17/20 **b)** 3 : 17 **c)** 17 : 3

15. a) 1/28,561 **b)** 1 : 28,560 **c)** $57,120

17. a) 2/5 **b)** 3/5 **c)** $15

19. a) 0.99994 **b)** 3:49,997 **c)** 49,997:3

Section 4.4

7. 0.92 **9.** 0.18

15. 2/5 **17.** 1/4

19. 17/20 **21.** 13/20

23. 17/20 **25.** 0.10

27. 0.064 **29.** 0.13824

31. 0.784 **33.** 0.28

35. 0.3456

37. b) 14/100 **c)** 5/100 **d)** 79/100

39. b) 23/450 **c)** 62/450 **d)** 412/450

41. 185/730 **43.** 150/730

45. 364/730 **47.** 0.2601

49. 0.4998 **51.** 0.7599

Section 4.5

7. 0.2222 **9.** 0.0672

11. 0.18, 0.3214 **13.** 8/15

15. 2/9 **17.** 234/1300

19. 163/397 **21. b)** 0.25

23. b) 2/7 **c)** 2/16

25. b) 39/412 **c)** 373/388

27. 85/185 **29.** 62/219

31. 0.51 **33.** 0.49

Section 4.6

9. $\approx -\$0.3333$; lose: $33.33

11. $20.00

13. a) $\approx \$0.225$ **b)** No **c)** 22.5¢

15. 10 **17.** −$0.1407

19. $\approx -\$0.0526$

21. $1.00 Ticket: E.V. $\approx -\$0.266$
$2.00 Ticket: E.V. $\approx -\$0.533$
$3.00 Ticket: E.V. $\approx -\$0.799$

23. Second

25. a) $\approx -\$0.28$ **b)** $50.00

CHAPTER 4 REVIEW

Review Section 4.1

1. 7/19 **2.** 5/36

3. Empirical **4.** 0.976

Review Section 4.2

5. 1/56 **6.** $0.0006 \approx \frac{1}{1641}$

7. $0.095 \approx \frac{1}{10.5}$ **8.** $0.029 \approx \frac{1}{35}$

9. $0.00019 \approx \frac{1}{5267}$ **10.** $0.00014 \approx \frac{1}{7384.5}$

Review Section 4.3

12. 1:3 **13.** 1/13

14. 1:11 **15.** 54,144:1

Review Section 4.4

17. 14/57 **18.** 43/57

19. 0.12

Review Section 4.5

20. 11/18 **21.** 12/19

22. 0.2877 **23.** 82/204

Review Section 4.6

25. −$0.31 **26.** No; $1,644,729

27. 0.03

CHAPTER 4 TEST

1. 1/16 **3.** 1/100

4. 0.2601 **5.** 0.386 or $\frac{132}{342}$

6. 0.877 **7.** 1/720; 1:719

8. a) $\frac{4}{C_{52,7}}$ b) $\frac{48}{C_{52,7}}$ c) $\frac{624}{C_{52,7}}$

9. E.V. ≈ 0.78

10. a) E.V. ≈ 1.4375 b) 14 points

11. a) 1/326; 1:325 b) 325/326; 325:1

c) −$0.23

12. a) 365^{20} b) $P_{365,20}$

c) 0.589

CHAPTER 5

Section 5.1

9. a)

Class	Frequency
1–3	3
4–6	4
7–9	3

b)

30%
30%
40%
1–3
4–6
7–9

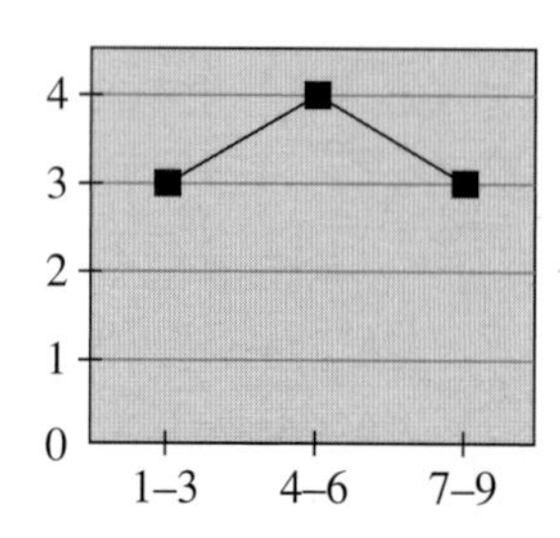

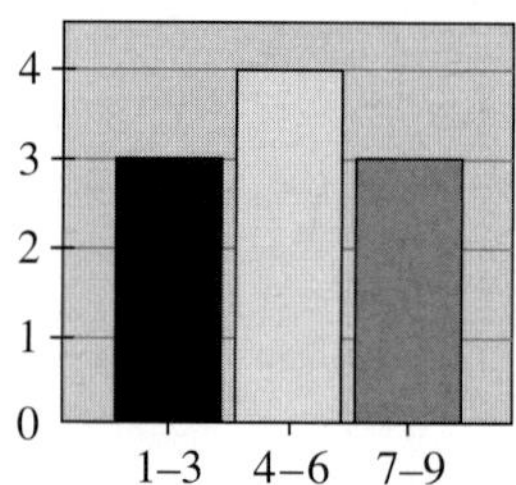

11. a)

Class	Frequency
10–11	1
12–13	4
14–15	2
16–17	4
18–19	1

b)

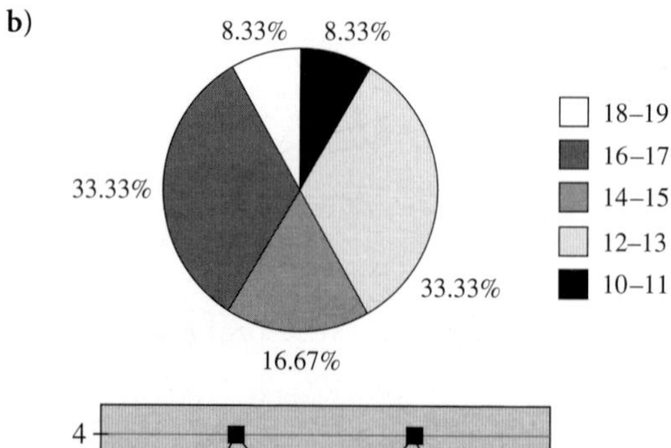

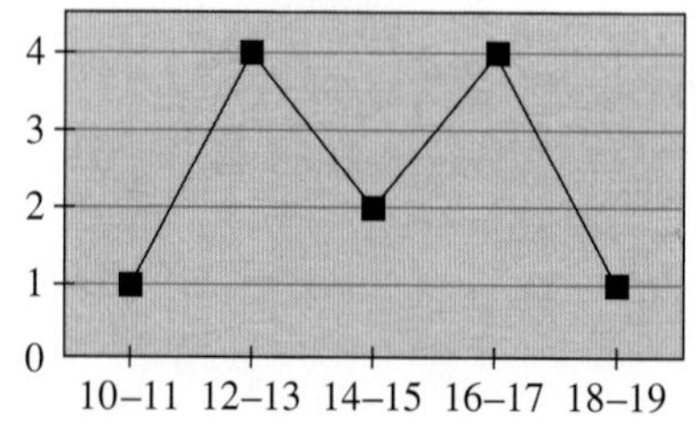

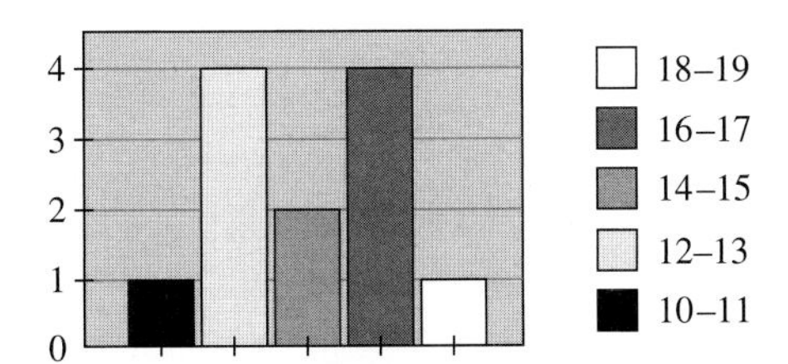

13. a)

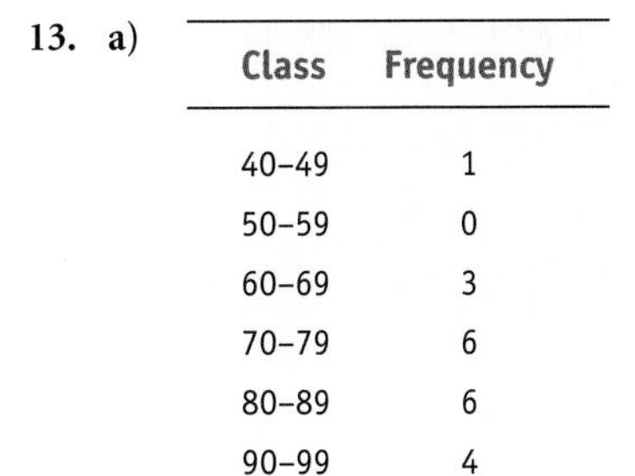

Class	Frequency
40–49	1
50–59	0
60–69	3
70–79	6
80–89	6
90–99	4

b)

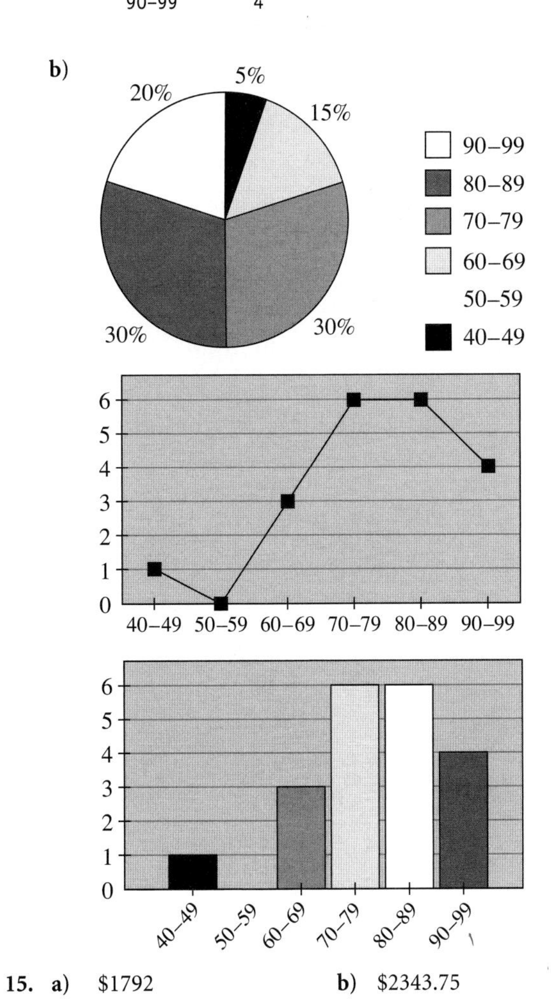

15. a) $1792 b) $2343.75

17. a)

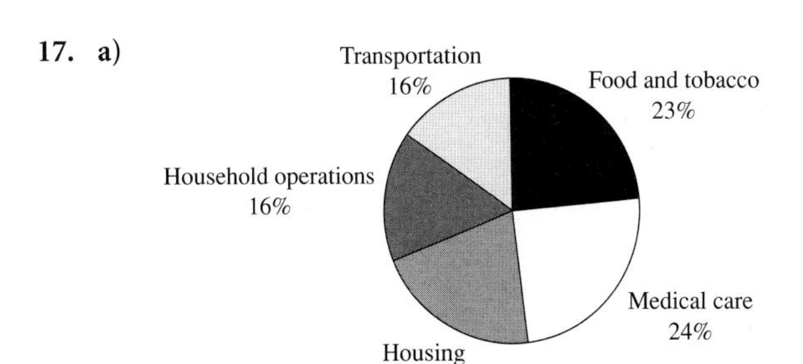

b)

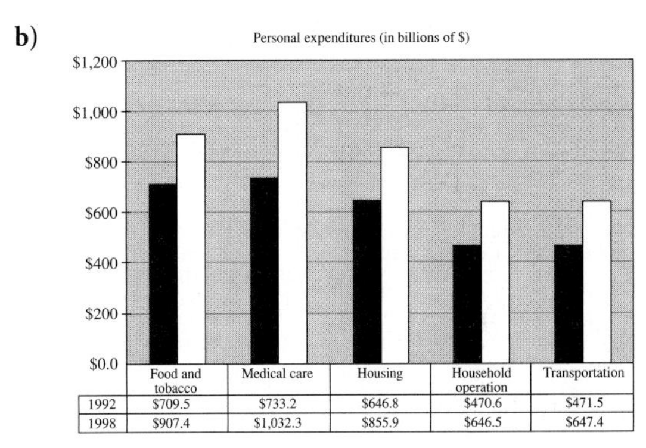

	Food and tobacco	Medical care	Housing	Household operation	Transportation
1992	$709.5	$733.2	$646.8	$470.6	$471.5
1998	$907.4	$1,032.3	$855.9	$646.5	$647.4

c) Medical care has increased by the largest amount ($299.1 billion) and its percentage of personal expenditures has increased from 24% to 25%.

19.

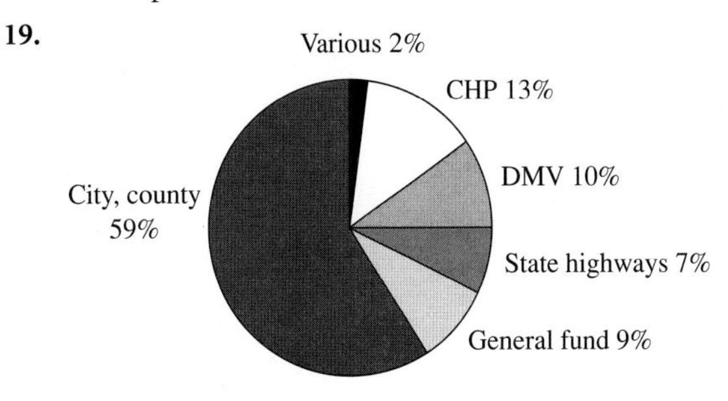

Section 5.2

11. a) $(2 + 4 + 7 + 2 + 1 + 8 + 9 + 10 + 9 + 6)/10 = 5.8$

b) Median = 6.5

c) The data is bimodal with modes 2 and 9.

13. a) $(20 + 29 + 2 \times 21 + 2 \times 28 + 3 \times 23 + 3 \times 26 + 5 \times 25)/17 = 24.65$

b) Median = 25

c) Mode = 25

15. a) At McDonald's

Calories	mean = 355 cal	median = 335 cal
Sodium	mean = 800 mg	median = 860 mg

b) At Wendy's

Calories	mean = 363 cal	median = 360 cal
Sodium	mean = 763.3 mg	median = 790 mg

17. Mean = 81, median = 90, mode = 94
The median

19. a)

	NFL	NBA	MLB	NHL
Mean	$6,006,300	$14,732,000	$12,455,200	$8,254,000
Median	$6,137,500	$14,900,000	$12,107,000	$8,250,000

b) Mean = $10,361,875, median = $10,680,000

c) The median salary for the entire group is not an effective measure of the average because it does not accurately reflect the salary of the top player salaries in any particular sport.

d) The mean salary for the entire group is not an effective measure of the average because it does not accurately reflect the salary of the top player salaries in any particular sport.

Section 5.3

11. a) 12
b) $\mu = 9, \sigma = 4.38$
c) Standard deviation is smaller

13. a) 8 b) $\mu = 4, \sigma = 2.45$

15. a) 56 b) 12.74

17. 132.2 mg

19. a) $11,794,000 b) $10,680,000
c)

Salary (in millions)	# of players
$4–7.999	14
$8–11.999	9
$12–15.999	15
$16–19.999	2

d)

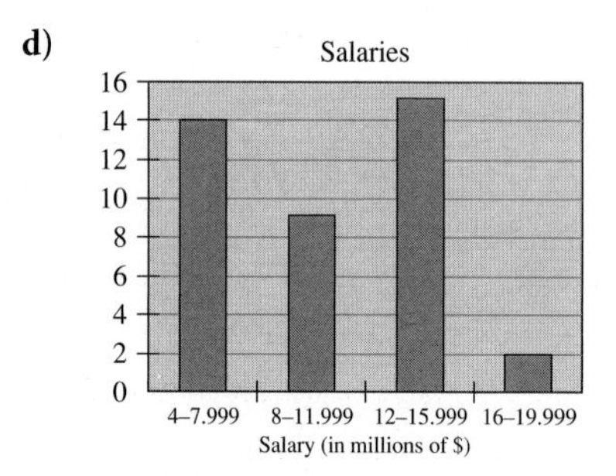

e) $10,499,500 f) $3,814,446

Section 5.4

11. 0.3413
13. 0.7692
15. 0.1156
17. 0.1335
19. −1.675
21. 0.845
23. −2.575
25. a) 13.25% b) 58.32%
27. a) 0.4090 b) 29.64% c) 11.58%
29. 48 cigarettes

Section 5.5

9. 1.28
11. 1.44
13. 59%–67%
15. 45.2%–50%
17. 4.2%
19. 2.3%
21. 99.97%
23. 95.22%
25. 3385
27. 456
29. a) 97.74% b) 339 c) 4145
31. a) 94.26% b) 37% to 43%
33. a) 97.56% b) 664 c) 792
35. a) 1068 b) 16,577

Section 5.6

5. a) $y = 1.84 + 1.05x$ b) When $x = 10, y = 12.34$

7. a) $y = 361.78 - 2.68x$ b) When $x = 70, y = 174.38$

9. a) $y = 9.72 - 0.24x$
b) When $x = 34, y = 9.72 - 0.24(34) = 1.56$.
c)

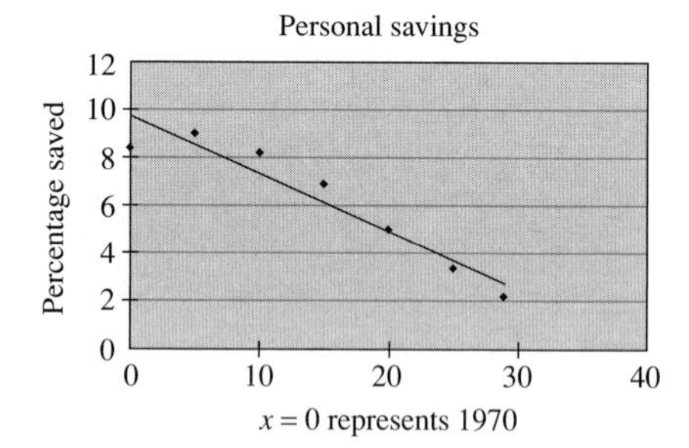

11. a) $y = 97.04 + 14.56x$.
b) When $x = 25, y = \$461.04$.
c)

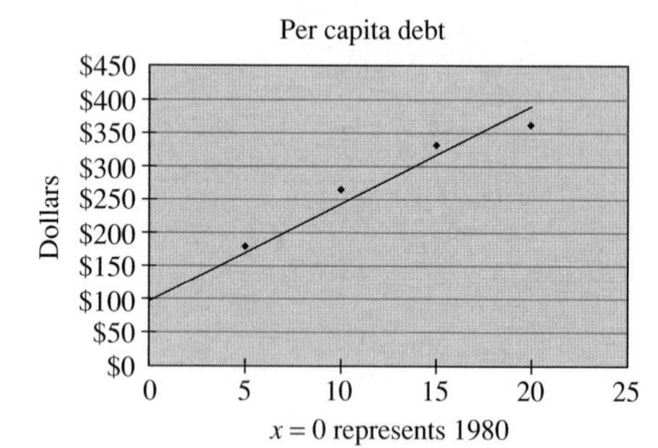

13. a) $y = 3.78 - 0.038x$.
b) When $x = 15$, $y = \$3.21$.
c)

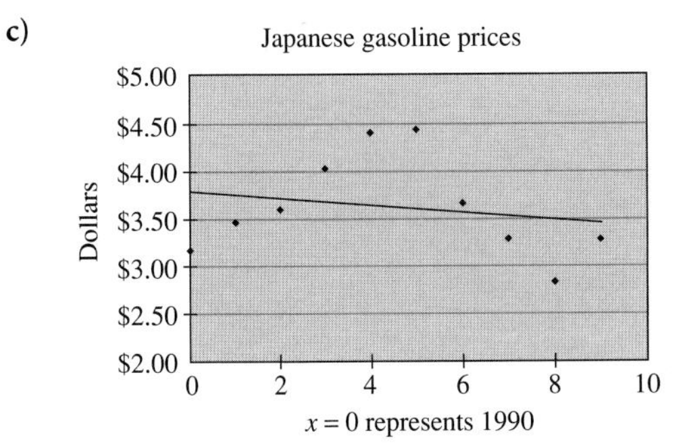

15. $y = -641.8 + 2748.2x$

CHAPTER 5 REVIEW

Review Section 5.1

1.

Class	Frequency
1–10	4
11–20	5
21–30	9
31–40	5
41–50	5

2.

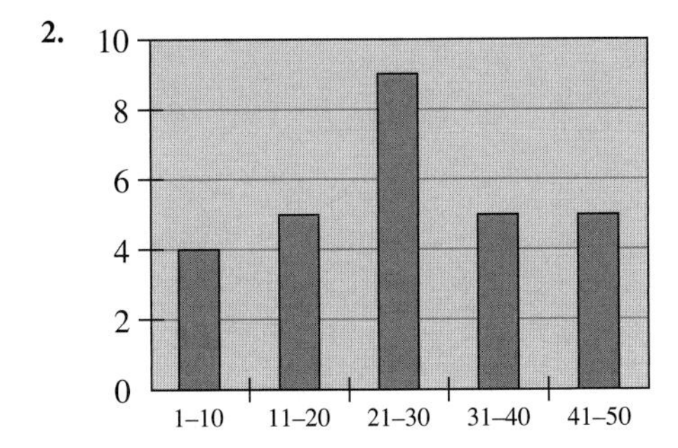

3.

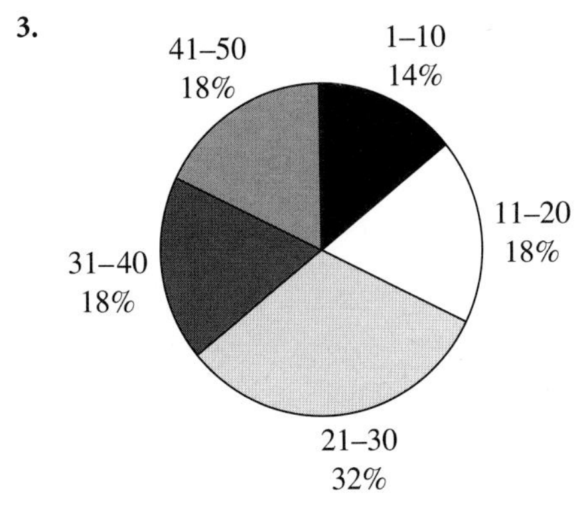

4.

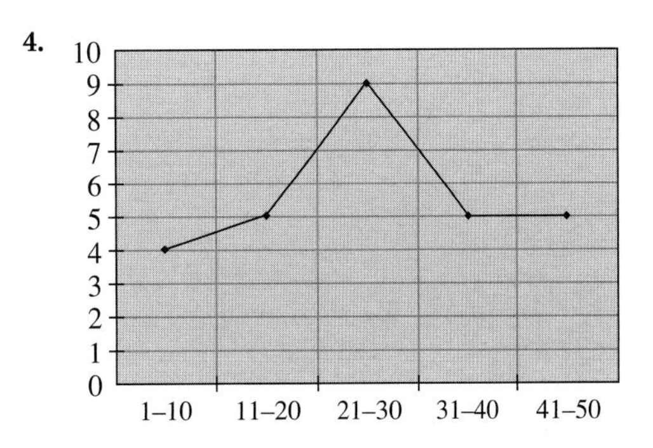

Review Section 5.2

5. 25.29, median = 26., mode = 23.
6. Mean = 26.21
7. The mode is the best measure of central tendency when you are trying to find the most frequently occurring value. For example, if you are trying to find the most popular type of car in a parking lot, you would count the number of Fords, Toyotas, etc., and the make that appears most frequently is the mode.
8. The median is the best measure of central tendency when you are trying to compute a numerical value but there may be a few extremely large or small values that would distort the mean. For example, if you had five house purchases in the town and the houses sold for $200,000, $225,000, $240,000, $250,000, and $2,500,000 the median would give the best measure of central tendency.

Review Section 5.3

9. 12.52 10. 12.8 11. 42
12. The standard deviation tells you how much the data are spread away from the mean.

Review Section 5.4

13. 0.0561 14. 1.16 15. 0.0307 16. 84

Review Section 5.5

17. The margin of error is the distance from the center of a confidence interval to each endpoint. It is how much error is expected in the results of a survey. The confidence level is a statement that specifies the probability that the confidence interval actually contains the value being estimated.
18. The margin of error should decrease as the sample size increases since a survey with a larger number of people should give a more accurate representation of the true opinions of the group. In terms of a formula, since $M = \dfrac{z}{2\sqrt{n}}$, if n increases, the denominator of the fraction increases, forcing the value of the function to decrease.
19. 6766 20. 2.3%

Review Section 5.6

21. $y = 6.69 + 1.81x$

22. If $x = 20$, $y = 42.89$.

23. 18.40

24.

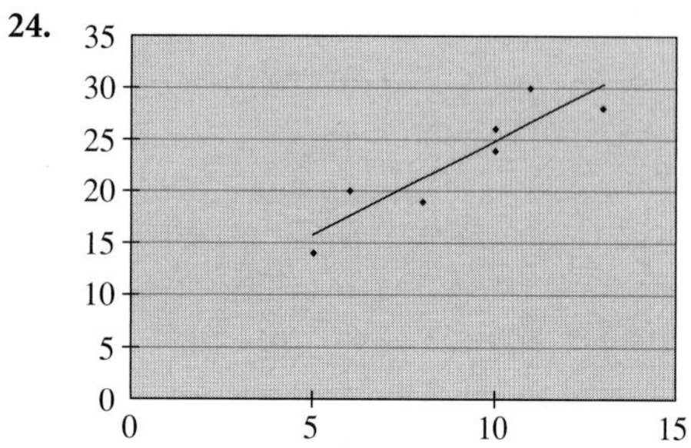

CHAPTER 5 TEST

1. a) 4.9

b) 5

c) Bimodal—the modes are 3 and 6.

d) 8

e) 2.55

2. a)

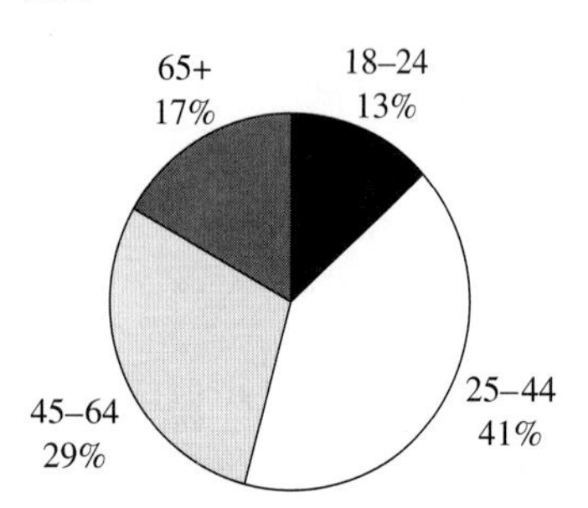

b)

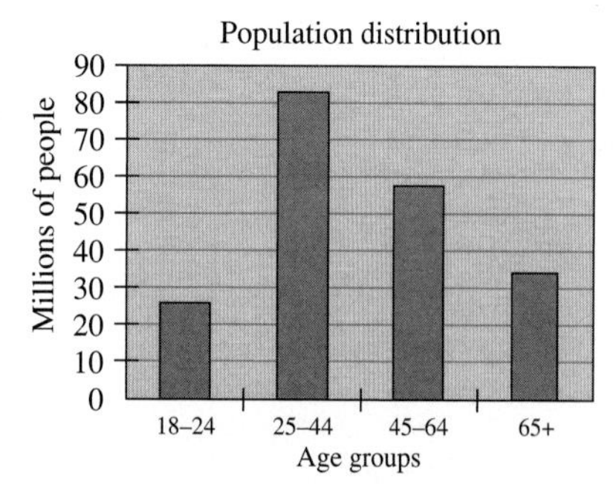

3. a)

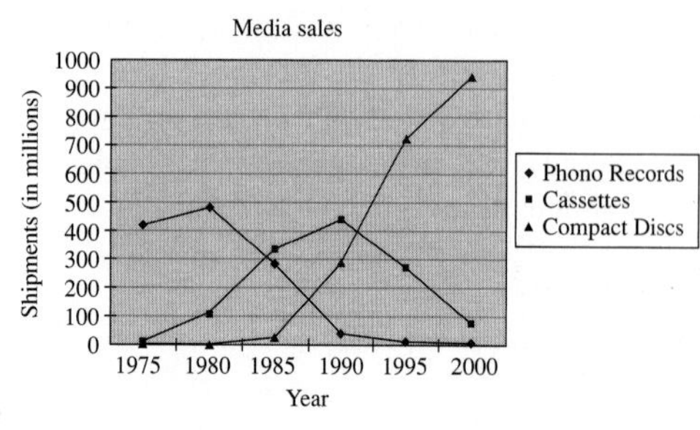

b) Some conclusions are:

Sales of phonograph records peaked in 1980 while the sales of cassettes peaked in 1990.

Sales of compact discs increased at a rapid and fairly consistent rate since 1985.

4. a) Mean Income
High School—$26,508
Some College—$33,778
Bachelors—$46,564
Advanced—$57,944

b) Greater education provides greater income.

5. Company A because of less variation in the product.

6. a) 36.24% **b)** 11.22% **c)** 0.6217 **d)** 0.2483

7. *A*'s 91.2–100
B's 80.4–91.19
C's 65.6–80.39
D's 54.8–65.59
F's 0–54.79

8. a) 94.26% **b)** 1842 **c)** 16,577

9. $y = 14 + 9.5x$; 156.5

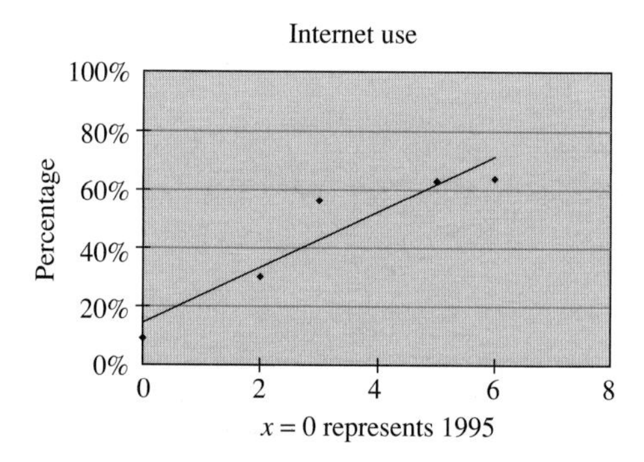

CHAPTER 6

Section 6.1

9.

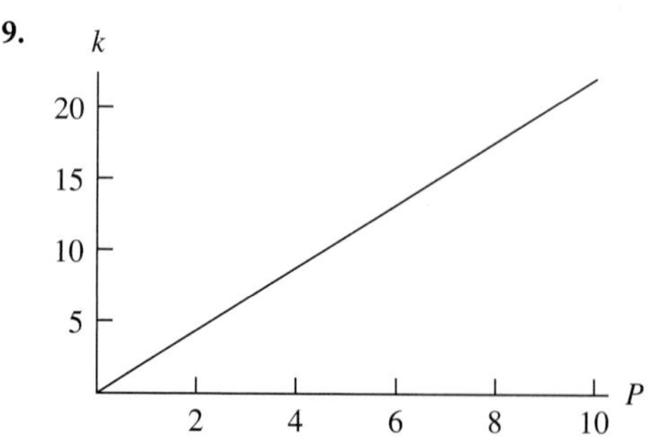

11.

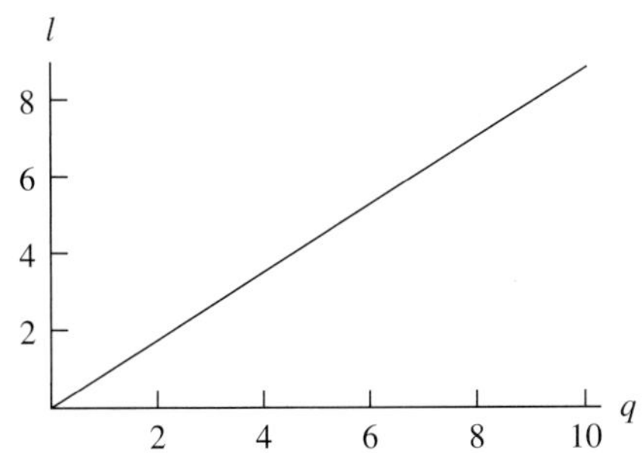

13.

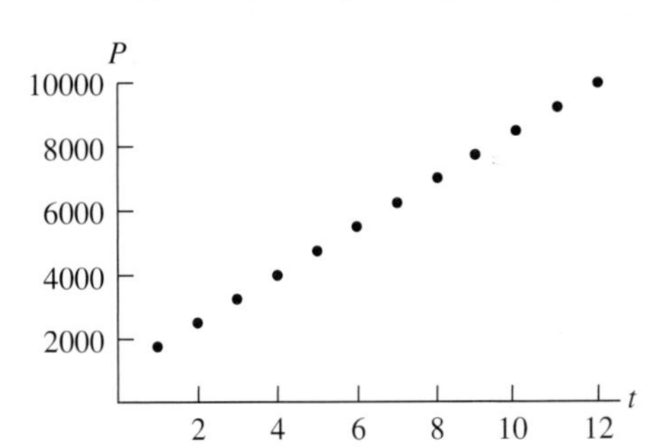

15.

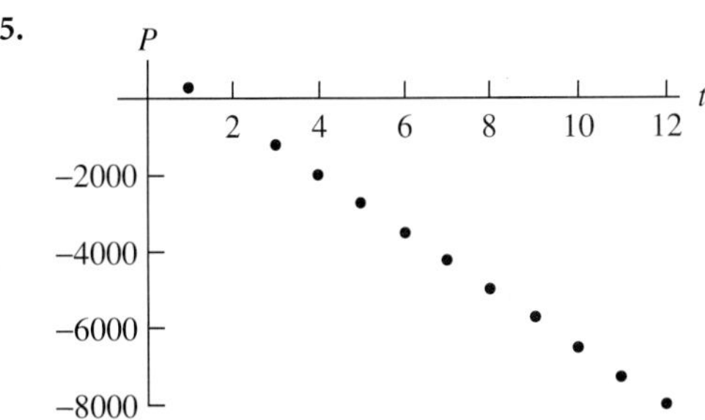

17. In Problem 13, the business is currently making a profit and expects the profit to increase. In Problem 14, the business is currently operating at a loss but expects to make a profit soon and expects the profit to increase. In Problem 15, the business is currently making a profit but, because of decreasing profits, expects to soon be operating at a loss. In Problem 16, the business is currently operating at a loss and expects that the losses will continue to increase.

19. $x = 1, y = 1$; $x = 6, y = -2$; $x = 11, y = -5$; $x = -4, y = 4$; $x = 16, y = -8$

21. a)

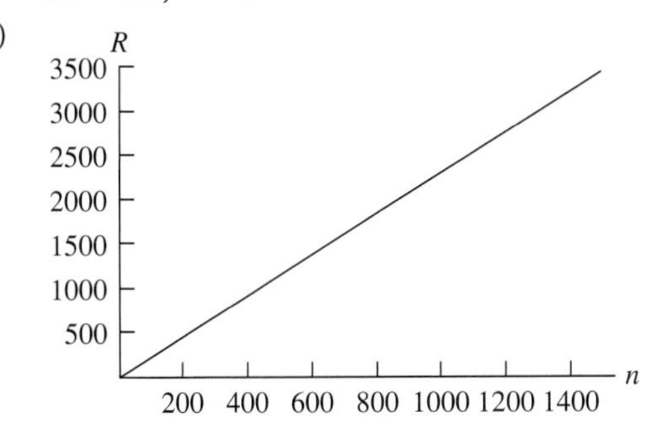

b) $m = 2.3$; for every additional customer, the revenue increases by $2.30.

c) $3450

23. a)

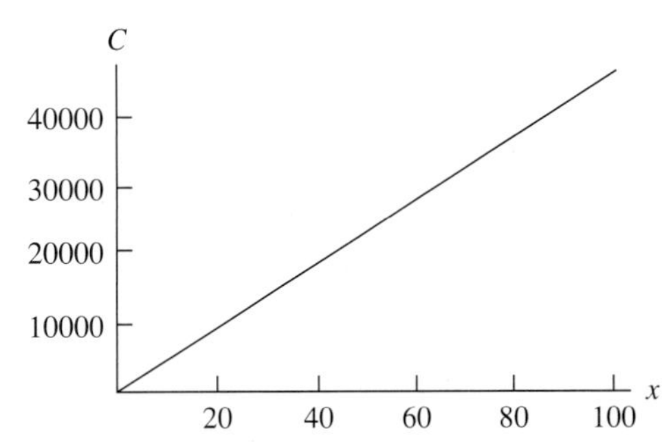

b) Since only a whole number of barrels may be produced, the line is actually a series of points at whole number values of x.

c) $m = 450$. This gives the cost of manufacturing one more wine barrel.

d) $46,200

e) 219 barrels

25. a) $t = -0.0035a + 59$

b) -25°F

27. a) $H = 8B + 10$

b)

H
60
50
40
30
20
10
1 2 3 4 5 6 B

c) $H = \frac{2}{3}B + \frac{5}{6}$

29. a) $T = -0.002a + 212$

b) 183.78°F

31. a) $T = 0.391x + 93{,}374$

b) $110,969

c) $365,447.19

33. a) For each increase in the height, there is a constant increase in the weight.

b) Women: $W = 4.5H - 148$; Men: $W = 4H - 128$

35. 3 ways

Section 6.2

5. Vertex: (0, 5), x-intercepts: $(-2.2, 0)$ and $(2.2, 0)$

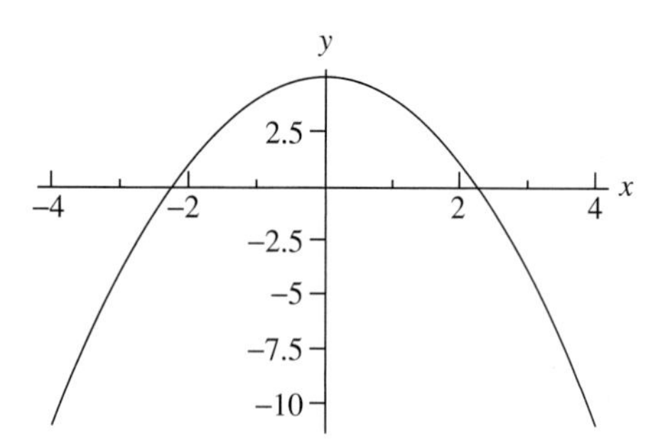

7. Vertex: $(-3, -4.7)$), x-intercepts: $(-7.9, 0)$ and $(1.0, 0)$

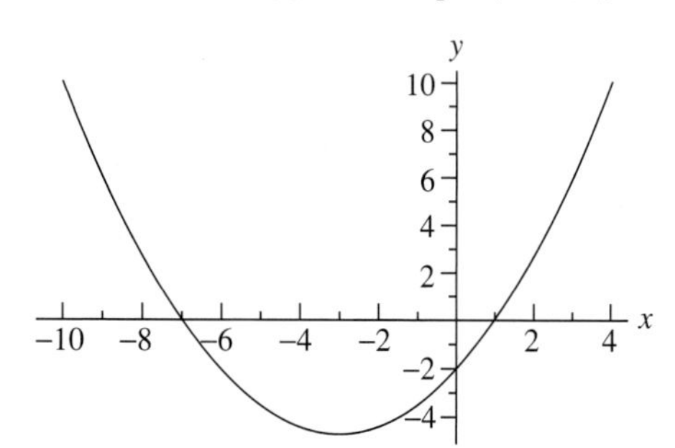

9. **a)** 148.8 ft **b)** 119.0 yards

11. **a)** 15 ft **b)** 208.7 ft **c)** 35.25 ft

13. **a)** $h = -0.039d^2 + 1.61d + 2$
    b) 18.6 ft **c)** 6.7 ft and 34.7 ft

15. **a)** $D = 0.0568s^2 + 1.06s$
    b) At 55 mph, 230 ft
    At 65 mph, 309 ft
    c) The formula for the stopping distance is accurate to within a few feet.
    d) 3249 ft

17. **a)** $V = \frac{1}{2}n^2 + \frac{1}{2}n$
    b) 5050

Section 6.3

7.

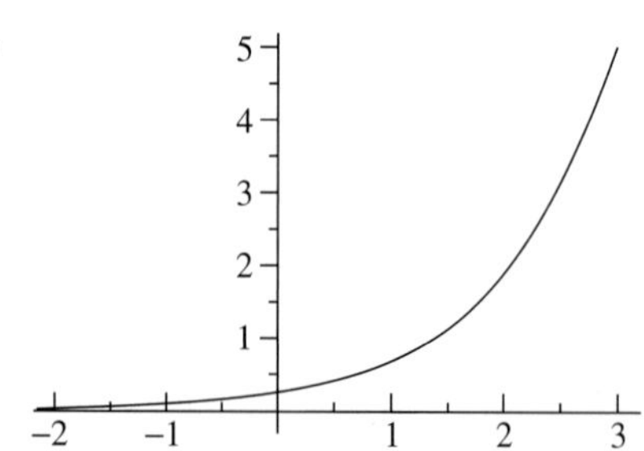

9.

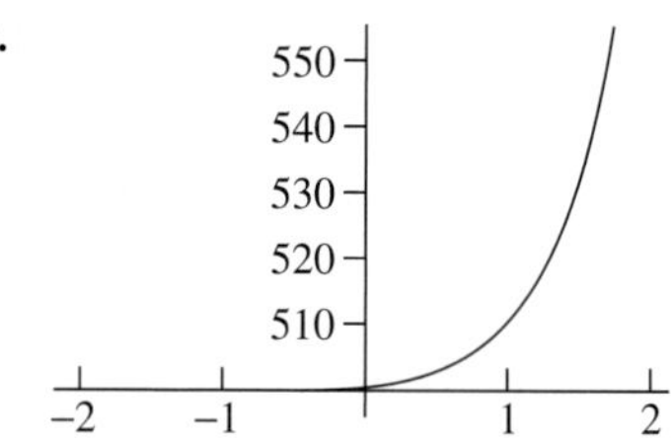

11.

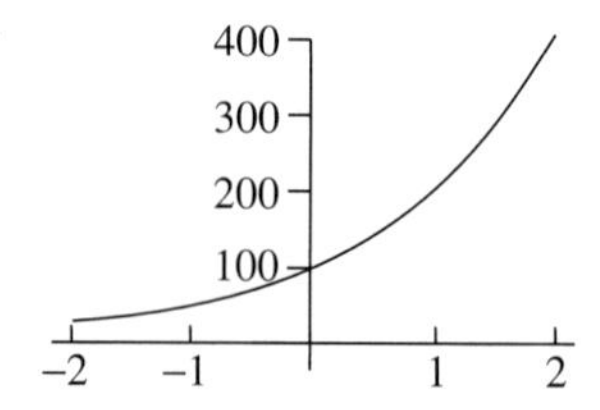

13. **a)**

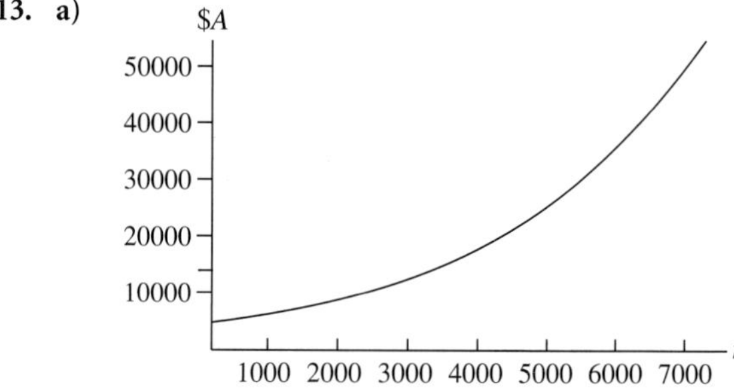

b) $5637.37 **c)** $55,094.10

15. **a)**

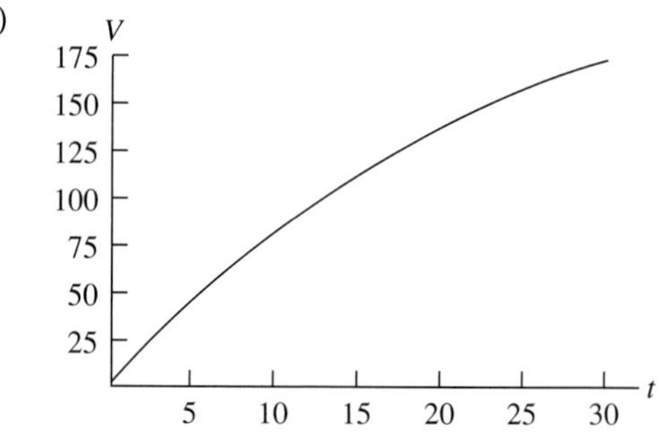

b) 9803 **c)** 107,198 **d)** 174,701

17. **a)**

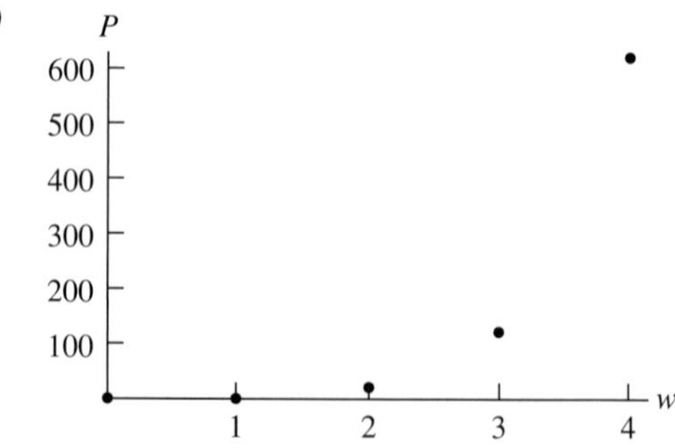

b) $p = 5^w$ **c)** 244,140,625

19. 332,534,400

21. 6.3 psi

23. $A = 407.5e^{0.1048t}$ and in 2004 ($t = 12$), $A = 1433.2$ million.

25. $A = 366.4e^{-0.1966t}$ and in 2004 ($t = 12$), $A = 34.6$ million.

27. 3.689×10^{19} kernels of corn or $\approx$ 10,541,000,000,000,000 pounds.

Section 6.4

7.

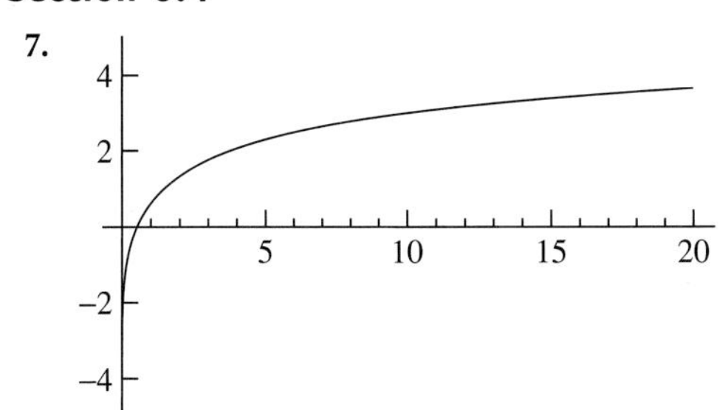

9.

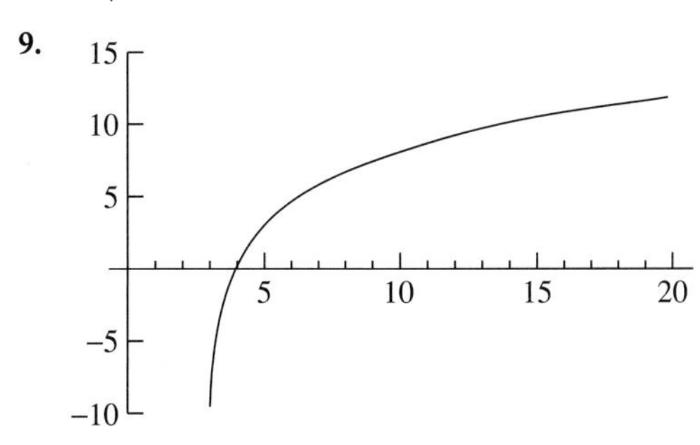

11.

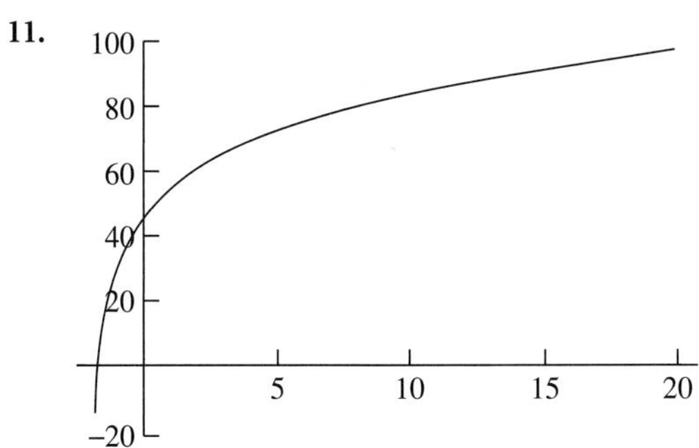

13.

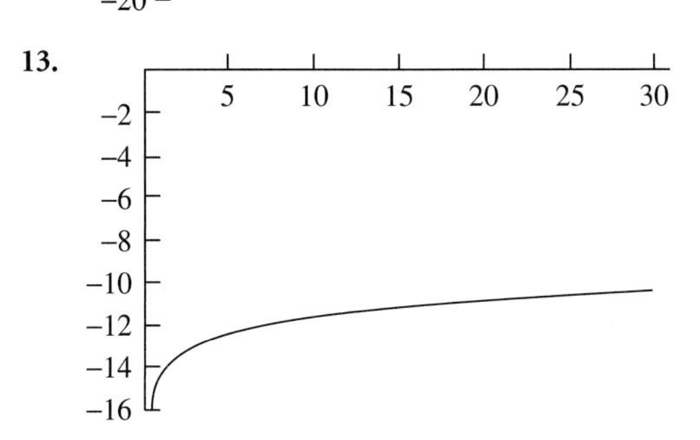

15. a) 89.2% **b)** 60.5 in.

17. a) 27.5 min **b)** 65.3 min

c)

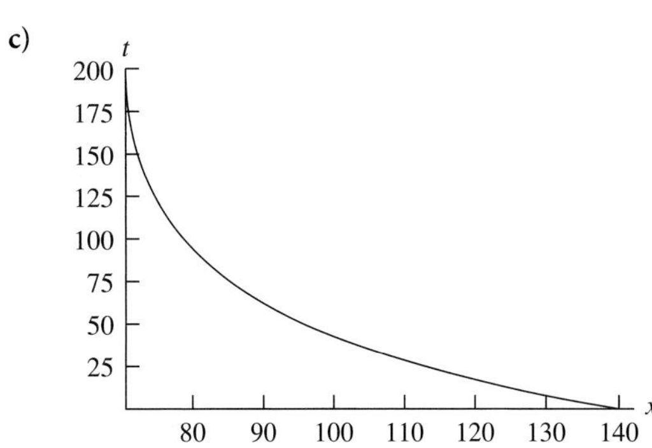

19. a)

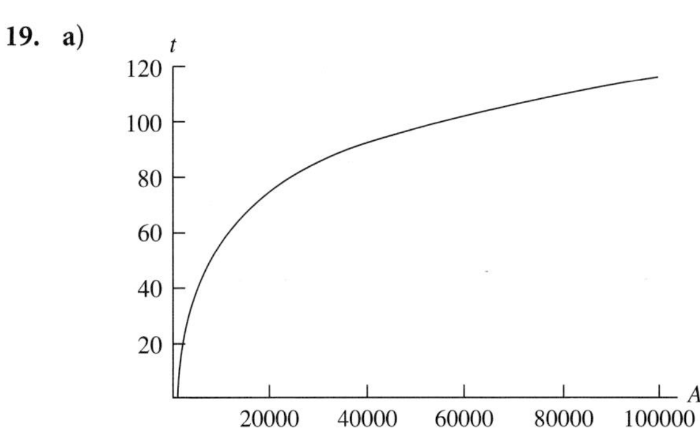

b) 172.7 hrs

21. a) 24,360 yrs **b)** 161,844 yrs

23. a) $y = 9778 + 371.7 \ln x$

b) 10,551

c) The model gives an answer that is 10 less than the actual amount.

d) 2013

CHAPTER 6 REVIEW

Review Section 6.1

1. a)

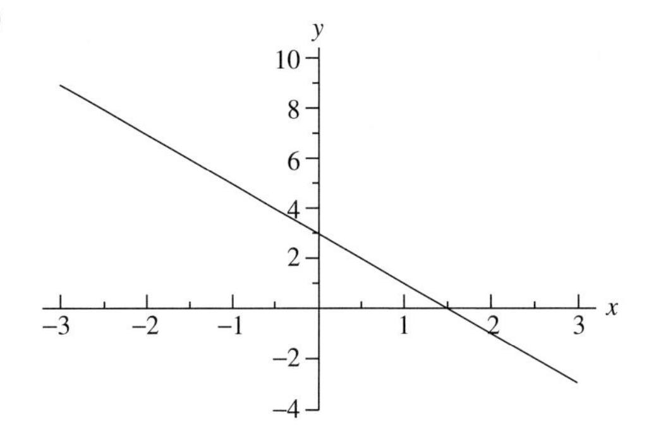

b)

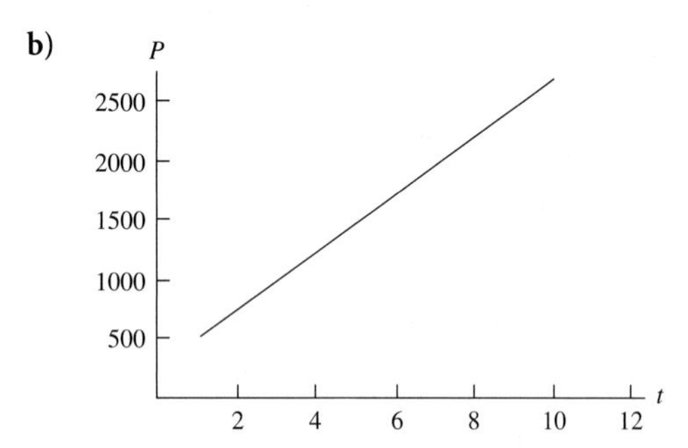

c)

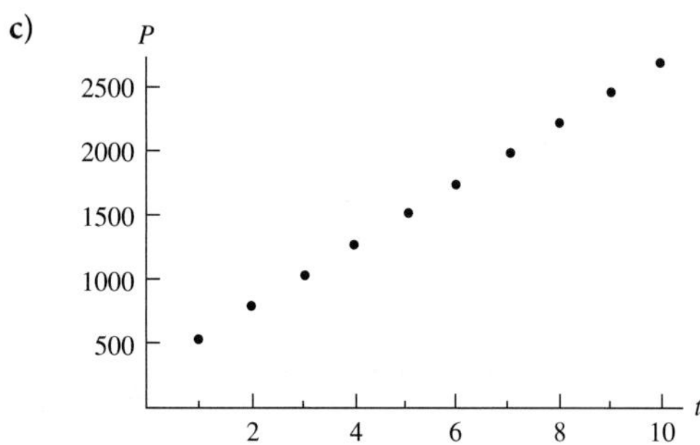

2. a) The revenue increases at a constant rate of \$0.77 for each can sold.

b) $P = 0.77x - 135$

c) \$604.20 **d)** 175

e)

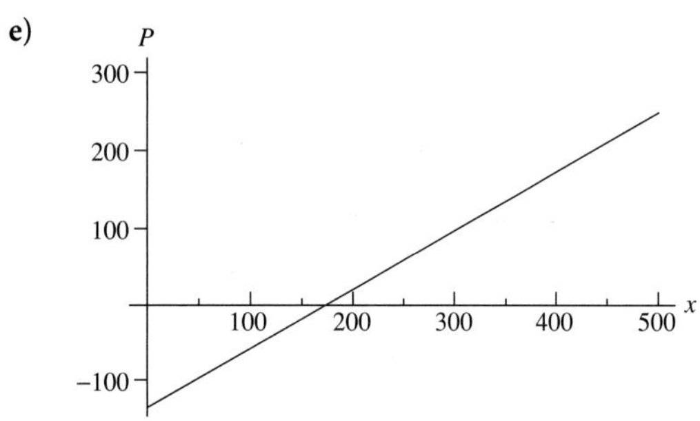

3. a) $s = 293a - 2535$, where a = Krista's age and s = the number of stamps

b) 2739 stamps **c)** 42.8 years old

4. $5x + 3y = 68$, five ways

Review Section 6.2

5. a) $(3, 9)$

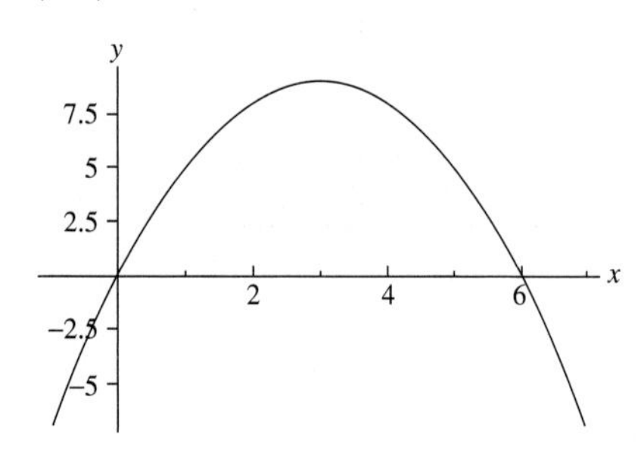

b) $(3, -1)$

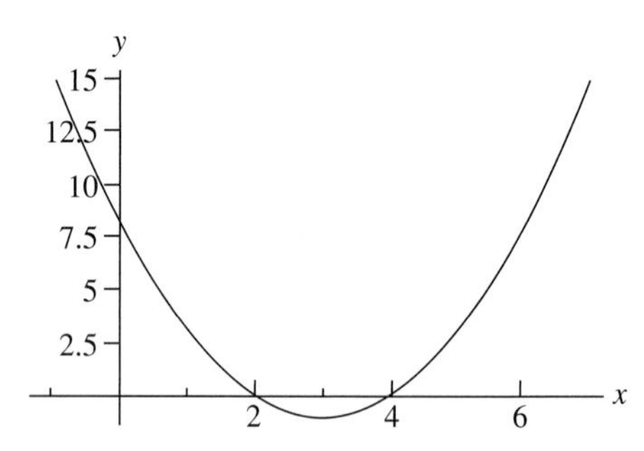

c) $(12, 27.6)$

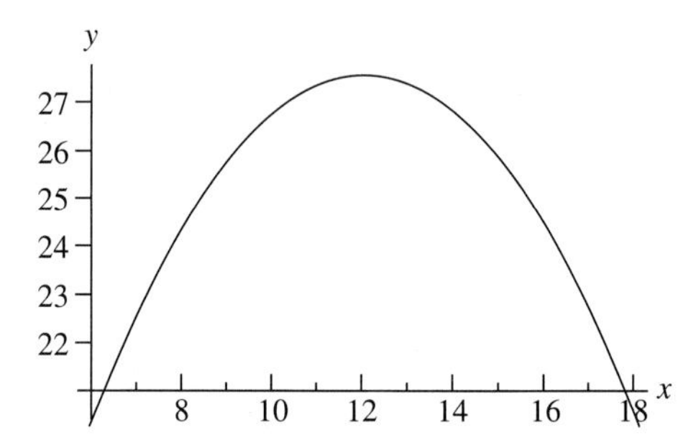

6. a) 7 ft **b)** 23 ft

c)

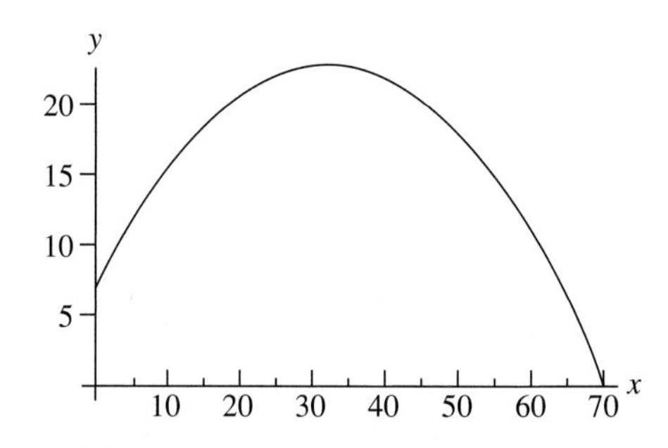

d) 70.4 ft

7. a) Coordinates measured in yards: (0, 0), (98, 3), (100, 0)

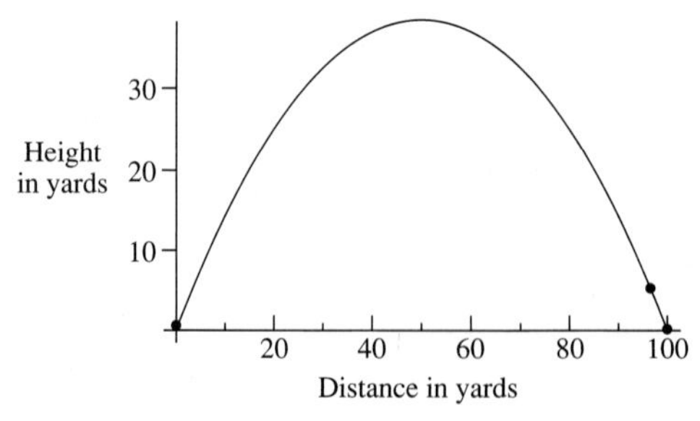

b) $y = -0.0153x^2 + 1.53x$

c) 38.25 yards

Review Section 6.3

8. a) 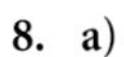

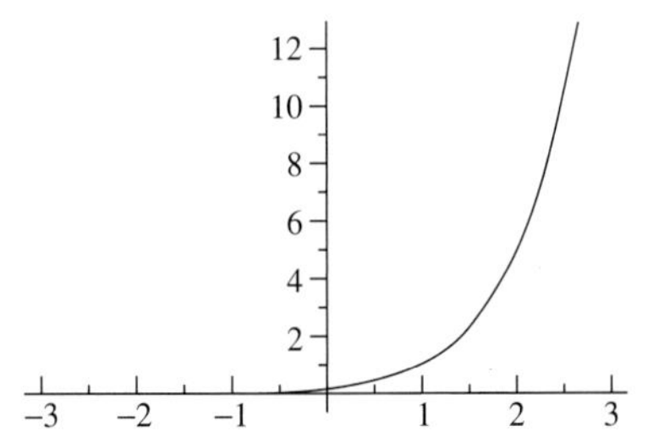

b)

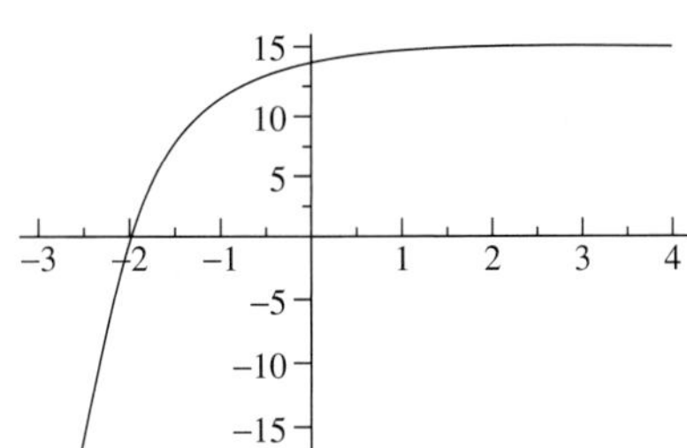

9. a) 60 calculators b) 71.8, 87.5, 90.0

c)

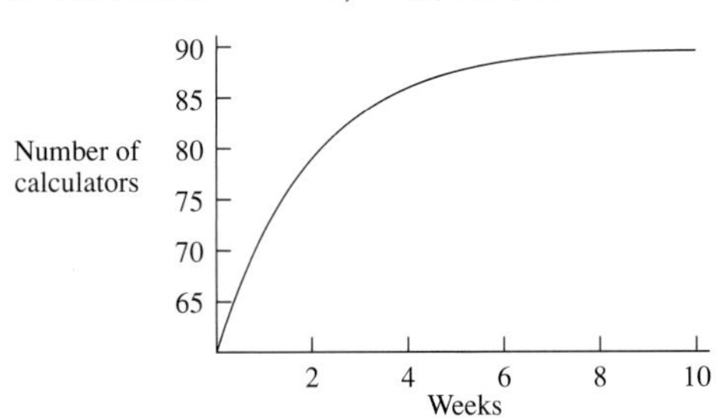

d) 90

10. a) The number of internet sites has been increasing more and more rapidly since 1996.

b) $A = 0.6e^{0.9242t}$

c) The data and the results of the model agree fairly well.

d) 24.2 million, 61.0 million

e) The exponential model was okay for 5 years but it does not seem to predict the number of internet sites after the year 2000. The number of sites has stopped growing exponentially.

Review Section 6.4

11. a)

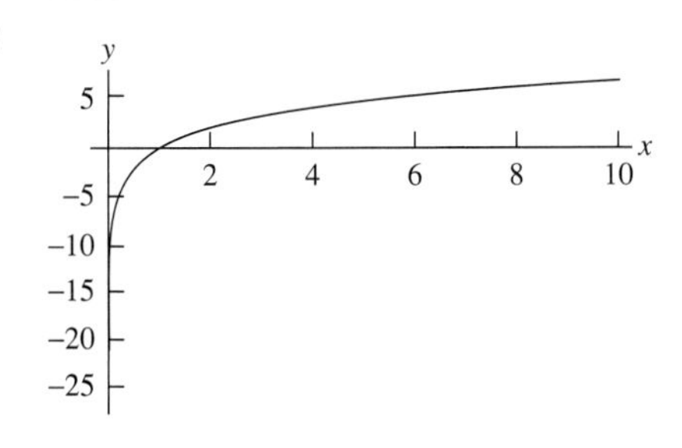

b)

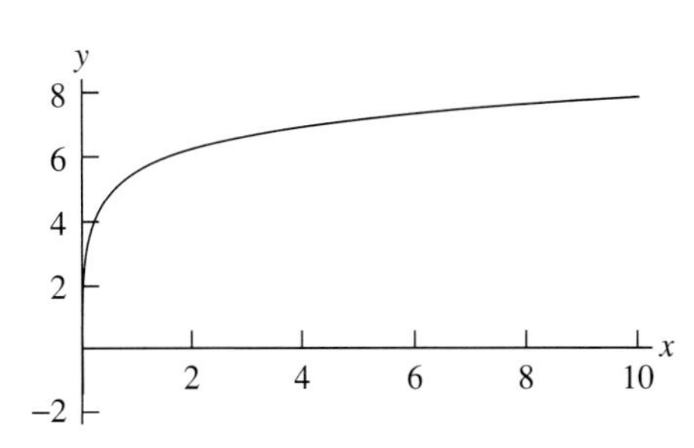

12. a) 97, 103, 107

b)

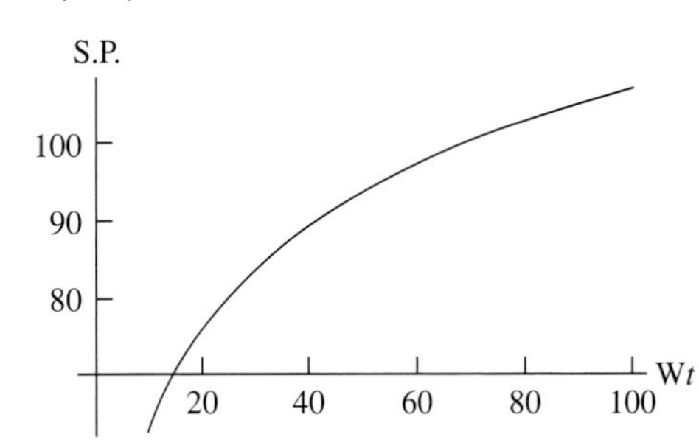

13. a) $y = 26.2 + 13.9\ln x$

b) 56.7%

c) 2030

CHAPTER 6 TEST

1. a)

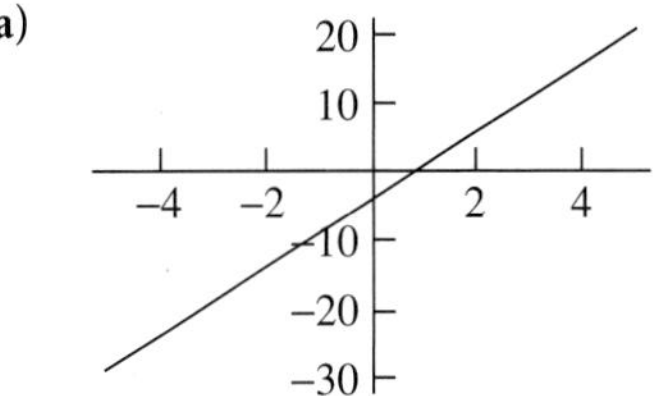

b)

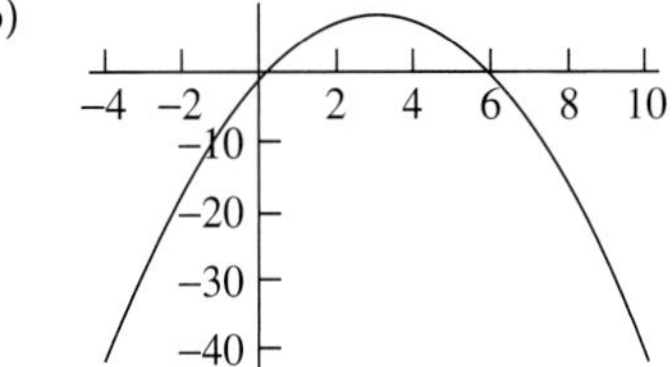

c)

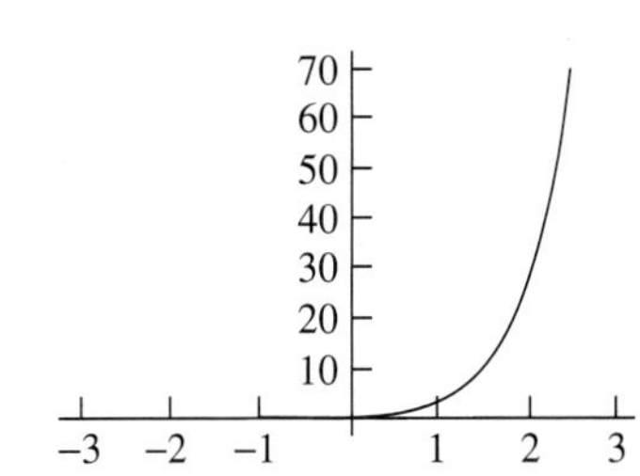

d)

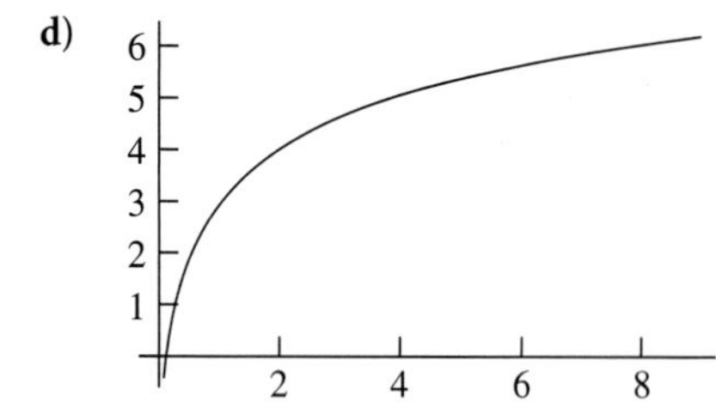

2. a)

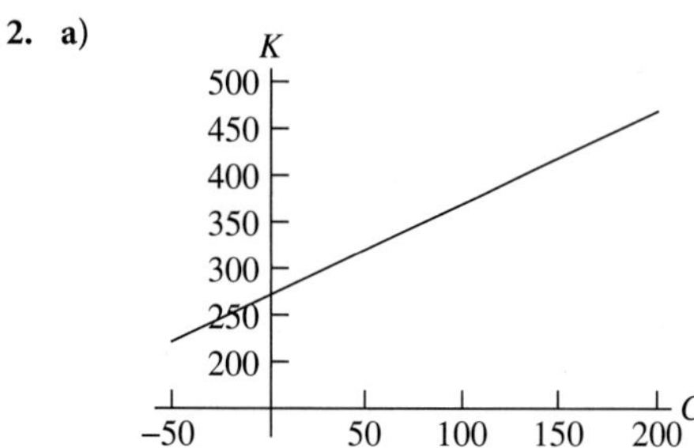

b) $-273°C$

c) $373°K$

3. a) A constant change in the weight causes a constant change in the calories.

b) $c = 2.727w + 140$

c) 576.3 cal

4. a) The number of logs in each row changes by a constant amount.

b) $L = -2r + 249$

c)

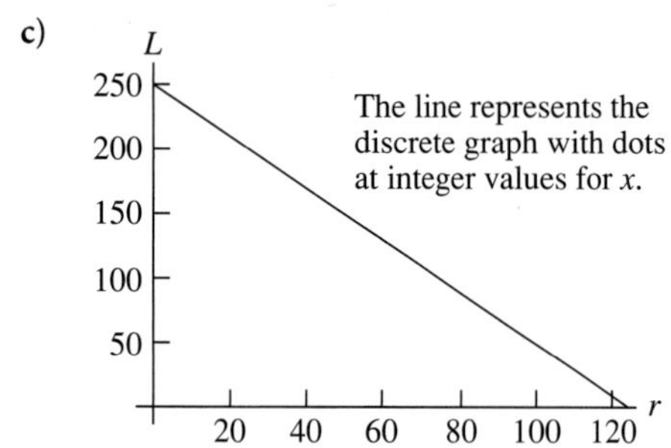

d) 149 logs

e) 124 rows

5. a)

b) 5121 ft c) 20.64 sec

6. $3x + y = 15$; 6 ways

7. a) $h = -0.001758x^2 + 0.616x$

b) 4′ 10.5″, 24′ 3.5″

8. a) $P = \frac{3}{2}t^2 - \frac{1}{2}t$

b)

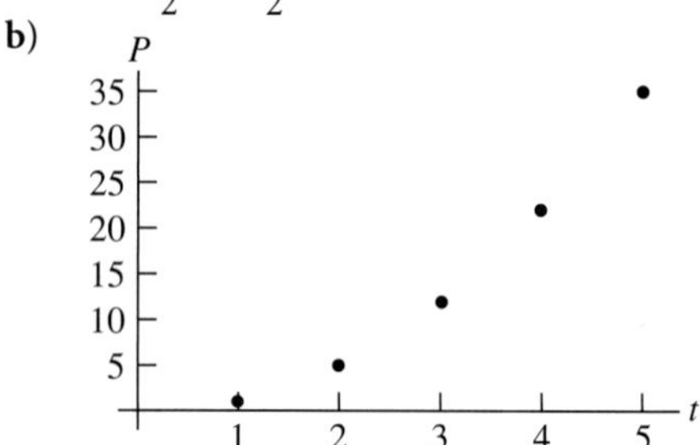

c) 14,950
Its shape consists of a pentagon with 100 dots on a side and 98 smaller pentagons contained inside the pentagon.

9. a)

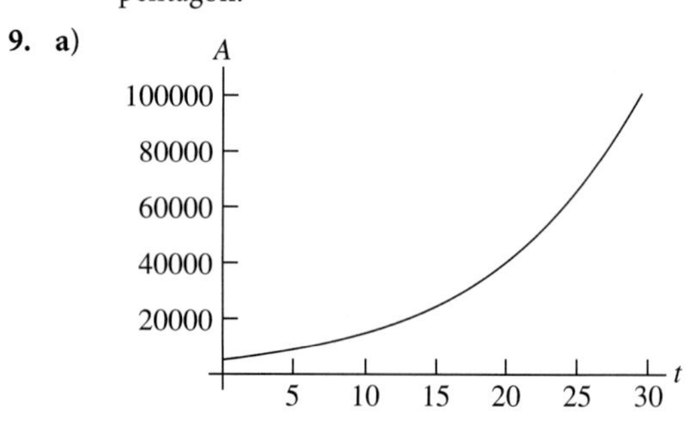

b) \$9,663.06, \$15,562.45, \$104,696.41

10. a)

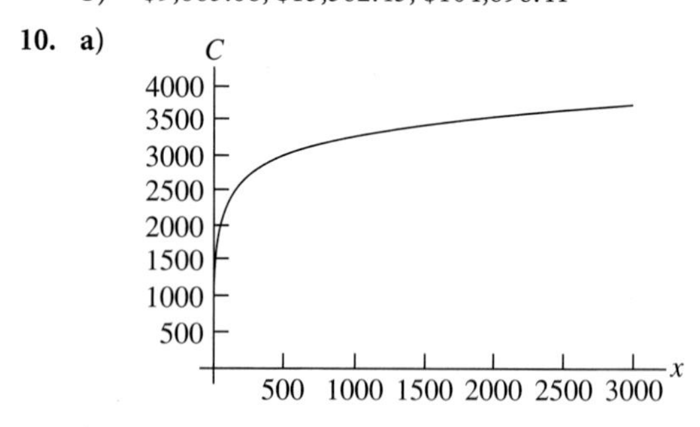

b) \$3,263.10, \$3,540.36

c) \$3.26; \$1.77

11. a) $y = 27.2 - 8.19 \ln t$ and in 2004 ($t = 7$), $y = 11.3$.

b) $y = 0.5e^{0.9435t}$; In 2004 ($t = 6$), $y = 143.7$ million DVDs.

12. a) $A = 3^s$, where s = the square on the Monopoly board and A = amount of money.

b) \$19,683; no

c) No, more than \$282,000,000,000 is needed.

d) \$12,157,665,459,056,928,801; 303,941,636 years

CHAPTER 7

Section 7.1

15.

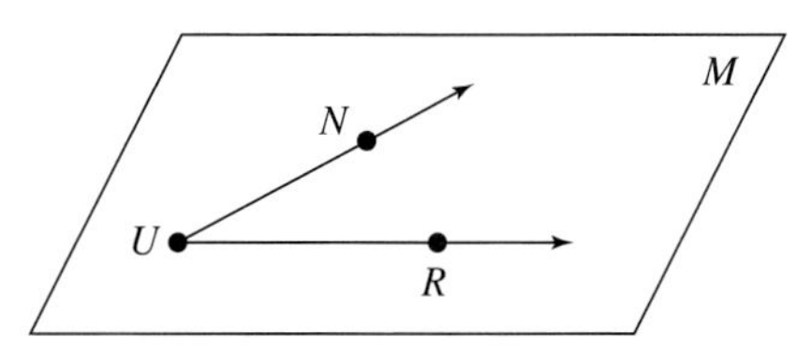

17.

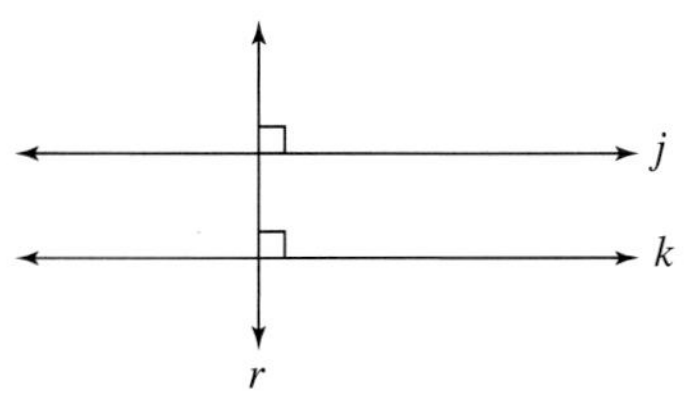

19.

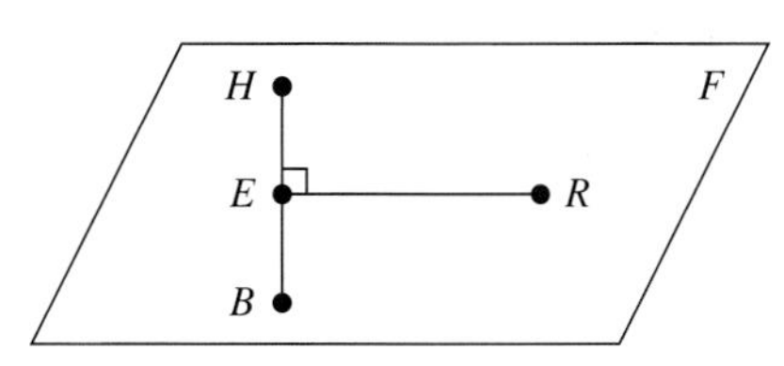

21.

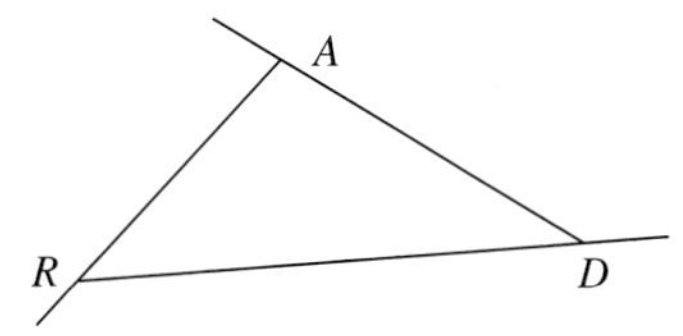

23. *Hypothesis:* m intersects n
Conclusion: m intersects n in only one point Q

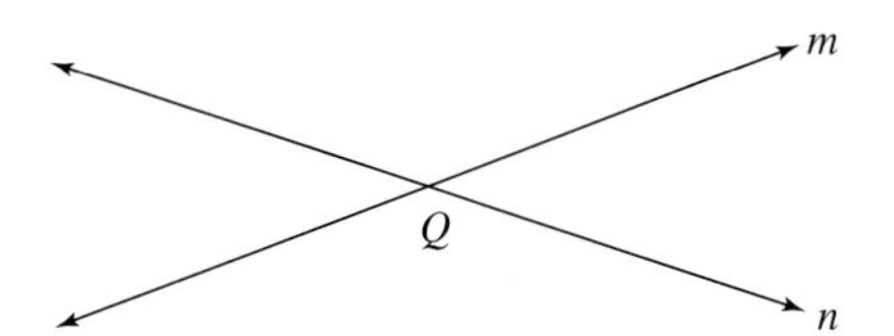

25. *Hypothesis:* $\angle 1$ is an exterior angle of ΔABC
Conclusion: $\angle 1 > \angle A$ and $\angle 1 > \angle C$

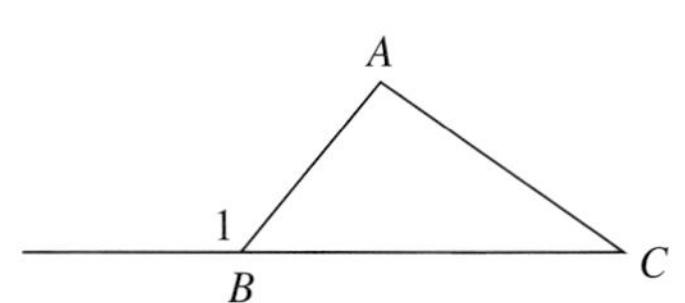

27.

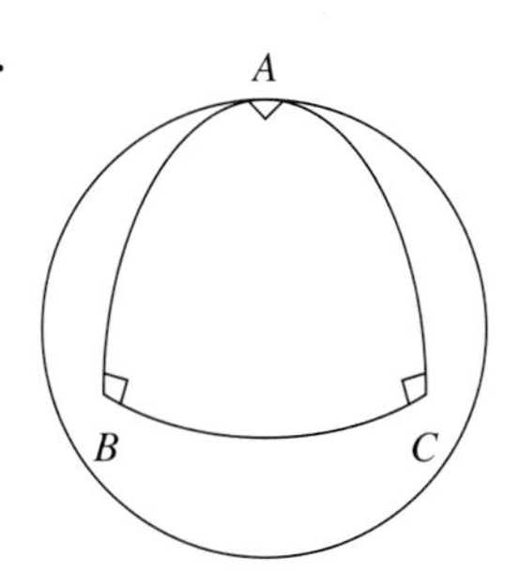

29.

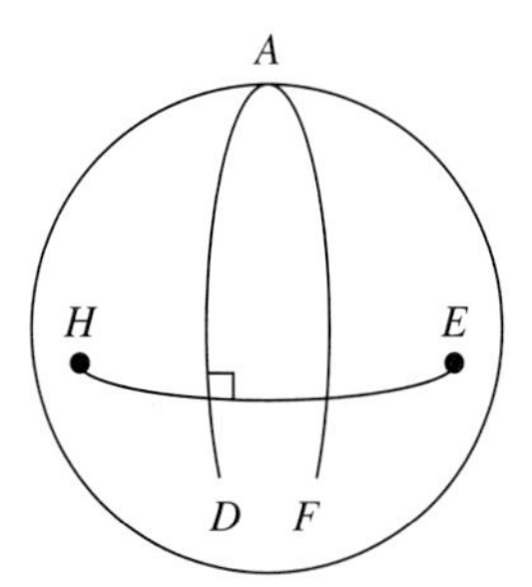

31.

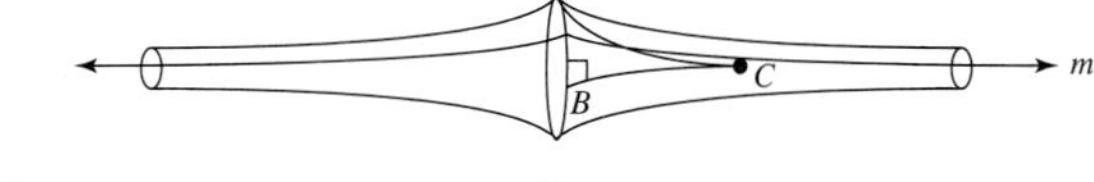

33.

35. It is not a good definition, since it does not distinguish a point from all other objects; for example, a moment in time also has no parts.

37. A possible sequence of definitions from a dictionary could be dimension $\rightarrow$ extent $\rightarrow$ length $\rightarrow$ dimension.

39.

$$
\begin{aligned}
5 + 6x &= 4x - 11 \\
5 + (-5) + 6x &= 4x - 11 + (-5) && \text{(A3)} \\
6x &= 4x - 16 \\
6x - 4x &= 4x - 4x - 16 && \text{(A3)} \\
2x &= -16 \\
x &= -8 && \text{(A4)}
\end{aligned}
$$

41. If a line intersects two parallel lines, the pairs of alternate interior angles have equal measures.

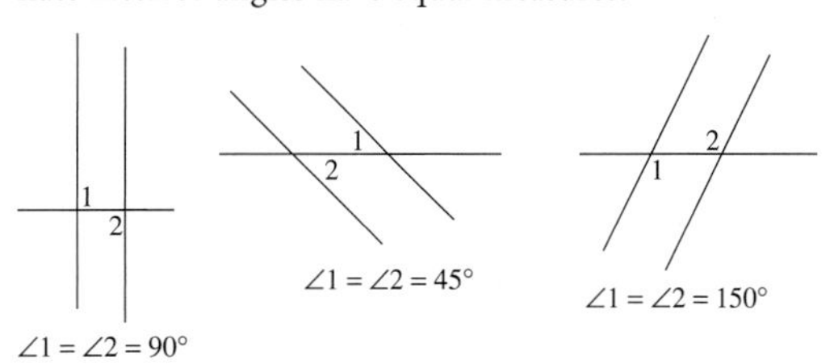

In all three cases, the alternating angles are equal.

43. Mathematics does not establish truths about the physical world. It can give only logical descriptions of the physical world based on beginning assumptions (postulates).

45. If you examine the set of parallel lines on the Lobachevskian model that follows, you will observe that $a \parallel b$, but as you move away from the center of the pseudosphere, the lines get closer together and are not the same distance apart.

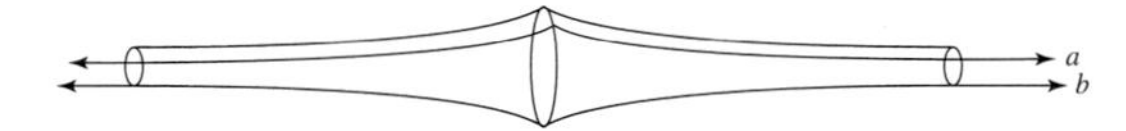

47. The fourth angle ($\angle A$) is obtuse.

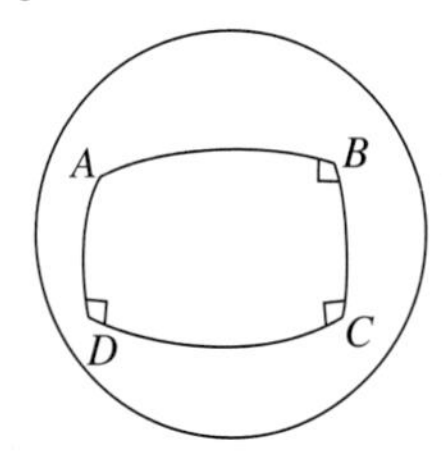

49. By drawing the diagonal of a four-sided plane figure, you get two triangles. Since in Lobachevskian geometry the sum of the angles of a triangle is less than 180°, the sum of the angles of the two triangles would be less than 360°. But a rectangle has four right angles, and the sum of its angles is exactly 360°. Thus, the Lobachevskian four-sided plane figure cannot have four right angles. Therefore, rectangles as defined do not exist in Lobachevskian geometry.

51. In equiangular ΔABC, $\angle A = \angle B = \angle C = x°$. In Lobachevskian geometry,

$$\angle A + \angle B + \angle C < 180°$$
$$x + x + x < 180°$$
$$3x < 180°$$
$$x < 60°$$

Thus, each angle is less than 60°.

53. In equiangular ΔABC, $\angle A = \angle B = \angle C = x°$. In Riemannian geometry,

$$\angle A + \angle B + \angle C > 180°$$
$$x + x + x > 180°$$
$$3x > 180°$$
$$x > 60°$$

Thus, each angle is greater than 60°.

55. In Riemannian geometry, the sum of the angles is greater than 180°. Thus, it is possible for two of the three angles to be equal to 90°.

Section 7.2

9. a)

b)

c)

11.

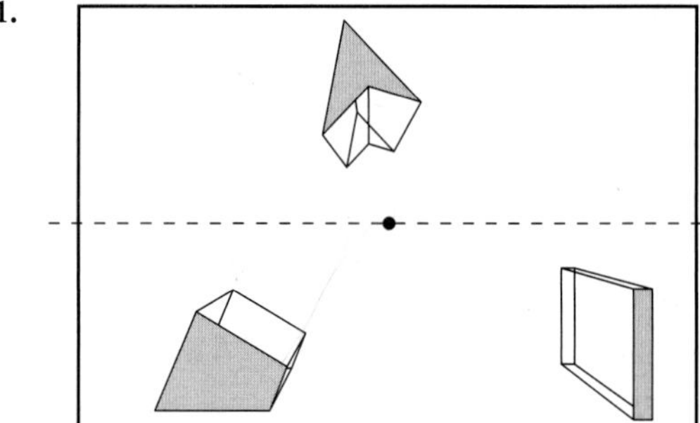

13.

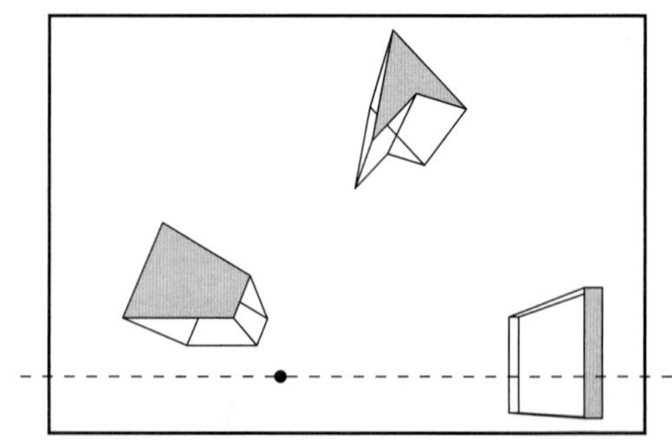

15. Use the four-step process described in Section 7.2 for equally spaced objects.

17.

19. $d = 12.5, c = 7.5$

21. $b \approx 2.2, c \approx 20$

23. $d = 15, b = 6$

25.

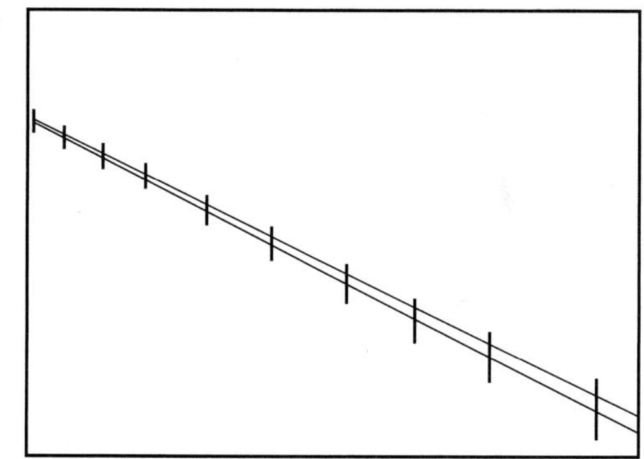

27. Overlapping shapes

29. One-point perspective

31. None of the methods is used. This is an example of a "flat painting."

33. 13 1/3 in.

35. The tracks seem to get closer together. This suggests a one-point perspective.

37. Answer depends on what work of art is examined.

Section 7.3

9. The point is 2.6 cm from one end of the segment.

11. The point is 3.8 cm from one end of the segment.

13. No

15. No

17. 37.3 ft, 14.2 ft

19. 92.0 m, 35.1 m

21. 14.6 cm, 5.6 cm

23. 58.3 in., 22.2 in.

25. $w = 9.9$ cm, $b = 9.9$ cm, $t = 6.1$ cm

27. Answers vary.

29. $w = 8.1$ in., $l = 13.1$ in.

31. $w = 5.6$ cm, $l = 14.6$ cm

33. $w = 14.6$ cm, $l = 23.7$ cm

35. Requires construction of a box with $h = 3$ in., $w = 4.9$ in., and $l = 7.9$ in.

37. The ratios of consecutive Fibonacci numbers approximate the Golden Ratio. Thus, three consecutive Fibonacci numbers give approximate Golden boxes.

39. Requires measurement of cans.

41. Requires a survey of ten people using different parallelograms.

43. $\dfrac{360}{3 + \sqrt{5}}$ or $90(3 - \sqrt{5})$

Section 7.4

21. $S = 540°$

a)

b)

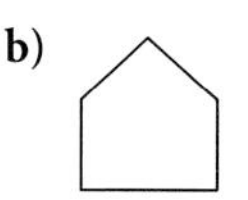

23. $S = 1980°$

a)

b)

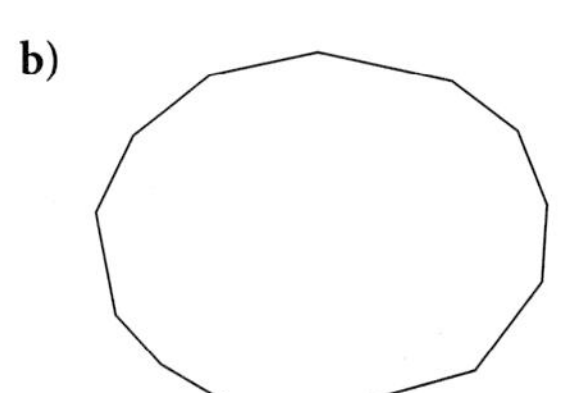

25. $S = 1440°$

27. Since the central angle is determined by dividing 360° by the number of points desired, we get the following:

Points	Angle
5	72°
6	60°
7	≈51.4°
8	45°
9	40°
10	36°
11	≈32.7°
12	30°
13	≈27.7°
14	≈25.7°
15	24°
16	22.5°
18	20°
20	18°
30	12°
36	10°

29. $A = 135°$

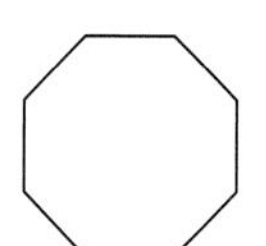

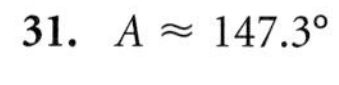

31. $A \approx 147.3°$

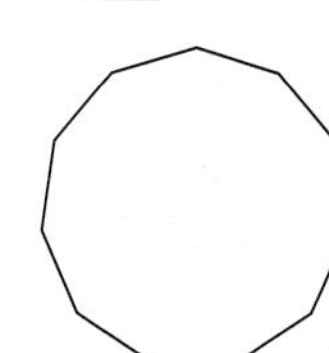

33. $A = 162°$

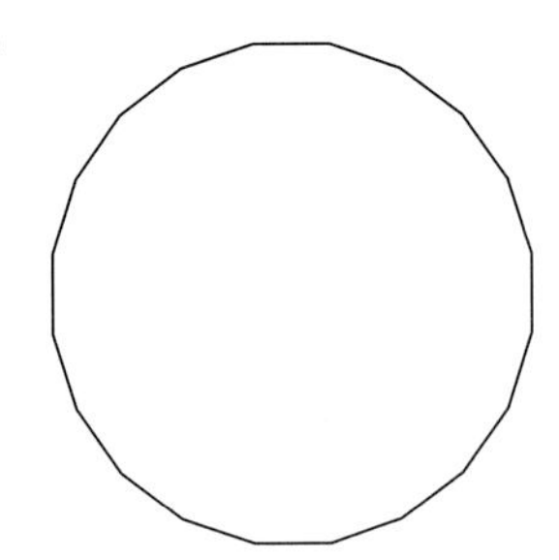

35. $A \approx 154.3°$

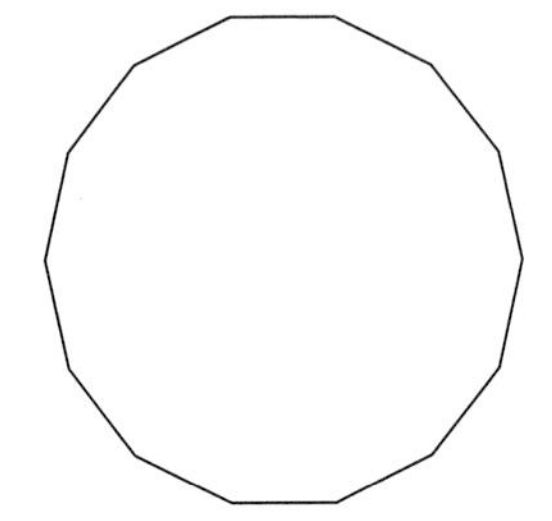

37.

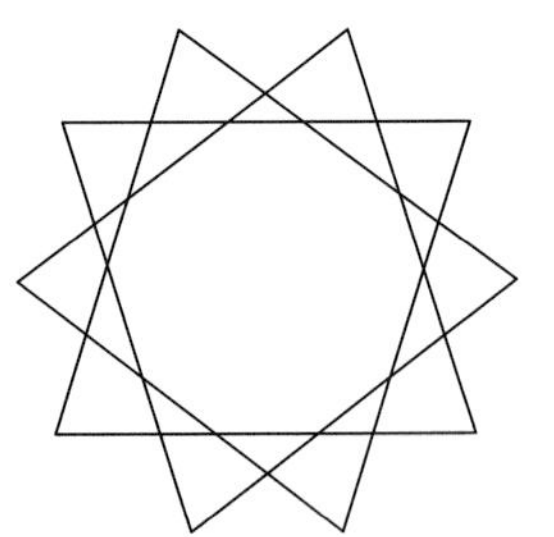

39.

41.

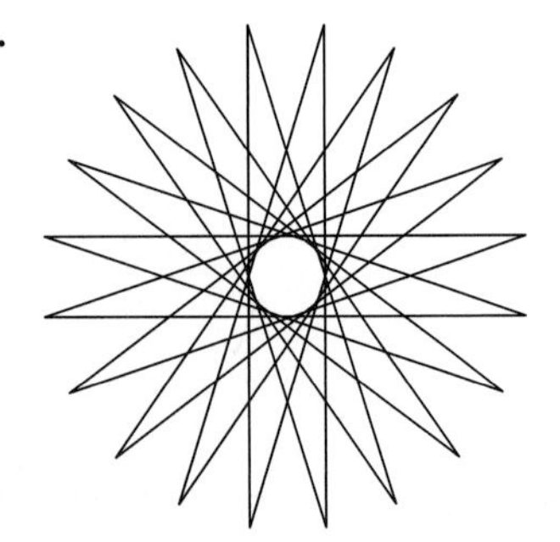

Section 7.5

11. A regular heptagon has angles of 128 4/7°. Since 360° is not evenly divisible by 128 4/7°, regular heptagons will not tessellate the plane.

13. A regular decagon has angles of 144°. Since 360° is not evenly divisible by 144°, a regular decagon will not tessellate the plane.

In the answers to Problems 15–25, each diagram is approximately 40% of actual size and shows four to seven tiles.

15.

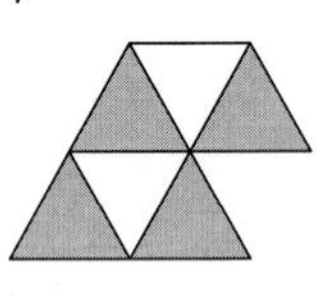

17.

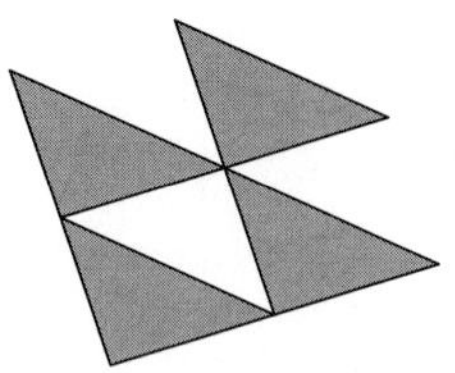

19.

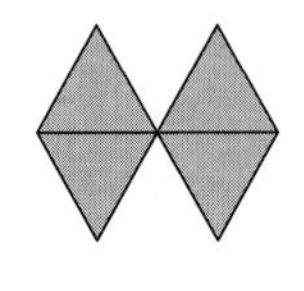

21.

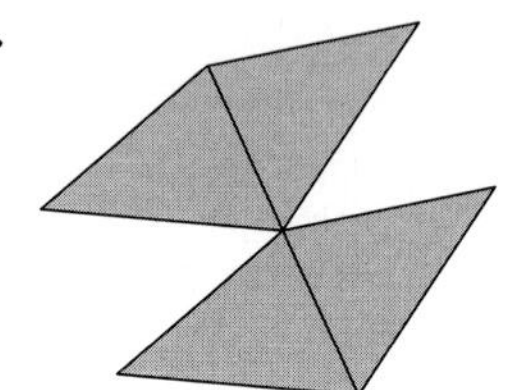

23.

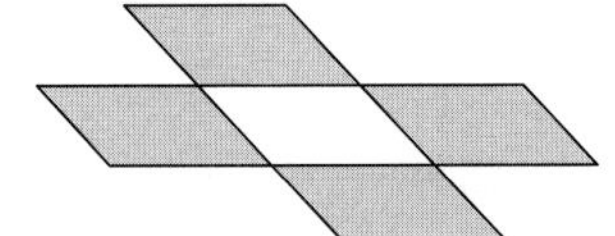

25.

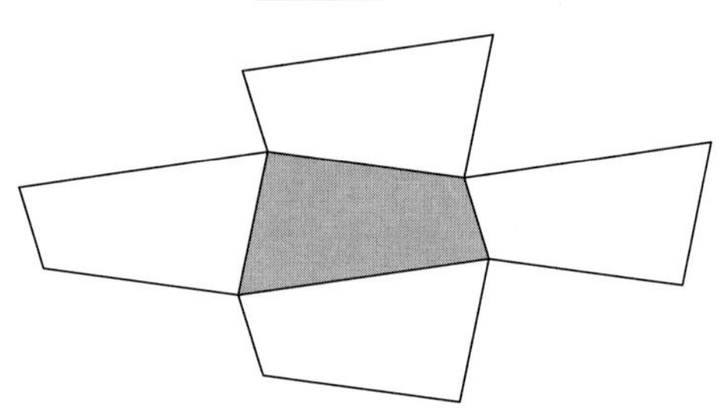

27. The sum of the angles for these three polygons does not add to 360°.

29.

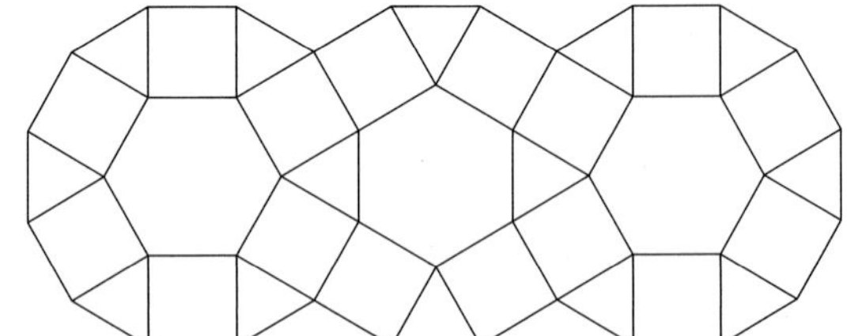

31. Two possible answers are to use two octagons and a square or two pentagons and a decagon.

33. Answers vary.

35. b)

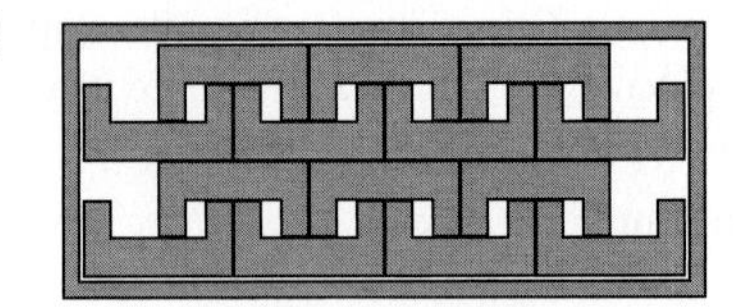

Section 7.6

9.

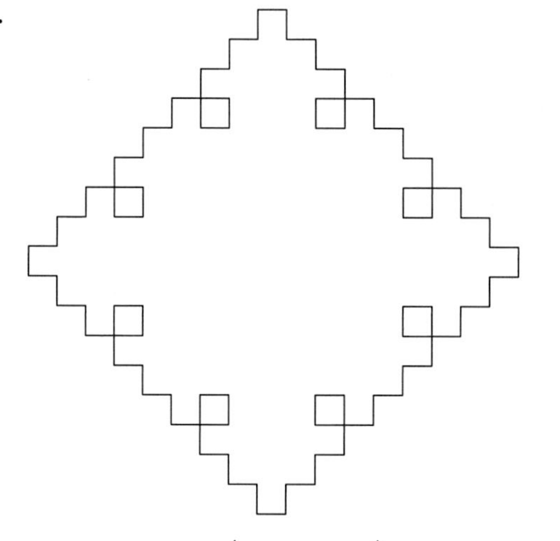

11.

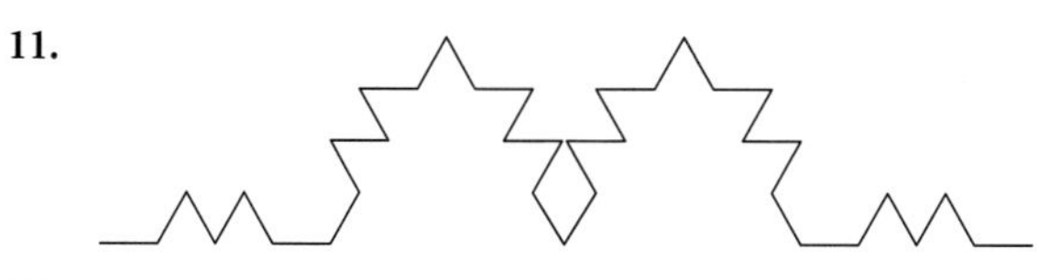

13.

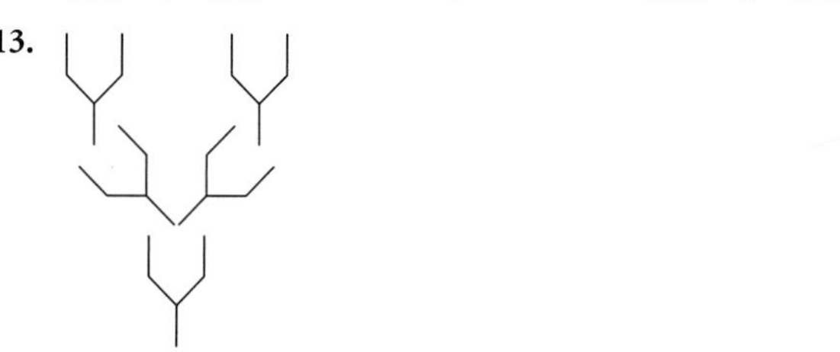

15. Answers vary.

17. 1.46 **19.** 1.29 **21.** 1.46 **23.** 0.68

CHAPTER 7 REVIEW

Review Section 7.1

1. It is necessary to have undefined terms and postulates in a deductive system of geometry because there must be something to start the system. Not all terms can be defined nor can all propositions be deduced.

2.
$$\begin{aligned} 6x + 5 &= 2x - 3 \\ 6x - 2x + 5 &= 2x - 2x - 3 \quad \text{(A3)} \\ 4x + 5 &= -3 \\ 4x + 5 - 5 &= -3 - 5 \quad \text{(A3)} \\ 4x &= -8 \\ x &= -2 \quad \text{(A4)} \end{aligned}$$

3. Euclidean

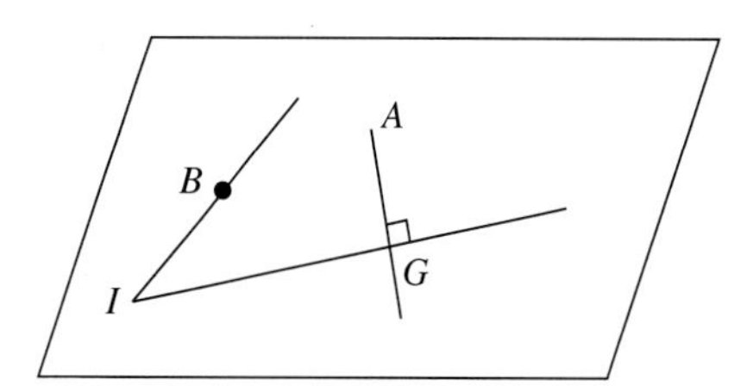

Lobachevskian

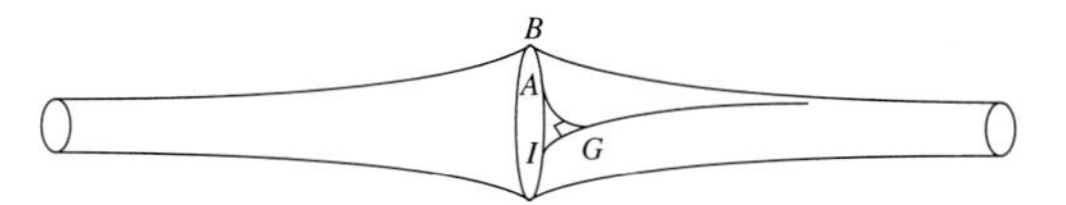

Riemannian

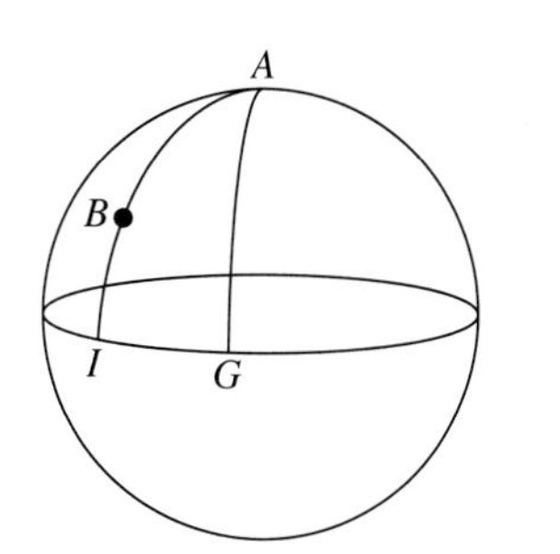

4. Euclidean

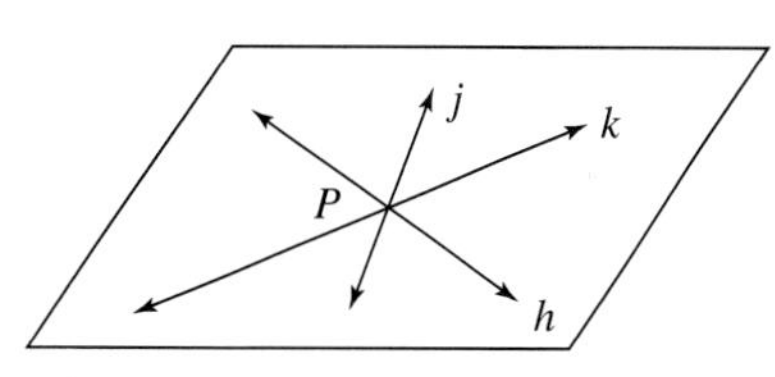

Lobachevskian

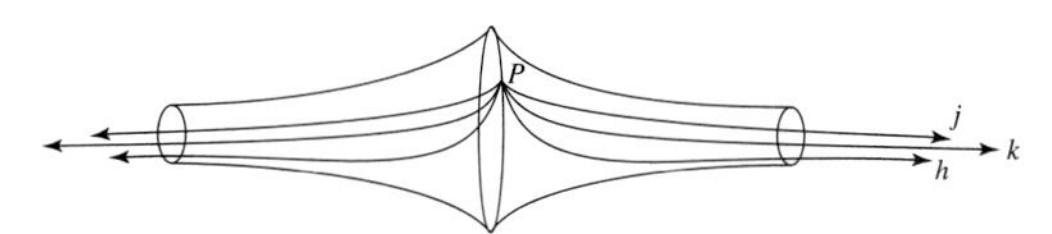

Riemannian

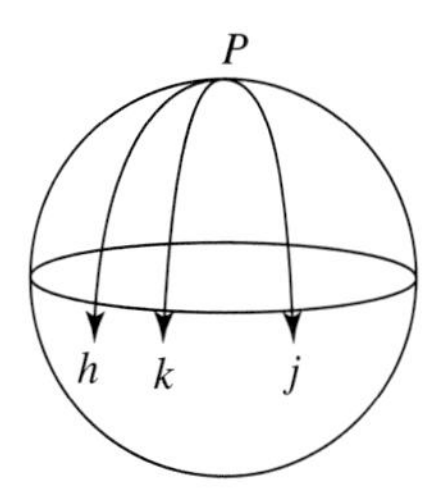

5. A square has an angle sum of 360° (four right angles). However, in Lobachevskian geometry, the angle sum of a quadrilateral is less than 360° and in Riemannian geometry, the sum is greater than 360°. Thus, squares do not exist.

Review Section 7.2

6. The artist uses overlapping shapes, diminishing sizes, and atmospheric perspective in a one-point perspective to create the inside of the building.
7. None of the techniques used in representing a three-dimensional scene on a two-dimensional canvas is used in the wall painting.
8. 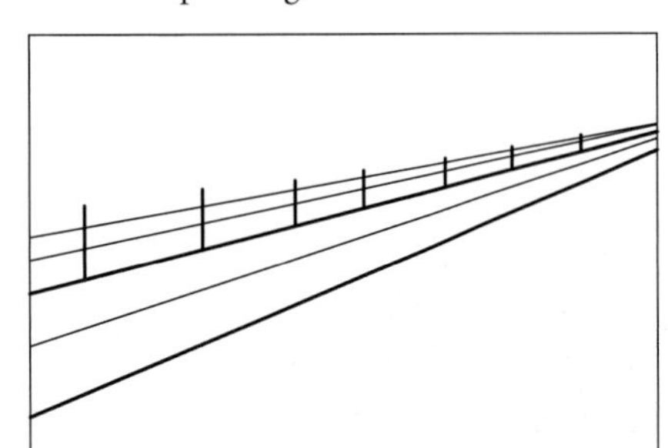
9. 1.8 in.

Review Section 7.3

10. The Greeks believed that if the ratio of distances equaled the Golden Ratio, it was pleasing to the eye. The Golden Ratio was used in Greek art and construction.
11. 13.6 ft from one end of the stage
12. 87.5″, 33.3″
13. Either the ratio of the diagonals or the ratio of the sides should be 1.62.

Review Section 7.4

14. 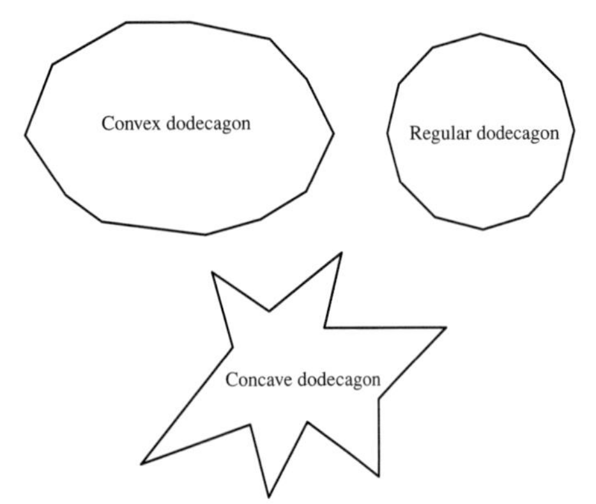

15. 1800°
16. 150°
17. 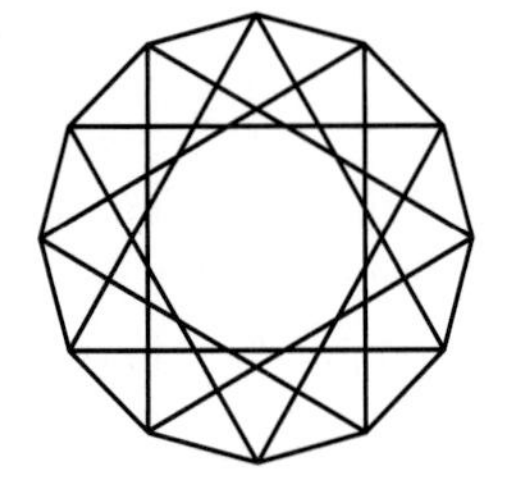

Review Section 7.5

18. From Problem 16, each angle of a regular dodecagon is 150°. No combination of 150° angles will add up to 360°. Therefore, regular dodecagons cannot tessellate a region.
19.
20. a)

b) 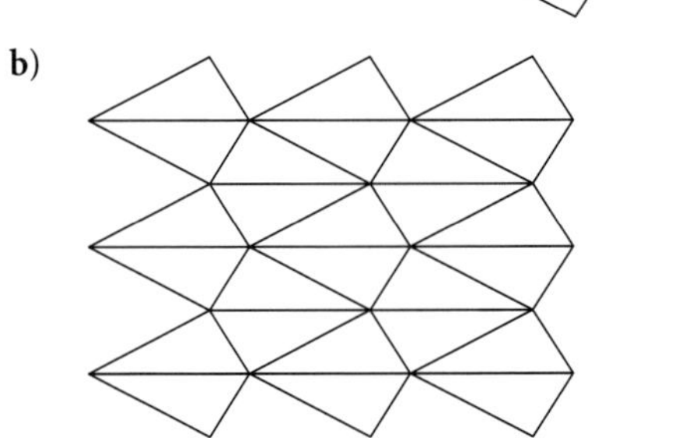

21. a)

b) 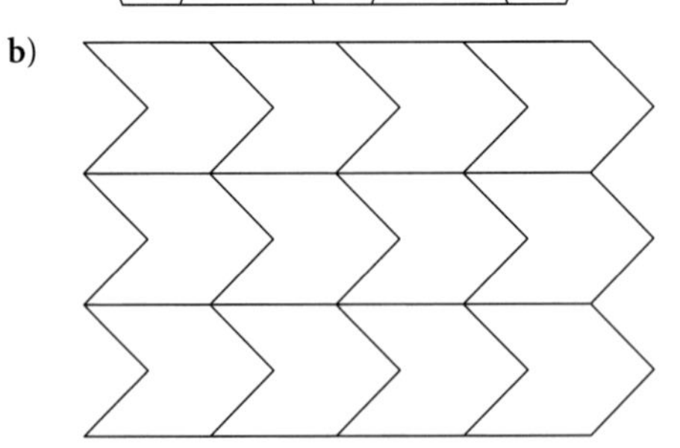

Review Section 7.6

22.
23. 1.77

24. A dimension of 1.62 indicates that objects exist between one-dimensional and two-dimensional space. Objects in this space are closer to two-dimensional objects than one-dimensional objects. This dimension would be "special," since 1.62 is the Golden Ratio. These objects might be very visually pleasing.

CHAPTER 7 TEST

1. a) Undefined terms are terms whose meanings are assumed to be known. Definitions are descriptions of terms using other known terms or undefined terms.

b) Axioms are the assumptions of algebra. Postulates are the assumptions of geometry. Theorems are propositions that can be proven using postulates, axioms, definitions, and previously proven theorems.

2. Euclidean

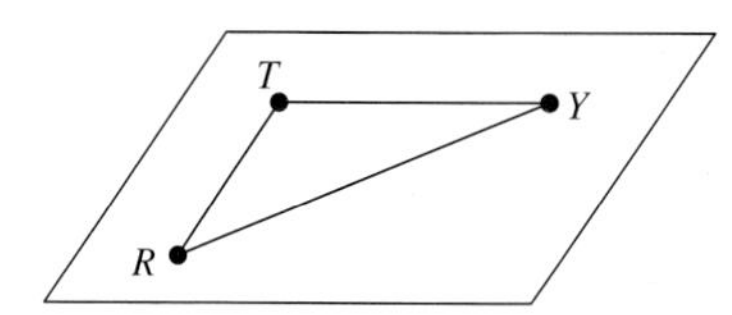

Lobachevskian

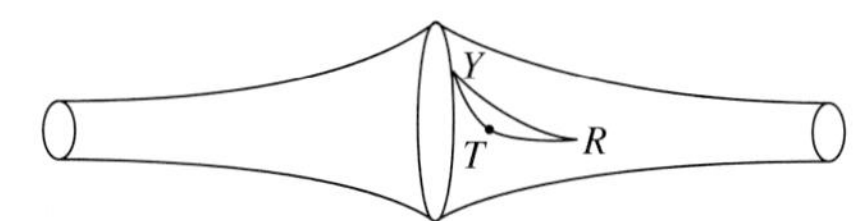

Riemannian

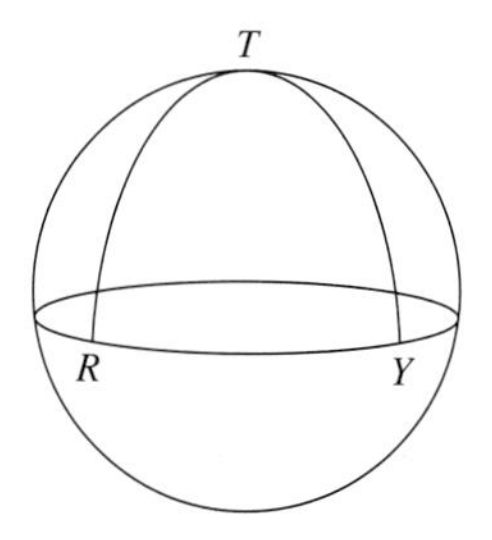

3. Mathematics does not give truth about the real world. It can give only logical descriptions of the real world based on beginning assumptions (postulates).

4. a)

b)

c)

5.

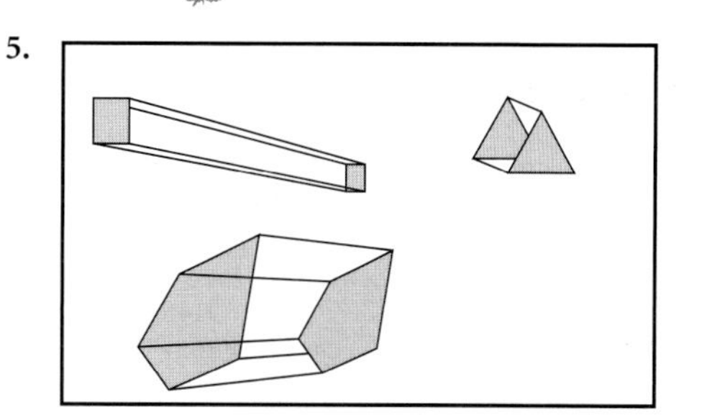

6. The point is 1.1 in. from one end of the segment.

7. 13.8 in.

8. a)

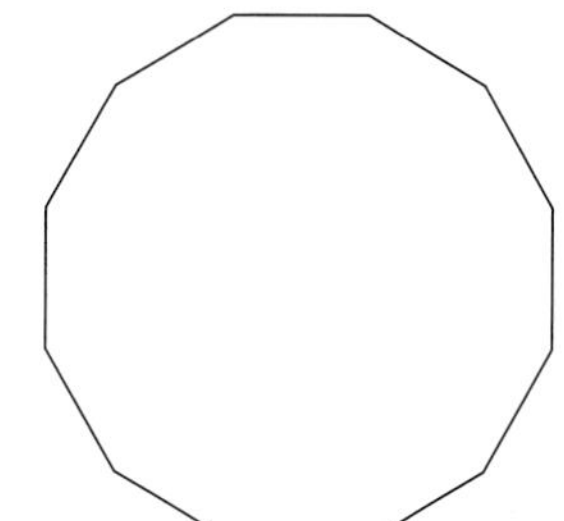

b)

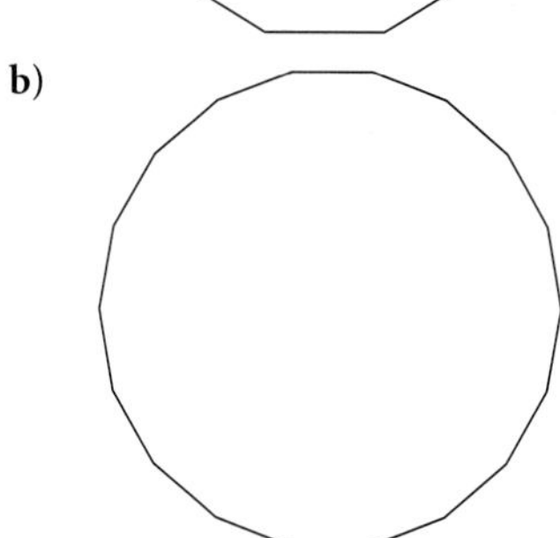

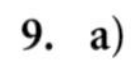

9. a)

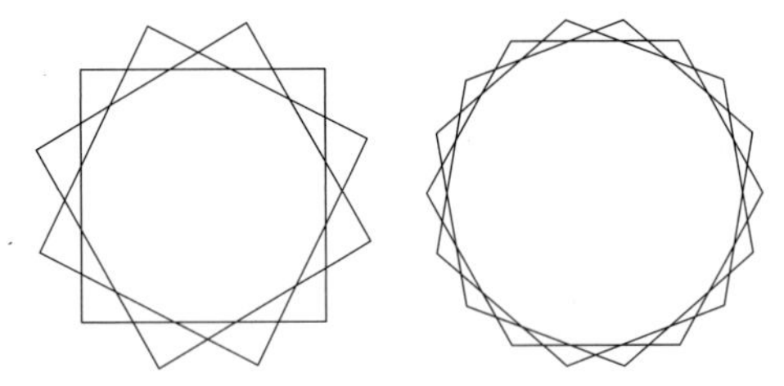

b)

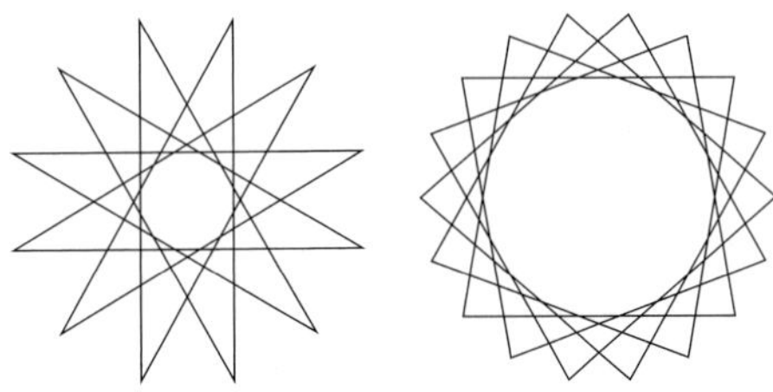

10. a)

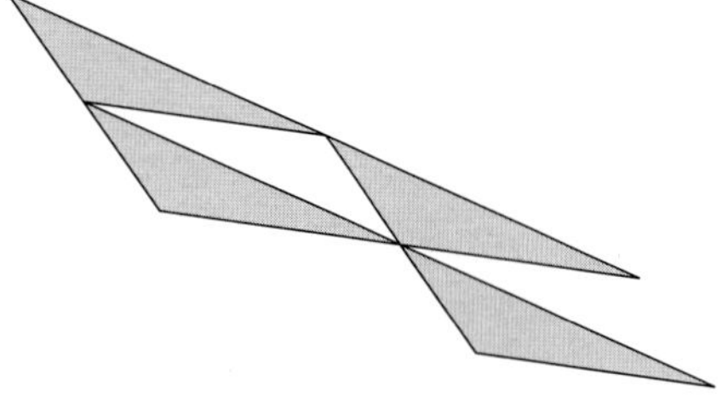

b)

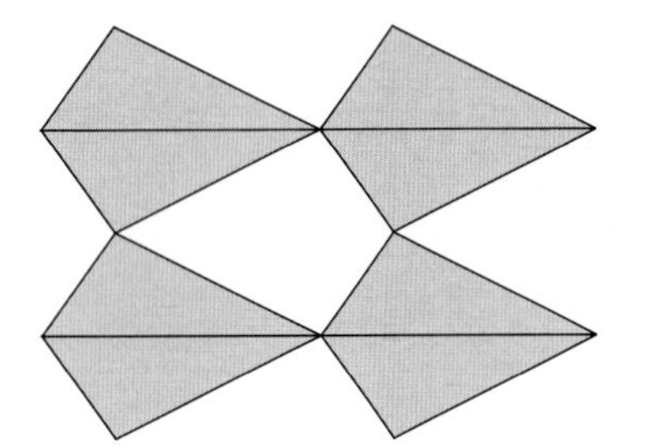

11. a)

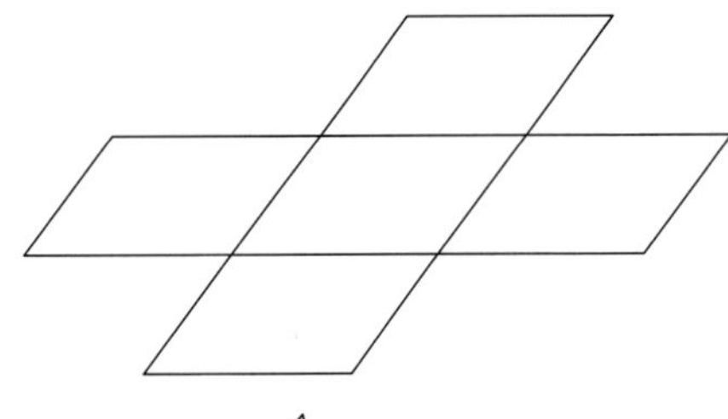

b)

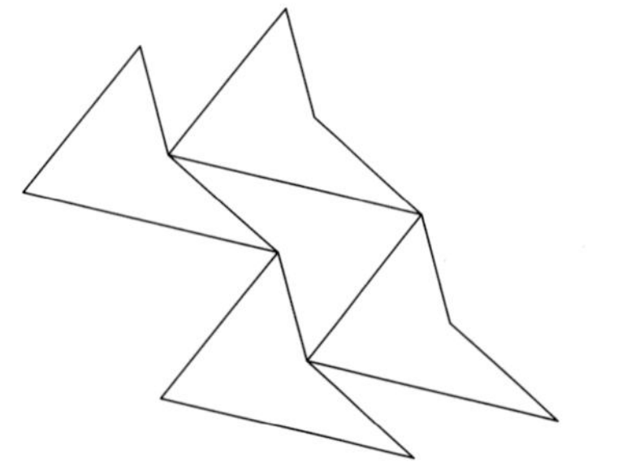

12. For a regular polygon to tessellate a region, the sum of the angles around any vertex must equal 360°. The angles of a square (90°), a regular hexagon (120°), and a regular dodecagon (150°) have a sum of 360°. Thus, they can tessellate a plane. The angles of an equilateral triangle (60°), a regular pentagon (108°), and a regular heptagon (128 4/7°) have a sum of 296 4/7°. Thus, they cannot tessellate a plane.

13.

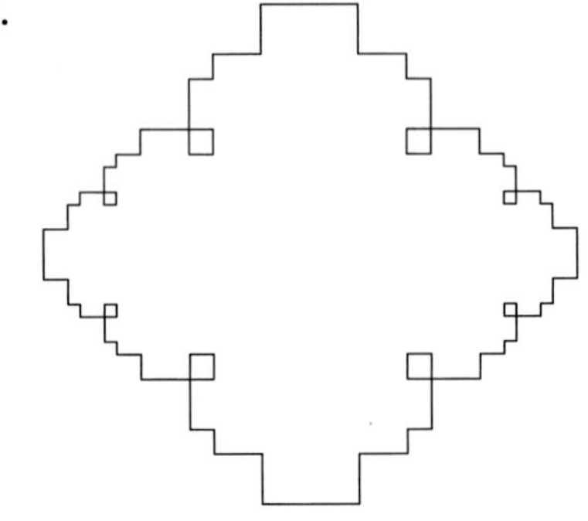

14. 1.29

CHAPTER 8

Section 8.1

11. legs: k, d; hypotenuse: y

13. (10) opposite: y; adjacent: d

(11) opposite: d; adjacent: k

(12) opposite: k; adjacent: y

15. 10.6 **17.** 67.2

19. $\sin A = 3/5$ $\sin B = 4/5$

$\cos A = 4/5$ $\cos B = 3/5$

$\tan A = 3/4$ $\tan B = 4/3$

21. $\sin A = 4/5$ $\sin B = 3/5$

$\cos A = 3/5$ $\cos B = 4/5$

$\tan A = 4/3$ $\tan B = 3/4$

23. 0.2181 **25.** 0.9947

27. 4.1022 **29.** 82.9°

31. 24.6° **33.** 73.7°

35. $36^2 + 48^2 = 60^2$

37. $33^2 + 56^2 = 65^2$

39. $5^2 + 12^2 = 13^2$

41. 36.9°, 53.1° **43.** 43.6°, 46.4°

45. 71.1°, 18.9°

47. $AC = 10$, $AD = 12.2$, $\angle DAC = 35.0°$

49. $AC = 7.6$, $AD = 9.2$, $\angle DAC = 34.4°$

Section 8.2

9. $\angle C = 53.1°$, $\angle A = 36.9°$, $b = 6.5$
11. $\angle A = 41°$, $b = 13.6$, $a = 11.8$
13. $\angle X = 73.3°$, $y = 39.0$, $z = 135.7$
15. $\angle X = 75.8°$, $\angle Y = 14.2°$, $z = 1004.6$
17. $\angle T = 2.2°$, $\angle P = 87.8°$, $p = 2600$
19. $\angle T = 74.6°$, $i = 783.1$, $t = 755.3$
21. $\angle P = 67°$, $i = 48.5$, $t = 18.9$
23. $h = 35.4$
25. $h = 92.4$
27. $x = 14.5$

Section 8.3

7. 2.6 mi
9. 231.8 m
11. pole: 50.0 ft, wire: 51.4 ft
13. 29.5 ft
15. 114.9 ft
17. 2192.9 ft
19. 65.3 sec
21. 14 ft, 12.1°
23. 29.5 ft

Section 8.4

7. $n = 26.1$
9. $u = 407.5$
11. $u = 41.0$
13. $n = 272.3$
15. $d = 32.6$
17. $g = 4.1$
19. $\angle O = 60.9°$
21. $\angle C = 78.6°$, $t = 23.0$, $a = 27.4$
23. $\angle C = 66.2°$, $\angle U = 41.7°$, $\angle P = 72.1°$
25. $e = 42.3$, $\angle G = 30.9°$, $\angle T = 81.6°$
27. $\angle P = 44.9°$, $\angle E = 54.2°$, $\angle T = 80.9°$
29. $AD = 42.9$ and $RD = 47.3$.
31. Impossible

Section 8.5

7. 460.0 ft
9. 27.2 ft
11. 33.0 ft
13. 1603.5 m
15. 438.6 mi
17. 12.0 ft
19. 49.6°, 44.5°, 85.9°
21. Methods may vary.
23. 1494.9 ft

Section 8.6

7. Initial: $x = 0$, $y = 58$ At $t = 5$, $x = 181.2$, $y = 334.1$.

9.

t	x	y
0	0	58
2	72.5	264.5
4	144.9	342.9
6	217.4	293.4
8	289.9	115.8

11. 1878.0 ft
13. 55.4 ft
15. 3.9 sec, 242.0 ft
17. 3.8 sec, 317.3 ft
19. Yes. The ball travels 523.5 ft in the horizontal direction.
21. 2.2 sec, 32.0 mph
23. 479.4 ft

CHAPTER 8 REVIEW

Review Section 8.1

1. 53 in.
2. 18.4 in.
4. a) 0.9205 b) 0.9759
5. a) 13.6° b) 88.4°

Review Section 8.2

6. $a = 35$, $\angle T = 36.9°$, $\angle X = 53.1°$
7. $\angle U = 67°$, $u = 86.9$, $s = 94.4$
8. $\angle P = 16°$, $u = 98.5$, $p = 28.2$
9. 90°, 43.6°, 46.4°

Review Section 8.3

11. 117.8 ft
12. 4979.0 ft

Review Section 8.4

13. $\angle A = 83°$, $a = 7.2$, $s = 4.2$
14. $\angle M = 46°$, $\angle T = 46°$, $a = 19.5$
15. $\angle P = 80.7$, $c = 44.4$, $a = 99.8$
16. $\angle N = 65.0°$, $\angle A = 31.2°$, $\angle P = 83.8°$

Review Section 8.5

17. 3.7 mi
18. 213.0 ft
19. 22.2 mi, 13.7 mi, 12.1 mi
20. 2950.8 yd, 2987.6 yd

Review Section 8.6

21. At $t = 0, x = 0, y = 10$.
At $t = 4, x = 201.9, y = 42.3$.

22. 67.6 ft **23.** 44.0 ft/sec

CHAPTER 8 TEST

1. $b = 51, \angle A = 28.1°, \angle C = 61.9°$

2. $\angle N = 84°, r = 20.3, n = 29.0$

3. $a = 38.0, \angle M = 73.1°, \angle Y = 44.4°$

4. $a = 45.7$ **5.** $\angle C = 84.9°$ **6.** 41.8°

7. a) 17.0 ft, 176.2 ft **b)** 48.0 ft, 178.9 ft

8. 851.8 mi **9.** N11.2°E, N48.2°W

10. 319.0 ft **11.** 46.9 ft/sec

CHAPTER 9

Section 9.1

11. a) 0.05 **b)** 0.067 **c)** 0.09125 **d)** 2.34 **e)** 0.0003

13. a) 35% **b)** 6% **c)** 0.25% **d)** 201.5% **e)** 75%

15. a) 71.4% **b)** 66.7% **c)** 55.2% **d)** 0.034 = 3.4% **e)** 7.8%

17. a) $2.38 **b)** $8.82 **c)** $47.31

19. a) $20.59 **b)** $29.17 **c)** $70.00

21. a) $27.00 **b)** $61.56 **c)** $148.49

23. a) $13.60 **b)** $20.40 **c)** $23.99

25. a) 13.4% **b)** 49.9% **c)** 40.0%

27. $381.13

29. a) $77.00 **b)** $8.80

31. $23,225

33. a) Mitek lost 21.1% **b)** Yotek lost 27.5%

35. a) $357 **b)** $889.55 **c)** $1873.75

Section 9.2

7. $400.00 **9.** $56.25 **11.** $3720.00

13. $35,466.67 **15.** $1090.16 **17.** $2500

19. 2.5 years **21.** 25% per year **23.** $3542.80

25. 1.16 years **27.** 3.98% per year

29. 8.11% per year

31. 6% per year **33.** 47.1 years

35. a) $1025.00 **b)** $1050.63 **c)** $1076.90

37. $10.58

Section 9.3

9. $2540.98 **11.** $6970.93 **13.** $5323.11

15. $5568.53 **17.** 3.98 years **19.** 14.13 years

21. 19.52% **23.** 7.35% **25.** 6.17%

27. 13.86%

29. 7.8%, compounded daily

31. $20,808.41

33. a) $562.72 **b)** $594.39 **c)** $627.83

35. $9580.50 **37.** 47.4 total years

39. 5.78 years **41.** $11,628,024,100

43. $20.57 **45.** $10.60

Section 9.4

5. $13,954.01 **7.** $34,320.46

9. $409.11 **11.** $432.86

13. 123 months or 10.22 years

15. 205 months or 17.08 years

17. $192,646.74

19. a) $973,151.26 **b)** $57,600.00
c) $915,551.26 **d)** $778,181.07
e) $91,200.00 **f)** $686,981.07

21. a) $77,405.71 **b)** $1,166,691.60

23. 18 years **25.** $2088.45

27. a) $139.74 **b)** $47,504.66

Section 9.5

7. $10,345.11 **9.** $4087.86

11. $609.11 **13.** $3019.28

15. 186 monthly payments or 15.5 years

17. 68 quarters or 17 years **19.** $1204.88

21. a) $215.41 **b)** $10,339.68 **c)** $2339.68

23. a) $135.38
b) 23 months or 1 year and 11 months.
c) $748.24

25. $160,625.11

27. a) $1367.60, $1048.82
b) $246,168.00, $377,575.20
c) The first loan

29. a) The first year costs of the loan are estimated at $13,681.63.
One year of renting costs 1650 × 12 = $19.800.
b) The first year costs of the loan are estimated at $11,041.05.

CHAPTER 9 REVIEW

Review Section 9.1

1. a) 96.875% b) 0.3125
2. $233.33
3. $31,252.25
4. Bill: 6.45%, Jill: 5.68%.

Review Section 9.2

5. a) $2504 b) $1597.44
6. $12,400
7. 3.76%
8. $511.25

Review Section 9.3

9. a) $2570.93 b) $1555.85
10. 4378 days ≈ 12 years
11. 10.4%
12. 6.17%
13. $1254.58
14. $2366.51
15. $7.63

Review Section 9.4

16. $8156.20
17. $220.69
18. 17 years, 3 months
19. $304.28
20. 32 years, 7 months

Review Section 9.5

21. $6344.93
22. $283.69
23. 12 years, 7 months
24. 20 years, 6 months
25. a) $395.14 b) $17,225.04 c) $2225.04
26. $4543.75
27. The first year costs of the loan are estimated at $11,958.95
One year of renting costs 1500 × 12 = $18,000.

CHAPTER 9 TEST

1. $19.93
2. a) $1806.25 b) $1750
3. 33.33%
4. $1,669,090.91
5. 7.2%
6. $1104.71
7. $5846.10
8. a) $24,160.79 b) $8,160.79 c) $175,036.37
9. 27.25 years
10. About 18.85 years
11. $99,516.29
12. 6.03%
13. 6 years 3 months
14. The first year costs of the loan are estimated at $14,926.04.
One year of renting costs 1800 × 12 = $21,600.
15. $55.36

CHAPTER 10

Section 10.1

7. a) B, D, F b) A, E c) C, G
9. Between 0.5 and 2
11. 2
13. $-2x$
15. $9x^2 - 5$
17. $4x^3 + 3x^2 + 2x + 1$
19. $20x^4$
21. $20x^4 - 10x$
23. $5x^4 - 6x + 7$
25. No minimums or maximums

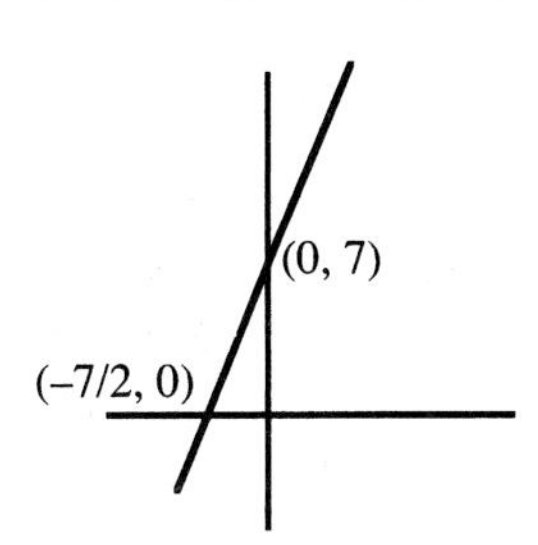

27. Maximum at (2, 4)

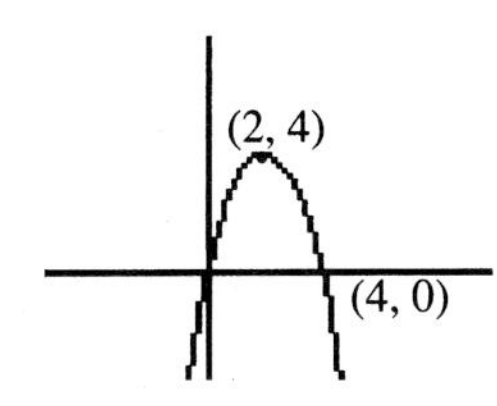

29. Maximum at (−1, 4); minimum at (1, −4)

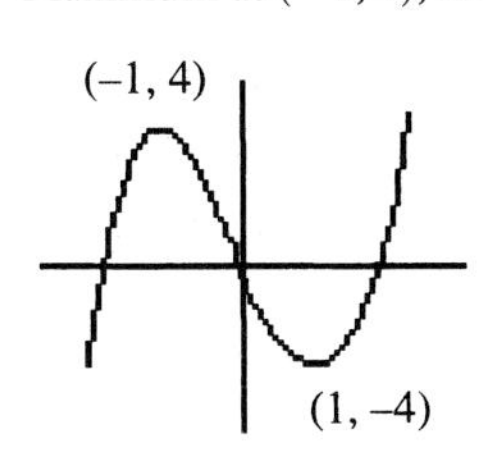

31. a) 42 in. b) 18 in.
33. 150 ft by 300 ft; area = 45,000 sq ft
35. 2 ft; 4 ft; $2\frac{2}{3}$ ft, $V = \frac{64}{3}$ cu ft
37. a) 20,000 b) $16.20

Section 10.2

5. 144
7. 96
9. 125/3
11. 288
13. 9750 sq ft
15. a) b) 1.35
17. a) 125.66 sq cm b) 124.56 sq cm

Section 10.3

13. $1 + 7x + 21x^2 + 35x^3 + 35x^4 + 21x^5 + 7x^6 + x^7$

15. $1 + 6x + 21x^2 + 50x^3 + 90x^4 + 126x^5 + 141x^6 + 126x^7 + 90x^8 + 50x^9 + 21x^{10} + 6x^{11} + x^{12}$

17. $1 + 4x + 10x^2 + 20x^3 + 31x^4 + 40x^5 + 44x^6 + 40x^7 + 31x^8 + 20x^9 + 10x^{10} + 4x^{11} + x^{12}$

19. 9 **21.** 15 **23.** 31

25. a) 1, 3, 6, 10, 15, 18, 19, 18, 15, 10, 6, 3, 1
1, 4, 10, 20, 35, 52, 68, 80, 85, 80, 68, 52, 35, 20, 10, 4, 1
1, 5, 15, 35, 70, 121, 185, 255, 320, 365, 381, 365, 320, 255, 185, 121, 70, 35, 15, 5, 1

Each entry in successive rows is obtained by adding five items from the previous row.

b) $1 + 3x + 6x^2 + 10x^3 + 15x^4 + 18x^5 + 19x^6 + 18x^7 + 15x^8 + 10x^9 + 6x^{10} + 3x^{11} + x^{12}$

c) $1 + 4x + 10x^2 + 20x^3 + 35x^4 + 52x^5 + 68x^6 + 80x^7 + 85x^8 + 80x^9 + 68x^{10} + 52x^{11} + 35x^{12} + 20x^{13} + 10x^{14} + 4x^{15} + x^{16}$

d) 80

e) 320

Section 10.4

5.

	Ann	Brenda	Cathy	Dawn	Elle	Francesca	Gloria
The Time Machine	1	2	1	2	4	4	2
The Mysterious Island	4	4	3	1	1	3	3
The Secret Garden	2	3	4	3	3	2	1
Wind in the Willows	3	1	2	4	2	1	4

Plurality
The Time Machine has 2 first place votes.
The Mysterious Island has 2 first place votes.
The Secret Garden has 1 first place vote.
The Wind in the Willows has 2 first place votes.
Therefore, *The Time Machine, The Mysterious Island,* and *The Wind in the Willows* are tied.

Plurality with Elimination
Removing *The Secret Garden* gives

	Ann	Brenda	Cathy	Dawn	Elle	Francesca	Gloria
The Time Machine	1	2	1	2	3	3	1
The Mysterious Island	3	3	3	1	1	2	2
Wind in the Willows	2	1	2	3	2	1	3

This results in *The Time Machine* winning with three first place votes.

Borda Count

	Points
The Time Machine	$(2 \times 4) + (3 \times 3) + (0 \times 2) + (2 \times 1) = 19$
The Mysterious Island	$(2 \times 4) + (0 \times 3) + (3 \times 2) + (2 \times 1) = 16$
The Secret Garden	$(1 \times 4) + (2 \times 3) + (3 \times 2) + (1 \times 1) = 17$
Wind in the Willows	$(2 \times 4) + (2 \times 3) + (1 \times 2) + (2 \times 1) = 18$

Therefore, *The Time Machine* wins.

Pairwise Comparison

Comparison	Winner
Time v. *Mysterious*	*Time*
Time v. *Secret*	*Time*
Time v. *Wind*	*Time*
Mysterious v. *Secret*	*Secret*
Mysterious v. *Wind*	*Wind*
Secret v. *Wind*	*Wind*

Therefore, *The Time Machine* wins.

7.

	1200	200	1000	1500	100
1st place	*A*	*A*	*B*	*C*	*B*
2nd place	*B*	*C*	*A*	*B*	*C*
3rd place	*C*	*B*	*C*	*A*	*A*

Plurality: A wins with 1400.
Plurality with Elimination: A wins with 1400.

Borda:

Candidate	Points
A	$(1400 \times 3) + (1000 \times 2) + (1600 \times 1) = 7800$
B	$(1100 \times 3) + (2700 \times 2) + (200 \times 1) = 8900$
C	$(1500 \times 3) + (300 \times 2) + (2200 \times 1) = 7300$

B wins.

Pairwise

Comparison	Winner
A v. *B*	*B* wins 2600 to 1400
A v. *C*	*A* wins 2400 to 1600
B v. *C*	*B* wins 2300 to 1700

Therefore, *B* wins.
Therefore, there is no clear winner.

9.

	13	10	8	7	2
Increase sales taxes	1	5	2	3	5
Increase income taxes	5	1	3	2	4
Decrease school funding	2	2	5	4	5
Decrease transportation funding	3	4	1	5	2
Decrease social service funding	4	3	2	1	1

The legislators agreed to use the Borda count method with the following weighting: first choice: 5 points; second choice 4 points, etc.

	Points
Sales tax	$(13 \times 5) + (8 \times 4) + (7 \times 3) + (0 \times 2) + (12 \times 1) = 130$
Income tax	$(10 \times 5) + (7 \times 4) + (8 \times 3) + (2 \times 2) + (13 \times 1) = 119$
School	$(0 \times 5) + (23 \times 4) + (0 \times 3) + (7 \times 2) + (10 \times 1) = 116$
Transportation	$(8 \times 5) + (2 \times 4) + (13 \times 3) + (10 \times 2) + (7 \times 1) = 114$
Social services	$(9 \times 5) + (8 \times 4) + (10 \times 3) + (13 \times 2) + (0 \times 1) = 133$

First: Decrease social service funding.
Second: Increase sales taxes.
Third: Increase income taxes.
Fourth: Decrease school funding.
Fifth: Decrease transportation funding.

11. First choice: 5 points; second choice: 5 points; third choice: 3 points; fourth choice: 2 points; and last choice: 1 point.

	Points
Sales tax	(13 × 6) + (8 × 6) + (7 × 3) + (0 × 2) + (12 × 1) = 159
Income tax	(10 × 6) + (7 × 6) + (8 × 3) + (2 × 2) + (13 × 1) = 143
School	(0 × 6) + (23 × 6) + (0 × 3) + (7 × 2) + (10 × 1) = 162
Transportation	(8 × 6) + (2 × 6) + (13 × 3) + (10 × 2) + (7 × 1) = 126
Social services	(9 × 6) + (8 × 6) + (10 × 3) + (13 × 2) + (0 × 1) = 158

First: Decrease school funding.
Second: Increase sales taxes.
Third: Decrease social service funding.
Fourth: Increase income taxes.
Fifth: Decrease transportation funding.

Section 10.5

9. Number of Seats Using Hamilton's Method

County	Population	Quota	Initial # of Seats	# of Seats
Brooke	48,859	4.87	4	5
Hopkins	161,135	16.06	16	16
Isaac	87,194	8.69	8	9
Haley	596,270	59.42	59	59
Wallace	110,006	10.96	10	11
Totals	1,003,464	100.00	97	100

11. Number of Seats Using Jefferson's Method

County	Population	Divisor = 10,035 Quota	# of Seats	Divisor = 9774 Pop/Divisor	# of Seats
Brooke	48,859	4.87	4	4.999	4
Hopkins	161,135	16.06	16	16.49	16
Isaac	87,194	8.69	8	8.92	8
Haley	596,270	59.42	59	61.01	61
Wallace	110,006	10.96	10	11.25	11
Totals	1,003,464	100.00	97		100

13. Number of Seats Using Webster's Method

County	Population	Divisor = 16,248 Quota	# of Seats
Carlisle	110,993	6.83	7
Newport	441,946	27.20	27
Sonoma	113,229	6.97	7
Totals	666,168	41.00	41

15. **Number of Courses Using Hamilton's Method**

Course	# Student	Quota	Initial # of Courses	# Courses
College Algebra	120	4.00	4	4
Calculus	80	2.67	2	3
Math for Business	235	7.83	7	8
Liberal Arts Math	165	5.50	5	5
Totals	600	20.00	18	20

17. **Number of Courses Using Jefferson's Method**

		Divisior = 30		Divisor = 27	
Course	# Students	Quota	# of Courses	New quota	# of Seats
College Algebra	120	4.00	4	4.44	4
Calculus	80	2.67	2	2.96	2
Math for Business	235	7.83	7	8.70	8
Liberal Arts Math	165	5.50	5	6.11	6
Totals	600	20.00	18		20

19. **Number of Trucks Using Webster's Method**

		Divisor = 544	
District	Population	Quota	# of Trucks
Bayview	25,456	46.80	47
Downtown	32,723	60.16	60
East End	27,568	50.69	51
Greenhaven	16,475	30.29	30
Ingleside	8,696	15.99	16
Riverview	11,458	21.07	21
Totals	122,376	225.00	225

21. **Number of Sites Using Hamilton's Method**

District	Population	Quota	Initial # of Sites	# of Sites
1	90,000	90.91	90	91
2	166,000	167.68	167	168
3	102,000	103.03	103	103
4	131,000	132.32	132	132
5	147,000	148.48	148	148
6	156,000	157.58	157	158
Totals	792,000	800.00	797	800

23. **Number of Sites Using Jefferson's Method**

District	Population	Pop/Divisor	# of Sites
1	90,000	91.19	91
2	166,000	168.19	168
3	102,000	103.34	103
4	131,000	132.73	132
5	147,000	148.94	148
6	156,000	158.05	158
Totals	792,000		800

25. **Number of Days Using Webster's Method**

District	Voters	Voters/Divisor	# of Days
1	167,000	26.02	26
2	116,000	18.07	18
3	237,000	36.93	37
4	231,000	35.99	36
5	215,000	33.50	33
Totals	966,000		150

Section 10.6

7. Maximum $P = 75$ at $(10, 9)$

9. Minimum $P = 27$ at $(0, 9)$

11. Maximum $P = 288$ at $(14, 12)$

13. Maximum $P = 116$ at $(9, 14)$; minimum $P = 25$ at $(2, 3)$

15. Maximum $P = 40$ at $(0, 8)$

17. Maximum $P = 88$ at $(0, 8)$

19. Minimum $C = 50$ at $(6, 4)$

21. Minimum $C = 95$ at $(5, 5)$

23. Minimum $P = 16$ at $(8, 0)$; maximum $P = 56$ at $(4, 12)$

25. 60 grams of mix A and 100 grams of mix B gives 4.8 grams of fat.

27. Use 400 pounds of each mix. Minimum cost is \$6000.

29. 18 mountain bikes and 2 touring bikes. Maximum $P = \$1460$.

CREDITS

This page constitutes an extension of the copyright page. We have made every effort to trace the ownership of all copyrighted material and to secure permission from copyright holders. In the event of any question arising as to the use of any material, we will be pleased to make the necessary corrections in future printings. Thanks are due to the following authors, artists, publishers, and agents for permission to use the material included.

Chapter 1

Page 1	© Art Resource, NY
Page 6	Egyptian Expedition of The Metropolitan Museum of Art, Rogers Fund, 1930. (30.4.44) Photograph ©1979 The Metropolitan Museum of Art
Page 18	The Metropolitan Museum of Art, Alfred Stieglitz Collection, 1949. (49.59.1) Photograph ©1986 The Metropolitan Museum of Art
Page 31	© Lester Lefkowitz/CORBIS
Page 39	Upper left, *o*, © David McLaughlin
Page 39	Upper right, Tom Marioni *Pi*, 1988 Woodcut printed in red on silk mounted on rag paper, 22 1/2″ × 23 1/4″, edition 30. Published by Crown Point Press
Page 39	Bottom left, *e*, © David McLaughlin
Page 39	Bottom right, *i*, © David McLaughlin

Chapter 2

Page 53	©Royalty-Free/CORBIS
Page 85	Rube Goldberg is the © of Rube Goldberg Inc.
Page 93	Magazine cover copyright © Dell Magazines, a division of Crosstown Publications. Used with permission.

Chapter 3

Page 109	© David McLaughlin
Page 110	The Metropolitan Museum of Art, Bequest of Stephen C. Clark, 1960. (61.101.1) Photograph ©1982 The Metropolitan Museum of Art
Page 113	© Philip Jaeger/CORBIS
Page 121	Henri de Toulouse-Lautrec, *The Jockey* (*Le Jockey*), Rosenwald Collection, Image © 2003 Board of Trustees, National Gallery of Art, Washington

Chapter 4

Page 151	Edouard Manet, *At the Races*, Widener Collection, Image © 2003 Board of Trustees, National Gallery of Art, Washington
Page 152	Lucas van Leyden, *The Card Players*, Samuel H. Kress Collection, Image © 2003 Board of Trustees, National Gallery of Art, Washington
Page 153	PEANUTS reprinted by permission of United Feature Syndicate, Inc.
Page 163	Courtesy of John Berggruen Gallery, San Francisco, CA

Chapter 5

Page 176	© Hilary Price. Reprinted with special permission of King Features Syndicate.
Page 211	© Forest J. Ackerman/CORBIS
Page 215	© Royalty-Free/CORBIS
Page 227	© Bob Daemmrich/CORBIS
Page 245	Peanuts reprinted by permission of United Feature Syndicate, Inc.
Page 257	© Bettmann/CORBIS
Page 264	FOXTROT © 2002 Bill Amend. Reprinted with permission of UNIVERSAL PRESS SYNDICATE. All rights reserved.

Chapter 6

Page 279 Joseph Stella, *The Brooklyn Bridge: Variation on an Old Theme, 1939.* Oil on canvas, 70 × 42 inc. (177.8 × 106.68 cm. Purchase 42.15. Photography by Geoffrey Clements. Whitney Museum of American Art, New York
Page 284 © Eric and David Hosking/CORBIS
Page 294 Left, © Royalty-Free/CORBIS
Page 294 Right, © Philip James Corwin/CORBIS
Page 316 CLOSE TO HOME © 1999 John McPherson. Reprinted with permission of UNIVERSAL PRESS SYNDICATE. All rights reserved.

Chapter 7

Page 333 Kristin Shinoda
Page 338 Courtesy of John Berggruen Gallery, San Francisco, CA
Page 352 Mary Cassatt, *The Letter,* Chester Dale Collection, Image © 2003 Board of Trustees, National Gallery of Art, Washington
Page 353 André Derain, *The Old Bridge,* Chester Dale Collection, Image © 2003 Board of Trustees, National Gallery of Art, Washington
Page 354 Martin Johnson Heade, *Cattleya Orchid and Three Brazilian Hummingbirds,* Gift of the Morris and Gwendolyn Cafritz Foundation, Image © 2003 Board of Trustees, National Gallery of Art, Washington
Page 357 Vincent van Gogh, *Flower Beds in Holland,* Collection of Mr. and Mrs. Paul Mellon, Image © 2003 Board of Trustees, National Gallery of Art, Washington
Page 360 Canaletto, *The Square of St. Mark's, Venice,* Gift of Mrs. Barbara Hutton, Image © 2003 Board of Trustees, National Gallery of Art, Washington
Page 361 © David McLaughlin
Page 365 Grant Wood, *New Road,* Gift of Mr. and Mrs. Irwin Strasburger, Image © 2003 Board of Trustees, National Gallery of Art, Washington. © Estate of Grant Wood/Licensed by VAGA, New York, NY
Page 366 Henri de Toulouse-Lautrec, *Quadrille at the Moulin Rouge,* Chester Dale Collection, Image © 2003 Board of Trustees, National Gallery of Art, Washington
Page 366 Marco Ricci, *Capriccio of Roman Ruins,* Ailsa Mellon Bruce Fund, Image © 2003 Board of Trustees, National Gallery of Art, Washington
Page 367 Hieronymus Bosch, *Death and the Miser,* Samuel H. Kress Collection, Image © 2003 Board of Trustees, National Gallery of Art, Washington
Page 367 © Wolfgang Kaehler/CORBIS
Page 373 © Lester Lefkowitz/CORBIS
Page 378 © Roman Soumar/CORBIS
Page 378 © Jim Sugar Photography/CORBIS
Page 390 © Anthony Cooper; Ecoscene/CORBIS
Page 390 © Royalty-Free/CORBIS
Page 402 © Craig Tuttle/CORBIS
Page 402 © Pat O'Hara/CORBIS
Page 410 Bob Bradshaw
Page 410 Bob Bradshaw
Page 411 Bob Bradshaw
Page 411 Bob Bradshaw
Page 412 Top, © Bill Ross/CORBIS
Page 412 Bottom, © Digital Art/CORBIS

Chapter 8

Page 423 Raoul Dufy, *Regatta at Cowes,* Ailsa Mellon Bruce Collection, Image © 2003 Board of Trustees, National Gallery of Art, Washington
Page 434 Georges Seurat, *The Lighthouse at Honfleur,* Collection of Mr. and Mrs. Paul Mellon, Image © 2003 Board of Trustees, National Gallery of Art, Washington
Page 441 Winslow Homer, *Eight Bells,* Gift of John W. Beatty, Jr., Image © 2003 Board of Trustees, National Gallery of Art, Washington
Page 458 © Keith Dannemiller/CORBIS SABA
Page 467 © Bettmann/CORBIS

Chapter 9

Page 479	© Erich Lessing/Art Resource, NY
Page 481	Giovanni Paolo Panini, *Interior of the Pantheon, Rome,* Samuel H. Kress Collection, Image © 2003 Board of Trustees, National Gallery of Art, Washington
Page 484	Jan Gossaert, *Portrait of a Merchant,* Ailsa Mellon Bruce Fund, Image © 2003 Board of Trustees, National Gallery of Art, Washington
Page 511	© Images.com/CORBIS

Chapter 10

Page 525	Gloria Guy
Page 539	Giovanni Paola Panini, *Interior of St Peter's, Rome,* Ailsa Mellon Bruce, Fund, Image © 2003 Board of Trustees, National Gallery of Art, Washington
Page 556	Digital Image © The Museum of Modern Art/Licensed by SCALA/ Art Resource, NY
Page 565	© Richard Nowitz/CORBIS
Page 574	Frank & Ernest reprinted by permission of Newspaper Enterprise Association, Inc.

INDEX

μ, 226
π, 45–46
~, 75
′, 20
$\vee$, 75
$\wedge$, 75
$\equiv$, 75
$\rightarrow$, 75
$\therefore$, 75

A

AAS. *See* Angle-angle-side
Abundant number, 44
Accelerating loan payments, 513–514
Achenwall, Gottfried, 212
Acute angle, 340
Acute triangle, 449, 460–462
Addition
ancient numeration systems using, 6–11
on computer, 35–37
Affirmative, 41
Ahmes, 280–281
Ahmes papyrus, 425
Al-Battâni, 425
Alberti, Leon, 336
Algebra
exponential models, 304–316
history of, 280–281
linear models, 284–294
logarithmic models, 316–325
quadratic models, 294–304
Al-Khwârizmi, 280, 281
Alphametric puzzles, 96–97
American Standard Code for Information Interchange. *See* ASCII
Amicable numbers, 44
Anaxagoras, 335
Ancient numeration systems
Attic Greek, 7
Babylonian, 11–12
Chinese, 8–9
Egyptian hieroglyphic, 6–7
Ionic Greek, 10–11
Mayan, 12–13
Roman, 7–8
And
compound events, probability of, 176–177
sets, translating, 123
symbol, 75
And statement, 78
Angle
Euclidean geometry, 340
of a polygon, sum of, 382
triangle, solving when one is known, 452–455
Angle-angle-side (AAS), 451
Angles, fifth postulate, 336
Angle-side-angle (ASA), 451
Annuities, 506–510
Antecedent, 58
Antecedent, affirming the, 68–69
Antiderivatives, 542–544
Apportionment
assumptions, 565–566
divisor methods, 567–568
Hamilton's method, 566, 568–570
Hill-Huntington method, 571
Jefferson's method, 567–568
rounding method, 566
Webster's method, 568

Aquinas, Thomas, 480–481
Arabia, 41–42
Archimedes of Syracuse, 45
Area, finding
  irregular shapes, 544–545
  rectangles, 533–535
Aristotle, 54, 67, 480
Arithmetic mean, 226–227
Ars Magna, 41
Art, featured
  The Brooklyn Bridge (Joseph Stella), 279
  The Card Players (Lucas van Leyden), 152
  The Card Players (Paul Cezanne), 110
  Cattleya Orchid and Three Brazilian Hummingbirds (Martin Johnson Heade), 354
  Death and the Miser (Hieronymus Bosch), 367
  Delphic Vision (David McLaughlin), 361
  Duet (Adolph Gottlieb), 113
  Eight Bells (Winslow Homer), 441
  Fissures in Sun-Baked Clay (Anthony Cooper), 390
  Flower Beds in Holland (Vincent van Gogh), 357
  Global Pathways (Lester Lefkowitz), 31
  Golden Venn (David McLaughlin), 109
  Horseshoe (Ron Davis), 163
  The Interior of the Pantheon, Rome (Giovanni Paolo Panini), 480
  I Saw the Figure 5 in Gold (Charles Henry Demuth), 18
  Le Penseur (Tami Whitt-Zenoble), 53
  The Letter (Mary Cassatt), 352
  Lighthouse at Honfleur (Georges Seurat), 434
  Mathematica (Bob Bradshaw), 548
  A modern representation of Van Gogh's Starry Night (Gloria Guy), 525
  O, e, i (David McLaughlin), 39
  The Old Bridge (André Derain), 353
  Portrait of a Merchant (Jan Gossaert), 484
  Quadrille at the Moulin Rouge (Henri de Toulouse-Lautrec), 366
  At the Races (Edouard Manet), 151
  Regatta at Cowes (Raoul Dufy), 423
  The Rendering of the Tribute Money (Masaccio), 479
  The Senate (William Gropper), 556
  The Square at St. Mark's (Canaletto), 360
  π (Tom Marioni), 39
  Urban Freeways (Wayne Thiebaud), 211
  Warriors and Dragons (Kristen Shinoda), 333
Articles of faith, 66
Aryabhata, 45, 425
ASA. *See* Angle-side-angle
ASCII (American Standard Code for Information Interchange), 32
Assumptions
  apportionment, 565–566
  reasoning, 66
Atmospheric perspective, 353–354
Atmospheric pressure, 310
At the Races (Edouard Manet), 151
Attic Greek system, 3
Attic numeration system, 7
Average, 226
Axioms, Euclidean geometry, 341–342

B

Babylon
  ancient numeration systems, 11–12
  fractions, 19–20
  geometry, 335
  interest, 480
  number system, 23, 41
  π, 45
  trigonometry, 425
Banking. *See* Finance
Banks, 481
Bar graph, 216–217, 219–220
Base 8, 32
Base 10 logarithms, 317
Base 16, 32
Bases
  from decimal, converting, 25–27
  to decimal, converting, 24–25
  fractions in other, 27–28
  place-value system for any, 23–24
Basic Counting Law, 136–137
Basimal point, 27
Bayes, Thomas, 213
Bearing, 460–462
Bell curve. *See* Normal distribution
Beltrami, Eugenio, 344
Ben Ezra, Rabbi, 152
Bhãskara, 152, 480
Biconditional statement, 60
Binary digits, 31–32
Binary numbers, converting, 33–35
Binary system, 3

BINGO, 162–163
Binomial expansion, Pascal-Yang Hui Triangle, 549
Bit, 31
Boëthius, 54, 152
Bolyai, Janós, 344
Bolzano, Bernhard, 110
Bombelli, Rafael, 42
Boole, George, 55
Borda count voting method, 558–559, 561
Bosch, Hieronymus (*Death and the Miser*), 367
Bradshaw, Bob (*Mathematica*), 548
Brahmagupta, 42, 281
Broken line graph. *See* Line graph
The Brooklyn Bridge (Joseph Stella), 279
Brunelleschi, Filippo, 336
Buffon, Georges, 163
Burley, Walter, 54

C

Calculator use
  correlation coefficient, 270
  logarithm, 317
  standard deviation of grouped data, 238
  trigonometric functions, 430–431
Calculus
  differential, 527–539
  integral, 527, 539–547
Canaletto (*The Square at St. Mark's*), 360
Cantor, Georg, 110–111, 117, 118, 403
Capriccio of Roman Ruins (Marco Ricci), 366
Cardano, Girolamo, 41, 42, 153
Cardinality of a union, 125
Cardinal number of sets, 114–115
The Card Players (Lucas van Leyden), 152
The Card Players (Paul Cezanne), 110
Cards, 164–167
Cassatt, Mary (*The Letter*), 352
Cattleya Orchid and Three Brazilian Hummingbirds (Martin Johnson Heade), 354
Celsius, converting to Fahrenheit, 285
Central tendency, measures of
  mean, 226–228
  median, 229
  mode, 230–231
Cezanne, Paul (*Card Players*), 110
Ch'ang Ts'ang, 335
Charts. *See* Flowcharts
Children, height of, 318–319
China
  ancient numeration system, 8–9
  combinatorics, 152
  geometry, 335
  Pascal-Yang Hui Triangle, 549
  trigonometry, 425
  zero and negative numbers, 41–42
Chinese, 41
Ch'in Kiu-shao, 281
Chrysippus, 54
Chu Shih-Chieh, 281
Circle graph. *See* Pie chart
Classes, 215–216
Class midpoint, 228
Coin toss, 157
Combinations
  counting, 137–143
  Pascal-Yang Hui Triangle, 550–551
Combinatorics, 152–153
Common logarithms, 317
Complementary events, probability, 158–159
Complement of a set, 121–122
Complex number, 42
Composite number, 43
Compound events probability, 176–177
Compound interest
  credit cards, 501
  defined, 497–498
  effective rate, 502
  exponential model, 307–308
  goal, time to reach, 499–500
  period, 498
  present value, 499
  rate, computing, 500–501
Computer use
  base of numeration system, 23
  binary digits, 31–32
  calculations, 35–37
Concave polygon, 379
Conclusion, 58
Conditional probability
  formula, 189–191
  symbols, 187
  tables, 188–189
  tree diagrams, 187–188, 191
  Venn diagrams, 190–191

Conditional statement
  logic, 58–60
  truth tables, 78–79
Conditional symbol, 75
Confidence interval, 257–258
Confidence level, 258
Conjunction, 75
Conjunction statement, truth tables, 78
Consequent
  denying the, 69–71
  in logic, 58
Constraint, linear programming, 575
Converting
  bases to decimal numerals, 24–25
  binary, octal, and hexadecimal numerals, 33–35
  Celsius to Fahrenheit, 285
  decimal numerals to other bases, 25–27
  odds and probability, 172
Convex polygon, 379
Cooling, Newton's Law of, 319–320
Cooper, Anthony (*Fissures in Sun-Baked Clay*), 390
Correlation coefficient, 270
Cosine (cos)
  described, 428–429
  Law of, 452–455
Countably infinite sets, 118
Counting
  Basic Counting Law, 136–137
  combinations, 137–138
  by listing possible outcomes, 135–136
  with permutations and combinations, 138–143
  tools, 137
Credit cards
  compound interest, 501
  simple interest, 493–494
Cross number puzzles, 97–98

D

Dante, 153
Data
  exponential models, 311–312
  information, arranging statistical, 215–218
Da Vinci, Leonardo, 336, 354
Davis, Ron (*Horseshoe*), 163
Death and the Miser (Hieronymus Bosch), 367
Decagon, 379
Decimal fractions, 18–19
Decimal numerals, converting
  from other bases, 24–25
  to other bases, 25–27
Decimal system. *See* Hindu-Arabic system
Decision symbol, 86
Decision theory, 198–199
Decrease, percentage, 488–489
Deduction
  antecedent, affirming the, 68–69
  consequent, denying the, 69–71
  described, 66–67
  hypothetical syllogism, 67–68
De Fermat, Pierre, 336
Deficient number, 44
Definite integrals
  area, calculating, 544–545
  integral calculus, 542–543
Definitions, 66
Degree mode, calculator, 430–431
Delphic Vision (David McLaughlin), 361
De Morgan, Augustus, 55
De Pisa, Leonardo. *See* Fibonacci, Leonardo
Deposit accounts
  annuities, 506–510
  compound interest, 497–501
  simple interest, 492–493
Depression, angles of, 443–446
De puritate artis logicae, 54
Derain, André (*The Old Bridge*), 353
De ratiociniis in ludo aleane, 153
Derivation, Law of Sines, 450–452
Derivatives
  applications, 530–535
  described, 527–530
  graphing functions with, 530–532
  reverse, 542–544
  shortcuts for finding, 530
Descartes, René, 42, 44, 280, 281, 336
Descriptive method, 114
Deviation. *See* Standard deviation
Dice problems
  four-sided, 553
  multiple tosses, 164
  Pascal-Yang Hui Triangle, 551–553
  single toss, 157–158
  two-sided, 551–552
Differential calculus, 527–539
Dimension
  of an object, 409
  of fractals, 407–408
  of rectangular region, 533–535
Diminishing sizes, 352–353
Diophantine equation, 289
Diophantus, 41, 281, 289

Disjoint sets, 125
Disjunction statement, 77–78
Dispersion, measures of
  range, 236–237
  standard deviation, 237–240
Divina porportione, 372
Divisor apportionment methods, 567–568
Dodecagon, 380
Dodgson, Charles, 111
Duet (Adolph Gottlieb), 113
Dufy, Raoul (*Regatta at Cowes*), 423
Dungeons and Dragons, 161–162
Duodecimal Society of America, 23
Dürer, Albrecht, 336

E

Each angle of a regular polygon, 382
Effective interest rate, 502
Egypt
  fractions, 19
  geometry, 110, 335
  numeration system, 3, 6–7
  trigonometry, 425
Eight Bells (Winslow Homer), 441
Element, 113
Elements, 335, 336
Elevation, angles of, 443–446
Elimination, plurality method with, 557–558, 560
Empty sets, 116
English, translating symbols to, 75–77
Equal sets, 116
Equation of a line, 284
Equivalent sets, 116
Equivalent statements, truth tables, 79–80
Eratosthenes, 43
Euclid, 334, 335, 336
Euclidean geometry
  acute angle, 340
  angle, 340
  axioms, 341–342
  exterior angle, 340
  line segment, 339
  obtuse angle, 340
  parallel rays and lines, 341
  perpendicular rays and lines, 341
  postulates, 341, 342–343
  ray, 339
  right angle, 340
  straight angle, 340
  theorems, 343–344
  triangle, 340
  undefined terms, 338–339
Euler, Leonhard, 42, 44, 55, 64, 111
Evenly spaced objects, 358
$\overline{E}$, complement of an event, 158
Expected value, 195–204
Experimental probability, 156–158
Exponential models
  compound interest, 307–308
  data, 311–312
  functions, 304–307
  growth, 308–309
  population growth, 309–310
Exterior angle, 340
Eye level line, 356–357

F

Factorials, 137
Factors, 43–46
Facts, commonly accepted, 66
Fahrenheit, converting from Celsius, 285
Fairness, 174, 197
Fatou, Pierre, 404
Fechner, Gustav, 368, 377
Fermat, Pierre de, 44, 153
Fibonacci, Leonardo, 281, 372, 425
Fibonacci numbers, 372
Fifth postulate, 336
Finance
  annuities, 506–510
  interest
    compound, 497–505
    history of, 480–481
    simple, 492–496
  loans, 511–519
  percents
    defined, 485–486
    increase and decrease, 488–489
    markups and markdowns, 487–488
    of a number, finding, 486–487
Finite sets, 113–120
Fissures in Sun-Baked Clay (Anthony Cooper), 390
Flowcharts
  constructing, 86–88
  organizational, 88–89
  symbols, 85–86
Flower Beds in Holland (Vincent van Gogh), 357
Flow line symbol in flowcharts, 86

Forecasting, 264–271
Formula
  conditional probability, 189–191
  probability of a union, 179
  quadratic, 296–301
  standard deviation, 237
Four-sided dice problems, 553
Fractals
  described, 402–403
  dimension of, 407–408
  history, 403–404
  Mandelbrot set, 410–413
  Sierpinski carpet, 406–407, 408
  von Koch snowflake, 404–406, 408
Fractions
  Hindu-Arabic system, 17–23
  in other bases, 27–28
Franceschi, Pietro, 336
Frequency, 216
Frequency polygon. *See* Line graph
Functions
  exponential, 304–307
  graphing with derivative, 530–532
  linear, 288–289
  logarithmic, 316–318
  objective with constraints, 575
  quadratic, 295–296
  trigonometric, 428–431
Fundamental Theorem of Calculus, 543, 544
Fundamental Theorem of Linear Programming, 576

G

Games and gambling. *See* Probability
Gauss, Carl, 344–345
Gaussian distribution. *See* Normal distribution
Geometric shape basis of numbers, 42–43
Geometry
  Euclidean, 338–344
  fractals, 402–413
  golden ratios and rectangles, 368–377
  history of, 110, 334–335
  non-Euclidean, 344–348
  perspective, 351–368
  polygons and stars, 378–389
  tesselations, 390–401
Geometry of Art and Life, 371
Ghyka, Matila, 371
Global Pathways (Lester Lefkowitz), 31
Goal, savings, 499–500
Golden ratio, 109, 368–377
Golden Venn (David McLaughlin), 109
Gossaert, Jan (*Portrait of a Merchant*), 484
Gottlieb, Adolph (*Duet*), 113
Graphing
  functions with derivatives, 530–532
  parabolas, 530
Graphs
  bar, 219–220
  line graph, 221
  pie chart, 220–221
  type, choosing, 218
Graunt, John, 212
Great Pyramid of Gizeh, 45, 335
Greece
  algebra, 281
  ancient numeration systems, 10–11
  Attic Greek numerals system, 3
  fractions, 20
  geometry, 110, 335
  logical theory in, 54
  trigonometry, 424–425
Gropper, William (*The Senate*), 556
Grouped data, standard deviation, 238
Growth, exponential model, 308–309
Guy, Gloria (*A modern representation of Van Gogh's Starry Night*), 525

H

Hamilton's apportionment method, 566, 568–570
Hanging Gardens of Babylon, 335
Heade, Martin Johnson (*Cattleya Orchid and Three Brazilian Hummingbirds*), 354
Height of children model, 318–319
Heptagon, 379
Hexadecimal numbers, converting to binary, 33–35
Hexagon, 379, 391
Hilbert, David, 403
Hilbert's curve, 403
Hill-Huntington apportionment method, 571
Hindu, 152
Hindu-Arabic system
  advantages, 28
  Babylonian fractions, 19–20
  decimal fractions, 18–19
  described, 17–18
  Egyptian hieroglyphic fractions, 19

Ionic Greek fractions, 20
origins, 3, 13–14
Roman fractions, 20
Hipparchus of Nicea, 425
Homer, Winslow (*Eight Bells*), 441
Horseshoe (Ron Davis), 163
House buying loans, 515–517
House odds, 173–174
Houthakker, Henrik, 412
Human face, perfect, 371
Huygens, Christiann, 153
Hypotenuse of right triangle, 427
Hypothesis, 58
Hypothetical syllogism, 67–68

I

I-Ching, 152
If . . . then statement, 78
If . . . then symbol, 75
I (imaginary numbers), 42
Ilm al-jabr walmuqabalah, 281
Inclusive disjunction symbol, 75
Increase percentage, 488–489
India, 41–42, 54
Inductive reasoning, 64–66, 71
Inferential statistics, 257–258
Infinite sets, 113–117
Information, arranging statistical, 215–226
Insurance risk calculation, 198
Integers, 39, 117
Integral calculus
antiderivatives, 542–544
differential versus, 527
method of rectangles, 539–542
Interest
compound, 497–505
history of, 480–481
simple, 492–496
Interest rate
computing, 500–501
effective, 502
The Interior of the Pantheon, Rome (Giovanni Paolo Panini), 480
Intersection, 122, 177
Intuitive probability, 155–163, 188–189
Irrational numbers, 39
Irregularly shaped area, 544–545
I Saw the Figure 5 in Gold (Charles Henry Demuth), 18

J

Jefferson's apportionment method, 567–568
Jews, 481
Jones, William, 45
Julia, Gaston, 404

K

Kanada, Y., 45
Keno, 167
Kepler, Johann, 372
Khayyam, Omar, 281
K'iu-ch-ang Suan-shu (Arithmetic in Nine Sections), 335
Klein, Felix, 344
Kowa, Seki, 54

L

La Geometrie, 281, 336
Lambert, Johann, 42, 44, 45
Large numbers, law of, 157
Laws
of Cosines, 452–455
of Large Numbers, 157
Newton's Law of Cooling, 319–320
in reasoning, 66
of Sines, 450–452
Legs of right triangle, 427
Leibniz, Gottfried, 54–55, 527
Le Penseur (Tami Whitt-Zenoble), 53
The Letter (Mary Cassatt), 352
Liber de ludo aleane, 153
Lighthouse at Honfleur (Georges Seurat), 434
Linear functions, 288–289
Linear models, 284–294
Linear programming
applications, 576–580
Fundamental Theorem of Linear Programming, 576
objective function and constraints, 575
variables, 574–575
Line graph, 217–218, 221
Lines
fifth postulate, 336
parallel, 336–337, 341
perpendicular, 341
slope of a, 527–530
Line segment, 339
Listing method, 114

Liu Hui, 45
LN button, 317
Loans
  accelerating payments, 513–514
  buying a house, 515–517
  described, 511–513
  refinancing, 514–515
Lobachevskian Parallel Postulate, 344
Lobachevskian Triangle-Sum Theorem, 347
Lobachevsky, Nicolai, 344
Logarithmic models
  height of children, 318–319
  logarithmic functions, 316–318
  Newton's Law of Cooling, 319–320
Logarithmic spiral, 372–373
LOG button, 317
Logic
  defined, 56
  flowcharts, 85–93
  history of, 54–55
  inductive and deductive reasoning, 64–74
  puzzles, 93–98
  statements, 56–60
  symbolic and truth tables, 74–85
Logical equivalency symbol, 75
Logically equivalent statements, 79–80
Logistic curve, 283
Lotteries, 167–168
Lukasiewicz, Jan, 55
Lulle, Raymond, 153
Luther, Martin, 480–481

M

Magic square, 95–96
Magic triangles, 94
Malthusian population model, 309–310
Malthus, Thomas Robert, 309–310
Mandelbrot, Benoit, 402–403
Manet, Edouard (*At the Races*), 151
Margin of error, 258, 259–260
Marioni, Tom (π), 39
Markups and markdowns, 487–488
Masaccio (*The Rendering of the Tribute Money*), 479
Mathematica (Bob Bradshaw), 548
Mayan numeration systems, 3, 12–13, 23, 41
McLaughlin, David
  Delphic Vision, 361
  Golden Venn, 109
  O, e, i, 39
Mean, 226–228
Measures of central tendency
  mean, 226–228
  median, 229
  mode, 230–231
Measures of dispersion
  range, 236–237
  standard deviation, 237–240
Median, 229
Megarian school, 54
Members of a set, 113–114
Méré, Chevalier de, 153
Method of rectangles, 539–542
Mill, John Stuart, 55
Minus, 41
Mode, 230–231
A modern representation of Van Gogh's Starry Night (Gloria Guy), 525
Money matters. *See* Finance
Mortgage loans, 515–517
Motion of a projectile, 467–471
Multiplication
  ancient numeration systems using, 8–11
  computer calculations, 35–37

N

N!, 137
Nasir ed-din al-Tûsi, 425
Natural logarithms, 317
Natural numbers, 39, 116, 125
Navigation problems, 460–462
Negation, 57–58, 75
Negative numbers, 41–42
Neolithic period, 2
Newton, Sir Isaac, 527
Newton's Law of Cooling, 319–320
Nominal interest rate, 502
Nonagon, 379
Non-Euclidean geometry
  described, 344–345
  Lobachevskian model, 346–348
  Riemannian model, 345–346, 347–348
Normal distribution
  polls and margin of error, 258–259
  statistics, 245–256
Notation, set, 113–114
Not symbol, 75

Numbers
  ancient systems, 6–17
  bases, other, 23–30
  complex, 42
  factors, based on, 43–46
  geometric shape basis, 42–43
  Hindu-Arabic system and fractions, 17–23
  history, 2
  percent of, finding, 486–487
  real, 39–41
  system of numeration, 3–4
  of technology, 31–39
  types, 39–47
  zero and negative, 41–42
Numerals, 3
Nunes, Pedro, 424
Nyaya schools, 54

O

Object, dimension of, 409
Objective function and constraints, 575
Obtuse angle, 340
Obtuse triangle, 449
Octagon, 379
Octal numbers, converting, 33–35
Odds, 172–174
O, e, i (David McLaughlin), 39
The Old Bridge (André Derain), 353
One-point perspective
  evenly spaced objects, 358
  eye level line, 356–357
  proportions, 358–359
One-to-one relationship, 117
Operations
  sets, 121–122
  translating words into, 123
ϕ (phi), 368
Or
  compound events probability, 176–177
  sets, translating into operations, 123
  symbol, 75
Order of objects, irrelevant. *See* Combinations
Organizational charts, 88–89
Organon, 54
Otho, Valentin, 45
Ounces, 20
Outcomes, listing possible, 135–136
Oval symbol, 85
Overlapping shapes, 352–353

P

Pacioli, Luca, 372
Paganini, Nicolo, 44
Pairwise comparison voting method, 559–560, 561–562
Paleolithic period, 2
Panini, Giovanni Paolo (*The Interior of the Pantheon, Rome*), 480
Parabola, graphing, 294–304, 530
Parallel lines or rays, 336–337, 341
Parallel postulate, 343
Parthenon, 371
Pascal, Blaise, 153, 549
Pascal-Yang Hui Triangle
  binomial expansion, 549
  combinations, 550–551
  dice problems, 551–553
  history, 549
  illustrated, 548
  polynomial expansion, 549–550
Payments, accelerating loan, 513–514
Peano, Giuseppe, 403
Pentagon, 379
Pentagonal numbers, 43
Pentagram, 371
Percents
  defined, 485–486
  increase and decrease, 488–489
  markups and markdowns, 487–488
  of a number, finding, 486–487
Perfect number, 44
Period, compound interest, 498
Permutations, 137, 138–143
Perpendicular rays or lines, 341
Perspective
  atmospheric, 353–354
  one-point, 355–359
  overlapping shapes, 352–353
  three-point, 361
  two-point, 360
Petty, William, 212
Phidias, 368
Phi (ϕ), 368
Pie chart, 218, 220–221
Pinochle, 162
Pi, symbol for, 45–46
Pitiscus, Bartholomaus, 425
Place value
  numbers systems, 11–14, 28
  system for any bases, 23–24

Plato, 335, 480
Playfair, John, 336–337
Plurality voting method, 557–558, 560
Poincaré, Henri, 344
Points, mortgage, 516
Polls
  confidence levels and normal distribution, 258–259
  inferential statistics and confidence intervals, 257–258
  margin of error and surveys, 259–260
Polygon
  angles, 381–382
  described, 378–381
  drawing regular, 383–384
  frequency, 217–218, 221
  tesselations
    with more than one, 398–399
    with other, 397
    with regular, 391–392
Polygonal numbers, 42
Polynomial expansion, Pascal-Yang Hui Triangle, 549–550
Population-density map, 215
Population growth model, 309–310
Portrait of a Merchant (Jan Gossaert), 484
Positive, 41
Post, Emil, 55
Postulate
  Euclidean geometry, 341, 342–343
  fifth, 336
  parallel, 343
  Riemannian Parallel, 345
Pound, 20
Practica Geometriae, 425
Present value, 499
Prime (′), 20
Prime number, 43
Primitive tribes, 23
Principia Mathematica, 55
Privative, 41
Probability
  calculating, 163–172
  complementary events, 158–159
  compound events, 176–187
  conditional, 187–195
  expected value, 195–204
  experimental, 156–157
  history, 152–153
  intuitive concepts, 155–163
  odds, 172–176
  restrictions, 156
  single events, 157–158, 158
  tables, reading, 178–179
  theoretical, 156–157, 157–158
  tree diagrams, 181–183
  of a union, formula for, 179
  Venn diagrams, 179–181
Projectile, motion of, 467–471
Proof of a theorem, 343
Proper factor of a natural number, 43
Proper subsets, 115–116
Proportion, one-point perspective, 358–359
Ptolemy, 45
Puzzles
  alphametric, 96–97
  cross number, 97–98
  magic squares, 95–96
  magic triangles, 94
Pythagoras, 44, 424, 427
Pythagorean theorem
  history, 335
  right angles, sine, cosine, tangent, 427–428
Pythagorean triple, 433

Q

Quadratic models
  formula, 296–301
  function, 295–296, 305
Quadrilateral, 379, 392–397
Quadrille at the Moulin Rouge (Henri de Toulouse-Lautrec), 366
Quetelet, Adolphe, 212–213

R

Range, 236–237
Ratio, golden, 109, 368–377
Rational numbers, 39, 117
Rays
  described, 339
  parallel, 336–337, 341
  perpendicular, 341
  symbol, 86
Real numbers, 39–41
Rectangles
  dimension of, 533–535
  golden, 368–377
  method of, 539–542
  symbol, 85

Refinancing loans, 514–515
Reflection
  defined, 394
  tesselations with triangle using translations, 395–397
Regatta at Cowes (Raoul Dufy), 423
Regression, statistical, 264–271
Regular polygons, 379
Relative maximum, 531–532
The Rendering of the Tribute Money (Masaccio), 479
Replacement, listing outcomes with or without, 135–136
Restrictions, probability, 156
Reverse derivative, 542–544
Rhind papyrus, 1, 280–281
Rhombus symbol, 86
Ricci, Marco (*Capriccio of Roman Ruins*), 366
Riemann, Georg, 345–346
Riemannian model, 345–346
Riemannian Parallel Postulate, 345
Riemannian Triangle-Sum Theorem, 346
Right angle
  Euclidean geometry, 340
  Pythagorean theorem, 427–428
Right triangle
  described, 427
  elevation and depression, angles of, 443–446
  solving, 434–440
Roberval, Gilles Persone de, 539
Roles, inductive and deductive reasoning, 71
Roman Catholic Church, 481
Roman numerals, 3, 13
Rome
  ancient numeration system, 7–8
  fractions, 20
  interest, 480
  logical theory in, 54
Roulette wheel, 158–159
Rounding method, 566
Round-off rules, 434
Rules, 66
Russell, Bertrand, 55, 111

S

Saccheri, Girolamo, 344
Sectio divina, 372
The Senate (William Gropper), 556
Set builder notation, 114
Set of items, arrangement of. *See* Permutations
Sets
  applications, 128–134
  cardinal number of, 114–115
  empty, 116
  equal versus equivalent, 116
  finite and infinite, 113–120
  history, 110–111
  notation and members of, 113–114
  operations, 121–122
  sizes, 117–118
  subsets and proper subsets, 115–116
  translating words into operations, 123
  universal, 116
  Venn diagrams, 123–125
Seurat, Georges (*Lighthouse at Honfleur)*, 434
Shape basis, geometric, 42–43
Shinoda, Kristen (*Warriors and Dragons*), 333
Sides
  and angles of triangle, relationship between, 450–452
  polygons, 378–379
  triangle, solving when known, 452–455
Sierpinski carpet, 406–407
Simple interest, 492–496
Sine (sin)
  functions, described, 428–429
  law of, 450–452
Single event probability, 157–158
SIN key, 430–431
Sizes
  diminishing, 352–353
  sets, 117–118
Skeeball, 204–205
Slope of a line, 284, 527–530
Sound, speed at certain temperature, 287–288
Spatial relationships, 334–335
Sports
  events, 159
  players' salaries, 234–236, 242–244
  scoring, 288–289
Square array of numbers. *See* Magic square
The Square at St. Mark's (Canaletto), 360
Square numbers, 42
Squares
  magic, 95–96
  tesselations, 391
S-shaped curve. *See* Logistic curve
Standard deviation, 237–240
Stars, 384–386
Start/stop symbol in flowcharts, 85

Statement
conditional, 58–60
defined, 56
negation of, 57–58
symbol in flowcharts, 85
Statistics
central tendency, measures of, 226–236
dispersion, measures of, 236–244
history, 212–213
information, arranging, 215–226
normal distribution, 245–256
polls and margin of error, 257–263
regression and forecasting, 264–271
Stella, Joseph (*The Brooklyn Bridge*), 279
Stifel, Michael, 41
Stoic school, 54
Straight angle, 340
Subsets, 115–116
Subtraction, 7–8
Sumeria, 480
Summa Logicae, 54
Sum of the angles of a polygon, 382
Surveys, 259–260
Sûrya Siddhanta, 425
Syllogism
hypothetical, 67–68
truth tables, verifying, 80–82
Symbol
conditional probability, 187
English, translating to, 75–77
flowcharts, 85–86
mean (μ), 226
standard deviation (Σ), 237
Symbolic logic, 74–77

T

Tables
conditional probability, 188–189
probability, reading, 178–179
reading, 128–129
Tally, 2–3
Tangent (tan), 428–429
Technology numbers
binary digits, 31–32
computer calculations, 35–37
converting binary, octal, and hexadecimal numerals, 33–35
Temperature
Celsius, converting to Fahrenheit, 285
speed of sound at certain, 287–288
Tesselations
in artwork, 333
described, 390–391
with more than one polygon, 398–399
with other polygons, 397
with quadrilateral or triangle, 392–397
with reflection, translations, and triangle, 395–397
with regular polygon, 391–392
with translations and triangle, 394–395
Theorems
Euclidean geometry, 343–344
Fundamental Theorem of Calculus, 543, 544
Fundamental Theorem of Linear Programming, 576
Pythagorean, 335, 427–428
Triangle-Sum, 344, 346, 347
Theoretical probability, 156–158
Therefore symbol, 75
Thiebaud, Wayne (*Urban Freeways*), 211
Three-dimensional space, illusion of. *See* Perspective
Three-point perspective, 361
Time to reach goal, 499–500
Tomb Hieroglyphs at the Valley of the Queen, 367
π (Tom Marioni), 39
Tools, counting, 137
Toulouse-Lautrec, Henri de (*Quadrille at the Moulin Rouge*), 366
Traditional Chinese system, 3
Translations, tesselations
with reflection and a triangle, 395–397
with triangle, 394–395
Tree diagrams, 181–183, 187–188, 191
Triangle
acute, 449
Euclidean geometry, 340
magic, 94
obtuse, 449
polygons, 381
right, 427, 434–440
round-off rules, 434
sides, 379
tesselations, 391, 392–397
Triangle-Sum Theorem, 344, 346, 347
Triangular numbers, 42, 304
Trigonometric functions
described, 428–429
finding with calculator, 430–431

Trigonometry
acute triangle applications, 458–466
history, 110, 424–425
motion of a projectile, 467–471
right angles, sine, cosine, and tangent, 427–433
right triangle applications, 441–449
sines and cosines, laws of, 449–458
True/false statements in logic, 56
Truth tables
conditional statement, 78–79
conjunction statement, 78
defined, 77
disjunction statement, 77–78
equivalent statements, 79–80
syllogisms, verifying, 80–82
Two-point perspective, 360
Two-sided "dice" problems, 551–552

U

Undecagon, 380
Union
of compound events, 177
probability formula, 179
of a set, 122
Universal sets, 116
Urban Freeways (Wayne Thiebaud), 211

V

Van Gogh, Vincent (*Flower Beds in Holland*), 357
Vanishing point, 356–357
Van Leyden, Lucas (*Card Players*), 152
Varâhamihira, 425
Variables, linear programming, 574–575
Venn diagrams
conditional probability, 190–191
history, 110
problems, solving, 129–131
sets, 123–125
union, probability of, 179–181
Venn, John, 55, 111, 123
Vertices (vertex), polygon, 378–379
Viewfinder, 355–359
Von Koch snowflake, 404–406
Voting systems
Borda count method, 558–559, 561
comparing, 562
defined, 556
elimination with plurality method, 557–558, 560
pairwise comparison method, 559–560, 561–562
plurality method, 557, 560

W

Warriors and Dragons (Kristen Shinoda), 333
Webster's apportionment method, 568
Weighted average, 228
Weights, 20
Whitehead, Alfred North, 55
Whole numbers, 39, 117
William of Ockham, 54
Wittgenstein, Ludwig, 55
Words, translating into operations, 123

Y

Yang Hui. *See* Pascal-Yang Hui Triangle

Z

Zero
first use of, 12
numbers, 41–42

IMPORTANT SYMBOLS AND FORMULAS

Egyptian Hieroglyphics

1	𓏤
10	𓎆
100	𓍢
1000	𓆼
10,000	𓂭
100,000	𓆐
1,000,000	𓁨
10,000,000	𓍶

Mayan

0	𝋠	10	𝋪
1	𝋡	11	𝋫
2	𝋢	12	𝋬
3	𝋣	13	𝋭
4	𝋤	14	𝋮
5	𝋥	15	𝋯
6	𝋦	16	𝋰
7	𝋧	17	𝋱
8	𝋨	18	𝋲
9	𝋩	19	𝋳

Attic Greek

1	Ι
5	Π
10	Δ
50	𐅄
100	Η
500	𐅅
1000	Χ
5000	𐅆
10,000	Μ
50,000	𐅇

Babylonian

1	𒁹
10	𒌋

Ionic Greek

1	α	10	ι	100	ρ
2	β	20	κ	200	σ
3	γ	30	λ	300	τ
4	δ	40	μ	400	υ
5	ε	50	ν	500	φ
6	ς	60	ξ	600	χ
7	ζ	70	ο	700	ψ
8	η	80	π	800	ω
9	θ	90	Q	900	T

Traditional Chinese

1	一
2	二
3	三
4	四
5	五
6	六
7	七
8	八
9	九
10	十
100	百
1000	千
10,000	万

Roman

1	I
5	V
10	X
50	L
100	C
500	D
1000	M
5000	$\overline{V}$
10,000	$\overline{X}$
50,000	$\overline{L}$
100,000	$\overline{C}$
500,000	$\overline{D}$
1,000,000	$\overline{M}$